HANDBOOK OF CELLULAR MANUFACTURING SYSTEMS

HANDBOOK OF CELLULAR MANUFACTURING SYSTEMS

Edited by

Shahrukh A. Irani

A WILEY-INTERSCIENCE PUBLICATION

JOHN WILEY & SONS, INC.

New York / Chichester / Weinheim / Brisbane / Singapore / Toronto

Copyright © 1999 by John Wiley & Sons. All rights reserved.

Published simultaneously in Canada.

This publication is designed to provide accurate and authoritative information in regard to the subject matter covered. It is sold with the understanding that the publisher is not engaged in rendering professional services. If professional advice or other expert assistance is required, the services of a competent professional person should be sought.

Library of Congress Cataloging-in-Publication Data:

Handbook of cellular manufacturing systems / edited by Shahrukh A. Irani.
  p. cm.
  "A Wiley-Interscience publication."
  Includes bibliographical references and index.
  ISBN 0-471-12139-8 (alk. paper)
  1. Group technology. 2. Manufacturing cells. 3. Manufacturing cells—Case studies.
 I. Irani, Shahrukh.
 TS155.H28134 1999
 670.42—dc21
                        98–29970

CONTENTS

Preface   ix

Contributors   xi

Introduction to Cellular Manufacturing Systems   1
Shahrukh A. Irani, Sanjay Subramanian, and Yosef S. Allam

1   **Investigation of Cellular Manufacturing Practices**   25
Ronald G. Askin and Steve Estrada

2   **Machine–Component Matrix Clustering Methods for Cell Formation**   35
T. T. Narendran and G. Srinivasan

3   **Cell Formation Using Production Flow Analysis**   69
Roland De Guio and Marc Barth

4   **Layout Design for Cellular Manufacturing**   111
Massoud Bazargan-Lari

5   **Scheduling Cellular Manufacturing Systems**   141
Robert A. Ruben and Farzad Mahmoodi

6   **Setup Time Reduction to Enhance Cell Performance and Flexibility**   155
Jerry W. Claunch

7   **Framework for Cellular Manufacturing Evaluation Process Using Analytical and Simulation Techniques**   179
Farzad Mahmoodi, Charles T. Mosier, and Anthony T. Burroughs

v

8 Operating Manufacturing Cells with Labor Constraints 195

Anand Iyer and Ronald G. Askin

9 Benchmarking Measures for Performance Analysis of Cells 225

Barrie G. Dale

10 Flexibility Considerations in Cell Design 249

Asoo J. Vakharia, Ronald G. Askin, and Hassan M. Selim

11 Quality Control in Cellular Manufacturing 275

Tapas K. Das and William A. Miller

12 Organizational and Human Issues in Cellular Manufacturing 319

Al Miller

13 Economic Justification of Cellular Manufacturing 389

Peter L. Primrose

14 Project Management and Implementation of Cellular Manufacturing 413

Bopaya Bidanda, Rona Colosimo Warner, Paul J. Warner, and Richard E. Billo

15 Design and Implementation of Lean Manufacturing Systems and Cells 453

J T. Black

16 Industrial Implementation of Production Flow Analysis 497

Marc Barth and Roland De Guio

17 Dishwashing Machine Assembly: A Case Study 527

Quarterman Lee

18 Cell Formation in a Coil Forging Shop: An Implementation Case 555

Marco Perona

19 Cases in Cellular Manufacturing **589**

William Wrennall and Frank C. Kerns

20 Classroom Tutorial on the Design of a Cellular Manufacturing System **613**

Lifang Yan and Shahrukh A. Irani

21 Systematic Redesign of a Manufacturing Cell **661**

Yosef S. Allam and Shahrukh A. Irani

22 Plantwide Conversion to Cellular Manufacturing **681**

Quarterman Lee

23 Conversion to Cellular Manufacturing at Sheet Metal Products **707**

Danny J. Johnson

Index **749**

PREFACE

The idea for this edited book was conceived as early as 1993. This was the year when I first taught a course—IEOR5990—on Cellular Manufacturing Systems. Later that year, using the course materials, I developed and taught an industry workshop on Cellular Manufacturing Systems, primarily as a mechanism to attract research funding from industry. Both teaching experiences proved to be excellent learning experiences. They were instrumental in convincing me that there was a need for a book that simultaneously achieved the two seemingly conflicting goals of (a) presenting a practice-oriented description of the subject to faculty and students on "soft" issues such as project management, lean manufacturing, human relations, and setup reduction that are critical when designing and implementing cells in industry and (b) presenting a methods and analysis-oriented description of the subject to industry practitioners on the analytical models and computer methods for automated solution of activities such as formation, layout, scheduling, and performance evaluation of cells that they usually undertake to do by hand.

Having identified the need for the book and received the blessings of John Wiley & Sons, the next task was to identify the best potential contributors to write the different chapters. I wish to take this opportunity to thank all of my friends and colleagues in academia and industry for having contributed their time and expertise to write chapters that are a unique blend of theory and practice. In particular, I thank them for their patience as they undertook the revisions that I suggested. This book is a tribute to all to them.

There are two people at Wiley whom I wish to thank profusely and who truly deserve my gratitude—Bob Argentieri, Senior Editor, and his Editorial Assistant, Akemi Takada. It suffices to say that, had it not been for their supreme patience, support, and guidance throughout this project period, this book would not have seen the light of day. To Bob and Akemi, I say, "Thank you *very* much!"

Lastly, I wish to dedicate this book to my family. Dad's spirit always hovered over me in full support, with Mom, my brother, Vispi, and my wife, Gulnar, being my morale boosters here on earth. They are testimony to the fact that family means everything.

Shahrukh A. Irani
The Ohio State University
Columbus, OH 43210

CONTRIBUTORS

Yosef S. Allam, 3903 Autumn Drive, Huron, Ohio 44839

Ronald G. Askin, Systems and Industrial Engineering Department, College of Engineering and Mines, University of Arizona, Tucson, Arizona 85721

Marc Barth, c/o Roland De Guio, Laboratoire de Recherche en Productique de Strasbourg, Ecole Nationale Supérieure des Arts et Industries de Strasbourg, 24 Boulevard de la Victoire, F-67084 Strasbourg Cedex, France

Massoud Bazargan-Lari, Department of Business Administration, Embry-Riddle Aeronautical University, 600 South Clyde Boulevard, Daytona Beach, Florida 32114-3900

Bopaya Bidanda, Department of Industrial Engineering, University of Pittsburgh, Pittsburgh, Pennsylvania 15261

Richard E. Billo, Department of Industrial Engineering, University of Pittsburgh, Pittsburgh, Pennsylvania 15261

J T. Black, Department of Industrial & Systems Engineering, 307 Dunstan Hall, Auburn University, Alabama 36849

Anthony T. Burroughs, Xerox Corporation, 800 Phillips Road, Webster, NY 14580

Jerry W. Claunch, Claunch & Associates Inc., 9123 N. Military Trail, Suite 104, Palm Beach Gardens, Florida 33410

Barrie G. Dale, Manchester School of Management, UMIST, PO Box 88, Manchester, M60 1QD United Kingdom

Tapas K. Das, Department of Industrial and Management Systems Engineering, University of South Florida, Tampa, Florida 33620

Roland De Guio, Laboratoire de Recherche en Productique de Strasbourg, Ecole Nationale Supérieure des Arts et Industries de Strasbourg, 24 Boulevard de la Victoire, F-67084 Strasbourg Cedex, France

Steve Estrada, c/o R. G. Askin, Systems and Industrial Engineering Department, College of Engineering and Mines, University of Arizona, Tucson, Arizona 85721

Shahrukh A. Irani, Department of Industrial, Welding and Systems Engineering, The Ohio State University, 210 Baker Systems Engineering, 1971 Neil Avenue, Columbus, Ohio 43210

Anand Iyer, i2 Technologies Inc., 909 E. Las Colinas Blvd., Suite 1600, Irving, Texas 75039

Danny J. Johnson, College of Business, Iowa State University, 300 Carver Hall, Ames, Iowa 50011

Frank C. Kerns, c/o William Wrennall, The Leawood Group, Ltd., 4701 College Boulevard, Suite 212, Leawood, Kansas 66211

Quarterman Lee, Strategos, Inc., 3916 Wyandotte, Kansas City, Missouri 64111

Farzad Mahmoodi, School of Business, Clarkson University, Potsdam, New York 13699–5770

Al Miller, Manufacturing Consultant, 14871 45th Avenue North, Plymouth, Minesota 55446

W. A. Miller, Department of Industrial and Management Systems Engineering, University of South Florida, Tampa, Florida 33620

Charles T. Mosier, School of Business, Clarkson University, Potsdam, NY 13699–5770

T. T. Narendran, Division of Industrial Management, Department of Humanities and Social Sciences, Indian Institute of Technology, Madras 600036 India

Marco Perona, Dipartimento di Ingegneria Meccanica, Universita di Brescia, Via Branze 38, 25123 Brescia, Italy, and Dipartimento d'Economia e Produzione, Politecnico di Milano, Piazza Leonardo da Vinci, 20133 Milano, Italy

Peter L. Primrose, Total Technology Department, UMIST, PO Box 88, Manchester, United Kingdom M60 1QD

Robert A. Ruben, Department of Management, New Mexico Institute of Mining and Technology, 801 Leroy Place, Socorro, New Mexico 87801

Hassan M. Selim, Business Administration Department, College of Management and Economics, United Arab Emirates University, Al-Ain POB 17555, Al-Ain, United Arab Emirates

G. Srinivasan, Division of Industrial Management, Department of Humanitites and Social Sciences, Indian Institute of Technology, Madras 600036 India

Sanjay Subramanian, c/o Shahrukh A. Irani, Department of Industrial, Welding and Systems Engineering, The Ohio State University, 210 Baker Systems Engineering, 1971 Neil Avenue, Columbus, Ohio 43210

Asoo J. Vakharia, Department of Decision and Imformation Sciences, University of Florida, 343 Bus, PO Box 117169, Gainesville, Florida 32611-7169

Paul J. Warner, Department of Industrial Engineering, University of Pittsburgh, Pittsburgh, Pennsylvania 15261

Rona Colosimo Warner, Department of Industrial Engineering, University of Pittsburgh, Pittsburgh, Pennsylvania 15261

William Wrennall, The Leawood Group, Ltd., 4701 College Boulevard, Suite 212, Leawood, Kansas 66211

Lifang Yan, MVE, Inc., 407 Seventh Street NW, P.O. Box 234, New Prague, Minnesota 56071-0234

INTRODUCTION TO CELLULAR MANUFACTURING SYSTEMS

Shahrukh A. Irani, Sanjay Subramanian, and Yosef S. Allam

Department of Industrial, Welding and Systems Engineering
The Ohio State University
Columbus, Ohio 43210

1. DEFINITIONS

Group Technology (GT) is a manufacturing concept that seeks to identify and group similar parts to take advantage of their similarities in manufacturing and design. Group Technology has been practiced around the world for many years as part of good engineering practice and scientific management. The concept of GT was originally proposed by Mitrofanov (1996) and Burbidge (1975). Mitrofanov (1966) defined GT as "a method of manufacturing piece parts by the classification of these parts into groups and subsequently applying to each group similar technological operations." A modern definition of GT is "the realization that many problems are similar, and that by grouping them, a single solution can be found to a set of problems, thus saving time and effort" (Shunk, 1987). This definition captures the true essence

Handbook of Cellular Manufacturing Systems, edited by Shahrukh A. Irani
ISBN 0-471-12139-8 © 1999 John Wiley & Sons, Inc.

of GT that the population of entities or activities in a manufacturing system, or subsystem, can be replaced by a smaller number of families. However, the most general definition of GT is that " it is a manufacturing philosophy which identifies and exploits the underlying proximity of parts and manufacturing processes" (Ham et al., 1985).

Cellular Manufacturing (CM) is an application of the GT concepts to factory reconfiguration and shop floor layout design. Cellular Manufacturing Systems (CMS), as illustrated in Figure 1, have been proposed as an alternative to job shops since they provide the operational benefits of flow line production. Cellular Manufacturing involves processing a collection of similar parts on a dedicated group of machines or manufacturing processes. A manufacturing cell can be defined (Ham et al., 1985) as "an independent group of functionally dissimilar machines, located together on the floor, dedicated to the manufacture of a family of similar parts." Furthermore, a part family can be defined (Ham et al., 1985) as "a collection of parts which are similar either because of geometric shape and size or because similar processing steps are required to manufacture them." Usually it is preferable that a cell be dedicated to a single part family, that each part family be preferably produced completely within its cell, and that cells in a CMS have minimum interaction with each other. In summary, a CMS is essentially a set of manufacturing and/or assembly cells, each dedicated to the manufacture or assembly of a part family or group of products, respectively.

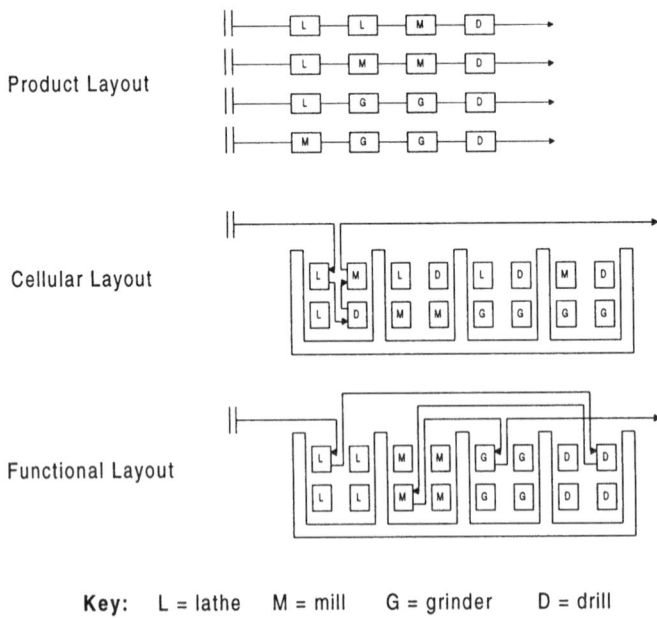

Key: L = lathe   M = mill   G = grinder   D = drill

Figure 1. Traditional types of manufacturing facility layouts.

2. PART FAMILY SELECTION TO DETERMINE THE MANUFACTURING FOCUS OF A CELL

The manufacturing focus of a cell is determined by the attributes of the family of parts that will be produced in it. Several criteria could be used to define the manufacturing focus of a cell, as determined primarily by the variety of processes included in the cell and/or the variety of parts assigned to the cell. For example, a cell can be defined by the variety of processes it contains (e.g., machining, grinding, inspection, assembly) that appear in the routings of the parts. Or, a cell could be defined by a part family derived from a single product or several products. Other criteria such as production volume, market segment, and degree of automation can also define a cell. A cell could be dedicated to frequently ordered parts or products, regardless of variety of tooling, material, processes, and work-holding requirements. Furthermore, considering market segments, a cell could be dedicated to some companies that order the same products. Considering the degree of automation, facilities with a common control and communication system are eligible to form a cell. For example, a group of machines served by a common material handling system can be considered as a cell. Table 1 gives a comprehensive classification of these criteria for defining the manufacturing focus of a cell.

3. EXAMPLES OF CELLS BASED ON DEGREE OF AUTOMATION

Within a cell, a variety of activities such as processing, handling, inspection, cleaning, packaging, loading/unloading, and setup changes need to be performed. The frequency with which these activities are performed, the number of stations involved, etc., could vary with the variety of parts and the level of operator skill required in the cell. Some representative types of cells based on degree of automation and variety, of the activities are briefly described next.

Machining Center   A machining center is a single automated machine tool. A machining center combines the processing capabilities of dissimilar conventional machine tools and can process a family of parts in random order. Modern machining centers are capable of automatically changing tools, workpieces, and cutting parameters. However, certain machining centers may require an operator to perform loading/unloading of workpieces and part programming functions. A machining center performs many different machining operations in a single setup. A single machining center capable of producing a family of parts is the simplest type of cell.

Single Operator, Multiple Machines   In this cell, the machines are usually all single-cycle automatics so they can complete the machining cycle untended, turning off automatically when finished with a cycle.

Single-Robot Automated Cell   To enhance the degree of automation of a cell operated by a single operator, it is possible to design an unmanned cell where a robot

Table 1. Criteria for defining the manufacturing focus of a cell

Variety of processes included
    Casting, forging, plating, etc.
    Machining
    Mechanical assembly
    Electronic assembly
    Inspection
    Heat treatment, welding, etc.
Production volumes
    Low volume, variable demand pattern
    High volume, steady demand pattern
Variety of parts included
    Single part
    Family of parts based on common raw material properties (e.g., size, shape, material)
    Family of parts based on similarity of geometry and/or operation sequences
    Family of parts from different products with dissimilar geometry and/or operation sequences
    Family of parts with common tools and fixtures that can be produced on a machining or turning center
    Identical parts in a range of sizes
    A single assembled product
    A family of products
Market segments
    Single customer or group of customers with same Standard Industrial Classification (SIC) code
    Products within a geographic region
    Products or a set of products within a size or capacity range
    Refurbishing returned products or reworking damaged parts for a specific industry (e.g., automotive or aerospace)
    Spare parts for a specific industry (e.g., valves)
Scope of application
    Factory level
    Department or shop level
    Group of workstations
    Single workstation
    Vendors
    Prototype manufacturing
Degree of automation of activities in the cell
    Manufacturing and assembly
    Material handling
    Inspection
    Loading/unloading
    Scheduling
    Tool changing
    Cleaning and packaging
Part attributes
    Size
    Shape
    Raw material composition, shape, dimensions, etc.
    Raw material properties, e.g., weight, cost, durability, etc.

replaces the single operator. A typical example of such a cell would be where one robot and three Computer Numerical Control (CNC) machines constitute the cell. The robot is employed for material handling, and the machines in the cell have adaptive control capability to allow autonomous operation of the cell.

Multiple Machines, Multiple Operators   For a nonautomated cell, several operators may be required if the operations of tool changing, inspection, workpiece loading/unloading, and the like are all performed manually. A larger cell with several operators provides more flexibility and capacity to produce a larger variety and/or volume of parts.

Multiple Robots, Automated Cell   Several dissimilar conventional machine tools and/or workstations could be arranged in a U-shape layout to allow greater accessibility and variety in the part family. These machines produce a family of parts. The operators are replaced by robots that do the work of tool changing, inspection, workpiece loading/unloading, material handling between machines, and so forth. A typical example of an automated cell would be where two robots, an AGV, and an automated loading system are integrated to form the cell.

Assembly Cell   An assembly cell is composed of several workstations placed in line based on the assembly sequence. An assembly cell is designed for assembling a product or a subassembly. Each workstation may consist of one or more machines and human workers to complete the assembly tasks.

Transfer Line   The transfer line is a product-oriented flow line cell where the machines are arranged according to the product's sequence of operations. The line is organized by the processing sequence needed to make a single product or similar variants of the same product. In a transfer line, the setup times to change from one product to another are often long and complicated. The investment costs of specialized machines and tooling are also quite high.

4. DOMAINS OF APPLICATION OF CELLS IN INDUSTRY

There is a spectrum of possible applications of the cell concept in manufacturing and assembly. Perhaps the primary requirement is the discreteness of the parts and products being produced using multistep process plans. Table 2 lists some typical sectors of industry for which cells are ideally suited.

Cells can be implemented in any manufacturing environment where one or more products may require the same combinations of processes and resources. Typical conditions favorable for the introduction of cells are shown in Table 3.

5. CHARACTERISTICS OF A REPRESENTATIVE CELL

A typical cell can have the characteristics listed in Table 4. However, this list of characteristics can easily change, depending on the company culture, level of

Table 2. Domains of application of cells in industry

Machinery and machine tools
Agricultural and construction equipment
Hospital and medical equipment
Defense products
Automobiles and engines
Piece parts and components
Electronic products
Chemical equipment
Packaging industries

Table 3. Conditions favorable for cellular manufacturing

Small variety of parts in product assemblies
Clearly defined and stable part families
Small number of parts that do not fit any of the families identified
Steady demand pattern and stable delivery schedules for assemblies or parts
Operations sequences of parts have minimum multimachine or nonmachining process
  requirements that cannot be performed within the cell
Subcontractors and raw material suppliers have a reliable delivery capability with small lead
  times
Flexible manufacturing or assemply equipment with low setup-times
Maintenance force is efficient and manufacturing equipment has low downtimes
Skilled and flexible workforce
Reliable and up-to-date part and assembly routings and equipment capabilities
Top management convinced that cellularization is a future direction for the company

automation, demand stability, and other operating or strategic factors unique to the business environment of the company.

6. METHODS FOR CELL DESIGN

The biggest single design problem in changing over to a cellular layout from an existing layout is the problem of grouping parts into families. There are three general methods for solving this problem. All of these methods are time-consuming and involve the analysis of much data by properly trained personnel. These methods are:

- Eyeballing
- Production Flow Analysis (PFA)
- Parts Classification and Coding (C&C)

6.1. Eyeballing

The eyeballing method (also called visual inspection method) is the simplest and least expensive method. It involves the classification of parts into families by looking

Table 4. Characteristics of a representative cell

Cell has a range of values for the number of people (1–6) and workstations (1–10) and is concnetrated in its own bay, in a distinct area with a specific entrance and exit.

Machines are laid out in a U-layout to simplify and minimize material movement.

Most work is completed in the cell but some work is sent to outside vendors or departments within the factory. In some cases, especially job shops, work may be moved in from other sections of the factory particularly at slack times.

Cell has materials, and production control documentation for each job and its own storage area for tools and fixtures.

Cell has assembly or production sequences and routes for all components. All relevant engineering, design work, and support is done by the group of operators in the cell.

Cell does its own inspection and work scheduling.

Assembly and production targets and manpower levels are set by management in consultation with cell operators.

Recruiting of all members is done by management in consultation with cell operators.

Cell leader is appointed by management in consultation with cell operators.

Cell produces a family of similar components that requires similar machines, or workstations, in the cell.

Members are paid on a group incentive scheme.

---

either at the parts themselves or their drawings and arranging them into groups based on general criteria. This method is very limited in scope when dealing with a large number of parts. The accuracy of grouping obtained by this method is generally considered to be the least among the three methods.

6.2. Production Flow Analysis

Production Flow Analysis was first introduced by Burbidge (1961, 1963) and is a method for forming part families and machine groups by analyzing the production process data listed in the route sheets of parts produced in a factory. It groups together the parts that have similar operation sequences. This method requires reliable and well-documented route sheets. Therefore, a drawback of PFA is that it assumes the accuracy of existing route sheets, with no consideration given to whether those process plans are up-to-date or optimal with respect to the existing mix of machines. For a single factory, the classical method of PFA consists of four nested stages where each stage addresses a smaller portion of the entire factory. These four stages are Factory Flow Analysis, Group Analysis, Line Analysis, and Tooling Analysis.

Stage 1: Factory Flow Analysis (FFA)   In this stage, flows due to parts that repeat-edly move to and from between shops (machine shop, forge, foundry, welding, press, assembly, heat treatment) are eliminated by a minor redeployment of equipment. This is done by graphically representing the material flows between the shops as a travel chart to show how the shops are connected. Often FFA may be redundant in a company that has essentially a single large manufacturing facility in a building.

Stage 2: Group Analysis (GA)   In this stage, the flow in each of the shops identified by FFA is analyzed. Since the number of machines in a shop is larger than the number of shops in a factory, a greater variety of flow patterns can be expected. Group Analysis considers the flow interactions among the machines due to the operation sequences of the parts in order to identify manufacturing cells. Loads are calculated for each part family to obtain the equipment requirements for each cell. Each cell usually contains all the equipment necessary to satisfy the complete manufacturing requirements of its part family. Due to equipment sharing problems, some intercell material flows may arise. With GA, there is a significant reduction in material flows within the shop.

Stage 3: Line Analysis (LA)   In this stage, a layout is designed for the equipment assigned to a particular cell. Some typical types of cellular layouts are given in Table 5. Line Analysis considers the operation frequencies and sequences of the parts and plans a cell layout that allows efficient transport to reduce travel times. The layout shape must also encourage multimachine tending by an operator. With LA, a further reduction in material flow is achieved within the cell.

Stage 4: Tooling Analysis (TA)   In this stage, the principles of GA and LA are integrated with C&C data on the shape, size, material, tooling, and setup attributes of the parts. The fixturing, tooling, and setup attributes of the machine tools are correlated with attributes such as the size, shape, material type, and tolerances for the parts. Tooling Analysis deals with the cell scheduling problem for parts with similar operation sequences, tooling, setup, and inspection requirements on the machines in the cell. Its aim is to sequence the parts on each machine tool and to schedule all the parts in the cell such that the throughput times of the parts and machine idle times caused by the transportation of parts and setup changes on the machines are minimized. Hence, with TA there is a further reduction in material flow at the machine level.

6.3.   Parts Classification and Coding

Parts Classification and Coding is a highly time-consuming and complicated activity. This method attempts to group parts with identical or similar design and manufacturing attributes into families. Attributes of a part such as dimensions, shape features,

---

Table 5. Types of cellular layouts

For a single cell
   U-shape layout
   L-shape layout
   S-shape layout
   W-shape layout
For multiple cells
   Cascading cells
   Cells with remainder cells
   Cells with common facilities
   Virtual cells
   Linked cells
   Hybrid flowshops

auxiliary holes, or gear teeth are captured in a code number. The code number for each part provides a compact and consistent description of the attributes of each part. Such numerically processable information serves as a basis for sorting and grouping the parts into families. The cell for each family is identified by matching the attributes of the parts in each family to machine capabilities and available capacity on those machines. Hence, cell formation using C&C requires the analyses used for GA and LA when using PFA to make machine allocation and cell composition decisions.

7. ADVANTAGES OF CELLS

As stated earlier, a manufacturing cell is a group of dissimilar machines located in close proximity and dedicated to the manufacture of a family of parts. The significant benefits gained from the use of cells have encouraged many companies to implement manufacturing cells. The advantages of manufacturing cells can be classified into two categories, as shown in Table 6.

The tangible advantages of manufacturing cells are due mainly to the proximity of all machines required to make a family of parts (Wemmerlöv and Hyer, 1989). This reduces the total distance that must be traveled by the batches of parts in the family. The reduction in lead times between successive operations reduces the average work-in-process levels of the parts as well as the assemblies in which they are used. If the material flow patterns in the cell can be simplified, an efficient layout and suitable handling system can be implemented for the cell. Parts in the family having identical or similar operation sequences can be sequenced consecutively through the cell to minimize setup changeover times. Quality control is simplified since operators at adjacent machines can provide feedback to each other about defective parts. Machine capabilities and the complexity of the parts spectrum that can be handled by the cell are easily defined. This information, when combined with the forecasts of available machine capacities, helps to accept or reject batches of new parts. Cells comprised of conventional machine tools are ideal for replacement by multifunction Computer Numerical Control machining centers if the workload due to the parts using that group of machines is economically justifiable. Hence, manufacturing cells have the capability to minimize the manufacturing throughput times of entire families of parts by reducing the material flow times, machine setup times, and queuing times. Furthermore, the creation of manufacturing cells allows the decomposition of a large manufacturing system into a set of smaller and more manageable subsystems. The material flows of these subsystems can be significantly simplified. Finally, manufacturing cells can be regarded as a first step in the implementation of new manufacturing technologies, such as Computer-Aided Design (CAD), Computer-Aided Manufacturing (CAM), Flexible Manufacturing Systems (FMS), Computer-Aided Process Planning (CAPP), and Concurrent Engineering (CE). This is because cells are a natural form of autonomous manufacturing system in a larger factory that can be automated and controlled by computer.

Manufacturing cells also have several intangible advantages (Vakharia and Kaku, 1993). First, the implementation of manufacturing cells can be regarded as a manage-

Table 6. Advantages of cellular manufacturing

Strategic advantages
   On-time delivery
   Improved response
   Reduced inventory
   Improved quality
   Improved work flow
   Increased operational flexibility
   Improvements in company culture
   Accountability
   Better equipment utilization
   Better use of skilled labor
   Job satisfaction
   Improved information flow
Shop floor advantages
   Speeds throughput in parts manufacturing and assembly
   Reduces work-in-process and finished goods inventory levels
   Reduces material handling and travel distances for parts
   Eliminates non-value-added operations (storage, inspection, and handling)
   Increases capacity by reducing setup times and encouraging group scheduling of parts
   Performs accurate machine and manpower requirements analysis
   Improves quality by reducing scrap and process variations due to better monitoring of operations
   Creates autonomous teams of workers who manage their own cells
   Creates sensible cost centers around each cell's activities and outputs
   Simplifies production and assembly planning and control, scheduling, load balancing, and capacity requirements analysis
   Helps to introduce just-in-time manufacturing to minimize lateness in delivery schedules
   Provides manufacturing capacity and schedule information promptly to sales and marketing personnel

ment philosophy that enforces the pursuit of successful teamwork, use of state-or-the-art manufacturing technology, and promotes the competitiveness of the organization. Second, manufacturing cells can lead to higher motivation and morale due to greater job satisfaction. Third, the conversion to a cellular system has the intangible benefit that it prepares the way for automation. With computer control for the entire CMS, the management and production activities can be accomplished in a more automatic and economical manner.

8. DISADVANTAGES OF CELLS

The implementation of cells could, however, have some disadvantages as compared to the traditional functional and product layouts. The disadvantages can arise from the underlying design characteristics of cells and the limitations of the methods used to design and evaluate the cells. These disadvantages are summarized in Table 7.

As presented in Table 7, cells have some inherent disadvantages. First, their implementation often leads to an increase in investment, as certain machines may need to be duplicated to create independent cells. Consequently, in implementing manufacturing cells, companies weigh the operational benefits such as reduced Work-In-Progress (WIP) inventory and throughput times against the costs of increased investment (Vakharia, 1986). Second, it can be argued that a CMS is less flexible than a functional or process layout. A CMS may lack the ability to completely process a new part within any single cell due to the nonavailability of equipment. Hence, it is generally true that a CMS fails to deal with long-term changes in part demand. Third, another potential disadvantage of manufacturing cells is its lower machine and labor utilization compared to a functional or process layout. To create completely independent cells, some machine types must be duplicated among several cells. This may result in low utilization rates on certain machine types. Sometimes, a load imbalance may occur when some machines in a cell are more utilized than others. It is not possible, however, to operate manufacturing systems where all machines and operators are equally utilized (Butbidge, 1971). Finally, when machine breakdowns occur, the production rate of a cell may be hindered because it lacks more machines of similar function to replace the disabled machine.

Some other disadvantages arise due to the methods available for design and evaluation of cells. First, a primary problem is the absence of a flexible yet simple and comprehensive cell formation method that can rapidly and cheaply identify manufacturing cells for a multiproduct facility. Existing methods, though powerful in their use of analytical tools, have inadequacies such as large data requirements, multiple stages of analysis and indirect or incomplete analysis in critical decision areas. This is complicated by the dynamically changing structure of the manufacturing environment. Next, collection and verification of data, selection of suitable data analysis methods, and interpreting or adapting these results

Table 7. Disadvantages of implementing manufacturing cells

Disadvantages of cells
    Need for high investment in machine installation and relayout
    Lack of flexibility in handling demand changes, product mix changes, infrequent ordering of parts, variable lot sizes, changes in part designs and process plans, improvements in manufacturing technology, etc.
    Imbalance of utilization of machines and labor
    Potential for crippling delays due to machine breakdowns, worker absenteeism, etc.
Implementation problems associated with methods used for cell formation
    Lack of a comprehensive cell formation method
    Data collection and analysis is time-consuming
    Significant difficulty in incorporating the impact of dynamic operational factors into the cell design process

can be time-consuming. Even a pilot study can prove costly. The data retrieved from company archives are often found to be either inaccurate or outdated at later stages of analysis. Finally, most cell formation methods assume that the population of parts and the manufacturing facilities used as input are stable in time. Unfortunately, several dynamic operational factors can influence the machine allocations, part family assignments, and layout for the cells (Wemmerlöv and Hyer, 1987). The cell and family compositions could change on a short- or long-term basis. Breakdowns and labor absenteeism are typical short-term disruptions that can force parts to be moved to machines external to their host cells. Examples of long-term changes are changes in the production technology, modifications in the part designs, rationalization of product designs to eliminate parts previously produced in a cell, and changes in process plans of the parts forced by machine breakdowns.

9. CONCLUSION

Cellular Manufacturing Systems have become a foundation on which a large number of Original Equipment Manufacturers (OEMs) and job shops introduce programs such as setup reduction, just in time, and design for manufacture to improve productivity and operational efficiency. A variety of criteria can be used to define the part family and the manufacturing/assembly focus of a cell. Cells have been implemented in various sectors of industry ranging from furniture manufacturing and welding fabrication to automotive assembly and shipbuilding. While cells provide advantages such as reduced lead times and WIP, they have several disadvantages, especially the inflexibility to react to product mix and volume changes and machine breakdowns. In addition to providing a basis for factory reorganization, the Cellular Manufacturing philosophy and design concepts can be extended to workforce training, equipment design, strategic planning for a manufacturing organization, and problem solving through decomposition.

APPENDIX 1: BOOKS ON GROUP TECHNOLOGY AND CELLULAR MANUFACTURING

Arn, E. A. (1975). *Group Technology: An Integrated Planning and Implementation Concept for Small and Medium Batch Production,* Springer, New York.

Black, J. T. (1996). *The Design of the Factory with a Future,* McGraw-Hill, New York.

Brandon, J. A. (1996). *Cellular Manufacturing: Integrating Technology and Management,* Wiley, New York.

Burbidge, J. L. (1975). *The Introduction of Group Technology,* Wiley, New York.

Burbidge, J. L. (1979). *Group Technology in the Engineering Industry,* Mechanical Engineering Publications, London.

Burbidge, J.L. (1989). *Production Flow Analysis for Planning Group Technology,* Oxford University Press, Oxford, UK.

Demyanyuk, F. S. (1963). *The Technological Principles of Flow Line and Automated Production, Volumes I & II* (English translation), Pergamon Press/ MacMillan, New York.

Dhavale, D. G. (1996). *Management Accounting Issues in Cellular Manufacturing Systems and Focused-Factory Systems,* IMA Foundation for Applied Research, Montvaler, NJ.

Edwards, G. A. B. (1971). *Readings in Group Technology,* Machinery Publications, London.

Gallagher, C. C., and Knight, W. A. (1973). *Group Technology,* Butterworth, London.

Gallagher, C. C. and Knight, W. A. (1986). *Group Technology Production Methods in Manufacture,* Wiley, New York.

Ham, I., Hitomi, K., and Yoshida, T. (1985). *Group Technology: Applications to Production Management,* Kluwer-Nijhoff, Hingham, MA.

Hyde, W. F. (1981). *Improving Productivity by Classification, Coding and Database Standardization.* Marcel Dekker, New York.

Hyer, N. L. (Ed.) (1984). *Group Technology at Work,* Society of Manufacturing Engineers, Dearborn, MI.

Hyer, N. L. (Ed.) (1987). *Capabilities of Group Technology,* Society of Manufacturing Engineers, Dearborn, MI.

Ivanov, E. K. (1963). *Group Production Organization and Technology* (English translation), Business Publications, London.

Kamrani, A. K., Parsaei, H. R., and Liles, D. H. (Ed.) (1995). *Planning, Design and Analysis of Cellular Manufacturing Systems,* Elsevier Science, New York.

Knox, C. S. (1987). *Organizing Data for CIM Applications,* Marcel Dekker, New York.

Lee, R. N. (Ed.) (1992) *Making Manufacturing Cells Work.* Society of Manufacturing Engineers, Dearborn, MI.

Mitrofanov, S. P. (1966). *Scientific Principles of Group Technology, Volumes I, II & III* (English translation), National Lending Library for Science and Technology, Yorkshire, UK.

Moodie, C., Uzsoy, R., and Yih, Y. (Eds.) (1995). *Manufacturing Cells: A Systems Engineering View,* Taylor & Fracis, London.

Muther, R. A., Fillmore, W. E., and Rome, C. P. (1996). *Simplified Systematic Planning of Manufacturing Cells,* Management and Industrial Research Publications, Kansas City, MO.

Petrov, V. A. (1968). *Flowline Group Production Planning* (English translation), Business Publications, London.

Ranson, G. M. (1972). *Group Technology,* McGraw-Hill, London.

Sekine, K. (1990). *One-Piece Flow: Cell Design for Transforming the Production Process,* Productivity Press, Cambrige, MA.

Singh, N., and Rajamani, D. (1996). *Cellular Manufacturing Systems,* Chapman & Hall, New York.

Snead, C. S. (1989). *Group Technology: Foundation for Competitive Manufacturing,* Van Nostrand Reinhold, New York.

Studel, H. J., and Desruelle, P. (1992). *Manufacturing in the Nineties: How to Become a Mean, Lean World-Class Competitor,* Van Nostrand Reinhold, New York.

Suresh, N. C., and Kay, J. M. (1998). *Group Technology and Cellular Manufacturing: State-of-the-Art Synthesis of Research and Practice,* Kluwer, Boston.

APPENDIX 2: COMMERCIAL SOFTWARE FOR DESIGN AND EVALUATION OF CELLS

Group Technology

1. CADIS-PMX
   CADIS, Inc.
   World Headquarters
   5700 Flatiron Parkway
   Boulder, CO 80301
   Telephone: (303) 440-4363
   Fax: (303) 440-5309
   http://www.cadis.com/

2. DCLASS/LOGICTREE
   Cam Software Inc.
   Westpark Bldg. Suit 208
   750 North 200 West
   Provo, UT 84601

3. GTS
   Southwest Research Institute
   6220 Culebra Road
   P.O. Drawer 28510
   San Antonio, TX 78228-0510
   Telephone: (210) 6845111
   http://www.swri.org

Cell Formation

1. SAS
   SAS Institute Inc.
   SAS Campus Drive
   Cary, NC 27513-2414
   Telephone: (919) 677-8000
   Fax: (919) 677-4444
   http://www.sas.com/

2. STORM
   Storm Software, Inc.
   24100 Chagrin Boulevard, Suite 300
   Cleveland, OH 44122-5535
   Telephone: (216) 464-1209
   Fax: +1 (216) 464-4222

Layout Design

1. STORM
   Storm Software, Inc.
   24100 Chagrin Boulevard, Suite 300
   Cleveland, OH 44122-5535
   Telephone: (216)464-1209
   Fax: +1 (216) 464-4222

2. CRIMFLO
   765, rue Notre-Dame Est–Victoriaville
   Québec, Canada G6P 6S4
   Tél.: (819) 758-8219
   Téléc.: (819) 758-5486
   http://www.cgpvicto.qc.ca/
   crimbo/services.htm

3. SPIRAL
   Marc Goetschalckx
   4031 Bradbury Drive
   Marietta, GA 30062-6165
   Telephone: (770) 565-3370
   Fax: (770) 578-6148
   http://www.isye.gatech.edu/
   mgoetsch/index.html

4. MALAGA
   AESOP Corporation
   233 South Wacker Drive
   Sears Tower, Suite 9604
   Chicago, IL 60606
   Telephone: (312) 559-9880
   Fax: (312) 559-9910
   http://www.aesop.de

5. FACTORY FLOW/FACTORY OPT
    EAI/CIMTechnologies Corporation
    2501 North Loop Drive, Suite 700
    Ames, Iowa 50010-8285
    Telephone: (515) 296-9914
    Fax: (515) 296-9909
    http://www.cimtech.com/

Performance Analysis

1. MPX
    Network Dynamics Incorporated
    128 Wheeler Road
    Burlington, MA 01803
    Telephone: (617) 270-4120
    Fax: (617) 270-4119

2. PROMODEL
    PROMODEL Corporation
    1875 S. State Street
    Suite 3400
    Orem, UT 84097
    Telephone: (801) 223-4600
    Fax: (801) 226-6046
    http://www.promodel.com/

3. ARENA
    Systems Modeling Corporation
    504 Beaver Street
    Sewickley, PA 15143
    Telephone: (412) 741-3727
    Fax: (412) 741-5635
    http://www.sm.com/

4. PROVISA
    LANNER GROUP INC.
    Suite 680
    11000 Richmond Avenue
    Houston, TX 77042
    Telephone: (713) 532-8008
    Fax: (713) 532-3732
    http://www.lanner.com/

5. CELLMASTER
    CMS Research Inc.
    627 Bayshare Drive
    Oshkosh, WI 54901
    Telephone: (920) 235-3356
    Fax: (920) 235-3816
    http://156.46.191.15/ cmsves/

products.html

6. AUTOSCHED
    AutoSimulations, Inc.
    655 E. Medical Drive
    Suite 200
    Bountiful, UT 84010
    Telephone: (801) 298-1398
    Fax: (801) 298-8186
    http://www.autosim.com/

7. MATFLOW
    LANNER GROUP INC
    Suite 680
    11000 Richmond Avenue
    Houston, TX 77042
    Telephone: (713) 532 8008
    Fax: (713) 532-3732
    http://www.lanner.com/

8. FACTOR/AIM
    Pritsker Corporation
    8910 Purdue Road, Suite 600
    Indianapolis, IN 46268-1170
    Telephone: (800) 428-7636
    http://www.pritsker.com

9. SIMUL8
    Visual Thinking
    55A Part Street East
    Missisauga, Ontario L56 4PG
    Canada
    Info@visualt.com
    http://www.visualT.com

Scheduling

1. PREACTOR
   Preactor Inc.
   18 Blooms Corners Road
   Warwick, NY 10990
   Telephone: (888) PREACT1
   http://www.preactor.com

2. TEMPO
   Systems Modeling Corporation
   504 Beaver Street
   Sewickley, PA 15143
   Telephone: (412) 741-3727
   Fax: (412) 741-5635
   http://www.sm.com/

3. MIXED MODEL SCHEDULER
   Kiran and Associates
   4411 Morena Blvd., Suite 105
   San Diego, CA 92117
   Telephone: (619) 270-9950
   http://www.kiran.com

Production Control

1. THRUPUT
   Thru-Put Technologies
   2115 O'Nel Drive
   San Jose, CA 95131
   Telephone: (408) 920-9711
   Fax: (408) 920-9715
   http://www.thru-put.com/

2. SHIVA
   455 Phipps McKinnon Building
   10020 101A Avenue,
   Edmonton, Alberta
   Canada T5J 3G2
   Telephone: 403-496-7495
   FAX: 403-423-0414
   http://www.shivasoft.com/

3. AWESIM
   Pritsker Corporation
   8910 Purdue Road, Suite 600
   Indianapolis, IN 46268-1170
   Telephone: (800) 428-7636
   http://www.pritsker.com

APPENDIX 3: SUPPORTING ACTIVITIES FOR THE INTRODUCTION OF CELLS

Product Design

- Standardize part geometries, tolerances, dimensions, etc. to reduce or eliminate tools, fixtures, gauges, setup adjustments, etc.
- Introduce new designs for parts such that all their operations can be performed in any one cell.

- Perform ABC analysis to subcontract or purchase parts that have low value or infrequent demand or belong to low-value products.
- Use principles of design for manufacture and assembly to design products with reduced parts count.
- Modify part designs to eliminate or combine features that would require special operations.
- Strive to design new products in the future with modularity in mind whereby each module could be fabricated and assembled in a single cell.
- Maximize use of common parts in different products as much as possible.
- Modify product designs to permit assembly without tools and operations such as riveting, screwing, joining, etc.

Manufacturing Engineering

- Revise process plans to eliminate exception operations or combine them with operations already done within a cell.
- Revise process plans to reduce the number of different workstations that a part must visit.
- Change machine selections in process plans to use machines located in any one cell.
- Acquire multifunction automated machines to replace a set of two or more existing conventional machines.
- Pick new manufacturing technology, ex. Net Shape Manufacturing (NSM) processes, to eliminate or reduce the number of operations for a part, to absorb processes external to a cell into the cell, etc.
- Evaluate and optimize manufacturing process parameters on all workstations to reduce operation times, operator skill, setup times, etc.
- Select manufacturing equipment, inspection equipment, fixtures, etc. so as to avoid too many variations in make, model, and/or supplier, thereby simplifying maintenance and operation in different cells.
- Select manufacturing and test equipment and fixtures to support a setup reduction program concurrent with the cell implementation project.
- Analyze process plans to eliminate operations made necessary by previous operations, e.g., deburring, rust removal, degreasing, scale removal, etc.
- Analyze the process plans for similar parts to check for special operations made necessary because of the requirements of only one or a few products (or customers).
- Reduce setup times on key machines in order that they can be run for shorter periods to make smaller batches of parts to match the demand profiles.

Plant Layout

- Determine how much of the shop floor will be converted to cells.
- Allocate areas on the shop floor to each group so as to reduce relocation of heavy equipment.
- Plan relative positions of cells and the configuration of aisles connecting them to link service areas such as heat treatment, welding, and inspection in order to encourage smooth and rapid material flows between the cells.

- Evaluate the existing material handling system for its ability to support rapid intercell flows of multiple small batches.
- Plan the overall layout based on those cells that share machines and service areas.
- Adjust the shapes of the cells to conform to the existing shop floor boundaries and intercell flow patterns.
- Locate the storage areas, tool cabinets, buffer areas, inspection tables, etc. in a cell for ease of material flow in the cell.
- Coincide the replacement of existing equipment with the purchase of multifunction manufacturing equipment to reduce the cost and time for relayout of the existing facility layout.
- Plan the layout with future growth and/changes in demands as well as changes in part mix.

Materials Handling

- Schedule the supply of raw materials, tools, gauges, etc. to each cell prior to each shift.
- Schedule to-and-from trips between cells giving priority to parts requiring intercell flows.
- Plan for the rapid removal of finished batches of parts to the shipping department to prevent clutter inside a cell.
- Select suitable handling equipment for intermachine transfers of parts within a cell or between cells.
- Select devices for automatic machine loading and unloading to allow a single operator to tend to several machines.
- Determine whether WIP storage will be stored within the cells or in a central area.
- If the cells are designed to make and assemble a product or subassembly, link these cells to conform to the product bill of material and assembly precedence diagram.
- Develop flowcharts for all material handling operations required for all parts to detect wasteful or idle times in their cycle times.
- Combine operations to eliminate material handling at a single workstation, in a cell, and between cells.
- For loading/unloading and handling of large parts, evaluate the use of mechanical handling devices such as hand trucks, electric trucks, overhead traveling crane, and waist-high portable benches or carts.
- Design special racks or trays to permit easier material handling, to minimize damage to parts during storage or handling, to regulate work-in-process inventory buffers between workstations, etc.
- Eliminate storage space external to a cell and have the parts delivered *directly* to the cell.
- Select racks and shelving to allow vertical storage and stacking of material with different shapes or sizes to save floor space and to encourage smooth material flows and operator movements within the cell.
- Purchase materials or deliver materials in quatitiies suitable for manual handling in the cell.
- Design the capacity of the storage containers based on production cycle times, safety stock levels for when machines break down, etc.
- Use lights or bells to notify material handlers that parts are ready for any intercell moves.

- Use several forklift trucks or a tractor-trailer (or AGV) with a variable schedule for any intercell moves.

Quality Control and Inspection

- Train cell personnel to prepare and analyze Statistical Process Control (*SPC*) charts for key machines whereby they inspect parts that they make at the same station.
- Perform inspection tasks inside the cells instead of sending parts to a central inspection department external to the cell.
- Develop standard procedures to troubleshoot any quality problems indicated by *SPC* charts.
- Determine sampling plans and develop reliable processes instead of resorting to 100% inspection.
- Train all cell operators to be aware of the critical operations for each and every part assigned to the cell.
- Analyze the operation sequence for each part using tolerance charting techniques to measure the effect of the quality of each operation on all succeeding operations.
- Seek to improve the quality of key operations by using modern processes and online gauging.
- Reduce the number of unique gauges and inspection operations required for all parts produced in the cell.
- Eliminate "special" inspection operations required by a particular customer, if possible.
- Find the cause of a defect and correct it immediately.
- Maintain process flowcharts for inspection and modification of process settings.

Production Control and Scheduling

- Make periodic short-term forecasts by product type.
- Ensure reliable and accurate forecasts for parts and subassemblies.
- Encourage customers to accept a steady delivery schedule.
- Try to combine due dates and batches of several customers to reduce setup changeover losses and obtain quantity discounts, respectively.
- Reject orders that require manufacturing capabilities that are not available in any one cell.
- Determine the set of parts to be made on each day based on finite capacity constraints.
- Schedule machines, operators, and intercell flows for on-time delivery.
- Sequence jobs in the queue at each machine using correct dispatching rules.
- Account for machine breakdowns and changes in part mix or order quantities in the daily cell schedule.
- Use simulation or Rapid Modeling Techniques (RMT) software to evaluate, fine tune, and plan the performance of each cell and intercell flows.
- Utilize idle times in the cell schedule to produce orders that are not requested frequently.
- Change the machines on which some operations are performed to utilize the underloaded machines better and to reduce queue sizes, WIP inventory, etc.
- Produce all items required in a family together, if possible.

- Design part families within product BOM's (BOM - Bill of Material) and not across product BOM's.
- Utilize Material Requirements Planning (*MRP*) for forecasting (PUSH) and just-in-time (JIT) for actual cell scheduling (PULL).

Preventive Maintenance

- Ensure that cell operations experience least disruptions due to interruptions in supply of cutting oil, lubricants, gas, compressed air, water, electricity, etc.
- Study the possibility of reducing the number of makes and models of equipment to reduce difficulties arising from nonavailablity of spare parts, differences in maintenance procedures, etc.
- Train operators to perform preventive maintenance in the cell.
- Create documentation for preventive maintenance tasks.

Setup Reduction

- Analyze the changeover procedures on key machines in the cell to identify and eliminate wasted times.
- Convert internal setups (process must be stopped to perform the changeover) into external setups (changeovers while process is running).
- Use preset tooling, quick change flexible fixtures, group tooling, etc. to eliminate machine stoppages.
- Sequence jobs with similar setups consecutively to reduce setup changeovers.

Purchasing and Subcontracting

- Plan for on-time deliveries by vendors.
- Ensure that parts are delivered in desired quantities only.
- Ensure acceptable quality of incoming batches of parts or raw materials.
- Ensure flexibility of vendors to accommodate variety in raw material types, order quantities and frequency of deliveries.
- Reduce the number of unique new materials and parts through standardization.
- Encourage vendors to perform work upon raw materials supplied to the cells to eliminate preparatory operations.
- Order raw materials in amounts and sizes that permit their utilization by more than one cell to minimize scrap and cutting stock waste.
- Reduce the number of vendors and ensure steady business for the selected ones to ensure consistent incoming quality of material to the cells.

Cell Team Development

- Select a group of cell operators who have compatible personalities and who will promote a cooperative working atmosphere.
- Let cell personnel have a say in selection of new operators.

- Select cell foreman with multimachine technical and maintenance skills and leadership qualities.
- Schedule a cell to balance operator work loads.
- Assign operators to machines based on skill levels determined by part tolerances, level of equipment automation, desired production rates, etc.
- Maintain flexibility in assignment of operators to machines depending on absenteeism, proximity of machines, overlaps in operation cycles, and complexity of parts in queue at each machine.
- Train and orient new members to cells in advance if there is significant labor turnover.
- Document quality problems, absenteeism, preventive maintenance, time standards, order cost estimates, and cell performance on a continual basis.
- If the work culture is appropriate, install a gain sharing plan or group incentive scheme that rewards the employees for helping to meet the cell's objectives.
- Recruitment and training of cell operators and foremen should seek to develop a flexible and cross-trained workforce with an emphasis on job enrichment.

Housekeeping within a Cell

- How are jobs assigned to the operator?
- Are schedules such that an operator is ever without a job to do?
- How are instructions imparted to the operators?
- How is material stored?
- How are drawings and tools stored?
- How are times at which a job is started and finished checked?
- What delays occur when releasing jobs to the cell?
- If the operator does his/her own setup, would economies be gained by providing special setup people?
- Is the layout of tool cabinets and storage racks orderly so that no time is lost searching for tools or equipment?
- Are the setup instructions and material handling equipment provided to the operators adequate?
- Are machines set up properly?
- Are machines adjusted for proper feeds and speeds?
- Are machines in good repair?
- Does the workplace layout conform to the principles of effective workplace layout design?
- Are preset tools, gauges, parts feeding, and workholding devices adequate?
- Are the first few pieces produced on any machine checked for correctness by anyone other than the operator?
- If an operation is performed continuously, are preliminary operations of a preparatory nature necessary at the beginning of the shift?
- Are adjustments to equipment for a continuous operation made by the operator?
- How is the material supply replenished?

- If a number of miscellaneous jobs are done, can similar jobs be grouped to eliminate certain setup elements?
- How are partial setups handled?
- Are benches or tables of proper design?
- Who maintains the machines?
- Is adequate provision made for transfer of responsibility for tools, setup, etc. from one operator to another between shifts without undue loss of time?
- Does variation in quantities ordered justify different setups on the machines?
- Are charts of feeds, speeds, etc. for machines available in the cell?
- Can a trip to return tools to the tool crib be combined with a trip to get tools for the next job?
- What is done with scrap, or defective parts?

Top Management Support

- One or more members of management must become conversant with Cellular Manufacturing to lead and coordinate the cell design and implementation project.
- Clear reasons, goals, and objectives for implementation of cells must be identified.
- Support to the team members of the cell design and implementation team must be shown by way of release hours, inviting consultants for in-house training, granting autonomy to team members to make decisions, etc.
- Establish mechanisms for two-way communication between management and the workforce.
- Encourage suggestions from employees.
- Display key progress milestones achieved by the cell design and implementation team/s.
- Mount notice boards at strategic locations in the company on various topics such as Total Productive Maintenance, Statistical Quality Control, ISO 9000, Just-In-Time, Total Quality Management, Continuous Improvement, etc.
- Seek to revamp the organizational structure of the company to eliminate departmental barriers and multiple levels of information flow.
- Encourage team appraoches to projects.

REFERENCES

Burbidge, J. L. (1961). The new approach to production, *Production Engineer,* **40**(12), 769-793.

Burbidge, J. L. (1963). Production flow analysis, *Production Engineer,* **42**(12), 742-752.

Burbidge, J. L. (1971). Production flow analysis, *Production Engineer,* **50**(4), 139-152.

Burbidge, J. L. (1975). *The Introduction of Group Technology,* Wiley, New York.

Ham I., Hitomi, K., and Yoshida, T. (1985). "Layout Planning for Group Technology," in *Group Technology: Applications to Production Management,* Kluwer-Nijhoff, Hingham, MA, pp. 153-169.

Mitrofanov, S. P. (1966). *The Scientific Principles of Group Technology,* National Lending Library for Science and Technology, Boston Spa, Yorkshire, England.

Shunk, D. L. (1987). "Computer Integrated Manufacturing" in *Manufacturing High Technology Handbook,* D. Tijunelis and K.E. McKee (Eds.), Marcel Dekker, New York, pp. 83-100.

Vakharia, A. J. (1986). Methods of cell formation in group technology–A framework for evaluation, *Journal of Operations Management,* **6**(3), 257-272.

Vakharia, A. J., and Kaku, B.K., (1993). Redesigning a cellular manufacturing system to handle long-term demand changes: A methodology and investigation, *Decision Sciences,* **24**(5), 909-930.

Wemmerlov, U., and Hyer, N. L. (1987). Research issues in cellular manufacturing, *International Journal of Production Research,* **25**(3), 413-431.

Wemmerlov, U., and Hyer, N. L. (1989). Cellular manufacturing in the U.S. industry: A survey of users. *International Journal of Production Research,* **27**(9), 1511-1530.

1

INVESTIGATION OF CELLULAR MANUFACTURING PRACTICES

Ronald G. Askin and Steve Estrada

Systems and Industrial Engineering Department
College of Engineering and Mines
University of Arizona
Tucson, Arizona 85721

1. INTRODUCTION

Group Technology (GT) implements the general philosophy that similar things should be performed in a similar manner (Askin and Standridge, 1993). Lessons learned from one activity should be exploited for their value in improving the efficiency of related activities. When lessons can be shared among multiple activities, the fixed cost per unit of investing in improved methodologies is reduced. Thus, it becomes feasible to invest resources to develop and implement optimal operating procedures with low variable cost and time requirements. Applied to manufacturing, at the design stage GT leads to the use of common, modular components in final products, similar process plans for similar components, and similar assembly sequences for products. Group Technology breaks down the product set into composite product families. Within each family, products are manufactured using similar processes and procedures. On the production floor, manufacturing cells composed of heterogeneous machines are constructed for these product families. Cellular Manufacturing (CM) naturally results

Handbook of Cellular Manufacturing Systems, edited by Shahrukh A. Irani
ISBN 0-471-12139-8 © 1999 John Wiley & Sons, Inc.

from the shop floor implementation of GT. Cellular Manufacturing attempts to lower part inventories, reduce material handling, simplify employee training, and reduce setup times. The simplified operations planning requirements of CM systems leads to improved operating efficiency.

Group Technology has proven to be very successful when implemented properly. Prior studies (Pullen, 1976; Houtzeel and Brown, 1984; and Wemmerlov and Hyer, 1989) have shown the following dramatic improvements:

1. Throughput time (5–90%)
2. Work-in-process inventory (8–80%)
3. Materials handling (10–83%)
4. Job satisfaction (15–50%)
5. Fixtures (10–85%)
6. Setup time (2–95%)
7. Space needed (1–85%)
8. Quality (5–90%)
9. Finished goods (10–75%)

The above advantages essentially lower manufacturing cost and produce a higher quality product. This is what makes GT so attractive. Group Technology is intended for medium variety and medium volume production environments. However, manufacturing cells can be used whenever short sequences of processing steps are found with sufficient demand volume to justify dedicated equipment.

This chapter documents the findings of a recent survey on CM practices. A variety of companies that have implemented manufacturing cells were surveyed. The project was designed to update and extend previous empirical studies of cellular manufacturing effectiveness. In addition to providing an update on the performance of cells, the project was designed to learn the motivational factors behind CM implementation, the common practices used in CM, and to identify the needs for future research and development.

2. SURVEY DESIGN AND ANALYSIS

The survey instrument evolved through preliminary discussions with potential respondents. The intent was to discover how cells are being used, how well they are working, and what we still need to learn in order to make cells work better. The survey was designed to obtain information on (1) motivational factors for CM implementation; (2) benefits and adverse effects obtained by implementing CM; (3) the size, flow pattern, and level of automation in cells; (4) problems found in trying to implement cells; (5) team autonomy of the cells; and (6) remaining open research questions. It is our hope that the results of this survey will aid practitioners in deciding whether to construct cells and in improving the performance of their existing cells and guide researchers in their definition of problems to be addressed.

Over 50 companies were identified that had implemented cells by scouring the literature and interviewing researchers in the area. Each company was contacted by mail and with several follow-up phone calls to request their participation. Overall, we were able to receive 30 responses from companies throughout the United States. The companies that took part in this survey ranged in size from small shops to large plants of multinational firms. They represent a wide range of industries but with particular representation from machine tool, defense, and automotive industries. In the remainder of this section we analyze the data collected in each section of the survey and describe the results.

2.1. Motivation for Cells

The survey contained a list of what we believed, based on the general literature and prior surveys, to be the top 15 motivations for implementing CM. Respondents ranked each of these on a scale of 1 to 5, with 5 being very important and 1 being not important. The list of factors along with the average score and standard deviation across all 30 companies is shown in Table 1. The factors are ordered based on the number of companies that assigned a value of 4 or 5 to that factor, thus indicating a high level of importance. Using the larger standard deviation of 1.45, a 95% confidence interval can be estimated by adding or subtracting 0.5 from the values in the table. Thus, a significant difference exists between the factors appearing at the top, middle, and bottom of the table. The exact number of 4 and 5 responses out of 30 companies is shown in Figure 1 for each factor. From the table and figure it is clear that reducing cycle time is a primary motivator along with improving product quality, reducing Work-In-Progress (WIP), and reducing material handling. Twenty-five of the 30 companies assigned high importance to reducing cycle time. Using a binomial model, this gives a proportion of $\hat{p} = 0.833$ with a two standard deviation confidence interval

Table 1. Motivating factors for implementing cells

Motivation	Average Score	Standard Deviation
1. Reduce cycle time	4.40	1.00
2. Improve product quality	4.17	1.18
3. Reduce WIP	3.90	1.24
4. Reduce material handling	3.86	1.25
5. Improve shop floor control	3.41	1.38
6. Improve team work	3.38	1.27
7. Reduce scrap/rework	3.57	1.37
8. Reduce labor cost	3.38	1.40
9. Organize activities by final product	3.29	1.36
10. Reduce indirect cost	3.43	1.43
11. Reduce setup time	3.22	1.22
12. Simplify production planning	3.18	1.42
13. Flexibility to produce more	3.24	1.45
14. Facilitate product variation	2.59	1.34
15. Simplify supervision	2.79	1.24

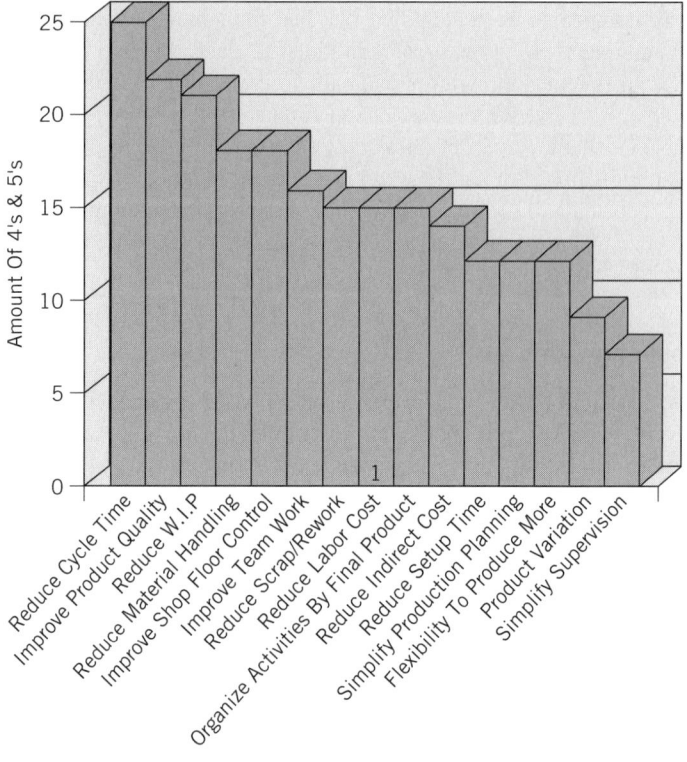

Categories For Motivation

Figure 1. Motivation for cells.

of $\hat{p} \pm 2\sqrt{\hat{p}(1 - \hat{p})/30} = 0.833 \pm 0.136$. Thus, we may conclude with 95% confidence that at least 70% of firms find reducing cycle time to be a primary justification for considering cells. The ability to reduce setup times through generic family setups has been widely discussed in the literature as an advantage of GT/CM (Shafer and Rogers, 1991). However, setup time reduction was viewed as being of less importance to the industrial respondents. Problems due to managing a complex system seem to be relatively unimportant motivators with simplify production planning, production variation complexities, and simplify supervision all appearing at the bottom of the list. Only six companies chose cells in order to simplify supervision of the shop. This translates to an upper 95% confidence limit of 35% of companies.

In examining the responses, we noted some correlation between motivations. Companies concerned with reducing scrap/rework seemed to have a prior belief that this could be achieved by better teamwork. Likewise, these companies also wanted to improve overall product quality. Companies desiring to increase their flexibility to produce more also wanted to improve their shop floor control and to increase their product variety.

2.2. Benefits and Adverse Effects

This next section of the survey investigated the observed benefits and adverse effects experienced in 12 areas when implementing CM. The survey allowed for the respondents to indicate the percentage increase or decrease in each area. The factors and a summary of results are provided in Table 2. The results can be seen in the last four columns indicating the percent of companies experiencing an increase and decrease, and the conditional average percentage change for those companies. In the area of tooling cost, for example, the table indicates that 69% of the population witnessed an increase, and the average increase was about 17%. On the other hand 31% witnessed a decrease in tooling, and the average decrease was about 10%. In some cases, companies were only able to indicate an increase or decrease without providing percentage changes. The values reported are based on all data reported.

Table 2 documents that companies implementing cellular manufacturing have a very high probablility of obtaining improvements in human resource management (labor cost, worker satisfaction, and turnover), quality, setup time, cycle time, and WIP. Tooling cost showed a minor increase in approximately two-thirds of the companies. Machine utilization and subcontracting showed mixed results, but over one-half of the companies were able to obtain an average reduction in subcontracting of 50% and an increase in machine utilization of 33%. Eighty percent of companies found an average cycle time reduction of 40%. As this was the primary motivator for CM, implementation would appear to be a success. Setup time and WIP inventory showed similar improvements. For WIP, this should be expected as a consequence of Little's law (Little, 1961). For setup time, this should be expected due to the economically induced relationship between setup time and batch size, and the subsequent impact of batch size on total processing time and batch arrival rate. While the reductions in cycle time possibly occurred through direct attention since this was the number one motivator for cells, the improvements in setup may well be endemic to cells

Table 2. Impacts experienced from implementing cells

Measure	Percent Experiencing Increase	Average Percentage Increase	Percentage Experiencing Decrease	Average Percentage Decrease
Tooling cost	69	17	31	10
Labor cost	9	25	91	33
Setup time	16	32	84	53
Cycle time	16	30	84	40
Machine utilization	53	33	47	20
Subcontracting	43	10	57	50
Product quality	90	31	10	15
Worker satisfaction	95	36	5	—
Space utilization	83	40	17	25
WIP inventory	13	20	87	58
Labor turnover/absenteeism	0	—	100	50
Variable production cost	7	10	93	18

since these improvements came despite not being a high priority. Improved product quality was a key motivator and, indeed, 90% of companies saw an improvement. However, the 31% improvement was not as substantial as that experienced in some other areas. Finally, the chance to reorganize seems to have been useful for improving space utilization in most cases. Comparing these results to those of other studies, we see that our results tend to concur with those of earlier studies. In summary, we note that when implementing CM, benefits are likely, significant, and spread across many areas, but clearly do vary from company to company.

Sample correlations were computed from the data to determine the tendency for savings in one area to have spillover effects in other areas. (Note that based on a sample of size 30, a 95% confidence interval for a correlation would be ± 0.36 approximately. However, not all companies provided numerical values for all categories, sometimes opting to only indicate the direction of change. Values were computed only using those companies providing quantitative estimates.) As expected, a strong correlation (91%) was found between setup time and cycle time reduction. Clearly the data is indicating that setup time reduction is a key for reducing cycle time as has been widely claimed in the literature. To the extent that the formation of cells facilitates setup time reduction, cycle time benefits will also accrue. As would be expected from Little's law, cycle time also showed a high correlation (91%) with WIP inventory. Surprisingly, as labor cost was reduced, worker satisfaction increased. On the other hand, product quality seemed unrelated to worker satisfaction (an insignificant 12% sample correlation). The data was less definitive due to incomplete responses, but total variable production cost had a correlation coefficient above 50% with every other factor except tooling. The highest correlations were with setup time, machine utilization, and product quality.

2.3. General Cell Information

The types of cells used in CM will depend greatly on a number of issues such as level and stationary of demand, similarity of processing sequences and relative utilizations of part types, ability to share tooling, and economics of automation. Cell structures can be used in manual assembly and automated parts fabrication. In general, we envision a family of manufactured part types with common processing sequences, tooling, and relative utilizations having sufficient demand to warrant a specially designed flow line cell. However, the term cell is used in other contexts as well, including a small automated cell wherein a robot transfers parts between Computer Numerical Control (CNC) machines and the case of manual assembly and light manufacturing environments where workers tradeoff tasks such as sewing and stitching in the textiles industry.

To better understand the types of cells used in practice, data was collected on the types of cells used by the respondents. Cells were classified based on material flow patterns and level of automation. Data was obtained on a total of 392 cells. The total number of cells falling into each combined class is shown in Table 3. Manual cells with flexible capability for general material flow between machines is the most common cell type representing 28% of the total. Overall, 55% of the cells were manually operated, labor-intensive cells. At the other end, only 5.6% of the cells were highly automated flexible cells with integrated material handling. Forty-seven percent of cells allowed general flow patters, 44% of cells had unidirectional flow paths, and just under 10% of cells were logical in structure with spatially separated machines. A chi-square goodness-of-fit test fails to reject the hypothesis that the level of automation and flow pattern are uncorrelated.

2.4. Cell Design and Implementation Problems

The cell design part of the survey was to determine what issues in the implementation of CM created the most severe problems. In the survey we listed what we believed to be the top 10 implementation problems in CM. As with the prior lists of options, these were based on our reading of the general literature in cellular manufacturing and our own experience in talking with industrial companies. The 10 roadblocks to implementing cells were rated based on the severity of the problem with 5 being very severe and 1 being not important. Graphing the responses (Fig. 2) in a similar fashion as motivation for cells, we were able to come up with the list that gave companies the greatest problems. Average scores are shown in Table 4. The factors are ordered based on the number of companies that assigned a value of 4 or 5 to that factor, thus indicating a high level of importance.

First, note that relative to motivation, problems proved to be less significant in that the highest rated factors had lower scores. Whereas the top four motivators received average responses of 3.8 or above, the most significant problem rated only a 3.5 on average. Over half of the companies assigned high importance to preexisting machine location. Using a binomial model, this gives a proportion of $\hat{p} = 0.524$ with a two standard deviation confidence interval of $\hat{p} \pm 2\sqrt{\hat{p}(1-\hat{p})/21} = 0.524 \pm 0.218$. Thus, we may conclude with 95% confidence that at least 30% of firms find preexisting machine location to be a primary implementation problem for CM. In a relatively tight group following preexisting machine locations (no significant difference between their average importance values) lies the need to upgrade human skills, accounting, and scheduling systems, economically equating cell capacity with product demand requirements, and resistance to change among line workers. Perhaps the most profound finding of the study is the relative unimportance of problems in identifying part families and machine groups. Researchers should clearly receive the message that while the vast majority of past research has dealt with forming these part families and machine groups, either this problem is now adequately solved or it never really existed in the first place. It seems the time to move on to procedures

Table 3. Summary of existing cells

Flow\Automation	Manual	Automated Machines	Automated and and Integrated
Flow line	88 (22%)	69 (18%)	14 (3.6%)
General flow	108 (28%)	72 (18%)	4 (1.0%)
Logical cell	21 (5.4%)	12 (3.1%)	4 (1.0%)

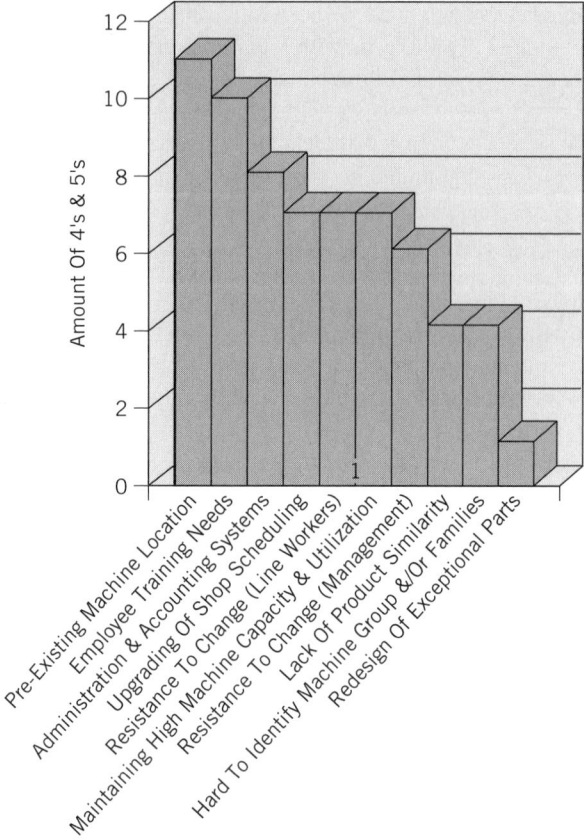

Categories Of Problems

Figure 2. Severity of problem.

that accept initial configurations and account for existing accounting and scheduling systems is here.

As expected, companies that faced a lack of product similarity, had trouble with identifying machine groups and part families. However, this did not appear to measureably impact the benefits obtained. Instead, these companies resorted to increased subcontracting.

2.5. Level of Team Autonomy

It has often been stated (see, e.g., Burbidge, 1975) that CM groups or teams should have a high level of autonomy. Teams should be responsible for their performance and have the authority to effect performance. In this section of the survey we wanted to know how much freedom was actually given to teams. The respondent was to indicate the responsibility level for each activity listed. The allowable responses for

Table 4. Severity of problems in cell implementation

Problem	Average Score	Standard Deviation
Preexisting machine location	3.50	1.28
Employee training needs	3.15	1.04
Administrative and accounting systems	2.95	1.47
Upgrading of shop scheduling	2.95	1.47
Maintaining high machine utilization	2.85	1.39
Resistance to change (line workers)	2.81	1.21
Resistance to change (management)	2.55	1.43
Identify machine groups and part families	1.93	1.21
Lack of product similarity	2.30	1.34
Redesign of exceptional parts	1.89	1.36

responsibility levels were full, partial, and no responsibility. Table 5 summarizes our findings showing the percentage of respondents assigning the three levels of responsibility to the cell team for each activity.

Worker control and empowerment has been assumed to be a key aspect of CM. It has been widely stated in the management literature that cells should have the authority to set there own policies and not be constrained from achieving their goals by management or other organizational entities. In practice, it appears that cell workers are in general collectively responsible for quality and productivity (worker assignment, scheduling, and methods improvement) but not hiring or purchasing. In addition, group incentives have not been used extensively. Referring back to Table 2, we see that this semiautonomy strategy has been effective.

2.6. Issues on Research

The final section of the survey was a free response in order to see where companies would like future research directed. Respondents indicated a desire for better coding systems, software packages for identifying common part routings, algorithms for

Table 5. Percentage of respondents assigning responsibility to team

	Full Responsibility (%)	Partial Responsibility (%)	No Responsibility (%)
Hiring	0.0	26.3	73.7
Scheduling	26.3	26.3	47.4
Worker assignment	31.6	47.3	21.1
Methods improvement	68.4	26.3	5.3
Job sequencing	22.2	66.7	11.1
Quality assurance	88.9	11.1	0.0
Purchasing	0.0	44.4	55.6
Salary and bonus based on team performance	0.0	22.2	88.9

scheduling parts, and methods to form worker teams. It was not clear from the responses to what degree the companies had taken advantage of scheduling and cell formation algorithms available in the literature.

3. SUMMARY

What seems to motivate companies to implement CM is that they want a high-quality product produced quicker at a lower manufacturing cost. Cellular Manufacturing can aid in accomplishing these objectives if implemented correctly. The survey has shown that CM can successfully reduce manufacturing cost while increasing quality through efficient manufacturing techniques. Cells were successfully formed to combat these problems without making machine utilization the center of attention. It was also found that the major research issue of the past, namely, cell formation, should be replaced by methods for cell scheduling, improving information and administrative systems, better part coding systems, and dealing with human resource issues.

Acknowledgments The authors wish to express their appreciation to Dr. Shah-rukh Irani for his assistance in identifying companies to take part in this survey. This material is based on work supported by the National Science Foundation under grant No. DDM 92–15432.

REFERENCES

Askin, R. G., and Standridge, C. R. (1993). *Modeling and Analysis of Manufacturing Systems*, Wiley, New York.

Burbidge, J. L. (1975). *An Introduction to Group Technology*, Wiley, New York.

Houtzeel, A., and Brown, C. S. (1984) "A Management Overview of Group Technology," in *Group Technology At Work*, N. L. Hyer (Ed.), Society of Manufacturing Engineers, Detroit, MI, pp. 3–16.

Little, J. D. C. (1961). "A Proof for the Queueing Formula L=xW," *Operations Research*, **9**(3), 383–385.

Pullen, R. D. (1976). A survey of cellular manufacturing cells, *The Production Engineer,* **55**(9), 451–454.

Shafer, S. M., and Rogers, D. F. (1991). A goal programming approach to the cell formation problem, *Journal of Operations Management*, **10**, 28–42.

Wemmerlov, U., and N. L. Hyer (1989). Cellular manufacturing practices, *Manufacturing Engineering*, **102**(3), 79–82.

2

MACHINE–COMPONENT MATRIX CLUSTERING METHODS FOR CELL FORMATION

T.T. Narendran and G. Srinivasan

Division of Industrial Management
Department of Humanities and Social Sciences
Indian Institute of Technology
Madras 600036 India

1. INTRODUCTION TO THE CELL FORMATION PROBLEM

Cellular Manufacturing (CM) is a strategy that divides a production system into small groups or cells such that each cell can produce a family of parts or components completely. It is an application of a well-known philosophy called Group Technology (GT) (Burbidge, 1971). *Group Technology* has been defined as *a realization that it is possible to divide a large problem into manageable groups and solve it efficiently.*

In CM, we divide the machines and parts (or components) into groups such that all processes to complete the parts in a family are available in the machine cell. In situations where such grouping is not possible, parts have to visit more than one machine cell resulting in intercell moves. When an existing process layout is being reorganized to a cellular system, the objective is to minimize the intercell moves because these represent the additional capacity (machines) required in cells to achieve independence. If we are building a cellular system from the beginning, product-based

Handbook of Cellular Manufacturing Systems, edited by Shahrukh A. Irani
ISBN 0-471-12139-8 © 1999 John Wiley & Sons, Inc.

cells are usually created so that the output from a cell is a complete product or assembly and not a set of related parts.

Reorganizing a functional layout to a cellular system is done using the route cards of components. From this, the machines visited by each component are known and can be read in the form of a matrix called *machine–component incidence matrix*. An example of such a matrix is shown in Figure 1. Usually rows represent machines, columns represent parts, and a "1" indicates that a component requires the machine for processing. Here information such as the actual sequence of operations and the number of machines available is not used. These will be considered in the analysis after the initial groups are obtained. The terms parts and components are used interchangeably in this chapter. Machine cells and part families are formed using this matrix using a variety of techniques listed in the next sections.

2. CELL FORMATION USING ZERO–ONE MATRICES

2.1. Block Diagonalization Problems

From the machine–component incidence matrix, machine cells and part families are found. One of the approaches is to block diagonalize the matrix where the rows and columns are interchanged so as to bring all the 1's to the diagonals. A machine component incidence matrix is shown in Figure 1(a) and its block-diagonal structure is shown in Figure 1(b). Here, the machine cells and part families are visible. There are three machine cells and part families.

Consider the machine–component incidence matrix shown in Figure 2(a) and a block-diagonal structure shown in Figure 2(b). Here, the machine cells and part families are not visible as in Figure 1(b). There are intercell moves, and the exact number of groups are also not clear. We will get different solutions depending on where we partition the rows and columns.

In fact, the problem of partitioning from the block-diagonal structure is as hard a

```
          11111111112                       112     1111    1111
       12345678901234567890              14956023581479670238

 1    1  1    1     11    1         1     111111
 2     11 1  1  1  1  1 1           4     111111
 3         11  1 11     1           7     111111
 4    1  1    1     11    1         2          11111111
 5         11  1 11     1           6          11111111
 6     11 1  1  1  1  1 1           9          11111111
 7    1  1    1     11    1         3                  111111
 8         11  1 11     1           5                  111111
 9     11 1  1  1  1  1 1           8                  111111
10         11  1 11     1          10                  111111

          (a)                                    (b)
```

Figure 1. (a) Machine–component incidence matrix. (b) Block-diagonal structure.

problem as the grouping problem itself. It is very important to provide the machine groups and part families along with the block-diagonal structure.

Utility of the Block Diagonalization Approach   The block-diagonal structure shown in Figures 1(*b*) and 2(*b*) help the planner in the following ways (Shargal et al., 1995):

```
                    1111111111122222222223333333333344444
                    12345678901234567890123456789012345678901234567890123

1                                                          1    1
2         1         1                     1    1    11 1 1
3              1              1                      111
4            1  11      1      1 1 1          1
5            1  11      111    1 1 1          1    1           1 1
6         11    111     111    1 1      1          111    1 11 11
7         1                 1           1
8         111     11 11   1   111 11   11   1         11   1 1
9          1 1        1        1              1    1    11 1 1
10        1              11            11    1         1
11         1       1            1    1 1 1
12                  1           1 1  1 1
13        1    1  1          1                1    11    1
14         1                        1
15        1  1            1      1 1              1
16           1            1      1 1          1           1 1
```

(*a*)

```
            11233344   1333    11111222344 12233 122223
            24088278026774565892456913931313563193102470

1                 1  1
2         1 1 111111
9         1111111111
16        1 11 111 1 1
3                    11111
14        1          1 1 1
4                        1 1 1111111
5                    111 1111111111
6         1     11 111111    1 11  1 1 1 111      1
8         1     1 11         111 1 111   111    1 111 11
15                          1    1   11   111
7                                          111
10                      1              111111
11                      1                     1 1 111
12                                                1 1111
13                                                1   1
```

(*b*)

Figure 2. (*a*) Machine–component incidence matrix (from King and Nakornchai, 1982). (*b*) Block-diagonal structure for the matrix shown in (*a*).

1. **Range of Number of Cells.** The block-diagonal structure of Figure 2(*b*) can be partitioned into four or five groups. The number of blocks defines a lower limit for the minimum number of cells.

2. **Machine and Part Composition.** Each block shown in the figure matches a machine cluster (group) to a part cluster (group).

3. **Complexity of the Cell.** This is indicated by the number of machine types and parts with nonmatching machine requirements that are included in the blocks.

4. **Bottleneck Machines.** Machine types shared by two or more part families, for example, machines 6 and 8 in Figure 2(*b*). Decisions to duplicate the machines can be made depending on cost and availability considerations.

5. **Bottleneck Parts.** Parts that require machines from more than two cells, for example, part 2 in Figure 2(*b*).

6. **Exceptional Operations.** Parts that require only one extra operation in another cell, for example, part 9 in Figure 2(*b*).

7. **Feasibility of Cell Formation.** Whether it is possible to identify the cells from the blocks and if so, whether it is practically feasible. For example, it is possible to easily identify cells from Figure 1(*b*) while it is a little difficult in Figure 2(*b*). In Figure 2(*b*), the biggest cell has five machines (machines 4, 5, 6, 8, 15). Is it acceptable to have six machines in a cell?

8. **Alternative Cell Configurations for the Same Number of Cells.** From Figure 2(*b*), we can identify two configurations with five cells. One configuration is {1, 2, 9, 16}, {3, 14}, {4, 5, 6, 8, 15}, {7, 10}, and {11, 12, 13} while another could be {1, 2, 6, 8, 9, 16}, {3, 14}, {4, 5, 8}, {7, 10}, and {11, 12, 13}.

9. **Alternative Cell Configurations with Different Number of Groups.** We can consider four cells with the following configuration {1, 2, 3, 9, 14, 16}, {4, 5, 6, 8, 15}, {7, 10}, and {11, 12, 13}. Choice among these alternatives could be finally resolved through a more involved cost analysis.

10. **Classification of Parts.** Parts can be further classified into three categories: (a) parts that use nonbottleneck machines only, (b) parts that use both bottleneck and nonbottleneck machines (requires a cost model that considers machine duplication, intercell flow, layout design, and subcontracting), and (c) parts that use only bottleneck machines (involving a more detailed cost analysis to choose among candidate cells).

11. **Shop Layout.** Cells that correspond to blocks requiring bottleneck operations and parts have to be laid close to each other.

12. **Cell Stability.** It may be possible to generate block-diagonal structures for different part mixes to verify whether the layouts are stable over a period of time or to create a stable layout for various part mixes.

13. **Improvement Strategies.** Strategies such as setup reduction and acquisition of new machines can be focused on bottleneck machines and cells.

2.2. Mathematical Formulation

Let us consider two mathematical formulations of the grouping problem from the machine–component incidence matrix. The first is the *p*-median problem (Kusiak,

1987) constructed with the objective of maximizing the sum of similarities among machines. The second is a linear cell formation that minimizes the number of intercell moves for a given number of groups.

The p-Median Formulation   The *p*-median formulation seeks to form a fixed number of groups by grouping either the machines or components into a fixed number of groups. A similarity matrix for machines is constructed from the machine–component incidence matrix using

$$S_{ij} = \sum_k d_k \tag{1}$$

$$d_k = \begin{cases} 1; & \text{if } a_{ik} = a_{jk} \\ 0; & \text{otherwise.} \end{cases}$$

This matrix is square, symmetric, and satisfies triangular inequality. The similarity between two machines *i* and *j* represents the number of components requiring processing on both the machines plus the number of components not requiring processing on either of them. Machines are to be grouped so as to maximize the similarity measure.

Let $x_{ij} = 1$; if machine *i* is allotted to a cell with nucleus machine (or median) *j*. Let $x_{ij} = 0$; otherwise.

$$\text{Maximize } z = \sum_i \sum_j s_{ij} x_{ij}, \tag{2}$$

$$\sum_i x_{ij} = 1, \forall j \tag{3}$$

$$\sum_j x_{jj} = p, \tag{4}$$

$$x_{ij} \leq x_{jj}, \forall i, j \tag{5}$$

$$x_{ij} = 0 \text{ or } 1.$$

The objective function [Eq. (2)] tries to maximize the sum of similarities that, in turn, is expected to group machines that have components with similar processing requirements.

The first constraint [Eq. (3)] assigns each machine to only one group. The second constraint [Eq. (4)] fixes the number of groups. There is a group with machine *j* as a median if $x_{jj} = 1$. The third constraint [Eq. (5)] ensures that a machine is assigned to a cell with median *j* only if such a cell exists.

The *p*-median problem is a known hard problem and does not have algorithms that guarantee optimal solutions in polynomial time.

Instead of maximizing similarities, the same problem can be solved to minimize a dissimilarity (or distance) given by

$$d_{ij} = \sum_k |a_{ik} - a_{jk}|. \tag{6}$$

It is seen that $d_{ij} = n - s_{ij}$; where n is the number of parts.

Once the machines are grouped, parts are assigned to groups that visit maximum number of machines. Ties are broken by assigning to the smaller group.

Linear Cell Formation Problem   The objective of this problem is to minimize the intercell moves for a given number of groups. Let

$$x_{ik} = \begin{cases} 1; & \text{if machine } i \text{ is in group } k \\ 0; & \text{otherwise} \end{cases}$$

$$y_{jk} = \begin{cases} 1; & \text{if part } j \text{ is in group } k \\ 0; & \text{otherwise} \end{cases}$$

$$\text{Minimize } Z = \sum_i \sum_j \sum_k a_{ij} \left| x_{ik} - y_{jk} \right| \tag{7}$$

$$\sum_k x_{ik} = 1 \ \forall i, \tag{8}$$

$$\sum_k y_{jk} = 1 \ \forall j, \tag{9}$$

$$x_{ik}, y_{jk} = 0 \text{ or } 1.$$

The objective function [Eq. (7)] represents the sum of the intercell moves. An intercell move occurs whenever a part j in one group requires a machine i in some other group. The term within the modulus takes a value 1 if machine i and part j belong to different groups. Equations (8) and (9) ensure that each machine and component is attached to only one group.

This is also a proven hard problem and cannot be solved optimally in polynomial time.

2.3. Approaches to Cell Formation

The p-median formulation and the linear cell formulation assume that the number of groups are known and provide solutions for a fixed number of groups. One of the earliest beliefs of group technology is that machine cells exist naturally, and the task of the researcher or manager is to discover them (Burbidge, 1971). It is seen that neither model considers this aspect of grouping while formulating the problem.

If the number of groups are not known *a priori*, these formulations have to be solved for number of groups ranging from 1 to m (there can be a maximum of m groups with each machine constituting a group). If it is assumed that a group should have at least two machines, there is a maximum of $m/2$ groups when m is even and $(m - 1)/2$ groups if m is odd. It is difficult to solve such problems for large matrices even once. Therefore, researchers resort to heuristic procedures that are quick and provide good or near-optimal solutions to these problems.

Most of the heuristics developed for obtaining machine cells and part families from the incidence matrix can be viewed as procedures to block diagonalize the matrix. These approaches can be classified as array-based methods, clustering-based methods, mathematical-programming-based methods, graph-theory-based methods, and others. In the following sections, a few important algorithms from the above categories are described using the matrix shown in Figure 2(*a*) for illustration.

3. ALGORITHMS FOR CELL FORMATION

3.1. Production Flow Analysis (PFA)

This approach was developed by Burbidge (1971) who later incorporated it in a manual method (Burbidge, 1977). A description of the manual method of PFA and an illustration of the algorithm using the matrix in Figure 2(*a*) are presented here.

From the machine–component incidence matrix, the number of components visiting each machine (frequency) is computed (column 2, Table1). The machine, which has least frequency, is chosen as the nucleus machine to form a module. Here, the nucleus is machine 1 with a frequency of 2. The module is shown in Table 2, with machine 1 in row 1. All the components, visiting this machine are listed (components 37 and 42) as shown in Table 2. We have exactly two components as indicated by the frequency.

The other machines required for making components 37 and 42 are added to the module (column 1 of Table 2). They are machines 2, 6, 8, 9, and 16. An asterix (*) is used to indicate that the component requires the machine in the module (Table 2). All machines, except 8, have two such markings while 8 has one because component 42 does not require machine 8. The number of markings against each machine is subtracted from the values in Table 1 (column 2) to result in column 3. Here, against machines 1, 2, 6, 9, and 16, the values are decreased by 2, while machine 8 has its value reduced by 1; others remain as they were.

Table 1. Frequencies of machines (shown up to seven modules)

Machines	1	2	3	4	5	6	7
1	2	0	0	0	0	0	0
2	8	6	6	6	6	6	6
3	5	5	5	5	5	5	5
4	7	7	7	7	7	6	6
5	13	13	13	13	13	12	12
6	19	17	17	15	13	13	13
7	3	3	3	0	0	0	0
8	20	19	17	16	14	11	10
9	10	8	8	8	8	8	8
10	7	7	7	4	0	0	0
11	6	6	4	4	4	0	0
12	5	5	4	4	4	2	0
13	2	2	0	0	0	0	0
14	4	4	4	4	4	4	4
15	7	7	7	7	7	7	7
16	8	6	6	6	6	6	6

Table 2. Module formation (module 1 with machine 1 as nucleus)

Machines	37	42
1	*	*
2	*	*
6	*	*
8	*	
9	*	*
16	*	*

The next module is created using column 3 of Table 1. Machine 7 is chosen as the nucleus because it has the smallest nonzero value of the frequency. Here ties have been broken arbitrarily. The module is shown in Table 3. This procedure of forming modules is continued until it is not possible to identify a nucleus machine for the next module. Fourteen modules can be found in our example and the machines and components in these are shown in Table 4.

Table 3. Module formation (module 2 with machine 13 as nucleus)

Machines	3	24
13	*	*
8	*	*
11	*	*
12		*

Table 4. Modules obtained for the matrix in Figure 2(a)

Modules	Machines	Components
1	1 2 6 8 9 16	37 42
2	13 8 11 12	3 24
3	7 6 8 10	1 13 25
4	10 6 8	12 26 31 39
5	11 4 5 8 12	9 20 27 30
6	12 8	11 22
7	14 2 6 8 9 16 3	2 6 17 35
8	3 6 16	7 34 36
9	16 2 9 6 8	10 8 32 38
10	2 8 9 6	28 40
11	9	4
12	4 5 15 6 8	5 14 19 21 23 29
13	6 5 8 15	8 33 43
14	15 5 8	41
15	8 5	15
16	5	16

We observe that there are several modules where one has a set of machines, which is a subset of another. Such modules are merged. For example, modules 1, 2, 8, and 9 can be merged and so on to result in six shown in Table 5. Here machines 6 and 8 appear in six out of the seven modules and other machines appear in more than one module. We need a total of 34 machines in the five cells shown in Table 5. However, if we make a few changes in the allocations, we can reduce the solution to five groups shown in Table 6 with three intercell moves (component 7 visits machine 16, component 9 visits machine 11, and component 2 visits machine 14). These changes are based on the judgment and expertise of the cell designer, and there is no specific algorithm to do these. At best, a few guidelines can be given to the cell designer if he desires to have them.

The initial machine–component incidence matrix and the block-diagonal structure are shown in Figures 3(a) and 3(b), respectively.

Production Flow Analysis Algorithm

Step 1. Read the machine–component incidence matrix and compute the frequency for each machine.

Step 2. Choose the machine with the lowest nonzero frequency as the nucleus. Form modules by including the components that require the nucleus machine and the other machines needed to make these components.

Step 3. Repeat step 2 until no nucleus machine can be found.

Step 4. Merge modules if a machine set for one is a subset of another.

Modify the solution in step 4 to suitably reduce the number of groups requiring intercell moves.

Table 5. Modules synthesis from Table 4

Module	Machines	Components
1	1 2 6 8 9 16	4 28 40 10 18 32 38 37 42
2	8 11 13 12	11 22 3 24
3	6 7 8 10	12 26 31 39 1 13 25
4	4 5 8 11 12	9 20 27 30
5	2 6 8 9 14 16 3	7 34 36 2 6 17 35
6	4 5 15 6 8	16 15 41 8 33 43 5 14 19 21 23 29

Table 6. Final machine groups and component families for the matrix in Figure 2(a)

Module	Machines	Components
1	1 2 6 8 9 16	2 4 10 18 28 32 37 38 40 42
2	8 11 12 13	3 11 20 22 24 27 30
3	6 7 8 10	1 12 13 25 26 31 39
4	3 6 14	6 7 17 34 35 36
5	4 5 6 8 15	5 8 9 14 15 16 19 21 23 29 33 41 43

```
                    1111111111222222222233333333334444
                    1234567890123456789012345678901234567890123

 1                                                  1        1
 2          1           1                1    1   11 1 1
 3            1                1                  111
 4            1  11      1      1 1 1        1
 5            1  11      111    1 1 1        1    1           1 1
 6          11   111     111    1 1   1           111   1 11  11
 7          1                1              1
 8          111      11 11   1   111 11   11   1       11   1 1
 9           1 1         1           1           1     11 1 1
10          1           11               11      1         1
11            1       1              1   1  1  1
12                  1                1   1  1  1
13          1     1  1            1              1     11    1
14           1                        1
15          1   1             1                      1
16             1           1        1 1           1         1 1
```

                              (a)

```
            11233344 122223 112233  1333    1111222344
            24088278023102470123561967745658945 69139313

 1                 1  1
 2          1 1 111111
 6          1     11 11
 8          1   1 11
 9          1111111111
16          1 11 111 1                1
 8                       111 1 1
11                       1 1 111              1
12                       1 1111
13                       1   1
 6                            1 11  1
 7                            1 11
 8                            111 1
10                            1111111
 3                                   11111
 6                                   111111
14          1                        1 1 1
 4                                          1 1 1111111
 5                                          1111111111111
 6                                          1 1  1 1 1 1
 8                                          11 1 111   11
15                                          1  1  11    111
```

                              (b)

Figure 3. (a) Machine–component incidence matrix (from King and Nakornchai, 1982). (b) Block-diagonal structure for the matrix shown in Figure 2(a).

This algorithm comes under the group analysis of PFA proposed by Burbidge. This seems to be the most widely used algorithm for cell formation in the industry.

3.2. Array-Based Algorithm

The array-based algorithm explained in this section is the Rank Order Clustering (ROC) algorithm (King, 1980; King and Nakornchai, 1982). Rank-Order Clustering is essentially a sorting algorithm where the machines are arranged in decreasing order of their weights. Column j (from the left) is given a weight $w_j = 2^{(n-j)}$, where n is the number of columns and for each row (machine), the weight $x_{ij} \leq x_{jj}$ is calculated. Row j is given a weight $w_j = 2^{(m-j)}$, where m is the number of machines and for each column (component), weight $W_i = \sum_j a_{ij} w_j$ is calculated. The rows and columns are now rearranged with decreasing order of row and column weights. This constitutes two iterations of the method. The original matrix and the rearranged matrix are shown in Figures 4(a) and 4(b), respectively.

With the matrix in Figure 4 (b), two more iterations are carried out, one for the rows and one for the columns. The rearranged matrix is shown in Figure 5.

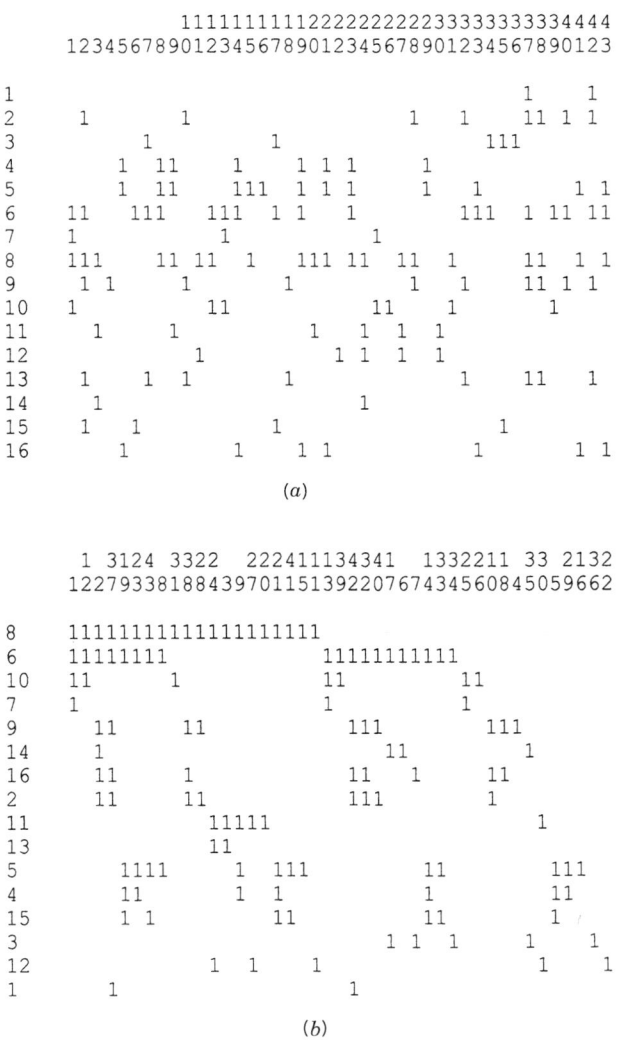

Figure 4. (a) Machine–component incidence matrix (from King and Nakornchai, 1982). (b) Rearranged matrix after two iterations of ROC algorithm.

From the solution shown in Figure 5, four groups are possible if machines 6 and 8 are either ignored or suitably duplicated. Since these machines have parts that are present in almost all the part families, the authors propose a way to eliminate the intercell moves by creating a row for every intercell move and iterating again as done previously. Twenty rows are created for machine 8 with each element of the matrix in each row and 19 rows are created for row 6. After two iterations and merging after merging rows of machines 8 and machine 6 in the corresponding machine cells, a solution as is shown in Figure 6 is obtained.

A better implementation suggested by King and Nakornchai (1982) is to treat the rows and columns as binary words and arrange them in the decreasing order of their binary value.

Rank-Order Clustering Algorithm

Step 1. Rearrange the rows of the incidence matrix in decreasing order of binary value.

Step 2. Rearrange the columns of the matrix after step 1 in decreasing order of binary value.

Efficient implementation of ROC for well-structured (perfectly block diagonalizable) matrices was reported by Kusiak and Chow (1987). Chandrasekharan and Rajagopalan (1986), who observed that ROC was dependent on the initial disposition of the matrix, pointed out the limitations of the algorithm. Rearranging the columns of the initial matrix can result in a different solution depending on the arrangement of the matrix. In spite of these shortcomings, ROC remains the most simple and popular algorithm for cell formation.

```
         121332 34331241   13 121   22443 112 2321233
         135219627282080847756914593931385 6437001246

   8     1  11   11 1   1          11   11 11 11 111 11
   6     11 1 1  111 1   1   11 11 1   1 1 11            1
  10     1111111
   7     111
   9            1111111111
   2            11111111
  16            111111  1 1
  14            1               111
   1              11
   5                            1111111111111
   4                            1111111
  15                            1111    111
  11                                1             11111
  13                                              11
  12                                              1 11 11
   3                            111                        11
```

Figure 5. ROC solution after four iterations.

```
                121332 33431241  13 33112 2 2434 112 22312
                1352916278220808477564694153993318564370012

  1        1111111
  7        111
  6        11 11
  8        1  1 1
  1                1 1
  9                1111111111
  2                11111111
 16                111111  1 1
  6                11 11  1  1
  8                111   1
 14                1             111
  3                              111 11
  5                                  1111111111111
  4                                  1111111
 15                                  1111    111
  8                                  1 1 11 1 111
  6                                  11  1   11 1
 11                                   1             11111
  8                                                 1111 1
 13                                                 11
 12                                                 1 1 111
```

Figure 6. Final block-diagonal structure using ROC algorithm.

3.3. Similarity-Based Clustering Algorithm

A similarity-based algorithm developed by McAuley (1972) is explained in this section. The author was the first to use a zero–one incidence matrix. Burbidge used tick marks instead of 1's. The algorithm uses a similarity measure different from the one used in the p-median model. It is called the Jaccard's similarity coefficient and is given by

$$ S_{ij} = \left(\sum_{k} a_{ij} \times a_{jk} \right) \bigg/ \left(\sum_{k} d_k \right), \qquad (10) $$

where

$$ d_k = \begin{cases} 1; & \text{if } a_{ik} = a_{jk} \\ 0; & \text{otherwise.} \end{cases} $$

This similarity coefficient is a number between 0 and 1 and has a wider dispersion than the previous measure. The similarity measures for pairs of machines in the current example is shown in Table 7 where all pairs with positive values of the coefficient are listed. All other pairs of machines have a similarity coefficient of zero.

Let us form machine cells by grouping the machines with maximum similarity. Machines 2 and 9 form the first group since they have a similarity of 0.8. Machine 16 gets added to this group since it has a similarity of 0.636 with machine 9. Machines

Table 7. Similarity matrix of machines from example from Figure 2(a)

	1	2	3	4	5	6	7	8	9	10	11	12	13	14	15	16
1	0.00	0.25	0.00	0.00	0.00	0.11	0.00	0.05	0.20	0.00	0.00	0.00	0.00	0.00	0.00	0.25
2	0.25	0.00	0.00	0.00	0.00	0.23	0.00	0.17	0.80	0.00	0.00	0.00	0.00	0.09	0.00	0.60
3	0.00	0.00	0.00	0.00	0.00	0.14	0.00	0.00	0.00	0.00	0.00	0.00	0.00	0.29	0.00	0.08
4	0.00	0.00	0.00	0.00	0.54	0.13	0.00	0.17	0.00	0.00	0.08	0.00	0.00	0.00	0.40	0.00
5	0.00	0.00	0.00	0.54	0.00	0.23	0.00	0.32	0.00	0.00	0.06	0.00	0.00	0.00	0.54	0.00
6	0.11	0.23	0.14	0.13	0.23	0.00	0.10	0.26	0.21	0.18	0.00	0.00	0.00	0.15	0.18	0.23
7	0.00	0.00	0.00	0.00	0.00	0.10	0.00	0.05	0.00	0.43	0.00	0.00	0.00	0.00	0.00	0.00
8	0.05	0.17	0.00	0.17	0.32	0.26	0.05	0.00	0.15	0.12	0.24	0.14	0.10	0.04	0.17	0.12
9	0.20	0.80	0.00	0.00	0.00	0.21	0.00	0.15	0.00	0.00	0.00	0.00	0.00	0.08	0.00	0.64
10	0.00	0.00	0.00	0.00	0.00	0.18	0.43	0.12	0.00	0.00	0.00	0.00	0.00	0.00	0.00	0.00
11	0.00	0.00	0.00	0.08	0.06	0.00	0.00	0.24	0.00	0.00	0.00	0.38	0.33	0.00	0.00	0.00
12	0.00	0.00	0.00	0.00	0.00	0.00	0.00	0.14	0.00	0.00	0.38	0.00	0.17	0.00	0.00	0.00
13	0.00	0.00	0.00	0.00	0.00	0.00	0.00	0.10	0.00	0.00	0.33	0.17	0.00	0.00	0.00	0.00
14	0.00	0.09	0.29	0.00	0.00	0.15	0.00	0.04	0.08	0.00	0.00	0.00	0.00	0.00	0.00	0.09
15	0.00	0.00	0.00	0.40	0.54	0.18	0.00	0.17	0.00	0.00	0.00	0.00	0.00	0.00	0.00	0.00
16	0.25	0.60	0.08	0.00	0.00	0.23	0.00	0.12	0.64	0.00	0.00	0.00	0.00	0.09	0.00	0.00

4, 5, and 15 form the next group with a similarity of 0.538. Machines 7 and 10 form the next group with a similarity of 0.429. Machines 11 and 12 form the next group with a similarity of 0.375, and machine 13 gets added to the group with a similarity of 0.333. Machine 8 is added to group {4,5,15} at a similarity level of 0.32. Machines 3 and 14 form a group at a similarity level of 0.286, and machine 1 is added to the group {2,9,16} at a similarity level of 0.25. Finally machine 6 is included in group {4,5,8,15} at a similarity level of 0.231.

We obtain 5 groups {1,2,9,16}, {4,5,6,8,15}, {7,10}, {11,12,13}, and {3,14}. The machine group formation is shown in the dendrogram in Figure 7. A similar procedure can be followed to obtain component groups. It is also possible to obtain component families by assigning each component to groups where they visit the maximum the number of machines. This gives the same solution as in Figure 2(b).

The Algorithm

Step 1. Compute the similarity coefficients for the all pairs of machines.

Step 2. Form machine cells by grouping machines with high similarity. Add machines to existing groups when an unassigned machine has a high similarity coefficient with, at least, one of the machines in the group.

Step 3. Form component families by assigning components to cells where they visit the maximum number of machines.

Permutation Generation Approach to Cell Formation   We illustrate the permutation generation approach to cell formation using the same example in Figure 2(a). The similarity coefficient for machines shown in Table 7 is used in this approach.

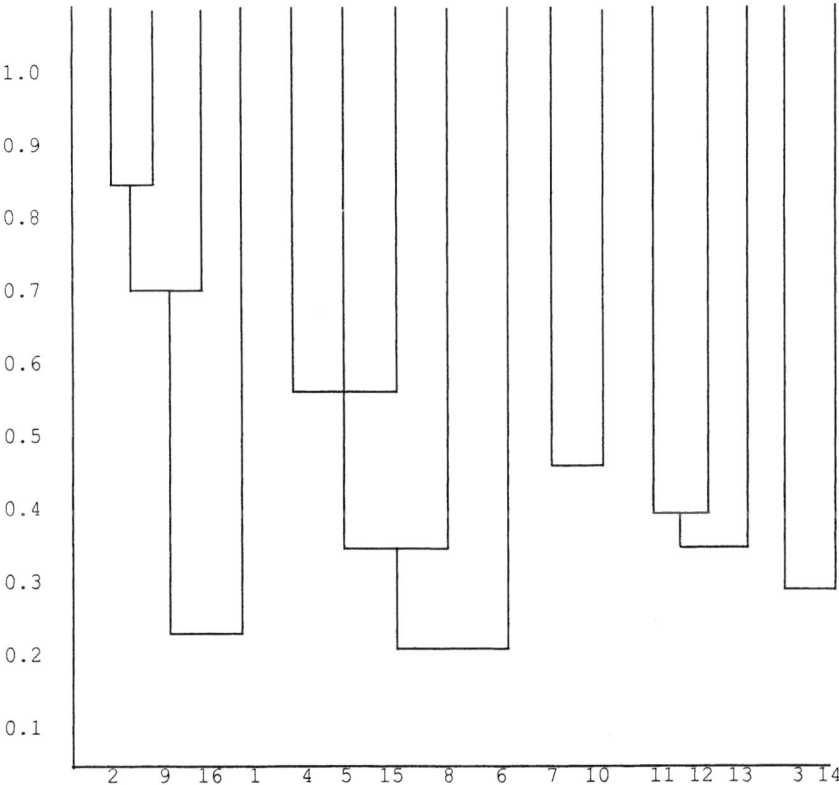

Figure 7. Dendrogram showing the machine group formation.

Initially a machine is chosen at random. Let this be machine 6. The machine with the highest similarity (machine 8) is added to it. This process of adding the machine (among the unassigned machines) with the highest similarity is continued until no more additions are possible. In our example when we reach the chain 6–8–5–4–15 we cannot add a machine to it because the rest of the machines have a zero similarity with machine 15. We choose a machine randomly (say, machine 11) and continue the process. We reach the chain 6–8–5–4–15–11–12–13 when we are unable to add a machine. We pick a machine randomly (say, machine 7) and continue until we get the final chain 6–8–5–4–15–11–12–13–7–10–1–2–9–16–14–3. A similar procedure can be used for the components to form a component chain from which the block-diagonal structure can be obtained.

The chain can be broken at the weak links to yield the desired number of groups. If we require five groups, the chain can be broken at the links 15–11, 13–7, 10–1, where the similarity is zero, and 16–14, where the similarity is 0.09, to yield five groups. The component families are formed by assigning each component to the group where it visits a maximum number of machines. The initial machine–component matrix and the block-diagonal form using the permutation approach are shown in Figures 8(a), and 8(b), respectively.

The permutation approach can result in a different solution depending on the choice of the first machine. Alternatively, the first two machines can be identified from the similarity matrix as those with the maximum similarity coefficient. Random choice of the first machine can provide different solutions for better understanding and analysis.

3.4. ZODIAC—A Nonhierarchical Clustering Algorithm

The ZODIAC (Zero–One Data–Ideal seed Algorithm for Clustering) was proposed by Chandrasekharan and Rajagopalan (1987). This is a nonhierarchical clustering

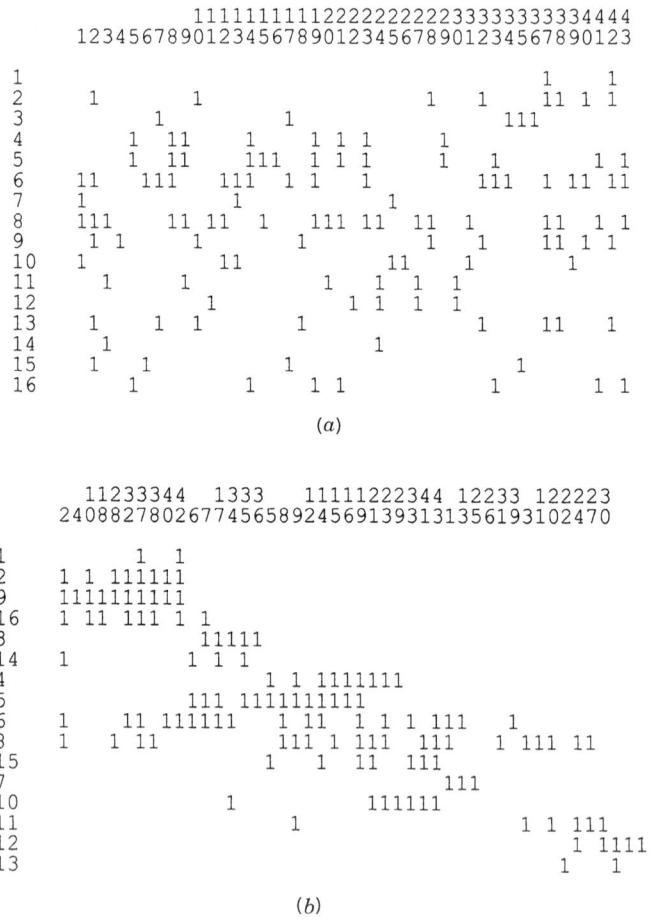

Figure 8. (a) Machine–component incidence matrix (from King and Nakornchai, 1982). (b) Machine component incidence matrix after permutation approach (from King and Nakornchai, 1982).

algorithm, which obtains a fixed number of groups. Here each machine vector (row) is treated as a point in a higher dimensional zero–one space and clustered around some fixed seed points, which may be among these points themselves. In this section we explain the natural seed clustering version of the algorithm followed by the ideal seed clustering algorithm.

The natural seed algorithm attempts to find the most natural machine cells from the data. It uses Jaccard's (Sukal and Sneath, 1963) similarity matrix to create initial seed points around which the machines are to be grouped. The seed points should be as far away from each other, that is, as dissimilar as possible so that the points clustered around them are similar and the clusters (groups) themselves are dissimilar.

The average of the Jaccard similarity coefficient for machines from Table 7 is found to be 0.0855. Since the matrix is symmetric, only the upper triangular values are considered. The matrix is scanned to obtain a pair of machines whose similarity is less than the average. Machines 1 and 3 have a value of zero and are chosen as the first two seed points. Machine 4 is chosen as the third seed point since its similarity with all the seeds is less than the average. Machines 7 and 11 are the fourth and fifth seed points. It is to be noted that any machine that has a similarity value more than the average even with one existing seed does not qualify to be a seed.

The machines are clustered around the seed points. The five seed points are

Seed 1: [00000000000000000000000000000000001000010]
Seed 2: [00000010000000001000000000000000001110000000]
Seed 3: [00001000100001000010101000001000000000000000]
Seed 4: [10000000000010000000000010000000000000000000]
Seed 5: [00100000100000000001000100100100000000000000]

Each machine vector is assigned to that seed with which the distance [dissimilarity measure given in Eq. (6)] is minimum. Machine 1 is assigned to seed 1, machine 2 to seed 1, and so on. The machine groups are

Group 1: {1,2,6,9,13,16}
Group 2: {3,14}
Group 3: {4,5,15}
Group 4: {7,10}
Group 5: {8,11,12}

These machine groups obtained using natural seeds will be used to create seed points for ideal seed clustering later. A similar procedure is performed for the components using the transpose of the matrix in Figure 8(*a*). Five component groups are obtained. They are

Group 1: {1,8,12,13,25,31,39}
Group 2: {2,3,4,10,15,16,18,20,22,26,28,32,37,38,40,42}

Group 3: {5,9,14,19,21,23,29,33,41,43}
Group 4: {11,22,24,27,30}
Group 5: {6,7,17,34,35,36}

Though five machine groups and component groups are available, the part family corresponding to the machine cells are not known. This is done using a procedure called *diagonalization.*

Each family can be assigned one machine cell. The number of 1's is computed for each of the 25 possible pairs of machine cells and part families. The combinations with the maximum number of 1's are paired together. This procedure is continued until each machine cell has a part family assigned to it. In the present example, the assignments are

Machine group 1 to part family 2
Machine group 2 to part family 5
Machine group 3 to part family 3
Machine group 4 to part family 1
Machine group 5 to part family 4

The goodness of the solution is given by a measure called *grouping efficiency* given by

$$\eta = q * a/b + (1 - q) * c/d, \tag{11}$$

where a = actual number of 1's in the diagonal blocks
b = maximum number of 1's that can be accommodated in the diagonal blocks
c = actual number of 0's in the off-diagonal area
d = maximum number of 0's that can be accommodated in the off-diagonal area.

Here zeros are replaced by blank spaces, and q usually takes a value of 0.5.

The current solution has 41 intercell moves and a grouping efficiency of 72.51%. This solution is improved using the ideal seed algorithm. Five seeds are obtained from the five component groups. They are

Seed 1: [1000000100011000000000001000001000000010000]
Seed 2: [0111000001000011010101000101000100001101010]
Seed 3: [0000100010000100001010100000100010000000101]
Seed 4: [0000000000110000000000001001001000000000000]
Seed 5: [0000011000000000100000000000000001110000000]

Seed 1 has been created from part family 1 and has 1's in locations {1,8,12,13,25,31, and 39} representing the parts grouped in family 1. The other seeds are created similarly. Each seed is a vector with 43 elements. The machine vectors are clustered using these seed points and the distance measure [Eq. (6)]. The machine groups are

Group 1: {6,7,10}
Group 2: {2,9,16}
Group 3: {4,5,8,15}
Group 4: {1,11,12,13}
Group 5: {3,14}

From machine–component groups obtained using natural seeds earlier, seed points are created to cluster components. The machine group {1,2,6,9,13,16} yields seed 1 with a 1 in positions 1, 2, 6, 9, 13, and 16. The seeds thus obtained from five machine groups are

Seed 1: [1100010010001001]
Seed 2: [0010000000000100]
Seed 3: [0001100000000010]
Seed 4: [0000001001000000]
Seed 5: [0000000100110000]

Each component is attached to a seed with which its distance is minimum. The component groups after clustering are

Group 1: {2,4,10,18,28,32,37,38,40,42}
Group 2: {6,7,17,34,35,36}
Group 3: {5,8,9,11,14,15,16,19,20,21,23,29,33,41,43}
Group 4: {1,12,13,25,26,31,39}
Group 5: {3,22,14,27,30}

Finally, part groups are assigned to machine groups. Using the diagonalization procedure, explained earlier, the assignments are

Machine group 1 to part family 4
Machine group 2 to part family 1
Machine group 3 to part family 3
Machine group 4 to part family 5
Machine group 5 to part family 2

This solution has 32 intercell moves and a grouping efficiency of 79.93%. Since our objective is to find a solution with maximum grouping efficiency, we continue the ideal seeding procedure. The solution obtained in the next iteration is shown in Table 8. This has 33 intercell moves and a grouping efficiency of 80.20%

Finding that further iterations do not improve the grouping efficiency, the algorithm stops here. Intermediate solutions from the algorithm may yield unequal numbers of machine and part groups. In such cases the solutions are not evaluated. The number

Table 8. Solution using the ZODIAC algorithm

Group	Machines	Components
1	1,11,12,13	3,24,27,30
2	2,9,16	2,4,10,18,28,32,37,38,40,42
3	3,6,14	6,7,17,34,35,36
4	4,5,8,15	5,8,9,11,14,15,16,18,19,20,21,23,29,33,41,43
5	7,10	1,12,13,22,25,26,31,39

of seed points for machines and parts are made equal by eliminating a few seed points (corresponding to small or singleton groups) and the iterations continued.

The machine–component incidence matrix and the block-diagonal form obtained by the ZODIAC algorithm are shown in Figures 9(a) and 9(b), respectively.

As in this example, it is possible to obtain two solutions, one with fewer intercell moves but with a lower value of grouping efficiency than another. Since the objective is to maximize grouping efficiency, the solution in Table 8 is chosen.

ZODIAC Algorithm

Step 1. Compute the Jaccard similarity coefficient matrix for machines (rows) and parts (columns).

Step 2. Choose seed points for natural seed clustering. Perform row and column clustering. Compute the grouping efficiency if the number of machine groups and part groups are equal. Else, eliminate small groups to have an equal number of groups and then evaluate the grouping efficiency.

Step 3. Perform ideal seed clustering. Evaluate grouping efficiency if the numbers of machine and part groups are equal. Update the solution if found better. Go to step 5.

Step 4. If the number of groups are different, eliminate small groups to have equal number. Go to step 3.

Step 5. Repeat step 3 until no better solution is found.

3.5. Assignment Model for Cell Formation

The assignment model was proposed by Srinivasan et al. (1990). It is a heuristic solution to the p-median problem where the number of groups is not fixed. It solves an assignment problem using the distance matrix for machines and parts and forms initial groups using the solutions. These are progressively merged using a set of rules to form the final machine cells and part families.

Let us apply the model to the incidence matrix in Figure 2(a). A matrix of dissimilarities (distances) is first computed for machines using Eq. (6). This matrix has a large value (M) along the diagonal in order to avoid assignment to these elements. The sixteen assignments obtained as the solution to the problem are:

```
                  1111111111222222222233333333334444
                  1234567890123456789012345678901234567890123
   1                                              1       1
   2      1         1                     1    1    11 1 1
   3         1           1                      111
   4        1 11    1      1 1 1         1
   5        1 11     111   1 1 1        1   1          1 1
   6     11   111   111  1 1    1            111  1 11 11
   7     1             1            1
   8     111    11 11  1    111 11  11  1         11   1 1
   9      1 1      1         1          1    1    11 1 1
  10     1         11            11    1        1
  11      1      1        1   1 1 1
  12           1          1 1 1 1
  13     1    1 1      1             1      11    1
  14      1               1
  15     1  1         1                 1
  16      1           1     1 1           1       1 1
```

(*a*)

```
              223   11233344   1333    111112222344 1122233
              3470240882780267745658914569013931312325619
   1                        1 1
  11      1111                  1        1
  12      111                   1                    1
  13      11
   2        1 1 111111
   9        1111111111              1
  16        1 11 111 1 1            1
   3                    11111
   6        1     11 111111   1  1   1  1 1 1111
  14        1        1 1 1
   4                       1 1 1   1 111
   5                       111 111 1 111111
   8     111 1   1 11       111 1  1111  1111     1
  15                        1   1   1 1  111
   7                                    1 1 1
   1                                    111 111
```

(*b*)

Figure 9. (*a*) Machine–component incidence matrix (from King and Nakornchai, 1982). (*b*) Block-diagonal structure of the solution obtained using ZODIAC algorithm.

$$X(1,13) = X(13,1) = 1$$
$$X(2,9) = X(9,16) = X(16,2) = 1$$
$$X(3,14) = X(14,3) = 1$$
$$X(4,5) = X(5,15) = X(15,4) = 1$$
$$X(6,8) = X(8,6) = 1$$
$$X(7,10) = X(10,7) = 1$$
$$X(11,12) = X(12,11) = 1$$

From the assignments, we can obtain seven preliminary groups. From $X(1,13) = X(13,1) = 1$ we obtain a group $\{1,13\}$. Similarly, we obtain the remaining six machine groups $\{2,9,16\}$, $\{3,14\}$, $\{4,5,15\}$, $\{6,8\}$, $\{7,10\}$, and $\{11,12\}$.

We construct the distance matrix for the components (columns) and solve another assignment problem (avoiding assignments along the diagonal) to minimize the total distance. From the solution, the following component groups are obtained:

$\{1,12\}$, $\{2,42,37\}$, $\{3,24\}$, $\{4,10,18\}$, $\{5,14,19,23,8,15,16,29\}$, $\{6,17\}$, $\{7,34\}$, $\{9,20\}$, $\{11,22\}$, $\{13,25\}$, $\{21,41\}$, $\{26,31,39\}$, $\{27,30\}$, $\{28,38\}$, $\{32,40\}$, $\{33,43\}$, and $\{35,36\}$

We have more component groups than machine groups and have to assign component groups to machine groups. We consider the first group $\{1,12\}$ and assign it to the machine group $\{6,8\}$ because components 1 and 12 have more visits to the machine group $\{6,8\}$ than to any other machine group. Each component group is assigned to the machine group where the number of visits (1's) is a maximum. The machine and components thus obtained are given in Table 9.

From Table 9 it is seen that machine group $\{1,13\}$ is empty and hence has to be assigned to a component group. It is assigned to the group $\{2,9,16\}$ because the components attached to this group are the ones that have maximum visits on $\{1,13\}$ compared to other component groups. The resultant solution is shown in Table 10.

Two machine and component groups are merged if the number of intercell moves saved by the merger is more than the number of blanks created by the merger. Here, we do not merge any further. The solution in Table 10 is the best with a grouping efficiency

Table 9. Machine cells and part families using assignment method (intermediate solution)

Group	Machines	Components
1	1,13	
2	2,9,16	2,42,37,4,10,18,28,38,32,40
3	3,14	6,17,7,34,35,36
4	4,5,15	5,14,19,23,8,15,16,29,9,20,21,41,33,43
5	6,8	1,12
6	7,10	13,15,16,31,39
7	11,12	3,24,11,22,27,3

Table 10. Final solution obtained using the assignment model

Group	Machines	Components
1	2,9,16,1,13	2,42,37,4,10,18,28,38,32,40
2	3,14	6,17,7,34,35,36
3	4,5,15	5,14,19,23,8,15,16,29,9,20,21,41,33,43
4	6,8	1,12
5	7,10	13,15,16,31,39
6	11,12	3,24,11,22,27,3

```
              1111111111222222222233333333334444
              1234567890123456789012345678901234567890123
 1                                              1    1
 2       1        1                    1   1    11 1 1
 3          1              1                111
 4        1 11      1      1 1 1         1
 5        1 11      111    1 1 1         1   1        1 1
 6       11   111   111   1 1    1          111  1 11 11
 7       1             1            1
 8       111    11 11  1    111 11   11  1      11  1 1
 9        1 1     1         1          1   1    11 1 1
10       1          11              11    1     1
11        1     1            1   1 1 1
12              1               1 1  1 1
13       1    1 1         1              1     11   1
14        1                       1
15       1   1          1                  1
16          1           1      1 1         1        1 1
```

(a)

```
              43 112334 1 333 112 112 22434 111133 21223
              2274088820677456549385699011331235519341270
 2       111 1 1111
 9       1111111111
16       111 11 11    1
 1         11
13                                                   11
 3                      11111
14       1             11  1
 4                          1111    11 1
 5                          111111111 1111    11
15                          111      1111
 6       111    111111    1111          11111   1
 8       1 1    11          1111    1111 111 1 1 111 1
 7                                        1 1
 1                                        111 11
11                          11                 11  11
12                                                 11111
```

(b)

Figure 10. (a) Machine–component incidence matrix (from King and Nakornchai, 1982). (b) Block-diagonal structure using the assignment model.

of 77.60%. The machine–component matrix and the block-diagonal structure obtained using the assignment algorithm are shown in Figures 10(a) and 10(b), respectively.

Assignment Model for Machine Cell and Part Family Formation

Step 1. Construct a distance matrix for machines and parts [Eq. (10)].

Step 2. Solve an assignment problem to minimize distances for machines and obtain the initial machine groups from the solution.

Step 3. Solve an assignment problem for components and obtain the component groups.

Step 4. Assign component groups to machine groups so that the number of visits is maximized. If a machine group is idle, assign it to the next best component group.

Step 5. Merge two machine and component groups if the number of intercell moves saved by the merger is more than the blanks created by the merger.

Extension of the Assignment Model   An extension of the assignment model is the GRAFICS algorithm proposed by Srinivasan and Narendran (1991). This algorithm gives better results than the assignment model. Let us apply this algorithm the same example.

The groups obtained after solving an assignment problem for machines are as follows: {1,13}, {2,9,16}, {3,14}, {4,5,15}, {6,8}, {7,10}, and {11,12}.

Each component is assigned to the machine group in which it visits the maximum number of machines. Ties are broken by assigning to the smallest group. For example, component 1, which requires machines 6, 7, 8, and 10, is assigned to the group {6,8} since it visits the maximum number of machines. The corresponding groups obtained after assigning all the components are:

{3}, {2,4,10,18,28,32,37,38,40,42}, {6,7,17,34,35,36}, {5,9,14,16,19,21,29,33,41}, {1,8,11,12,15,20,23,31,39,43}, {13,25,26}, and {22,24,27,30}

The first group has only one component and is reassigned to the group {1,8,11,12, 15,20,23,31,39,43} since it is the next best. We have seven machine groups and six component groups. We again form machine groups so that each machine is attached to the group from which the maximum components visit it. The six corresponding groups are: {1,2,9,16}, {3,14}, {4,5,15}, {6,8,10}, {7}, and {11,12,13}.

Here {7} is a singleton group and is assigned to the group {6,8,10} because the largest number of components attached to this group visit machine 7. We have five machine groups. Each component is now allotted machine groups where it visits the maximum number of machines. The machine and component groups are shown in Table 11.

This solution has 31 intercell moves and a grouping efficiency of 79.90%. Using the component groups, machines are assigned to groups such that the maximum number of components visits the machine. The procedure of assigning components to fixed machine groups and assigning machines to fixed component groups is repeated until

Table 11. Intermediate solution using the GRAFICS algorithm

Group	Machines	Components
1	1,2,9,16	2,4,10,18,28,32,37,38,40,42
2	3,14	6,7,17,34,35,36
3	4,5,15	5,9,14,15,16,19,21,23,29,33,41,43
4	6,7,8,10	1,8,12,13,25,26,31,39
5	11,12,13	3,11,20,22,24,27,30

Table 12. Final solutions obtained using the GRAFICS algorithm

Group	Machines	Components
1	1,2,9,16	2,4,10,18,28,32,37,38,40,42
2	3,14	6,7,17,34,35,36
3	4,5,6,8,15	5,8,9,12,14,15,16,19,21,23,29,33,41,43
4	7,10	1,13,25,26,31,39
5	11,12,13	3,11,20,22,24,27,30

no further improvement in grouping efficiency is found. The final solution is given in Table 12, which has 27 intercell moves with a grouping efficiency of 79.40%

The machine–component incidence matrix and the block-diagonal structure obtained by the GRAFICS algorithm are shown in Figures 11(*a*) and 11(*b*), respectively.

The Algorithm

Step 1. Construct a distance matrix for machines.

Step 2. Solve an assignment problem for machines using a machine distance matrix. Form initial groups.

Step 3. Assign each component to the group in which it visits the maximum number of machines. Break ties by choosing the smallest group. Replace singleton clusters by assigning them to the next best group.

Step 4. If the number of machine and component groups are same, evaluate grouping efficiency. Update if better solutions are found. Stop if no better solution is found.

Step 5. If machines were clustered last, obtain component groups from the machine clusters. If components were clustered last, obtain machine groups from these. Go to step 3.

4. CELL FORMATION USING RECENT SEARCH TECHNIQUES

This section considers a few search techniques, which start with an initial solution and try to progress to the best solution. In the improvement process, these techniques sometimes accept inferior solutions at intermediate stages with the hope of finding a better solution in the end. The methods differ in the improvement procedures and the approach adopted in accepting inferior solutions at intermediate stages.

These procedures are usually generic and can be applied to a variety of problems by making suitable modifications to suit the problem and objective. Some of the popular techniques are simulated annealing, tabu search, genetic algorithms, and neural networks. Only a few research studies report the application of these techniques to cell formation problems.

4.1. Simulated Annealing

This approach has been applied successfully to solve several combinatorial optimization problems such as the traveling salesman problem and the flow shop scheduling

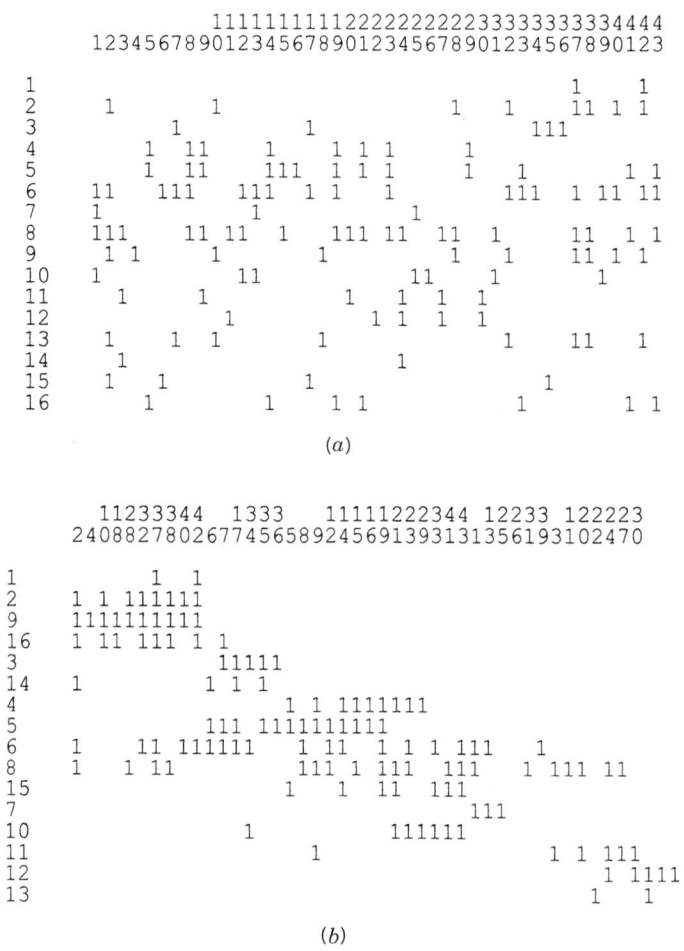

Figure 11. (a) Machine–component incidence matrix (from King and Nakornchai, 1982). (b) Block-diagonal structure using the GRAFICS algorithm.

problems. This uses properties from statistical mechanics in the process of improving a given solution. This algorithm is similar to an annealing process where the energy level is brought down gradually by a slow cooling process.

Kirkpatrick et al. (1983) proposed this technique to solve problems in networks and in traveling salesmen problems. Boctor (1991) applied this technique to solve cell formation problems. Chen and Srivastava (1994) have also proposed an algorithm based on simulated annealing for cell formation. The objective function represents the energy level of the system while accepting that an inferior solution is equivalent to a temporary increase in the energy state.

The algorithm for cell formation starts with an initial solution from which neighbor solutions are generated. If the neighbor is superior to the initial solution, it is accepted.

If the neighbor is inferior, it may still be accepted as the current solution with a certain probability. Neighbors are generated from the current solutions. The algorithm searches for the best solution assuming that it will be possible to find it by carefully accepting inferior solutions, in order to avoid being trapped in local optima.

The simulated annealing approach is illustrated using the matrix in Figure 2(*a*). Assume that we require two groups with a maximum of eight machines in each group.

Let us define an initial solution: $a = [2122112111222211]$.

The vector has 16 entries where $a(i)$ represents the group to which machine i belongs. We assign components such that each component is assigned to the group where it visits a maximum number of machines. The number of intercell moves is 23.

Using a pairwise interchange algorithm to improve upon the grouping, we get a solution

$a = [2121112112222211]$ with 19 intercell moves. The solution does not improve further.

If we follow an adjacent pairwise interchange and accept all the intermediate solutions, we obtain the solutions shown in Table 13. The machine–component incidence matrix and the block-diagonal structure are shown in Figures 12 (*a*) and 12 (*b*), respectively.

Note that inferior solutions have been accepted at the second and third stages in order to obtain better solutions without being trapped in local minima.

Boctor's (1991) algorithm for simulated annealing is as follows:

Step 1. Initialize $c_{max} = 320$; $r_{max} = 64$; $T = 32$; $a = 0.5$; $b = 0.8$.

Step 2. Generate an initial solution and store the intercell moves as $\mathbf{z}_{opt}$.

Step 3. Repeat until $c = c_{max}$, $R = r_{max}$, and $T = aT$. Repeat until $r = R$. Generate a neighbor solution and evaluate the intercell moves. Let z represent the difference between the intercell moves of the current solution and the neighbor. If $z \leq 0$, accept the neighbor. If $z \leq z_{opt}$ store the current solution as the best and update z_{opt}. If $z \geq 0$, Generate a random number between 0 and 1. If $z \leq \exp(-z/T)$ accept the inferior solution as the current solution: $r = bR$. Generate another neighbor and repeat the step.

Step 4. Terminate the algorithm when $T = 1$.

Boctor's initial solution is created as follows: Let M represent the number of machines and m the maximum machines allowable in a group. Find $h =$ higher integer

Table 13. A few feasible solutions to the matrix in Figure 3

Solution	Intercell Moves
[2122112111222211]	23
[2121122111222211]	27
[2112122111222211]	37
[2112212111222211]	29
[2112211211222211]	25
[2112211211222121]	15

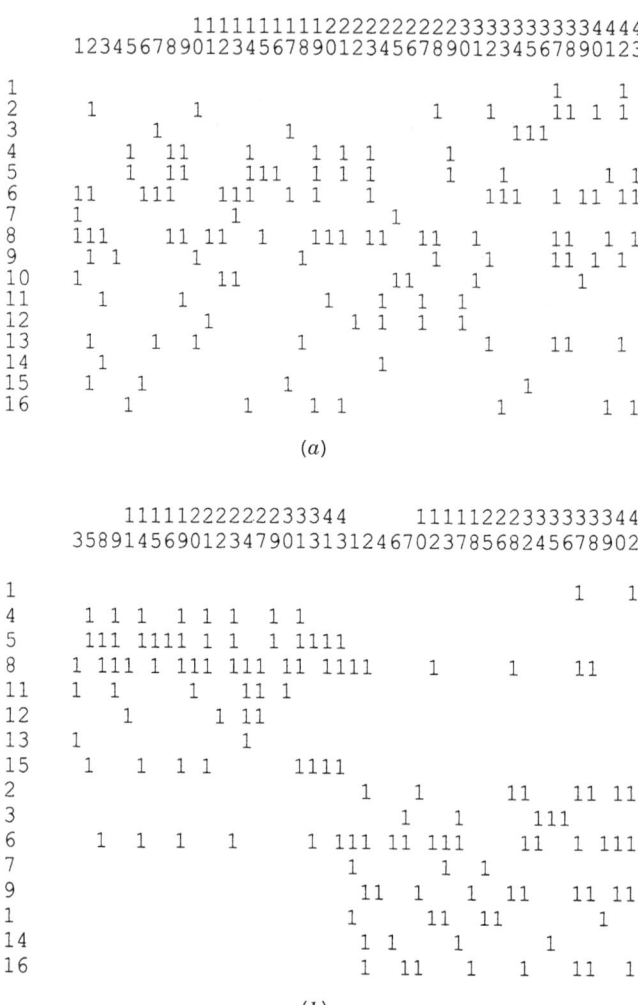

Figure 12. (a) Machine–component incidence matrix (from King and Nakornchai, 1982). (b) Block-diagonal structure for the matrix in (a).

of M/m. Allot the first h machines to group 1, the next h machines to group 2 and so on until all machines are allotted. Allot components to groups that visit the maximum number of machines. Reassign machines to components such that each machine is in the group where maximum components visit it. If the number of machines exceeds m, reassign machines to other groups such that the minimum increase in intercell moves is minimum.

The neighbor is generated as follows: Randomly pick a machine and assign it to an arbitrarily chosen group. If the upper limit of the new group capacity is exceeded, reassign any other machine from that group, ensuring minimum increase in intercell moves until a feasible solution is obtained.

Boctor's initial solution is the same as the first solution in Table 13, and his algorithm results in the last solution of Table 13 for two groups with eight machines in each. The intermediate solutions are different.

4.2. Genetic Algorithms

Genetic Algorithms (GAs) are based on the ideas of natural evolution of chromosomes. An initially chosen chromosome leads to several new chromosomes through operators such as crossover, mutation, and inversion. These are incorporated in the cell formation problem by considering each feasible solution to be a chromosome. There is a fitness function, which is evaluated and made progressively better by accepting superior chromosomes.

Venugopal and Narendran (1993) applied GAs for cell formation considering two objectives, namely minimizing intercell moves and maximizing grouping efficiency. We illustrate the use of ideas from GAs for the example in Figure 2(a).

Suppose two groups are to be formed with eight machines in each group. Consider an initial solution $a1 = [2122112111222211]$, which results in 23 intercell moves. Here $a(i)$ represents the group to which machine i is assigned. The vector **a** has 16 locations, one for each machine. Components are assigned to groups where they visit a maximum number of machines.

Generate two random numbers between 1 and 16, say, 3 and 15. Invert the values of locations 3 and 15 in **a**, resulting in a new solution. The old and new solutions are given below with values of locations 3 to 15 shown in bold:

Old solution ($a1$) = [21**2211211122221**1]
New solution ($a2$) = [21**1222211121122**1]

This operation is called *inversion*. The solution $a2$ has 31 intercell moves.

Consider the two solutions $a1$ and $a2$. Generate two random numbers between 1 and 16 say, 6 and 12. Create two more solutions $a3$ and $a4$ such that $a3$ is obtained from $a1$ with locations 6 to 12 taken from $a2$ and $a4$ from $a2$ with locations 6 to 12 taken from $a1$. The solutions $a3$ and $a4$ along with $a1$ and $a2$ are shown below with locations 6 to 12 shown in bold:

$a1 = [21221$**1211112**$22211]$
$a2 = [21122$**2211121**$1221]$
$a3 = [21221$**2211121**$2211]$
$a4 = [21122$**1211112**$21221]$

Solution $a4$ has 28 intercell moves. The operation that led to the evolution of two more feasible solutions (chromosomes) is called *two-point crossover*.

Consider solution $a4$ and an inversion operator between locations 7 and 10 to yield a solution $a5 = [2112211$**112**$2221221]$ with 31 intercell moves.

Consider solutions $a4$ and $a5$ and randomly choose a number between 1 and 16, say, 10. Create two new solutions $a6$ and $a7$ such that $a6$ has all elements of $a4$ up to location 10 and elements from $a5$ for locations 11 to 16. Similarly $a7$ has locations 1 to 10 from $a5$ and locations 11 to 16 from $a4$. These are shown below with locations 11 to 16 in bold:

$a6 = [2112212111\mathbf{221221}]$

$a7 = [2112211112\mathbf{221221}]$

The operator that led to solutions $a6$ and $a7$ is called *single-point crossover operator.*

Consider $a6$ and randomly generate two positions where there is a 1 and there is a 2. Suppose these are 8 and 7. Interchange the values to result in a solution $a8 = [2112211211221221]$. This operation is called *mutation.*

Consider $a8$ and perform mutation for locations 13 and 14 to obtain a solution $a9 = [2112211211222121]$. This solution has 15 intercell moves.

Thus, we can obtain the best solution using a careful combination of the genetic operators, namely single-point and two-point crossovers, inversion, and mutation.

Genetic algorithms perform these operations according to a prespecified number and accept better solutions progressively. The steps in a genetic algorithm for cell formation are as follows:

Step 1. Initialize population size (P) and size of mating pool (M).

Step 2. Generate as many feasible solutions as the population size. This constitutes a population. Compute the mean and variance of the fitness values (intercell moves) of the solutions in the population.

Step 3. Pick any solution randomly from the population and calculate $z = ($fitness $-$ mean$)$/standard deviation. Pick a random number r between 0 and 1. If abs(z) $\leq r$, accept the solution into the mating pool. Repeat this procedure to create a mating pool of size M.

Step 4. Perform genetic operations such as crossover, mutation, and inversion to create new offspring (feasible solutions). Accept the solution into the population if the fitness value of the offspring is less than the mean of the previous population. Create a population of size P.

Step 5. Perform steps 3 and 4 until the differences between the averages of two successive populations is less than a prespecified small value (say 0.001).

4.3. Neural Networks

A neural network is made up of a number of simple, highly interconnected, processing elements (neurons) that process information by their dynamic state response to external inputs. Neural networks mimic the human brain and show good performance with high computation rate using a parallel processing feature. A simple artificial neural network model is illustrated for the example in Figure 2(a).

Consider the component grouping problem. Since there are 16 machines, there can be a maximum of 16 groups. A 16×16 matrix of 1's is created, representing the 16 input nodes. Another 16×16 matrix with all values fixed at $1/(m + 1)$ called the weight matrix is created (for the example it is 1/17).

Each column of the incidence matrix is chosen and its similarity with each input node is calculated. The similarity is the dot product of the component vector and the column of the weight matrix corresponding to the input node. The input node with maximum similarity is chosen. Ties are broken by choosing the first among the tied nodes.

An acceptance ratio defined as the ratio between the number of 1's in the chosen input node and the number of matching 1's between the input node and the chosen column is computed. If this ratio is more than a prespecified value, the chosen column is attached to the input node. Otherwise, the next best-input node is chosen for consideration. This procedure is repeated until the component is attached to a node.

The input node is replaced by the chosen column and the weights corresponding to the input node are updated. This procedure is repeated until all columns are grouped. Input nodes that have not attracted any column are ignored. The component families are obtained from the input nodes that have attracted components. The machines required to produce the family of components are grouped to form the corresponding machine cells.

Kaparthi and Suresh (1992) applied this model to the matrix shown in Figure 3 and reported five machine groups {1,2,9,16}, {3,14}, {4,5,15}, {7,10}, and {11,12,13}. They left out machines 6 and 8 from the analysis since they are required by many components. If we allocate machines 6 and 8 to the third group where most components require them, we get a solution with 27 intercell moves.

With this algorithm, it is possible to obtain different solutions by changing the acceptance level and by changing the initial weights.

The Algorithm

Step 1. Create a matrix of input nodes with 1 in all the locations and a matrix of weights, W, with a fixed constant weight.

Step 2. Consider a column from the incidence matrix X_i. Compute the sum, $\sum_j (W_{ij} \times X_i)$. Choose node j such that the sum is a minimum. Break ties by choosing the first among them.

Step 3. Compute acceptance ratio r. If r is greater than a predefined R, accept input node and go to step 4. Otherwise, choose the next best node for consideration.

Step 4. Replace input node by the column chosen in step 2. Replace the weights corresponding to the chosen input node.

Step 5. Repeat steps 2–4 for all components and identify component families. Form machine cells by including all the machines required for producing the components. Reduce the number of machines to be duplicated by providing for intercell moves.

5. A BRIEF SURVEY OF RELATED ALGORITHMS FOR CELL DESIGN USING ZERO–ONE MATRICES

The algorithms presented in this chapter use the machine–component incidence matrix to form machine cells and component families. These can be programmed on a computer and can provide quick and efficient solutions. Software packages for some of these methods are available.

Algorithms have been developed in the last couple of decades for cell formation using zero–one incidence data. Owing to limitations of space and number, we have not described each of them in detail, though all of them provide good solutions to the cell formation problem. Some of these include the graph-theoretic approach developed by Rajagopalan and Batra (1975), MACE by Waghodekar and Sahu (1984), and Wei and Kern (1989).

Algorithms that use operation sequences and those that consider alternate routings for components have also been proposed. Rajamani et al. (1990) and Kusiak (1987) have developed algorithms that use alternate process plans. Vakharia and Wemmerlov (1990) and Harhalakis et al. (1990) have developed algorithms to design a cellular manufacturing system using operation sequences. Gunasingh and Lashkari (1989) have developed algorithms that trade off between intercell and intracell movements and include restrictions on cell size.

Askin and Subramanian (1987), Choobineh (1988), and Shafer and Rogers (1991) have developed models treating the cell formation problem as a cost minimization problem. Vannelli and Kumar (1986, 1987) have proposed algorithms that consider machine duplication and part subcontracting strategies while designing cells.

Chu (1989), Singh (1993), and Wemmerlov and Hyer (1989) have reported surveys of cellular manufacturing algorithms and applications. The first two reports survey algorithms for cell formation and identify areas for further research in this area. The third article reports the findings of a survey of 53 U.S.-based industries that practice cellular manufacturing. It indicated that most of these industries were aware of production flow analysis, rank-order clustering, and similarity-based methods as cell formation procedures and some of them have used these in their cell design.

6. CONCLUSIONS

This chapter presents a few cell formation algorithms using zero–one incidence data of machines and components. Each algorithm in this chapter has been explained using a numerical illustration that is widely used by researchers in this field. Algorithms using processing times, production volumes, alternative routes, and the like are discussed in the later chapters of this book.

All the algorithms discussed in this chapter are quantitative in their approach and involve computation of similarity coefficients and intercell moves. They can be done manually only for small problems. Software is available for these algorithms, some on a commercial basis for the ready use by practitioners. Alternatively all of these can also be easily coded into computer programs that can be used to obtain initial configuration of cells.

When applied to real-life problems, these algorithms can sometimes yield solutions with too many groups or too few groups, which are very large, since most of the methods do not consider limitations on cell size or number of cells. The solutions obtained using these models provide block-diagonal forms that are useful to obtain initial groups that can be modified to meet user specifications, such as number of cells, cell size, permissible number of intercell moves, and cell capacity. These solutions also help in understanding and analyzing the various possible cell configurations before finalizing the actual layout of cells.

REFERENCES

Askin, R. G., and Subramanian, S. P. (1987). A cost-based heuristic for group technology configuration, *International Journal of Production Research*, **25**, 101–113.

Boctor, F. F. (1991). A linear formulation of the machine-part cell formation problem, *International Journal of Production Research*, **29**, 343–356.

Burbidge, J. L. (1971). Production flow analysis, *Production Engineer*, **50**, 139–152.

Burbidge, J. L. (1977). A manual method for production flow analysis, *Production Engineer*, **56**, 34–38.

Chandrasekharan, M. P., and Rajagopalan, R. (1986). MODROC—An extension of rank order clustering for group technology, *International Journal of Production Research*, **24**, 1221–1233.

Chandrasekharan, M. P., and Rajagopalan, R. (1987). ZODIAC—An algorithm for concurrent formation of part families and machine cells, *International Journal of Production Research*, **25**, 835–850.

Chen, W. H., and Srivastava, (1994). Simulated annealing procedures for forming machine cells in group technology, *European Journal of Operational Research*, **75**, 100–111.

Choobineh, F. (1988). A framework for the design of cellular manufacturing systems, *International Journal of Production Research*, **26**, 1161–1172.

Chu, C. H. (1989). Cluster analysis in manufacturing cell formation, *Omega*, **17**, 289–295.

Gunasingh, K. R., and Lashkari, R. S. (1989). Machine grouping problems in cellular manufacturing systems—an integer programming approach, *International Journal of Production Research*, **27**, 1465–1473.

Harhalakis, G., Nagi, R., and Proth, J. M. (1990). An efficient heuristic in manufacturing cell formation for group tehnology applications, *International Journal of Production Research*, **28**, 185–198.

Hyer, N. L., and Wemmerlov, U. (1989). Group technology in the US manufacturing industry—A survey of current practices, *International Journal of Production Research*, **27**, 1287–1304.

Kaparthi, S., and Suresh, N. C. (1992). Machine–component cell formation in group technology—a neural network approach, *International Journal of Production Research*, **30**, 1353–1367.

King, J. R. (1980). Machine–component grouping in Production Flow Analysis: An approach using a rank order clustering algorithm. *International Journal of Production Research*, **18** 213–232.

King, J. R., and Nakornchai, V. (1982). Machine-component group formation in group technology: Review and extensions, *International Journal of Production Research*, **20**, 117–133.

Kirkpatrick, S., Gellat, Jr., and Vecchi, M. P. (1983). Optimization by simulated annealing, *Science*, **220**, 671–680.

Kumar, K. R., and Vannelli, A. (1987). Strategic subcontracting for efficient disintegrated manufacturing, *International Journal of Production Research*, **25**, 1715–1728.

Kusiak, A. (1987). The generalized group technology concept, *International Journal of Production Research*, **25**, 561–569.

Kusiak, A., and Chow, W. S., (1987). Efficient solving of group technology problem, *Journal of Manufacturing Systems*, **6**, 117–124.

McAuley, J. (1972). Machine grouping for efficient production, *Production Engineer*, **51**, 53–57.

Rajagopalan, R., and Batra, J. L. (1975). Design of cellular production systems: A graph theoretic approach, *International Journal of Production Research*, **13**, 567–569.

Rajamani, D., Singh, N., and Aneja, Y. P. (1990). Integrated design of cellular manufacturing systems in the presence of alternative process plans, *International Journal of Production Research*, **28**, 1541–1554.

Shafer, S. M., and Rogers, D. F. (1991). A goal programming approach to the cell formation. *Journal of Operations Management*, **10**, 28–43.

Shargal, M., Shekhar, S., and Irani, S. A. (1995). Evaluation of search algorithms and clustering efficiency measures for machine-part matrix clustering, *IIE Transactions*, **27**, 43–59.

Singh, N. (1993). Design of cellular manufacturing system—An invited review, *European Journal of Operational Research*, **69**, 284–291.

Sokal, R. R., and Sneath, P. H. A. (1963). *Principles of Numerical Taxonomy*, Freeman, San Francisco.

Srinivasan, G., and Narendran, T. T. (1991). GRAFICS—a nonhierarchical clustering method for the group technology problem, *International Journal of Production Research*, **29**, 463–478.

Srinivasan, G., Narendran, T. T., and Mahadevan, B. (1990). An assignment model for the part families problem in group technology, *International Journal of Production Research*, **28**, 145–152.

Vakharia, A. J., and Wemmerlov, U. (1990). Designing a cellular manufacturing system: A material flow approach based on operation sequences, *IIE Transactions*, **22**, 84–97.

Vannelli, A., and Kumar, K. R. (1986). A method for finding minimum bottlenecks for grouping part-machine families, *International Journal of Production Research*, **24**, 387–400.

Venugopal, V., and Narendran, T. T. (1993). A genetic algorithm approach to the machinecomponent grouping problem with multiple objectives, *Computers and Industrial Engineering*, **22**(4), 469–480.

Venugopal, V., and Narendran, T. T. (1994). Machinecell formation through neural network models, *International Journal of Production Research*, **32**(9), 2105–2116.

Waghodekar, P. H., and Sahu, S. (1984). MACE—Machine-component cell formation in group technology, *International Journal of Production Research*, **22**, 937–948.

Wemmerlov, U., and Hyer, N. L. (1989). Cellular manufacturing in the US industry—A survey of users, *International Journal of Production Research*, **27**, 1530–1551.

Wei, J. C., and Kern, G. M. (1989). Commonality analysis: A linear cell clustering algorithm for group technology, *International Journal of Production Research*, **27**, 2053–2062.

3

CELL FORMATION USING PRODUCTION FLOW ANALYSIS

Roland De Guio and Marc Barth

Laboratoire de Recherche en Productique de Strasbourg
Ecole Nationale Supérieure des Arts et Industries de Strasbourg
F-67084 Strasbourg Cedex, France

1. INTRODUCTION

The design of a Cellular Manufacturing System (CMS) is a complex task. The practitioner who is knowledgeable about the workshop is not necessarily aware of the operations research techniques and may wish to comprehend all the theoretical aspects before using sophisticated analytical techniques to design the manufacturing system.

Section 2 gives a general view of the main problems facing the cell designer starting with knowing how

- To represent the different aspects of the functioning of a shop using matrices representing the manufacturing routings
- To formalize in an objective manner the dilemmas linked with the functioning of a workshop organized into cells

Section 3 permits the deepening of the concepts introduced in Section 2 and concretely shows how to gather the data useful for the various routing analyses. Section 4 is a synthesis section that shows a manufacturing cell designing method

Handbook of Cellular Manufacturing Systems, edited by Shahrukh A. Irani
ISBN 0-471-12139-8 © 1999 John Wiley & Sons, Inc.

from the concepts developed in the previous sections. This method can also be integrated into a more general workshop designing method, which is based on the value engineering method, described in another chapter of this handbook.

Before beginning Section 2, let us introduce the terminology we shall use throughout this chapter: The terms *parts* and *machines* are generally used instead of, and as well as, the more general terms *items* and *workcenters,* except for Section 3 where their use seemed inescapable. Nevertheless, the method introduced can clearly be extended to any kind of items and workcenters in other manufacturing systems. It also seems necessary to clarify the concept of cells. A cell is a group of machines associated with a family of parts, the machine of which are geographically gathered together in a shop. It may or may not be an autonomous production unit.

The example in Figure 1 illustrates this definition. It represents the production flows of 10 machines within a shop divided into three subsets of machines, namely (1,2,3), (4,5,6), and (7,8,9,10), which are associated, respectively, with the families of parts (A,B), (C,D), and (E,F). The arrow labels define the parts traveling through the machines. Thus, part B successively travels through machines 1, 4, and 3. The sets of machines (1,2,3) and (4,5,6) are cells. Indeed, machines (1,2,3) of cell 1 essentially produce A and B, but machine 4 is required in order to produce B. Hence, cell 1 is dependent on cell 2. Conversely, cell 2 entirely controls the production of the family of parts (C,D) associated with it. It is nevertheless dependent on cell 1, since one of its machines intervenes in the manufacturing of part B carried out by cell 1. The third set of machines (7,8,9,10) is an autonomous cell, first because parts E and F are entirely produced within it and second because its machines are of no use for the production of other parts in the shop.

2. OVERVIEW OF CELL FORMATION

The cell identification process comprises three main steps:

1. Group analysis (Section 2.2)
2. Comparison of solutions obtained with Production Flow Analysis (PFA) (Section 2.3)
3. Final cell formation process (Section 2.4)

The first step consists in analyzing the existing routings from several point of views. At the end of this step, the splitting up of the shop into cells is obtained from the existing routings. During this stage, a great number of breakups of the machine set can be reached. Nevertheless, whatever the partitioning is, some machines are systematically or very often associated with the same parts. These similarities between solutions are identified and analyzed during the second step. The last step consists in reducing intercell flows by acting on many parameters of the manufacturing system whereas previous steps mainly analyzed and reorganized production flows resulting from the existing routings. Before developing these three steps, we describe the representations traditionally used to implement this process in Section 2.1.

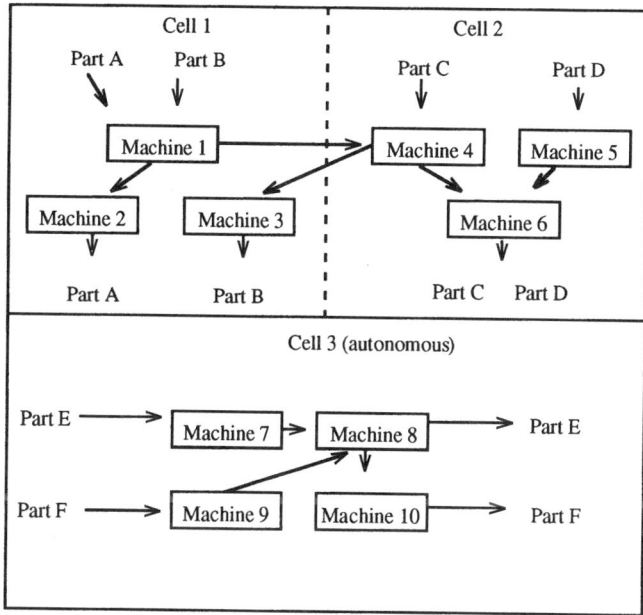

Figure 1. Layout into three cells including an autonomous cell.

2.1. Representation of the Problem

Let us consider an imaginary shop in which three types of parts, denoted by PA, PB, and PC, are manufactured. The shop is laid out in a job shop according to the diagram in Figure 2. The machines are gathered together according to similar type of processing technology. The first section gathers two lathes, noted L1 and L2; the second section comprises a milling machine, noted M1; the third one is made up of three drilling machines noted D1, D2, and D3. The routings in Figure 3 describe the production flows between the machines.

In industrial situations, the number of parts and machines is much greater, and the graphical representation of the flows as in Figure 2 is not feasible to permit an easy cell search. In order to do so, the routings are represented in the form of a matrix similar to the one in Figure 4. This representation leads to an easy automation of the reasoning we shall later describe. Many methods presented in this very book make use of it.

In this matrix, the part routings are listed in the rows. Henceforth, for part PA, the number 1 at the intersection of row PA and column L1 indicates that part PA makes use of machine L1 for the first operation of the routing, of machine D2 for the second routing, and of machine L2 for the third one. Thus, in this matrix, there are nine squares filled in corresponding to the nine operations to be carried out on PA, PB, and PC. The analysis of the table following the columns allows us to identify the

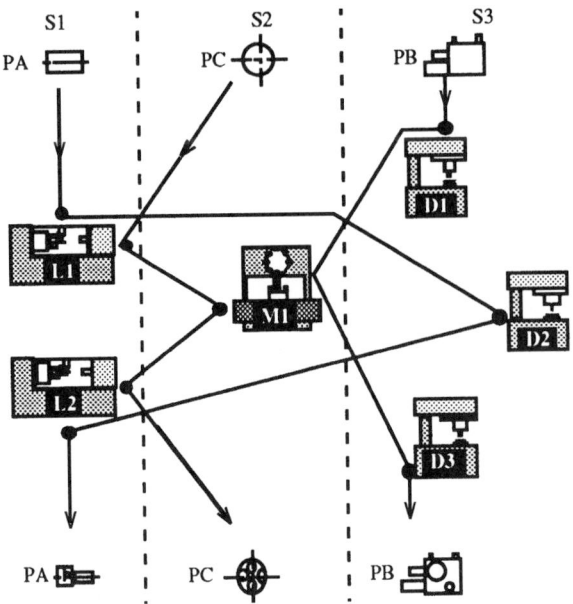

Figure 2. Job shop layout (flows).

Operation	PA	PB	PC
1	Turning on workcenter L1	Drilling on workcenter D1	Turning on workcenter L1
2	Drilling on workcenter D2	Milling on workcenter M1	Milling on workcenter M1
3	Turning on workcenter L2	Drilling on workcenter D3	Turning on workcenter L2

Figure 3. Routings.

	D1	D2	D3	L1	L2	M1
PA		2		1	3	
PB	1		3			2
PC				1	3	2

Figure 4. Matrix representation of the routings of the shop in Figure 2.

routings that make use of the machine corresponding to the column. In the example shown in Figure 4, we can read that machine L1 will be utilized by parts PA and PC.

2.2. Group Analysis

Different criteria can be used to create cells that lead to solutions that may not be the same. Each criterion is a way to analyze the production flow from a different angle. We shall now present the principal criteria we are using, their flow and matrix representation.

Search for Resource-Independent Parts   When two parts never use the same machines, they are said to be resource independent. The aim of this stage is to identify families of resource-independent parts that will lead to a first partitioning of the set of machines. The knowledge of the order of use of the workcenters in the routings is useless in detecting these sets of parts. The search is done as follows: It is customary to utilize a simplified representation of the table in Figure 4, also called part—machine matrix. To achieve this, we start with the matrix in Figure 4 and replace each nonzero value by a 1 and fill the other squares with 0's. We then have the binary matrix in Figure 5 in which the digit 1 represents the utilization of a machine by a part, whereas the digit 0 means that the part does not utilize the machine. We denote this matrix the operation part—machine matrix. In order to simplify the reading of the matrices, the 0's will not be included in the later representations of the matrices.

Looking at the Figure 5, it appears that the routing of part PA uses machines D2, L1, and L2, and the routing of part PB uses machines D1, D3, and M1. As these two parts never use the same machines, they are resource independent. On the contrary, part PC shares machines L1 and L2 with part PA, and machine M1 with part PB. The routing of PC is not independent of the routings of PA and PB. These findings can be visualized on the operation matrix by permuting the rows and columns of Figure 5. Figure 6 shows the resulting matrix where two potential cells can be identified. The

	D1	D2	D3	L1	L2	M1
PA	0	1	0	1	1	0
PB	1	0	1	0	0	1
PC	0	0	0	1	1	1

Figure 5. Binary representation of the part–machine matrix.

			Reference of the machine's group					
			1	1	1	2	2	2
			D2	L1	L2	D1	D3	M1
Reference of	1	PA	1	1	1			
the part's	2	PB				1	1	1
family	?	PC		1	1			1

Figure 6. Permuted binary matrix.

first one includes part PA and machines D2, L1, and L2; the second one comprises part PB and machines D1, D3, and M1. The shaded squares show the coupling of a part and a machine within the same cell and underline the independence of routings PA and PB. At this stage, part PC has not been linked with any group of machines; hence, no square of the row corresponding to part PC has been shaded.

The permutation, by arranging the machines and the parts in ascending cell number, inevitably leads to shaded squares forming diagonal blocks. In Figure 7, the machines of each previous identified cells have been geographically gathered together in order to point out the resource-independent parts. Notice that the flow organization in Figure 7 looks clearer than the one in Figure 2. Contrary to the job shops, the cells consist of machines with different technologies.

If part PC has to be manufactured, the question of the allocation of part PC to one of the cells arises. In Figure 8, this problem is graphically represented. We find that part PC uses machines (L1,L2) of cell 1 as well as machine M1 of cell 2. Whichever solution is chosen, the flow will stay unchanged. Some allocation criteria of part PC to the cell are described in the remainder of this section.

Minimization of Extracellular Operations This criterion is aimed at minimizing the number of operations to be carried out external to the cells to which the parts are assigned. That is the reason why we call this criterion the operation criterion. The number of operations to be carried out outside the cells is measured on the operation matrix by the number of 1's outside the diagonal blocks. If it was decided not to manufacture part PC in the example's shop, the flows could be represented by

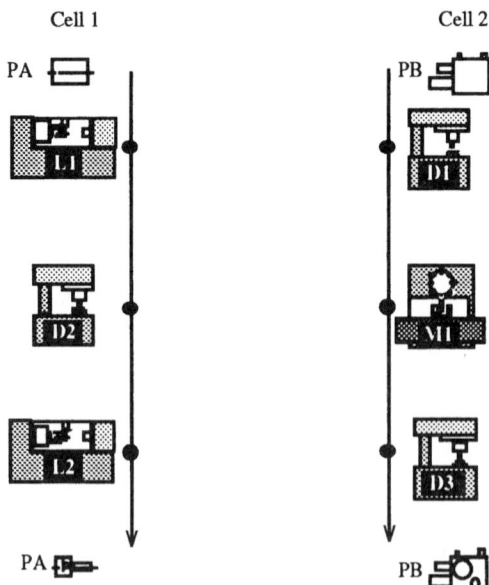

Figure 7. Resource-independent flows.

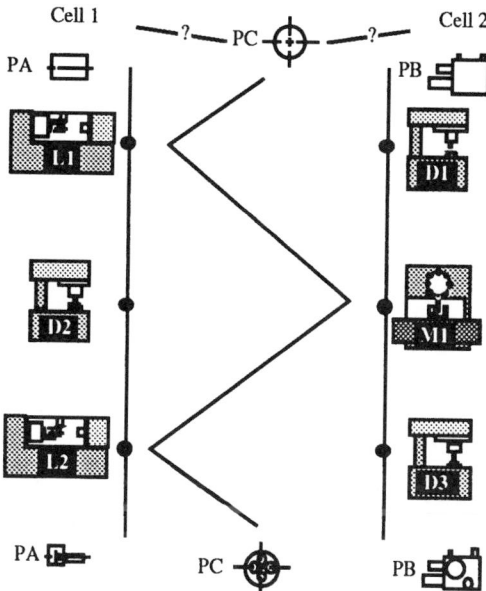

Figure 8. In which cell should part PC be classified?

Figures 9 and 7, and the number of extra cellular operations would be naught. Should part PC be manufactured in the shop, the purpose should be to allocate it to a cell in such a way that it minimizes the number of extracell operations. Let us explore, for instance, the degree of independence of the cells, according to the allocation of part PC.

If part PC is classified with part PB within cell 2 (Figure 10), there are two extracell operations that are the turning operations on lathes L1 and L2. If part PC is within the same cell as part PA (Figure 11), the only operation executed on the outside of the cells is the milling, on machine M1. This solution, graphically represented in Figure 12, minimizes the number of extracell operations. Each dot beside a machine represents an operation on this machine. The cells, the intracell and extracell operations, as well as the intercell and intracell moves (flows) can clearly be seen.

Minimization of the Extracellular Machines' Load   A machine load is the

		1	1	1	2	2	2
		D2	L1	L2	D1	D3	M1
1	PA	1	1	1			
1	PB				1	1	1

Figure 9

		1	1	1	2	2	2
		D2	L1	L2	D1	D3	M1
1	PA	1	1	1			
2	PB				1	1	1
2	PC		1	1			1

Figure 10

		1	1	1	2	2	2
		D2	L1	L2	D1	D3	M1
1	PA	1	1	1			
1	PC		1	1			1
2	PB				1	1	1

Figure 11

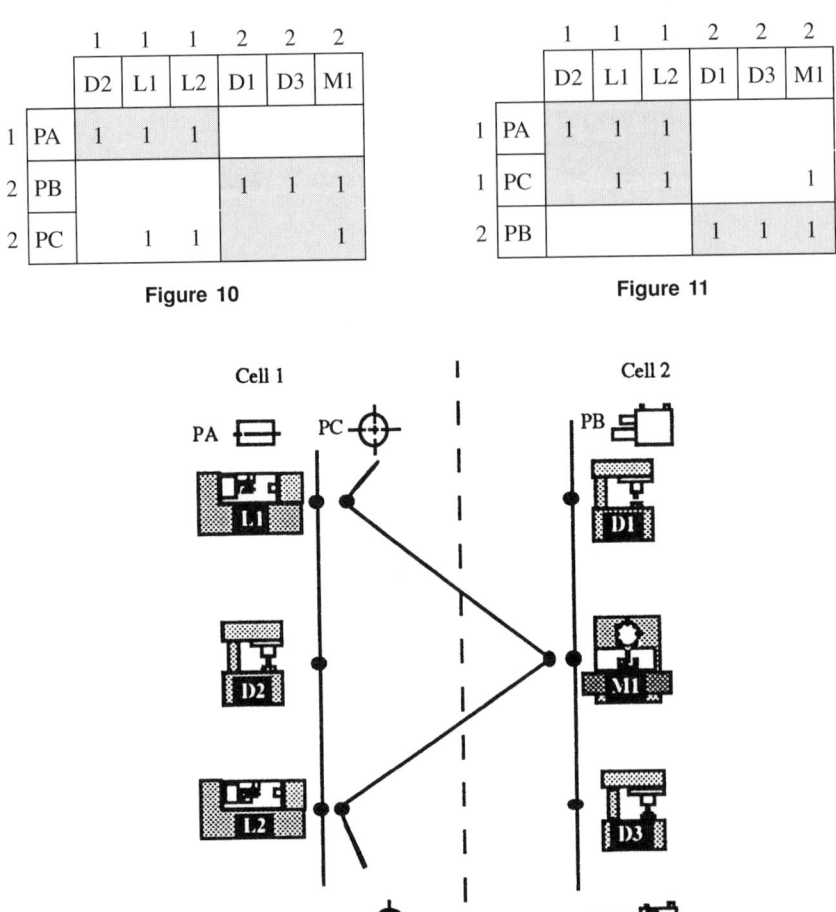

Figure 12. Cells according to the operation criterion.

amount of time required to produce a given number of parts on that machine. The purpose of the classification of the parts and machines according to the load criterion is to minimize the machine load for the entire set of extracell operations. In our example, part PA turning on lathe L1 lasts 4 hours, which means that machine L1's load is 4 hours for part PA. This is expressed by replacing digit 1 in the (PA,L1) square of the operation matrix (Figure 5) by digit 4. Proceeding that way for all the operations leads to a part—machine matrix whose generic element corresponds to the machine load for a part. As in the case of the operation minimization, two assignments of part PC can be considered, as shown in Figures 13 and 14. If part PC is allocated to cell 1 (Figure 13), the milling operation, which corresponds to a load of 50 hours for the milling machine M1, is executed outside the cell. If part PC is allocated to cell 2 (Figure 14), two operations corresponding to a total load of 12 hours $(2+10)$ are executed outside the cell. Hence the solution represented in Figure 14 best satisfies the machine load criterion. Also, note that this solution does not optimize the operation criterion. The solution which optimizes the load criterion is represented in Figure 15 on which the cells, the operations, the extracell and intracell loads, as well as the flows can clearly be seen.

Minimization of the Intercells Production Flow   Both preceding classifications make use of the routings' similarity criteria but do not take into account the flows of parts among the machines. The classification according to the flow criterion is based on a measure of the part moves between machines and allows us to gather in a same cell the machines between which the demand for labor is high. Figure 16 keeps the same organization into cells as in Figure 15, but the arrows and their labels

		1	1	1	2	2	2
		D2	L1	L2	D1	D3	M1
1	PA	2	4	3			
1	PC		2	10			50
2	PB				3	8	10

Figure 13

		1	1	1	2	2	2
		D2	L1	L2	D1	D3	M1
1	PA	2	4	3			
2	PC		2	10			50
2	PB				3	8	10

Figure 14

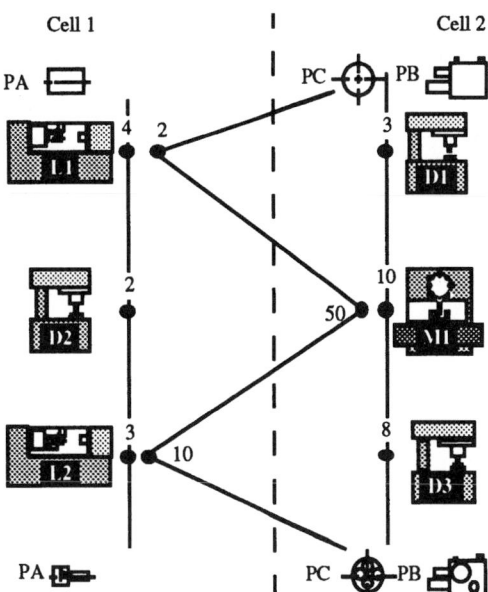

Figure 15. Cells obtained with the load criterion.

indicate the number of manufacturing orders transiting through the shop. Parts PA and PB are manufactured, respectively, with 3 and 2 production orders. Part PC is manufactured with 50 production orders. To underline this major difference, the flow corresponding to the PC routing is represented with a larger arrow. Seen from this new point of view, machines L1, M1, and L2 would intuitively be placed in the same cell.

This observation can also be seen on the matrix in Figure 17 where the loads of Figure 14 are replaced by the number of production orders of each part. The intercell flow (50+50 = 100) is larger than the intracell flow (3+3+3+2+2+2+50 = 65).

The organization into cells proposed in Figure 18 describes a solution that minimizes the intercell flow (50+50+50+3+2+2 = 157 intracell against 3+3+2 = 8 intercell). This solution, satisfactory with respect to the intercell moves, is graphically

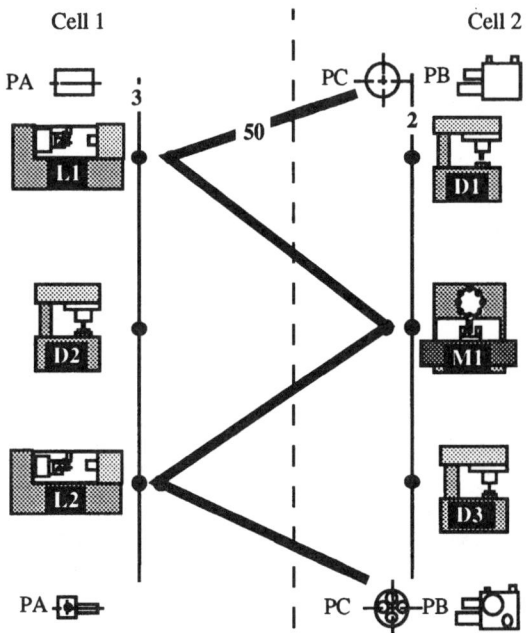

Figure 16. Flow between the machines.

		1	1	1	2	2	2
		D2	L1	L2	D1	D3	M1
1	PA	3	3	3			
2	PC		50	50			50
2	PB				2	2	2

Figure 17. Flow matrix and cells minimizing the extracells load.

		1	1	1	2	2	2
		L1	L2	M1	D2	D1	D3
1	PC	50	50	50			
2	PA	3	3		3		
2	PB			2		2	2

Figure 18. Cells minimizing the intercellular flow.

represented in Figure 19, where the flow corresponding to the largest arrow stays within cell 1. It is worth noting that in this example, the solution that best fulfills the flow criterion is a solution to none of the criteria previously studied. Let us remark that in this example we took the production orders as a measure of the production flow; and depending on the problem other measures can approximate the production flow.

2.3. Analysis of Basic Production Flow Analysis Similarities

We have just shown that the solutions are not always identical in terms of part families or group of machines, depending on whether the criterion aims at minimizing the number of extracell operations, the load of the machines due to the extracell operations, or the intercell flow. Figures 20, 21, and 22 remind us of the three solutions minimizing the studied criteria (Figures 11, 14, and 18). None of the cells among these solutions is identical to another solution's cell. Nevertheless, some similarities can be found. Some machines, such as L1 and L2, are systematically placed in the same cell, whatever criteria is used; whereas other machines, such as D1 and L1, are never assigned to the same cell. We call this convergence of points of view total agreements. The cell designers, taking into account the standpoint reflected by the three criteria, know that pairs of machines (L1,L2) and (D1,D3) are robust cell cores. On the contrary, D2 is classified with L1 and L2 (Figures 20 and 21) and then with D1 and D3 (Figure 22); and the cell designers who want to take into account the three preceding criteria has to find a consensus about machine D2 allocation.

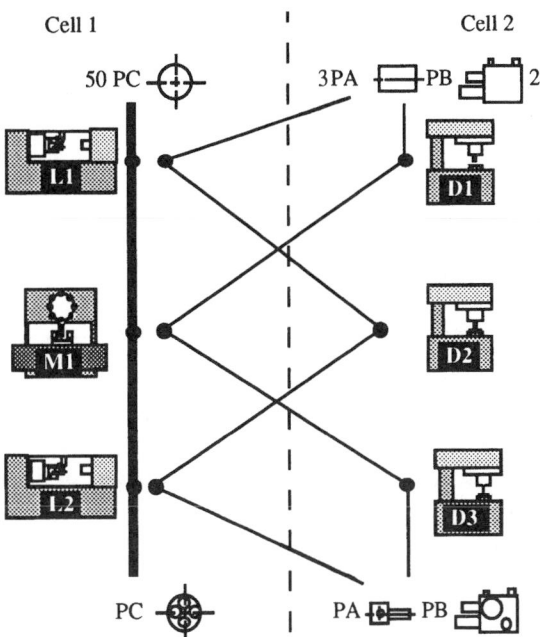

Figure 19. Cells obtained from minimizing the intercellular flows.

		1	1	1	2	2	2
		D2	L1	L2	D1	D3	M1
1	PA	1	1	1			
1	PC		1	1			1
2	PB				1	1	1

Figure 20. Cells based on the operation criterion.

		1	1	1	2	2	2
		D2	L1	L2	D1	D3	M1
1	PA	2	4	3			
2	PC		2	10			50
2	PB				3	8	10

Figure 21. Cells based on load criterion.

		1	1	1	2	2	2
		L1	L2	M1	D2	D1	D3
1	PC	50	50	50			
2	PA	3	3		3		
2	PB			2		2	2

Figure 22. Cells based on flow criterion.

	D1	D2	D3	L1	L2	M1
D1	3	1	3	0	0	2
D2	1	3	1	2	2	0
D3	3	1	3	0	0	2
L1	0	2	0	3	3	1
L2	0	2	0	3	3	1
M1	2	0	2	1	1	3

Figure 23. PFA similarity matrix.

	D2	D3	L1	L2	M1
D1	–	3	0	0	–
D2		–	–	–	0
D3			0	0	–
L1				3	–
L2					–
M1					

Figure 24. Total agreement matrix.

The identification of robust cell cores and the search for consensus's is greatly facilitated by the use of a machine—machine matrix similar to the one in Figure 23 and automatic clustering techniques. The content of any of this matrix's squares indicates the number of times two workcenters are classified together in the same cell, all criteria being considered. We name this matrix the PFA similarity matrix. Digit 3, row D1, column D3 indicates that these two machines are classified three times together in the same family, as can be noticed on the above matrices. It also corresponds to the number of studied solutions. The digit 0 at the intersection of row D1 and column L1 indicates that these two machines are never grouped together in the same cell, whichever cells are proposed. The values 3 and 0 therefore indicate total agreements between solutions. The other values comprised between 0 and 3 indicate disagreements, which can be understood as partial agreements. Indeed, digit 2 at the intersection of row D2 and column L1 indicates that these two machines are in the same cell twice out of the three proposed solutions (Figures 20 and 21). The matrix in Figure 23 is a symmetric matrix, and its diagonal elements are always equal to 3; they therefore do not give any useful information. The total agreements thus obtained are represented in the upper semimatrix of Figure 24 in which only the 3's and 0's are kept. Figure 25 graphically represents the questions facing the cell designer in the example's case after identifying the two robust cell cores (L1,L2) and (D1,D3). The designers will have to allocate the parts as well as machines D2 and M1 to the cell cores. Contrary to our example, the number of total agreements -0 and 3 values- are more numerous than partials agreements in most industrial cases. The consensus's are usually quickly and easily found case by case.

The concept of PFA similarities can be applied to the parts, to the whole set of parts plus machines, as well as to the machines. It is herein developed in relation to the machines, but the reader can easily transpose it to the parts or the set of parts plus machines. It is interesting to note the lack of total agreement concerning the parts in our example. But the use of the parts plus machine matrix would reduce the number of questions the designers have to answer. Indeed, part PB is always in the same cell as the machines of the cell core (D1,D3), so assignment of part PB is not a problem. Moreover, part PA is always classified in the same cell as machine D2, thus part PA and machine P2 assignment will be done simultaneously.

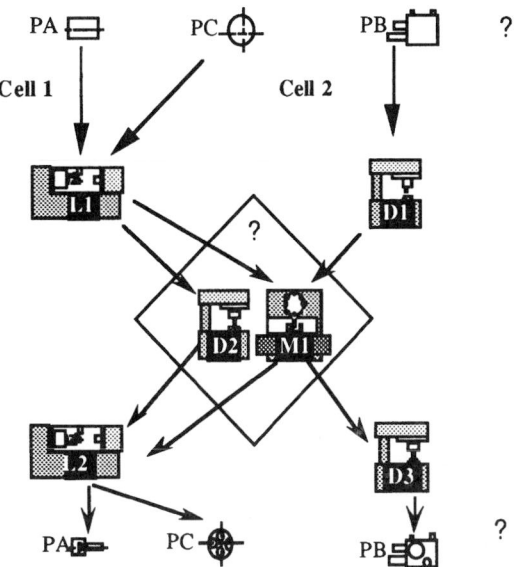

Figure 25. Visual representation of the assignment to perform.

2.4. Final Cell Formation Process

Whatever the solutions proposed by the previous routing analyses, moves between cells are noticed in any case. These moves prevent any decentralized cell management. In some cases, the elimination of the intercellular moves can be considered. Numerous solutions are developed in Section 4. As an example, it is shown how to solve this problem through an investment in machines. If we use the classification obtained from the operation criterion in Figure 26, we notice that one of the operations for part PC is executed outside of cell 1 since it requires machine M1. If the firm can justify the purchase of a second machine, noted M1A, executing the same tasks as M1, its allocation to cell 1 leads to the elimination of this intercellular move. The matrix in Figure 27 shows the elimination of all 1's outside the diagonal blocks, and thus of all intercellular moves. Two independent cells have just been created. Figure 28 graphically represents this solution.

3. BASICS FOR PRODUCTION FLOW ANALYSIS PRACTICE

Section 2 has given an overview of the complexity of designing cells by

- Showing the multiplicity of the points of view and the way to represent it through matrices
- Suggesting the way to analyze the matrices representing the points of view
- Presenting a way to synthetize the points of view by means of matrices to be analyzed

The discussion in this section describes tools to implement the previous concepts. It is first shown how to extract from a production control software the data necessary for the routing analysis (Section 3.1) and the way to construct the various matrices to be analyzed (Section 3.2). The set of matrices to be analyzed—operation matrix, load matrix, and the like—depends strongly on the firm's objectives; Section 3.3 states a few links between the objectives and these matrices. Whenever the matrices are made up of many rows or columns, it is advisable to make use of automatic classification methods to analyze them. Section 3.4 sets out a few general rules, which should

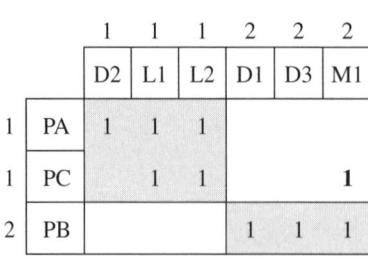

		1 D2	1 L1	1 L2	2 D1	2 D3	2 M1
1	PA	1	1	1			
1	PC		1	1			1
2	PB				1	1	1

Figure 26. Initial solution with the operation criterion.

		1 D2	1 L1	1 L2	2 M1 A	2 D1	2 D3	2 M1
1	PA	1	1	1				
1	PC		1	1	1			
2	PB					1	1	1

Figure 27. Independent cells obtained by purchasing of a machine.

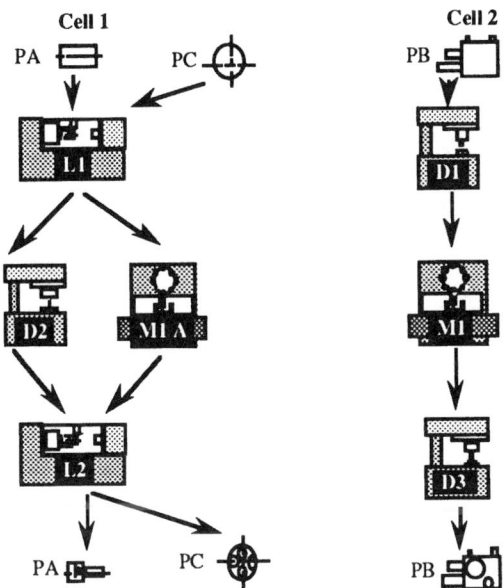

Figure 28. Example of final solution with two independent cells.

be kept in mind in order to properly use the routing analysis softwares. Section 3.5 shows how to analyze consensus. Section 3.6 is about visualization and performance measures, which permit the comparison of the various solutions.

3.1. Data Required

Production Flow Analysis makes use of the information concerning the items, the workcenters, and the routings. These data are traditionally easy to extract from the databases of the production control software. The three essential data files are the items, which are not necessarily parts, workcenters (machine), and routings files. The useful fields of these files are specified below. The fields in italics are not included in the database but can be deduced from it. It is shown how to obtain them. Extracts from an industrial example illustrate the definitions. It concerns a firm manufacturing bakery ovens. The items, workcenters, and routings files contain, respectively, 3000, 30, and 8500 entries. In order to be concise, mere extracts of these files are presented.

The Items File   An item is an article stocked in a warehouse and ready for use by another shop or by a customer. It can consist of raw materials, an elementary part, a subassembly, or a finished product. An item is distinguished by the characteristics listed in the items file (Figure 29).

Item Number. Coded item's denomination

Designation. Text describing the item

Item Number	Designation	Annual Quantity	Lot Size	Transfer Batch	Number of Production Orders	Number of Transfer Orders	*Annual Load*
00023641	TO PERF AG3+T2	33000	3300	3300	10	10	*1125.7*
05096401	JOINT KLINGERIT	61000	6100	1525	10	40	*4274.1*
30111101	RENF FERMETURE	24000	120	120	200	200	*1808.0*
58012901	TRAVERSEE SONDE	30000	3000	750	10	40	*757.5*
58013101	TRAVERSE	70000	3500	875	20	80	*4210.0*
....	....	....	....	....	....	....	....

Figure 29. The Item file.

Annual Quantity. Amount produced on the cell designing study horizon, corresponding herein to one year, that is, 240 open days

Lot Size. Amount produced between two line changes

Transfer Batch. Amount transferred between two workcenters during the handling. The transfer batch is less or equal to the lot size.

Number of Production Orders. Annual quantity to lot size ratio. Concerning item 30111101, the number of production orders is 24000/120 = 200.

Number of Transfer Orders. Annual quantity to transfer lot size ratio. Concerning item 580012901, the number of transfer orders is 30000/750 = 40.

Annual load. Summation of the annual item's routing operation load. Concerning item 5096401, annual load = $611.60_{\text{routings}} + 3662.50_{\text{routings}} = 4274.1$.

The indices indicate the files in which the numerical values are found.

Workcenters File A workcenter can be made up of one or more machines and/or an operator. Its useful characteristics are listed in the workcenter file (Figure 30).

Workcenter Number. Coded workcenter's denomination

Designation. Text describing the workcenter

Workcenter Number	Designation	Number	Theoretical Annual Capacity	Loss of Capacity	Total Actual Capacity	*Annual Load*	*Capacity Utilization*
4103	Bombled 28 J shears	1	1920	10%	1728	*1372.60*	*79.4%*
4104	Bombled 9 F shears	3	1920	15%	4896	*3796.20*	*77.5%*
4105	Behrens 9 F shears	1	1920	15%	1632	*992.00*	*60.8%*
4121	Behrens punch	1	1920	10%	1728	*1125.00*	*65.1%*
4151	Languepin welder	3	1920	10%	5184	*4889.50*	*94.3%*
....	....	....	....	....	....	....	....

Figure 30. The workcenter file.

Number. Number of available workcenters of the same kind

Theoretical Annual Capacity. Workcenter's availability over a year. In this example, the capacity is 8 hours a day for 240 working days, that is, 1920 hours for each post.

Loss of Capacity. Percentage of theoretical capacity loss due to breakdowns and other production cuts

Total Actual Capacity. Workcenter's actual availability over a year. It is equal to theoretical annual capacity per workstation multiplied by the number of workcenter's minus the capacity losses. Concerning workcenter 4104, the total actual capacity is equal to $1920 \times 3 \times (1 - 0.15) = 4896$ hours.

Annual Load. Summation of the annual operations load of the routing using workcenter. For workcenter 4104: Annual load $= 133.20_{routing} + 3662.5_{routing} = 3796.20$ hours.

Capacity Utilization. Ratio of the load to the actual capacity of workcenter 4104: capacity utilization $= 3796.20/4896 = 77.5\%$.

Routings File   An item's routing represents all the operations required for its production. An example of a routings file is shown in Figure 31.

Item. Item number

Operation. Operation's serial number. Item 0023641 requires 2 operations. The first one makes use of workcenter 4104 and the second one uses workcenter 4105.

Workcenter. Workcenter number

Setup Time. Time necessary for the workcenter's adjusting before the production of a lot

Run Time. Time necessary for an item operation to be executed

Item	Operation	Workcenter	Setup Time	Run Time	Operation's Annual Load
00023641	1	4104	0.17	0.004	*133.70*
00023641	2	4105	0.20	0.030	*992.00*
05096401	1	4103	0.16	0.010	*611.60*
05096401	2	4104	0.25	0.060	*3662.50*
30111101	1	4103	0.16	0.001	*56.00*
30111101	2	4121	0.20	0.020	*520.00*
30111101	3	4151	0.16	0.050	*1232.00*
58012901	1	4121	0.50	0.020	*605.00*
58012901	2	4151	0.25	0.005	*152.50*
58013101	1	4103	0.25	0.010	*705.00*
58013101	2	4151	0.25	0.050	*3505.00*
. . . .	. . . .	. . . .	. . . .	. . . .	. . . .

Figure 31. The routing file

Operation's Annual Load. Time necessary for this load corresponds to operation's run time multiplied by the item's annual quantity added to the setup time multiplied by the number of manufacturing orders. Concerning operation 2 of item 05096401, its annual load is equal to: $0.060_{routing} \times 61000_{item} + 0.25_{routing} \times 10 = 3662.50$ hours.

3.2. Data Representation Mode

Section 2 showed how to represent and handle the routings with matrices in order to design cells according to some criteria. We now explain how to create these matrices and some others from the above item, workcenter, and routing files. Each matrix can be used to analyze the routings from a standpoint. The link between each matrix and the point of view it represents is discussed in Section 3.3. We shall present separately the formation of the item–workcenter and workcenter–workcenter matrices.

The item–workcenter matrices contain the information linking the items to the workcenters. In these matrices, a row corresponds to an item and a column to a workcenter. Each element within the matrix informs about an item's routing operation. The matrix's element may be binary (Figure 32) or weighted by the operation's load (Figure 33), by the number of manufacturing orders or by the number of transfer lots (Figure 34). The matrices in Figures 32 and 33 are deduced from the sole routing file, whereas the creation of the matrix in Figure 34 requires the item file too. Other weightings, such as the weight of the items to be moved from a workcenter to another, are sometimes also used. Depending on the classification criteria used, the number of extracell operations (Figure 35), the machines' extra cellular load (Figure 36) or the number of transfer orders involving the carrying of items between cells (Figure 37), will be minimized.

The workcenter–workcenter matrices, which are not described in Section 2, link the workcenters to each other. Each matrix's element measures a global flow between two workcenters, not taking into account the direction of the flow. These are symmetric matrices. The binary operation matrix in Figure 38 is obtained from the sole routing file. Digit 1 tying workcenters 4103 and 4104 means that at least one sequence of operations successively using these workcenters exists. The computing of the matrix in Figure 39 makes use of the routing and item files. Thus, the value 240 of the square (4151,4121) is obtained from the following calculation: the existence of the sequence (4121,4151) corresponding to items 30111101 and 58012901 is detected on the routings file. The number 240 ($240 = 200 + 40$) is the summation of the number of these items' transfer orders defined in the item file. The classification of the binary workcenter–workcenter matrix minimizes the moves between cells without taking into account the quantities moved (Figure 40). The classification of the weighted workcenter–workcenter matrix minimizes the moves between cells and takes into account the transferred amounts (Figure 41). Other weightings, such as the weight to be handled, are sometimes used.

3.3. Link between Objectives and Type of Representation

Various types of information and criteria, which can affect the analysis of the routings, have been described in the previous sections. It is not always necessary, possible, or

		Workcenter				
		4103	4104	4105	4121	4151
	23641		1	1		
	5096401	1	1			
Item	30111101	1			1	1
	58012901				1	1
	58013101	1				1

Figure 32. Operation item–workcenter matrix.

		1	1	2	2	2
		4104	4105	4103	4121	4151
1	23641	1	1			
1	5096401	1		1		
2	30111101			1	1	1
2	58013101			1		1
2	58012901				1	1

Figure 35. Classified Figure 32's matrix.

		Workcenter				
		4103	4104	4105	4121	4151
	23641		133,7	992		
	5096401	611,6	3662,5			
Item	30111101	56			520	1232
	58012901				605	152,5
	58013101	705				3505

Figure 33. Load item–workcenter matrix.

		1	1	2	2	2
		4104	4105	4103	4121	4151
1	23641	133,7	992			
1	5096401	3662,5		611,6		
2	30111101			56	520	1232
2	58013101			705		3505
2	58012901				605	152.5

Figure 36. Classified Figure 33's matrix.

		Workcenter				
		4103	4104	4105	4121	4151
	23641		10	10		
	5096401	40	40			
Item	30111101	200			200	200
	58012901				40	40
	58013101	80				80

Figure 34. Flow item–workcenter matrix.

		1	1	1	2	2
		4103	4121	4151	4104	4105
1	30111101	200	200	200		
1	58012901		40	40		
1	58013101	80		80		
2	23641				10	10
2	5096401	40			40	

Figure 37. Classified Figure 34's matrix.

advisable to analyze all the matrices presented here. When all the data are available, the cell designers must select the data to be analyzed according to the objectives of the analysis. Some remarks are made here on various scenarios that could arise.

Annual Quantities To Be Produced Are Not Known   Some analyses are therefore impossible. Indeed, neither the load of each operation nor the number of production orders can be calculated, and only the item–workcenter or workcenter–workcenter binary matrices are available.

Figure 38. Binary workcenter–workcenter matrix.

Figure 40. Classified Figure 38's matrix.

Figure 39. Flow workcenter–workcenter matrix.

Figure 41. Classified Figure 39's matrix.

Production Forecasts Are Too Imprecise The operation criteria appears useful when the production forecasts are too uncertain or vague. Studying the structure of the routings independently from the future production quantities might thereby be preferred. Other analyses should, however, not be ruled out; the combination of the operation criterion with other weighted criteria can be studied by means of the concept of consensus.

Weighted Matrices Should Generally Be Preferred to Binary Matrices

Weighted matrices more precisely echo the reality of the firm. Industrial cases have established that the load, the number of production orders, the annual quantities, or the costs of the items often have a Pareto distribution. The main part of the machines' workload is dominated by a small number of items. Likewise, the majority of part flows concerns only a limited number of workcenters. In that case, the results of the analysis of the binary and weighted matrices differ greatly, since the classification-weighted criteria naturally give more weight to the most loaded machines or to the workcenters between which the flows are numerous (see Figures 21 and 22).

Analyzing the Production Lead Times The reduction of the production lead times is the main objective mentioned by industrialists faced with the laying out of a shop. The analysis of an item's production lead time shows that a great part of this time is due to the queue time at its workcenters. The number of transfer orders represents, on the study horizon, the number of queues for an operation of the routing. From the use of the number of transfer orders are obtained cells for which the number of queues of extracellular parts is minimal. A production manager can thereby have within his/her cell the greater part of the queues of which he/she is in charge. This production manager can thus more easily curb or even reduce an item's production lead time. In the example introduced in Section 3.2, the classification minimizing the number of queues is given by Figure 37. The production manager of cell 1 has to handle 40 queues of item 0509640 linked to cell 2, but entirely handles the 900 queues of cell 1's items ($900 = 3 \times 200 + 2 \times 40 + 2 \times 80 + 2 \times 10 + 1 \times 40$).

Item–Workcenter and Workcenter–Workcenter Matrices   The type of matrix, item–workcenter or workcenter–workcenter, has little effect on the results of the workcenters' classification. The item–workcenter matrices indicate the operations bringing about intercellular moves, without stating which upstream or downstream workcenter causes these moves. The workcenter–workcenter matrices give information about the workcenters between which moves exist, without indicating which items originate these moves. Henceforth, the choice of the use of one of these two matrix types is not really a problem. Since they are complementary, both types of matrix are useful to study industrial cases. Nevertheless, if the matrix's size is considerable, a representation on paper medium turns out to be impossible. The number of pages for a matrix of these type, comprising 10,000 items and about a hundred workcenters, is close to 100. In that case, the representation of the workcenter–workcenter matrix is preferred, for its size is always visually reasonable.

3.4. Algorithms for Production Data Analysis

In industrial cases, the amount of data justifies the use of classification algorithms. Two examples illustrate the complexity of the classification problem. In a firm producing farming tractors, the workshop includes 250 items and 130 workcenters, and the routing file comprises 1800 operations. In another firm producing storage cupboards for electrical equipment, the sheetmetal workshop produces 10,345 items with the help of 81 workcenters and 36,207 operations. The manual analysis of such an amount of data is tedious, costly, or even impossible. Recent developments in the field of classification allow the designer to make use of fast and reliable algorithms. These algorithms are not supposed to give the best possible solutions; they simply help the workshop designer who can easily, using a personal computer, calculate and visualize a set of solutions into cells. The use of such algorithms constitutes an assistance to creativity for the search for cells. Section 4 integrates the use of such algorithms in a cells searching method. Whatever the classification criteria are, the cell calculus algorithms follow almost the same principle. It is a question of forming item and workcenter families by minimizing the quantities outside the diagonal blocks. Nevertheless, this formulation is incomplete and gives us the opportunity to introduce the problem of the workcenter and item families' sizes. If the previous formulation was used, the self-evident solution of a cell including all items and workcenters would clearly be justified. More generally, the smaller the cell's size aimed at, the more numerous the intercell moves and extracell operations. This intuitive result, illustrated in Figures 42 and 43, is mathematically established.

The previous formulation has thus to be complemented with the specification of the objective. Several approaches are possible. It can be decided, for instance, that at least two cells are necessary. If the solution thus obtained comprises two oversized cells, a solution comprising at least three cells will be searched for and so on. It is preferred to fix the maximal rate of the studied quantities—operations, transfer orders, number of queues, and the like—outside the cells. It can be shown that the use of one of the previous criteria leads to a finite number of solutions. In practice, for the last criteria and for a given matrix, this number is seldom greater than 20 and nearly always less than 10 in industrial cases.

In actual fact, the cell formation algorithms present a whole collection of different solutions depending on the rate. Figure 44 represents the evolution of the number of cells according to the percentage of extracellular operations obtained from the operation item–workcenter matrix of a production shop of a firm producing farming tractors. This shop comprises 125 machines and manufactures 250 types of parts. In this example, only eight solutions noted S1 to S8 are suggested. In this figure, it is confirmed that the number of cells and the extracellular operation rate increase concurrently. Solution S8 includes item and workcenter families of a size larger than in the other solutions.

Choosing the size of a cell is a complex issue. It is verified, in industrial firms, that the number of machines per cell varies between 2 and about 30. The criteria involved in this choice are mainly the number of operators, the number of machines, the cell's surface, and the civil engineering constraints. If the cell is too large, its visual and mental mastery by a production manager or a foreman is impossible. If too small, the cell requires too many annex resources. An arbitrage is necessary.

		1	1	2	2	3
		M1	M2	M3	M4	M5
1	P1	1	1			
1	P2	1	1		1	
2	P3			1	1	
2	P4			1	1	
3	P5				1	1
3	P6	1			1	1

Extra cell operations = 4

Figure 42. Great number of cells.

		1	1	2	2	2
		M1	M2	M3	M4	M5
1	P1	1	1			
1	P2	1	1		1	
2	P3			1	1	
2	P4			1	1	
2	P5			1		1
2	P6	1		1		1

Extra cell operations = 2

Figure 43. Small number of cells.

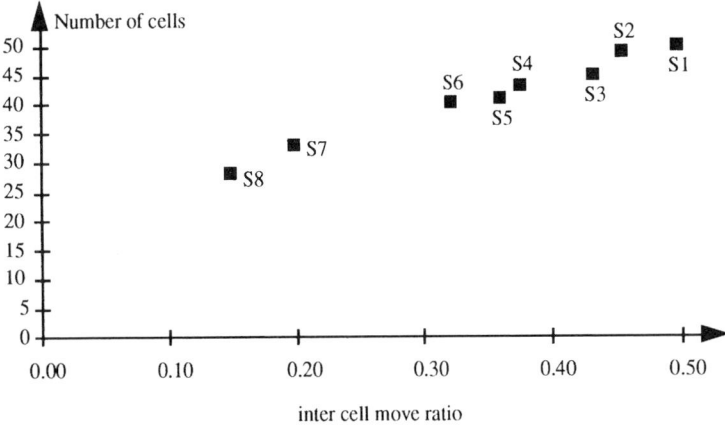

Figure 44. Variation of the number of cells with respect to the intercell moves.

Nowadays, literature on the subject abounds in proper cell formation algorithms. These algorithms lead to approximately identical solutions and selecting the best algorithm is not an actual problem. In an industrial case study, it is essential to make use of an algorithm authorizing the analysis of all the points of view mentioned in this chapter, and suggesting not one but several solutions. These algorithms must first of all be employed as creativity and exploration tools, since all the technical, economic, social, and organizational constraints inherent in cell formation cannot be modeled simultaneously and solved in one go by any method.

3.5 Algorithm to Search Consensus among Production Flow Analysis Solutions

Whichever criteria is used, several workcenters and items often or even always are allocated to the same families throughout the different classification processes. Other workcenters seldom or even never are gathered together in the same cell. This is what we named a total agreement in Section 2.3. The knowledge of this result is important for the implementation of the project since it facilitates the activities of a working party and increases its efficiency. Indeed, if the working party identifies the total agreements, it can focus on finding consensus about the disagreement points, which are generally less numerous than the agreement points. Algorithms for searching for consensus promptly lead to the analysis of all the solutions generated by the classification algorithms. It has been shown in Section 2.4 how to conceive a PFA similarity matrix; in this section, a method is suggested to analyze such a matrix. The explanations are based on the study case given in Figure 45. The elements of the main diagonal of the matrix indicate the number of solutions studied in order to conceive the PFA similarity matrix. In our example, the matrix summarizes the solutions of 3 PFAs. Some pairs of workcenters, such as (M6,M8), always are placed in a common cell independently of the solution, and other pairs of workcenters, such as (M6,M9),

	M1	M2	M3	M4	M5	M6	M7	M8	M9	M10
M1	3		1		2				3	
M2		3	2	1		1	3	1		3
M3	1	2	3	1		1	2	1	1	2
M4		1	1	3		2	1	2		1
M5	2				3				2	
M6		1	1	2		3	1	3		1
M7		3	2	1		1	3	1		3
M8		1	1	2		3	1	3		1
M9	3		1		2				3	
M10		3	2	1		1	3	1		3

Figure 45. Solution similarity matrix.

never are. The agreements on pairs (M6,M8) and (M6,M9) are said to be total. Other workcenter pairs, such as (M2,M3), do not systematically belong to the same cell; the agreements on pair (M2,M3) is called a partial agreement or disagreement.

Study of the Total Agreements.

Let us consider the matrix (Figure 46) obtained after having suppressed each partial agreement and replaced it with a dash. In our case, there are 42 dashes; the working party hence unanimously agrees on 58% of the matrix's cases [20 digit 3 and 38 digit 0; 58% = 58/(10 × 10)]. An easy permutation of this matrix's rows and columns sheds light on groups of machines that are gathered within the same cell independently of the solutions (Figure 47). Groups (M1,M9), (M6,M8), and (M2,M7,M10) are cores of cells, the gathering of which is not questionable. The groups made up of one machine only are machines the allocation of which has to be further studied.

Groups of machine number 1 to 6 are the equivalence classes of the binary relation "is always assigned in the same cell than" defined for each pair of machines. As a consequence there are many fast and easy ways to get them. If a graph toolbox is available to the designer, these groups can be found automatically, otherwise the small number of machines—less than 300—allows to perform this research manually or more conveniently with the help of a spreadsheet. When using a graph toolbox, the groups are the connected components of the graph of the previous defined binary relation. In our example the adjacency matrix of this graph is obtained by replacing 3 values of Figure 45 matrix with 1's and other by 0's.

	M1	M2	M3	M4	M5	M6	M7	M8	M9	M10
M1	3		–		–				3	
M2		3	–	–		–	3	–		3
M3	–	–	3	–		–	–	–	–	–
M4		–	–	3		–	–	–		–
M5	–				3				–	
M6		–	–	–		3	–	3		–
M7		3	–	–		–	3	–		3
M8		–	–	–		3	–	3		–
M9	3		–		–				3	
M10		3	–	–		–	3	–		3

Figure 46. Total agreement matrix.

		1	1	2	2	3	3	3	4	5	6
		M1	M9	M6	M8	M2	M7	M10	M3	M4	M5
1	M1	3	3						–		–
1	M9	3	3						–		–
2	M6			3	3	–	–	–	–	–	
2	M8			3	3	–	–	–	–	–	
3	M2			–	–	3	3	3	–	–	
3	M7			–	–	3	3	3	–	–	
3	M10			–	–	3	3	3	–	–	
4	M3	–	–	–	–	–	–	–	3	–	
5	M4			–	–	–	–	–	–	3	
6	M5	–	–								3

Figure 47. Reorganized total agreement matrix showing the indivisible groups of machines.

Study of the Partial Agreements Families   The working party may consider that the groups of machines that are at least twice out of three times within the same cell are reasonable future consensual groups. In order to identify these groups, a matrix including solely the similarity greater or equal to 2 and equal to 0 is conceived. These groups are obtained from the classification algorithms used for the cell identification, slightly modified to take into account the matrix's dashes.

This leads to the matrix in Figure 48. The cell cores of this matrix clearly show three groups of machines: (M1,M5,M9), (M6,M8,M4), and (M2,M3,M7,M10). Figure 48 shows that the group of machines (M1,M5,M9) is often gathered within the same cell and that, furthermore, this group's machines are hardly ever gathered with machines from the two other groups. This group is probably a proper cell core.

We have already pointed out that 38% of the matrix's elements are 0's. This information should be exploited. For instance, the analysis of the PFA similarity matrix can be undertaken by trying to get rid of the machines that are never gathered together in the same cell. The grouping illustrated in Figure 49 is thus obtained. This figure separates the machines into two main groups of machines that, with the exception of M3, are never together within the same cell. Once it has ruled on M3, the working party can split its work and deal independently with the cells of machines (M1,M5,M9) and (M6,M8,M4,M2,M7,M10).

The few analyses just presented tally each other and are sometimes redundant. Our purpose is to give the reader means of analysis of the PFA similarity matrix and tracks to facilitate the working party's progress. The person conducting the working party selects the strategy of exploitation of the PFA similarity matrix according to the simplicity of the results or the preferences of the group.

For more details about the clustering algorithms and existing softwares you can contact the authors.

3.6. Representation of the Results

The representation of the results is a crucial stage in cell formation (Figure 50). The ability of the designer to understand and devise different types of solutions corresponding to their objectives is dependant on the quality of the representation of the results. In many industrial case studies, it is observed that visual and calculated results objectively shed light on facts sometimes unknown *a priori*, but conceivable *a posteriori*. There exist numerous performance measures. Nevertheless, they can be categorized into two types: the visual qualitative indicators and the quantitative indicators.

Numerous studies have shown that vision is a powerful tool for the communication and representation of abstract or concrete phenomena. The matrices introduced since the beginning of this chapter are powerful working tools. Their representation on paper sheets seems inevitable. If the item–workcenter matrix is too large, the workcenter–workcenter matrix or the PFA similarity matrix should be judiciously used. The analysis of the PFA similarity matrix often leads to the reduction in the number of rows and columns of the workcenter–workcenter or item–workcenter matrices. For example, a single row or column can serve as a substitute for the rows or columns

		1	1	1	2	2	2	3	3	3	3
		M1	M5	M9	M6	M8	M4	M2	M3	M7	M10
1	M1	3	2	3					–		
1	M5	2	3	2							
1	M9	3	2	3					–		
2	M6				3	3	2	–	–	–	–
2	M8				3	3	2	–	–	–	–
2	M4				2	2	3	–	–	–	–
3	M2				–	–	–	3	2	3	3
3	M3	–		–	–	–	–	2	3	2	2
3	M7				–	–	–	3	2	3	3
3	M10				–	–	–	3	2	3	3

Figure 48. Partial consensus matrix.

		1	1	1	2	2	2	2	2	2	2
		M1	M9	M5	M6	M8	M4	M2	M3	M7	M10
1	M1	3	3	2					1		
1	M9	3	3	2					1		
1	M5	2	2	3							
2	M6				3	3	2	1	1	1	1
2	M8				3	3	2	1	1	1	1
2	M4				2	2	3	1	1	1	1
2	M2				1	1	1	3	2	3	3
2	M3	1	1		1	1	1	2	3	2	2
2	M7				1	1	1	3	2	3	3
2	M10				1	1	1	3	2	3	3

Figure 49. Partial consensus matrix.

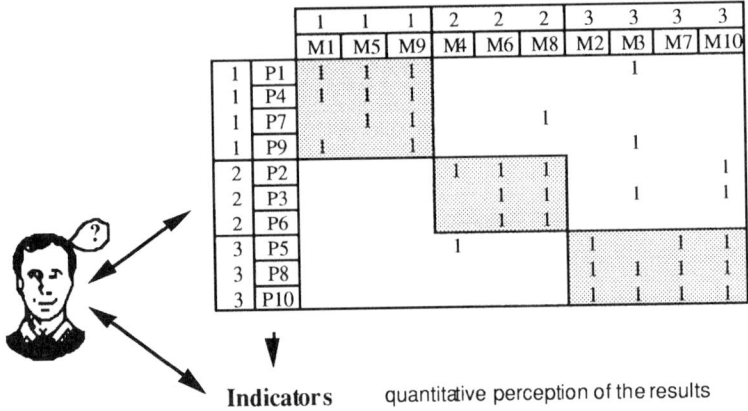

		1	1	1	2	2	2	3	3	3	3
		M1	M5	M9	M4	M6	M8	M2	M3	M7	M10
1	P1	1	1	1				1			
1	P4	1	1	1							
1	P7		1	1			1				
1	P9	1		1					1		
2	P2				1	1	1				1
2	P3					1	1	1			1
2	P6					1	1				
3	P5				1			1		1	1
3	P8							1	1	1	1
3	P10							1	1	1	1

Qualitative and visual perception of the results.

Indicators   quantitative perception of the results

Figure 50. Designer and the two types of representation of the results.

of a total agreements' family. Thus, the designer has a global and visual perception of the intercellular transfer volume and of the cells' sizes, as shown in Figure 51.

In any case, quantitative indicators must complete the analysis. It seems utopian to exhaustively list the quantitative indicators useful for the results' analysis, since from each case study arise specific problems. However, some indicators cannot be bypassed and are categorized into two types. The following examples illustrate the computation of these indicators. The matrices in Figures 51, 52, and 53 give the indicators of, respectively, operation, load, and flow matrix for each item and workcenter. The table in Figure 54 represents the indicators linked with the cells. We do not specify the indicators' meaning, which evidently varies according to the matrix used.

Quantitative Indicators Linked to Items: I1–I9

I1, I4, I7 row summation in the operation, load and flow cases, respectively.
Ex.: for P01, $I1 = \text{"}1_{M01}\text{"} + \text{"}1_{M05}\text{"} + \text{"}1_{M09}\text{"} + \text{"}1_{M03}\text{"} = 4, I4 = 31 + 40 + 45 + 34 = 150, I7 = 10 + 10 + 10 + 10 = 40$.

I2, I5, I8 row summation of the extracellular operations in, respectively, the operation, load, and flow cases.
Ex.: for P01, $I2 = \text{"}1_{M03}\text{"} = 1, I5 = 34, I8 = 10$.

I3, I6, I9 row percentage of extracellular operations in, respectively, the operation, load, and flow cases.

$I3 = I2 / I1$; $I6 = I5 / I4$; $I9 = I8 / I7$.
Ex.: for P01, $I3 = 1/4 = 25\%, I6 = 34/150 = 23\%, I9 = 10/40 = 25\%$.

Quantitative Indicators Linked to Items: I10–I18

I10, I13, I16 column summation in, respectively, the operation, load and flow cases.

		Cell										**I1** Total "1"	**I2** Extra "1"	**I3** % Extra "1"
		1			2			3						
	I24 Nb.W.C./Cell	3			3			4						
	I26 Nb.W.S.	3	3	2	1	3	3	2	5	2	3			
		M01	M05	M09	M04	M06	M08	M02	M03	M07	M10			
C e l l 1	P01	1	1	1					1			4	1	25
	P04	1	1	1								3	0	0
	P07	1		1			1					3	1	33
	P09		1	1							1	3	1	33
C e l l 2	P02				1	1	1		1			4	1	25
	P03					1	1		1		1	4	2	50
	P06					1	1					2	0	0
C e l l 3	P05				1			1		1	1	4	1	25
	P08							1	1	1	1	4	0	0
	P10							1	1	1	1	4	0	0
	I10 Total "1"	3	3	4	2	3	4	3	5	3	5			
	I11 Extra "1"	0	0	0	1	0	1	0	3	0	2			
	I12 % Extra "1"	0	0	0	50	0	25	0	60	0	40			

Figure 51. Indicators of the operation matrix.

97

Figure 52. Indicators of the load matrix.

			Cell										**I4**	**I5**	**I6**
			1	1	1	2	2	2	3	3	3	3	Total load	Extra load	% Extra load
			M01	M05	M09	M04	M06	M08	M02	M03	M07	M10			
C e l l	1	P01	31	40	45					34			150	34	23
	1	P04	21	10	2								33	0	0
		P07		40	12			28					80	28	35
		P09	32		11					12			55	12	22
	2	P02				14	46	8				14	82	14	17
	1	P03					24	27		3		6	60	9	15
		P06					38	47					85	0	0
	3	P05				20			16	37	24	12	72	20	28
		P08							9	35	11	3	60	0	0
		P10							12	35	5	49	101	0	0
I13 Total load			84	90	70	34	108	110	37	121	40	84			
I14 Extra load			0	0	0	20	0	28	0	49	0	20			
I15 % extra load			0	0	0	59	0	25	0	40	0	24			
I19 Capacity			108	96	72	36	114	114	72	180	72	96			
I20 L/C %			78	94	97	94	95	96	51	67	56	88			

		Cell											**I7**	**I8**	**I9**
		1			2			3					Total order	Inter order	% Inter order
		M01	M05	M09	M04	M06	M08	M02	M03	M07	M10			
C e l l	1	P01	10	10	10					10			40	10	25
		P04	1	1	1								3	0	0
		P07		4	4			4					12	4	33
		P09	10		10					10			30	10	33
	2	P02				2	2	2				2	8	2	25
		P03					5	5		5		5	20	10	50
		P06					10	10					20	0	0
	3	P05				8			8		8	8	32	8	25
		P08							1	1	1	1	4	0	0
		P10							5	5	5	5	20	0	0
I16 Total order			21	15	25	10	17	21	14	31	14	21			
I17 Inter order			0	0	0	8	0	4	0	25	0	7			
I18 % inter order			0	0	0	80	0	19	0	81	0	33			

Figure 53. Indicators of the flow matrix.

	Binary (Fig. 51)			Load (Fig. 52)	Flow (Fig. 53)
I21	Num. of cells	3			
I22	Num. of item/cell	4	3	3	
I23	Num. I.U.O.C.W.M.	3	2	1	
I24	Num. of W.C./Cell	3	3	4	
I25	Num. W.C.U.O.C.I.	0	2	2	
I26	Num. of W.S.	8	7	12	

	Workcenter group			Workcenter group			Workcenter group		
Item families	Cell 1	Cell 2	Cell 3	Cell 1	Cell 2	Cell 3	Cell 1	Cell 2	Cell 3
I27 Cell 1	10	1	2	244	28	46	61	4	20
I28 Cell 2		7	3		204	23		36	12
I29 Cell 3		1	11		20	213		8	48
I27 Cell 1 %	29	3	6	31	4	6	32	2	11
I28 Cell 2 %		20	9		26	3		19	6
I29 Cell 3 %			31		3	27		4	25

I30	Extracell operations Intercell moves (flows)	7	117	44
I31	Intracell operations or moves	28	661	145
I32	Total operations or moves	35	778	189
I33	% extracell operations Intercell moves (flows)	20%	15%	23%

Figure 54. Indicators linked with the cells.

Ex.: for M10, I10 = "1_{P02}" + "1_{P03}" + "1_{P05}" + "1_{P08}" + "1_{P10}" = 5, I13 = 14 + 6 + 12 + 3 + 49 = 84, I16 = 2 + 5 + 8 + 1 + 5 = 21.

I11, I14, I17 column summation of the extracellular operations in, the operation, load, and flow cases, respectively
    Ex.: for M10, I11 = "1_{P02}" + "1_{P03}" = 2, I14 = 14 + 6 = 20, I17 = 5 + 2 = 7.

I12, I15, I18 column percentage of extracellular operations in the operation, load, and flow cases, respectively.
    I12 = I11 / I10 ; I15 = I14 / I13 ; I18 = I17 / I16.
    Ex.: for M10, I12 = 2/5 = 40%, I15 = 20/84 = 24%, I18 = 7/21=33%.

I19 capacity of the workcenter. For M01, the capacity is 108 hours/week for the three work centers.
    Ex.: for M10, I19 = 96 hours per week.

I20 capacity utilization of the workcenter. I20 = I13 / I19.
    Ex.: for M10, I20 = 84/96 = 88%.

Quantitative Indicators Linked with Items: I21-I33

Indicators I21 to I26 are independent of the operation, load, and flow criteria.

I21 number of cells: 3, noted cell 1, cell 2, cell 3.

I22 number of items per cell.
    Cell 1 I22=4, P01, P04, P07, P09.
    Cell 2 I22=3, P02, P03, P06.
    Cell 3 I22=3, P05, P08, P10.

I23 number of items of a cell utilizing other cells' workcenters (IUOCWC). These items generate intercellular moves.
    Cell 1 I23=3, P01, P07, P09.

I24 number of workcenters per cells (Fig. 51).
    Cell 1 I24=3, M01, M05, M09.
    Cell 2 I24=3, M04, M06, M08.
    Cell 3 I24=4, M02, M03, M07, M10.

I25 number of workcenters of a cell utilized by other cells' items (WCUOCI). These items generate intercellular moves.
    Cell 2 I25=2, M04, M08.

I26 number of workstations (ws) of a cell (Figure 51).
    Cell 1 I26 = 8 i.e., workcenter M01, is made of 3 ws, workcenter M05 is made of 3 ws, and workcenter M09 is made of 2 ws. This quantity is calculated with the assistance of the workcenter file which is not given in this example.

I27, I28, I29: These indicators summarize the item–workcenter matrices by summing the quantities in each block of the matrix stemming from the crossing of an item family with a group of workcenters.
    I27, Ex.: Total workcenter load of cell 1 for the items of cell 1.
    I27 (load,cell1,cell1)=31+40+45+21+10+2+40+12+32+11=244.

I28, Ex.: number of queues within cell 3 generated by the items of cell 2 (i.e., in front of the workstations of cell 3). This quantity is calculated from the flow matrix. I28 (flow,cell2,cell3)=5+5+2=12.

I29, Ex.: number of queues generated within cell 2 by the items of cell 3. I29(flow,cell3,cell2)=8.

I30,I31,I32: These indicators are more global than the indicators I27—I29 from which they are deduced.

I30 sums up the quantities outside the diagonal blocks of a matrix. For the operation matrix I30(operation)=1+2+3+1=7. This number is the total number of extracellular operations executed within the workshop and corresponding to this solution.

I31 sums up the quantities within the diagonal blocks of a matrix. For the operation matrix I31(operation)=10+7+11=28.

I32 shop's load, flow, or operations summation. For the operation matrix I32(operation)=7+28=35.

I33 = I30 / I32. For the operation matrix I33=7/35=20%.

Other indicators can be of great use for an analysis of the data. It is verified, in the majority of industrial cases, that the annual quantities, the workcenters' load, the number of manufacturing orders have a Pareto distribution. The classification of these variables provides useful information.

4. CELL FORMATION METHOD

In Section 3, the data and techniques required for cell formation were described. This section describes the sequence of stages necessary for the cell design process. There are six of them: data gathering, data consistency checking, data screening and modification, suggestion of cells, consensus searching, and final cell formation process. These stages are detailed later on. The shop designing stages generally follow one another in the order previously mentioned. However, we note that it is always difficult to formulate from the very beginning the objectives or constraints of each stage. The production system designer sometimes has to return to a stage, as is illustrated in Figure 55. The design process is never linear; it directs the designer toward a more explicit grasp of the production system. It modifies the designer's perception of the system. The designer obtains qualitative and quantitative information allowing him to clarify or redefine the objectives and constraints for cell formation.

4.1. Data Gathering

The search for cells starts with the gathering of the data. The cell designer must entirely know the data representative of the items, workcenters, and routings described in Section 3.1. Some specific data, such as the cost price or the parts' weight or size, are sometimes required. In most cases, these data are extracted from databases used for

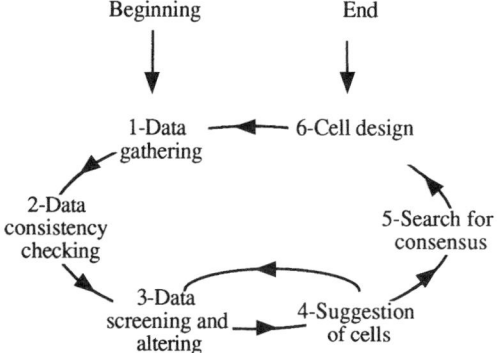

Figure 55. Cell formation method: the sequence of stages.

production control. ASCII files, which can be read on any computer, are often used. The size of these files varies between a few hundred kilobytes and several megabytes. If these data do not exist, they must be manually collected; this is a weighty and tedious task but quite necessary for an elaborate study.

4.2. Data Consistency Checking

Before analyzing the items, workcenters, or routings, it is always necessary to verify the consistency of the files. Some data can be false or incomplete. Some often inconsistent data in the item, workcenter, and routing files are listed. This list is obtained from mere requests in the previous three files. These data must be validated by the designer before their screening. Examples for data inconsistencies in the different files follow:

- Item file
  Item without operation in the routing file
  Annual quantities or lot size or number of orders or annual loads equal to nought
  Transfer lot size larger than production lot size
  Production lot size larger than the annual quantity
- Workcenter file
  Workcenter unused in the routings
  Load, capacity, or number of workcenters equal to nought
- Routing file
  Item undeclared in the Item file
  Workcenter undeclared in the Workcenter file
  Setup time, run time, or loads equal to naught
  These possibly inconsistent data are validated, modified, or eliminated by the designer

4.3. Data Screening

During the PFA, it is not necessary to study all the items, workcenters, or routings. Data screening eliminates from the study the items and workcenters that are not representative of our problem.

Items that should be ruled out of the study are

- Items with a light annual load
- Lightly loaded workcenters
- Subcontracting operations in the routing file
- Subcontracted items
- Obsolete items

In the same way, some workcenters might be eliminated from the study. Traditionally, several cases occur, as described next:

- The moving cost of some machines requiring foundations makes any displacement impossible. It is the case with large presses, forges, painting, or surface treatment/surfacing lines. These machines are seldom integrated within cells.
- Other machines are noisy or require an environment protecting them from dust or excessive temperatures. These machines are often laid out in isolated premises and cannot be integrated within cells. It is the case with the high precision adjusting machines, dimensional control machines.
- The control operations declared in the routings can easily be integrated in future cells. There is no use in taking them into account during the analysis.
- The firm purchases, in the near future, production means replacing the existing ones. In that case, the new workcenters must be stored in the workcenter file, and the old ones removed. The routing operations utilizing the old workcenters must be modified.
- Finally, some machines are systematically utilized by every item. It is the case in particular for some washing plants in mechanical engineering. Taking into account such workcenters makes it difficult to form cells. The exclusion of the workcenters from the routings makes it possible to consider solutions that are not linked to these constraints.

Other data have to be created. These concern the items or workcenters appearing in a future close enough to be compatible with the designer's study horizon. In most cases, the production quantities are based on sales forecasts. The data validation and screening are important stages.

4.4. Suggestion of Cells

Once data are consistent, the different matrix can be created and group analysis can be managed using automatic clustering algorithms. We shall give here some tips on how to carry through this step successfully.

Sections 3.3 and 3.4 described the multiplicity of data, optimization criteria, and objectives used in the algorithms for searching for cells. The analyzed data can be binary or weighted by the load or the number of manufacturing orders. According to the agreed intercellular move ratio, the number and size of the cells can vary. Although there is a link between the type of data and the analysis objectives, it is advised to search for a great number of solutions in accordance with several optimization criteria for at least two reasons: First of all, the computing time of today's algorithms on large matrices does not constitute a check to explore numerous solutions (on a personal computer: a few minutes for several thousands items and hundreds of workcenters). Furthermore, the search for a great number of solutions promotes the working party's creativity. The calculation method we recommend is the following:

1. Analyze the item–workcenter or workcenter–workcenter matrix according to the operation, load, and flow criteria.
2. For each of the three criteria, search for a set of solutions by changing the cell's size while keeping the intercellular move ratio below 50%. Furthermore, the intercellular moves are more important than the moves within the cell. In practice, the number of different solutions is between 3 and 8. The total number of solutions thus found varies between 9 and 24 (3 criteria per 3 to 8 solutions).

Each solution is then summarized by indicators defined in Section 3.6. The results are qualitatively and quantitatively analyzed. This analysis must imperatively be carried out by a working party mastering various skills. The analysis of the results is a particular stage in cell designing. It is at this very moment that is felt the group's creativity. New questions are asked: What would happen with new production quantities? Machine X involves many intercellular moves, can it be distributed? These questions often require new analysis. Screening the data as described in Section 4.3 can initiate new propositions of cells.

After the search and analysis of the solutions, the working party does not always select all of them. The reasons for rejection can vary . The following list of reasons is not exhaustive:

- The cells comprise too many workcenters.
- The number of workcenters per cell is judged inadequate.
- The intercellular moves are too numerous.
- The cell sizes are inadequately proportioned, some very large and some very small.
- Some cells' production loads are not significant enough to justify allocating machines to them.

At the end of this stage, the designer selects a limited set of solutions per criteria.

4.5. Search for Robust and Consensual Cell Cores

This stage consists in comparing the solutions developed in Section 4.4 and defining a few robust and consensual solutions. They are obtained with the assistance of the methodology introduced in Section 3.5. The search for consensus is an alternative way to represent and analyze PFA solutions. At this level, the cell designer can identify the sources of agreement or disagreement between solutions. During a case study, the item or workcenter total agreements are often more numerous than the disagreements; this is an important piece of information. The similarity measure between PFA solutions is a way of expressing the solutions' robustness. And the designer is faced with a clearly set problem. The belonging of an item or a workcenter to a cell depends on the classification optimization criterion and the cell's size. Decisions still have to be made, but with full knowledge of the facts. The consensus search directs the designer or working party toward further queries. The screening and proposition of new cells sequences is sometimes necessary. At the end of this step, a few solutions emerge that must be studied in-depth. That is the purpose of the next step.

4.6. Final Cell Formation Process

The propositions of cells were made by analyzing existing data and without modifying the set of workcenters, items to be manufactured in the workshop and routings. These propositions contain often intercell flows. It is not necessary to accept as-is the noninteger machine requirements that are calculated for the shared workcenters. Usual operation management techniques, tooling analysis, and computer-aided process planning can be used to reduce the machine requirements through redesign of parts, standardization of features and tooling, design of group fixtures, optimization of machining parameters, and so forth. Hence instead of planning around noninteger machine requirements for the cells, a practical strategy is to reduce machine requirements causing intercell flows. There always exists actions to eliminate the sharing problem that forces intercell flows; however, the extra study, investment, or operating costs may make the elimination of the intercellular moves expensive. A compromise between financial criteria and physical flow criteria must then be reached. We shall briefly comment ways to reach the elimination of intercell moves.

Aggregation of the Small Cells The elimination of the moves between two cells can be obtained from the gathering of these cells within a sole cell. The new cell thus formed has a larger size. In numerous industrial cases, this particularly simple technique is applied to consensus families that comprise few workcenters. Indeed, many cells' sizes are to small. A list of cell couples arranged in descending order with regard to the transfers eases the analysis of simple cases. In less obvious cases, the designer can construct a matrix intersecting the cells with themselves. Each square of this symmetric matrix contains the intercellular transfer quantities. The diagonal elements of this table are neutralized, as it has been done to the nontotal agreements of the PFA similarity matrix—dashes in Figures 46–49. The aggregation of cells is obtained using the same automatic clustering technique as for the partial agreement analysis in Section 3.5. Two remarks concerning the interest of this method for forming cells seem indispensable. The first one is that the use of this sole method is not sufficient to form flow-independent cells, since the automatic clustering method used to manage group analysis would have detected this possibility. The second remark is that this method can be used earlier during the search for consensus.

Distribution of the Existing Workcenters   Some items utilize other workcenters external to the cell to which they are assigned. If these workcenters are made up of several workstations, their distribution among the cells may contribute to lower the number of intercellular moves. In the example of Section 3.6 we can see in Figure 51 that the workcenter M08, which has one extracellular operation, is made of three workstations. The capacity of the workcenter is 114 hours; the load external to the cell is 28 hours (Figure 52). Let us suppose an equal capacity to the workcenters (114/3 = 38 hours); then assigning a workstation of the workcenter M08 to cell 1 and two of them to cell 2 eliminates intercell flows. Although the distribution of the workcenters reduces the number of intercell moves, the benefits from this distribution might be lessened by other constraints. Indeed, with the new distribution, the capacity utilization of the workstation allotted to cell 1 is 74% (I20 = 28/38 = 74%) while the capacity utilization of the set of workstations allotted to cell 2 becomes 108%. The designer is then faced with the problem of decreasing the capacity utilization of some workstations. The following steps could then be taken to decrease the capacity utilization:

1. Decrease the load of the workstations by decreasing their setup and run times or eliminate operations in the routings of the workpieces.
   a. Reduce tool changing times with preset tooling, automatic tool changers, combination tooling, cutting tool materials having longer tool life, large tool magazines, replaceable inserts, and the like.
   b. Reduce the number of tools by applying variety reduction techniques to standardize workpiece dimensions, tolerances, fillet radii, chamfer angles, thread types, hole radii, and so forth.
   c. Perform group scheduling based on sequence dependencies and similarity of process plans.
   d. Optimize process parameters to reduce operation times.
   e. Plan with lower scrap levels.
   f. Decrease the number of setup changes by increasing batch sizes.
2. Increase the capacity of the workstation.
   a. Apply Total Productive Maintenance (TPM) concepts: Apply preventive maintenance on the critical machines by operators in the cell, eliminate idle times, reduce labor absenteeism in cells, and so forth.
   b. Train the workers to be multifunctional.
   c. Increase the number of shift or do overtime.
   d. Adopt an intracell layout that allows one operator to monitor several machines.

To maintain the workstation capacity utilization at an acceptable rate is one of the most important constraint when distributing the workstations. Some algorithms, which automatically distribute the workcenters, can be found in the appropriate literature; nevertheless, we must attract the reader's attention on the fact that these general algorithms constitute merely an assistance. Indeed it is not a profitable operation to create an algorithm that takes into account the specific constraints of each workshop as illustrated in the following examples:

1. Although described as identical in the workcenter file, some workstations are older than others. In that case, the designer may prefer to allocate the old machines to the rough operations and the new ones to the finishing operations. This difference is not formalized in the input data of the algorithm, which will generally not propose an unacceptable solution in this case.

2. An investment in tooling and handling means is sometimes indispensable in order to obtain autonomous distributed workcenters. It may be sometimes better to subcontract noncritical parts that distribute the machines.

3. The utilization of some workcenters requires an expertise mastered at the moment by a sole operator. The distribution of these workcenters thus implies the training of other operators.

Investment in Production Means   Investing in production means reduces the number of intercellular moves and increases capacity of the machines:

- Purchase extra machines of the same type as the machines of the workcenter whose load is shared by several cells.
- Purchase higher efficiency Computer Numerical Control (CNC) center and dedicate the existing machines to a few key parts only.

Once again, a compromise between the reduction in the number of intercellular moves, the investment and operating costs, and the capacity utilization has to be reached—underloaded machines gives bad capacity utilization too.

Subcontracting of the Items and the Elimination of the Items   The items utilizing other workcenters' cells may be subcontracted in order to lower the number of intercellular moves. An economic evaluation in terms of cost price and subcontracting required time is necessary. In the case of noncritical items involving low cost prices, subcontracting might be profitable, since the extension of the allotted time thus caused can be offset by security inventories. The case of the other items must be discussed case by case. Numerous other criteria are involved, such as the protection of a knowhow, the weakness in transport, and, of course, the effect on the shared machine loads. Subcontracting items decreases the load of several machines and therefore their capacity utilization.

Modification of the Routings   The parts causing intercell moves utilize other cells' workcenters. The modification of the routings can be executed several ways :

1. The workcenter causing the move is replaced by a workcenter in the item's cell.
2. The routing is revise to use solely the workcenters in the item's cell.
3. Eliminate final inspection of a part by in-process dimensional gauging.

The modification of the production process is generally executed by members of the Process Engineering and Production Department, which have good knowledge of the production means' technology. These modifications also affect the costs, the allowed times, and capacity utilization of the machines.

Redesign of the Parts   The redesign of a part can be twofold:

- Redesign parts to eliminate the feature(s) requiring the exceptional operation external to the cell.
- Redesign the parts requiring workcenters that could be distributed after eliminating those features requiring operations on these workcenters.

Contrary to the other approaches, this technique requires a close collaboration of the Process Engineering and the Research Departments. It is probably the most efficient technique in the long term, but also the most difficult to implement.

Concluding Remarks about the Final Process   As just stated, the elimination of intercell moves is always technically feasible, but a set of technical and financial criteria defines the limits of such a transformation. The steps to eliminate intercell moves described above can be combined to obtain independent cells.

5. CONCLUSION

In this chapter we have shown that the identification of potential cells is a simple and fast operation when using existing clustering techniques and usual data of the production control databases. Nevertheless, the high number of parameters involved in the design of manufacturing cells makes, on the one hand, the elimination of intercell moves a little more difficult; on the other hand, it gives the cell designer a large degree of freedom. In order to make sure that the most important parameters have been taken into account, it is advisable to integrate the method proposed in this chapter in a more general workshop designing method, described in another chapter of this handbook.

4

LAYOUT DESIGN FOR CELLULAR MANUFACTURING

Massoud Bazargan-Lari

Department of Business Administration
Embry-Riddle Aeronautical University
Daytona Beach, Florida 32114-3900

1. CELLULAR MANUFACTURING

John Burbidge (1975), who introduced the concept of Group Technology (GT) to the western countries, considers that the traditional type of organization for manufacturing is process organization, in which organizational units each specialize in a particular process. This is gradually being replaced by product organization, in the form of either continuous line flow or GT. Group Technology is a manufacturing philosophy whose idea is to capitalize on similar, recurrent activities. It is a philosophy with broad applicability, potentially affecting all areas of a manufacturing organization. Cellular Manufacturing (CM), an application of GT, utilizes the concept of divide and conquer and involves the grouping of machines, processes, and people into cells responsible for manufacturing or assembly of similar parts or products. Cellular Manufacturing is now an established international practice to integrate equipment, people, and systems into "focused factories," "minibusinesses," or "cells" with clear customers, responsibilities, and boundaries. It has its origins in Russian and German "group technology" in the early part of this century.

The main difference between the conventional type of manufacturing and CM is the way the machinery is arranged on the shop floor. There are three basic ways to

Handbook of Cellular Manufacturing Systems, edited by Shahrukh A. Irani
ISBN 0-471-12139-8 © 1999 John Wiley & Sons, Inc.

arrange machines in a factory: by line, by function, and by group (Bedworth et al. 1991). Figure 1 illustrates each of these layouts.

- *Line/Product Layout.* The machines and other workcenters are arranged in the sequence in which they are used to manufacture the product. This arrangement is suitable for high-volume/low variety products.
- *Function/Process Layout.* Machines of a specified type are grouped together. This layout can result in large amounts of material handling, a large amount of Work-In-Process (WIP) inventory, excessive setup times, and long manufacturing lead times.
- *Group/Cell Layout.* Machines are arranged as cells. Each cell is capable of performing manufacturing operations on one or more families of parts. Consequently, the capacity of a cell can be determined by considering only the families of parts that utilize that cell. As a result, this layout should be easier to manage. This is just one reason why a group layout may be more desirable.

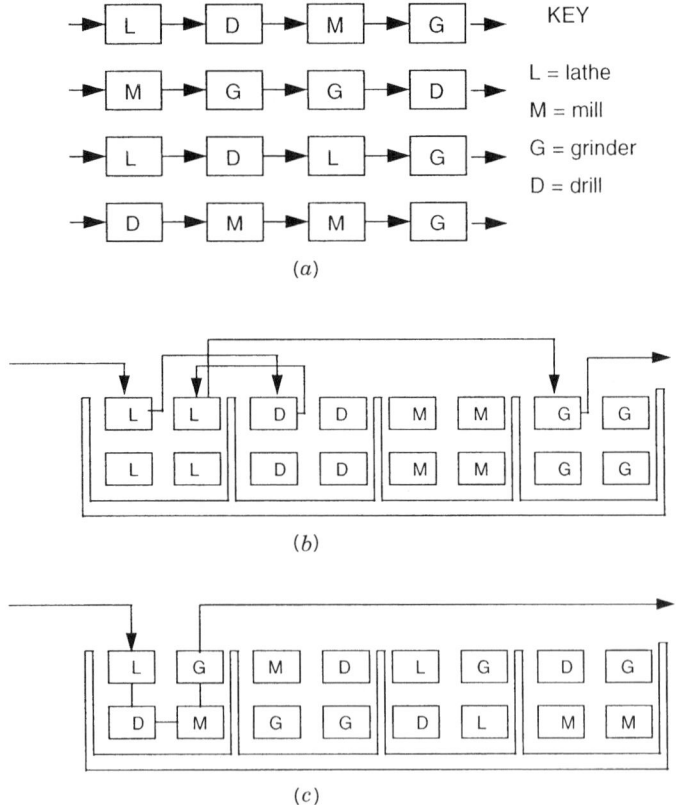

Figure 1. Three types of plant layout: (*a*) line layout, (*b*) functional layout, and (*c*) cell layout.

2. MACHINE LAYOUT DESIGNS

The emergence of new manufacturing technologies and philosophies such as flexible manufacturing system, GT, and just-in-time require new models and solution methodologies to address their design and operations, of which one important activity is determining machine arrangement on the facility floor (Hassan, 1995). The locations and arrangements of the machines and workcenters contribute in a large measure to the manner in which a company is operating. A key element to exploit the benefits of CM is efficient layout designs. A poor physical layout will offset some or all of the benefits expected from the company.

The *right* solution to plant layout problems is important because material handling costs comprise anywhere from 30 to 75% (usually nonadded value) of total manufacturing costs. Any savings in material handling realized through a better arrangement of the machines is a direct contribution to the improvement of the overall efficiency of the operation (Sule 1994).

Apple (1977), Hales (1984), Anon (1986), Francis et al. (1992), Meyers (1993), Sule (1994), Tompkins et al. (1996), and Heragu (1997) are among the authors who provide general and introductory readings on the facility layout problem and the importance and significance of designing efficient layout configurations.

2.1. Need for a New Layout Design

The growth of manufacturing companies and introduction of new machinery and workcenters over the years is the major cause resulting in an inefficient/unsafe shop floor layout. This difficulty arises almost with every growing company and worsens over time. These companies normally start with one or two workcenters. As they grow, they introduce new workcenters on the shop floor without considering the impact of material movements or other layout considerations such as safety and ease of material handling on the total operation of the company. In many cases the location of a new workcenter is dictated by other workcenters that already exist on the shop floor.

The need for a new plant layout may also arise due to

- Relocation of the plant
- Automation
- Increase or decrease in the demand for a product
- Changes in the design of a process
- Replacement of one or more pieces of equipment
- Adoption of new standards, technologies, and strategies
- Organizational changes within the company

Research conducted in leading industrial countries concludes that the average life of an efficient plant layout is not more than 5 years.

3. MATHEMATICAL MODELS FOR MACHINE LAYOUT DESIGN

The Machine Layout Problem (MLP) has been studied under a broader term called Facility Layout Problem (FLP). Although it is closely related to FLP, certain restrictions prevent direct application of FLP algorithms to MLP. For example, many FLP algorithms assume that the facilities are all of the same size and/or the locations of the facilities to be known *in advance*. This assumption is usually not valid for MLP. Although some FLP methods, such as CRAFT (Armour et al., 1964), consider different facility areas using grids, the shape of the facility may have to be altered in the final solution, which is undesirable in MLP. Recent continuous models have proved to provide better solutions than discrete models in overcoming this problem by considering rectangular departments with fixed boundaries that be located anywhere in the continuous plane (Bazargan-Lari and Kaebernick, 1997; Lacksonen, 1994).

The following provides a brief outline on the machine layout methodology developed by Bazargan-Lari and Kaebernick (1997). This method is of particular interest because of the following:

- It has been used to generate several real-world industrial machine layout designs.
- It is the base method for addressing the various stages of CM design described later in this chapter.

We skip the complete mathematical background to this methodology and refer the interested readers to Bazargan-Lari and Kaebernick (1997, 1996a, 1996b).

The proposed mathematical model adopts a continuous-plane approach. In this model the machines are considered to be rectangular blocks with known dimensions in a continuous plane. Three variables are associated with each block, namely the coordinates of the bottom left-hand corner and the aspect ratio, which is the height over the width of the block and is used to determine the orientation of block. The orientation of a machine is considered horizontal if the aspect ratio is less than 1 and vertical otherwise (see Figure 2).

Width and height of a block are the dimensions of the smallest rectangle enclosing the machine parallel to the *x* axis and *y* axis, respectively. Irregular-shaped machines can be broken into two or more blocks, and then certain restrictions are imposed on the positioning of these blocks so that the shape of the machine is maintained. The task of this model is to determine the coordinates and aspect ratio (orientation) of each machine in the continuum plane by achieving and satisfying certain goals and constraints.

3.1. Problem Constraints and Objectives

The following sections present the development of the constraints and objective functions of the proposed mathematical model. As goal programming is used as part of the solution methodology (described later in Section 3.2) to solve the mathematical model, the distinction between the objective function and the constraints is not as

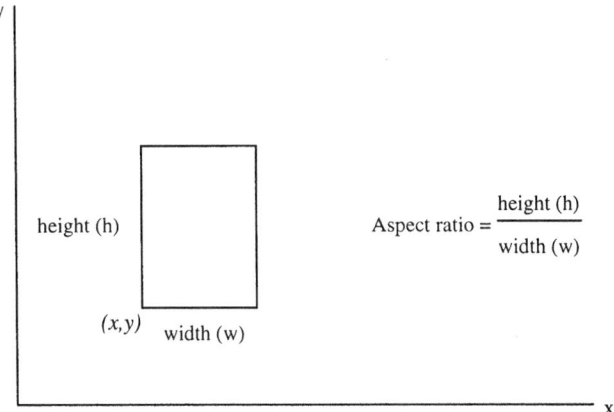

Figure 2. Definition of the coordinates and aspect ratio of a rectangular block.

clear-cut as it is the case with conventional mathematical programming such as linear programming.

- *Shop Floor Boundaries.* For a feasible solution each product must be fully positioned inside the shop floor dimensions (see Figure 3).
- *Nonoverlapping Conditions.* This constraint ensures that no two rectangular blocks overlap (see Figure 3).
- *Closeness Relationships.* In some cases of machine layout planning it may be desirable to position certain machines adjacent or nonadjacent to each other. In this model adjacencies or nonadjacencies between machines are expressed in terms of their desired physical distances (see Figure 4). This restriction may

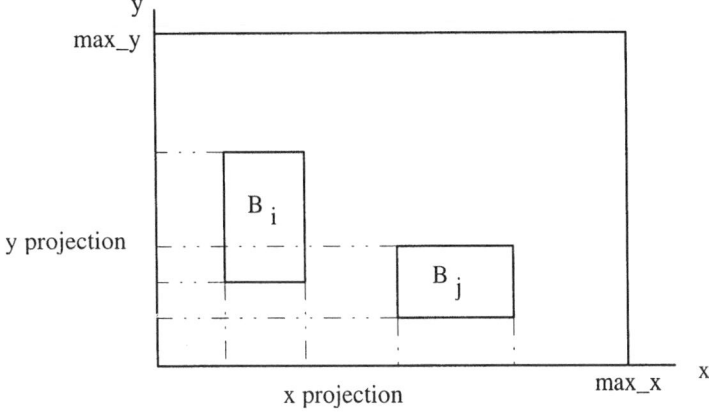

Figure 3. Definition of *x* and *y* projection for blocks B_i and B_j.

arise due to operational conditions, safety, multitasking workers, sharing the same utilities, and the like. The distance between the two machines (blocks) is measured by the nearest pair of points, each on the perimeter of one block.

- *Location Restrictions.* The model assumes a continuous space for the machine layout where every block can be positioned anywhere within the boundaries of the shop floor. In every shop floor, however, there are spaces that must not be used for placing the machinery. Examples include the aisles, offices, washrooms, reserved spaces, and so forth. This constraint restricts the location of any block within such forbidden spaces (see Figure 5).

- *Location Preferences.* It may be desirable that one or more machines be located in a specifically defined space within the shop floor because of several

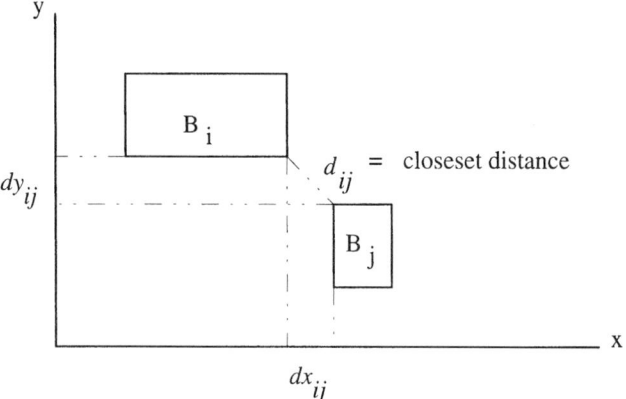

Figure 4. Closest distance between two blocks.

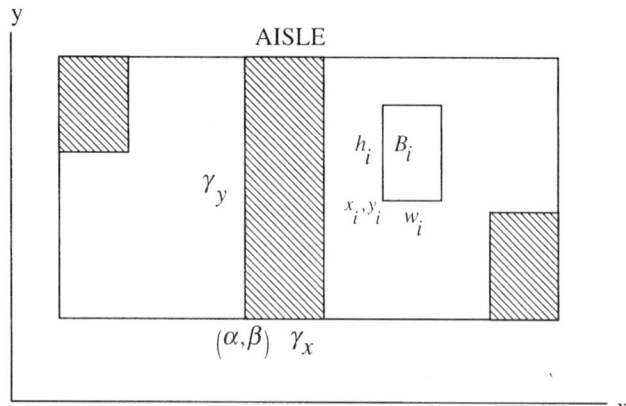

Figure 5. Utilizing location restrictions to address aisles and irregular shape sites (shaded areas).

reasons such as special storage environment for machines or where a machine is already installed and its relocation would be economically unjustified. This set of constraints enforces the desired machines to be positioned within such predefined areas.

- *Orientation Constraints.* In real-world layout problems the decision maker may express a certain preference regarding the orientations of machines. For example, it may be desired that certain machines be horizontal or vertical in the generated layout for purposes such as safety and/or multitasking workers. To address these restrictions, the aspect ratios are used in a manner to satisfy the requirements.

- *Material Handling Cost.* In most industrial shop floors, travel occurs along a set of aisles, arranged in a rectangular pattern parallel to the walls of the building. In such a situation the distance traveled is referred to as rectilinear distance. The traveling distance is calculated as the sum of the flow of materials between any two machines (in number of trips) times the rectilinear distances between the centroids of these two blocks (Figure 6).

3.2. Solution Methodology

A combination (hybrid) of both goal programming and simulated annealing is used to solve the above mathematical model. For a complete description of each method and the merge between these two optimization methodologies, the interested readers are referred to Bazargan-Lari and Kaebernick (1997). This hybrid methodology was primarily developed to overcome some of the difficulties of the existing methods as the mathematical model becomes large and complex.

Goal programming, a rewarding approach in handling multiple objectives, was adopted to address the various practical and often conflicting objectives in modeling real-world layout problems. In goal programming, the decision maker provides the information about the desired levels of achievement (targets) for the various criteria. Prior to solving the goal programming problem, the decision maker provides an

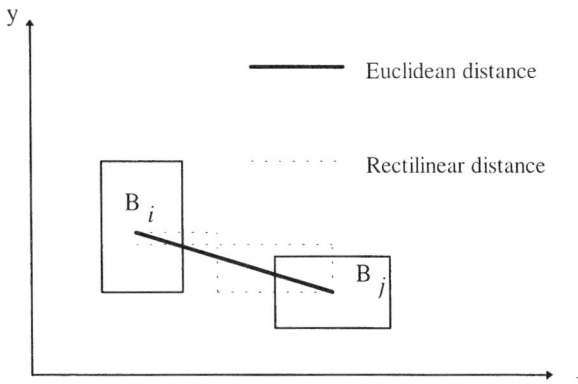

Figure 6. Definition and measure of Euclidean and rectilinear distances.

ordinal or cardinal ranking of the criteria. The goal programming method tries to find a solution that comes as close as possible to the prescribed set of targets in the order of priorities specified. The objectives/constraints are ranked based on preemptive priority order such that satisfaction of higher order goals is always preferred over lower order goals. In this formulation, the objective functions are also expressed as constraints with some desired values (goals) as the right-hand side of the constraints. We then distinguish all the constraints of the model as real (absolute) or goal constraints. The real or absolute constraints represent the constraints that absolutely should be satisfied and any violations to these constraint(s) will lead the solution(s) to be infeasible. The goal constraints on the other hand are less rigid in the sense that they can be violated and are ranked according to decision maker's preference.

Simulated annealing was selected as part of the solution methodology to widen the search horizon for better solutions. This technique has been found to be useful in solving combinatorial problems. The name comes from a useful analogy between statistical mechanics (the behavior of systems with many degrees of freedom in thermal equilibrium at a finite temperature) and combinatorial optimization (Wilhelm and Ward, 1987). It was initially developed for statistical mechanics by Metropolis et al. (1953), who used Monte Carlo random numbers to simulate the achieving of thermal equilibrium. Local search methods, often called "greedy algorithms," start the search from a given initial solution and then continually try to improve the quality of the solution. These methods stop when such improvement is no longer possible. The main drawback of such algorithms is that the solution gets trapped in a local optimum, and there is no way for the algorithm to exit this local optimum. Thus, the quality of the final solution depends very much on the selection of the initial solution. In simulated annealing, on the other hand, we occasionally accept a worse solution in the hope that the solution exits a local optimum and a better solution is obtained at later stages (Heragu and Alfa 1992).

A major capability of this hybrid method is that the user is not required to provide an initial feasible layout solution. The methodology generates multiple nondominant solutions. It constructs the layouts and improves the quality of the solutions using both goal programming and simulated annealing.

3.3. Machine Layout Examples

This section presents the application of the above model to two examples with realistic features.

Six-Machine Test Problem   This problem is extracted from the mechanical room layout proposed by Eastman (1972). Figure 7 shows a plan of a mechanical room. Inside the room must be located the six machines shown. You are to arrange the machines in the room so as to satisfy the following rules:

1. Machines 1 and 2 have control lights on their long face. Their long face must be visible from the door.
2. Machine 3 gives off sparks. Its long side must not be visible from the door.

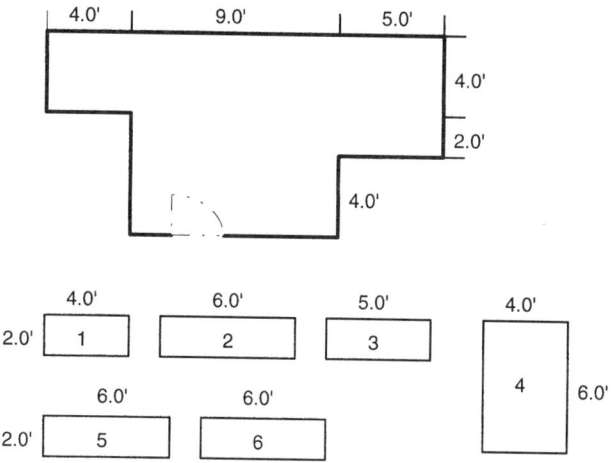

Figure 7. Eastman's mechanical room layout problem.

3. Machine 1 and 3 are connected and must be less than 1 ft apart.
4. Machine 4 passes hot water to machines 2 and 5. Machine 4 must be adjacent to them.
5. Machines 2 and 6 are connected by a flexible cable and must be less than 2 ft 6 in. apart.
6. All machines must be accessible from the door with the pathway being 2 ft wide.
7. All machines shall be located with a long side against a wall.

This problem is of interest because it involves preference satisfaction as the sole criterion and no consideration is given to the traveling cost.

The following procedure explains how each rule was addressed by the proposed model to solve the problem. Rules 1, 6, and 7 are accounted for by defining multiple location preferences for possible locations of each machine. Rule 2 is addressed by defining an area that will serve as location restriction for machine 3 (shaded area in Figure 8).

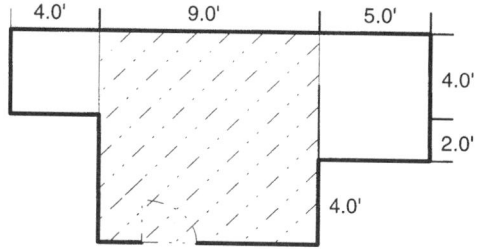

Figure 8. Location restriction area (shaded) for machine 3.

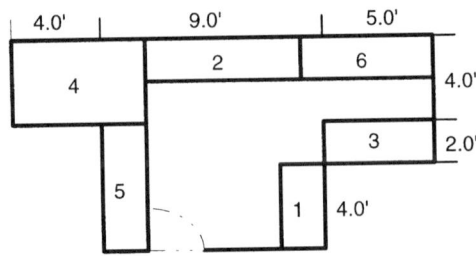

Figure 9. Solution to Eastman's six-machine layout problem by the proposed model.

Rules 3, 4, and 5 can be addressed by closeness relationships as was described in Section 3.1. Since all the rules must be met, they were all assigned to priority level 1. Figure 9 presents the final layout generated by the proposed model.

Eight-Machine Test Problem   This example is extracted from Bazargan-Lari and Kaebernick (1997). There are eight machines to be located in a cell of the size 10 × 8 m. The dimensions of the machines are shown in Table 1.

The frequency of the trips made between machines in 1000's of trips for all parts in one year is given by the matrix presented in Table 2 [similar to work flow for the eight-facility problem in Fortenberry and Cox (1985)].

The constraint/objectives are

- At least 1.5-m space placed between any two machines
- Machines 3 and 6 at least 5 m apart from each other
- Machines 4 and 8 at least 6 m apart
- An aisle is considered for material transport from and to the cell defined by (0,2) as bottom left-hand corner with 10 m width and 1 m length. This area serves as location restriction for machines,

Since certain machines are operated by one operator, it is desirable to have these machines close to each other and on the same side of the aisle as follows:

Table 1. Machine dimensions in centimeters for the layout problem example

Machine	Dimension (cm)
1	200 × 200
2	100 × 100
3	150 × 150
4	100 × 100
5	150 × 200
6	150 × 250
7	100 × 100
8	100 × 150

Table 2. Total frequencies of trips made between machines (in 1000s) in one year

	1	2	3	4	5	6	7	8
1	0	6	1	1	8	2	4	4
2	6	0	1	2	3	3	6	2
3	1	1	0	5	2	3	1	10
4	1	2	5	0	2	8	3	3
5	8	3	2	2	0	4	10	10
6	2	3	3	8	4	0	8	8
7	4	6	1	3	10	8	0	2
8	4	2	10	3	10	8	2	0

- Machines 2 and 7 should be 1.5 m apart.
- Machines 3 and 4 should be 1.5 m apart.
- Machine 6 is already installed at coordinate (3 ,0) in the cell and it incurs a certain cost to relocate it.
- It is preferred to locate machine 4 in an area denoted by (5,3) as bottom left-hand corner with 3 m width and 3 m length.
- Machines 2 and 5 are preferred to be located horizontally and vertically, respectively,
- Machines 6 and 8 should have the same orientation.
- Minimize total traveling cost.

In order to solve this problem, it was assumed that all the above requirements except the traveling cost are absolute (rigid) constraints and consequently were set as priority level 1. The traveling cost was assigned to priority level 2. This problem was solved 10 times. All the solutions were feasible; that is, they all satisfied the absolute constraints assigned to priority level 1. Figure 10 shows the best solution obtained by the proposed model at a total traveling cost of 726.54 km per year. The average traveling cost for the 10 solutions generated was 774 km per year with standard deviation of 18.7 km. Each solution required an average of 18 CPU seconds.

In the eight-machine problem the location preferences for machines 4 and 6 were imposed by expressing these requirements as absolute constraints and assigning them to priority level 1. It is of interest, however, to perform a what-if analysis on this issue. That is, what are the advantages in terms of traveling cost if we permit limited movements (relocation) of these machines? To solve this problem, the absolute constraints corresponding to these requirements were removed from priority level 1. That means a solution that relocates machines 4 and 6 outside the designated areas is also considered feasible, but we prefer these relocations to be as small as possible.

Now there are two desirable objectives, the first one is to reduce the traveling cost and the second is to keep machines 4 and 6 as close as possible to their present locations. The traveling cost was assigned to priority level 2 and the (soft) constraints on machine 4 and 6 to priority level 3. Note that assigning the priority levels the other

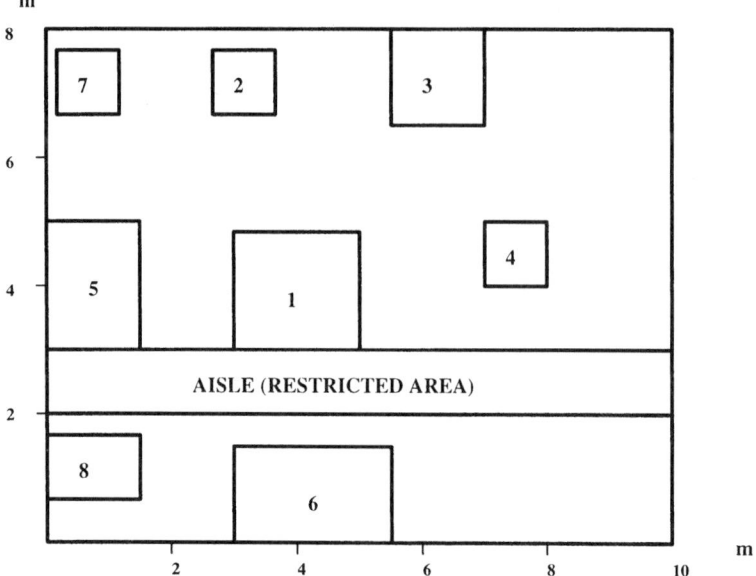

Figure 10. Solution for the eight-machine test problem with machine 6 fixed.

way round will force the positions of machines 4 and 6 to their present location, thus generating a solution similar to the previous layout. Since the constraints on machines 4 and 6 are of the same measure and scale, namely distances, grouping them under one priority level is justified. Figure 11 shows the solution generated by the proposed model at a total cost of 669.9 km per year. This represents a reduction of 56.6 km per year from the best previously found solution or about 8%. This solution helps the decision maker to decide whether the relocations are justified or not. Similar analysis can be performed by examining other criteria of this example.

4. INTRACELL AND INTERCELL LAYOUT DESIGNS

In order to fully achieve the benefits of CM, an efficient design of the physical layout of machines (intracell) and cells (intercell) is absolutely essential. This section introduces a new approach in integrating both intracell and intercell layout designs in an effort to generate multiple efficient alternatives and detailed layouts. The problem formulation, solution methodology, and further refinement of the methods introduced in the previous section provide a good foundation in order to solve intracell and intercell layout problems in a CM environment. In this proposed methodology nondominated intracell layouts are generated for each cell based on two criteria, namely the area allocated to the cell and the traveling cost. A filtering process is used to select the most different layout designs in an effort to handle information overload and to reduce the number of the nondominated solutions. Finally, these nondominated

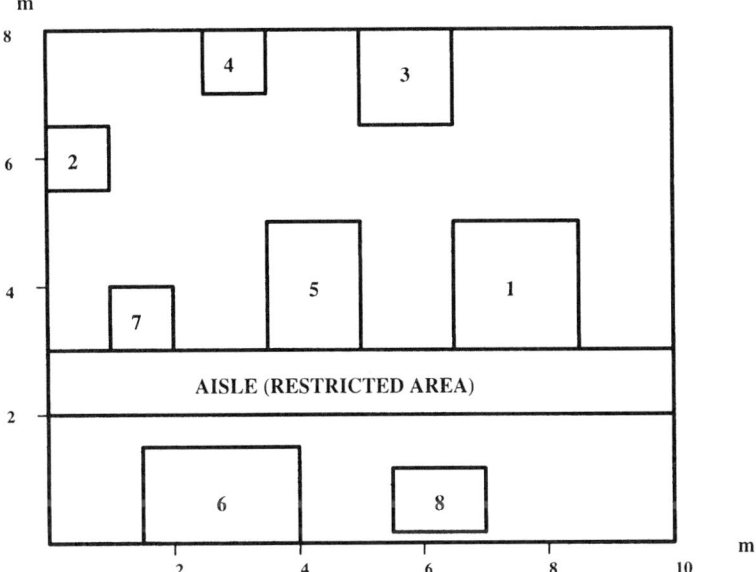

Figure 11. Solution for the eight-machine test problem with relocations of machines 4 and 6.

intracell layout designs are integrated to produce multiple efficient (nondominated) intercell layouts.

4.1. Intracell Layout Model

The intercell layout design is closely related to the machine layout model (Hassan, 1995). The machine layout algorithm presented in the last section is now utilized to address intracell designs. The main difference between the intracell layout and the machine layout problem relies on the cell/shop floor boundaries. In machine layout problems the size (width and height) of the shop floor is normally assumed to be known *a priori,* and the problem is to determine the location of the machines so that certain goals are achieved. In designing the intracell layout for CM, however, the objective is not only to identify the locations of the machines inside each cell, but also to determine the dimensions of the cells, which are not known, and their size is left to the layout designer to decide. Probably the only boundaries known in a manufacturing company planning to implement CM is the overall shop floor size that should accommodate all the cells.

The mathematical model is very similar to the machine layout problem with the following exceptions:

Cell Boundaries   In this model we aim at generating alternative efficient layouts for cells. Thus, the shape of the cell is left to be determined by the model as follows:

$$\max \_x = \max \{(x_1 + w_1), (x_2 + w_2), \ldots, (x_n + w_n)\}$$

$$\max\_y = \max\{(y_1 + w_1), (y_2 + w_2), \ldots, (y_n + w_n)\}$$

$Area$ of the cell $= \max\_x \times \max\_y$

$$x_i, y_i \geq 0, \qquad i = 1, 2, \ldots, n$$

where $\max\_x$ and $\max\_y$ are the width and height of the cell, w_i and h_i are width and height of block i, and n is the number of machines inside the cell, respectively (see Figure 12). The *Area of the cell* is now introduced to the problem formulation as a constraint/goal that we desire to minimize.

4.2. Intercell Layout Design Solution Methodology

Since the cell shapes and sizes are not known in advance and different areas and shapes lead to different layout configurations and costs, the generation of alternative nondominated layouts having different shapes and sizes is absolutely necessary and inevitable. A point within a set is considered to be efficient (nondominated) in that no other point is feasible at which the same or better performance could be achieved with respect to all criteria, with at least one being strictly better.

In generating machine and cell layout, three distinct sets of objectives have been considered. The first set concerns the feasibility of the solution. The second objective is to minimize the area of the cell and finally the third one is to minimize the traveling cost. Again a combination of goal programming and simulated annealing is adopted to solve the mathematical model (Bazargan-Lari and Kaebernick, 1997, 1996a, 1996b).

The program is first run by assigning minimization of the area of the cell and intracell traveling cost to priority levels 2 and 3, respectively. The solution thus generated has the lowest area and highest intracell cost (point z_1^* in Figure 13). Starting from this solution, the program is next run again only reversing the priority levels

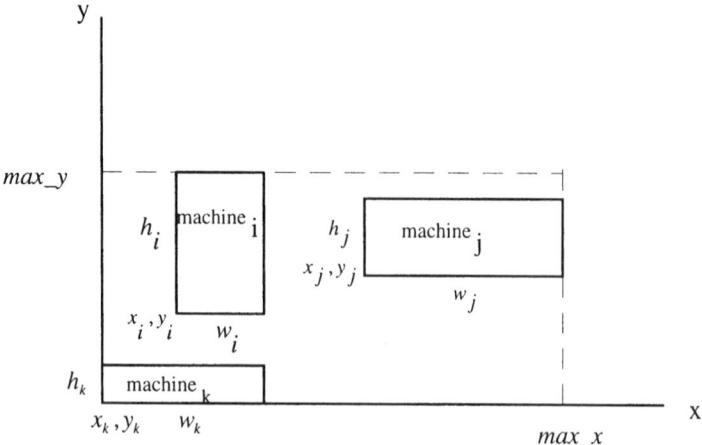

Figure 12. Definition of the cell boundaries.

of these two objectives. During this transition of going from one extreme solution having the smallest area and the largest cost to the other with the smallest cost and largest area (point z_2^* in Figure 13), a large number of solutions are generated of which many are efficient (nondominated) points, which are saved by the program. This phase of the model is repeated for all cells, resulting in a number of efficient solutions for each cell.

The number of nondominated solutions generated is normally large. Steuer (1986) suggests a filtering process to deal with such information overload. Filtering is a process of selecting subsets of points that are most different from others from a larger finite set of points.

The filtering process is performed according to the following procedure: A randomly efficient point among the set of nondominated solutions is selected as the seed point and is added to the top of the list of the retained points. Each nondominated point is then assessed with the seed point according to the desired distance relationship. Each point not satisfying the test is considered to be insignificantly different from the seed point and is excluded from further consideration. Any point, however, satisfying the test is added to the list of retained points. This process is continued, ensuring that each subsequent point is assessed against all the points retained in the filtering list until all the nondominated points are evaluated.

This process is called the method of the first point outside the neighborhood and graphically is shown in Figure 14. In this figure we have seven efficient points of which we would like to reduce it to the three most different points. Starting with x_1 as the seed point, points x_2 and x_3 are excluded from further consideration since their L_2-norm distances from x_1 is less than d. Point x_4, however, is the first point to be significantly different from x_1 and therefore is retained by the filtering process. The subsequent points should be assessed against both x_1 and x_4. Point x_5 is excluded since its distance with x_4 is insignificant. Similarly point x_6 is retained and x_7 excluded from the list. Thus, the filtering process reduces the initial size of seven nondominated points to x_1, x_4, and x_6. It is possible to select the value for d so that the number of selected points is less or more than these three points.

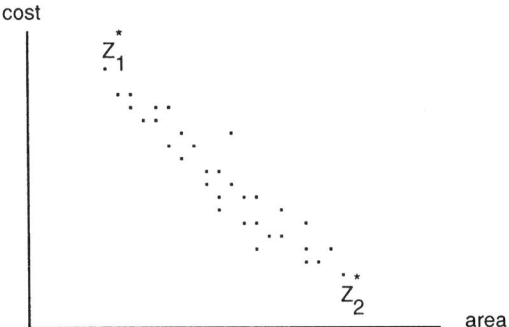

Figure 13. Generating the intermediate solutions between the two extreme points.

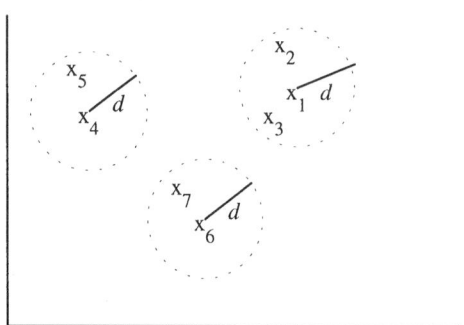

Figure 14. Method of first point outside the neighborhoods.

4.3. Intracell Layout Example

A five-cell layout problem is extracted from the literature and solved using the proposed model. The cell descriptions that are used as test cases are as follows: Cells 1 and 2 are the 6- and 12- machine test problems presented in Welgama and Gibson, (1993). Cells 3 and 4 are the 3- and 9- machine test problems introduced in Heragu and Kakuturi (1997), and finally cell 5 is the 8- machine test problem introduced in the previous section.

To produce realistic layouts, the following restrictions are also imposed:

- To facilitate smooth material flow, a minimum 1.5-m distance should be placed between any two machines in any cell.

Certain relationships between machines are required for cell 5, such as:

- Machines 3 and 6 should be at least 5 m apart from each other.
- Machines 4 and 8 should be at least 6 m apart.
- Machines 2 and 7 should be 1.5 m apart.
- Machines 3 and 4 should be 1.5 m apart.
- Machines 2 and 5 are preferred to be located horizontally and vertically respectively.
- Machines 6 and 8 should have the same orientation.

Table 3 shows the efficient intracell layouts generated for each cell after the filtering process. As the table implies, each shape according to width and height of the cell results in different intracell traveling cost. The filtering process was very useful in reducing the number of these solutions. For example, for cell 1 more than 38 solutions were generated, which after filtering were reduced to 3.

Figures 15–19 present the alternative layout designs for each cell. Tables 4–8 present area and cost information. It is, of course, possible to increase or decrease the above number of alternative layout designs for each cell by changing the values of test-distance parameters (d).

Table 3. Results for alternative efficient intracell layout

Cell	Alternative	Width (m)	Height (m)	Area (m²)	Cost (m)
	a	4.00	4.70	18.80	225.97
1	b	3.55	6.62	23.50	208.42
	c	5.28	5.57	29.42	207.04
	a	7.98	8.64	68.89	1956.01
2	b	10.98	6.98	76.67	1854.50
	c	11.47	9.47	108.60	1710.63
	d	11.31	7.99	90.30	1806.90
	a	3.98	4.5	17.90	8.50
3	b	4.50	3.75	16.86	9.01
	c	4.50	4.10	18.45	8.4
	a	7.13	8	57.07	470.84
4	b	12.93	4.5	58.19	364.33
	c	8.97	8.5	76.23	328.09
	d	8.47	8.49	71.99	353.74
5	a	7.88	8.00	63.07	723.18
	b	7.69	8.00	61.55	729.09

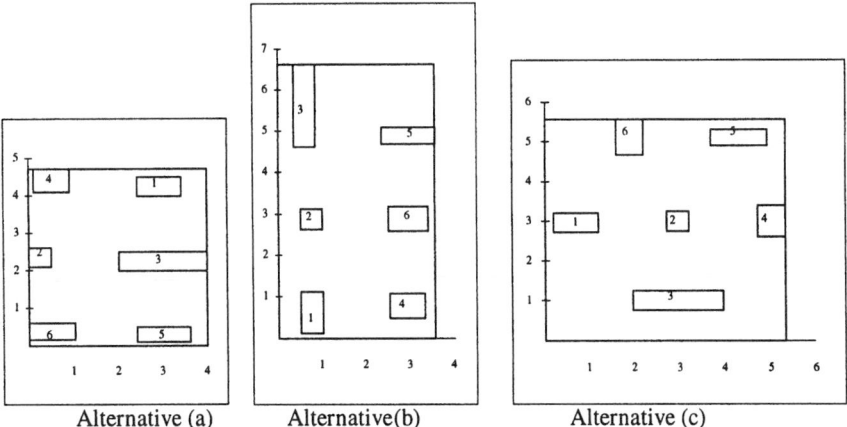

Alternative (a)   Alternative (b)   Alternative (c)

Figure 15. Alternative nondominated physical layouts for cell 1 with six machines.

Table 4. Area and cost for three alternatives in cell 1

Alternative	Area (m²)	Cost (m)
a	18.80	225.97
b	23.50	208.42
c	29.42	207.04

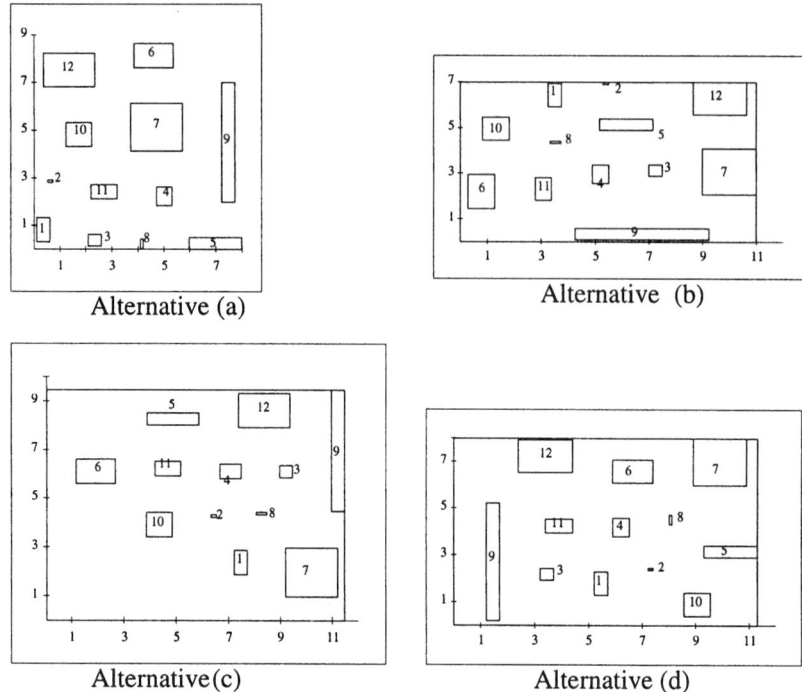

Figure 16. Alternative nondominated layout designs for cell 2 with 12 machines.

Table 5. Area and cost for four alternatives in cell 2

Alternative	Area (m²)	Cost (m)
a	68.89	1956.1
b	76.67	1854.50
c	108.60	1710.63
d	90.30	1806.90

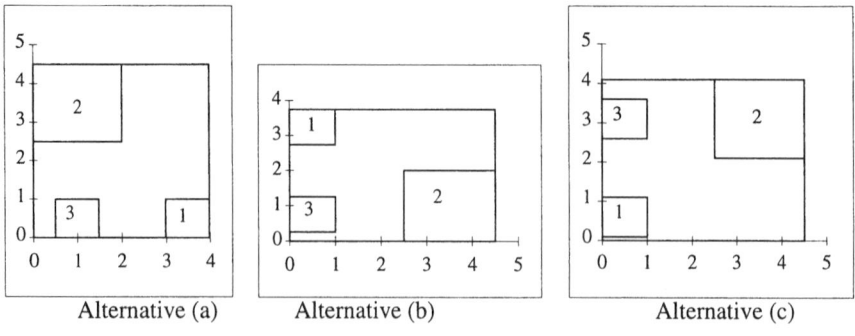

Figure 17. Alternative nondominated layouts for cell 3 with three machines.

Table 6. Area and cost of three alternatives for cell 3

Alternative	Area (m^2)	Cost (m)
a	17.90	8.50
b	16.86	9.01
c	18.45	8.4

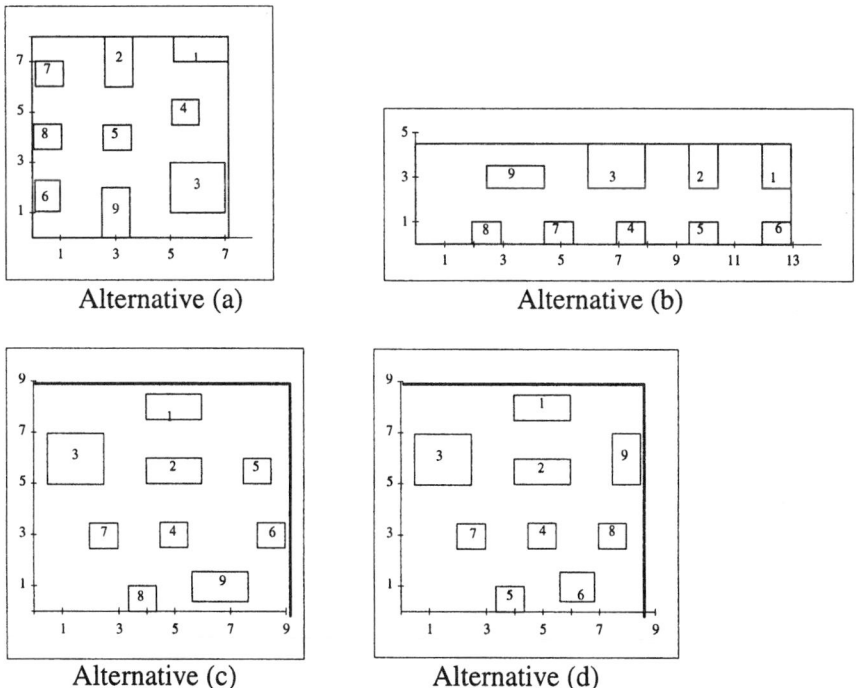

Figure 18. Alternative layouts for cell 4 with nine machines.

Table 7. Area and cost of four alternatives for cell 4

Alternative	Area (m^2)	Cost (m)
a	57.07	470.84
b	58.19	364.33
c	76.23	328.09
d	71.99	353.74

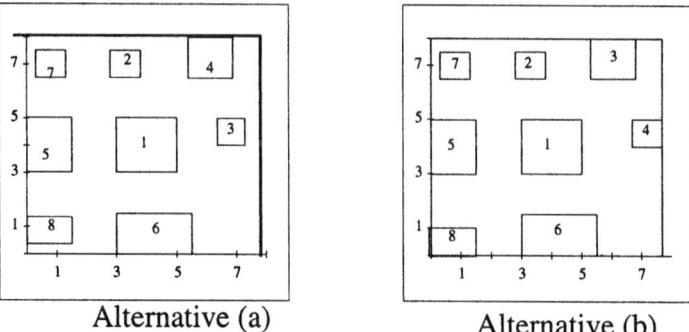

Alternative (a)   Alternative (b)

Figure 19. Alternative layouts for cell 5 with eight machines.

Table 8. Area and cost of two alternatives for cell 5

Alternative	Area (m^2)	Cost (m)
a	63.07	723.18
b	61.55	729.09

4.4. Intercell Layout Model and Integration with Intracell Layout Model

The intercell layout design concerns the location of the cells with respect to each other on the shop floor. The variables and constraints adopted for this model are similar to those presented for intracell layout model. The procedure to produce efficient intercell layout designs is to integrate all the possible combinations of efficient intracell layout configurations and filtering the nondominant solutions thus generated.

In order to demonstrate this heuristic, the five-cell example introduced in Section 4.3 is used again. The aim is to produce efficient intercell layout designs based on two criteria, namely the total area required to accommodate all the cells and the corresponding traveling cost associated with this area. Note that in this methodology the total cost is comprised of intracell cost (depending on which intracell designs are used) and the intercell cost (the cost associated with material handling between the cells).

The flow between the cells (intercell flow) is taken from the five-facility test problem in Nugent et al. (1968). However, to allow for comparable values between inter and intracell flows, the data pertaining to this five-facility test problem in Nugent et al. (1968) were magnified by a factor of 10.

Considering the number of alternative intracell layout designs for each cell, there are 288 ($3 \times 4 \times 3 \times 4 \times 2$) possible cases that the model should evaluate. The CPU time to run each case was less than 3 sec taking a total of about 15 min of CPU time. Table 9 and Figure 20 present the information regarding the area and cost for

Table 9. Results for alternative efficient intercell layout designs

Alternative	Width (m)	Height (m)	Area (m²)	Cost (m)
1	12.50	20.07	250.88	5496.99
2	17.32	14.82	256.68	5329.93
3	20.42	13.14	268.35	5269.56
4	20.47	13.37	273.55	5145.01
5	23.97	13.33	319.57	4868.71
6	27.92	12.93	360.98	4788.79

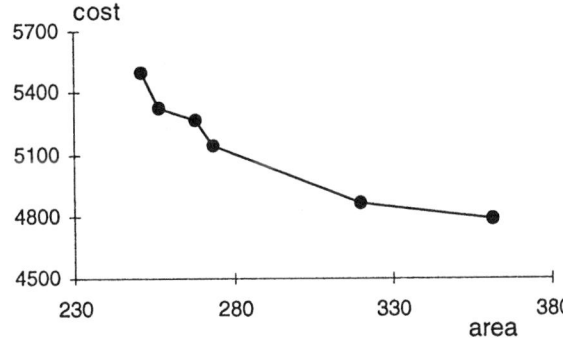

Figure 20. Alternative solutions for final intercell layout designs

six efficient (most different) intercell solutions that were retained after the filtering process.

Figures 21–26 show the six efficient layouts. As certain objectives are unquantifiable and cannot be expressed numerically, the integrated approach enables the decision maker to have wider choices and assess each layout design against these unquantifiable objectives. It is of interest to mention that to generate the lowest cost intercell layout design, it does not mean that we have to include all the lowest cost intracell layout configurations. This approach supports the idea that the most efficient solution for a system is not necessarily generated by the sum of all most efficient components of that system. For example, in alternative 6 above (the lowest intercell traveling cost), for cells 1, 2, and 4 alternatives a, d, and b are used, respectively, while the least-cost alternatives for each of these cells is alternative c. The intercell solutions generated by including all the lowest cost intracell alternative were found to be dominated by alternatives 4, 5, or 6. Other experiments with different interflow data, such as the original five-facility flow data in Nugent et al. (1968) (without the magnification factor), also supported the hypothesis that the final lowest cost intercell alternative does not necessarily include all the lowest cost intracell designs. This example demonstrated the need for generating alternative layout designs.

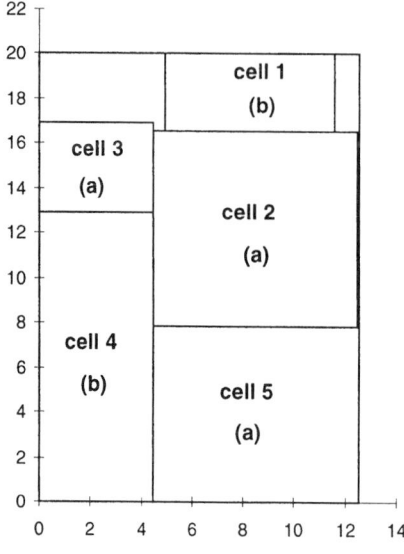

Figure 21. Alternative (1) for intercell layout design (area = 250.88 m², total intra and intercell traveling cost = 5496.99 m).

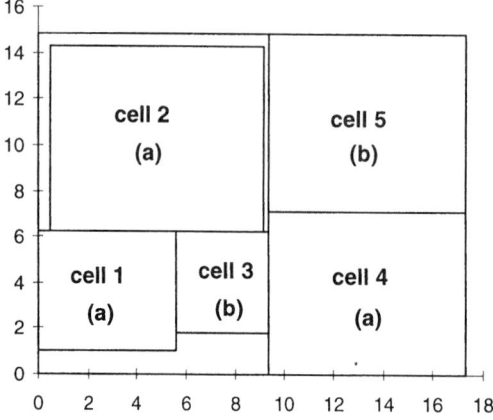

Figure 22. Alternative (2) for intercell layout design (area = 256.68 m², total intra and intercell traveling cost = 5329.93 m).

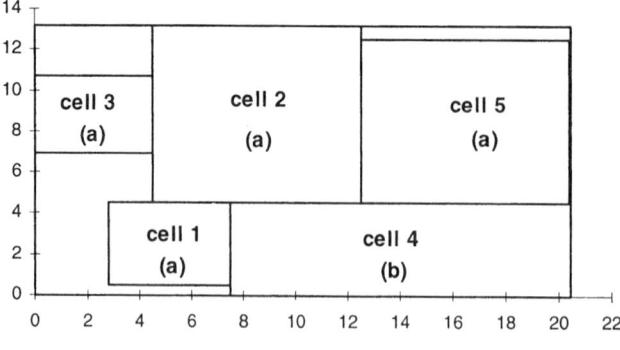

Figure 23. Alternative (3) for intercell layout design (area = 268.35 m², total intra and intercell traveling cost = 5269.56 m).

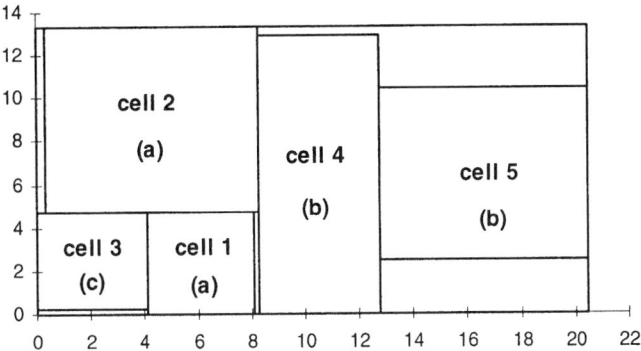

Figure 24. Alternative (4) for intercell layout design (area = 273.55 m^2, total intra and intercell traveling cost = 5145.01 m).

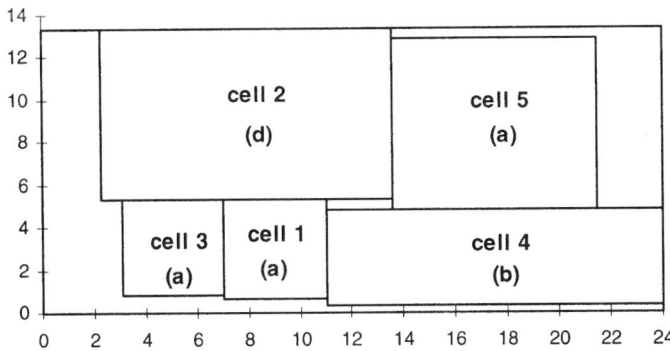

Figure 25. Alternative (5) for intercell layout design (area = 319.57 m^2, total intra and intercell traveling cost = 4868.71 m).

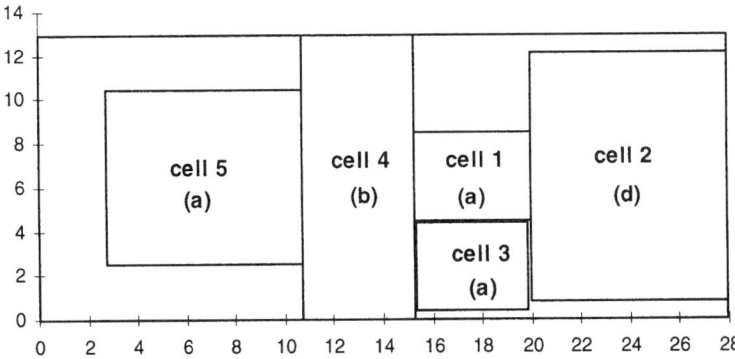

Figure 26. Alternative (6) for intercell layout design (area = 360.98 m^2, total intra and intercell traveling cost = 4788.79 m).

5. INDUSTRIAL APPLICATIONS

The following presents the application of the methodologies proposed to two manufacturing companies.

5.1. Case Study

A medium-sized "white-goods" manufacturing company that produces electric household goods, started implementing CM in early 1994. The need for changing and improving the manufacturing system in the press-shop was long felt by both the management and shop floor employees. Among the problems expressed were excessively long lead times resulting in late deliveries or frequent backlogs, high volumes of work in process, confusion and difficulties over production planning, and ever-increasing material movements.

The company had adopted a process-based layout for the press-shop, where similar presses were grouped together. Cellular Manufacturing was considered to be an appropriate strategy for this company to overcome some of the difficulties described. This strategy was additionally favored by the management as the change did not require new capital investment and included the cost of relocating the machines and equipment and possibly employee reorganization and education.

The total number of parts produced in the press-shop are 153. The list of machinery and equipment used to produce these parts, the available number of presses, and the references used are presented in Table 10. Figure 27 shows the shop floor plan (a potion of a larger shop floor) assigned specifically for positioning of cells and departments with the corresponding dimensions.

Table 10. References used for departments and machines

Department/Machine	Numbers Available	Reference
Raw material store	1	STOR
Guillotine	1	GULT
Degreasers	1	DGRS
11-ton feeder	1	ORRI
Linisher	1	LNSH
150-ton feeder	1	AIDA
Brake press	1	BRAK
30-ton press	5	203A
200-ton press	5	200C
200-ton hydraulic Press	2	200H
250-ton press	2	250C
300-ton press	2	300C
80-ton press	3	207A
200-ton wide	2	200W
110-ton press	1	PSCA
150-ton press	2	150C

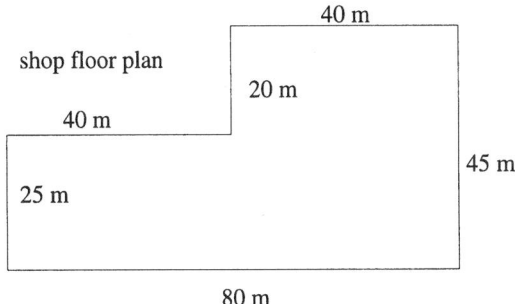

Figure 27. Shop floor plan to accommodate the machines and cells.

The company imposed several restrictions regarding the relocation of certain presses as they were very heavy and were installed on special foundations. Accordingly, any relocation in these machines incurred heavy costs and inconvenience. Furthermore some departments were desired to retain their current positions, for instance, the raw material store because of easy access to unloading docks. These fixed departments and machines are store (STOR), guillotine (GULT), linisher (LNSH), 150-ton feeder (AIDA), 110-ton feeder (ORRI), 300-ton press (300C), 250-ton press (250C), and brake press (BRAK).

The information regarding the parts routings, annual demand, transfer quantities, machine sizes, aisles, and clearance spaces was gathered. Table 11 shows the grouping of the machines (with corresponding number of duplicates in brackets) and the number of parts per cell.

As a test for the proposed layout methodology, the company independently developed a five-cell layout based on traditional in-house expertise. Figures 28 and 29 present the intercell layout design generated by the company and one of the alternative solutions produced by the proposed method, respectively. The solution generated by the proposed algorithm proved to be not only feasible by addressing all the constraints expressed by the company, but also provided a 30% lower traveling cost. This cost reduction attains larger magnitude when translated into other resources such as time and money. In these layout plans RSVD denotes the reserved areas assigned to offices, washrooms, utilities, die racks, and the like.

It is of interest to mention that the intracell and intercell layout designs generated

Table 11. Assignment of parts and machines into five cells

Cells	Number of Parts	Machines Assigned to Cells
Cell 1 (C1)	80	203A (5), 207A(3)
Cell 2 (C2)	18	200C (2), PSCA, 150C (2)
Cell 3 (C3)	20	150C (1), 200W(1), 200H (1), 200C (3)
Cell 4 (C4)	17	BRAK, 250C (1), 300C (1), 200H(2)
Cell 5 (C5)	18	300C (1), 200W(1), 250C (1)

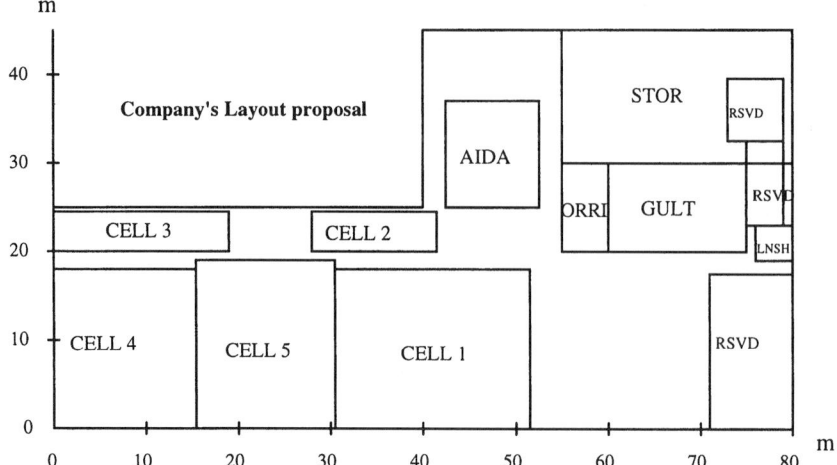

Figure 28. Layout configuration based on the company's proposal plan with a total traveling cost of 486.7 km/year.

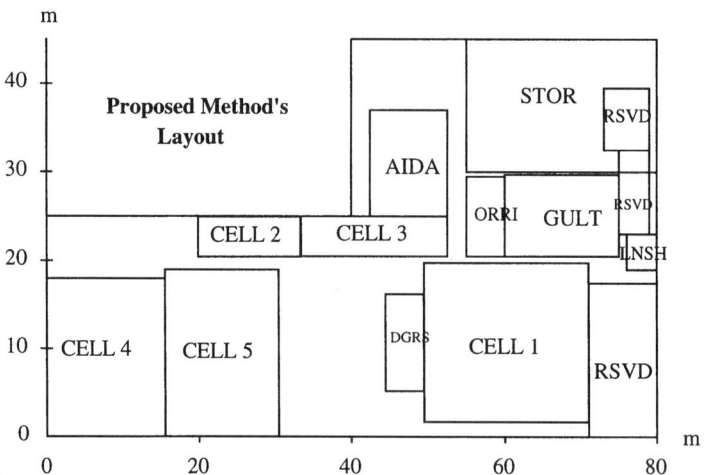

Figure 29. Layout configuration based on the hybrid proposed method with a total traveling cost of 340.3 km/year.

by the company were developed by an employee over several weeks based on his several years working experience at the company, while the results obtained by the hybrid model were generated in less than 15 min. This statement by no means is intended to claim that the proposed method can fully and completely replace the human expertise but should be looked at as an intelligent tool to assist the designer (employee) to generate better and more efficient layouts. The proposed method not only enables the designer to simultaneously target feasible and cost-efficient solutions but also perform what-if analysis.

5.2. Industrial Application

The industrial application for the proposed method involved another medium sized manufacturing plant. By the time the information was collected, the company had already implemented CM and the layout design generated by the company was in practice. The company was, however, interested in evaluating the performance of its layout design with other possible alternatives that the proposed method may produce.

Because of the technical constraints, such as a special and dedicated material handling systems, the shop floor was divided into 25 (consisting of 5×5) equal size zones each having a dimension of 5 by 4 m (see Figure 30). The overall manufacturing system was partitioned into 11 cells (C1 to C11), a store (S), and a forwarding area (F), each occupying one zone. The unassigned zones were considered candidates for tool storage, administrative desk, inspection bench, and so forth.

The information regarding the flow between the cells, store, and forwarding areas was collected. The interzone travel was considered to be "chessboard" (rectilinear), and the distances between the zones are measured from center to center. The materials are transferred at an average speed of 14 m/min.

The company also wished to impose the following constraints:

- The parts/machines allocated to the 11 cells should not be changed.
- The store (S) has to be located in zone 3.
- Cell 11 should be retained in zone 22.
- The forwarding area (F) should be placed in zone 24.

The company was also interested in determining the possible gains by relaxing the constraint on the fixed position of cell 11.

Figures 31 and 32 show the intercell layout designs generated by the company (currently in practice) and the proposed method, respectively.

1	2	3	4	5
6	7	8	9	10
11	12	13	14	15
16	17	18	19	20
21	22	23	24	25

Figure 30. Shop floor consisting of 25 zones

C-9	C-8	S	C-6	C-4
C-2	C-7	C-3		C-10
		C-5	C-1	
	C-11		F	

Figure 31. Company's intercell layout solution.

	C-1	S		
	C-10	C-5		
	C-9	C-7	C-8	
	C-4	C-3	C-6	
	C-11	C-2	F	

Figure 32. Proposed method's intercell layout solution.

Table 12 presents the performance of these two configurations based on three different measures, namely traveling distances in kilometers and traveling time in hours. As the table implies, the hybrid model produces a layout design with 23% lower cost than the company's existing design. The monetary (cost) saving of the proposed method was not considered to be significant by the company. The time saving, however, was received with high interest as any savings in this regard represents reduced lead time and accordingly increased productivity.

Figure 33 and Table 13 provide the alternative solution by relaxing the constraint on the fixed position of cell 11. The relocation of cell 11 represents a 27 and 6% improvement in traveling cost over the solutions generated by the company and the proposed methodology, respectively.

Table 12. Cost comparison between the solution generated by the company and the proposed method with S, C11, and F fixed

Method	Cost in Traveling Distances (km)	Cost in Time (hours)
Company	263.4	313.6
Proposed method	203.5	242.3

C-1	S	C-5	
C-11	C-7	C-3	
C-4	C-9	C-8	
C-2	C-10	C-6	
		F	

Figure 33. Proposed solution with S and F fixed but C-11 relocated.

Table 13. Cost for solution generated with S and F fixed but C-11 relocated

Cost in Traveling Distances (km)	Cost in Time (hours)
191.5	228.0

REFERENCES

Anon (1986). Layout principals that upgrade materials flow, *Modern Material Handling*, **41**, 25–27.

Apple, J. (1977). *Plant Layout and Material Handling*, Wiley, New York.

Armour, G. C., Buffa, E. S., and Vollman, T. E., (1964). Allocating facilities with CRAFT, *Harvard Business Research*, **42**, 136–158.

Bazargan-Lari, M., and Kaebernick, H. (1996a). An efficient hybrid method to solve equal and unequal-size facility layout problems, *International Journal of Industrial Engineering—Applications and Practice*, **3**(1), 51–63.

Bazargan-Lari, M., and Kaebernick, H. (1996b). Intra-cell and inter-cell layout designs for cellular manufacturing, *International Journal of Industrial Engineering—Applications and Practice*, **3**(2), 139–150.

Bazargan-Lari, M., and Kaebernick, H. (1997). An approach to the machine layout problem in a cellular manufacturing environment, *Production Planning & Control*, **3**, 139–150.

Bedworth, D. D., Henderson, M. R., and Wolfe, P. M. (1991). *Computer-Integrated Design and Manufacturing*, McGraw-Hill, New York.

Burbidge, J. L. (1975). *Introduction to Group Technology*, Academic, New York.

Eastman, C. M. (1972). Preliminary report on a system for general space planning, *Comm. ACM*, **15**(2), 76–87.

Fortenberry, J. C., and Cox, J.F. (1985). Multiple criterion approach to the facilities layout problem, *International Journal of Production Research*, **23**, 773–782.

Francis, R. L., McGinnis, L. F., and White, J. A. (1992) *Facilities Layout and Location: An Analytical Approach*, 2nd ed., Prentice Hall, Englewood Cliffs, NJ.

Hales, H. L. (1984). *Computer Aided Facilities Planning,* Marcel Dekker, New York.

Hassan, M. M. D. (1995). Layout design in group technology manufacturing, *International Journal of Production Economics*, **38**(2–3), 173–188.

Heragu, S. S. (1994). Group technology and cellular manufacturing, *IEEE Transactions on Systems, Man and Cybernetics*, **24**, 203–215.

Heragu, S. S., and Alfa, A. S. (1992). Experimental analysis of simulated annealing based algorithms for the layout problem, *European Journal of Operations Research*, **57**, 190–202.

Heragu, S. S., and Kakuturi, S. R. (1997). Grouping and placement of machine cells, *IIE Transactions*, **29**(7), 561–571.

Lacksonen, T. A. (1994). Static and dynamic layout problems with varying areas, *Journal of Operational Research Society*, **45**, 59–69.

Metrolpolis, N., Rosenbluth, A., Teller, A., and Teller, E. (1953). Equation of state calculations by fast computing machine, *Journal of Chemical Physics*, **21**, 1087–1092.

Meyers, F. E. (1993). *Plant Layout and Material Handling*, Regents/Prentice Hall, Englewood Cliffs, NJ.

Nugent, C. E., Vollmann, T. E., Ruml, J. (1968). An experimental comparison of techniques for the assignment of facilities to locations, *Operations Research*, **16**, 150–173.

Steuer, R. (1986). *Multiple Criteria Optimization: Theory, Computation and Applications*, Wiley, New York.

Sule, D. R. (1994). *Manufacturing Facilities: Location, Planning and Design*, PWS Publishing, Boston.

Tompkins, J. A., White, J. A., Bozer, Y. A., Frazelle, E. H., Tanchoco, J. M. A., and Trevino, J. (1996). *Facilities Planning*, 2nd ed., Wiley, New York.

Welgama, P. S., and Gibson, P. R. (1993). A construction algorithm for the machine layout with fixed pick-up and drop-off points, *International Journal of Production Research*, **31**, 2575–2590.

Wilhelm, M. R., and Ward, T. L. (1987). Solving quadratic assignment problems by simulated annealing, *IIE Transactions*, **19**(1), 107–119.

5

SCHEDULING CELLULAR MANUFACTURING SYSTEMS

Robert A. Ruben

Department of Management
New Mexico Institute of Mining and Technology
Socorro, New Mexico 87801

Farzad Mahmoodi

School of Business
Clarkson University
Potsdam, New York. 13699–5770

1. INTRODUCTION

The scheduling of work in a manufacturing cell has long been recognized as critical to the successful implementation of Group Technology (GT) (Mosier, et al., 1984; Flynn, 1987; Wemmerlov and Hyer, 1989). Accordingly, a great deal of research has been devoted to the issue in recent years. A basic assumption of this research and the characteristic that differentiates it from traditional scheduling research is that there exists, within the family of parts produced in the cell, subsets of parts that have similar setup requirements. We shall refer to these subsets of parts as *subfamilies*. The basis for this assumption in the context of GT may have originated in an article by Vaithianathan and McRoberts (1982). However, the notion that setup similarities among parts exist and can be exploited to produce more efficient production schedules was recognized much earlier (e.g., Maxwell, 1961). While formal methods have

Handbook of Cellular Manufacturing Systems, edited by Shahrukh A. Irani
ISBN 0-471-12139-8 © 1999 John Wiley & Sons, Inc.

been proposed to identify subfamilies (White and Wilson, 1977; Vaithianathan and McRoberts, 1982), they are more commonly the result of the standardization of tooling, fixtures, numerically controlled (NC) tapes, and the like, which are readily attained given the similarities in processing requirements of the parts assigned to the cell (Suresh and Meredith, 1985. Wemmerlov, 1992).

The setup times incurred when processing successive parts belonging to a particular subfamily are, by definition, minor, and researchers have typically assumed that they are included in the individual processing time of parts. The setup time incurred when processing successive parts from different subfamilies are more significant, and minimizing their occurrence and in some cases their duration has been the focus of much of the research devoted to scheduling in cellular manufacturing (CM).

The key to setup avoidance is the incorporation of each part's subfamily membership into the logic of the scheduling heuristics. To this end, the queue of parts waiting processing is partitioned on the basis of subfamily membership. While a physical partitioning is typically not required, conceptually, a single queue is replaced by multiple queues each dedicated to parts belonging to a single subfamily. We will hereafter refer to these conceptual queues as *subfamily queues*. The scheduling decision then consists of two stages. In the first stage the parts within each subfamily are ordered on the basis of the desired criteria, for example, processing time. In the second stage the choice of subfamily is made.

In the vast majority of applications, the complexity of the problem precludes deriving an optimal solution. Therefore, heuristics are employed. The criteria used to make the decision at each stage of the decision process define the heuristic. These two-stage heuristics are referred to as *group scheduling heuristics*. A number of such heuristics have been proposed and many are discussed in this chapter. Group scheduling heuristics can be categorized according to the frequency in which the second stage of the decision process is invoked.

Exhaustive group scheduling heuristics consider a change in subfamily only when the subfamily queue being processed is emptied of work. The goal of heuristics of this type is to reduce the number of setups performed and, consequently, increase the time devoted to actual production. As a result, queues are shorter and the average time to produce a part (flow time) is reduced. A potential weakness of exhaustive group scheduling heuristics is that by reducing the opportunity for parts from all subfamilies to be considered in each scheduling decision, parts may be delayed for long periods of time and the effectiveness of the system in maintaining critical priorities can be compromised.

In an effort to address this concern *nonexhaustive* group scheduling heuristics have been proposed. Heuristics of this type consider the subfamily choice upon completing the processing of each part. The focus remains on setup reduction; however, the status of all parts is regularly monitored in an effort to maintain system effectiveness.

In many applications the setup time between subfamilies is dependent on the sequence in which the subfamilies are processed. For example, some subfamilies requiring processing on a turning center might require different tool magazines and NC tapes but use a self-centering three-jaw chuck to hold the work. Other subfamilies might use different holding devices but require similar tooling. Thus, the setup time required for processing a subfamily is dependent on the subfamily that it follows in the schedule. In such cases setup times are termed *sequence dependent*. The term *sequence independent* setup times is used to describe applications where the setup time required for a subfamily is constant regardless of its' place in the schedule. When sequence-dependent setup times are present, heuristics that consider the sequence dependency in the choice of subfamily can result in reductions in the duration of individual setups in addition to reducing the total number of setups.

A large body of research exists on group scheduling (e.g., Mahmoodi and Dooley 1992; Shambu et al., 1996). A variety of shop configurations, environmental factors, performance measures, and assumptions have been incorporated in these research studies. Generally, two types of cell configuration have been studied: flow line cells and job shop cells. All parts in a flow line cell share a common routing through the work centers in the cell. Thus, the work centers in a flow line cell are usually arranged sequentially to achieve an unidirectional flow of work through the cell. Flow line cells are considered the ideal configuration and are particularly amenable to Just-In-Time (JIT) production systems (Wemmerlov and Vakharia, 1991). However, as the number of part families assigned to the cell is increased, the number of work centers assigned to the cell often increases as well. In such cases the flow line configuration becomes increasingly difficult to maintain. A job shop cell refers to the case where there are a variety of routing requirements among the parts processed in the cell, creating a jumbled flow of work through the cell.

In subsequent sections we describe a number of heuristics for use in scheduling work in a manufacturing cell. Our choice of heuristics to highlight is influenced by their performance in research studies. Since the conditions under which scheduling decisions are made in industry are not homogeneous, the influence of important environmental factors on the performance of scheduling heuristics will be discussed as well.

We begin with a discussion of scheduling in flow line cells. A number of studies have been conducted under the assumption of a single workcenter. Since a single workcenter cell is a special case of a flow line cell, results from these studies will be included in the discussion. We then turn to a discussion of the literature devoted to scheduling in a job shop cell. Finally, conclusions are presented.

2. SCHEDULING IN FLOW LINE CELLS

Heuristics for scheduling jobs in a flow line cell have been primarily devoted to the development of *permutation schedules*. A permutation schedule refers to a schedule that is determined at the first station of the flow shop and maintained throughout. Thus, the scheduling heuristic operates only on the queue at the first station and the First-Come-First-Served (FCFS) discipline is used to determine the production sequence at all remaining stages. An obvious advantage of a permutation schedule is ease of implementation. However, an optimal permutation schedule is not necessarily optimal in a global sense (Vakharia and Chang, 1990).

A second distinction among studies devoted to scheduling flow line cells is the nature in which jobs arrive for processing. Specifically, *static arrivals* refers to the case where all jobs to be scheduled are present at the time the schedule is generated. When jobs arrive for processing in the cell intermittently, over time *dynamic arrivals* are said to occur. The nature of job arrival is largely dependent on the production planning and control system that is in place. Period batch control (PBC) is a single-cycle system where all parts have identical planned starting times and due dates (Hyer and Wemmerlov, 1982). Static arrivals may be a more appropriate assumption when PBC is in place. Multicycle production planning and control systems such as Materials Requirements Planning (MRP) are more accurately modeled under the assumption of dynamic arrivals.

Finally, the criteria used to gauge schedule performance are an important consideration. Some heuristics have been shown to perform well on efficiency-based measures such as average flow time and Work-In-Process (WIP) inventory while others have shown good performance on due-date-related measures. Companies that produce parts to stock may be more interested in efficiency measures while make-to-order firms may favor effectiveness (i.e., due date oriented) measures. Assemble-to-order firms may strive for efficiency in the production of components but require effectiveness in their assembly operations. Thus, it is important to consider a number of characteristics of the production environment when choosing a group scheduling heuristic. We now present group scheduling heuristics for use in a variety of environments.

2.1. Static Arrivals and Sequence Independent Setup Times

The Campbell, Dudeck, and Smith (CDS) heuristic was originally proposed for scheduling jobs in a pure flow shop (Campbell et al., 1970). Wemmerlov and Vakharia (1991) subsequently modified the procedure for use in scheduling a flow line cell. Sequence-independent setup times are assumed. The objective of the procedure is to produce a permutation schedule that minimizes makespan, that is, the total time required to complete the processing of all parts scheduled.

The essence of the CDS procedure is to partition an M-stage flow shop problem into M-1 artificial two-stage subproblems. Each two-stage problem is then solved with Johnson's algorithm (Johnson, 1954). Johnson's algorithm produces a sequence of jobs with the optimal makespan for a two-stage flowshop. Johnson's algorithm is applied as follows:

Step 1. List all jobs and their total production time (setup and runtime) at each stage.

Step 2. Select the job with the lowest total production time. If the lowest total production time is at the first stage, place the job in the earliest, unassigned position in the sequence. If the lowest total processing time is at the last stage, place the job in the latest, unassigned position in the sequence. Ties can be broken arbitrarily. Remove the job from further consideration.

Step 3. Repeat step 2 until all jobs are assigned a position in the sequence.

As discussed previously, group scheduling requires producing a sequence of sub-families as well as a sequence of parts within each subfamily. Thus, the modification suggested by Wemmerlov and Vakharia (1991) simply applies the CDS procedure in two stages. In the first stage the sequence of subfamilies is produced by considering the total processing time to produce each subfamily at each stage in the flow line, that is, the sum of the subfamily setup time and the processing times for each job in the subfamily. Once the sequence of subfamilies is constructed, the CDS procedure is then reapplied to produce a sequence of jobs within each subfamily. Note that this procedure is exhaustive in that once a subfamily is chosen, all jobs in the subfamily are scheduled before another subfamily is considered.

In their original work Wemmerlov and Vakharia (1991) the procedure was applied intermittently under the assumption of dynamic arrivals with limited success. Indeed a very simple group scheduling heuristic, FC/FCFS (discussed below), yielded similar performance on average flow time and lateness measures. However, studies by Vakharia and Chang (1990) and Skorin-Kapov and Vakharia (1993) indicate under static arrivals the procedure performs quite well both as a stand-alone procedure and as a starting point for stochastic search techniques when the performance criteria is makespan.

The research of Vakharia and Chang (1990) and Skorin-Kapov and Vakharia (1993) applied simulated annealing and tabu search, respectively, to the flow line group scheduling problem. The schedules that result are generally not permutation schedules and thus are more difficult to implement. However, their results indicate that in many cases near optimal schedules in terms of makespan are possible with these methods. Essentially, these stochastic techniques search the solution space by generating and evaluating a limited number of alternate solutions. Both studies use simple pairwise exchanges to generate alternate solutions. If an alternate solution yields lower makespan, it is retained and pairwise exchanges are again applied. In addition, there is a nonzero probability that an inferior solution will be retained that may allow the procedure to escape local optima. While these procedures require a careful choice of parameters, the results of the research indicate the increase in solution quality may well be worth the effort.

Each of these studies indicates that the magnitude of the subfamily setup times relative to the processing times of individual jobs has a significant impact on the performance of group scheduling heuristics. Since the focus of group scheduling is setup avoidance, the impact of group scheduling increases with the magnitude of subfamily setup times.

Finally, a group of Japanese scholars has developed procedures for finding optimal solutions on a number of criteria. Many of these procedures are based on branch and bound techniques and are feasible for only relatively small problems. The interested reader is referred to Ham et al. (1985) for details.

2.2. Static Arrivals Sequence-Dependent Setup Times

When setup times are sequence dependent, the scheduling problem becomes more complex (e.g., Gupta and Darrow, 1986; Irani et al., 1988). If the objective is to minimize the total setup time, the problem can be cast as a traveling salesman problem. Simply stated, the traveling salesman problem is to determine the minimum length route from a departure point through a given number of locations and back to the departure point, visiting each location exactly once. In solving the scheduling problem, the sequence-dependent setup times between families are analogous to the distances between locations in a traveling salesman problem. Branch and bound, dynamic programming, and heuristic approaches to the traveling salesman problem are available in standard operations research texts (e.g., Winston, 1994). It must be noted, however, that schedules that are optimal for total setup time are not necessarily optimal in terms of other performance measures.

Foo and Wager (1983) note that the acyclic version of the traveling salesman problem where the salesman does not return to the departure point may be a more appropriate formulation for the group scheduling problem. In the acyclic version of the problem the selection of the initial subfamily to process can have an impact on the solution quality. The authors suggest a simple technique for choosing the first subfamily that involves including a dummy row and column into the setup time matrix. The problem is then solved as a cyclic traveling salesman problem.

2.3. Dynamic Arrivals and Sequence-Independent Setup Times

We are aware of only two studies in a multistage flow line cell where dynamic arrivals and sequence-independent setup times are assumed. One is the aforementioned study by Wemmerlov and Vakharia (1991). The group scheduling heuristic that provided the best overall performance was the FC/FCFS heuristic (designated FC-FAM in their study). This exhaustive heuristic produces permutation schedules and operates as follows: In the first stage, jobs within each subfamily are ordered on the basis of their time of arrival to the cell, earliest to latest. In the second stage, the subfamily with the job that arrived the earliest is chosen.

While providing good performance in this study, FC/FCFS was dominated by a number of heuristics in the study by Russell and Philipoom (1991), which utilized a similar shop model. All group scheduling heuristics tested produced permutation schedules, and their performance was largely dependent on the method used to set the due dates. A large number of group scheduling heuristics were tested. We will not attempt to reproduce them all in this chapter. Rather, we highlight three of the most promising as indicated by the results:

EDD/EDD   This exhaustive group scheduling heuristic operates in a similar manner to FC/FCFS discussed. The difference is that jobs in each subfamily are ordered by their due date. The subfamily containing the job with the earliest due date is then chosen. This heuristic provided good performance on average tardiness, particularly when the magnitude of the family setup times was high.

APT/SPT   This is an exhaustive group scheduling heuristic that provided superior performance on average flow time across a wide range of experimental conditions. Jobs within subfamilies are ordered on the basis of processing time. The family with the least average total processing time is chosen. The average total processing time is calculated by adding the subfamily setup time to the sum of the processing time for all jobs in the queue. This quantity is then divided by the number of jobs in the queue.

EDD/T   This heuristic is nonexhaustive and was shown to provide good performance on due date measures when the subfamily setup times were low. The heuristic operates somewhat differently than the two-stage approach described above; however, the effect is the same. All jobs in the queue are ordered by their due date; however, those jobs that do not belong to the subfamily currently being processed have a positive constant added to their due date. Thus, a switch in subfamilies only occurs if the queue is exhausted of the current subfamily or there is a job from another subfamily with a severely critical due date. The choice of the constant will affect the number of subfamily switches and must be selected carefully.

Wemmerlov (1992) assumes dynamic arrivals and sequence-independent setup times in his study of a single workcenter cell. While no statistical analysis was conducted, the results do provide insights into the environmental conditions under which the two-stage group scheduling approach yields the greatest advantage over traditional single-stage scheduling heuristics. Specifically, the results suggest that high utilization rates, fewer part families, and high setup times increase the importance of the setup avoidance mechanism and, hence, favor the group scheduling approach. In addition, the results suggest that as randomness is removed from the system in the form of more stable interarrival times and processing times, the advantage of group scheduling decreases.

Finally, the effect of subfamily mix was examined. Previous studies have made the assumption that each subfamily accounts for an equivalent amount of the processing requirements of the cell. Wemmerlov (1992) modified this assumption and examined the case where 25% of the subfamilies accounted for 90% of the processing requirements. The number of part families was varied from 4 to 32. As expected, flow times were reduced and the advantage of group scheduling diminished as the demand distribution was skewed in favor of a small number of subfamilies.

2.4. Dynamic Arrivals and Sequence-Dependent Setup Times

An early study by Sawiki (1973) introduced a number of nonexhaustive group scheduling heuristics under the assumption of dynamic arrivals and sequence-dependent setup times. The study assumed a single workcenter. A potential pitfall of using nonexhaustive heuristics is the excessive number of setups that may result. To determine the potential time available for setups at a workstation, one must first estimate the rate at which jobs arrive to the workstation and the rate at which jobs can be processed at the workstation. The ratio of these two estimates provides an estimate of workstation utilization (generally represented as (ρ). For example, if, on average,

5 jobs arrive to a workcenter in an hour and the workcenter has capacity to process 10 jobs per hour, then the utilization of the workcenter is 50%. The percent of time available for setups is then $1 - \rho$. In our example 50% of the time (30 min/hour) is available for use in performing setups. If the time devoted to setups exceeds this amount then, the rate at which jobs arrive at the workcenter will exceed the rate at which jobs are processed on the workcenter resulting in ever growing queues unless capacity is increased (e.g., overtime).

Sawiki's (1973) heuristics rely on a simple procedure for determining the timing of subfamily setups, which ensures that the capacity of the workcenter is not exceeded. The procedure sets a target processing time for each subfamily as follows:

$$P = K(\rho/1 - \rho)(\text{SSN}/N) - \Delta$$

where P = target amount of accumulated processing time before a switch is considered

K = parameter of the rule; ≥ 1

ρ = System utilization, i.e., the arrival rate/service rate

SSN = Sum of the setup time and idle time over the last N switches

N = parameter of the rule; ≥ 1

Δ = Net amount by which the target processing times has been exceeded up to the present time

The procedure operates as follows: each time a new subfamily is chosen, the target processing time, P, is calculated. Jobs from the subfamily are processed until the total processing time meets or exceeds P, at which point a switch is considered. If the total processing time exceeds P, then the difference (i.e., Δ) is subtracted from the target processing time for the next subfamily. If the subfamily is exhausted before P is reached, then more time is made available for the next subfamily. Any criteria can be used to choose the next subfamily and to sequence the jobs within each subfamily. However, Sawiki based both choices on the earliest due date, similar to EDD/EDD discussed above.

The choice of the K and N parameters had a significant effect on the performance of this heuristic. The K parameter affects the number of switches. Note that when $K = \infty$, the rule becomes exhaustive. The term (SSN/N) is the average time spent on setup and idle time over the last N subfamily switches. Thus, larger N values provide a greater smoothing effect. The procedure yielded impressive performance on average tardiness in the original study. The study by Russell and Philipoom (1991) discussed previously tested a number of different versions of Sawicki's heuristic in a flow line cell assuming sequence-independent setup times with mixed results.

A study by Mahmoodi et al. (1992) compared the performance of a number of heuristics in a balanced (no bottlenecks) five-stage flow line cell. Setup times were sequence dependent and permutation schedules were employed. Two group scheduling heuristics emerged as superior:

MS/SPT   This rule yielded good performance on average flow time and the percent of jobs tardy. Jobs within each subfamily are ordered on the basis of processing time. The subfamily with the minimum setup time is chosen. Note that this is equivalent to dynamically applying the nearest-neighbor traveling salesman heuristic to the setup matrix.

DD/SI[x]   This heuristic uses the SI[x] dispatching heuristic (Oral and Malouin, 1973) to order jobs within each subfamily. Initially, the slack for each job is calculated as follows:

$$S_i = \text{DDATE}_i - \text{TP}_i - \text{TNOW}$$

where, S_i = Slack time of job i
   DDATE_I = Due date of job i
   TP_I = Sum of the expected remaining processing time for i
   TNOW   = Present time

Two classes of jobs are then formed on the basis of slack. Those jobs with negative slack are placed in one class while jobs with positive slack are placed in the other. Within each class, jobs are ordered on the basis of processing time. The class of jobs with negative slack is then given priority. The subfamily that has the job with the minimum due date is chosen. This heuristic dominated all other heuristics on average tardiness and percent tardy while yielding good flow time performance.

The study also examined two simple single-stage dispatching rules, that is, Shortest Processing Time (SPT) and First-Come-First-Served (FCFS). Data was gathered under two levels of due date tightness, shop load, Setup-to-Runtime (S/R) ratio, and interarrival time distribution. The exponential distribution (high variability) and uniform distribution (low variability) were used to model interarrival times.

The results indicated that the performance of the single-stage heuristics (relative to the group scheduling heuristics) improved significantly when the interarrival times were less variable. Further, conditions that tend to increase queue lengths (i.e., high shop load and long setup times) tend to magnify the advantage of group scheduling heuristics. The group scheduling heuristics were also found to be very robust with respect to all of the experimental factors studied. Finally, the results indicated that unlike the single-stage rules, the performance of the group scheduling heuristics is very similar, especially in stable, underutilized manufacturing environments. Thus, the most easily implemented group scheduling heuristics should be selected. This result is similar to those of Frazier (1996) who showed that more complex heuristics result in poorer performance than simpler heuristics, under a variety of experimental conditions.

More recently, Mahmoodi and Martin (1997) developed an efficiency-oriented queue selection heuristic with the objective of minimizing total setup times. This heuristic dynamically assesses variation in a subfamily's arrival rate that is critical in highly transient-state environments. Results indicated favorable flow time and due date performance under a variety of experimental conditions.

3. SCHEDULING JOB SHOP CELLS

A number of studies have examined the group scheduling problem in a job shop cell (e.g., Mosier et al., 1984; Flynn, 1987; Mahmoodi et al., 1990a, 1990b; Mahmoodi and Dooley, 1991; Ruben et al., 1993; Suresh and Meredith, 1994; Kannan and Lyman, 1994). Recall that a job shop cell is characterized by a jumbled flow of jobs through the workcenters in the cell. Thus, the easily applied permutation schedules used to schedule jobs in a flow line cell cannot be employed in a job shop cell. Instead, scheduling heuristics must be applied to the queue of jobs at each workcenter.

An early study by Mosier et al. (1984) tested three methods for subfamily selection in combination with five job selection rules yielding 15 group scheduling heuristics. The procedures were applied to a job shop cell with sequence-independent setup times. Two subfamily selection rules emerged as dominant and each has been studied extensively in subsequent research.

WORK   This rule is exhaustive. The subfamily that has the greatest total expected processing time is chosen. This rule exhibited good overall performance across a variety of performance measures.

ECON   This rule is nonexhaustive. The formulation of this rule has been modified from the original in a subsequent study (Ruben et al., 1993). The result is a more general formulation that affords the user more control over the number of setups performed. It is this formulation that is described here. For each of the j subfamilies not currently being processed, a priority index is calculated as follows:

$$P_j = n_j / E(S_j),$$

where, P_j = Priority index for subfamily j

n_j = Number of jobs belonging to subfamily j that is currently in the queue

$E(S_j)$ = Expected setup time for subfamily j

Note that the priority index has the intuitive appeal of requiring more jobs to accumulate for subfamilies with higher expected setup times. A switch is made to the candidate subfamily with the highest value of P_j if the following condition is met:

$$(n_j - n_i - 1)/n_j > Z,$$

where n_i is the number of jobs in the subfamily currently being processed and Z is a parameter of the rule. Note that when $Z = \infty$, the rule becomes exhaustive; hence, larger Z values tend to reduce the number of setups incurred. This rule has produced mixed results. It performed quite poorly in the study by Mahmoodi et al. (1992) and the study by Mahmoodi and Dooley (1991). Both studies used the original formulation. The results of Ruben et al. (1993), however, indicate that it has potential for managing the tradeoff between due date performance and efficiency-based measures.

Mahmoodi et al. (1990a) tested many of the group scheduling heuristics and experimental factors discussed above with reference to the flow line cell study by Mahmoodi et al. (1992) in a job shop cell. Sequence-dependent setup times were assumed. Indeed the same setup time matrix was used in both studies. The similarity in the results indicates that the conclusions of that study are applicable under both cell configurations. Specifically, the DD/SI^x and MS/SPT heuristics dominated on average tardiness and average flow time, respectively. Further, these group scheduling heuristics were very robust to changes in the levels of due date tightness, shop load, and S/R ratio. Note that these group scheduling heuristics were robust with respect to these factors in a flow line cell as well (Mahmoodi et al., 1992).

Ruben et al. (1993) extended the work of Mahmoodi et al. (1990a). Many of the group scheduling heuristics proposed in the earlier study was examined along with the ECON/SI^x heuristic in the same job shop cell with sequence-dependent setups. The nonexhaustive ECON/SI^x heuristic coupled the modified ECON logic (described above) for switching between subfamilies with the SI^x job sequencing rule. The effect of the part family mix and variability of the interarrival time distribution were examined along with two levels of S/R and shop load.

The results indicated that while DD/SI^x and MS/SPT generally dominated on average tardiness and average flow time, respectively, ECON/SI^x was best at managing the competing priorities of effectiveness and efficiency. Furthermore, the effect of a skewed demand distribution and reduced variability in interarrival times was similar to that of Wemmerlov (1992). Specifically, when the demand for parts from a single subfamily accounted for 80% of the demand for all products, the difference in performance among various heuristics was reduced. A reduction in the variability of interarrival times led to a reduction in the performance differential among heuristics as well.

The effect of labor scheduling policies in a job shop cell was examined by Wirth et al. (1993). The shop model was identical to that used in the Ruben et al. (1993) study. A total of three laborers were available to service the five-workcenter cell. Shop load, interarrival time distribution, S/R ratio, and transfer-to-runtime ratio was each examined at two levels. A number of previously reported group scheduling heuristics were modeled in combination with a total of five labor scheduling rules. Two labor scheduling heuristics emerged as superior:

IDLE   A laborer is available for transfer only upon completion of all jobs at the workcenter, that is, when the workcenter becomes idle.

QCOMP   A laborer is available for transfer only upon the completion of all jobs in the current subfamily. Subfamily queues at the current workcenter as well as those at workcenters with no laborer assigned are considered. Thus, the laborer need not necessarily leave the current machine.

The average amount of time needed to transfer labor between workcenters had the greatest effect on the labor scheduling policies. Specifically, when labor transfer time was high the IDLE heuristic performed best. However, when labor transfer times were low QCOMP yielded the best performance. Thus, the level of flexibility in moving labor between machines is inversely related to the transfer time. In a cell, where workers are cross-trained and workcenters are in close proximity to one another, labor transfer times are small and, hence, flexibility will be high.

Finally, the MS/SPT group scheduling heuristic described above dominated all other heuristics, including DD/SI^x, on average flow time, average percent of jobs tardy, and average tardiness. This was true under both levels of labor transfer time. This indicates that when labor is a constrained resource, efficiency-oriented group scheduling heuristics that limit the buildup of WIP inventory in the shop are preferred.

4. CONCLUSIONS

In our discussion of scheduling CM systems, we have concentrated primarily on group scheduling heuristics. Accordingly, we have presented detailed descriptions of a number of these heuristics. Many have been suggested in the literature and more will be forthcoming in the coming years. When choosing among the many heuristics available, a number of factors should be considered. First, the nature of arrivals needs to be determined. For example, the modified CDS heuristic is an easily implemented heuristic that will provide schedules, which perform well on the makespan criteria when arrivals are static. If dynamic arrivals are a more accurate reflection of reality, group scheduling heuristics such as DD/SI^x and MS/SPT have been shown to provide good performance in both flow line and job shop cells. Practitioners that are more concerned about meeting due dates should consider DD/SI^x, while MS/SPT should be the choice if efficiency is the primary goal. MS/SPT is particularly appropriate when setup times are sequence dependent. In addition, a number of nonexhaustive heuristics have been suggested that hold the promise of managing the tradeoff between efficiency and effectiveness.

Research has also shown that the impact of group scheduling is enhanced when conditions are present that magnify the importance of setup avoidance. Specifically, increased variability in interarrival times and/or processing times, high machine utilization, lengthy setup times, and a low number of part families, have all been shown to increase the performance differential of group scheduling over single-stage scheduling. These are important considerations and management can directly influence many. Finally, research has shown that, simpler heuristics perform as well as more complex heuristics, under a variety of experimental conditions. Group scheduling heuristics can be expected to perform at least as well as single-stage heuristics under most conditions. Furthermore, the performance of group scheduling heuristics is much more robust with respect to different experimental factors than those of the single-stage heuristics. Differences in performance among the group scheduling heuristics are much narrower than those of single-stage heuristics. Thus, selecting an inappropriate single-stage heuristic can result in a more serious deterioration of shop performance than choosing an inappropriate group scheduling heuristic.

REFERENCES

Campbell, H. G., Dudeck, R. A., and Smith, M. L. (1970). A heuristic algorithm for the n job m machine sequencing problem, *Management Science*, **16**, 630–637.

Flynn, B. B. (1987). Repetitive lots: The use of a sequence-dependent setup time scheduling procedure in group technology and traditional shops, *Journal of Operations Management*, **7**, 203–216.

Foo, F. C., and Wager, J. G. (1983). Setup times in cyclic and acyclic group technology scheduling systems, *International Journal of Production Research*, **21**, 63–73.

Frazier, G. V. (1996). Evaluation of group scheduling heuristics in a flow-line manufacturing cell, *International Journal of Production Research*, **34**, 959–976.

Gupta, J. N. D., and Darrow, W. P. (1986). The two machine sequence dependent flowshop scheduling problem, *European Journal of Operational Research*, **24**, 439–446.

Ham, I., Hitomi, K., and Yoshida, T. (1985). *Group Technology*, Kluwer-Nijhoff, Boston.

Hyer, N. L., and Wemmerlov, U. (1982). MRP/GT: A framework for production planning and control of cellular manufacturing, *Decision Sciences*, **13**, 681–701.

Irani, S. A., Gunasena, U., Davachi, A., and Enscore, E. E. (1988). Single machine setup-dependent sequencing using a setup complexity ranking scheme, *Journal of Manufacturing Systems*, **7**, 11–23.

Johnson, S. M. (1954). Optimal two and three stage production schedules with setup times, *Naval Research Logistics Quarterly*, **1**, 61–68.

Kannan, V. R., and Lyman S. B. (1994). Impact of family-based scheduling on transfer batches in a job shop manufacturing cell, *International Journal of Production Research*, **32**, 2777–2794.

Mahmoodi, F., and Dooley, K. J. (1991). A comparison of exhaustive and non-exhaustive group scheduling heuristics in a job shop cell, *International Journal of Production Research*, **29**, 1923–1939.

Mahmoodi, F., and Dooley, K. J. (1992). Group scheduling and order releasing: Review and foundations for research, *International Journal of Production Planning and Control*, **3**, 70–80.

Mahmoodi, F., and Martin, G. E. (1997). A new shop-based and predictive scheduling heuristic for cellular manufacturing, *International Journal of Production Research*, **35**, 313–326

Mahmoodi, F., Dooley, K. J., and Starr, P. J. (1990a). An investigation of dynamic group scheduling heuristics in a job shop manufacturing cell, *International Journal of Production Research*, **28**, 1695–1711.

Mahmmodi, F., Dooley, K. J., and Starr, P. J. (1990b). An evaluation of order releasing and due date assignment heuristics in a job shop manufacturing system, *Journal of Operations Management*, **9**, 548–573.

Mahmoodi, F., Tierney, E. J., and Mosier, C. T. (1992). Dynamic group scheduling heuristics in flow-through cell environment, *Decision Sciences*, **23**, 61–85.

Maxwell, W. L. (1961). An investigation of multi-product, single machine scheduling and inventory problems, unpublished Doctoral Dissertation, Cornell University.

Mosier, C. T., Elvers, D. A., and Kelly, D. (1984). Analysis of group scheduling heuristics, *International Journal of Production Research*, **22**, 857–875.

Oral, M., and Malouin, J. L. (1973). Evaluation of shortest processing time scheduling rule with truncation process, *AIIE Transactions*, **5**, 357–363.

Ruben, R. A., Mosier, C. T., and Mahmoodi, F. (1993). A comprehensive analysis of group scheduling heuristics in a job shop cell, *International Journal of Production Research*, **31**, 1343–1369.

Russell, G. R., and Philipoom, P. R. (1991). Sequencing rules and due date setting procedures in flow line cells with family setups, *Journal of Operations Management*, **10**, 524–545.

Sawicki, J. D. (1973). The problem of tardiness and saturation in a multi-class queue with sequence-dependent setups, *AIIE Transactions*, **5**, 250–255.

Shambu, G., Suresh, N. C., and Pegels, C. C. (1996). Performance evaluation of cellular manufacturing systems: A taxonomy and review of research, *International Journal of Operations and Production Management*, **16**, 81–103.

Skorin-Kapov, J, and Vakharia, A. J. (1993). Scheduling a flow-line manufacturing cell: A tabu search approach, *International Journal of Production Research*, **31**, 1721–1734.

Suresh, N. C., and Meredith, J. R. (1985). Achieving factory automation through group technology principles, *Journal of Operations Management*, **5**, 151–167.

Suresh, N. C., and Meredith, J. R. (1994). Coping with loss of pooling synergy in cellular manufacturing systems, *Management Science*, **40**, 466–483.

Vakharia, A. J., and Chang, Y. (1990). A simulated annealing approach to scheduling a manufacturing cell, *Naval Research Logistics Quarterly*, **37**, 559–577.

Viathianathan, R., and McRoberts, K. L. (1982). On scheduling in a GT environment, *Journal of Manufacturing Systems*, **1**, 149–155.

Wemmerlov, U. (1992). Fundamental insights into part family scheduling: The single machine case, *Decision Sciences*, **23**, 565–595

Wemmerlov, U., and Hyer, N. L. (1989). Cellular manufacturing in U.S. industry: A survey of users, *International Journal of Production Research*, **27**, 1511–1530.

Wemmerlov, U., and Vakharia, A. J. (1991). Job and family scheduling in a flow-line manufacturing cell, *IIE Transactions*, **23**, 383–393.

White, C. H., and Wilson, R. C. (1977). Sequence dependent setup times and job sequencing, *International Journal of Production Research*, **15**, 191–202.

Winston, W. L. (1994). *Operations Research: Applications and Algorithms*, 3rd ed., Duxbury Press, Belmont, CA.

Wirth, G. T., Mahmoodi, F., and Mosier, C. T. (1993). An investigation of scheduling policies in a dual-constrained job shop cell, *Decision Sciences*, **24**, 761–788.

6

SETUP TIME REDUCTION TO ENHANCE CELL PERFORMANCE AND FLEXIBILITY

Jerry W. Claunch

Claunch & Associates, Inc.
Palm Beach Gardens, FL 33410

Once you have achieved the design, organization, and implementation of your cell, you now have the opportunity to make even further improvements. While setup time reduction does not require cells, it certainly contributes to continuous improvement, especially in the areas of increasing capacity, reducing lot size, and working capital investment. Your goal should be to reduce setup time, then reduce the lot size until you get to "make to order." Cells require a focus on lot sizing and a change in mind-set toward zero batching. The overall measures of capacity, efficiency, and productivity should be drivers to the improvements necessary.

1. WHAT IS SETUP TIME?

Setup time is defined as the time the machine is stopped while the operator is changing items to make the machine capable of producing the next job. This includes removing and replacing items like fixtures, tools, gages, paperwork, gears, wheels, rollers, programs, and all those adjustments necessary when changing jobs on a machine. It also includes any quality checks and first article approvals. The time it currently takes

Handbook of Cellular Manufacturing Systems, edited by Shahrukh A. Irani
ISBN 0-471-12139-8 © 1999 John Wiley & Sons, Inc.

today to change over can be greatly reduced if you take the time to follow the steps outlined in this chapter.

2. WHY IS SETUP TIME REDUCTION IMPORTANT TO CELLS?

You installed the cell to be responsive to demand, reduce lead time, and to reduce lot sizes. You obtained many benefits by installing the cell already. Now it is time to make even greater gains in the areas listed above. Flexibility is the key word, being able to accommodate a changing part mix, produce small batches without holding excess inventory, increasing the capacity of the machines to produce, and greatly reducing the idle time on the machine during setups. Less emergency interruptions that may cause scrap, rework, mistakes, and delays are all reasons in favor of becoming flexible.

3. SPEED

The reason for forming a cell is to increase the speed at which we can respond to a demand. The initial improvements made by installing the cell are now the norm, and it is time to make further improvements. Examination of the flow, quality problems, and other issues will surface along with the setup time. In order to become the quickest, achieving rush order production may provide the best opportunity.

4. SETUP TIME REDUCTION REGAINS LOST CAPACITY

We all realize that capacity that is lost due to the machine being down during setup can never be regained, it is lost forever. Your goal must be to reduce the setup time so that the capacity becomes available and is not lost in the future. Most companies experience 20% downtime due to setups. In extreme cases, this lost capacity may be as great as 60%. So what does that mean to your business?

5. DETERMINING IF SETUP TIME NEEDS TO BE REDUCED

It is safe to assume that every cell should reduce its setup time, but rather than assume that setup time is an issue, it may be well for you to measure and compare setup time to run time. Most companies find that the cell machines are idle in excess of 20% of the time. There are even recorded cases of machines being in setup greater than 60% of the time. It is recommended that you begin to monitor the time every machine in the cell is producing and the time the machine is down for setup, maintenance, lack of material, or no production requirements. This will allow you to calculate the percent of downtime for all these reasons. Supervision and management can then determine the percent of downtime due to setup. As a benchmark, you should consider that any percentage greater than 5 offers excellent opportunity for reduction in cost and lead times by reducing setup time. The more competitive companies are reducing their setup times in cells to less than 1% of the run time.

In addition to this time comparison, you can also determine if setup time reduction is needed in a cell by listening to the cell operators. Complaints about availability of tools or other change parts, constant searching by cell employees for items needed during setups, and borrowing from other cells during setup are general indications that setup time needs to be reduced. Talk to the cell employees about the importance of setup time reduction, and they will provide you valuable insight into this critical issue. Some of them may be confused that you are concerned about setup time since most companies simply accept it as part of manufacturing. You may want to consider installing rush order production into the cell.

6. RUSH ORDER PRODUCTION

One major problem we face today are those annoying rush orders. When ordered to happen, rush orders mess up all the other jobs going through the shop. The good things about rush orders is that they won't get canceled if we meet the delivery requirement, and the product ships immediately, speeding up the accounts receivable process. Consider what it would be like if you could setup quickly enough that every order goes through your cell as quickly as a rush order does today. If you could get to that point, product would be flowing quickly, in small lots, and there would be no more rush orders. Rush order production is possible through setup time reduction provided you reduce the lot size as you reduce the setup time. Today's rush orders provide valuable information on how responsive we need to become in order to provide the service our customers deserve.

6.1. Developing a Strategic Plan in the Cell

It is important that you understand the rush order. What is the typical quantity, delivery time, and cause of the rush order? With this information, you can develop a strategic plan in the cell to reduce setup time and make rush order production a reality. It would be a shame to invest in cells and not get the real benefit of reduced inventory, continuous flow, and cycle time reduction. First let's look at where to start.

6.2. Questions That Allow You to Develop a Strategic Plan for the Cell

Many of these questions could be answered casually. Don't do that. Conduct surveys, and keep seeking the answers. Never assume that you know the answers. Answering the following questions provides data on which you can develop your strategic plan.

Question 1. Do We Have Scrap and/or Rework as a Result of Setups in Our Cell?   This alone may not create rush orders, but it will keep you from being able to respond quickly and economically. Never assume that your assembly or packaging area will notify you of rework or problems with the product you produce in the cell. Go talk to them and encourage them to notify you whenever they find a quality problem from your cell. When they do notify you of a quality problem, welcome them and appreciate their feedback. Before any other improvements are made, you should address this issue first. Implementing statistical methods and reducing variation are the keys to elimination of scrap. In addition capability studies should be done to determine problems with each machine in the cell. The problems may include improper preventive maintenance in the cell, measurement error, insufficient quality education of cell operators, raw material variation, and so on.

Question 2. Why Do We Have Rush Orders?   The primary cause may be that your lead times are too long. This may be disguised because you want to blame the customer for not planning ahead. You then fail to address the things you can do to reduce your lead times while your customer seeks another supplier that has shorter lead times. Before you jump on this issue, the third question should be asked.

Question 3. Why Does Management Trigger the Rush Order?   Don't jump to the conclusion that it is simply because the customer called one of the senior managers or that the manager simply wanted to take the easy way out and launch the rush order. Your competitors may have shorter lead times or have the product on hand, and you may lose the order if you fail to deliver.

Question 4. What Special Steps Are Taken to Get a Rush Order Out? Normally, there is exceptional communication and tracking involved in a rush order. Supervision communicates about rush orders and follows it through the process closely. Updates are required from your cell often, and progress reports are given to sales and the customer frequently.

Question 5. What Is the Average Time, Start to Completion, for a Rush Order?   The answer to this question gives you a good idea of the cycle time that you can achieve in your cell. The start time is when the raw material is available and the finish is when all quality checks or approvals are done and the product moves out of the cell.

Question 6. How Many Rush Orders Do We Have Monthly?   Don't think that this question stands alone. The amount of rush orders may be the tip of the iceberg, and the next question must be asked, and the quantity of those orders should be added to the quantity identified in this question to obtain the magnitude of the opportunity.

Question 7. How Many Orders Do We Lose Because of Our Lead Times? Sales, Marketing, and Customer Service should be able to give you this data. You may need to give them time to track this information since records may not be kept presently on this. Be patient; this information will greatly assist in your implementation.

Question 8. How Much Does a Rush Order Cost the Cell?   There are lots of intangibles involved such as other product that is not produced. The cost to break into a current run to do the rush order is one example. This is where you need to get the Finance or Accounting Departments involved to help develop the real costs. Intangible costs can be identified, but not given an absolute cost if it cannot be substantiated.

After all these questions are answered, you now have facts that will scream for attention. Your cell will recognize the importance of making improvements to the cell and be able to set its goals accordingly. Knowing why improvements are necessary keeps the cell employees focused with tangible goals.

6.3. Now Develop a Strategic Plan

Based on the above information, you can develop your plan of improvements for the cell. There is no doubt that you will begin with quality/setup time reduction. Your strategic plan should include creation of quality awareness, training, participation, support, and pressure for results. What, who, and when are the keys to a good strategic plan.

7. ORGANIZATIONAL AND PEOPLE ISSUES ASSOCIATED WITH SETUP TIME REDUCTION

You need to be aware that there will be issues that need to be addressed prior to setup time reduction in the cell. These issues begin with management. If the management of the cell is not committed to setup time reduction, don't ask for the involvement of the operators in the cell. Commitment on the part of management and supervision must come first! Too often, companies have started efforts in setup time reduction only to have them stop due to lack of long-term commitment by the company. Don't start setup time reduction until there is a commitment by the company to finish the task and hold the gains. Once that commitment is in place, present setup time reduction to the cell employees. The following responses are common:

Criticism: "Why should we change, what about other departments?" "Everybody knows you always have setup time." "This is just a ploy to get us to work harder." "Why are you all of a sudden interested in setup time? We've been telling our supervisor of the problems we have due to setup time for years and nothing has been done!"

Concern: "Is this going to eliminate jobs?" "How can we reduce setup time?" "Will setup time reduction work in our cell?"

Support: "It makes sense to me, our setups take way too long."

The key culture change needed to support setup time reduction is that management and supervision recognize the need to reduce setup time and that they encourage the cell employees daily in setup time reduction. Make this reduction part of the daily culture (work on it, talk about it, and share the successes daily), and it will be successful. Involve all shifts, supervisors, and support management (planning, scheduling, purchasing, engineering, tooling, stores, maintenance, etc.) in the effort along with the cell employees and your success will be great.

8. QUALITY IS FIRST! Couldn't Be Said Any Better

Without fail, improving the product quality in the cell is most important. Setup time reduction is important but any scrap or rework generated as a result of setup in the cell must be eliminated. Scrap and rework that is ignored means that making product faster, generates scrap or rework faster and doesn't get the results you deserve.

8.1. Setup Time Reduction Should Immediately Follow Quality on the Priority List

Don't relegate setup time reduction to the second or later task in importance. Next to quality, nothing is more important than setup time reduction. Start with quality improvement, then begin reducing setup time immediately.

9. STARTING SETUP TIME REDUCTION IN THE CELL

Applying setup time reduction in cells requires that you focus on the machine rather than the parts produced. Your goal is to make each machine quick to setup so that any part required can be produced quickly, and the machine is then ready to produce the next part order. Always remember, your goal is rush order production, not to just produce more product that goes into stock. Your first step is to organize the cell for quick setups.

9.1. Organizing the Cell for Setup Time Reduction

As soon as your cell is in place, the first step is to organize the cell for faster setups. This organization will normally reduce your setup time by 30% and provide a basis on which to obtain significant reductions in setup time. Implementing the organization requires the involvement of other support departments such as tool crib, maintenance, stores, safety, and purchasing to name a few. Make sure you have the support before you begin the cell organization. Once the support is available, most of the cell organization can be done concurrently. Following these principles will make the organization process simple and straightforward.

Hand Tools   Make sure all the necessary hand tools needed for setups are *always* available, prior to starting the setup. Hanging the hand tools on a rack and utilizing shadows on the rack will clearly identify missing tools. This is a simple and effective way to know if hand tools are available prior to starting the setup.

Fasteners   It seems every setup requires fasteners. Until you can get to the next stage and eliminate the usage of fasteners during the setup, you should ensure that there are plenty of every fastener at every machine. Since fasteners get lost, damaged, or stripped, a good rule of thumb is to have five of each fastener required at every machine. Making it simple to replace them as they get damaged, lost, or stripped is also important. If you make every effort to have five sets of every fastener, and establish a replacement system when they become damaged or lost, you should find that no setup is delayed for lack of a fastener or its associated hand tool. The best system for replacement is that whenever an operator needs a new fastener, he or she turn the old one into the crib and the replacement is immediately given with no hassles or complaints. If the fastener is lost, the operator notifies the crib and the fastener is replaced immediately with no hassles or complaints. Far too often, people get a lot of grief over a lost fastener, which is extremely cheap to replace. If the fastener is not replaced, your setup times will increase, costing a great deal more than the lost fastener.

Perishable Tools   In most cases, there are perishable tools that must be replaced during setup. Getting an adequate supply of the perishable tools at each machine is a must. In addition, you should install a replenishment program to ensure these perishable supplies are always on hand and available during setup. A kanban replenishment system is recommended for your perishable tools in a cell. In order to determine the kanban size, determine how many perishable parts are used in a shift. That number usually makes an ideal kanban quantity. Then store 1.5 times the number of kanbans used during the lead time to replenish. If the supplier has a lead time of three days, and you have a two-shift operation, the recommended number of kanbans is 9. This will keep your inventories to the lowest while eliminating the possibility of stockouts due to unexpected delays in replenishment.

Location and Organization of Setup Parts   Moving all setup parts to the machine is an important step in reducing setup time. If the setup parts can be permanently located in the cell close to the machine, this task becomes simple. If the configuration or size of the cell prohibits locating the setup parts in the cell, the setup parts should be brought to the cell prior to the start of the setup either by using kits that are put together while the previous job is running or put into carts that contain all setup parts that can be brought to the cell just prior to the start of the setup. All parts requiring cleaning should have duplicate sets and be cleaned after the new job is setup and running.

Maintenance of Change Parts, Tools, and Fixtures   None of these items used to set up should require any maintenance during setup. Establishing a procedure that ensures that items needing maintenance work are sent to the appropriate area and maintained before the setup begins should be a requirement. No setup parts should be missing due to maintenance or not working properly due to lack of maintenance. Queues in maintenance are just as costly as queues in production. A system for scheduling of maintenance based on the average time between failures is important to a setup time reduction program. Your goal should be to eliminate all queues in the facility, be it paperwork, production, or maintenance.

Lubricants, Chemicals, and Solvents   Within the cell, all lubricants, chemicals, and solvents needed for the setup should be available when needed. Material

Safety Data Sheets (MSDS) should be available in the cell and all of these items stored in proper storage containers with the correct labels.

Guard Removal   Many machines require the removal of guards in order to complete the setup. The guards then have to be replaced prior to the run. Guard removal typically requires the removal of fasteners, which is time consuming, and they get damaged and lost. Installing "one-quarter-turn" fasteners on all guards or some other quick disconnect method will simplify this part of the setup. Figure 1 shows how guard removal can be simplified using toggle clamps and dowel pins.

Scrap and Rework   In no case should scrap or rework resulting from a setup be accepted. Install process control to ensure that the cause of scrap and rework is identified and eliminated. During setup, nominal of the specification should be the target. This will allow the process to operate within statistical control limits and reduce or eliminate adjustments early during the run.

Standardize Setup Procedures   Many companies do not have standardized procedures to be followed during setup. Most setup experts use methods they have developed over the years. Typically, there are some very good habits that need to be explained to others doing setup and some bad habits that need to be eliminated in everyone's setup method. The keys to a standard method is organization and sequencing of the setup to make it efficient and simple to achieve. Many setups are haphazard in that there is no logic to starting at one end of the machine and working sequentially to the other end while doing setups. Utilizing setup documentation will provide consistency from operator to operator, shift to shift, or crew to crew. Documentation of a current setup is the initial step; then as you make changes to the setup procedure, the documentation should also be updated. A recommended documentation of the setup is provided in Figure 2. Documenting the current setup will probably demonstrate the need to make significant improvements because of difficulty, lack of information, and wasted time.

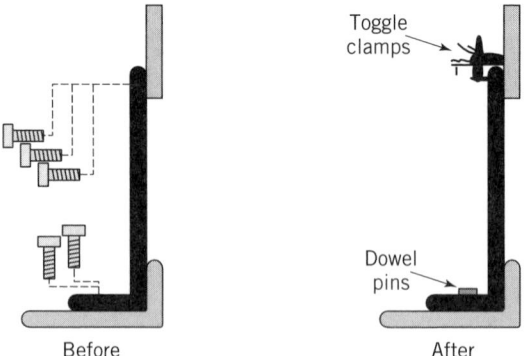

Figure 1. Guard removal can be simplified using toggle clamps and dowel pins or one-quarter-turn fasteners.

Machine Set Up Sheet

Hand tools required:	3/4" socket
	1" socket
	3/8" hex wrench

Change parts required:	CNC tape #12-436B
	Cutting tool #3456
	Cutting tool #2434
	Cutting tool #5465
	4 Hydraulic clamps #334

| Fixture required: | #567-98767 |

| Gages required: | Air gage #345 |
| | Dial bore #454 |

Detailed set up steps

Element:	Time:
Install fixture on machine table at location A-6	1 minute
Clamp fixture using hydraulic clamps	2 minutes
Load CNC tape	1 minute
Install tool #3456 into location 1 turret A and clamp	1.5 minutes
Etc.	

Figure 2. Documenting the setup procedure.

Deliver the Tooling in Kits   Putting all tooling in kits before the setup begins and delivering those tools to the cell provides a tremendous opportunity to reduce setup time. You need to ensure that all kits are complete and any presetting of tools, gages, or fixtures is done with 100% quality prior to delivery. The kit should consist of all the items identified as needed on the setup documentation and be delivered before the completion of the previous job. This way the items needed for setup will be immediately available to the operator before the setup begins. Many companies use a cart to deliver the kit to the cell.

Standard Preset Tooling   If your setups today use nonstandard tooling, you should standardize all tooling. Standardizing the tooling means you eliminate as many single-application tools as possible. It is common for 80% of all tooling to be standardized. Standardizing of tooling greatly reduces setup time as well as the inventory required. Once tools are standardized, presetting prior to delivery to the cell should be instituted. Typically when tools are installed into machines, there is a lot of time spent setting the length or determining the location of the tool in the spindle in reference to the part in the fixture. Standardizing the lengths and presetting them prior to delivery to the cell will greatly reduce your setup time. This effort should be led by your manufacturing engineering department and should not be limited to the cell.

Standard Locations   In machining and turning centers especially, standard locations for tooling is the next step after standardizing the tooling. If tool changers are used, standardize the location of the tool in the magazine as well. This eliminates much of the time and effort required to install tooling, and to edit the program in Computer Numerical Control (CNC) machines as well.

Standard Routing   Much like standard tooling, standard routings may apply to at least 80% of the product produced in the cell. These standard routings provide for fewer setups and simplify the documentation required for the setup. Standard routings always follow the same sequence and simplify the setup primarily through tool location in the tool changer or turret. An example would be that operation 10 is always rough turning, operation 20 may be end finish. Operation 30 may always be finish turning and so forth. Standard routings just like standard tooling will reduce your setup times because they always follow the same sequence.

No-Fault Reporting of Problems   Many companies, in an effort to control expenditures and eliminate wasteful practices, make it difficult for employees to be completely open when a problem occurs. A machine crash, lost hand tools, damaged fasteners go unreported in many cases because the employees are more concerned with being "blamed" than in getting the problem resolved and then instituting corrective action so that the occurrence never happens again.

No-fault reporting of problems or occurrences is the first step to eliminate problems. No one in the cell should hesitate to report a problem. If you have not instituted no-fault reporting, it is recommended that you so do immediately, communicate it to all employees, and ensure that it works on a daily basis.

A checklist for organizing the cell for setup time reduction is provided in Figure 3.

10. MAKING SIGNIFICANT REDUCTIONS IN SETUP TIME IN THE CELL

Once you have completed the cell organization, you are now ready to make significant improvements to setup time. It may be simplest to start with the machine that takes the longest to set up, but that may not provide the benefit you expect. You may also find it difficult to clearly identify the setup time. By running jobs by "families," you incur only partial setups. At other times there are full setups when you go to another product family. Identifying the average setup time for both full and partial setups for each machine is a good starting point. In Figure 4 we have a new cell layout that we can use as an example. In this example, the company produces covers in both aluminum and stainless steel. Prior to the cell implementation, the cycle time for an order of covers was 28 days. With the cell, the cycle time for an order is 7 hours. Now the team needs to reduce its lot size.

10.1. Full and Partial Setups

You may meet resistance to reducing setup time because employees may state that some setups take only a few minutes. This is likely because they confuse full and partial setups. Full setups occur when more than 80% of the change parts must be

Hand tools: Make sure they are always available, prior to starting the setup. Shadows on the rack that clearly identifies missing tools is a simple and effective way to know if tools are available.

Fasteners: Get plenty of every fastener at every machine that uses them during setup.

Perishable tools: Get plenty of the perishable tools at each machine and install a replenishment stocking program.

Forming dies, tools, tool holders, fixtures, dies, faceplates, printing cylinders, forming tools, chuck jaws, and other change parts: Move all change parts to the machine externally either by using kits that are put together externally or carts that contain all change parts. All parts requiring cleaning should be cleaned externally.

Maintenance of change parts, tools, and fixtures: None of these items used to changeover should require any maintenance during set up. Establish a procedure that ensures that items needing maintenance work are completed before the setup begins.

Lubricants, chemicals, and solvents: Like all other items needed to complete a changeover, these should be available when needed and be in proper storage container.

Guard removal: Typically requires the removal of fasteners, which is time consuming, and they get damaged and lost. Install one-quarter-turn fasteners on all guards.

Scrap and rework: Should not be accepted during setup. Install process control to ensure that the cause of scrap and rework is identified and eliminated.

Standardize setup procedures: Most companies do not have standardized procedures to be followed during setup. Most setup experts use methods they have developed over the years. Typically there are some very good habits that need to be explained to others doing setup and some bad things that need to be eliminated in everyone's method. Utilizing the setup documentation method will provide documentation of a current setup, and the documentation can be changed as your team makes improvements to its documented setup.

Deliver the change parts in kits: Putting all change parts in kits external to the setup provides a tremendous opportunity to reduce setup time. Ensure that all kits are complete, and any presetting of tools, gages, or fixtures is done with 100% quality.

Standard preset tooling: Instead of nonstandard tooling. Normally at least 80% of all tooling could be standardized, which may eliminate some steps in your current setup.

Standard locations: Is the next step after standard tooling. If tool changers are used, standardize the location of the tool in the magazine as well.

Standard routing is much like standard tooling in that 80% of the product or more could be manufactured utilizing a standardized routing.

Implementing these changes should result in at least a 30% reduction in your current setup time, which makes them well worth the effort.

Figure 3. Checklist to organizing the cell for setup time reduction.

replaced during the setup. This includes the fixture(s), tools gages, and other change parts. Partial setups typically require 20% or less of the change parts to be replaced. An analysis of the documentation you do, as shown in Figure 5, will keep anyone from confusing the importance of setup time reduction.

10.2. Determining Where to Start Setup Time Reduction in the Cell

Using the chart in Figure 5 provides cell members a sequential listing of the machines in the cell with their appropriate setup time for full and partial setups, and the average

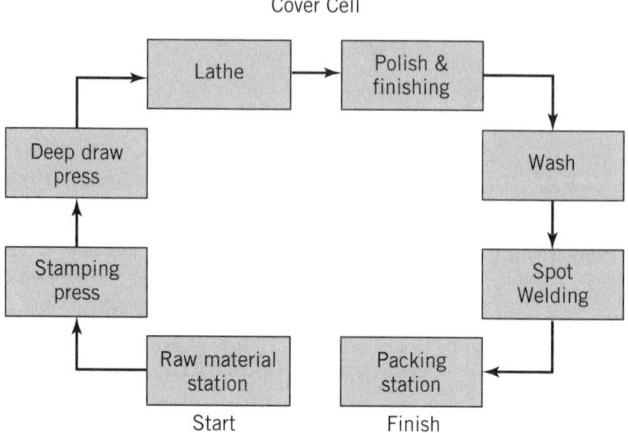

Figure 4. Layout of the cover cell.

Operation/Machine	Average Partial Setup Time	Average Full Setup Time	Comments	Average Value-Added Time per Piece
Raw material staging	N/A	20 min		0.1 min
Operation 10, stamping	12 min	40 min	Product ID stamp	0.15 min
Operation 20, stamping	12 min	40 min	Customer ID stamp	0.15 min
Operation 30, deep draw	N/A	70 min		0.75 min
Operation 40, lathe turn	10 min	35 min		1.5 min
Operation 50, finishing	N/A	1 min		3.5 min
Operation 60, wash	0	0		0.75 min
Operation 70, drying	0	0		7 min
Operation 80, spot welding	10 min	25 min		1.5 min
Operation 90, packaging	N/A	20 min		1 min

Figure 5. Sequential listing of machines in cell with setup and run times.

value-added time per piece. This information provides insight where to begin the setup time reduction efforts. The deep draw operation is the longest average full setup time, but the stamping operation has two changeovers (operations 10 and 20). If the stamping operations can be combined by remaking the die, that should be considered first. If combining is not possible, the team should begin by reducing the full setup time on the stamping operation. Once that is completed, the team should reduce the setup time on the draw operation. The next project should be reducing the value-added time for the finishing operation.

11. SETUP TIME REDUCTION STEPS

Being able to identify the improvements made in setup time will encourage everyone in the cell to work on setup time reduction. Figure 6 is a wall chart that can be displayed in the cell , thus communicating the old method of setup and the improved setup. Not only does this provide documentation of the new setup, it demonstrates the improvement rate. Time can be placed on the chart using the average time it takes to complete an element of the setup. This chart also communicates to anyone walking past the cell. They can easily see what the operations are and which ones require more setup time. A reduced copy of this chart can be used by the operators as they do setups.

12. DETERMINING COST AND BENEFIT FOR EXPENDITURES

In your efforts to reduce setup time, there will be a need to spend funds to purchase and implement setup improvements. It is recommended that the cell consider low-cost ideas for improvement first. Low cost means that the payback is 2.5 months or less. If your cell requests large sums of money that require many months or years to payback, they are probably over looking many simple opportunities for improvement. Figure 7 is a form to determine justification for expenditures. The cell members can use this document to determine if an expenditure is justified, and then according to the rules of thumb at the bottom determine where to get the funding. It is always recommended that management provide funds to a cell to reduce setup time. This can be done as a lump sum, petty cash, or a simplified allocation when low-cost funding is required. Management should periodically review the status of the available funds and ensure that low-cost improvements are always available to the cell. Figure 8 shows a timeline typical of a cell in its efforts to reduce setup time. Low cost are those improvements that result in a payback of 2.5 months or less (20% payback in Figure 7). These are the items of which a team should have funding provided without hesitation by management.

13. GETTING MANAGEMENT TO BUY IN

Utilizing the cost—benefit analysis in Figure 7 will help greatly in proving the worth of setup time reduction. Most managers and supervisors have simply accepted setup time as a necessary task and have not taken the time to understand its impact. Figure 9 shows how you can analyze the possibility of setup time reduction. Management should then realize the possible improvement through reducing the setup time. All you need to do is determine the amount of time each machine is in setup per year. Imagine a cell that has five machines with an average of 1 hour 15 minutes of setup time. If they normally do two setups per shift, have a two-shift operation, and work an average 220 days per year, the annual setup hours are 5500. If you then multiply those hours by 60 minutes you get 330,000 minutes of setup annually. Now multiply

Old Procedure

Element	Log onto setup at computer terminal	Get print, routing sheet from file	Make list of tools needed and get from tool crib	Remove old tools from tool changer	Remove old fixture and return it to storage shelf	Remove new fixture from storage shelf and take to machine	Install new fixture on machine	Remove old coolant lines	etc.
Time (min and sec)	1:23	3:06	5:24	6:45	4:15	5:23	6.54	2:15	**Total time: 2:40**

New Procedure

Element	All tools and change parts delivered to machine prior to setup on a cart	Log onto setup at machine CRT	Remove old tools from tool changer using air wrench	Remove old fixture and set on cart	Get fixture from cart and install on machine	Install new tools	Adjust collant lines	Put part in machine and run	etc.
Time (min and sec)		:23	3:10	2:40	3:15	3:16	:24	2:18	**Total time: :40**

Figure 6. Setup wall chart demonstrating the difference between the previous setup and the improved setup.

Determining Payback for Expenditures

A. Identify the Overall Value of Setup Time Reduction
1. Forecasted annual cost of goods sold from the cell = \$_____.
2. Divided by the number of primary machine tools in the cell _____ = _____.
3. Divide by the number of working days per year _____ = _____.
4. Divide by working hours per day _____ = _____.
5. Times the average % of downtime for setups __ . _____ (%) = \$_____.
6. Divide by 60 (minutes per hour) =   \$ _____.
                                                                     (One-minute value)

B. Identifying the Total Purchase Cost
7. Identify each item to be purchased (material and equipment)

Item	Delivered price	
_____	\$ _____	
_____	\$ _____	
_____	\$ _____	etc. (use additional pages)
Total purchase price \$ _____		

8. Add design costs
Internal \$ _____ + External \$ _____ = Total design cost \$_____
9. Identify labor costs to install
Internal \$ _____ + External \$ _____ = Total labor costs \$ _____
10. Identify training costs to implement
Development \$ _____ + Conducting \$ _____ = Total training costs \$ _____
11. Add all costs   Total purchase cost: \$ _____

C. Identify the Value of the Purchase
12. Identify number of minutes reduced on each setup = _____.
13. Times the number of setups per year _____ = _____.
14. Times the one-minute value \$ _____ =   \$ _____
                              (one-minute value)   (annual benefit)

D. Identifying the Payback
15. Total purchase cost = \$ _____.
                         (Line 11)
16. Divided by annual benefit \$ _____ = _____ % payback.
                             (Line 14)

> *Payback Rules of thumb:*
> 20% or less should be immediate purchase (2.5 month or less payback)
> 20% to 100% should be available from budget (1 year payback maximum)
> Over 100% needs capital justification (greater than 1 year payback)

Figure 7. Justifying funds for setup time reduction.

the minutes by the 1 minute value determined in Figure 7, and you have the amount that setup costs the cell. This should get the attention of management. Imagine if 75% or even 90% of that number were added to the profit from the cell. Now imagine all the cells in your company achieving this result.

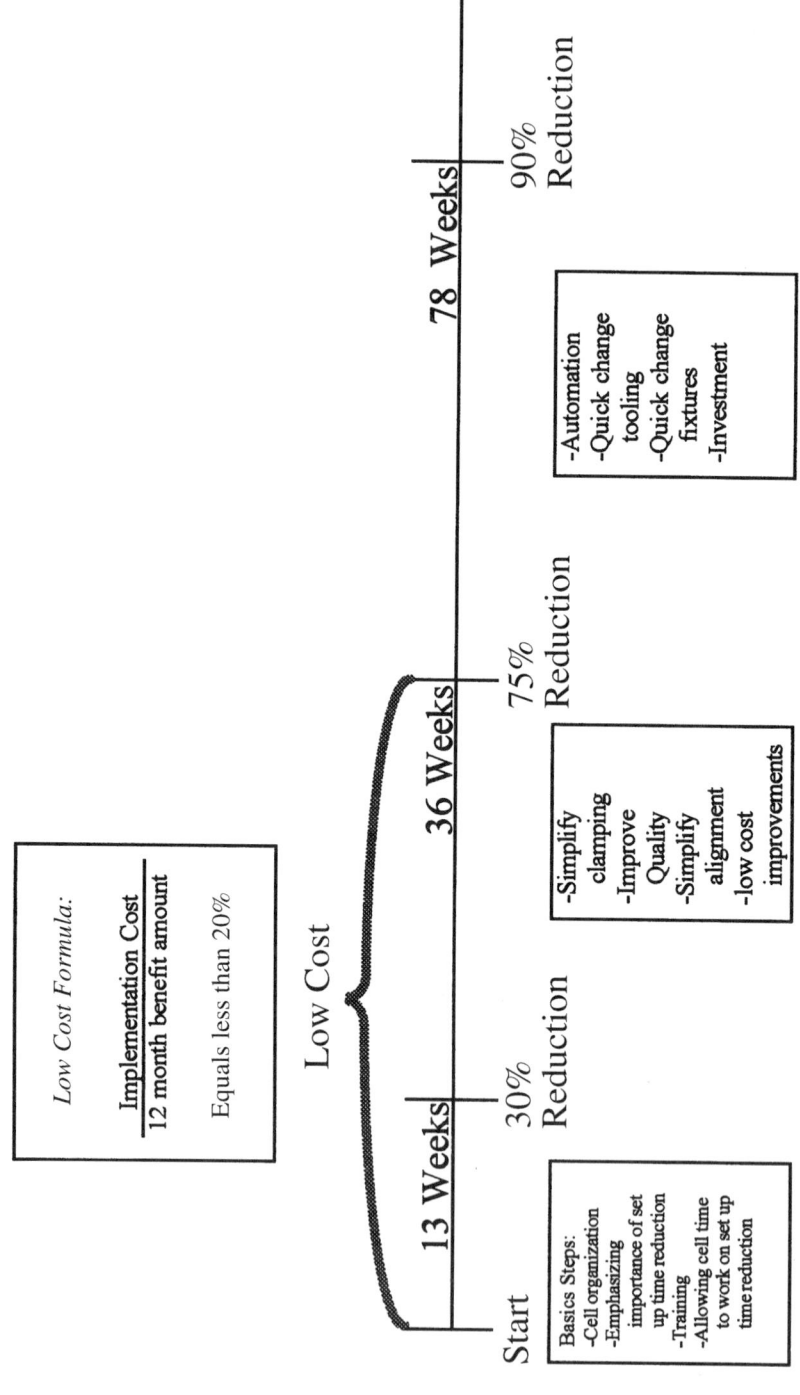

Figure 8. Cell setup time reduction timeline.

Identify the Overall Value of Setup Time Reduction

1. Forecasted annual cost of goods sold from cell = $12,000,000.00
2. Divided by the number of primary machine tools 5 = $2,400,000.00
3. Divide by the number of working days per year 220 = $10,909.00
4. Divide by working hours per day 16 = $681.00
5. Times the average % of downtime for setups 37.5 (%) = $ $255.38
6. Divide by 60 (minutes per hour) = $_____$4.26_____
                                    (one-minute value)

Now multiply the one-minute value times the annual minutes of setup in the cell (330,000) = **$1,405,800.00 Annual cost of setup in the cell.**

Figure 9. Annual cost of setup time in a cell example.

13.1. Installing, Clamping, and Unclamping of Setup Parts

The installation and removal of setup parts provides many opportunities for improvement. Eliminating the use of fasteners during setup should be one of the earliest improvements made. Fasteners get lost, damaged, and stripped and they require hand tools to install. They are time consuming and cannot be tolerated during setup. Some fastener elimination methods are simple and inexpensive, while others require more funds. All parts changed during the setup should have true positioning. Hard stops, dowel pins, and fixed locations are important in making the setup quick and typically do not require great funds. Quick attachment and clamping is the next step. In order to accomplish this, it is necessary to identify stress force direction. If engineering support is not available in the cell, management may need to provide that support to help determine the stress forces that need to be overcome during the operation. Any "overkill" of stress forces can be eliminated before the improvements that eliminate fasteners. Now the cell can consider quick clamp methods in place of fasteners. Figures 10–16 provide examples of improvements that have been made to eliminate fasteners in setup.

13.2. Transporting

Transporting of any setup parts that cannot be kept in the cell can be simplified by the use of carts. The parts can be left in the cart as their permanent location or the parts can be put in the cart while the previous job is running. It seems that every company in every industry that wants to reduce setup time eventually will purchase carts to achieve setup time reduction. It is recommended that your cell purchase these carts early in its development.

13.3. Locating to Eliminate Further Adjustments and Changing of Settings

Many times after a setup is completed, the operator makes adjustments or changes settings during the running of the part. This practice needs to be eliminated, and

Clamp System

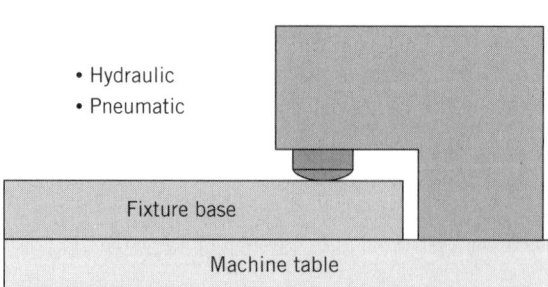

Figure 10. Using hydraulic or pneumatic clamps in place of fasteners.

Standard Heights & Standard Clamping
- Die height standardization eliminates shut height adjustment
- Blocks standardize clamping

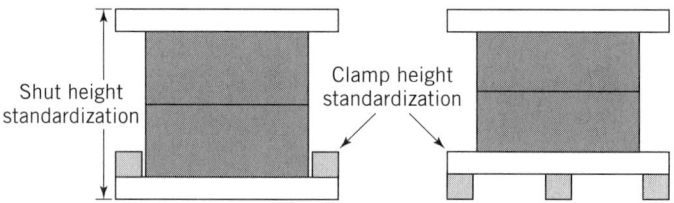

Figure 11. Standardizing overall height and clamping height.

Sub-bases

Standard sub-base with different size fixtures.

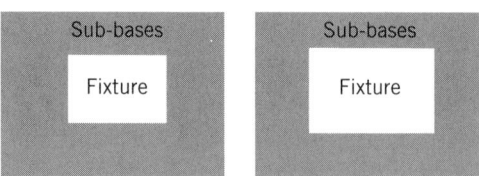

Figure 12. Utilizing subbases to standardize clamping positions with different size fixtures or dies.

the cells efforts to reduce setup time should include the elimination of adjustments or changing of settings after the setup is complete. The cell should determine the adjustments or settings that are finally made, record them in the setup documentation, and everyone should make those adjustments or settings properly during future setups. Adjustments typically apply to the location of the fixture, die, or other part changed during setup. Settings typically apply to speeds, feeds, dwell time, temperature, flow volume, and so forth.

Utilizing Base Plates

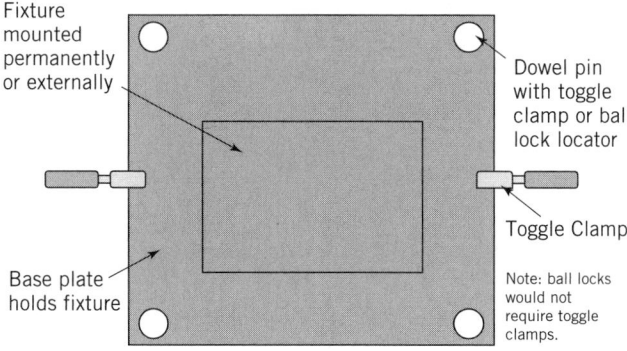

Figure 13. Locating and clamping with baseplates.

Using Toggle Clamps for Holding

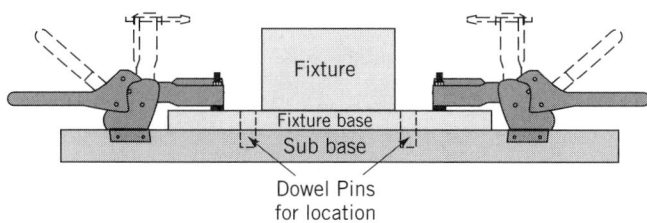

Figure 14. Utilizing toggle clamps for holding and dowel pins for location.

13.4. Reducing the Time to Make Connections (Air, Water, Hydraulic, Vacuum, or Electrical Lines)

The key to reducing connection time is to use quick connect/disconnect technology. All connections made in the cell during setup should be identified and all should be changed to quick disconnect. A method to code the different lines may also be in order. Color coding may be the simplest, provided every cell member can see the colors used. There is nothing wrong with employees that are color blind, it simply means that another method may need to be employed to ensure the connections are quick and easy to identify and make.

13.5. Cleanup

Any parts needing cleaning during setup should have duplicate sets purchased so the cleaning can be done when the machine is running. No cleanup should be done during setup since it is time consuming and typically messy.

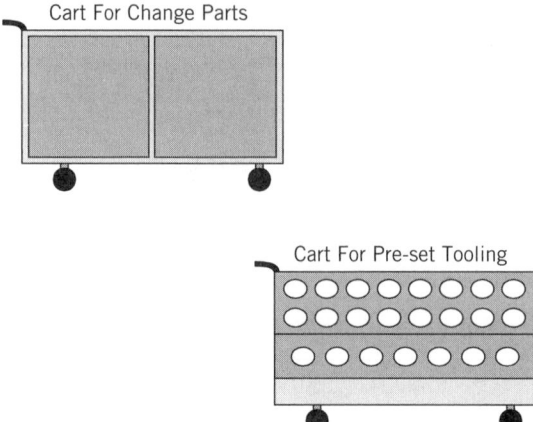

Figure 15. Carts can be used to reduce transportation of setup parts to the cell.

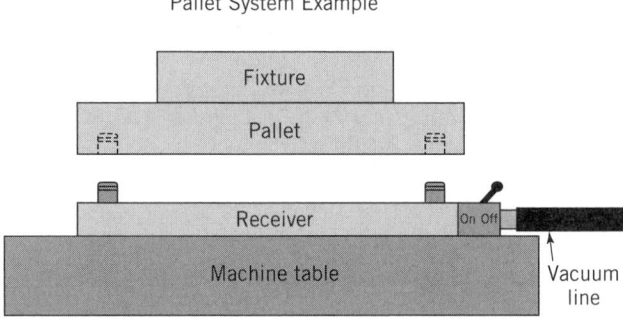

Figure 16. Automating setups with pallet systems.

13.6. Trial Runs

The cell employees should always question why they make any trial runs and go about eliminating the trial runs. If the cause is scrap or rework, the trials may be easy to eliminate by documentation that provides the correct settings and eliminates adjustment for locations. Many cells have determined that trial runs are unnecessary and have successfully eliminated this time-consuming practice.

13.7. Inspection

Typically inspection by the quality department is part of the setup. Examine the inspections that can be made by the operator during setup and don't have inspection duplicate those checks. In most companies a study of the inspection reveals that the inspections by the operator during setup are always correct when the inspector checks the product as well. As was stated earlier in this chapter, the setup should always target nominal of the specification. If this is done, no further inspections should be required. The only inspections that may be required are positional checks, which typically can't be done at the machine. These are usually done on a Coordinate Measuring Machine (CMM). Functional gaging may be used to eliminate the need for CMM inspection. One Fortune 500 company found that when they began researching inspection during setup that the inspectors never had any problem with the dimensional checks made by the operators during setup. They immediately eliminated all except positional inspection by the inspection department and reduced an average of 12 minutes on every setup. The cost savings in machine time alone was $2,400,000.00 per year in their cells.

13.8. Data Entry

Always question any data entry, be it manual or computerized. If it is necessary, then simplify as much as possible. Whatever data entry cannot be eliminated, make sure it is done while the machine is running. Work with the computer programmers to simplify any data recording required during setup.

13.9 Automation

There are many products on the market that will assist in the automation of setup time. Quick change tooling, quick change systems for stamping, forging, and injection molding are all available in the market, and the suppliers can typically be found in the manufacturers registers. While more expensive, these systems can greatly reduce setup time and make your cell flexible. Always remember, the goal of setup time reduction is to more quickly respond to known demand, and doing so with setup time reduction provides that benefit at the lowest cost.

14. SETUP TIME REDUCTION ACTIVITIES SCHEDULE

Based on everything presented in this chapter, here is a project schedule of activities that need to take place in your cell to achieve setup time reduction.

- Train the cell employees and support area employees, including management and supervision.
- Implement the 30% reduction steps discussed in this chapter.
- Determine where to start setup time reduction by machine.
- Identify the setup steps today, and empower the cell employees to reduce the steps and the time to do setups.
- Track the progress and post the results in the cell.

15. CASE STUDIES OF SETUP TIME REDUCTION IN CELLS

A few years ago, it was recommended to a cell in Arizona that it could reduce setup time. The employees were given a three-day training session on setup time reduction.

One employee in the cell advised the instructor on the first day that he thought setup time reduction was just a bunch of bull. He was asked by the instructor to "give it a chance," which he did. Three months later, the instructor had the opportunity to visit the cell, and, when he approached the operator, the operator immediately pointed to the clock on the wall and said "That's my enemy! I now understand that I need to do every setup faster than the previous time." The employees of the cell now understood that setup time reduction is not about job reduction, it is about becoming competitive and staying in business. Flexibility to customer needs, reduced lead time, and increasing profits are the keys to survival in manufacturing.

As this cell began reducing setup time, the members realized that many of the things they began implementing were needed in the other cells in the plant. They recommended to management that a "housekeeping" team be established to work in every cell to accomplish the cell organization outlined earlier in this chapter. The opinion of the cell employees was that there were not enough of the tools needed during setup. These tools included boring rings, cutting tools, wrenches, fasteners, and clamps. Once they organized the cell, they were able to determine which tools they had plenty of and which ones needed to be purchased. Had they not organized the cell, they would have wasted a lot of money purchasing tools that in fact were not needed. This caused the credibility of the cell to rise greatly in the eyes of management.

The team then decided to reduce the setup time on the CNC milling machine in the cell first since it was the "bottleneck" workcenter. Parts normally waited for the milling operation most often. The team did a Pareto chart on time spent during setup. They found that a lot of time was spent getting the fixture located on the machine and decided to put fixtures on a subbase while the machine was running, and using dowel pins and hydraulic clamping reduced the locating and clamping time by over 90%. Next they determined that their normal procedure of making a cut (adding a feature) and then checking the feature against the print tolerance was very time consuming since 5–10 features were added to each part at this operation. The team evaluated the results over the previous 6 months and determined that there was little risk in adding all the features to the part, then checking all the features at one time. The time to complete this part of the setup was reduced by over 85%.

The team then evaluated the time spent getting all the cutting tools organized and ready to put in the machine. They determined that the operator should not have to go get tools; the tools should be delivered in kits by the tool crib. Working with manufacturing engineering to identity on the routing what tools were needed and what the preset should be, the operators now have that information delivered to the tool crib when the previous job is running and the kits are assembled with preset tools and delivered to the cell just prior to the setup.

The team asked the supervisor to have a white board installed in the cell so the scheduler could identify the next job coming to the cell and so the operators could identify the status of the job(s) running in the cell. The operators were then able to anticipate the next job to be setup and could communicate with the rest of the company on the status of the jobs running presently. The scheduler joined the team and became a valued team member. Instead of giving a weekly schedule as previously done, she simply listed the next job. This simplified making changes to the schedule and allowed the kits to be delivered in time by the tool crib.

The team next evaluated the time spent waiting for first article approval after a setup. It was typical for the machines to sit idle for an average of 17 minutes waiting for the first article approval from the quality inspector. Working with inspection, the team determined that feature checks done by the operator were never found to be incorrect, yet every feature was checked by inspection. Inspection had a Coordinate Measuring Machine that allowed them to do positional checks that the operator was unable to do in the cell. The team recommended and was given approval to eliminate all checks by inspection except for positional checks. This resulted in the first article approval time to be reduced from an average of 17 minutes to an average of 4 minutes. Part quality has not gone down at all in over 4 years.

The team next looked at the time spent during setup and run going to the tool crib for carbide inserts. They evaluated the time spent in every other cell as well and received approval for a two-step improvement. They first installed lights in every cell, which when turned on indicated that there was a tooling need, and the "runner" from the tool crib would come to the cell, find out the need, and bring the necessary inserts to the cell. The second phase used two-way radio communication. When the team determined a tooling need, they could notify the runner without him coming to the cell and reduce the time even more. When this team started, the average setup time was 1 hour 22 minutes. In 4 months time, the setup time was reduced to under 12 minutes. This equates to an 85% reduction in setup time. All of this was done on a low-cost basis.

The cell outlined in Figure 4 was implemented in Wisconsin. In addition to the improvement in lead times, there were many other improvements realized because of the cell and setup time reduction. The covers produced were made from both aluminum and stainless steel. In the past, if there were stains in the raw material, the polishing operator would scrap the covers, and all the value added at the stamping press, deep draw press, and turning operation was wasted. As soon as the cell started operation, the finishing operator went over to the stamping press operator and showed him how to determine if a stain in the material would cause scrap at his operation. After that, raw material with stains was not allowed to enter the process in the cell, and scrap due to stains was eliminated immediately simply due to improved communication. The cell employees decided to start setup time reduction on the stamping press since it was the first operation, and then went to each subsequent operation in the cell and reduced setup time. The improvements were immediate as the cell implemented simple, low-cost solutions to the time spent in setup. Each operation has a reduction of over 70% of the setup time when the cell began 2 years ago. Setup time reduction doesn't require great sums of money nor does it require superior intelligence. Good old common sense with a focus on reducing setup time is all it takes. If you could simply imagine your cell with setup times reduced drastically, you may then be on the road to success. Take a few minutes and see yourself in the cell with setups that take a few minutes or seconds, and you'll have the vision of where you need to get.

16. IMAGINE YOURSELF THERE

There are so many important reasons for installing cells in most companies, and there are just as many reasons for reducing the setup time. In many cases the reasons are the same. The primary reason is to "delight the customer." Delighting the customer is so easy, yet we make it so difficult. Customer delight requires 100% quality, on-time delivery (in shorter and shorter lead times), and correct quantities of parts. Setup time reduction coupled with rush order production will enable your cell to delight its customers. All it takes is for you to have the vision of customer delight and take your cell to that vision. I recently saw a paper pinned to a cubicle wall that gives us the right understanding of where we are:

> "Imagination is more important than knowledge. For knowledge is limited, whereas imagination embraces the entire world." Albert Einstein

My advise to you is to "imagine yourself there!" Imagine your cell is in place, setups take a few seconds, every part is made to order quickly, and you have 100% delighted customers. Imagine this many times a day and never lose focus of that image. Repeat it often to the other cell members, and support departments and achieve greatness in your career.

7

FRAMEWORK FOR CELLULAR MANUFACTURING EVALUATION PROCESS USING ANALYTICAL AND SIMULATION TECHNIQUES

Farzad Mahmoodi

School of Business
Clarkson University
Potsdam, New York 13699-5770

Charles T. Mosier

School of Business
Clarkson University
Potsdam, New York 13699-5770

Anthony T. Burroughs

Xerox Corporation
800 Phillips Road
Webster, New York 14580

Handbook of Cellular Manufacturing Systems, edited by Shahrukh A. Irani
ISBN 0-471-12139-8 © 1999 John Wiley & Sons, Inc.

1. INTRODUCTION

Cellular Manufacturing (CM) has received a significant amount of attention in recent years by demonstrating great potential for productivity improvements in batch manufacturing. Often batch facilities produce items from a population of thousands of active part types using hundreds of machines organized in functional layouts. The cell design problem is to reconfigure the existing machines into smaller groups (or cells) of dissimilar machines with each cell dedicated to a family of parts, where, ideally, all parts in a given family are produced in a single cell. However, the typical case is that dedicated cells and families cannot be created to accommodate the whole population of parts and all machines, and, often a sizable "remainder" cell is left to serve those parts that do not fit into any families. The cell design process can be lengthy, requiring substantial effort to evaluate candidate cells according to a diversity of operational and economical performance measures. Thus, the general cell formation process can be viewed as having two phases:

1. Creation of candidate cells (i.e., identification of part families and machine groups)
2. Evaluation of the performance of candidate cells

The explosive combinatorics can result in many possible configurations to be developed and evaluated.

There has been much current research devoted to the development of cell formation procedures (i.e., techniques that identify part families and machine groups). However, the performance evaluation of manufacturing systems in general, and manufacturing cells, in particular, has received less attention.

Operationally, the cell evaluation process may involve a number of phases of decision making, ranging from an analysis of initial feasibility to detailed redesign of the facility layout. Due to the high degree of interconnection and integration in modern manufacturing, often seemingly "local" decisions have "global" impacts. Likewise, the decisions made concerning the configuration of cells in most manufacturing facilities are interrelated, that is, decisions concerning the makeup and operational characteristics of a particular cell impact configuration of the other cells.

Alternatively, the primary benefits of CM are directly related to the operational "independence" of each cell. The "best" cellular reconfiguration, in this context, is the one where the impacts of local decision making are restricted to within the cell, while still retaining the productive capabilities required for the assigned manufacturing tasks. Even experienced manufacturing personnel will have difficulty in predicting the impact of a proposed cellular reorganization. There is copious anecdotal literature describing specific manufacturing evaluation approaches, but very little describing integrated efforts progressing from a specific mix of machines and parts and resulting in a set of cells having the required operational features. Thus, it is of interest to consider an evaluation framework using a variety of analytical and simulation techniques to support analysis throughout the cell evaluation process.

More specifically, a framework for evaluation of cellular designs is of interest for several reasons:

1. Most cell design techniques do not have explicit objectives tied to them, so the cell evaluation is often independent of the formation of the cell (e.g., McAuley, 1972; King and Nakornchai, 1982; Mosier, 1989; Arvindh and Irani 1994).
2. Some cell design techniques have one-dimensional objectives, such as minimization of makespan, that do not necessarily produce good performance for a system of cells (Shafer and Rogers, 1991; Song and Hitomi, 1992; Srinivasan et al., 1990).
3. The cell design process has a truly integrated impact, since altering the makeup of one cell can affect others in a variety of ways. To incorporate this into the design process, evaluation techniques are needed that provide sensitivity information, to direct the designer to favorable variations in machine groupings and part families.

2. FRAMEWORK FOR CELLULAR MANUFACTURING EVALUATION

Our proposed framework for CM evaluation consists of three steps. Each step differs in the level of detail required for its use and in the evaluation information that it provides. Note that it is assumed that several candidate cell configurations are available for evaluation. The three evaluation steps are:

1. Static evaluation using computational schemes
2. Stochastic evaluation using queuing theory techniques
3. Dynamic evaluation using computer simulation

As shown in Figure 1, the steps are applied sequentially, with the intention of identifying poor cell configurations, and subsequently reducing the number of candidate cell configurations for further evaluation. The term "candidate" is used since there are many ways to arrange the machines into cells. Each candidate cell configuration would consist of a collection of cells with each cell having a dedicated part family, and a remnant cell to accommodate all those parts that do not fall into one of the families. The three evaluation steps are discussed in detail in the following sections.

2.1. Static Evaluation

The static evaluation is the simplest and the most basic evaluation method. This analysis consists of utilizing a spreadsheet software tool to compare the work allocated to each resource and its available time. Static evaluation models assume a deterministic environment (e.g., predictable schedules, no breakdowns, and part availability) and ignore all dynamics, interactions, and uncertainties typically seen in manufacturing systems (Suri and Diehl, 1986). However, they provide a rough estimate of the system

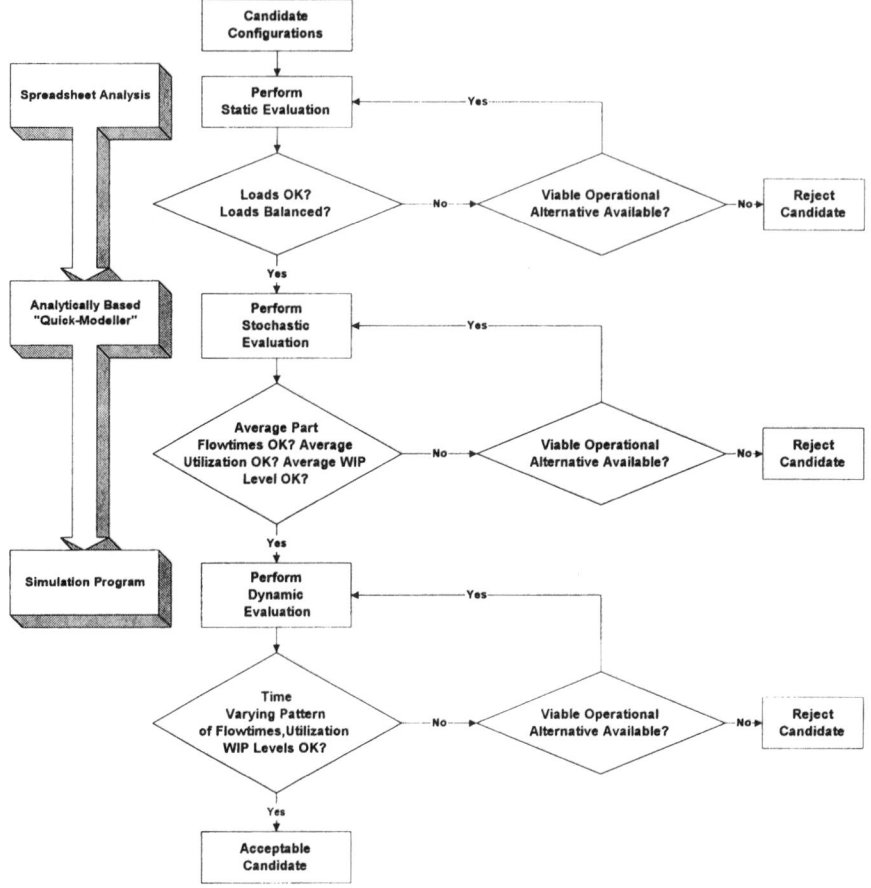

Figure 1. Framework operation.

performance. For example, the "rough-cut capacity planning" module of Materials Requirements Planning (MRP) systems use similar approaches (Suri et al., 1993). However, part flow times through the system and Work-In-Process (WIP) inventory cannot be estimated using this analysis.

This method is useful in the very early planning and design stage to identify system bottlenecks. Thus, design considerations can be evaluated to address the bottlenecks. This analysis can also be used in an operating plant to determine where the bottleneck operations will move to if system modifications or improvements are made. Therefore, early design alternatives can be explored quickly, and eliminated if they do not support capacity targets. Finally, what-if analysis can be performed by adjusting the inputs and evaluating the impact on the cell performance. The static evaluation results in an initial selection of promising candidates for further evaluation. The information required and provided by the static analysis is outlined next.

Information Required

- Production rate and yield for each product
- Uptime for each workstation
- Production run sequence
- Average changeover/setup time
- Number of production runs for each product
- Shift configuration (e.g., 3 shifts, and 5 days a week)
- Hours per shift (e.g., 4, 8, and 12)
- Number of workdays available per year
- Number of operators required for each workstation
- Quantity of each product to be produced per year

Information Provided

- Aggregate weighted average rate, yield, and uptime at each workstation
- Annual capacity for each workstation based on yielded input from the previous workstation
- Bottleneck workstation(s)

2.2. Stochastic Evaluation

The next level of evaluation uses stochastic models based on queuing network theory, which can give estimates of job flow time and WIP inventory in each cell in a candidate configuration (Rehman and Diehl, 1993). While these methods are more complex and require more information than the static methods, they do account for some of the dynamics, interactions, and uncertainties in the system. However, the performance measures estimated are often only the aggregate steady-state averages. Furthermore, the queuing analysis requires assumptions of steady-state operations (assuming moderate-to-large size production runs), certain probability distributions for arrival and processing times, and assumptions as to how multiple parts are managed in the cell. On the other hand, with recent advances these models tend to give reasonable estimates of a system's performance and (relative to computer simulation) are very efficient (Suri et al., 1993).

Candidate cells can be compared based on the estimated throughput times and workstation utilization levels. Thus, this method can be used to determine which design alternatives are worth simulating. Furthermore, warehouse and other storage design considerations can be addressed. Finally, what-if analysis can be performed to determine the impact of various changes on cell performance. In recent years there has been growing use of queuing network theory in manufacturing (e.g., Chen et al., 1988; Brown, 1988; Suri, 1989), particularly with the availability of commercially supported software packages such as Manuplan (Suri et al., 1986) and MPX (Suri and De Treville, 1991).

Information Required In addition to the information required for the static evaluation, the following information is required:

- Changeover/setup time
  - Average time required to set up for the next product or production run
  - Total time at equipment and labor time at equipment
  - Setup time variability factor
- Manpower levels
- Percent time unavailable due to breaks, lunches, meetings, training, etc.
- Percent overtime allowed
- Maximum utilization possible
- Equipment reliability: mean time to failure and mean time to repair
- Variability in arrival to each workstation, labor, and equipment
- Percent of the equipment assigned at each workstation

Information Provided

- Equipment and labor utilization
- WIP inventory at each workstation
- Flow time of each product
- Number of setups performed at each workstation

2.3. Dynamic Evaluation

The dynamic evaluation employs computer simulation to trace the status of each product and workstation over time, thereby identifying the actual loading patterns, idle times, system bottlenecks, and queue lengths. Obviously, this method is the most complex and expensive and requires an enormous amount of detail. For industrial related problems, this method requires one to three orders of magnitude more effort than the stochastic models, as shown by actual case studies (Anderson, 1987; Brown, 1988). On the other hand, these models can mimic real systems in as much detail as required. Thus, they can be very accurate. They can also be combined with visual animation to better communicate the results of the analysis (Suri et al., 1993).

Dynamic models are especially helpful if multiple parts are scheduled to be in the cell simultaneously, material handling systems and/or dynamic routings are considered, breakdowns of machines are expected during production, and measures of lateness are desired. Simulation models allow the study of detailed interactions of operations and enable detailed decisions to be made about a system and then judge the impact of the decision (Suri and Diehl, 1986; Kamrani et al., 1995).

Computer simulation has been extensively used to evaluate the performance of various manufacturing cells (e.g., Ang and Willey, 1984; Dale and Dewhurst, 1984; Steudel, 1986; Sassani, 1990; Suresh, 1992). Furthermore, several studies have used simulation analysis to investigate the performance of cellular verses functional layouts under a variety of shop operating conditions (e.g., Flynn and Jacobs, 1987; Morris and Tersine, 1990; Shafer and Charnes, 1993). Simulation modeling has a long application history in the study of scheduling, in general (e.g., Aigbedo and Monden, 1996; Dawood, 1996; Lyu and Gunasekaran, 1997; Jernigan et al., 1997), and scheduling in the CM environment, in particular (e.g., Vaithianathan and McRoberts, 1982; Mosier et al., 1984; Mahmoodi et al., 1990; Ruben et al., 1993; Wirth et al., 1993; Kelly et al., 1997; Mahmoodi and Martin, 1997).

Computer simulation is a tool that involves significant software and implementation costs (*IIE Solutions,* 1996). Acquiring the expertise to do high-quality simulation modeling and experimentation is an expensive proposition whether this expertise is developed in-house or through outside consultants. Additionally, the accuracy of simulation modeling results is always in question. Analytical models are able to provide accurate answers to very limited set of problems while simulation analysis is able to provide solution estimates to an almost unlimited problem domain.

Lastly, the most critical failing of simulation analysis is its potential for misuse. Here the notion of misuse might be better considered as "amoral" rather than "immoral." If the analyst has a vested interest in the accuracy of his or her model or in one of the candidate policies, configurations, or layouts being considered, it is possible to manipulate the execution and subsequent analyses to justify any result desired. More insidious is the fact that this may be an unconscious manipulation. Oftentimes the target audience is simply too gullible. Many in upper management are more convinced by slick simulation animation than by the results of a rationally designed experiment comparing well thought out alternatives. The opposite is sometimes true as well. Sometimes those at the corporate level do not trust the results of any simulation modeling effort because of the intrinsic complexity.

Alternatively, given the expense and the intrinsic weaknesses of simulation modeling, there is no alternative technique for modeling manufacturing systems with any degree of complexity. In the real-world manufacturing, systems are complex, often involving complex structural elements, complex and diverse production paths, and stochastic aspects that are certainly not amenable to analytical modeling. The use of simulation modeling in solving complex manufacturing problems is commonplace (e.g., Miller, 1990; Thompson, 1996).

Information Required

- Changeover/setup time
  - Average time required to setup for the next product or production run
  - Setup time distribution
- Detailed work schedule (shifts/day, days/year)
- Number of operators required for each workstation
  - Schedule of breaks, lunch, meetings, training, etc.
- Equipment data
  - Downtime distribution

- Distribution of yields parameters
- Order schedule

Information Provided

- Time-varying patterns of equipment utilization
- Lot schedules at each workstation
- Number of setups and the average setup time performed at each workstation
- Off-shift: number of occurrences and the average length of time
- Breakdowns: number of occurrences and the average length of time
- Waiting: number of occurrences and the average length of time

Note: Other information that relates to the cell operation can be provided as required. The following case study illustrates the application of the discussed methods to the evaluation of an existing manufacturing cell in an attempt to increase its productivity.

3. OFFICE AUTOMATION COMPANY COLOR TONER PLANT CASE[*]

Opened in the late 1960s, the plant previously produced black toner for the Canadian market. The changes in copier technology and the replacement of an older generation of copiers made the facility uneconomical to operate. Even with modifications to improve the productivity of the plant, it was still not feasible to keep the plant running. The problem for the plant was the booming popularity of color photocopiers— the demand for toners in blue, yellow, and magenta colors, which was beyond the capabilities of the plant. With plants scattered around the world, color production could have been located anywhere. The Office Automation Company considered four locations to produce color toner: two locations in the United States and two locations in Canada. One of the Canadian plants won the internal bidding for the color business because of its reputation for high-quality production, inexpensive electric power in Ontario and the Canada—U.S. free-trade agreement.

The Canadian plant is currently the sole source for the color products for the Office Automation Company. In 1991, there were less than 25 people working at the plant when it reopened its doors. Today that number has increased to greater than 50 employees. The plant operates three, 8-hour shifts a day, five days per week. Color toner is manufactured for several copier product families. Each product family has 3 toner colors, except for one that has 4. Thus, more than 10 different toners are manufactured. Changeovers between colors are a major activity for the plant given the current design.

---

[*]The information presented in the analyses associated with this case study has been *sanitized* to protect the client firm.

3.1. Canada Plant Manufacturing Cell Process

The plant uses a manufacturing cell to produce copier toner. Toner is a fine dry powder ranging from 5 to 40 microns in size. The major constituent is a low-melting, low-molecular-weight thermoplastic polymer (resin) or polymer blend, with colorants or pigments incorporated into the blend making up the bulk of the additives. Components are sourced from commercially available raw materials suppliers.

The process to manufacture toner involves a number of steps, categorized as follows:

- *Preblend/Melt Mix*   Polymers, resins, pigments, and other minor additives are blended in a mixing vessel to homogeneity.
- The blend is then melt-mixed in a process resulting in a material in the form of round pellets.
- *Grind/Classify*   The polymer composite pellets are crushed and ground to a fine powder via fluid energy milling.
- The ground powder is sized, separated, and classified via an aerodynamic sieving and inertial separation process.
- *Blend/Screen*   The powder is blended with external (functionality related) additives.
- The powder is "screened" to remove oversized materials and contaminants.
- *Packaging*   The final product is packaged via an auger-type filling system into containers in preparation for shipping.

Figure 2 provides a process map for the production process.

A changeover occurs after each run. This will allow the equipment to be totally cleaned before the next color product is run. Each production area requires a certain changeover time, as indicated in Table 1. The changeover times reduce the available production time.

To avoid confusion, the following terms are defined:

- Rates (lb/hr): The rate at which product flows through each production area.
- Yields (%): The ratio of the amount of "good" material leaving a workstation to the amount of material that enters the workstation; losses are due to scrap, samples, contamination losses, etc.
- Uptime (%): The percentage of the time that production is occurring; downtime is due to maintenance, breaks, equipment problems, poor operational practices, etc.

3.2. Management Experiences and Objectives

The volatile color copier market has been frustrating for the plant management. Fluctuations and uncertainty in the projected volumes make it difficult to develop accurate plans. The present MRP system does not allow for quick what-if analysis.

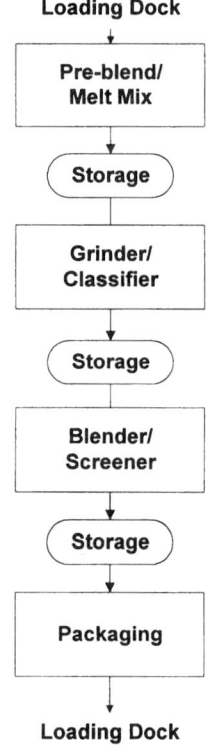

Figure 2. Toner Production Process.

Table 1. Changeover times

Production Area	Changeover Time (hr)
Preblend/melt mix	32
Grinder/classifier	80
Blender/screener	32
Packaging	32

Productivity improvement projects to reduce changeover times, increase capacity, reduce WIP inventory, and identify bottlenecks cannot be adequately evaluated by the current plant systems. Quick testing of different production rates, yields, and uptimes to determine their impacts on plant operations are not easily achievable.

Continuous evaluation of the manufacturing cell would aid the plant management to respond to the evolving environment effectively. For example, currently business needs require a 15% increase in output of the manufacturing cell for the upcoming year. This would require the annual output to increase to 1,308,161 pounds. The evaluation methods used would need to be simple to allow the plant to implement them as working tools.

3.3. Application of the Evaluation Methods

We apply the proposed evaluation methods to the plant's manufacturing cell with the objective of meeting the increased forecasted demand in the most efficient manner. Static evaluation was conducted by utilizing a Lotus spreadsheet (Lotus, 1994). This analysis required information on the machine rates, yields, and uptimes. Lotus formulas, functions, and macros were used for the analysis. The spreadsheet level of complexity was medium. That is, while a spreadsheet can be manipulated for very complex and elaborate calculations, this one was developed by an engineer with a B.S. degree and no formal training on spreadsheets, except for the tutorial that comes with the software.

The stochastic evaluation was conducted by utilizing the MPX modeling software package (Network Dynamics Inc., 1992). Based on Rapid Modeling Technology (RMT), MPX allows models to be built in a matter of hours and each what-if takes only a few seconds on a 386-class PC. MPX predicted average machine utilization, product flow times, and WIP inventory. No programming was needed. The package involved multiple input screens, which required detailed data about the manufacturing cell. Specifically, MPX required information on the numbers and types of machines, downtime statistics (i.e., rough estimates of the average time between failures and average time to repair), and estimates of the percent yield. The MPX level of complexity was medium. The model was constructed by an engineer who had 6 hours of classroom training on RMT and the MPX package.

The dynamic evaluation utilized the SIMAN V simulation software language (Banks et al., 1995). Many hours of programming time were required to develop the simulation model. In addition, detailed data about the systems operation was required. The analysis's level of complexity was high. The simulation model required several inputs as described in the previous sections.

What-If Analysis Using Static Evaluation   The spreadsheet model was used to perform what-if analysis and develop a variety of alternatives to increase the output to the desired level. The following three options were determined by management as practical and economically viable.

Option A
   Step 1. Increase grinder/classifier yield from 80 to 92% by stabilizing the process using statistical process control procedures.
Observations: Output increased to 1,292,034 pounds. However, the blend/screen is now the bottleneck operation. Further improvements in yield will not impact the output.
   Step 2. Increase blend/screen rate from 353 to 358 pounds per hour by altering the process control program.
Observations: The bottleneck has moved back to the beginning of the manufacturing cell. Output increased to 1,308,161 pounds.

Option B

Step 1. Increase preblend/melt mix rate from 360 to 399 pounds per hour by installing a larger feed hopper to supply material to the workstation.
Observations: Output increased to 1,255,373 pounds. The bottleneck has now moved to the grinder/classifier.

Step 2. Increase blend/screen yield from 96 to 98% by implementing a preventative maintenance program.
Observations: Output increased to 1,281,526 pounds. The bottleneck operation is still the grinder/classifier. It is being ignored in this option.

Step 3. Increase packaging yield from 96 to 98% by reducing inspection frequency and sample size due to improved process stability.
Observations: The bottleneck operation is still the grinder/classifier. Output increased to 1,308,225 pounds.

Option C

Step 1. Increase grinder/classifier yield from 80 to 92% by stabilizing the process using statistical process control procedures.
Observations: Same as step 1 of option A.

Step 2. Decrease blender/screener changeover time from 32 hours to 16 hours by installing quick disconnects and clamps that will allow the equipment to be disassembled and assembled quickly (low capital investment required).
Observation: Output increased to 1,308,161 pounds.

In this case, the static analysis identified the impact of the decisions on the cell's output for each option. In addition, it identified the bottleneck workstation in each case.

What-If Analysis Using Stochastic Evaluation   To narrow down the available alternatives, stochastic evaluation was utilized to compare the average WIP inventory and product flow times under the three options. It was determined that option A was infeasible due to insufficient capacity to produce the desired output. However, options B and C were feasible. The results are summarized in Table 2. In this example the

Table 2. Results of stochastic evaluation

Product	Option B		Option C	
	Average WIP (lb)	Average Flow Time (days)	Average WIP (lb)	Average Flow Time (days)
1	227,868	82.4	159,509	54.9
2	175,793	80.5	107,620	55.4
3	110,545	75.7	70,152	47.0
Total WIP	514,206		337,281	

stochastic analysis identified the impact of the decisions on the cell's performance for each option.

What-If Analysis Using Dynamic Evaluation   To further narrow down the available alternatives, dynamic evaluation was utilized to examine time-varying patterns (e.g., maximum levels) of product flow times and WIP inventory. The order input and production mix were determined using the previous year's demand, adjusted to attain the desired level of overall production. Future variations in this order and mix will induce system behavior that will be somewhat different than indicated by these results. The sequence of orders used for this simulation analysis is illustrated in Table 3. Furthermore, the setup requirements for each of the products are not constant; rather, it varies depending upon the color and product family. The simulation results are summarized in Table 4. The order size indicates the amount of raw materials needed to achieve the desired output considering the yield at each station.

Option B is infeasible due to lack of sufficient WIP inventory storage space. The maximum storage capacity is 360,000 pounds (product density varies very little, so poundage translates directly into volume). Thus, option C is the preferable alternative in this case.

4. SUMMARY OF THE PROPOSED FRAMEWORK

Each evaluation method within the proposed hierarchical evaluation framework has its benefits and drawbacks. However, the different methods compliment each other. The static analysis can be used as an initial alternative eliminator and a valuable benchmark. The stochastic evaluation can be used to obtain WIP inventory and product flow time estimates. The simulation modeling can be used for determining such performance measures as time-varying patterns of product flow times, WIP inventory, and workcenter utilization levels by considering specific scheduling schemes, lot sizes, machine breakdown patterns, and the like. These methods complement

Table 3. Order input stream

Sequence	Product	Sequence	Product
1	3	11	2
2	1	12	3
3	1	13	1
4	1	14	1
5	3	15	1
6	3	16	3
7	3	17	3
8	2	18	3
9	2	19	2
10	1	20	2
		21	2

Table 4. Results of dynamic evaluation

	Option B			Option C		
Product	Average WIP (lb)	Maximum WIP (lb)	Average Flow Time (days)	Average WIP (lb)	Maximum WIP (lb)	Average Flow Time (days)
1	94,739	284,340	57.8	83,627	252,410	56.3
2	69,938	240,240	88.8	61,941	213,260	85.5
3	77,752	227,560	64.8	69,572	201,220	62.5
Total WIP	242,430	388,240		215,140	352,220	
	Production (lb)			Production (lb)		
1		496,749			496,758	
2		356,303			356,309	
3		455,112			455,120	
Total		1,308,164			1,308,187	
Order Size		1,791,863			1,590,626	

each other quite well as the modeling and analyses progress in studying design and operation of a manufacturing cell (Shimizu and Zoest, 1988). The methods could also be used to perform what-if analysis at each stage of the evaluation process by adjusting the inputs and evaluating the impact on cell performance.

REFERENCES

Aigbedo, H., and Monden, Y. (1996). A simulation analysis for two-level sequence-scheduling for just-in-time (JIT) mixed-model assembly lines, *International Journal of Production Research, 34*, 3107–3145.

Anderson, K. R. (1987). "A Method for Planning Analysis and Design Simulation of CIM systems," *Proceedings of the 1987 Winter Simulation Conference*, IEEE Computer Society Press, Los Alamitos, CA, 715–720.

Ang, C. L., and Willey, P. C. T. (1984). A comparative study of the performance of pure and hybrid group technology manufacturing systems using computer simulation techniques, *International Journal of Production Research, 22*, 193–233.

Arvindh, B., and Irani, S. A. (1994). Principal component analysis for evaluating the feasibility of cellular manufacturing without initial machine-part matrix clustering, *International Journal of Production Research, 32*, 1909–1938.

Banks, J., Burnette, B., Kozloski, H., and Rose, J. (1995). *Introduction to SIMAN V and to CINEMA V*, Wiley, New York.

Brown, E. (1988). IBM combines rapid modeling technique and simulation to design PCB factory-of-the-future, *Industrial Engineering, 20*, 23–26.

Chen, H., Harrison, J. M., Mandelbaum, A., Van Ackere A., and Wein L. M., (1988). Empirical evaluation of a queuing network model for semiconductor wafer fabrication, *Operations Research, 36*, 202–215.

Dale, B. G., and Dewhurst, F. (1984). Simulation of a group technology product cell, *Engineering Costs and Production Economics*, **8**, 45–54.

Dawood, N. N. (1996). A simulation model for eliciting scheduling knowledge: An application to the precast manufacturing process, *Advances in Engineering Software*, **25**, 215–224.

Flynn, B. B., and Jacobs, F. R. (1987). An experimental comparison of cellular (group technology) layout with process layout, *Decision Sciences*, **18**, 562–581.

IIE Solutions, Buyer's Guide (1996). (May), 54–63.

Jernigan, S. R., Ramaswamy, S. A., and Barber, K.S. (1997). Distributed search and simulation method for job flow scheduling, *Simulation*, **68**, 377–402.

Kamrani, A. K., Parsaei, H. R., and Leep, H.R. (1995). A Simulation Approach for Cellular Manufacturing System Design and Analysis," in *Planning, Design, and Analysis of Cellular Manufacturing Systems*, A.K. Kamrani, H.R. Parsaei, and D.H. Liles, (Eds.), Elsevier Science, Amsterdam, 351–381.

Kelly, C., Mosier, C. T., and Mahmoodi, F. (1997). An analysis of maintenance policies in a group technology environment, *International Journal of Production Research*, **35**, 767–787.

King, J. R., and Nakornchai, V. (1982). Machine-component group formation in group technology: review and extension, *International Journal of Production Research*, **18**, 117–133.

Lotus Development Corporation (1994). *Lotus 123 Release 5 User's Guide*, Cambridge, MA.

Lyu, J., and Gunasekaran, A. (1997). An intelligent simulation model to evaluate scheduling strategies in a steel company, *International Journal of Systems Science*, **28**, 611–616.

Mahmoodi, F., and Martin, G. E. (1997). A new shop-based and predictive scheduling heuristic for cellular manufacturing, *International Journal of Production Research*, **35**, 313–326.

Mahmoodi, F., Dooley, K. J., and Starr, P. J. (1990). An investigation of dynamic group scheduling heuristics in a job shop manufacturing cell, *International Journal of Production Research*, **28**, 1695–1711.

McAuley, J. (1972). Machine grouping for efficient production, *Production Engineering*, **51**, 51–57.

Miller, D. (1990). Simulation of a semiconductor manufacturing line, *Communications of the ACM*, **33**, 98–108.

Morris, J. S., and Tersine, R. J. (1990). A simulation analysis of factors influencing the attractiveness of group technology cellular layouts, *Management Science*, **36**, 1567–1578.

Mosier, C. T. (1989). An experiment investigating the application of clustering procedures and similarity coefficients to the GT machine cell formation problem, *International Journal of Production Research*, **27**, 1811–1835.

Mosier, C. T., Elvers, D. A., and Kelly, D. (1984). Analysis of group technology scheduling heuristics, *International Journal of Production Research*, **22**, 857–875.

Network Dynamics, Inc. (1992). *MPX Rapid Modeling Software Manual*, Burlington, MA.

Rehman, A., and Diehl, M. B. (1993). Rapid modeling helps focus setup reduction at Ingersoll, *Industrial Engineering*, November, **25**, 52–55.

Ruben, R. A., Mosier, C. T., and Mahmoodi, F. (1993). A Comprehensive Analysis of Group Scheduling Heuristics in a Job Shop Cell, *International Journal of Production Research*, **31**, 1343–1369.

Sassani, F. (1990). A simulation study on performance improvement of group technology cells, *International Journal of Production Research*, **28**, 293–300.

Shafer, S. M., and Charnes, J. M. (1993). Cellular verses functional layouts under a variety of shop operating conditions, *Decision Sciences*, **24**, 665–681.

Shafer, S. M., and Rogers, D. F. (1991). A goal programming approach to the cell formation problem, *Journal of Operations Management*, **10**, 28–43.

Shimizu, M., and Zoest, D. V. (1988). Analysis of a factory of the future using an integrated set of software for manufacturing systems modeling, Technical Report, College of Engineering, University of Wisconsin-Madison.

Song, S., and Hitomi, K. (1992). GT cell formation for minimizing the intercell parts flow, *International Journal of Production Research*, **30**, 2737–2753.

Srinivasan, G., Narendran, T. T., and Mahadevan, B. (1990). An assignment model for the part-families problem in group technology, *International Journal of Production Research*, **28**, 145–152.

Steudel, H. J. (1986). SIMSHOP: A job shop/cellular manufacturing Simulator, *Journal of Manufacturing Systems*, **5**, 181–190.

Suresh, N. C. (1992). Partitioning workcenters for group technology: analytical extensions and shop level simulation investigations, *Decision Sciences*, **23**, 267–290.

Suri, R. (1989). Lead time reduction through rapid modeling, *Manufacturing Systems*, **7**, 66–68.

Suri, R., and DeTreville, S. (1991). Full speed ahead: A look at rapid modeling technology in operations management, *OR/MS Today*, **18**, 34–42.

Suri, R., and Diehl, G. W. (1986). Manuplan: A precursor to simulation for complex manufacturing systems, *Proceedings of the Winter Simulation Conference* JEEE, Piscataway, NJ, pp. 411–420.

Suri, R., Diehl, G. W., and Dean, R. (1986). Quick and easy manufacturing systems analysis using MANUPLAN, *Spring IIE Conference*, Dallas, TX, IIE, Norcross, GA, 195–205.

Suri, R., Sanders, J. L., and Kamath, M. (1993). "Performance Evaluation of Production Networks," in *Handbooks in OR & MS*, Vol. **4**, Elsevier Science, Amsterdam, Chapter 5.

Thompson, M. (1996). Simulation-based scheduling meeting the semiconductor wafer fabrication challenge, *IIE Solutions* (May), 30–34.

Vaithianathan, R., and McRoberts, K. L. (1982). On scheduling in a GT environment, *Journal of Manufacturing Systems*, **1**, 149–155.

Wirth, G. T., Mahmoodi, F., and Mosier, C. T. (1993). An investigation of scheduling policies in a dual-constrained manufacturing cell, *Decision Sciences*, **24**, 761–788.

8

OPERATING MANUFACTURING CELLS WITH LABOR CONSTRAINTS

Anand Iyer

i2 Technologies, Inc.
Irving, Texas 75039

Ronald G. Askin

Systems and Industrial Engineering Department
College of Engineering and Mines
University of Arizona
Tucson, Arizona 85721

1. INTRODUCTION

Cellular Manufacturing (CM), the production implementation of a philosophy called Group Technology (GT), is a strategy that attempts to identify and economically exploit similarities in product design and processing. There has been a great deal of recent interest in GT as a scheme for parts grouping, machine dedication, and shop layout. Among the benefits cited are a reduction in material handling and setup times, lower Work-In-Process (WIP), and shorter throughput times. Conventionally, in batch-type manufacturing for multiple products, each product was treated as unique from design through manufacturing. Group Technology is based on the philosophy that by grouping similar parts into part families based on some defined attributes (usually processing similarities) it is possible to increase productivity through more effective manufacturing standardization.

Handbook of Cellular Manufacturing Systems, edited by Shahrukh A. Irani
ISBN 0-471-12139-8 © 1999 John Wiley & Sons, Inc.

The concept of GT was first popularized in the early 1960s by Burbidge (1963). However, only recently has it begun attracting attention from U.S. manufacturing firms. In a bid to remain competitive by taking advantage of the benefits of GT, more U.S. firms are implementing GT on their manufacturing floors. Although much of the research effort has been devoted to the technical and algorithmic aspects of the cell formation problem, published and anecdotal evidence (Brandon, 1993, 1995) indicates that at least part of the success in implementing cells can be attributed to changes in organization and policy. Indeed, it has been suggested that some of the earlier failures in implementing manufacturing cells stemmed from treating the problem as a purely technological one. Traditional organizational demarcations based on function were significant impediments to realizing the benefits of cells.

There is a growing realization that the full value of cells can be realized only through a combination of technical and social factors. Sociologists and management experts alike (Brandon, 1996) have observed that work is a group activity and that groups evolve naturally. It is, therefore, in the best interests of management to create self-managing teams and thereby enhance manufacturing efficiency. It is natural to extend these concepts to encompass group ownership of a single product or a small family of products, which satisfies the natural desire of people to belong to goal-seeking groups. Since the 1960s, several studies in the literature (Brandon, 1996) have provided insights into the design of such teams. They suggest that this group structure coupled with job enrichment and enlargement play a significant role in improving the productivity and effectiveness of cells.

Thus, many successful implementations can be expected to consist of an empowered team of cross-trained workers and the resources required to produce a family of related products. It will also not be atypical for the machines to outnumber workers. Such cells will require significant changes in organizational policies for personnel, wage administration, accounting, and scheduling. Since a discussion of incentive schemes and accounting policies is beyond the scope of this chapter and the expertise of this author, this chapter will focus on structuring operating policies for cells. We define an operating policy to be a consistent specification of lot sizes, material movement rules, cell WIP limits, machine queue dispatching rules, and worker assignment rules. The definition of an operating policy clearly indicates that its various components are closely linked. This fact usually makes analysis using traditional industrial engineering/operations research tools very difficult. Unfortunately, very little is known about operating policies in general or tailoring operating policies for specific cells. In this chapter, we will focus on some lessons learnt from various studies on operating policies for manufacturing cells. It will become quickly apparent that it is very difficult to provide ready-made recipes that ensure success. As such, we will limit ourselves to outlining some general rules of thumb in defining operating policies. Finally, we will discuss the use of tools and techniques to translate these general insights into schedules that can be implemented on a shop floor.

2. LITERATURE REVIEW

Much of the scheduling research in the area of CM has concentrated on machine queue dispatching rules (Mahmoodi and Dooley, 1991; Morris and Tersine, 1989; Mosiers et al., 1984; Sundaram, 1983; Vaithianathan and McRoberts, 1982) and usually makes the assumption that cells are not labor constrained. Other aspects of operating policies such as transfer batches have been studied in other contexts as well (Goyal, 1976; Szendrovits, 1975, 1976; Baker and Pyke, 1990).

As mentioned earlier, much of the early work on dispatching rules implicitly made the assumption that cells were not labor constrained. Harris (discussed in Nelson, 1967), in 1967, pointed out that operating conditions in most job shops did not warrant ignoring labor constraints. Harris's conclusion was based on studying operating data from job shops over a 7 month period. Subsequently, several simulation studies of dual resource constrained systems appeared in the literature (Nelson, 1967, 1970; Weeks and Fryer, 1976; Hogg et al., 1975a, 1975b; Huang et al., 1984). Most of these studies focused on the interaction between job structure and labor assignment rules as well as the efficacy of various assignment rules. Nelson (1967, 1970) concluded that the choice of queue disciplines at each workstation produced a large effect on the mean and variance in the system for jobs. However, there was a negative interaction between the mean and variance in that an improvement in one measure was obtained at the expense of the other. The labor assignment rule, on the other hand, had a smaller effect on the performance measures. However, in this case there were no tradeoffs between the mean and variance of the time in system. He also concluded that the mean and variance of the time in system for a job shop with a highly flexible workforce was much lower than the corresponding measures for a job shop with an inflexible workforce. Weeks and Fryer (1976) also confirmed Nelson's conclusion that making a worker available for reassignment after every operation reduced the mean flow time of jobs through the system. The alternative reassignment policy (called decentralized control) was to allow workers to switch only when they became idle, that is, when the queue ahead of them emptied. A recent study by Malhotra and Kher (1994) concluded that when transfer delays, (i.e., delays incurred when workers are switched between machines) are significant, centralized control caused the performance of the job shop to deteriorate. In the case where transfer delays are insignificant, they confirmed the findings of Nelson (1967, 1970) and Weeks and Fryer (1976). They also concluded that with a heterogenous workforce (not all workers are equally efficient at each machine), the timing of the workers assignment (WHEN) was not as important as the WHERE (the machine the worker is sent to). Hoopes et al. (1995) developed tabu search (Glover, 1989, 1990) based procedures that can be used for aggregate planning in the case where n assembly cells compete for labor resources to process multiple jobs. The processing time of a job in the cell depends on the number of labor resources allocated to the cell. The model is motivated by problems that arise in the production of printed circuit boards. The heuristics generate good solutions in a reasonable amount of time and demonstrate the advantages of a good labor allocation policy in situations where processing times are inversely proportional to the amount of allocated labor.

Russell et al. (1991) included labor flexibility as a factor in their study of dual resource-constrained CM systems. They concluded that there is a strong interaction between routing flexibility and labor flexibility (measured by the degree of cross training). Askin and Iyer (1993) compared traditional batch processing rules with multiproduct dedicated cells and a worker batch assignment strategy for machine-limited systems as well as machine and labor-limited systems. The multiproduct approach was shown to perform best for low to moderate utilization. The worker batch assignment strategy was also shown to perform well in labor-limited systems. This study assumed complete labor flexibility, that is, all workers were completely cross trained. Also, this study assumed that all part types had the same processing times at all machines.

Operating policies that find application in very specific manufacturing environments have been studied in some detail. One of these, labeled TSS (Toyota Sewn Products System), has been studied and described in some detail recently. The TSS policy is used later in this chapter as one of four families of policies that are compared and as such is described in some detail here. The TSS policy is used mainly in the apparel and garment manufacturing industries currently and uses flexible work assignments as a means of automatically balancing workloads among workers. The number of workers is usually smaller than the number of machines. Such systems are easy to control and run efficiently with very low levels of WIP. Workers are cross trained to operate all machines and work as a team. Typically, upon completing a task at a machine, workers use simple rules to decide which machine to proceed to next. Preemption rules may specify conditions under which a worker may "bump" another worker from a machine and take over the processing of the task. Bartholdi and Eisenstein (1993) studied the TSS policy for serial line where workers always maintain their ordering along the line. Each worker moves forward as much as possible and may be blocked at the subsequent machine. The last worker in the line triggers a series of preemptions when he or she reaches the last machine in the line and starts moving backward and bumps the first worker encountered or takes over his/her task. They showed that such a line is stable only if workers are sequenced slowest to fastest along the line. Bischak (1993) showed that reduction in processing time variability between machines and the presence of buffers are beneficial for such TSS lines. In the both studies, each worker was assumed to be equally efficient at all machines.

Although not directly applicable in the context of manufacturing cells, there is a growing body of work devoted to queueing models which are related to the operator scheduling problem. The interested reader is referred to Avi-Itzhak et al. (1965), Eisenberg (1971, 1972, 1978), Federgruen and Green (1984), Jacobson and Lazowska (1982), Silva and Muntz (1987), Srinivasan (1988), and Takacs (1968) and the references therein.

3. OPERATING MANUFACTURING CELLS

As is evident from the previous sections, analytical modeling of operating policies remains a difficult problem. The models discussed previously are applicable only in special cases. As such, simulation models are used here to compare policies across a broad spectrum of operating conditions. We assume a flow shop environment and investigate the impact of and the interaction between admittance policies, cross-training, levels and transfer batches.

4. SIMULATION EXPERIMENT

A simulation experiment was designed to compare the operating policies described later under flow shop conditions. Within this framework, we sought to address the following basic issues :

- Cross Training. Traditionally, the assumption made in many models (implicit or explicit) has been that workers are either completely cross trained or not at all. With fewer workers than machines, clearly cross training is a necessity not a luxury. However, it is not very practical to completely cross train all workers. In many environments, such as apparel manufacturing (Bartholdi and Eisenstein, 1993; Bischak, 1993), a high degree of cross training can be achieved. Also, workers can take over from some other worker in midtask quickly and efficiently. However, there are many environments where hand-offs cannot be accomplished in a seamless manner. In general, we explore the hypothesis that greater cross training, measured by both the number of machines that can be operated by the average worker and the average number of workers trained on each machine, improves cycle time and throughput.
- Admittance Policies. In recent years, many research studies extolling the virtues of CONstant Work-In-Process (CONWIP) systems (Buzacott and Shanthikumar, 1992; Spearman et al., 1990) have been published. Other previously published work (Askin and Iyer, 1993; Morris and Tersine, 1989) has reported on the good performance of policies that limit WIP in the cell in the hope of gaining faster flow times at the expense of waiting times at the cell dispatch queue. We examine, for various operating strategies, the interaction effect between the degree of cross training and the admittance policy.
- Process vs. Transfer Batches. Previous research on operating policies (Askin and Iyer, 1993, 1994) has shown that policies where process batches are moved in unit transfer batches perform very well under conditions of low to moderate utilization. In this study, we explore this phenomenon further. Variants of other policies were tested in which batch movement rules were modified to include the use of transfer batches.

The factors and levels chosen for study are given in Table 1, where σ_m^2 and σ_a^2 refer to the variability of processing times at machines and the variability of job

Table 1. Factors and levels for the simulation experiment

Levels ↓	Machines	ρ	σ_m^2	σ_a^2	Pattern	Cross Training	Workers
	5	0.85	High	High	Increasing	B	2
					Decreasing	U	4
		0.65	Low	Low	Random	C	5
	9	0.85	High	High	Increasing	B	4
					Decreasing	U	7
		0.65	Low	Low	Random	C	9

interarrival times to the cell respectively. High and low levels refer to the cases when the SCV (squared coefficient of variation) of the processing time or interarrival time distribution are 1.0 and 0.3, respectively. The term ρ refers to the targeted utilization levels. A high level refers to a maximum utilization level of 85% while a low level refers to a maximum utilization level of 65%. It should be noted that when there are fewer workers than machines, the utilization levels refer to worker utilizations. Pattern refers to the pattern of processing times along the line. Thus, an increasing pattern refers to the fact that the mean batch processing times increase from machine 1 to 5 (or 9). The cross-training factor is intended to reflect the pattern of worker assignments to machines. The three levels U, B, and C refer to unbalanced, balanced, and complete, respectively. Balanced and unbalanced refer to the degree to which the orbits of workers overlap. Thus, in the balanced case there is a higher degree of overlap between the orbits of workers than in the unbalanced case where workers are cross trained such that there is little or no overlap. Complete cross training refers to the case when all workers are completely cross trained. Batch sizes were set to 10 in all the experiments.

In order to understand the interactions that define the efficacy of operating policies, several variants of four families of policies, each embodying a different approach were studied. The four basic policies are described below. The first policy we study is called Machine Batch Loading (MBL). Modeled after traditional job shop operation, the MBL policy has no limit on cell WIP, and workers are stationed at machines until the machine queue is empty. At this point, workers switch to the idle machine with the largest queue. Batches are processed and moved as process batches. The second policy studied, called Dedicated Cell Loading (DCL), splits process batches into unit transfer batches. In addition, cell resources (workers and machines) are temporarily dedicated to one batch of parts at a time. Cell WIP is naturally limited by the pattern of job arrivals and resource requirements. Other batches wait in the cell dispatch queue until resources are released by the previous batch in the cell. The third policy called Worker Batch Assignment (WBA) is motivated by the Japanese practice of using limited WIP and cross-trained workers. Process batches are assigned to workers who then assume responsibility for that batch for the length of its sojourn through the cell. When blocked from using a machine, workers are allowed to switch to some other batch for which they are also responsible. The final policy we study is based on the Toyota Sewn Products System (TSS) policy described earlier. Workers move forward with batches until they reach the last machine in the process plan or are relieved by a downstream worker. This sets off a chain reaction of backward movements that eventually results in a new batch being admitted to the cell. Workers are blocked when they attempt to move forward and reach a busy machine.

The factors affecting the performance of each family of policies are identified and their behavior characterized by studying them under different operating conditions. In particular, we contrast the performance of the four basic policies under varying levels of maximum allowable WIP levels, worker cross training, processing time variability, worker utilization, workers per machine ratio, transfer batch size, and batch interarrival time variability.

The 12 policies can be categorized as belonging to one of two groups: policies with process batch flow and those with transfer batch flow. The policies belonging to the first group are

- MBLO. The original Machine Batch Loading policy as described earlier.
- MBL1. In this variant of the MBL policy, the maximum number of batches within the cell is restricted to be equal to the number of workers. Extra batches wait in queue outside the cell until space becomes available. Workers switch to the longest queue within their orbit.
- MBL4. This is similar to the MBL1 policy except that the number of batches within the cell is restricted to be equal to four times the number of workers.
- WBA2. This refers to the Worker Batch Assignment policy with the WIP being held to a maximum of two batches per worker. Whenever a worker reaches the limit of his or her orbit, responsibility for that part is transferred to an eligible worker with the least number of current responsibilities. An eligible worker is one who is trained to perform the next operation required by a batch.
- WBA4. This policy is similar to the WBA2 policy with the WIP being held to a maximum of four times the number of workers.
- TSSO. This is the original Toyota Sewn Products System policy with changes being made to the forward and backward movement rules for workers. This was done to account for the limited cross training of workers. Thus, a worker begins to implement the backward movement rules when he or she reaches the last machine in his or her orbit. Buffers are assumed to exist at the end of orbits so that upon processing a batch at the last machine in the orbit, a worker can drop off the batch at the next machine's input buffer. However, when a worker arrives at a busy machine in his or her orbit, he or she is blocked. Similarly, when a worker moves backward, if he or she reaches the first machine in the orbit and still does not have a batch to take over/work on, he or she becomes idle.
- TSS1. This policy differs from the TSSO policy only in the fact that workers are not blocked when they arrive at a busy machine. Instead, they drop off the batch and begin to move backward. Thus, a worker moving backward could arrive at an idle machine with a few batches waiting in the input queue. This worker would then start working on the first batch in the queue. The WIP is limited to be equal to the number of workers in the cell.
- TSS4. This policy is the same as TSS1 with the WIP being equal to four times the number of workers in the cell.

The policies belonging to the second group with unit part transfers are

- DCLO. This is the original Dedicated Cell Loading policy as described earlier.
- DCLI. In this variant of the DCL policy, there is no admittance control so that batches are admitted to the cell as soon as they arrive to the cell. However, priority is still given to the oldest batch in the cell with workers switching between parts only when they reach the limit of their orbits.

- DCL4. Unlike the DCLI policy, admittance control is exercised by limiting the maximum number of batches in the cell to be four times the number of workers.
- DSS1. In this "transfer batch" variant of the TSS1 policy, the upper limit on WIP is equal to the number of workers in the cell. However, the batch is moved in transfer batches of unit size. Thus, the worker also potentially switches machines after each part.

A complete factorial simulation experiment was run using the Cell Operating Policy simulator (described later) for 7500 hours of shop time with the initial 2500 hours being discarded to remove initialization bias. Three replicates were run for every combination of factor levels. Common random numbers were used across all cases to increase the precision of comparison.

5. SIMULATION INSIGHTS

In order to make meaningful comparisons, the observed average batch flow times were normalized by the ideal flow time. The ideal flow time is defined as the sum of mean batch processing times at all machines. All the numbers reported are in multiples of this quantity. All times are measured from the arrival of the batch to the system until completion of the last unit. Thus, if batch transfers are used and a batch has no waiting time from arrival to completion, flow time = 1.0. Since more than a few cases resulted in an unstable system, that is, the load imposed on the cell was much higher than its capacity, values larger than 4.0 were truncated at 4.0.

5.1. Instantaneous Move Times

The experiments described in the preceding section were run assuming instantaneous moves. Thus, all 384 cases were run for each of the 12 policies desribed previously. As a preliminary step, all the data collected from the simulation experiments was analyzed using the general linear models procedure using the software package, SAS. Initially, only two and three factor interaction effects were considered in addition to the main effect estimates. The output revealed that almost *all* effects were significant. With this many significant interactions, no significant insights can be gleaned from the data. However, the number of significant interactions does point to the necessity for choosing an appropriate policy with due consideration of the operating environment. As a first step in understanding the effects of choosing various policies, *Box and Whisker* plots were drawn for a few key factors. Figure 1 summarizes the effect of policies on the normalized flow time values. It is clear that the choice of policies has a marked effect on the flow times. Figure 1 also indicates that transfer-based policies such as DSS1 and DCL4 have much lower flow times than the process-batch-based policies. This finding is substantiated by Figure 2, which shows the effect of the batch transfer policy. In Figure 2, T stands for transfer batch while P stands for process batch. The effect of the cross-training level is shown in Figure 3. Figure 3 indicates that there is more variability in flow times when the cross-training level is C or U. Thus, a high degree of overlap in worker orbits (cross-training level B) appears to be a robust setting. Similarly, Figure 4 indicates that low levels of admittance control (higher levels of allowable WIP) result in lower flow times with less variability. In order to understand these effects better, the simulation was examined on a policy-by-policy basis.

The effect of cross training and the magnitude of the cross training–admittance control interaction are clearly revealed in Figure 5 where the normalized values of flow times are plotted against the simulated conditions by policy. The three policies being compared are DCLO, DCLI, and DCL4 only for the cases where σ_m^2, σ_a^2, and ρ are all set to their low values. The original DCL policy is the worst case under most circumstances. The exception is when workers are completely cross trained. In this case, the performance of the DCLO policy is comparable to the performance of other variants of the DCL policy, particularly when the utilization and processing

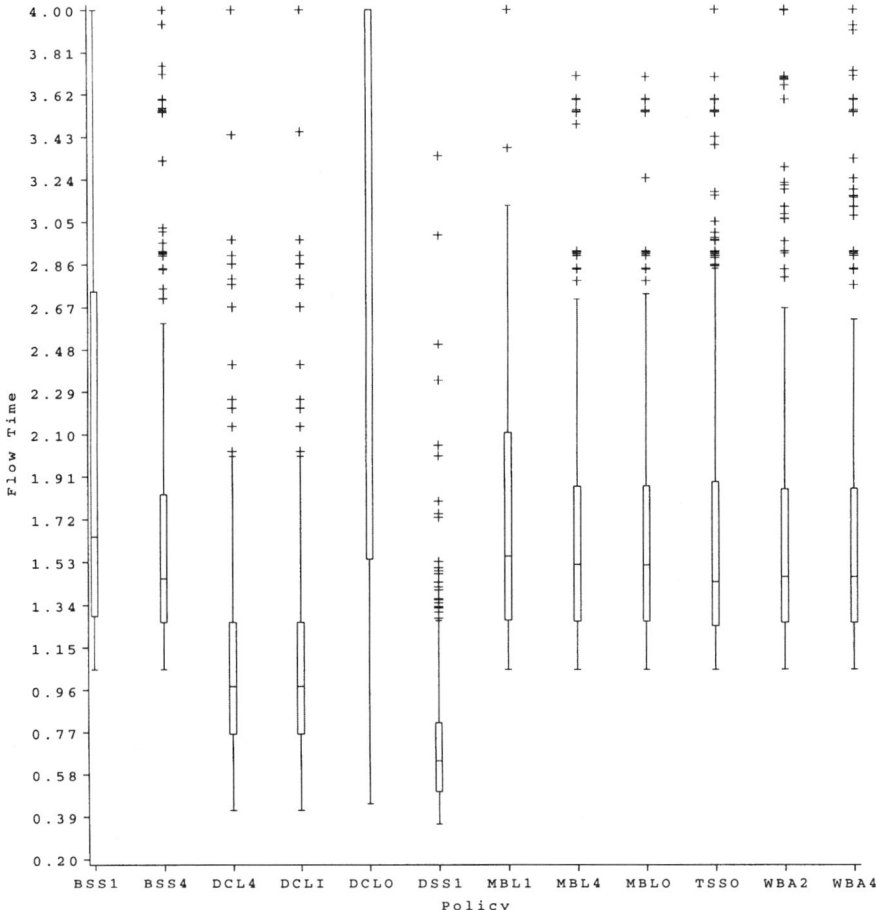

Figure 1. Impact of policies on flow times.

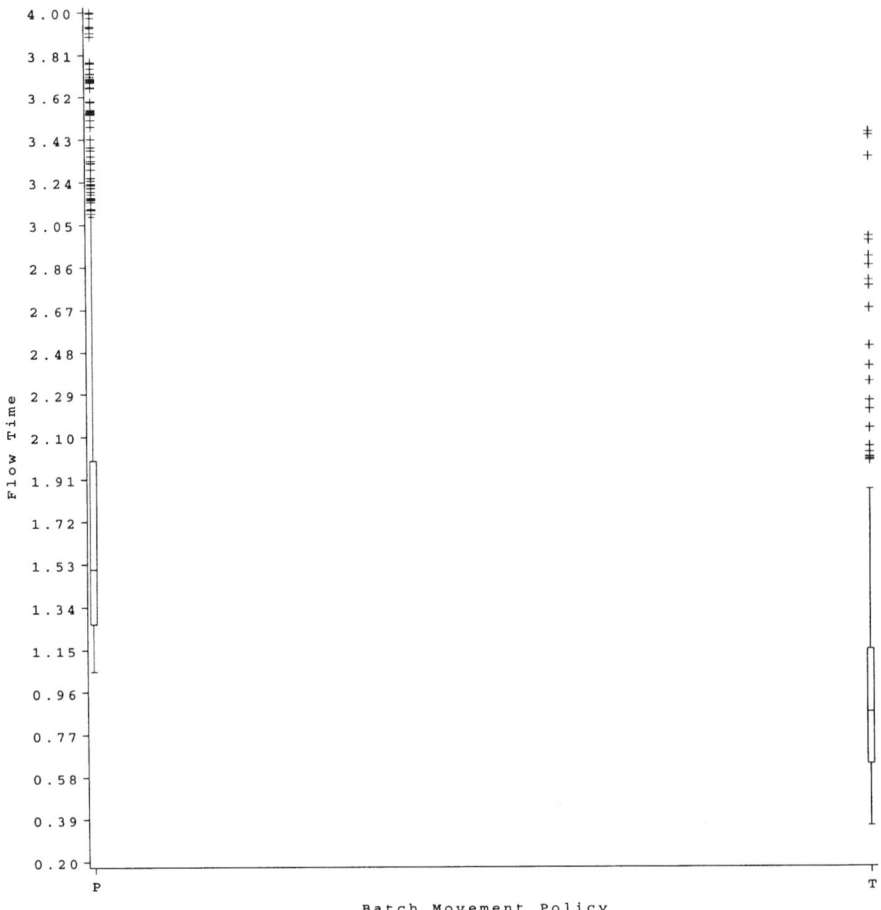

Figure 2. Impact of using transfer-batch-based policies on flow times.

time variability are low. Previous studies (Askin and Iyer, 1993, 1994; Morris and Tersine, 1989) demonstrated the superiority of the DCL policy under a wide range of conditions. In particular, low-to-moderate utilization and low processing time variability were shown to be conditions very favorable to the DCL policy. However, all those studies implicitly or explicitly assumed complete cross training on the part of the workers. The patterns seen in Figure 5 are visible for all other combinations of σ_m^2, σ_a^2, and ρ as well. The bad performance of the original DCL policy can be traced directly to the interaction between the cross-training level and the admittance policy. Recall that under the original DCL policy, a new batch is admitted to the cell when the previous batch has completed processing at the $(M - N + 1)$th machine in its sequence where M denotes the number of machines and N the number of workers. At that point, the worker who performed the first operation for the previous batch becomes free to work on the new batch. With limited cross training, the flow line has

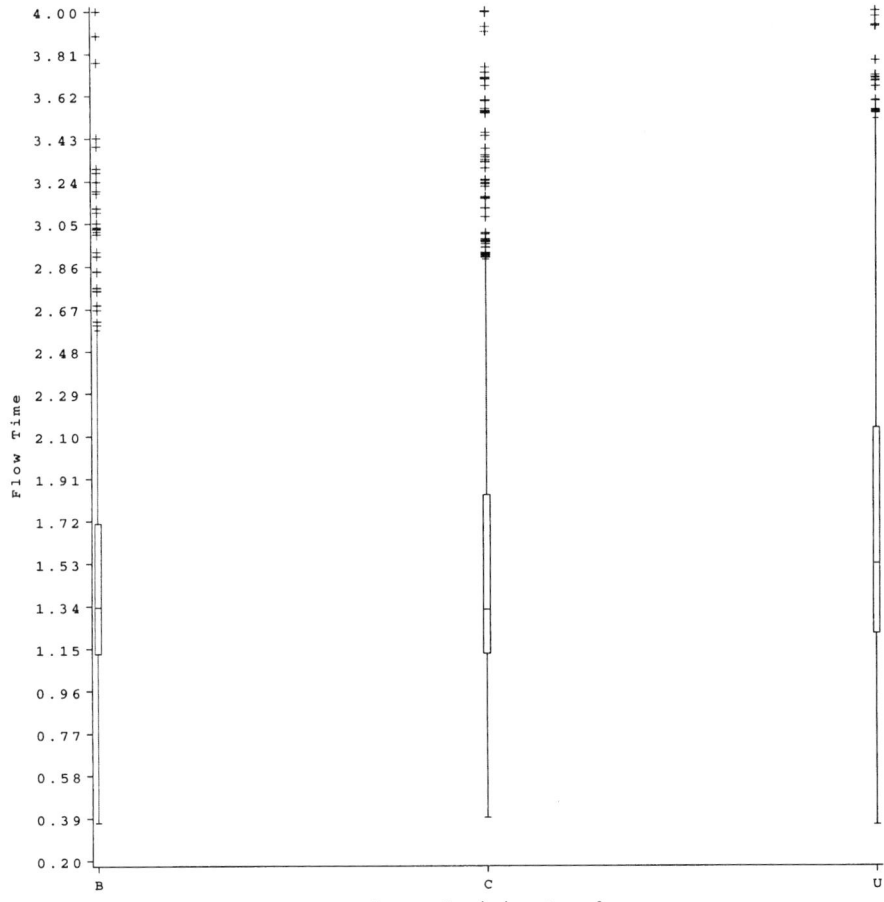

Figure 3. Impact of cross-training levels on flow times.

in fact been partitioned into sublines staffed according to the degree of cross training. The benefits of transfer batches are lost since workers may not be available to work on single units as they are moved to the next machine in their sequence. In addition, by adhering to the original admittance policy, workers are being kept idle even when they cannot be switched to machines required by the current batch due to limited cross training. If the cell were viewed as a single-server machine with the batch flow times being the service times at this machine, the queue at the machine becomes unstable since the arrival rate exceeds the service rate. One way of recovering lost capacity is to change the admittance policy. Thus, when the allowable WIP limits are increased, workers are no longer idle when they reach the limits of their orbits but can instead begin working on the oldest batch in their section of the cell.

Similar interactions can be seen in the case of the MBL family of policies as well. Machine Batch Loading and its variants are virtually indistinguishable at low

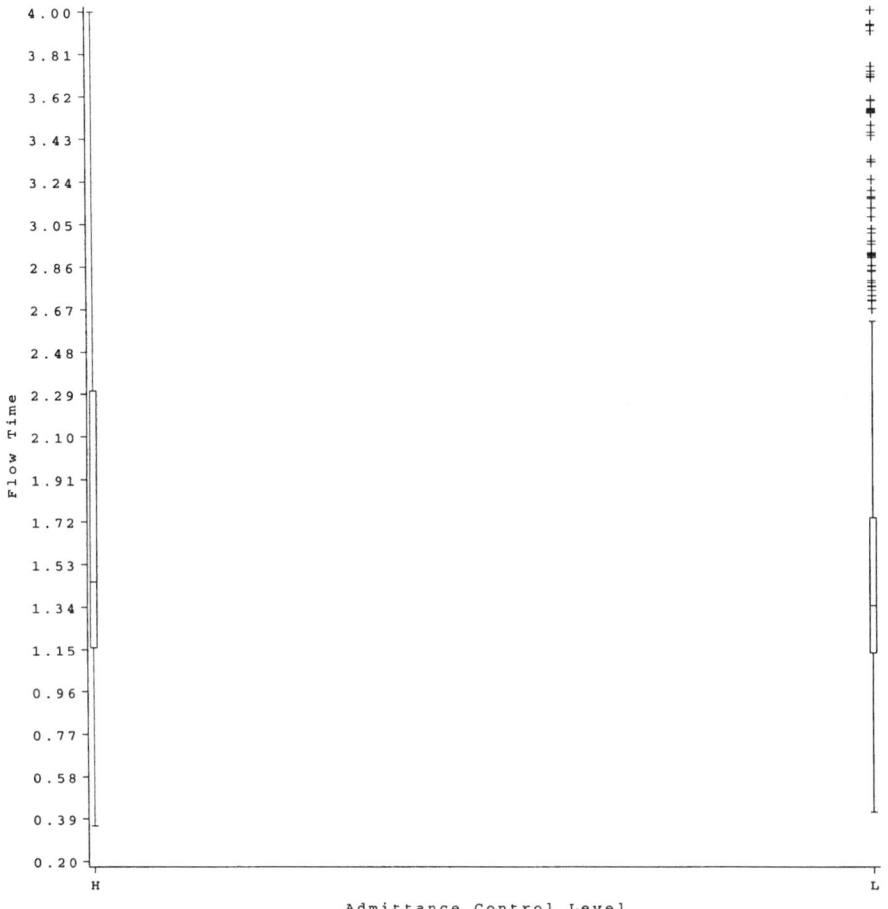

Figure 4. Impact of admittance control on flow times.

utilization levels and processing time variability. The effects of cross training are not felt as dramatically as in the case when both processing time variability and utilization are high (Figure 6). Though none of the policies perform as well as in the previous case, it is clear that MBL1 is the worst of the three. The effect of limiting WIP is exacerbated by the limited cross training, which results in workers being kept artificially idle. Relaxing admittance control alleviates this problem to some extent. As the number of workers in the line increases, the value of cross training diminishes.

Similar patterns are observed for the WBA family of policies as well. At low levels of utilization and processing time variability, there is not much difference between the three variants studied. Since workers hand off jobs when they reach the limit of their orbit, patterns and cross training levels determine how and when workers are blocked. With few workers, complete cross training is preferred since workers rarely block each other but must prevent the accumulation of work at the hand-off point. This

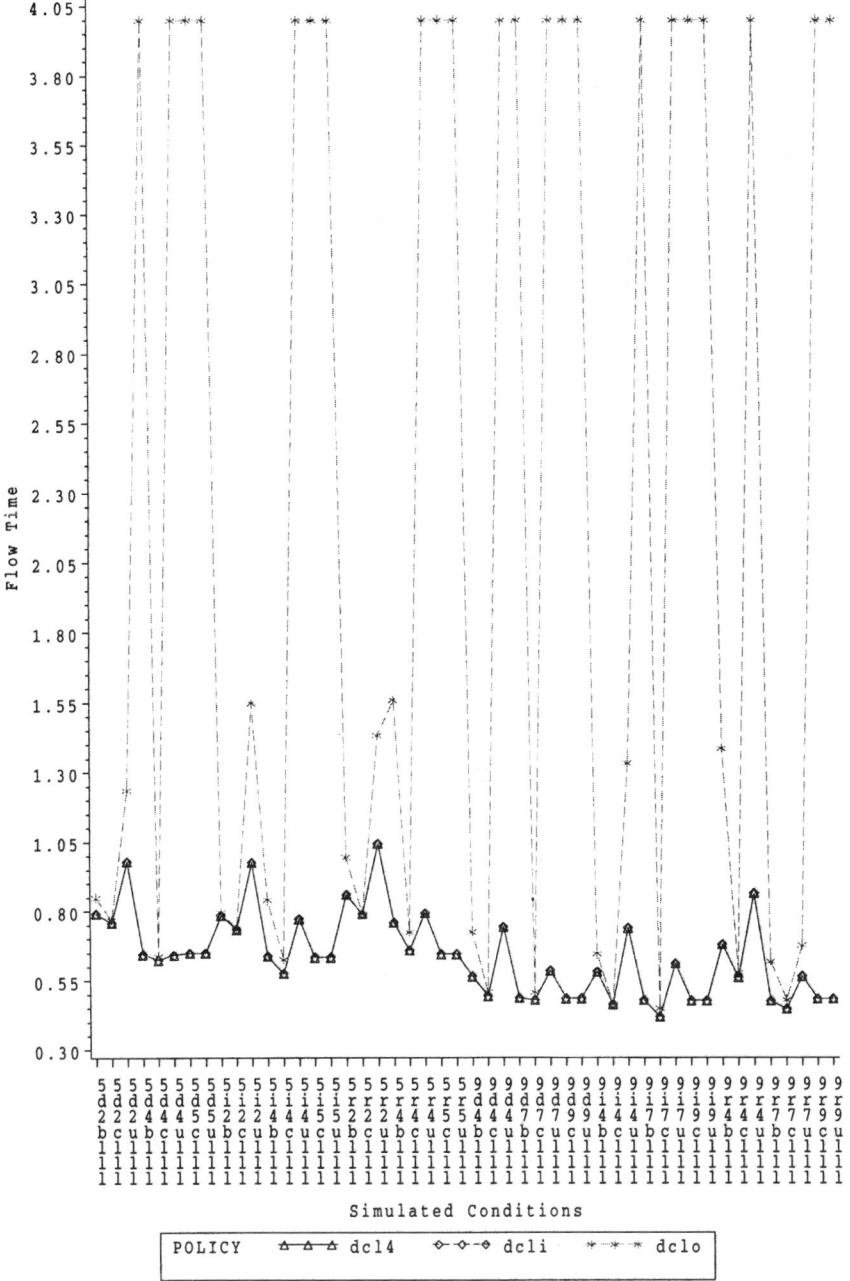

Figure 5. DCL and its variants.

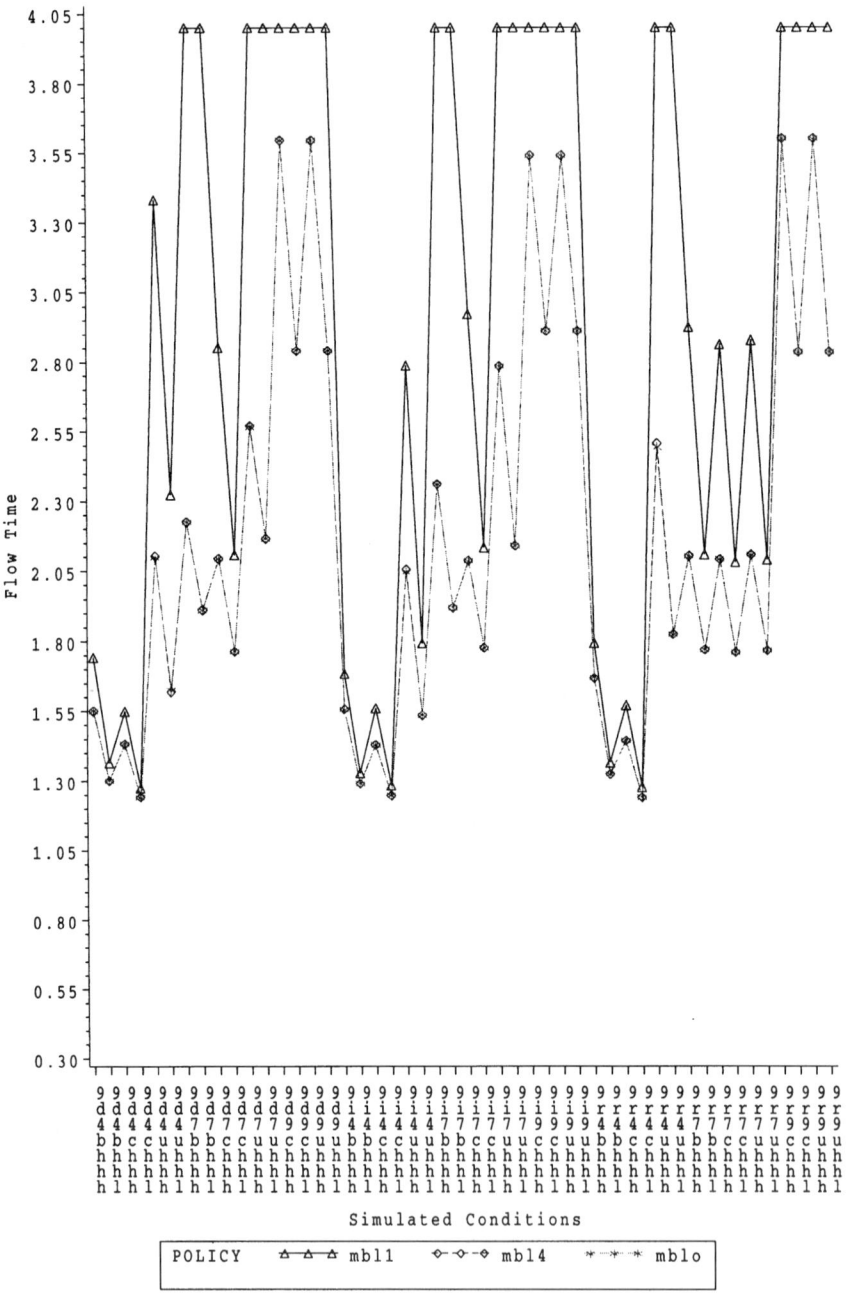

Figure 6. Comparing MBL and its variants ($M = 9$, $\sigma_m^2 = H$, $\rho = H$).

can become a severe problem with unbalanced worker assignments. As the number of workers increases, they tend to get blocked often. With minimal overlap in the skill sets of workers, blocking is prevented. With some overlap, workers get blocked but are able to switch to other jobs provided the admittance control policy allows for it. With tight admittance control, restricted orbits result in bad performance since there are no avenues for recourse.

Figures 7 and 8 exemplify the contrast between the variants of the TSS policy with batch transfers. At low levels of utilizations and processing time variability, there is not much to choose between the three policies (Figure 7). However, as both processing time variability and utilization increase, the superiority of BSS4 with increasing levels of WIP is apparent. While complete cross training is beneficial for most of the policies we have compared so far, the original TSS policy fails at high values of σ_m^2 and ρ even with complete cross training. This can be attributed to the fact that in the original TSS policy, workers are frequently blocked at the next machine as they move forward. Under the BSS4 policy, workers are allowed to drop off the batch they are moving with and start moving backward. This coupled with the admittance policy allows workers to switch to another batch and continue moving forward. This is the same strategy that allows the WBA strategy to work well, particularly when workers are allowed to be responsible for more than one batch at a time. The benefits of adopting this policy are apparent in Figure 8.

In order to simplify the analysis, further analysis is limited to six policies. The policies chosen are DSS1, DCL4, MBL4, TSSO, BSS4, and WBA4. DSS1 and DCL4 were chosen because they are always among the best two policies across all cases simulated. Of the policies with batch transfers, TSSO, BSS4, and WBA4 were chosen since they perform reasonably well and represent different approaches. MBL4 was chosen to represent traditional batch processing policies.

One of the most obvious patterns observed in all the simulations is the significant effect of processing time variability and utilization. The data also clearly demonstrates the superiority of the transfer batch policies over the other policies that use process batches. Careful analysis of the figures reveal several interactions that might account for these differences. It would appear that the relative superiority of transfer batch policies is magnified as the utilization and processing time variability increase. However, the magnitude of this difference is also a function of the number of machines and workers in the system. The disparity between transfer and process batch (TB and PB, respectively) based policies is greater when there are more machines and workers in the system. With few machines or few workers, the magnitude of the disparity decreases. This phenomenon is explained by considering the effect transfer batches strive to create. With transfer batches, several units of the same batch can be under production at different machines in the cell. Thus, processing (and setup) times at one machine are nested within the processing times at a previous machine. The actual number of units that can be under parallel production is given by min (M,W) assuming that the batch size is greater than the number of machines in the cell. Thus, when both M and W are high, the staggered processing time approach is most effective. When the number of workers is low, the dominance of the TB-based policies is muted by the fact that workers have to be shared between many machines. Given this explanation

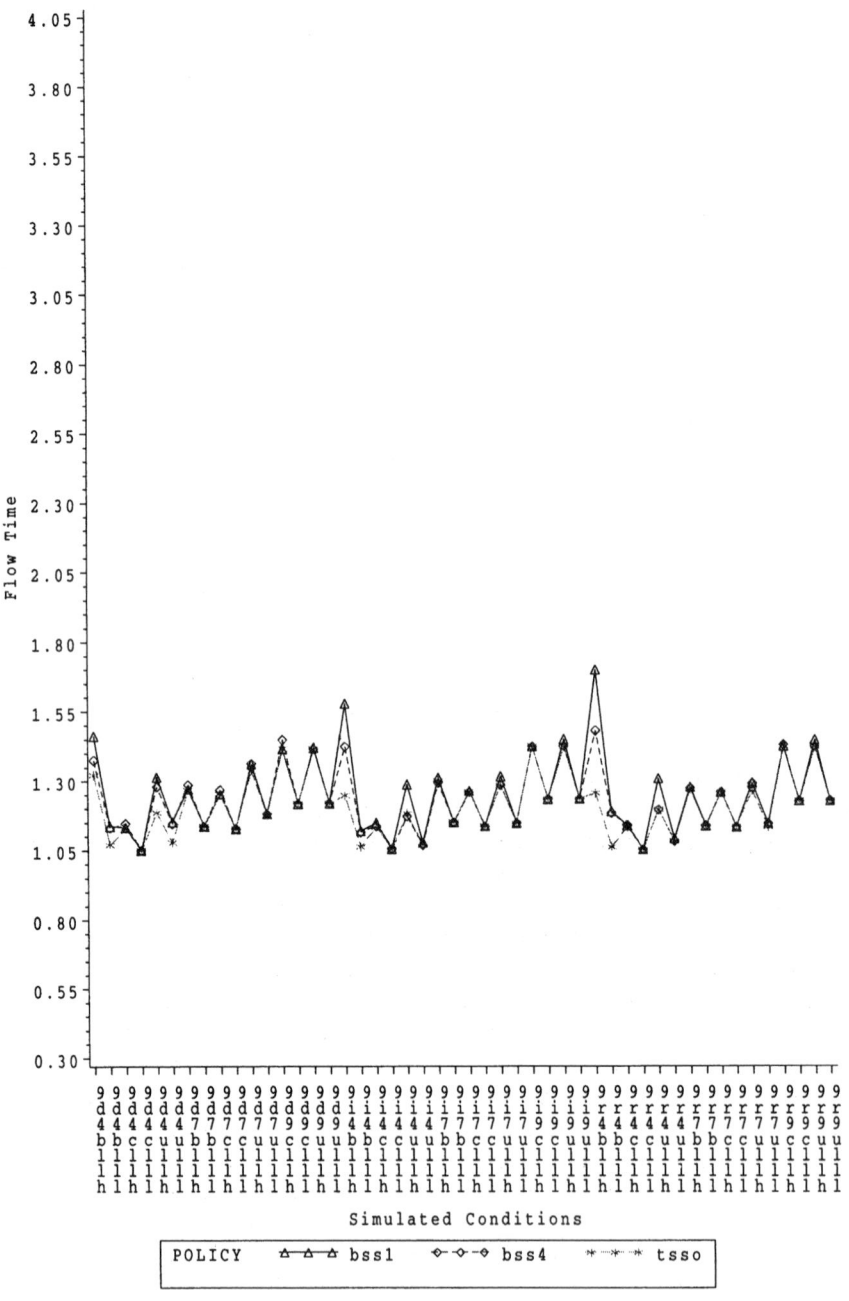

Figure 7. Comparing TSS and variants ($M = 9$, $\sigma_m^2 = L$, $\rho = L$).

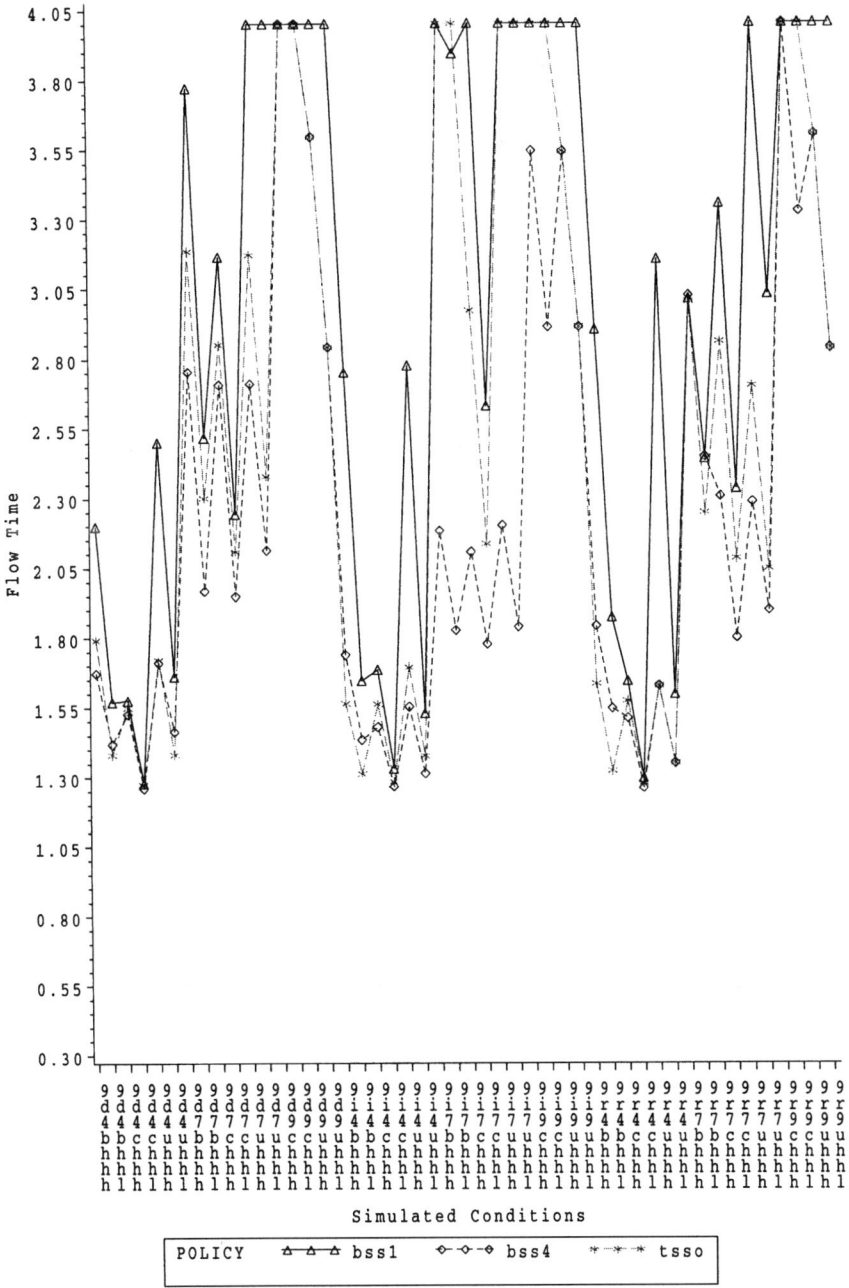

Figure 8. Comparing TSS and variants ($M = 9$, $\sigma_m^2 = H$, $\rho = H$).

for the dependence on the number of workers and machines, it is not surprising that the number of machines and workers interact with the levels of cross training as well. Cross-training levels have the effect of partitioning the flow line into smaller lines, the size of the line being determined by the number of workers and their cross-training level. When there is little overlap between the orbit sets of workers thus limiting the number of machines in their orbits, the intended effect of TB-based policies is better realized than if workers were allowed to range over all the machines in the system (complete cross training).

5.2. Positive Move Times

Impact of Move Times on Policy Performance   The simulation results presented so far have assumed negligible move times between machines. Implicitly, we were assuming that the layout was such that move times could be neglected. Under this scenario, DSS1 outperformed all variants of the DCL policy since the frequent moves required of workers by the DSS1 policy were not penalized. Our general understanding of the two policies would suggest that as move times between machines increased, DSS1 would be a less attractive option than the best DCL variant. We would also expect cross-training levels, processing time patterns, utilization level, and processing time variability to have a significant impact on the performance of DSS1 since these are the factors that limit the movement range of workers. In order to determine the effect of move times, DSS1 and DCL4 (termed MCL4 to denote the fact that move times are not zero) were tested for two levels of move times, one where move times were set to 0.1 times the average batch processing time and another where move times were equal to 0.25 times the average batch processing times. The results confirm our intuition. As move times increase, flow times increase for both policies. However, in the case of the DSS1 policy, the increases are much larger than those for MCL4. In fact, the DSS1 policy is unable to handle the workload in a number of cases. As the number of workers increases, DSS1 performs better since workers are often blocked from moving forward and hence drop off a unit at the next machine and continue backward. The simulation charged a move time delay whenever a worker or unit moved forward or when a worker moved backward. Move times were assumed to be independent of the machines the workers were switching between. The simulations also assume that workers can determine the status of machines they intend to switch to and thus would not try to switch to busy machines. Thus, in the DSS1 policy a worker might begin moving backward at machine 6 and eventually stop only at machine 3. However, the move time delay charged would be for one move only. Similarly, when the mean batch processing times increase along the line, the probability of encountering a worker in the forward direction increases, limiting the forward movement of a worker. With a few workers, complete cross training has a deleterious impact on flow times since not only can workers move over a larger range of machines, these moves are more expensive in terms of time. With high move times, the increase in flow times due to worker absences is the dominant effect and in these cases, the DSS policy usually fails. In contrast, MCL4 is relatively unaffected by these factors. Since workers move only after they have processed a batch completely at a machine, the number of moves made by a worker is relatively low. These findings agree with earlier results for dual resource constrained shops presented in Malhotra and Khen (1994), where the authors found that with significant move times, making the workers available for switching frequently caused the performance of the shop to deteriorate.

6. SOME GENERAL CONCLUSIONS

Some general conclusions can be drawn from the results of the simulation experiments. It is clear that more attention must be given to choice of operating policies as a whole and not just specific aspects of it. This should be apparent from the significant interactions between cross-training levels, the number of workers, processing time variability, and utilization levels. The importance of choosing an appropriate admittance control policy is also demonstrated by the impact of allowable WIP on the performance of policies. In general, low WIP combined with limited cross training tends to be bad. Relaxing the admittance control mitigates the problem to a certain extent. Regardless of the choice of policies, the importance of well-designed cells is illustrated by the importance of processing time variability (a synonym for diverse processing requirements) and the interaction with the pattern of processing times. With negligible move times, transfer-batch-based policies clearly outperform process-batch-based policies with batch flow time as the measurement criteria. Unit transfer batches are feasible as a result of the insignificant move times. When move times are significant, the choice of machines/parts to assign workers to acquires added significance. For example, it may be necessary to limit the orbits of workers explicitly. Transfer batch sizes must also be determined accordingly if workers are used as material handlers as well since large move times tend to starve the remainder of the process batch. Simply having completely cross-trained workers is not enough. Complete cross training must be managed effectively keeping in mind the various interactions between components of policies.

7. SOME TOOLS FOR STUDYING OPERATING POLICIES

As mentioned earlier, the intricate linkage between various components of an operating policy leads to many significant interactions that are often dependent upon the local manufacturing environment. This makes it very difficult to make all-encompassing statements on the structure of *good* operating policies and as a result translate these insights into schedules that can be implemented on a shop floor. Although, it would be nice to generate solutions that specify both the structure of the operating policy and its translation into a schedule (The solution *Worker A processes batch j completely at machines 1 and 2 and partially at machine 3 until it is taken over by by worker B* allows one to infer the operating policy), this is currently not possible with traditional prescriptive tools such as mathematical programming. As a result, insight into the structure of an appropriate policy must be developed locally. This knowledge then serves to provide constraints that the schedule builder must respect in developing schedules. In this section, we will discuss one approach to doing this using appropriately designed simulation models and commercially available schedule builders.

7.1. Simulating Operating Policies Using Object-Oriented Technology

The description of an operating policy (Iyer and Askin, 1998a) can be split into two parts: A static component that describes the relatively time-invariant components of the system state and a dynamic component that describes the evolutionary aspects of the system state. The static component of the state describes the design decisions that act as constraints on the real-time operation of the system such as machine capabilities and tooling, level of worker cross training, and process plans for parts. The dynamic component of the state consists of describing the positions (actual and potential) of workers and parts along with indicators of the status of machines. Thus, in general, operating policies can be described by a 5-tuple $(P,M,O,S,\mathscr{F})$. The sets P and M describe the static attributes of parts and machines, respectively. The set O represents the skill sets of workers. The set of possible dynamic configurations of the system is represented by S. This usually involves representing the states of machines, the composition of their queues, and worker positions. The quantity $\mathscr{F}$ represents the transition function that determines the sequence of system configurations visited during cell operation.

The cell operation protocol can be specified by describing the policies for batch admittance, part routing and batch movement, WIP limits, and worker assignment. The various parameters required to specify the physical and logical system to be modeled are described below. The reader will note that the specification of decision rules is done in terms of other static and dynamic state variables. We use k (worker) or j (part) as a superscript to indicate the principal entity being manipulated, with subscripts referring to other relevant parameters. A superscript of 0 will indicate the dispatch queue to the cell. Conditions are used to indicate the timing or prerequisite states for an action to occur. Rules dictate what action will be taken when the condition is met. Later, we will discuss how a discrete-event general-purpose simulator is obtained from this structured representation of policies.

- AC^0, is the Admittance Condition, which must be satisfied for a batch to be admitted to the cell. For example, in a CONWIP system (Spearman et al., 1990), the AC for a part is that the WIP in the cell be below a specified quantity.

- AR^0, is the Admittance Rule invoked when multiple batches are eligible to enter the cell and one must be chosen. The selection of the particular batch is dependent on the AR being used. One of the simplest rules used is the First-Come, First-Served (FCFS) discipline.

- SC, the set of Switching Conditions. Each element of the set, SC^j_{mpq}, represents the switching condition for the qth unit of a batch of type j at machine m after operation p has been performed. The switching condition for parts represents the condition that must be satisfied in order to invoke the routing policy for the part. The switching condition for parts allows the transfer batch size to be specified.

- SR, the set of Switching Rules. Each element of the set, SR^j_{mpq}, represents the switching rule for unit q of part j at machine m after operation p has been performed. The switching rule specifies the routing policy to be followed for a part at a machine when the conditions representing the switching condition have been realized. Consider a serial system where every operation can be performed on only one machine. Let $M(p^j)$ represent the set of machines that can perform operation p on part j. $\Gamma(p^j)$ is the set of operations that must follow p. The switching rule *route to the machine required for the next operation* can then be written as

$$SR^j_{mpq} : m' | m' \in M(p'^j), \, p' \in \Gamma(p^j)$$

- PSC, the set of Part Switching Conditions for workers. Each element of the set, PSC^k_{mjp}, represents the part switching condition for worker k at machine m performing operation p for part j. This represents the conditions that must hold before a worker is eligible for possible reassignment to another part in the system. In conventional queues, for example, the end of batch processing for the part at the head of the queue represents the part switching condition for the worker. Note that in some cases the switch may be to another batch of the same part type.

- PSR, the set of Part Switching Rules. Each element of the set, PSR^k_{mjp}, represents the part switching rule for worker k at machine m performing operation p for part j. The PSR is invoked when the conditions representing the PSC are true. Conceptually, the part switching rule can be thought of as a function that accepts as input information about the current worker (including current location) and the state of the system and outputs the part to which the worker is to be reassigned. This part then becomes the worker's current responsibility.

- MSC, the set of Machine Switching Conditions. Each element of the set, MSC^k_{mjp}, denotes the machine switching condition for worker k at machine m performing operation p on part j. It represents the conditions that must be satisfied for the worker to be reassigned to another machine. For example, in single-server polling systems where workers switch only after exhausting the queue at the current machine, the machine switching condition for the lone worker would be when the queue at the current machine is cleared.

- MSR, the set of Machine Switching Rules. Each element of the set, MSR^k_{mjp}, denotes the machine switching rule for worker k at machine m performing operation p on part j. It represents the rule used to dispatch workers to machines. In cyclical service systems with M machines and a single server, the MSR at machine n is to switch to machine $(n + 1) \bmod M$.

    With four switching rules in effect, it is entirely possible that at some instant in time all four could be invoked simultaneously. It then becomes necessary to specify the precedence relationships among the rules. We will uniformly assume that the switching rule for parts always takes precedence over the other three and is followed in the hierarchy by the admittance rule. For specific policies, the relationship between the machine and part switching rules for workers will have to be specified. The notation $\gg$ will be used to represent the priority of the rule on the left over the rule on the right.

- z is the limit on WIP in the cell counted in batches.

7.2. Example: Machine-Based Batch Loading with Truncated Shortest Processing Time

As an example of how specific policies can be characterized by assigning values to the policy descriptors, we consider the dual resource constrained system (fewer workers than machines) operating under a traditional batch loading policy with the truncated shortest processing time (SPT) queue discipline at each machine. Under this policy, workers stay at a machine until the queue at the machine is empty and then switch to the idle machine with the longest queue in front of it. Under the truncated SPT policy, high priority is given to the jobs with the shortest processing times unless there are other jobs in the queue that have waited more than a prespecified amount of time. Batches are processed in their entirety before being moved to their next destination. Jobs arrive to the cell and join the queue of the machine at which processing is required first.

The notation used to describe the relevant state variables is explained below. It must be noted that many descriptors of the state of the system are functions of time. The notation that follows supresses this time dependence for ease of explanation. The reader will note that all the framework parameters are described in terms of state variables:

- X_m, the ordered set representing the queue at machine m.
- L_m, length of the queue at machine m.
- Y_m, status indicator for machine m. A value of 0 indicates an idle machine, while 1 indicates a busy machine.
- $\Gamma(p^j)$, the *potential-successor* set for operation p of part j.
- $M(p^j)$, represents the set of machines that can perform operation p on part j, an idle but functional machine.
- s_{mpq}^j, remaining processing time for operation p on unit q of part type j at machine m. Q_j is used to represent the last unit in the batch. S_{mp}^j will be used to represent the remaining *batch* processing time for operation p of part type j at machine m. Thus, $S_{mp}^j = \sum_{q=1}^{Q_j} s_{mpq}^j$.
- W_m^j, waiting time of part j at machine m.

The parameter choices that correspond to this policy are

$$z : \infty, \tag{1}$$

$$AC^0 : \phi, \tag{2}$$

$$AR^0 : \phi, \tag{3}$$

$$SC_{mpq}^j : s_{mpQ_j}^j = 0, \tag{4}$$

$$SR_{mpq}^j : m'|m' \in M(p'^j), \, p' \in \Gamma(p^j), \tag{5}$$

$$PSC_{mjp}^k : s_{mpQ_j}^j = 0, \tag{6}$$

$$PSR_{mjp}^k : \begin{cases} j'|j' = \mathrm{Argmax}_{\hat{j} \in X_m} W_m^{\hat{j}} & \text{if } \exists \hat{j} \in X_m | W_m^{\hat{j}} \geq T, \\ j'|j' = \mathrm{Argmin}_{\hat{j} \in X_m} S_{mp}^{\hat{j}} & \text{else,} \end{cases} \tag{7}$$

$$MSC_{mjp}^k : s_{mpQ_j}^j = 0, \, L_m = 0, \tag{8}$$

$$MSR_{mjp}^k : m'|m' = \mathrm{Argmax}_m L_m, \, Y_{m'} = 0, \tag{9}$$

$$MSR \gg PSR. \tag{10}$$

Equation (1) indicates that the cell is being operated without WIP limits. As a result, there is no need for admittance control (Eqs. (2), and (3). The entire batch is transported together upon completion of batch processing at a machine [Eq. (4)]. Upon completion, batches are sent to a machine that is able to perform one of the operations in the successor set of the current operation [Eq. (5)]. After processing all batches at the current machine [Eq. (8)], workers switch to the idle machine with the longest queue in front of it [Eq. (9)]. Workers are allowed to start a new batch only after the processing an entire batch [Eq. (6)]. The next batch is selected based on the truncated SPT rule [Eq. (7)], that is, switch to the job that has waited the longest if its waiting time is longer than T time units [Eq. (7), part II]. If there are no jobs that have waited more than T time units, switch to the job with the shortest processing time. The machine switching rule takes precedence over the part switching rule. Therefore, if a worker finds the queue empty after processing a batch, the machine switching rule is invoked. More details of the framework and examples of its application are given in Iyer and Askin (1998a).

7.3. Simulating Operating Policies

The framework representation of a policy is well-suited to be the basis for developing simulation models for two reasons. First, the movement of part batches through the system can largely be described by the same event constructs that would be used for the simulation of a network of queues. The superposition of additional rules, which are invoked at *characterizable* (from the definition of switching conditions) points in time, completes the model. Second, the framework definition of policies lends itself naturally to an object-oriented translation that follows the rules of good Object-Oriented (OO) modeling and design (Rumbaugh et al., 1991). First among such principles is the maxim that simulation objects should correspond to typical elements of the manufacturing system such as products, machines, processes, and controllers in an obvious manner. This is in keeping with the general requirement that the object model represent a conceptual model of a class of systems by abstracting features of interest through the use of objects. Specific models of systems must be derivable from this conceptual model. Another principle states that decision objects must be separated from the objects that represent physical elements in the system. Yet another requirement is that the object definitions should allow the system to be modeled at different levels of resolution. This ensures that studying the system at different levels of detail requires minimal refinement.

As an illustration of the earlier assertion that the framework-based representation of policies can be used to develop simulation models, consider the event graph representation of a simulation model for a policy in Figure 9. Event graphs are used to represent the discrete-event simulation model by depicting the events (and associated state changes) as nodes with the arcs and the conditions on them representing the transitions between events. Figure 9 presents a simple event transition diagram to illustrate the idea of superimposing the framework on a regular event transition diagram. The event diagram in Figure 9 differs from the diagram for networks of queues only in the addition of the events and transition arcs represented by dotted lines. The letters of the alphabet on the arcs denote the conditions that must be true for those transitions to be realized. For example, (a) specifies the condition under which a new arrival is admitted to the cell and allowed to join a queue; (g) represents the testing for various switching conditions and event 7 is the action determination event. Though 7 is shown as one node, in reality it is a composition of events such as worker assignment, worker to machine, and part routing. However, the exact relationship between these events depends on the specific policy and hence the superevent representation. The dotted line arcs represent the possible transitions after appropriate actions have been determined. For example, if the end of service occurs at the last machine in a batch's sequence, in limited WIP policies this causes a new batch to enter the cell. In other cases, an end of service triggers a reassignment of workers and thus the beginning of service or setup. In some other cases, some end of service events trigger the cancellation of an event. Depending on the policy being simulated, different transition relationships could be modeled. The program for a discrete-event simulation can therefore be developed easily from the appropriate event transition diagrams. The transition arcs translate into placing an event structure with appropriate values for its various fields onto the future events list. The events and transition arcs would be implemented as methods of classes.

Several authors (Adiga, 1998; Adiga and Glassey, 1988; Glassey and Adiga, 1989; Govindaraj et al., 1990) have argued in favor of using the OO paradigm for developing manufacturing simulation software. Among the reasons cited have been the inability to separate the decision-making aspects of the system from the physical objects being modeled in traditional simulation languages and simulators, and the absence of an adequate reference model for the abstraction upon which the simulator is built. As a result, the abstractions built into the simulator are often inadequate, making rapid model development for complex systems a very difficult task. Also, these tools are not able to take into account the dynamic nature of modern manufacturing systems, forcing the modeler to implement the decision algorithms via the underlying standard programming language. In contrast, the OO paradigm, by its very structure, provides at least a partial reference model and allows the modeler to make a distinction between physical objects (e.g., machines, parts) and the control objects (e.g., operating policy). This provides a mechanism for representing the control system (or operating

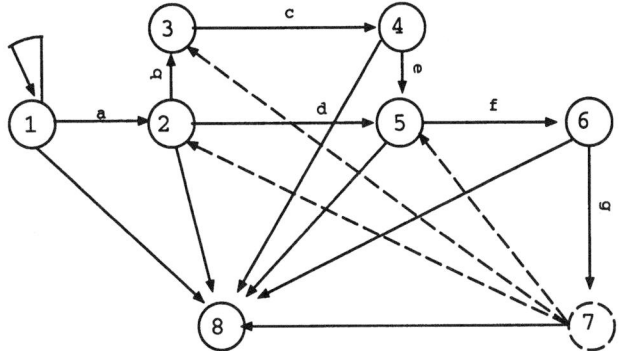

1	New_Arrival
2	Arrive_2_Queue
3	Begin_Setup
4	End_Setup
5	Begin_Service
6	End_Service
7	Det_Actions
8	Accounting

Figure 9. Event graph.

policy, in our case). Stated differently, object-oriented software is more robust as requirements and specifications change since it is based on the underlying framework of the application domain itself rather than the functional requirements of a single problem instance.

One prototype termed COPS (Cell Operating Policy Simulator) has been developed using the framework definition of policies. The reader is referred to Iyer and Askin (1998b) for details on design and implementation. The simulation results reported in this chapter were obtained using this simulator.

7.4. Developing Schedules for Execution

The outcome of a simulation study of the kind described earlier in this chapter should provide managers and planners with some knowledge of the ideal operating protocol.

For example, the simulation study might indicate that a worker should be limited to a certain range of machines or that a specific dispatching criterion works best at the bottleneck machine(s). These rules would be specified as hard or soft constraints to the schedule builder who develops schedules for execution. It must be emphasized that this kind of activity is undertaken infrequently when there is reason to believe that a change in the structure of the operating protocol is needed due to a change in the manufacturing environment.

Even when a number of such rules are known, the solution space for the scheduling problem might be large and, therefore, require a sophisticated schedule generator. Depending on the nature of the scheduling problem, one of several approaches/tools may be used.

- *Mathematical Programming.* The use of mathematical programming models to solve scheduling problems is well documented in the literature. While math programming approaches have worked very well for aggregate planning problems, they do not scale very well when it comes to developing scheduling solutions for large problems with finer granularity. Typically, most scheduling problems have several binary/integer variables that requires prohibitively expensive computational effort. Nonetheless, math programming models work very well for certain classes of problems.

- *Constraint Programming.* Several successful commercially available scheduling packages such as i2 Technologies' Rhythm Optimal Scheduler (i2 Technologies, 1997) use a constraint programming approach to generate scheduling solutions. The main advantage to using these packages is the wide variety of constraints that can be modeled. Constraints such as"if x is scheduled on a resource, every fifth task thereafter must be of type y," which cannot be easily expressed using math programming models, can be incorporated into the model easily. Optimal Scheduler, for example, has its own language with primitives for declaring decision variables, stating constraints, and classifying them as *soft* or *hard*. Constraint programming methods use a combination of domain reduction and local search to arrive at a solution. For more details on these methods, the reader is referred to Russell and Norwig (1995) and some of references mentioned therein.

8. CONCLUSIONS

Success in implementing manufacturing cells is attained by adopting a holistic approach that includes both technological and sociological factors. Consideration of these factors has led to cells with a group of dedicated cross-trained workers. Efficiently operating such cells requires managing and exploiting a variety of interactions some of which are not intuitively obvious. In this chapter, we have presented studies that highlight some significant interactions that determine the efficacy of an operating policy while making a case for a more broad-based approach that avoids myopic optimization. We also discuss the use of tools to help planners develop plans and schedules based on the specifics of their situation.

REFERENCES

Adiga, S. (1988). Software modelling of manufacturing systems: A case for an object-oriented programming approach, *Annals of Operations Research*, **17**, 363–378.

Adiga, S., and Glassey, C. R. (1988). Object-oriented simulation to support research in manufacturing, Technical Report, Engineering Systems Research Center, University of California, Berkeley, CA.

Askin, R. G., and Iyer, A., (1993). A comparison of scheduling philosophies for manufacturing cells, *European Journal of Operational Research*, **69**, 438–449.

Askin, R. G., and Iyer, A., (1994). Strategies for controlling the flow of material through manufacturing cells. Technical Report, Department of Systems and Industrial Engineering, University of Arizona, Tucson, AZ.

Avi-Itzhak, B., Maxwell, W. L., and Miller, L. W. (1965). Queueing with alternating priorities, *Operations Research*, **13**, 306–318.

Baker, K. R., and Pyke, D. F. (1990). Solution procedures for the lot streaming problem, *Decision Sciences*, **21**, 475–491.

Bartholdi III, J. J., and Eisenstein, D. D. (1993). A production line that balances itself, Technical Report, Georgia Institute of Technology, School of Industrial and Systems Engineering, Atlanta.

Bischak, D. P. (1993). Throughput of a manufacturing module with moving workers, Technical Report, Department of Business Administration, University of Alaska, Fairbanks.

Brandon, J. A. (1993). On the vulnerability of programmes of strategic change to functional interests: A case study, *Journal of Strategic Change*, **2**(3), 151–156.

Brandon, J. A. (1995). "Concurrent Engineering and Manufacturing Infrastructure," in *Fifth International Conference on Flexible Automation and Integrated Manufacturing*, R. D. Schraft et al. (Eds.), Begell House, New York.

Brandon, J. A. (1996). *Cellular Manufacturing: Integrating Technology and Management*, Research Studies Press, Taunton, Somerset, England.

Burbidge, J. L. (1963). Production flow analysis, *Production Engineer*, **42**, 742–752.

Buzacott, J. A., and Shanthikumar, J. G. (1992). A general approach for coordinating production in multiple-cell manufacturing systems, *Journal of Production and Operations Management*, **1**(1), 34–52.

Eisenberg, M. (1971). Two queues with changeover times, *Operations Research*, **19**, 386–401.

Eisenberg, M. (1972). Queues with periodic service and changeover times, *Operations Research*, **20**(2), 440–451.

Eisenberg, M. (1978). Two queues with alternating service, *Siam Journal of Applied Mathematics*, **36**(2), 287–303.

Federgruen, A., and Green, L. (1984). An M/G/c queue in which the number of servers required is random, *Journal of Applied Probability*, **21**, 583–601.

Glassey, C. R., and Adiga, S. (1989). Conceptual design of a software object library for simulation of semiconductor manufacturing systems, *Journal of Object-Oriented Programming*, **2**, 39–43.

Glover, F. (1989). Tabu search—Part I, *ORSA Journal of Computing*, **1**, 190–206.

Glover, F. (1990). Tabu search—Part II, *ORSA Journal of Computing*, **2**, 4–32.

Govindaraj, T., McGinnis, L. F., Mitchell, C. M., and Platzman, L. K. (1990). Manufacturing simulation using objects, Technical Report, School of Industrial and Systems Engineering, Georgia Institute of Technology, Atlanta.

Goyal, S. K. (1976). A note on manufacturing cycle time determination for a multi-stage economic production quantity model, *Management Science*, **23**, 332–333.

Hogg, G. L., Phillips, D. T., Maggard, M. J., and Lesso, W. G. (1975a). GERTS QR: A model for multi-resource constrained queueing systems: Part I, *AIIE Transactions*, **7**(2), 89–99.

Hogg, G. L., Phillips, D. T., Maggard, M. J., and Lesso, W. G. (1975b). GERTS QR: A model for multi-resource constrained queueing systems: Part II: An anlaysis of parallel-channel dual-resource constrained queueing systems with homogenous resources, *AIIE Transactions*, **7**(2), 100–109.

Hoopes, B. J., Daniels, R. L., and Mazzola, J. B. (1995). Scheduling parallel manufacturing cells with resource flexibility, Technical Report, Pamplin School of Business, Virginia Polytechnic Institute and State University, Falls Church, VA.

Huang, P. Y., Moore, L. J., and Russell, R. S. (1984). Workload versus scheduling policies in a dual resource constrained job shop, *Computers and Operations Research*, **11**(1), 37–48.

i2 Technologies, (1997). *Rhythm Optimal Scheduler: User Manual*, i2 Technologies, Inc. Cambridge, MA.

Iyer, A., and Askin, R. G. (1998a). A general framework for comparing operating policies in manufacturing cells, *Annals of Operations Research*, **77**, 23–50.

Iyer, A., and Askin, R. G. (1998b). Modeling and simulating operating policies for manufacturing cells, *IIE Transactions*, to appear.

Jacobson, P. A., and Lazowska, E. (1982). Analyzing queueing networks with simultaneous resource possession, *Communications of the ACM*, **25**(2), 142–151.

Mahmoodi, F., and Dooley, K. J. (1991). A comparison of exhaustive and non-exhaustive group scheduling heuristics in a manufacturing cell, *International Journal of Production Research*, **29**(9), 1923–1939.

Malhotra, M. K., and Kher, H. V. (1994). A comparison of worker assignment policies for dual resource constrained shops, *International Journal of Production Research*, **32**(5), 1087–1103.

Morris, J. S., and Tersine, R. J. (1989). A comparison of cell loading practices in group technology, *Journal of Operations Management*, **2**(4), 299–313.

Mosiers, C. T., Elviers, D. A., and Kelly, D. (1984). Analysis of group technology scheduling hueristics, *International Journal of Operations Research*, **22**, 857–875.

Nelson, R. T., (1967). Labor and machine limited production systems, *Management Science*, **13**(9), 648–671.

Nelson, R. T., (1970). A simulation of labor efficiency and centralized assignment in a production model, *Management Science*, **17**(2), B97–B106.

Rumbaugh, J., Blaha, M., Premerlani, W., Eddy, F., and Lorensen, W. (1991). *Objected-Oriented Modeling and Design*, Prentice Hall, Englewood Cliffs, NJ.

Russell, S., and Norwig, P. (1995). *Artificial Intelligence*, Pergamon, New York.

Russell, R. S., Huang, P. Y., and Leu, Y. (1991). A study of labor allocation strategies in cellular manufacturing, *Decision Sciences*, **22**, 594–610.

Silva, E., and Muntz, R. R., (1987). Approximate solutions for a class of non-product form queueing network models, *Performance Evaluation*, **7**, 221–242.

Spearman, M. L., Woodruff, D. L., and Hopp, W. J. (1990). CONWIP: A pull alternative to Kanban, *International Journal of Production Research*, **28**, 879–894.

Srinivasan, M. M. (1988). An approximation for mean waiting times in cyclic server systems with nonexhaustive service, *Performance Evaluation*, **9**, 17–33.

Sundaram, R. M. (1983). "Some Scheduling Rules for a Group Technology Manufacturing System," in *Computer Applications in Production and Engineering*, G. Ackerman (Ed.) pp. 765–772.

Szendrovits, A. Z. (1975). Manufacturing cycle time determination for a multi-stage economic production quantity model, *Management Science*, **22**, 293–308.

Szendrovits, A. Z. (1976). Manufacturing cycle time determination for a multi-stage economic production quantity model—A rejoinder, *Management Science*, **23**, 334–338.

Takacs, L. (1968). Two queues attended by a single server, *Operations Research*, **16**, 639–650.

Vaithianathan, R., and McRoberts, K. L. (1982). On scheduling in a GT environment, *Journal of Manufacturing Systems*, **1**, 149–155.

Weeks, J. K., and Fryer, J. S. (1976). A simulation study of operating policies in a hypothetical dual-constrained job shop, *Management Science*, **22**(12), 1362–1371.

9

BENCHMARKING MEASURES FOR PERFORMANCE ANALYSIS OF CELLS

Barrie G. Dale

Manchester School of Management
UMIST
Manchester, United Kingdom M60 1QD

1. INTRODUCTION

Group Technology (GT) is defined by Dale (1979, p. 125) as "an arrangement of machine tools and services to complete the manufacture of a defined range of similar components, subassemblies or products, generally under the control of one supervisor." Cellular Manufacturing (CM) is a special case or application of GT. The purpose of CM is to change the plant layout from a process-based to a product-based system. Cellular Manufacturing can be, and has been, adopted by many companies in jobbing and batch production so as to enable their production schedules to be achieved using "cells" of purpose-allocated machines, equipment, and operators.

Each cell contains all the machines and equipment required to complete the manufacture of a family of parts, subassemblies, or products. Cellular Manufacturing takes account of the fact that high utilization levels are more important for manpower than they are for machine tools. Consequently, the number of operatives in a cell is generally less than the number of machines. In general, within each cell the manpower can be used very flexibly, thereby reducing the dependence of the company on

Handbook of Cellular Manufacturing Systems, edited by Shahrukh A. Irani
ISBN 0-471-12139-8 © 1999 John Wiley & Sons, Inc.

specialized skills and developing the concept of multiskill teams and team leaders who develop and communicate a vision for the people under their control and then align and inspire them to this common vision. This is in contrast to the traditional supervisory approach where supervisors instruct and tell people what to do. For example, some members of the cell can, and do, operate more than one machine when required. The raw materials are scheduled to the appropriate cell, and loading, sequencing of parts, work flow, progressing, and intermachine transfers are all within the confines of the cell. Parts can be manufactured and inspected within the cell, only leaving it if specialized work needs to be carried out on them, such as painting, plating, testing, heat treatment, or when completed.

How do companies decide whether or not CM is appropriate to their components, products, and mode of operation? What are the probable benefits of using CM in preference to more conventional forms of organization? How do organizations predict the likely benefits should CM be introduced? What problems are likely to be faced? These are typical questions raised when the concept of CM is considered within an organization.

The most popular method of obtaining this type of data is a comparison with companies that are already using CM and are manufacturing similar products using similar equipment. This generally involves visits to these users of CM and holding discussions on a range of factors with management, engineers, production control staff, operators, and others. Some companies making products for which there is no direct comparison often argue that this sort of evaluation is of little value to them.

The many studies on CM outlining successful applications (e.g., Dale and Russell, 1983; Mosier and Taube, 1985; Burgess et al. 1993) and describing methods of cell design (e.g., Burbidge, 1982; Green and Sadowski, 1984) are also used as a basis for comparative analysis. Published material, in particular, those reporting benefits are often the principal reasons for organizations taking an initial interest in CM. An existing management experience of operating CM elsewhere can also be quite useful in convincing a company's management of its potential for CM. If a senior manager or director can say "This is how I operated CM in my last company—if it worked for them it can work for us," this is a very powerful argument.

Only a limited amount of work (e.g., Leonard and Rathmill, 1977a, 1977b; Dale and Willey, 1980; Burgess et al. 1993) has been done on providing general methods to help management decide whether it can benefit from using CM. Most companies, before deciding on such a big system change as CM involves, would like to have an indication of the benefits that the change is likely to bring. At a time when capital is scarce, committing a company to a major reorganization involving considerable expenditure is an act of faith of senior management if no guide and indication as to the expected results is available.

The method outlined in this chapter, which has been developed from a major research study (Dale, 1979), enables a company planning to introduce CM to consult the accumulated experience of other companies, and to determine for its particular characteristics the expected amount of improvement it might obtain from the introduction of CM.

2. DATA COLLECTION AND ANALYSIS

Data were collected on a before and after basis from 35 companies who have introduced CM. The sources of data accessed within each company included records, documents, and databases from the departments of Production Planning and Control, Stock Control and Storage, Dispatch, Production and Operations, Sales and Marketing, and Accounts. The time period over which the data was collected depended on the time that the organization had been operating CM concepts. The "Before CM" data reflected the last financial year before the reorganization to CM and the "After CM" data refers to the latest financial period. The type of questions used to collect data from each company is given in the Appendix. The questions were designed to cover all types of companies and manufacturing situations, some being applicable to particular companies, but not to others. The data was collected by the author during field visits to companies participating in the research.

This data has been measured for 42 different factors and ratios. A number of companies preferred to present information as ratios in order to ensure anonymity and to prevent absolute comparisons. Ratios, of course, eliminate the scale effects of the data collected from companies of differing sizes. Analyzing and measuring manufacturing performance tends to be easier with ratios and to produce more directly comparable results than absolute figures. Where financial data was requested, companies also supplied further information that enabled pre-CM and post-CM figures to be compared on a constant value basis independent of the effects of inflation.

An example of the type of data collected is shown in Tables 1 and 2. The average values of the 42 factors and ratios before and after the introduction of CM, along with the average percentage change, is shown in Table 3.

As a tool to help management determine whether or not its company had potential for CM, it was decided, after consultation with a number of industrialists, to present the information using a series of graphs. These graphs illustrate the relationship between the values of factors and ratios before the introduction of CM and the percentage changes in the same factors and ratios after CM had been introduced. An example of these graphs is shown in Figure 1. From an investigation using linear regression and correlation analysis, it was found that 17 out of the 42 relationships between the values of factors and ratios before CM were significantly correlated with their change at the 95% level of confidence. Linear regression and correlation analysis were used in the first instance since they were the simplest methods of handling and testing the data and, also more important, because it was thought to be more readily understood by potential users. This latter point was confirmed by validation and prototyping of the predictive method. A further examination of the data using curvilinear and exponential functions was found to represent the data no better than the linear relationships.

Using these relationships, including those that are not significantly correlated, any company planning to introduce CM can determine the expected amount of improvement it might obtain if, while certainly being different from all the other

Table 1. Actual product lead time (weeks) through manufacturing areas

Value before CM	Value after CM	Percentage Reduction after the Introduction of CM
3.0	2.0	33.3
3.1	2.1	32.3
3.9	2.0	48.7
5.0	4.0	20.0
5.7	2.2	61.4
6.0	1.2	80.0
7.0	3.0	57.1
8.0	5.0	37.5
8.0	2.9	63.8
8.0	4.0	50.0
8.0	4.0	50.0
8.2	4.1	50.0
8.6	3.4	60.5
8.7	4.4	49.4
9.0	4.0	55.6
9.6	4.5	53.1
11.9	4.3	63.9
12.0	2.0	83.3
12.0	4.0	66.6
12.0	4.0	66.6
12.0	5.0	58.3
12.0	8.0	33.3
12.3	6.4	48.0
13.0	7.0	46.2
13.1	4.3	67.4
15.0	4.0	73.3
15.0	6.0	60.0
20.5	10.2	50.2
25.6	14.6	43.0
26.0	3.1	88.1
39.4	13.2	66.5

companies, it nevertheless fits somewhere into the general pattern of companies that have previously adopted CM. Thus, a company planning to introduce CM is able to consult the accumulated experiences of other companies and to determine for its own particular characteristics the expected amount of improvement it might obtain from CM.

3. PREDICTING THE BENEFITS OF CELLULAR MANUFACTURING

Predictions have been carried out for 12 companies that wished to assess their potential for CM. These companies wanted answers to the following two questions:

Table 2. Proportion of orders delivered on time

Value before CM (%)	Value after CM (%)	Percentage Increase after the Introduction of CM
30.0	60.0	100.0
30.0	70.0	133.3
30.0	70.6	135.3
35.0	60.0	71.4
37.5	77.5	106.7
40.0	83.0	107.5
40.0	90.0	125.0
50.0	80.0	60.0
62.0	84.0	35.5
65.0	90.0	38.6
68.0	84.0	23.5
70.0	70.0	0
70.0	85.0	21.6
75.0	90.0	20.0
75.0	96.0	28.0
75.0	85.0	13.3
80.0	99.0	23.8

- Would a CM system be more efficient and beneficial than their existing functional layout?
- Could independent research support the detailed internal analysis already carried out for CM in terms of the potential benefits?

Both these aspects were found to be especially important when proposals for CM were being presented at the board level, even if long-term plans had not, at that stage, been decided upon. In one company, proposals for CM had been sent to the parent company in the United States for approval and had been returned by the American corporation with a request to justify the claims made.

In order for companies to be advised as to their CM potential, they had to provide their own company profile under the existing manufacturing system and on the product lines they were considering for CM organizations. For example, in a large multinational organization involved in the manufacture of capital equipment, the decision was taken to base the prediction for CM on the existing gears and shafts used in a range of their British products. Three main classes of data are required by the predictive tool; the full range of information is indicated in the Appendix:

- Financial information [e.g., value of annual sales, value of outstanding orders, value of work in progress (including semifinished stock), value of average monthly period of production completed, value of production scrap, rate of buying raw materials, value of tooling, jigs, fixtures, etc.]

Table 3. Specific factors and ratios measured before and after the introduction of CM

Factor or Ratio Measured	Average Value before CM	Average Value after CM	Number Pairs of Data	Average Percentage Change
Actual product lead time (weeks) from order receipt to dispatch	22.1	14.0	18	-41.8
Theoretical product lead time (weeks) from order receipt to dispatch	16.3	9.8	15	-40.4
Actual product lead time (weeks) through manufacturing areas	11.7	4.8	31	-55.4
Theoretical product lead time (weeks) through manufacturing areas	10.4	4.7	17	-50.9
Average number of operations per component	8.2	7.0	27	-11.4
Average total time (hours) spent on machines or equipment (including setting and processing per batch)	35.5	28.3	20	-15.2
Number of manufacturing employees	143.3	127.4	16	-4.1
Number of manufacturing supervisory staff	9.4	8.4	14	-6.2
Number of production control staff	9.5	7.2	13	-15.1
Number of progress chasers	3.9	1.7	14	-41.6
Number of production control staff including progress chasers	16.3	12.1	15	-21.9
Number of inspectors	13.8	8.8	12	-11.3
Proportion of orders delivered on time	54.9%	80.8%	17	+61.4
Average lateness (weeks) of late orders	5.6	2.4	17	-53.8
Average manufacturing batch size	1708.2	1697.4	24	-0.67
Average range of manufacturing batch sizes	38 645.6	38 644.9	20	-2.2
Number of different components manufactured	12 491.9	12 180.8	18	-3.3
Average batch traveling distance (meters)	713.9	130.1	17	-79.1
Setting time (hours) per batch	7.1	5.8	17	-16.5
Percentage scrap rate	5.7%	3.6%	18	-32.3
Average machine utilization	58.5%	63.3%	7	+16.7
Average labor utilization	83.2%	93.8%	5	+14.2
Labor turnover rate	9.1%	8.8%	3	-2.0
Labor absenteeism rate	6.2%	3.5%	4	-31.0
Value of annual sales to value of WIP (including semifinished stock)	3.8	7.5	14	+161.0
Value of annual sales to value of finished goods, part stocks, and subassemblies	3.3	4.9	11	+58.8
Value of annual sales to value of raw material stock	10.7	20.7	8	+100.2
Value of finished goods, part stocks, and subassemblies to value of average monthly period of production completed	4.6	3.0	9	-30.5

Table 3 Continued

Factor or Ratio Measured	Average Value before CM	Average Value after CM	Number Pairs of Data	Average Percentage Change
Value of raw material stock to rate of buying raw materials (value per year)	0.63	0.44	5	-33.1
Value of WIP (including semifinished stock) to value of average monthly period of production completed	4.2	2.1	9	-42.8
Value of average monthly period of production completed to value of monthly production scrap	22.8	40.7	8	+58.4
Actual product lead time (from order receipt to dispatch) to theoretical product lead time (from order receipt to dispatch)	1.12	1.02	17	-7.7
Actual product lead time through manufacturing areas to theoretical product lead time through manufacturing areas	1.18	1.02	18	-9.6
Actual product lead time through manufacturing areas to actual product lead time from order receipt to dispatch	0.61	0.50	20	-24.3
Actual product lead time (weeks) through manufacturing areas to average number of operations per component	1.47	0.71	28	-52.3
Actual product lead time through manufacturing areas to average total time spent on machines and equipment per batch	21.4	9.3	21	-50.8
Number of manufacturing employees to number of supervisory staff	16.5	15.8	16	-0.43
Number of manufacturing employees to number of production control staff	23.7	24.4	13	+8.5
Number of manufacturing employees to number of production control staff including progress chasers	14.4	17.1	14	+23.8
Number of production control staff to number of progress chasers	2.6	4.1	13	+48.7
Number of manufacturing employees to number of inspectors	19.4	19.1	13	+22.5
Value of outstanding orders to value of annual sales	0.1075	0.048	12	-60.2

- Production information (e.g., actual and theoretical product lead time through the manufacturing areas, proportion of orders delivered on time, average and range of lateness of late orders, average and range of batch sizes manufactured, setting time, average batch traveling distance, machine and labor utilization)

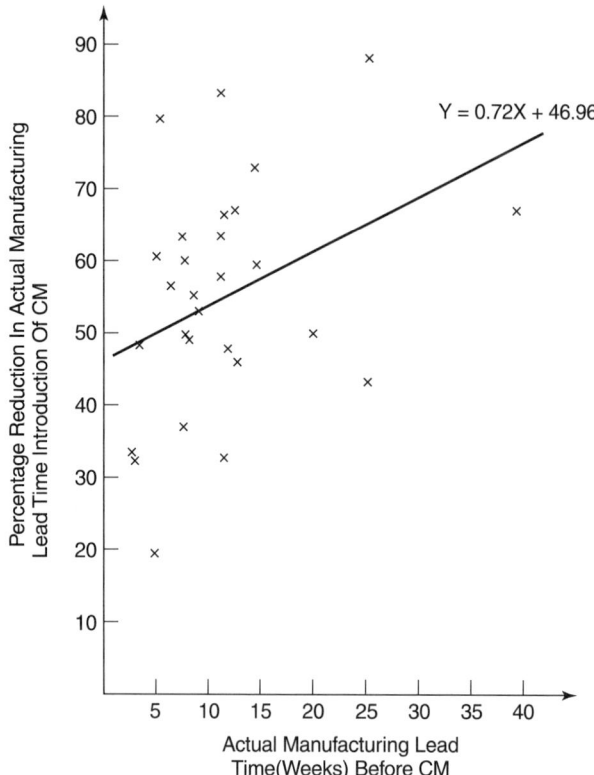

Figure 1. Actual product lead time through manufacturing areas.

- Staffing levels (e.g., number of manufacturing employees, supervisors, production control staff, progress chasers, inspectors)

Each individual piece of data provided by a company was applied to a graph of previously accumulated results so that the likely change could be calculated from the linear regression lines of previous data. For example, in Figure 2 it is seen that a company with an average delivery performance of 50% before CM, might have been expected to obtain an average increase in the percentage of orders delivered on time of 72%, which would be an improved on-time delivery performance of 86% after adopting CM, providing that the company conformed with the general pattern of companies that have previously adopted CM.

When this exercise was carried out for each piece of data, the company could obtain a useful indication of what it was likely to achieve as improvements and which aspects were likely to show no change, or even adverse change after the introduction of CM. It was not necessary for a company to provide data on each of the 42 relationships in order to predict the likely changes, since particular relationships might not have been applicable to the company in question.

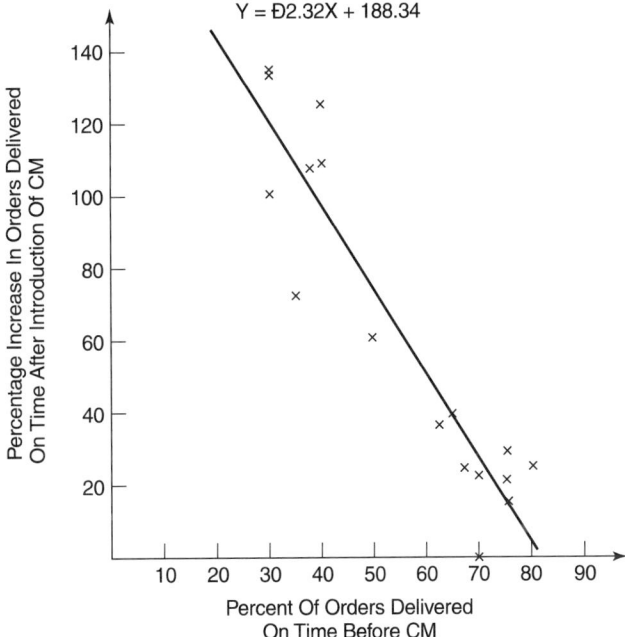

Figure 2. Orders delivered on time.

This prediction was carried out for 12 companies, and Table 4 shows samples of the predicted results. It is not possible to show a complete profile of just one company for reasons of confidentiality, so the sample information in Table 4 is taken from a selection of these 12 companies. The predictions were presented to each of the companies using the format shown in Table 4.

It can be seen that if columns 1 and 2 are compared, they show how similar the manufacturing characteristics and performance of one company are to those of other companies before introducing CM. Column 2 can be considered as the profile of the "average company" that has successfully applied CM. Therefore, when comparing a company with this profile, it is possible to see how near it is to the average company and also determine, before CM, in which areas it performed better or worse than the average company. Comparing columns 3 and 5 in Table 4 the company potential for improvement under CM can be seen, with reference to the average improvement. At the request of many of the companies investigated, column 4 was included. This shows the range, based on the 95% confidence interval, within which one may be 95% confident that the change in each factor or ratio will lie. Boards of companies are naturally interested in the range within which future results will lie and tend to use this information as a basis for decision making and planning.

In general, when these two comparisons are made, it is possible to see exactly where a company is more likely to obtain the greatest benefits from CM. For example, if using this method it were predicted that the company could make a substantial reduction in

Table 4. Measuring a company's potential for CM

1	2	3	4	5
Company A before CM	Average of all other companies before CM	Average Percentage Change in Companies with CM	Range of percentage change in these companies	Likely change after CM for company A
Actual product lead time through the manufacturing areas = 12 weeks	Actual product lead time through the manufacturing areas = 11.9 weeks	A reduction of 60.2%. If the company achieved this average change, the actual manufacturing lead time would be reduced to 6 weeks under CM.	A reduction of 90.4% to a reduction of 30.2%. Based on these average figures, the actual manufacturing lead time under CM could be expected to be between 1.15 and 8.4 weeks.	A reduction of 59.5%. The actual product lead time through the manufacturing areas would be reduced to 6.05 weeks.
The proportion of orders delivered on time = 75%	The proportion of orders delivered on time = 71.4%	An increase of 60.5%. If the company achieved this average change, the proportion of orders delivered on time would be increased to 100% under CM.	An increase of 149.3% to a reduction of 28.3%. Based on these average figures, the proportion of orders delivered on time under CM could be expected to be between 100 and 54%.	An increase of 14.5%. The proportion of orders delivered on time would be increased to 85.9%.
Machine setting time per batch = 6 hours	Machine setting time per batch = 9 hours	A reduction of 17.3%. If the company achieved this average change, the machine setting time per batch would be reduced to 5 hours under CM.	A reduction of 46.7% to an increase of 12.1%. Based on these average figures the machine setting time per batch under CM could be expected to be between 3.2 and 6.7 hours.	A reduction of 11.8%. The machine setting time would be reduced to 5.3 hours.

Value of annual sales/Value of WIP (including semifinished stock) = 6.49	Value of annual sales/Value of WIP (including semifinished stock) = 3.5	An increase of 174.4%. If the company achieved this average change, the value of this ratio would be increased to 17.8.	An increase of 564% to a reduction of 216%. Based on these average figures, the value of this ratio under CM could be expected to be between 43 and 0.	No change in the value of this ratio.
Value of annual sales/Value of finished goods, part stocks, and subassemblies = 5.87	Value of annual sales/Value of finished goods, part stocks, and subassemblies = 3.3	An increase of 56.5%. If the company achieved this average change, the value of this ratio would be increased to 9.2 under CM.	An increase of 121.3% to a reduction 8.3%. Based on these average figures the value of this ratio under CM could be expected to be between 12.9 and 5.38.	No change in the value of this ratio.
Actual product lead time through the manufacturing areas/Average number of operations during manufacture per component = 1.5 weeks/operation	Actual product lead time through the manufacturing areas/Average number of operations during manufacture per component = 1.5 weeks/operation	A reduction of 59.5%. If the company achieved this average change, the value of this ratio would be reduced to 0.61 weeks/operation under CM.	A reduction of 92.5% to a reduction of 25.5%. Based on these average figures, the value of this ratio under CM could be expected to be between 0.11 and 112.	A reduction of 59.5%. The value of this ratio would be reduced to 0.61 weeks/operation.
Number of manufacturing employees/Number of production control staff = 13.8	Number of manufacturing employees/ Number of production control staff = 21.8	An increase of 11.2%. If the company achieved this average change, the value of this ratio would be reduced to 15.3 under CM.	An increase of 47.64% to a reduction of 25.24%. Based on these average figures, the value of this ratio under CM could be expected to be between 20.3 and 10.3.	An increase of 15.2%. The value of this ratio would be increased to 15.9.

the number of production control staff with CM, the prediction would be important. First, it would act as an indication of potential and second as a warning given early enough for the company to initiate retraining schemes for the staff to do other jobs, or to start negotiations on the issues of organizational restructuring.

When these predictions were carried out, certain comments were made, drawing attention to factors that could affect the predictions made. Some typical comments follow:

- The machine tools to be transferred from the partner company in Germany could affect these predictions.
- The predicted reduction in manufacturing employees would be in direct conflict with the company's present manpower planning policies.
- The prediction that machine setting time will be reduced under CM is affected by the machine type population and production control policies.
- The prediction about changes in the staffing of the production control function will be influenced by the development of the Manufacturing Resources Planning (MRP II) system.
- In collecting the raw data for the model, when a company carried out an assembly operation as well as component processing, the average product lead time before CM included the assembly lead time. This is why the average product lead time in other companies is more than twice its own lead time average.
- The reduction in the average number of manufacturing operations per component is dependent on the machine tools used and on production engineering expertise.
- The lowest percentage internal scrap rate before the introduction of CM in any of the companies surveyed was 0.78%. It can be seen that the existing scrap rate is much lower than this figure, indicating that the company's approach to continuous improvement is progressive.
- The greatest proportion of orders delivered on time before the introduction of CM in any of the 35 companies surveyed was 80%. The unique manufacturing situation of supplying the internal assembly line with gears and shafts has resulted in an excellent delivery performance. Under CM it is not envisaged that this situation will improve, even though the model has predicted a small increase.

When the predictions had been carried out for each of the 12 companies, improvements were as expected for most of the factors and ratios. On average, there were only four factors or ratios where no improvement was predicted. These were staffing levels and ratios, number of operations per component, the ratio of value of annual sales to value of work in progress (including semifinished stock), and the ratio of value of annual sales to value of finished goods, part stocks, and subassemblies.

In four of the companies for whom a prediction was made, CM was implemented and new systems have been in operation for sufficient time to enable effective comparisons to be made between the changes as predicted and those actually achieved. The results for the four companies in relation to the predicted values compared to achieved results are shown in Table 5.

Table 5. Results for four companies in relation to predicted values

Company	Average Percentage Error in Prediction		
	Overestimated	Underestimated	Overall Trend
1	11.2%	41.2%	23.8% underestimated
2	17.7%	24.2%	11.8% underestimated
3	17.0%	38.0%	8.7% underestimated
4	24.1%	22.6%	7.3% underestimated
			Average error 13.7% underestimated

When the average error of prediction for each factor and ratio was examined, it was found as expected that the error of prediction was least in those relationships that were significantly correlated at the 95% confidence level. Even for those relationships that were not strongly correlated, the error of prediction was not excessive, and these relationships may still be used with some confidence by companies wishing to assess their suitability for CM. Where this sort of comparison has been made, the resulting average error in the predicted values compared to achieved results has been 13.7% underestimated.

4. FACTORS THAT INFLUENCE THE SUCCESS OF CELLULAR MANUFACTURING

A number of questions have been raised by various researchers (e.g., King and Nakornchai, 1982; Hyer, 1984) and practitioners about the types of products and manufacturing situations best suited to CM. Companies will often ask how does manufacturing system complexity, average (and range of) batch size, component variety, number of machine tools, and size of the organization affect achievement of the benefits of CM? An investigation has been carried out to determine the effects of these characteristics by comparing percentage changes in the 42 ratios and factors after companies have reorganized along CM lines. By using the operating experiences and performances of companies from which data has been collected, it is possible to discover something about the industrial environment that is most suitable for CM, serving as a guide to other potential users of CM.

4.1. Complexity of Manufacturing Systems

In order to study how manufacturing complexity affected the success of CM, four categories of manufacturing systems complexity were decided upon, as follows:

Type 1. Single-piece parts, relatively simple to manufacture (e.g., brake linings, twist drills). Twenty percent of companies investigated fell into this category.

Type 2. Single-piece parts, relatively complex to manufacture (e.g., frames, castings). Three percent of companies investigated fell into this category.

Type 3. End products that are simple assemblies (e.g., valves, pumps). Forty percent of companies investigated fell into this category.

Type 4. End products that are complex assemblies (e.g., machine tools, packaging machinery). Thirty-seven percent of companies investigated fell into this category.

Using these four categories of manufacturing system complexity the factors and ratios were analyzed on a before and after CM basis. The greatest changes in the factors and ratios were found most often for type 4 manufacturing system complexity, that is, on 39% of the occasions. The largest changes occurred next in type 1, for 25% of the ratios and factors, followed by types 2 and 3 with 18% each. Types 1, 2, 3, and 4 registered the smallest percentage changes in the ratios and factors in 11, 39, 32, and 18%, respectively, of the values measured. This evidence suggests that the greatest benefits of CM are obtained in companies manufacturing either very simple, single-piece products or those manufacturing parts going into complex assemblies of end products. Two explanations can be advanced for these findings, which are discussed in the following sections.

4.2. The More Complex the Product, the Greater the Benefits of Cellular Manufacturing

In a functional layout system, more effort is required from the manufacturing and associated control systems to produce complex products than simple ones. The main reasons for this are that the more complex products need a larger number of manufacturing operations and the batches of components visit more machines and departments. Numerous transfers of responsibility occur during the manufacturing process, and management and production control is both difficult and complex. The lead times, both in manufacture and from order receipt to dispatch, are large, resulting in high levels of Work-In-Process (WIP) and semifinished stocks. Production planning is often overtaken by changes in the situation.

With a reorganization to CM, these are just the type of features that tend to show improvements. With more simple products the same problem exists, but at a much lower level. The companies manufacturing simple single-piece products have, in general, the most favorable factors and ratios indicating greater efficiency before the introduction of CM. Thus, it seems that the introduction of CM brings greater benefits to those companies that are manufacturing more complex products. It is much easier, obviously, to improve delivery performance by 50% if the current level of proportion of orders delivered on time stands at 30% than to improve it by 50% if it is already at 60%.

4.3. The More Simple the Product, the Greater the Potential for Cellular Manufacturing

With a simple product there is more chance of reorganizing the existing functional layout into a total CM system. There are, however, exceptions to this. Where the product is simple, component families and associated groups of machines and equipment may be easier to identify. The components generally require fewer manufacturing operations, visit fewer departments, and are loaded onto fewer machines. There is also a tendency for simple products to be produced in larger batch quantities and the product and component variety tend to be less. So, with the introduction of CM, the greatest benefits occur in companies manufacturing simpler products, perhaps because a total CM system is easier to introduce, and the total benefits are correspondingly easier to achieve. It has been suggested by writers such as Malik et al. (1973), Edwards (1974), Burbidge (1975), and Burgess et al. (1993) that the full benefits of CM will only be achieved by those companies who attain a total CM system. The extent to which CM is able to replace functional layout systems is, therefore, likely to be a factor affecting success, and simpler products generally allow more complete conversion of the manufacturing system.

4.4. Component Variety

The relationship between component variety and percentage changes in the factors and ratios after the introduction of CM was examined. It was found that only 4 out of the 38 relationships tested were significantly correlated with their change after introducing CM at the 95% level of confidence. This finding would indicate that component variety is not as important to the success of CM as outlined by researchers such as Burgess et al. (1993).

It was found that about half of the factors and ratios measured under CM had exhibited the largest changes when associated with high levels of component variety, while the other half, which showed the largest changes, were for lower levels of varieties. There was no particular pattern between high and low variety, although the four relationships that were significant exhibited the greater changes in the factors and ratios at the higher levels of component variety. From discussions with industrialists, and from practical experiences of CM, component variety certainly is very important in cell design. In general, the lower the component variety, the easier it is to identify component families and to design manufacturing cells. With a high level of component variety it is generally found that the cell size, in terms of number of machines, has to be made larger in order to cater for the variations in product mix associated with higher levels of component variety.

4.5. Average Batch Size

The relationship between average batch size and percentage change in factors and ratios after the introduction of CM was investigated. There was not a single pair of data that showed significant correlation at the 95% level of confidence. Average batch size, therefore, cannot be shown, on the basis of the collected data, to be a mechanism that influenced the change to CM. Furthermore, there does not seem to be confirmable limits of average batch size that predetermine the application and success of CM in companies.

It was found that 55% of the larger changes in the factors and ratios were achieved in companies with small batch sizes and 45% by companies with the larger batch sizes. This is logical, since CM is applicable to companies with small average batch sizes. It is assumed that companies with larger batches would be operating at higher levels of efficiency. Therefore, if these types of companies were to apply CM, the benefits would be relatively lower than those manufacturing smaller batches.

The relationship between average range of batch sizes manufactured and percentage change in factors and ratios after introduction of CM was also investigated. Again, there was not a single pair of data that exhibited a significant correlation, and so it can be concluded that variations in batch sizes manufactured did not seem to be important mechanisms that influenced the changes that took place when companies introduced CM. There is also no observable limit to the range of batch sizes manufactured that predetermined the application or success of CM in companies.

The percentage size distribution was such that 66% of the larger changes in the factors and ratios were achieved in companies with smaller batch size ranges and 33% by companies with larger range of batch sizes. The same argument as applied to average batch sizes also applies here.

4.6. Number of Machine Tools in the Functional Layout

The relationship between the number of machine tools and the percentage change in the factors and ratios after the introduction of CM was examined. It was found that only one of the relationships tested was significantly correlated with its change after introducing CM at the 95% level of confidence. This indicates that the number of machine tools is not important to the success of CM.

Also of interest is whether, on average, the greatest percentage changes in the factors and ratios occur in companies with large or small numbers of machine tools. Using this method, 61% of the greater changes in the factors and ratios measured were found to occur in the companies with small numbers of machine tools, the remaining 39% being in companies with more machine tools. There seems to be no clear pattern in this distribution when the percentage changes are analyzed; but based on the collected data, companies with a smaller number of machines in a functional layout appear to be more likely to achieve the greatest changes in certain factors and ratios when they introduce CM. Companies with small numbers of machines and equipment in the functional layout system would, in general, find it easier to rearrange a greater proportion, or even all, of their production facilities along CM lines, thus achieving the greatest benefits. Companies with larger numbers of machine tools certainly find it much more difficult to rearrange all their production facilities along CM lines and so are more likely to be left operating a CM and a functional system in parallel. These companies tend to introduce one or two cells and go no further.

There are a number of reasons why companies with a large number of machines in the functional layout may find it more difficult to achieve total CM.

- It is more difficult to identify component families and associated machine groups where the population of components and machines is larger.
- The greater the number of machines present, the greater the number of CM cells that need to be created. The assumption is often made that the more cells required, the more additional machinery is likely to be needed. The prospect of additional capital expenditure on plant and machinery can act as a disincentive.
- In large organizations there tends to be greater resistance to change. It is much more difficult to convince every member of a large management team about the potential benefits of CM. In large organizations there are likely to be many different ideas of what CM entails. Burbidge (1975) suggests that the greater resistance to introducing CM into a company comes not from the shop floor but from management.

In a functional layout with a smaller number of machine tools, there already exists some of the benefits of CM such as small integrated group working, with the consequential social and production advantages of flexibility of labor. This should result in faster manufacturing lead times, lower levels of stock and WIP, and good delivery performances. Intuitively, there must be a limit to the number of machines above which the effect of small size no longer contributes. Consequently, when companies with few machines introduce CM, it might be thought that the benefits would not be as great as in companies having larger numbers of machines. A law of diminishing returns may operate, but the facts arising from this investigation show the contrary to be true.

4.7. Company Size

In their studies of organizational structure, Hickson et al. (1970) concluded that the features of organizations they tested and that were closely associated with the size of the organizations, as measured by the number of employees, showed that the smaller organizations were more affected by "technology" because everyone was "closer to the shop floor." This contrasts with the larger organizations, where technology affected only the organizational characteristics centered on work flow, such as the number of employees in technological functions. Introducing CM into a company causes changes in the manufacturing system from small batch manufacture to quasi-large-batch manufacture. It is, therefore, of interest to discover if smaller companies benefit more from the introduction of CM than larger ones.

The reasoning given earlier, outlining why a company with a small number of machine tools would benefit more from CM than a company with a larger number, also applies here since greater numbers of machine tools are usually present in the larger companies. The relationship between the company size before CM and percentage changes in the factors and ratios after the introduction of CM was studied. None of the factors were significantly correlated with the changes after CM at the 95% level of confidence. There was also no definite pattern in the percentage change distribution. Just over 50% of the larger changes in the factors and ratios were achieved in smaller size companies, and just under 50% by the larger size companies. So this evidence indicates that company size is not an important criterion for the success of CM.

4.8. Which Are the Most Important Variables?

Correlation testing was carried out for component variety, average batch size, average batch size range, number of machine tools, and company size against percentage changes in each of the factors and ratios measured. For these five variables it was found, respectively, that 4, 0, 0, 1, and 0 of the changes in the factors and ratios examined were found to be significantly correlated at the 95% level of confidence, with these five variables. Based on this analysis, none of these five variables can be considered to be either a constraint or a determinant of the success of CM.

Which of these five variables was most strongly correlated with changes in the 38 factors and ratios after introducing CM? This was tested by observing which five variables exhibited the largest algebraic values of correlation coefficient with the changes in factors and ratios examined. For 39% of the "after CM" changes in the factors and ratios, component variety was most strongly correlated, followed by the number of machine tools (25%), company size (15%), average batch size (12%), and batch size range (9%). The average correlation coefficient calculated from the 38 relationships measured was strongest for component variety (0.39) followed by a number of machine tools and company size (0.32), then average batch size and range (0.26). From this component, variety does seem to be more important to the success of CM than the other variables tested.

5. CONCLUSIONS

The benefits surrounding CM as reported in publications is the key factor that tends to convince Boards of Directors of the need to introduce the concept. The analysis described in this chapter has attempted to measure data before and after the introduction of CM, in order to study the changes that took place. Forty-two relationships have been established between a range of management factors and ratios before CM and their percentage changes after CM had been introduced. Seventeen of these relationships are significantly correlated at the 95% confidence level.

The research findings suggest that a number of companies that have implemented CM are not using it to best advantage. Companies that have already implemented CM are advised to consult these relationships in order to assess if they have achieved the benefits that have been gained by other organizations.

Actual and theoretical manufacturing lead times before CM and the percentage change in manufacturing lead times after CM were not expected to be correlated. The reason for this was that some data was obtained from companies operating only one or two cells; these results may have distorted the overall picture. Isolated cells in a functional organization may be regarded as operating under "artificial" CM conditions. The fact that these two relationships are significantly correlated indicates the potential that CM has for reducing manufacturing lead times regardless of the number of cells that are in operation.

It is possible to predict with some accuracy the changes that will take place when CM is implemented. The average error of prediction in the predicted values compared to achieved results has been 13.7% underestimated. Even those relationships, which are not strongly correlated, may still be used by companies wishing to assess their suitability for CM and may help to predict the likely changes that may take place if they reorganize along CM lines. The error of prediction is, in general, least for those relationships that are significantly correlated at the 95% level of confidence. As more data is collected and added to the data bank of CM case histories, the predictions will become more accurate.

From the study a tentative description of a company likely to achieve the greatest benefit from CM is as follows:

- Manufactures either a very simple single-piece product or parts going into a complex assembled product.
- Average batch sizes and the batch size range are both relatively small.
- The company is a relatively small organization with relatively small numbers of machine tools and manufacturing equipment.
- The company is not typified by either large or small component variety.

It is hoped that this chapter will encourage more applications-orientated research in Cellular Manufacturing.

APPENDIX: CM SURVEY SPECIMEN QUESTIONNAIRE

Part 1

(a) What event(s) or circumstances caused your company to become interested in CM principles?

(b) What improvements did you expect from CM?

(c) How did you apply CM?

(d) What were the main factors and considerations that convinced your Board of Directors of the potential benefits of CM?

(e) At the time you were thinking about CM, would it have helped to have been able to foresee the benefits arising from and the costs of introducing a CM system?

(f) If the answer to question (e) is *YES*, in what form would a "predictive indicator" of your company's suitability for CM have been most useful?

Part 2(a)

(The data requested is required in the before and after CM situation).

1. *Stock Turnover*
   (a) Value of annual sales
   (b) Value of work in progress (including semifinished stock)
   (c) Value of stock

(i) Finished parts   (ii) Finished goods

(iii) Subassemblies   (iv) Raw materials
   (d) Value of average (daily*/weekly*/monthly*) period of production completed (*Please state period)
   (e) Rate of buying raw materials (value per week*/month*/year*) (*Please state period)
   (f) Value of tooling, jigs, fixtures, etc.
   (g) Value of output of toolroom and purchase of tooling (value per week*/month*/year*) (*Please state period)
   (h) Value of production scrap (value per week*/month*/year*) (*Please state period)

2. *Lead Time*
   Please indicate the mean and range of variability (where present) by particular product, if more than one is involved. When indicating the order, product, or component lead time (from order receipt to dispatch), can you break down the time into activities such as design, procurement of materials, manufacturing, test and inspection, shipping, and dispatch?
   (a) Actual order, product, or component lead time (weeks) from order receipt to dispatch
   (b) Theoretical (scheduled, estimated, or promised) lead time (weeks) from order receipt to dispatch
   (c) Actual order, product, or component lead time (weeks) through the manufacturing areas
   (d) Theoretical (scheduled, estimated, or promised) lead time (weeks) through the manufacturing areas
   (e) Average number of operations during manufacture per order, product, or component
   (f) Average total time (hours) spent on machines or equipment (including setting and processing) per order, product, or component batch

3. *Staffing*
   (a) Number of manufacturing employees
   (b) Number of manufacturing supervisory staff
   (c) Number of production control staff
   (d) Number of progress chasers

   (e) Number of production control staff, including progress chasers

   (f) Number of inspectors

4. *Miscellaneous Information*

   (a) Value of outstanding orders

   (b) Proportion of orders delivered on time

   (c) Average and range of lateness of late orders

   (d) Average and range of batch sizes manufacture (components, subassemblies, finished products)

   (e) Number of different components manufactured

   (f) Average batch traveling distance (meters)

   (g) Setting times (hours) per batch

   (h) Percentage scrap rate

   (j) Average machine utilization

   (k) Average labor utilization

   (l) Labor turnover rate

   (m) Labor absenteeism rate

   (n) Proportion of bought-out finished parts that go into the assembly of your product(s)

Part 2(b)

(The ratios requested are required in the before and after CM situation).

1. *Stock Turnover Ratios*

   (a) Value of annual sales

      Value of work in progress (including semifinished stock)

   (b) Value of annual sales

      Value of finished goods, part stocks, and subassemblies

   (c) Value of annual sales

      Value of raw material stock

   (d) Value of finished goods, part stocks, and subassemblies

      Value of average (daily/weekly/monthly) period of production completed

   (e) Value of raw material stock

      Rate of buying raw materials (value per week/month/year)

   (f) Value of WIP (including semifinished stock)

      Value of average (daily/weekly/monthly) period of production completed

   (g) Value of tooling, jigs, fixtures, etc.

      Value of output of toolroom and purchase of tooling (value per week/month/year)

   (h) Value of average (daily/weekly/monthly) period of production completed

Value of production scrap (value per week/month/year)

2. *Lead Time Ratios*

   (a) Actual order, product, or component lead time from order receipt to dispatch

   Theoretical (scheduled, estimated, or promised) lead time from order receipt to dispatch

   (b) Actual order, product, or component lead time through manufacturing areas

   Theoretical (scheduled, estimated, or promised) lead time through manufacturing areas

   (c) Actual order, product, or component lead time through manufacturing areas

   Actual order, product, or component lead time from order receipt to dispatch

   (d) Actual order, product, or component lead time through manufacturing areas

   Average number of operations during manufacture per order, product, or component

   (e) Actual order, product, or component lead time through manufacturing areas

   Average total time spent on machines or equipment (including setting and processing) per order, product, or component batch

3. *Staffing Ratios*

   (a) Number of manufacturing employees
   Number of supervisory staff

   (b) Number of manufacturing employees
   Number of production control staff

   (c) Number of manufacturing employees
   Number of production control staff including progress chasers

   (d) Number of production control staff
   Number of progress chasers

   (e) Number of manufacturing employees
   Number of inspectors

4. *Miscellaneous Ratios*

   (a) Average value of outstanding orders
   Value of annual sales

REFERENCES

Burbidge, J. L. (1975). *The Introduction of Group Technology*, Wiley, New York.

Burbidge, J. L. (1982). The simplification of material flow systems, *International Journal of Production Research*, **20**(3), 339–347.

Burgess, A. G., Morgan, I., and Vollman, T. E. (1993). Cellular manufacturing: Its impact on the total factory, *International Journal of Production Research*, **31**(9), 2059–2077.

Dale, B. G. (1979). A General Method of Predicting the Performance of Group Technology Manufacturing Systems and Some Evaluations of Particular Applications, Ph.D. Thesis, University of Nottingham.

Dale, B. G., and Russell, D. (1983). Production control systems for small group production, *OMEGA*, **11**(2), 175–185.

Dale, B. G., and Willey, P. C. T. (1980). How to predict the benefits of group technology, *Production Engineer*, **59**(2), 51–54.

Edwards, G. A. B. (1974). Group technology—Technique, concept on new manufacturing system, *Chartered Mechanical Engineers*, **21**(2), 67–70.

Greene, T. J., and Sadowski, P. R. (1984). A review of cellular manufacturing assumptions: Advantages and design techniques, *Journal of Operations Management*, **4**(2), 85–97.

Hickson, D. J., Pugh, D., and Pheysey, D. (1970). Organization: Is technology the key? *Personnel Management*, February, 21–28.

Hyer, N. L. (1984). The potential of group technology for US manufacturing, *Journal of Operations Management*, **9**(3), 185–202.

King, J. R., and Nakornchai, K. (1982). Machine-component group formation in group technology: Review and extension, *International Journal of Production Research*, **20**(2), 117–133.

Leonard, R., and Rathmill, K. (1977a). The group technology myths, *Management Today*, January, 66–69.

Leonard, R., and Rathmill, K. (1977b). Group technology: A restricted manufacturing philosophy, *Chartered Mechanical Engineer*, **24**(10), 42–46.

Malik, M. Y., Connolly, R., and Sabberwall, A. J. P. (1973). Considerations for the formation of cells in group manufacture, *Proceedings of the 14th International MTDR Conference*, UMIST, Manchester, pp. 125–137.

Mosier, C. T., and Taube, L. (1985). The facets of group technology and their impacts on implementation: A state-of-the-art survey, *OMEGA*, **13**(5), 381–391.

10

FLEXIBILITY CONSIDERATIONS IN CELL DESIGN

Asoo J. Vakharia

Department of Decision and Information Sciences
University of Florida
Gainesville, Florida 32611-7169

Ronald G. Askin

Systems and Industrial Engineering Department
College of Engineering and Mines
University of Arizona
Tucson, Arizona 85721

Hassan M. Selim

Business Administration Department
College of Management and Economics
United Arab Emirates University
Al Ain, United Arab Emirates

1. INTRODUCTION

In a modern consumer society the now ubiquitous concept of customization is an accepted ingredient in many different commercial and industrial products, ranging from abstract products such as insurance policies to simple physical components such as rubber hoses or flexible couplings. At this level of understanding, the concept is commonplace and is about providing choice. To provide such products, a flexible production system is required, one with the capability of responding well to changing

Handbook of Cellular Manufacturing Systems, edited by Shahrukh A. Irani
ISBN 0-471-12139-8 © 1999 John Wiley & Sons, Inc.

requirements. In general conceptual terms the flexibility of an object is understood to be its ability to adapt to its constantly changing environment. Being flexible is an observable characteristic that says something about the internal structure or design properties of an object.

An interpretation of flexibility in a manufacturing context is exacerbated by the complexity of manufacturing. Flexibility can apply to different levels in a manufacturing organization, that is, to individual machines or manufacturing systems, to manufacturing functions, to individual products or groups, to the plant, and to the company. For flexibility to have meaning at a particular level, it must be expressed in the corresponding context. A substantial amount of literature dealing with manufacturing flexibility has accumulated over the last 50 years. The major part of this literature is devoted to defining various types of flexibility and identifying systems that exhibit one or more of these flexibility types. The literature makes one thing abundantly clear: Flexibility is a complex, multidimensional, and hard-to-capture concept. In practical terms, flexibility has been viewed as a tradeoff between efficiency in production and dependability in the marketplace (Abernathy, 1978; Wheelwright, 1981; Hayes and Wheelwright, 1984).

Flexibility of a manufacturing system can be defined as its adaptability to a wide range of possible environments that it may encounter (Sethi and Sethi, 1990). Thus, manufacturing flexibility can be regarded as the ability to reconfigure manufacturing resources so as to produce efficiently different products of acceptable quality. Jaikumar (1984) emphasizes the fact that flexibility in manufacturing is always constrained within a domain. Such a domain should be defined in terms of the portfolio of products, process, and procedures and should be well understood by product designers, manufacturing engineers, and software programmers. Moreover, other limitations on manufacturing flexibility are the speed and cost of responsiveness (Gustavsson, 1985; Garrett, 1986), the amount of required reinvestment, and the extent of interruptions in the existing system.

Uncertainties in a manufacturing system could stem from both internal and external forces (Garrett, 1986). The internal disturbances for which flexibility is useful include equipment breakdowns, variable task times, queuing delays, rejects, and rework. External forces are a function of the competitive environment and may reflect uncertainties in demand levels, product prices, product mix, and availability of resources. These could arise out of the actions of competitors, changing consumer preferences, technological innovations, new regulations, and the like. Thus, manufacturing flexibility clearly has major implications for a firm's competitive strength, and it plays a significant role in formulating a corporate strategy. In fact, as noted earlier, Hayes and Wheelwright (1984) consider flexibility as one of the dimensions of the competitive strategy of a business along with price, quality, and dependability. Management of manufacturing flexibility must invariably come to terms with the question of what are the "optimal" levels of various types of flexibilities. The answer to this question requires that the management identify and be able to measure the various flexibilities that the manufacturing system must have in order to gain maximum competitive advantage.

In the context of a Cellular Manufacturing (CM) system, flexibility could refer to the ability to respond to external (such as volume and mix) and internal (such as part design and machine capability) changes. However, it has been argued in the literature on CM that such systems are typically less flexible than the job shops that they replace (Vakharia, 1986). Hence, the focus of this chapter is to (i) overview generally accepted definitions of manufacturing flexibility, (ii) operationalize the concept of manufacturing flexibility as it relates to CM systems, (iii) develop a detailed procedure for designing flexible cells, and (iv) illustrate the procedure.

2. MANUFACTURING FLEXIBILITY

Flexibility from a manufacturing perspective can be classified as being related to the (a) shop floor, (b) manufacturing system, and/or (c) organization. Under each category, general types of flexibility can be defined as follows (Sethi and Sethi, 1990):

Shop Floor Flexibility

- Machine flexibility: This is one of the basic types of flexibility and refers to the ability of a machine to perform various types of operations without requiring a prohibitive effort to switch from one operation to another. Two critical issues with reference to this definition that must be kept in mind are (i) how do we define types of operations (e.g., are we simply talking about drilling holes of a different diameter or do different operations refer to a machine being capable of performing a drilling and grinding operation), and (ii) how do we operationalize the idea of prohibitive effort (i.e., is it in terms of time or cost or both, and regardless of how we assess it what is the threshold beyond which this effort becomes prohibitive).
- Material handling flexibility: This is the ability of a material handling system to move distinct part types to different machines for proper positioning and processing. In general, we can assume that the greater the set of part types that can be handled by a material handling system, the higher the flexibility of the system.
- Operation flexibility: This is the flexibility of a part type to be manufactured in different ways. For example, a part that can be manufacturing with several process plans is assumed to have a high operation flexibility. Similarly, in printed circuit board (PCB) manufacturing, if the order of chip placement on a PCB is immaterial, then we have high operation flexibility.

Manufacturing System Flexibility

- Part mix flexibility: This type of flexibility assesses the ability of the manufacturing system to process different sets of part types without major reconfiguration. In some cases, this is referred to as the process flexibility of system.
- Product flexibility: This is the ease with which new parts can be added or substituted for existing parts. Thus, product flexibility assesses the ability of a manufacturing system to respond to changes in the part mix.

- Routing flexibility: The greater the number of alternative manufatcuring routes that exist for a part type, the higher is the routing flexibility of the system. The primary distinction of this type of flexibility and operation flexibility is that the latter is part specific while this type is related to system components. Thus, a part with zero operation flexibility could be processed through multiple manuafacturing routes since there exist multiple machines that can be used to perform the same set of operations.
- Volume flexibility: The ability of a manufacturing system to be operated at different output levels. In general, this type of flexibility should be related to the slack capacity in the system.
- Expansion flexibility: This assesses the ease with which capacity and capability can be added as and when needed. While capacity focuses on increasing the output rate per unit time, capability refers to other process characteristics such as quality, technology type, and so forth.

Organization Flexibility

- Program flexibility: The ability of a system to run virtually untended for a long period of time. In general, completely authomated systems (which through design have limited variability) will have high levels of program flexibility.
- Production flexibility: The total number of part types that can be manufactured in a system without adding any capital equipment.
- Market flexibility: The ease with which a manufacturing environment can adapt to a changing market environment.

Given these different types of manufacturing flexibility, it should be obvious that the manufacturing system flexibility types are aggregate measurements of one or more shop floor system flexibility types, and, in turn, organization flexibility types reflect aggregation of one or more manufacturing system flexibility types. For example, production flexibility could be regarded as an aggregated version of part mix and product flexibility while routing flexibility could be an aggregated version of machine, material handling, and operation flexibility. Further, it should also be noted that there are interdependencies and conflicts between flexibility types. For example, we could design a system with high levels of routing flexibility, and this would probably result in higher levels of part mix flexibility. Also, an automated system (which would have high levels of program flexibility) would probably have low levels of production flexibility since by design the system would be able to produce a smaller set of part types. We now proceed to discuss the relevance of each of these flexibility types in CM.

3. FLEXIBILITY IN CELLULAR MANUFACTURING

A main feature of any manufacturing system, especially a CM system, is its capability to maintain a stable performance under changing conditions. This feature can be gained by designing and operating a "flexible" CM system. Flexibility, in the context of a CM system, can be defined as the ability of the system to adjust its resources to any changes in relevant factors such as product, process, loads, and machine failures. In this section, the dimensions and domain of flexibility in a CM system are described based on the assumption that manufacturing flexibility in CM stems from a combination of physical characteristics related to the system's design and control strategies.

There exists no rigorous method for identifying the domain of manufacturing flexibility in CM. The approach advocated here is a modified version of an approach developed by Gerwin (1987) and is based on the fact that CM systems facing changes utilize flexibility as an adaptive response. It is, therefore, necessary to examine the changes faced by CM systems in order to understand the type of manufacturing flexibility that needs to be built into the system. We classify the different types of changes as follows:

- Machine reliability changes and operator absenteeism, which could result in excess capacity or undercapacity within the current system.
- Machine processing capability changes could result in equipment becoming more specialized or more general-purpose.
- Part process plan and part design changes, which could result in capacity imbalances in the system as well as decreases in the desired levels of materials handling flexibility of the cell system.
- Part volume changes could result in increases/decreases in the total number of batches processed in the system.
- Part mix changes could result in new demands (in terms of new operation sets due to the introduction of new parts or modified operations sets resulting from part design changes) being placed on the system.

The latter two changes could stem from a variety of sources. For example, based on market costs and prices, shifting to subcontractors to supply key (or standard) parts rather than produce them in-house would result in changes in the part mix. On the other hand, if a portion of excessive demand is to be subcontracted rather than handled in-house, this would result in smaller changes to the existing part volume. Thus, any make versus buy analysis could result in changes in part demands.

Manufacturing cells should be designed with these types of changes in mind since most (if not all) of them will tend to result in poorer performance of the CM system. All these changes have several implications for CM systems. First, flow control within and between cells (in terms of materials handling and scheduling policies) is obviously affected when one or more of these changes occur. For example, changes in operation sets/operation sequences (due to part design and/or process plan changes) could result in nonunidirectional flow patterns between equipment within a cell and could also result in increased material flow between cells. Hence, the materials handling system as well as scheduling policies may need to be modified in light of such changes. In a similar manner, if the reliability of a particular machine declines over time and/or there is high labor turnover/absenteeism, there will be a necessity to reroute the load on that machine (or assigned to a particular operator) to another machine (or operator) within the same cell or another machine in another cell. Thus, flow control procedures may need to be modified.

Second, an obvious impact of these changes is on the utilization of equipment within the system. Due to increases in the volume requirements for certain parts as well as the introduction of new parts in the system, there may be insufficient capacity in the system, and, hence, additional investment in equipment may need to be contemplated. On the other hand, if a part is dropped from the current mix, certain types of equipment may be underutilized, and, hence, there may be little justification for including them in a particular cell.

Finally, another possible impact of these changes is on workload balance within and between cells. For example, changes in machine availability, machine processing capability, and the part mix could result in load imbalances that could impact the operational efficiency of the system. In the same vein, if a flexible automated machine is introduced in the system, then there will be a tendency to route work to that machine to justify its procurement. However, this could have a detrimental impact on the utilization of other equipment, leading to workload imbalances that could in turn increase average in-process inventories and cycle times.

3.1. Definitions of Flexibility for Cellular Manufacturing

In line with the types of changes that may impact a CM system, we define the following types of flexibility at the machine, cell, and system levels. To start with, at the machine level, we define the following type of flexibility:

- Machine flexibility: defined as the universe of operations that a machine is capable of performing with respect to the universe of operations that the system can perform.

In general, if machine flexibility is high, then there is a greater possibility that part design changes will be handled within the same cell. On the other hand, it is also likely that with high machine flexibility, the justification for procuring additional versatile machines is lower, and thus the possibility of machine processing capability changes is minimized.

Flexibility at the cell level depends on the elements composing the cell (i.e., part families and machines allocated to the cell) and represents the aggregated properties of a cell. Two types of flexibility defined at this level are:

- Cell comprehensiveness flexibility: the capability of the machines contained in each cell to carry out a variety of operations.
- Primary comprehensiveness flexibility: the ability to completely process individual part types within a single cell and multiple cells.

A high level of cell comprehensiveness flexibility indicates that part design changes as well as new product introductions can be accommodated more easily. The primary comprehensiveness flexibility depends on the acceptability of intercell moves in the system (i.e., the relative importance of identifying independent cells). If the CM system can process all part families without any need for intercell moves, then there is high primary comprehensiveness flexibility. Further, if there is more than one cell capable of carrying out all the operations for a part type, then this should facilitate real-time load balancing (since work from an overutilized cell can be offloaded to an underutilized cell).

Flexibility at the system level focuses on the responsiveness of the system to external changes that impact a CM. Two measures of flexibility at this level are as follows:

- Part volume flexibility: defined as the ability of the system to deal with volume changes in the current product mix.
- Part mix flexibility: defined as the ability of the CM system to handle different product mixes with minimum disruptions.

Hence, given high part mix and demand flexibilities, the impact of changes in the part mix and part volume can be minimized. To assess each of these types of flexibility, we propose measures for each type in the next section.

3.2. Measures of Flexibility in Cellular Manufacturing

Three issues are of relevance when developing measures to assess the flexibility of a cellular system. The first issue focuses on measure interpretation. To facilitate interpretation of the proposed measures we attempt to develop measures such that a "higher" value implies greater flexibility. Further, all except the aggregate measure are developed such that they are bounded (essentially each measure is bounded below by 0 and above by 1). Second, there is the issue of measure dependence and/or independence. We attempt to propose completely independent measures in order to assess all the different aspects of flexibility relevant in a CM system and also to avoid the "double-counting" of a particular type of flexibility. Once again, this is not necessarily true for the aggregate flexibility measure. Finally, the issue of measure validity needs to be addressed. All the measures developed in this chapter are based on the part and machine populations for a given data set. Hence, these measures cannot be compared across data sets. Given that the implementation/design of cells is context dependent, we contend that the comparison across data is not relevant. Thus, our measures facilitate the comparison of alternative cell configurations (existing or implemented).

The measures presented in this section can be grouped into four sets. These sets of measures are related to the definitions of flexibility in the context of CM described in the previous section as follows:

- Machine flexibility is assessed by the machine processing capability measures.
- Cell level flexibility is assessed by the cell, primary comprehensiveness, and routing flexibility measures.

- Part volume flexibility is assessed by the volume flexibility measure.
- Part mix flexibility is assessed by the mix flexibility set of measures.

The machine processing capability measures are a function of the machine populations in an existing or contemplated cellular system. On the other hand, all the other measures are dependent on the structure of the cell system designed/implemented. Further, the first two sets of measures focus explicitly on assessing the responsiveness of the system to the internal changes while the latter two sets of measures assess the system responsiveness to external changes. Finally, in addition to the above four sets of measures, we also present an aggregate measure that helps to assess the joint responsiveness of the system to all types of internal changes.

Notation

Operations and Parts An operation is defined as a transformation of material requiring inputs of labor, machine time, material, and/or tooling. Each part is characterized by a list of sequenced operations that must be performed on the part. Let Ψ denote the set of parts to be processed in the CM system and let $|\Psi|$ denote the cardinality of the set of parts (the number of parts). For each part $p \subset \Psi (p = 1, \ldots, |\Psi|)$, there is a set of operations denoted by $O_p \in \Omega$. D_p is the demand in batches per unit time for a part and Q_p is the anticipated batch size of the part.

Machines and Machine Cells Let M be the set of machine types in the CM system and let m be the machine type index, $m = 1, \ldots, |M|$. We assume that each machine of a type will be available a certain percentage of time, A_m. This percentage depends upon the machine reliability and maintenance. The machine availabilities are defined by the row vector $\tilde{A}$. The number of machines of each type available is N_m while c_m and F_m are the annual operating cost and annualized procurement cost for one machine of type m, respectively.

Machine flexibilities are a basic determinant of a CM system's capability given a generic operation set Ω. This is given by a machine–operation matrix $\tilde{X}_\Omega$, where each entry in the matrix X_{mj} is defined as

$$X_{mj} = \begin{cases} 1 & \text{if machine type } m \text{ is capable of performing operation } j, \\ 0 & \text{otherwise.} \end{cases}$$

Let j_p be the index of elements in O_p, $j_p = 1, \ldots, |O_p|$. Then the machine processing capabilities to perform the operations required by a part p is represented by a submatrix of $\tilde{X}_\Omega$ called $\tilde{X}_\Omega^p$ where each entry is defined as

$$X_{mj_p} = \begin{cases} 1 & \text{if machine type } m \text{ is capable of performing operation } j_p, \\ 0 & \text{otherwise.} \end{cases}$$

The run and setup times for each generic operation j on each machine type m are given in the matrices $\tilde{T}$ and $\tilde{S}$. Each entry in each matrix is defined as

t_{mj} = run time per unit of any part for generic operation j on machine type m,

s_{mj} = setup time for a batch of any part for generic operation j on machine type m.

Note that similar operations with different run and setup times are considered different generic operations.

The capacity and capability of a manufacturing cell is determined by the set of machines, tooling, and operating schedule assigned to it. In our analysis, we assume that sufficient manufacturing capacity is installed to meet anticipated maximum demand. The assignments of machine types to cells is given by K_{mc}, which is the number of type m machines assigned to cell c.

Machine Processing Capability Measures   We propose a set of machine Processing Capability (PC) measures that are all a function of the flexibility of the machines. Thus, they are independent of the actual cellular configuration. At an aggregate level, the machine PC of an individual machine type m is the proportion of total number of operations that the machine type can perform (i.e., $M_m^{PC} = [\sum_{j=1}^{|\Omega|} X_{mj}] / |\Omega|$; $0 \leq M_m^{PC} \leq 1$). Machine PC can also be measured relative to parts and operations required by those parts. Let M_{mp}^{PC} ($0 \leq M_{mp}^{PC} \leq 1$) denote the processing capability of machine type m relative to part p's operation set O_p. Then, this can be determined as the proportion of operations from the set O_p machine type m is capable of performing:

$$M_{mp}^{PC} = \frac{\sum_{j_p \in O_p} X_{mj_p}}{|O_p|}. \tag{1}$$

This machine PC measure can be modified to accommodate relative importance for each operation, machine type, or part. Assuming that the operation processing times on different machine types are a measure of the machine type processing efficiency, the Weighted Processing Capability (WPC) of machine type m relative to the jth operation of part $p \in \Omega$ is denoted by $M_{mj_p}^{WPC} = X_{mj_p} / t_{mj_p}$. Then, we determine the WPC of machine type m relative to the set O_p as ($0 \leq M_{mp}^{WPC} \leq 1$):

$$M_{mp}^{WPC} = \frac{\sum_{j_p \in O_p} M_{mj_p}^{WPC}}{|O_p|}. \tag{2}$$

The machine type WPC decreases with an increase in operation processing time. This measure is useful in comparing the Processing Capabilities of different machine types. For example, if machine type A can perform operation j in 10 seconds and machine type B can perform the same operation in 15 seconds, this measure indicates that machine type A has more Processing Capability than machine type B relative to performing operation j.

In terms of the responsiveness to potential changes, the machine PC measures are useful to assess if the machine population can handle part design changes. If M_m^{PC} is 1 for any one machine type, then we know that any part design changes or new parts introduced can always be accommodated by at least one machine type (assuming that no operations are added to the operation set Ω). Further, these measures can also be used to compare and evaluate the flexibility of different machine types in the system.

Primary Comprehensiveness Measures   If the cell comprehensiveness flexibility is 1 for a certain cell c, then the machines in this cell are capable of performing all the operations required by the current part mix. Hence, given a cell system consisting of C cells, the cell Comprehensiveness Flexibility (CF) of each individual cell c (CF_c^C; $c = 1, \ldots, C$) is determined as the percentage of total operations, which can be performed in the cell. This is computed as ($0 \leq CF_c^C \leq 1$):

$$CF_c^C = \frac{\sum_{m:K_{mc}>0} \sum_{j \in \Omega} X_{mj}}{|\Omega|}. \tag{3}$$

On the other hand, the primary CF is determined based on the operations of a certain part p, which can be carried out in any cell c. This primary comprehensiveness, A_{cp}^C, is a function of the availability of the machines allocated to the cell and is determined as ($0 \leq A_{cp}^C \leq 1$):

$$A_{cp}^C = \prod_{j_p \in O_p} A_{cj_p}^o, \tag{4}$$

where

$$A_{cj_p}^o = 1 - \prod_{m:X_{mj_p}=1} (1 - A_m)^{K_{mc}}.$$

Assuming independence of machine availabilities, A_{cp}^C is the proportion of time cell c has a set of machines operational that are capable of completely processing part p. This measure is useful to assess whether there exists a single cell to process all the operations on a part. Further, it can easily be used to compute the number of primary cells available to process a part. This would be useful when part volume changes cannot be handled in a single cell due to capacity limitations in that cell. Hence, an alternative cell could be used to handle potential increases in part volume requirements if slack capacity is available in that cell.

Routing Flexibility Measure   Once we have determined the primary comprehensiveness measure, we can extend it to assess the routing flexibility of a cell system as follows. Define $F_{cp} = 1$ is $A_{cp}^C > 0$ and 0 otherwise. Then the routing flexibility of a given cell system for a part p (F_p) is simply

$$F_p = \sum_{c=1}^{C} F_{cp}, \tag{5}$$

and the aggregate routing flexibility of the entire cell system is

$$F = \sum_{p=1}^{|\Psi|} F_p. \tag{6}$$

Volume Flexibility Measure   The volume flexibility of a given cell system is computed as the maximum *equal* percentage (defined as δ^*) increase in volume for all parts that can be processed without changing the system configuration (in terms of part and machine type assignments to cells, part operation assignment to machines, and the number of machines of each type). Hence, this measure is directly related to the slack capacity built into the system.

Mix Flexibility Measures   The flexibility of the cell system to respond to a change in part mix can be assessed depending upon the type of such change. A mix change occurs either when the relative volume requirements of the current part mix change or a new part with an associated operation set and volume requirement is introduced in the system. To assess the flexibility of a cell system to the first type of change (i.e., relative volume change), we compute the maximum percentage of demand for each current *individual* part (defined as δ_p^*), which can be accommodated within the current cellular configuration.

In order to assess the flexibility of the cell system to respond to the introduction of a new part (a new part is assumed to be one for which the process plan consists of a set of operations where the operation set is developed based on the statistical probabilities of operation commonality in the current part set), we need to consider two aspects: (i) the amount of slack capacity in the cell system to completely process all the demand requirements for the new part and (ii) whether there is a set of machines within any one cell with adequate slack capacity to completely process all the demand requirements for the new part. Obviously (ii) is preferred since the new part will be completely processed in a cell (leading to no increase in intercellular flows). However, if this is not possible, it may be possible to process the new part through multiple cells as long as adequate capacity is available in the system to do so. In order to assess these two aspects, we propose three measures.

The first measure is denoted by γ_1 and is defined as the percentage of new parts that can be accommodated in the current configuration without regard to intercellular flows. The second measure focuses on the percentage of new parts that can be completely processed within an existing cell without regard to capacity constraints (i.e., there exists at least one primary comprehensive cell to process the new part) and is denoted by γ_2. The final measure (γ_3) integrates the first two measures and assesses the percentage of new parts that can be completely processed within a single existing cell by explicitly considering the availability of capacity in the cell.

Aggregate Flexibility Measure   One of the major problems with developing flexibility measures for individual parts, machines, and/or cells is that it is difficult to compare these measures across different cellular configurations. Further, the number of parts, machines, and cells is data set dependent. Thus, there is a need to develop aggregate system measures. In this section, we present an aggregate measure that explicitly combines the machine availability, machine processing capability, and the primary comprehensiveness measures. Before doing so, we define

$$M_{pcm}^{UC} = \frac{D_p}{\sum_{p=1}^{|\Psi|} D_p} C_{cp}^{CP} M_{mp}^{WPC} A_m K_{mc},$$ (7)

where

$$C_{cp}^{CP} = \begin{cases} 1 & \text{if } A_{cp}^C > 0, \\ 0 & \text{otherwise.} \end{cases}$$

Here M_{pcm}^{UC} represents the "used" processing capability of a machine type m included in cell c to process part p. In addition, the measure explicitly incorporates the relative importance of part p (assessed in terms of proportion of total demand), the machine processing capability of machine type m in terms of operation set O_p for part p (i.e., M_{mp}^{WPC}), the availability of machine type m (i.e., A_m), and the availability of cell c in terms of machine types allocated (i.e., K_{mc}). Thus, Eq. (7), although focusing on each individual part, machine type, and cell, incorporates all the aspects of internal flexibility operationalized earlier. There are several properties of Eq. (7):

- It is an increasing function of the relative importance of part p (assuming that the relative demand reflects the importance of a part).
- It is an increasing function of the primary comprehensiveness (routing flexibility) of a cell c with respect to a part p.
- It is an increasing function of the efficiency of a machine type in processing the operations required by part p (i.e., the faster a machine type can process all the operations on a part, the greater is the value of the measures).
- It is an increasing function of the availability of machine type m.
- It is an increasing function of the number of type m machines included in cell c.

To develop an aggregated system flexibility measure that relates flexibility at lower levels to flexibility at higher levels, we draw upon the work of Shannon (1948) in developing an entropy measure in information economics. This has been used by several authors in formulating an objective theory of flexibility based on this measure (e.g., Kapur, 1986; Kapur et al., 1985; Kumar, 1986, 1987; and Yao, 1985). Based on this analysis, we propose the following aggregate entropic measure of system flexibility:

$$F_C^S = -\sum_{p\in\Psi}\sum_{c=1}^{C}\sum_{m\in M}\left[\left(\frac{M_{pcm}^{UC}}{S_C}\right)\ln\left(\frac{M_{pcm}^{UC}}{S_C}\right)\right],$$ (8)

where

$$S_C = \sum_{p\in\Psi}\sum_{c=1}^{C}\sum_{m\in M} M_{pcm}^{UC}.$$

In Eq. (8), the term $[(M_{pcm}^{UC}/S_C)\ln(M_{pcm}^{UC}/S_C)]$ represents the flexibility contribution of type m machines in cell c provided the cell is primary comprehensive with respect to part p. This measure can also be decomposed into the total cell flexibility and total machine type flexibility as shown in Selim et al. (1995). Further, this entropic measure also satisfies the essential and desirable axioms for developing aggregate flexibility measures that relate flexibility at the machine and cell levels to flexibility at the system level (Brill and Mandlebaum, 1989; Kumar, 1986, 1987).

This concludes our description of the measures related to the different flexibility types discussed in Section 2.1. In addition to measures that assess the flexibility of the machine population grouped into cells, we have also presented measures that assess flexibility of the cell system configuration. Further, to facilitate a comparison and evaluation of total system flexibility across cellular configurations, an aggregate system flexibility measure to assess the responsiveness of the system to internal changes has been proposed. We now proceed to describe a method for designing flexible cells.

4. DESIGNING FLEXIBLE CELLS

The proposed Flexible Cell Formation (FCF) method is operationalized in four phases. Phase I identifies the most economical set of machine types to process the required operations of the entire part set based on machine costs, capabilities, and capacities. Phase II assigns individual part-operations to individual machines. Phase III forms candidate cells that are then improved and evaluated in phase IV for added flexibility relative to the current part population. Each phase of the procedure allows user interaction in terms of parameter settings and once the cell design (with a fixed set of parameters) is generated, it can be evaluated using the measures described earlier. Based on such an evaluation, the user has the option of either completely regenerating the system design (i.e., start again at phase I) or simply changing parameter settings at some intermediate level to modify the cell system design generated. Details of each phase are described next.

4.1. Phase I: Assign Operations to Machine Types

Given a machine type population with different processing capabilities (in terms of operation types), phase I is concerned with assigning operation types to machine types.[*]

A comprehensive mathematical model for carrying out this assignment is as follows:

$$\text{Minimize } Z = \sum_{m \subset M}\sum_{j \subset \Omega} c_m u_{mj} x_{mj} + \sum_{m \subset M} F_m N_m \qquad (9)$$

---

[*]Each operation type represents a set of <u>identical</u> operations required by parts. Thus, if two parts with different features, geometry, and manufacturing requirements require the same operation, then this is defined as two distinct operations—one for each part.

subject to

$$\sum_{m \subset M} x_{mj} = 1, \qquad \forall j; \tag{10}$$

$$\sum_{j \subset \Omega} u_{mj} x_{mj} \leq U_m N_m, \qquad \forall m; \tag{11}$$

$$x_{mj} \in \{0, 1\}, \qquad \forall m \, \forall j; \tag{12}$$

$$N_m \geq 0 \text{ and integer}, \qquad \forall m. \tag{13}$$

In this model, the objective function minimizes the total annual operating cost of the operation–machine assignments and the total annualized procurement cost of machines. The first constraint ensures that each operation is assigned to a single machine type. The second constraint ensures that operation–machine type assignments are feasible at the system level by ensuring that we acquire adequate capacity. In this case u_{mj} is the fraction of machine type m required, assuming that all parts requiring operation j will be processed on machine type m and U_m ($0 \leq U_m \leq 1$) is a user-specified parameter that indicates the amount of slack capacity that should be built into the system. The user parameter U_m controls the demand flexibility of a cell system design. By setting lower values of U_m, we can build in more slack capacity into the system and, hence, allow the system to handle part volume changes without disruption. Finally, the technological constraints on the decision variables x_{mj} and N_m are enforced in the last two sets of constraints. Note that if we do not require that a part-operation always be performed on the same machine type, we can relax the integrality restrictions on the x_{mj}.

For a known set of N_m values, this model reduces to the Generalized Assignment Problem (GAP). In the context of our problem, we would know N_m if we were reconfiguring an existing system into a cellular system. On the other hand, when designing a completely new system, we propose that initial estimates of N_m be obtained by linearizing the fixed machine cost and then selecting the least expensive machine type for each operation. With this assignment, estimate the number of machines of each type by computing $\lceil (\sum_j u_{mj} x_{mj})/U_m \rceil$. Once these estimates of N_m have been obtained, we use the Martello and Toth (1981) enumerative algorithm for obtaining good solutions to the GAP model. Note that the final N_m values are based on the operation–machine type assignments resulting from applying this algorithm. A branch-and-bound strategy can be added to search for a better solution if desired. The output of this stage is the x_{mj} and N_m (if required) decision variables.

4.2. Phase II: Assign Part–Operations to Specific Machines

Given operation assignments to machine types, phase II assigns each part–operation to a specific machine of each type such that the similarity between part–operations assigned to a machine is maximized and the user-specified maximum machine usage (U_m) is not exceeded. The goal is to assign operations from potentially similar part types to the same machine. The process for this assignment is described below. The complete phase II problem is separable by machine type, and we solve this problem independently for each machine type that requires more than one machine.

In stage one, we first construct a graph. Each node in the graph represents a specific part–operation assigned to this machine type in phase I. The arc weight between each pair of nodes is 1 if both part–operations are required by the same part type. However, if a pair (say i and j) of nodes (part–operations) correspond to different parts (say p_i and p_j, respectively), then the arc weight is defined as

$$S_{p_i p_j} = \max \left\{ \frac{N_{p_i p_j}}{O_{p_i}}, \frac{N_{p_i p_j}}{O_{p_j}} \right\}, \tag{14}$$

where $N_{p_i p_j}$ is the number of common operations between part p_i and p_j. Note that $0 \le S_{p_i p_j} \le 1$. Thus, $S_{p_i p_j}$ represents the relative desirability for two operations to be assigned to the same machine.

In the second stage, an initial partition for this graph is obtained using the following heuristic. A part–operation pair with the lowest similarity is identified and each part–operation is assigned to two separate machines. This process is repeated iteratively until each machine has been assigned one part–operation. Then, we sequentially assign the remaining part–operations to machines in order to maximize the internal similarity of all part–operations assigned to the same machine subject to U_m. Once we get this initial partition, in the third and final stage, we improve on the solution using a variant of the Kernighan and Lin (1970) graph partitioning procedure (see Askin et al., 1997, for details). The output of this phase is the specification of the $X_{k_m j_p}$ decision variables.

4.3. Phase III: Identify Candidate Manufacturing Cells

Phase III of the FCF method involves clustering individual machines to identify candidate manufacturing cells. The individual part–operations carried out in each cell are a function of the individual machines assigned to a cell (phase II specifies the assignment of part–operations to individual machines). The algorithm used to create the candidate manufacturing cells is similar to that of phase II, except that we now consider partitioning the individual machine set into the number of desired cells. However, a major difference is in the manner by which the arc weights between individual machines is computed. The arc weight or desirability index $d_{n1,n2}$ between nodes $n1$ and $n2$ (each representing an individual machine of type $m1$ and $m2$, respectively) is defined as a convex combination of two parts:

$$d_{n1,n2} = \alpha W^s_{n1,n2} + (1 - \alpha) W^c_{n1,n2}, \qquad 0 \le d_{n1,n2} \le 1, \tag{15}$$

where

$$W^s_{n1,n2} = \max\{W_{n1,n2}, W_{n2,n1}\}$$

$W_{n1,n2}$ = proportion of machine $n1$ workload associated with parts that also use machine $n2$

$$W^c_{n1,n2} = \left[\frac{\sum_{j\in\Omega} X_{m1,j} + \sum_{j\in\Omega} X_{m2,j} - \sum_{j\in\Omega} X_{m1,j} X_{m2,j}}{|\Omega|} \right]$$

$$\cdot \left[\frac{\sum_{j1=1}^{|\Omega|-1} \sum_{j2=j1+1}^{|\Omega|} f_{j1j2} X^c_{j1j2}}{\sum_{j1=1}^{|\Omega|-1} \sum_{j2=j1+1}^{|\Omega|} X^c_{j1j2}} \right]$$

f_{j1j2} = proportion of current parts that require operations $j1$ and $j2$

X^c_{j1j2} = 1 if operation $j1$ can be performed by machine type $m1$ and operation $j2$ by machine type $m2$, or if operation $j2$ can be performed by machine type $m1$ and operation $j1$ by machine type $m2$; 0 otherwise

Here $W^s_{n1,n2}$ is computed as the maximum proportion of machine $n1$'s or $n2$'s work that involves parts that also visit the other machine given the current part–operation assignments; $W^c_{n1,n2}$ is a function of the combined processing capability of machine pair $n1$ (of type $m1$) and $n2$ (of type $m2$). The motivation for $W^c_{n1,n2}$ rests in the desirability, in a dynamic environment, to create cells that have the ability to process future parts that exhibit minor technological modifications or to handle minor changes in part mix and volume. Also, $W^c_{n1,n2}$ is the product of the proportion of part–operations that can be performed by the combined set of machines $n1$ and $n2$, and, for those operation pairs that require both machines, the weighted average that pair of operations will be needed by an arbitrary future part type. The parameter α is defined by the user. A high value of α indicates a user preference for designing cells based on the current part set and demand requirements. On the other hand, by specifying a lower value of α, the user can instead opt to focus on the *possible* set of operations that could be performed by both machines $n1$ and $n2$ rather than the current set that has been assigned. If the firm has additional technological forecasting information beyond that contained in the current part set, this information can be easily entered into the decision process by modifying the f_{j1j2} parameters accordingly.

Once the arc weights between all nodes have been defined, the user is required to specify the number of cells to be identified (i.e., C) and the variability to be permitted in cell sizes (i.e., V). Cell size is assessed in terms of the number of machines: C controls the number of partitions created while V controls the allowable range of the number of machines placed within each partition (cell). Based on these inputs, the procedure for initial partitioning and improvement described in phase II is used to partition the set of nodes, each representing a machine, into machine groups (cells). The output of phase III is the assignment of individual machines to cells.

The three user-defined parameters control the flexibility of cell system design in different ways; α allows the incorporation of machine level flexibility (in terms of possible operations that can be performed) during the cell design process. Thus, by varying α the user can investigate how responsive the current part–operation to machine assignments are in the context of the set of operations that can be performed by each machine. The number of cells (C) can be hypothesized to impact the flexibility of the cell design. By setting a low value of C, the size of each cell in terms of machines will be larger. Hence, we can argue that there will be more comprehensiveness flexibility built into the system as machine variety in each cell is likely to increase. Regarding the variability in cell size parameter (V), it can be hypothesized that by allowing cell sizes to be more variable, there is the likelihood that a "master" cell consisting of a large number of machine types will be identified and thus, comprehensiveness flexibility will once again be higher. Likewise a few highly specialized cells can be formed for current highly similar part families when appropriate. Note that both cases (i.e., small C and high V) will probably increase the part mix flexibility of the cell system configuration. This is primarily the result of there being a single cell that will contain multiple machine types and, hence, new parts can be processed in such a cell.

4.4. Phase IV: Improvement and Evaluation

In phase IV of the method, we first attempt to improve upon the preliminary cellular configuration. Depending upon the user input, we can improve the comprehensiveness flexibility and/or the volume flexibility of this preliminary system design. In order to improve upon the comprehensiveness flexibility of a given cell system design, we attempt to reassign parts, individual machines, and/or part–operations in that order (without changing the total number of machines in the complete system). The general procedure is as follows. Each part is considered for reassignment to another cell that contains all the necessary equipment to process the part, provided machine capacity is available in that cell. After we carry this out individually for every part, machine reassignment is considered if it will result in an increase in the number of cells to process one or more part types. Finally, the reassignment of part–operations is carried out for those part types that cannot be completely processed in one cell. The objective of such a reassignment is to check if we can identify another cell to completely process each part.

Another improvement algorithm available to the user at this phase of the FCF method focuses on volume flexibility of the system design. Given a cell system design, the volume flexibility is computed as the maximum *equal* percentage (defined as δ^*) increase in volume for all parts that can be handled without changing the system configuration. The improvement in volume flexibility is carried out as follows. First, we identify the bottleneck machine(s) in the system [i.e., the machine(s) that restrict the volume flexibility of the current design]. Next, we attempt to reroute some load from this machine to another machine in the same cell that can perform the same operation. If this is not possible, we attempt to reroute the workload to a machine of the same type (or a machine that can perform the same operation) in another cell. The procedure is iteratively repeated until no further rerouting of load is possible.

Once the cell design has been improved upon, we can evaluate it not only in terms of comprehensiveness and volume flexibility, which are explictly incorporated in the FCF method, but also in terms of other types of flexibility. For example, the flexibility of the cell system to respond to a change in part mix can be assessed depending upon the type of such change. A mix change occurs either when the relative volume requirements of the current part mix change or a new part with an associated operation set and volume requirement is introduced in the system. To assess the flexibility of a cell system to the first type of change (i.e., relative volume change), we compute the maximum percentage of demand for each current *individual* part (defined as δ_p^*), which can be accommodated within the current cellular configuration.

In order to assess the flexibility of the cell system to respond to the introduction of a new part,[*] we need to consider two aspects: (i) the amount of slack capacity in the cell system to completely process all the demand requirements for the new part and (ii) whether there is a set of machines within any one cell with adequate slack capacity to completely process all the demand requirements for the new part. Obviously, aspect (ii) is preferred since the new part will be completely processed in a cell (leading to no increase in intercellular flows). However, if this is not possible, it may be possible to process the new part through multiple cells as long as adequate capacity is available in the system to do so.

Other types of flexibility that may be relevant to assess at this point are those that relate to material handling, degree of worker cross training required to operate cells, and routing flexibility. Depending upon the interest of the user in these flexibility types, the measures proposed in Section 3.2 could be used.

5. ILLUSTRATION

We illustrate the FCF method and computation of the flexibility measures with an example. There are 10 different machine types in the system and the system produces 19 parts requiring 12 different operations. For each machine type, the procurement and operating costs and the availability of each machine were randomly generated. Similarly, for each part type, the part demand in batches per year, the batch size and part processing requirements matrix, and the run and setup times were also randomly generated (see Selim et al., 1994, for details). For this illustrative example, we assume that the operations are not part dependent. Thus, regardless of which part–operation is carried out on a machine type, run and setup times are identical. We assume that the shop in question operates 8 hours per day, 250 days a year. The weighted and unweighted machine processing capability measures for this data set are shown in Table 1.

At phase I, assume that U_m, the maximum allowable utilization of any machine is set at 0.90 for all m. We also assume the number of machines is not fixed; thus each operation is assigned to the machine type with minimum relaxed cost $u_{mj}(c_m + F_m/U_m)$. Based on the solution to phase I, individual operations are assigned to specific machine types. The number of machines (N_m) required of each type are (2,2,1,2,2,2,1,1,2,2).

Phase II is concerned with assigning each part–operation to a machine. Note that based on the N_m requirements from phase I, this is only relevant for machine types 1, 2, 4, 5, 6, 9, and 10. Based on these part–operation to machine assignments, in phase III we generate a candidate cell design using $\alpha = 1$ (to emphasize current part–operation to machine assignments), $C = 4$ (4 cell design) and $V = 0$ (i.e., cell size variability is not considered). Finally, in phase IV, an improvement of the phase III cell design is carried out. We search for improved solutions in terms of volume and comprehensiveness flexibility (in that order). The phase III and IV part–operation assignments to individual machines and the cells to which these machines

---

[*]A new part is assumed to be one for which the process plan consists of a set of operations. This operation set is developed based on the statistical probabilities of operation commonality in the current part set.

Table 1. Machine processing capability measures

Part (p)	Measure[a]	\multicolumn{10}{c}{Machine Type(m)}									
		1	2	3	4	5	6	7	8	9	10
1	U	0.20	0.20		0.40				0.20	0.20	
	W	0.10	0.06		0.20				0.08	0.04	
2	U	0.25			0.25	0.25		0.25			
	W	0.06			0.08	0.08		0.07			
3	U				0.25	0.25		0.25	0.25		
	W				0.08	0.08		0.07	0.10		
4	U	0.20	0.20		0.40			0.20	0.20	0.20	
	W	0.10	0.06		0.20			0.05	0.08	0.07	
5	U	0.25	0.25		0.25			0.25	0.25	0.25	
	W	0.13	0.07		0.17			0.06	0.10	0.08	
6	U					0.50	0.25			0.25	0.25
	W					0.15	0.10			0.05	0.10
7	U	0.20				0.40	0.20			0.20	0.20
	W	0.05				0.12	0.08			0.04	0.08
8	U	0.33		0.17	0.33	0.17		0.17			
	W	0.13		0.04	0.17	0.06		0.05			
9	U	0.40		0.20	0.40	0.20					
	W	0.15		0.04	0.20	0.07					
10	U	0.17			0.17	0.33	0.17			0.17	0.17
	W	0.04			0.06	0.10	0.07			0.03	0.07
11	U	0.25				0.50	0.25				0.25
	W	0.06				0.15	0.10				0.10
12	U		0.20		0.20		0.20	0.20	0.20	0.20	
	W		0.06		0.07		0.08	0.05	0.08	0.07	
13	U		0.33			0.33	0.33				0.33
	W		0.10			0.10	0.13				0.13
14	U		0.50			0.50					0.50
	W		0.14			0.14					0.20
15	U		0.50			0.50					0.50
	W		0.14			0.14					0.20
16	U		0.25		0.25		0.25		0.25		
	W		0.07		0.08		0.10		0.10		
17	U		0.25				0.25	0.25	0.25	0.25	
	W		0.07				0.10	0.06	0.10	0.08	
18	U		0.50			0.50					0.50
	W		0.14			0.14					0.20
19	U		0.50			0.50					0.50
	W		0.14			0.14					0.20

[a]U is the unweighted measure while W is the weighted measure.

are allocated is shown in Table 2. Table 3 shows the final individual cell compositions (i.e., after phase IV) in terms of machines and the average usage of each machine in each cell.

The final assignments of individual part–operations to machines and the cells to which individual machines are allocated is shown in Table 2. For example, part 2 requires operations (2,3,4,8). After phase IV, these operations are assigned to machine types (7,1,4,5), respectively, as shown in column 3; and individual machines to which the load for these part–operations is allocated are placed in cells (3,3,3,3), respectively. The cell compositions shown in Table 3 are the individual cell compositions after implementing phase IV of the FCF method. Although cell size variability was set at 0%, cell 4 (2 machines) is smaller than the remaining cells (each with 5 machines). This is due to the fact that in order to split up the 17 machines into four cells, the maximum size of a cell is set to $\lceil (17/4) \rceil = 5$.

An evaluation of the generated cell design in terms of flexibility measures is now carried out. In terms of the comprehensiveness flexibility measures, the cell comprehensiveness flexibility of each of the four cells is as follows. The machines allocated to cells 1 and 2 can perform 58.33% of the total number of operations (i.e., 7 of the 12 distinct operations); cell 3 can perform 66.67% of the total number of operations (i.e., 8 out of the 12 distinct operations); while only 16.67% of the total number of operations (i.e., 2 out of the 12 distinct operations) can be performed by the machines in cell 4. In terms of primary comprehensiveness flexibility, of the current design, we can see that all part types except 3, 10, and 13 have all operations assigned

Table 2. Final part–operation/machine–cell assignments

Part	Operation Set	Machine Assignment	Cell Assignment
1	(1,4,6,9,11)	(8,4,9,4,2)	(2,2,2,2,2)
2	(2,3,4,8)	(7,1,4,5)	(3,3,3,3)
3	(1,2,4,8)	(8,7,4,5)	(2,3,3,3)
4	(1,4,7,9,11)	(8,4,9,4,2)	(2,2,2,2,2)
5	(1,7,9,11)	(8,9,4,2)	(2,2,2,2)
6	(6,8,10,12)	(9,5,10,6)	(1,1,1,1)
7	(3,6,8,10,12)	(1,9,5,10,6)	(1,1,1,1,1)
8	(2,3,4,5,8,9)	(7,1,4,3,5,1)	(3,3,3,3,3,3)
9	(3,4,5,8,9)	(1,4,3,5,1)	(3,3,3,3,3)
10	(3,4,6,8,10,12)	(1,4,9,5,10,6)	(1,3,1,1,1,1)
11	(3,8,10,12)	(1,5,10,6)	(1,1,1,1)
12	(1,4,7,11,12)	(8,4,9,2,6)	(2,2,2,2,2)
13	(10,11,12)	(10,2,6)	(4,4,1)
14	(10,11)	(10,2)	(4,4)
15	(10,11)	(10,2)	(4,4)
16	(1,4,11,12)	(8,4,2,6)	(2,2,2,2)
17	(1,7,11,12)	(8,9,2,6)	(2,2,2,2)
18	(10,11)	(10,2)	(4,4)
19	(10,11)	(10,2)	(4,4)

Table 3. Final cell design

Cell Number	Machine Type Assigned	Number of Machines	Average Usage per machine[a]
1	1	1	0.576
	5	1	0.583
	6	1	0.532
	9	1	0.652
	10	1	0.485
2	2	1	0.756
	4	1	0.489
	6	1	0.331
	8	1	0.696
	9	1	0.465
3	1	1	0.470
	3	1	0.234
	4	1	0.639
	5	1	0.421
	7	1	0.447
4	2	1	0.621
	10	1	0.461

[a]This is computed as the average utilization per machine divided by the average availability of the machine of that type (i.e., A_m).

to machines included in a single cell. Thus, there is at least one cell to completely process these part types. The individual comprehensiveness flexibility of each cell c given the four-cell design with respect to each part p (i.e., A_{cp}^4) is shown in Table 4. Thus, although cell 2 contains all the machines required to completely process part 1, $A_{21}^4 = 0.62$ indicates that the machines required to perform the complete operation set on this part are only available 62% of the time to perform these operations.

In order to assess volume flexibility of the cellular system shown in Tables 2 (phase IV assignment) and 3, we increased the demand (in batches per year) for *every* part type by the same percentage until the utilization of one or more machines exceeded the availability. For this cell system, this occurred when the number of batches for all parts were increased beyond 145% of their current demand. Hence, the volume flexibility (δ) is 45% for the given cell system. In this case, the critical machine was machine type 2 in cell 2.

Regarding mix flexibility, the maximum percentage change in volume for an individual part (δ_p), which can be handled by the current system (keeping all other part volumes fixed), is shown in Table 5. Further, for each individual part, we identify the related bottleneck machine type. Thus, Table 5 shows that for part type 3, we can at most accommodate a 740% increase in batch volume before machine type 7 in cell 3 becomes overloaded.

Table 4. Cell/part primary comprehensiveness measures (A_{cp}^C)

Part Type	Cell 1	Cell 2	Cell 3	Cell 4
1	0.00	0.62[a]	0.00	0.00
2	0.00	0.00	0.43	0.00
3	0.00	0.00	0.00	0.00
4	0.00	0.62	0.00	0.00
5	0.00	0.69	0.00	0.00
6	0.58	0.00	0.00	0.00
7	0.50	0.00	0.00	0.00
8	0.00	0.00	0.25	0.00
9	0.00	0.00	0.36	0.00
10	0.00	0.00	0.00	0.00
11	0.52	0.00	0.00	0.00
12	0.00	0.55	0.00	0.00
13	0.00	0.00	0.00	0.00
14	0.00	0.00	0.00	0.72
15	0.00	0.00	0.00	0.72
16	0.00	0.58	0.00	0.00
17	0.00	0.62	0.00	0.00
18	0.00	0.00	0.00	0.72
19	0.00	0.00	0.00	0.72

[a]This indicates that although *all* the operations on part type 1 can be completed in cell 2, all the machines required to perform these operations are jointly available only 62% of the type.

In order to assess the flexibility of the system to respond to the introduction of new parts, we proceed as follows. First, we generate the demand (in batches) and batch size (in units) for a potential new part using a discrete uniform distribution with parameters (4,11) and (100,961), respectively. Second, for the part, we randomly generate the number of operations required for processing using a discrete uniform distribution with parameters (2,6). These parameter settings are chosen based on the minimum and maximum demand, batch sizes, and number of operations required for all parts. Third, we generate the individual operations required to process the part randomly from the *current* set of operations as follows. The probability (p_j) that a particular operation is selected first is based on the frequency (f_j) that operation is used for processing parts in the current mix and is computed as $f_j/(\sum_k f_k)$. Once the first operation has been selected, the second operation for the new part is selected conditioned on the first. Thus, the probability that operation $j1$ is selected as the second operation given that operation j is the first operation is computed as $f1_{j1}/(\sum_k f1_k)$. In this case $f1_{j1}$ is defined as the frequency that operation $j1$ appears along with operation j in the operation sequences for all parts. This sequential procedure is repeated until all the operations in the operation set for a new part have been generated.

Table 5. Mix flexibility of the cell design

Part (p)	Mix Flexibility (%) (δ_p)	Critical Machine
1	930	2
2	260	7
3	740	7
4	350	2
5	270	7
6	270	9
7	300	9
8	1960	7
9	370	4
10	220	9
11	390	1,5
12	280	2
13	790	2
14	380	2
15	340	2
16	290	2
17	260	2
18	360	2
19	560	2

Using the procedure described, we assessed the flexibility of the current system to respond to new part introduction. For the designed system, 68% of the 40 randomly generated new parts could be processed in the current system if we allowed intercellular flows of batches (i.e., $\gamma_1 = 68\%$); 48% of the new parts could possibly be processed within a single cell (without considering machine availability, i.e., $\gamma_2 = 48\%$); and 40% of the new parts could be processed within a single cell (i.e., without intercellular materials flows and explicitly considering machine availability, i.e., $\gamma_3 = 40\%$).

Finally, in order to illustrate the aggregate flexibility measure, we choose to illustrate it in the context of alternative cell structures. For the given data, we generated multiple cell systems consisting of 1–8 cells using the FCF method. For each cell system, the aggregate flexibility measure and its components are shown in Figure 1. As can be seen, there is a strong correlation between the aggregate measure (and its components) to the number of cells identified. As expected, the aggregate flexibility of the system declines as the number of cells increases. This is due to the fact that as the number of cells increases, the possibility that a part can be completely processed in a single cell decreases and, hence, the primary flexibility decreases. Further, we also see that as the number of cells increases, both the cell level and machine level flexibility components of the aggregate system measure also decrease. Given that both these reflect the flexibility of a cell and machine type to process a part completely, it is likely that they will be lower as the number of cells increases. This attests to the face validity of the aggregate system measure and its components.

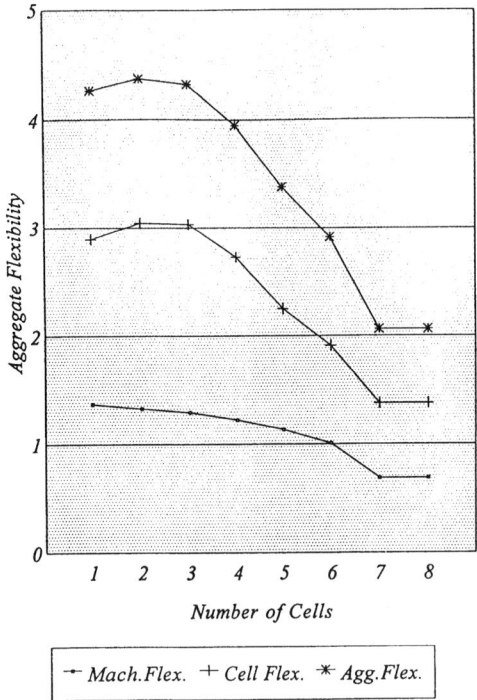

Figure 1. Aggregate flexibility measure and its components.

6. CONCLUSIONS AND IMPLICATIONS

In this chapter, we have identified the domain of flexibility in the context of CM systems, developed flexibility measures for CM systems, and have proposed and illustrated a procedure to design flexible cells. Using various user-defined inputs, the procedure can be used to enhance the flexibility of a cell system design. The reader interested in further details of this procedure and its applicability can contact any one of the authors or refer to Askin et al. (1997) where the procedure is described in more detail.

It is our view that flexibility is one of the most critical but ignored issues when designing, implementing, and operating CM systems. The reason underlying this perspective is that when cells were first implemented it was argued that by dedicating equipment to a select part set, we would always loose some degree of flexibility. In fact, the view that cells were always less flexibile than the job shops they replaced was a commonly accepted one in practice and research. However, current competitive pressures are forcing us to achieve "everything" simultaneously. In manufacturing, and in CM in particular, this translates to being efficient *and* flexible. Thus, our chapter provides a first thorough analysis of flexibility issues and how they can be accommodated when designing CM (which have been well established as being more efficient that other batch processing systems). Currently, we are interested in analyzing the cost versus flexibility tradeoff in designing and implementing CM systems. This, we believe, is the next critical area that needs to be addressed by managers and researchers who are interested in cellular manufacturing.

Acknowledgments

This work is supported in part through grant number DDM-92-15432 awarded by the National Science Foundation. We would like to acknowledge the comments provided by Professor Shahrukh Irani on a previous version of the chapter that, in our opinion, have significantly improved its value.

REFERENCES

Abernathy, W. J. (1978). *The Productivity Dilemma: Roadblock to Innovation in the Automobile Industry*, Johns Hopkins Press, Baltimore, MD.

Askin, R. G., Selim, H., and Vakharia, A. J. (1997). A methodology for designing flexible cellular manufacturing systems, *IIE Transactions*, **29**(7), 599–610.

Brill, P. H., and Mandlebaum, M. (1989). On measures of flexibility in manufacturing systems, *International Journal of Production Research*, **27**(5), 747–756.

Garrett, S. E. (1986). "Strategy First: A Case in FMS Justification," in *Proceedings*, Second Special Interest Conference on FMS, K. E. Stecke and R. Suri (Eds.) Ann Arbor, MI, pp. 17–30.

Gerwin, D. (1987). An agenda for research on the flexibility of manufacturing processes, *International Journal of Operations and Production Management*, **7**(1), 38–49.

Gustavsson, S. O. (1985). Flexibility and productivity in complex production processes, *International Journal of Production Research*, **22**(5), 801–808.

Hayes, R. H., and Wheelwright, S. C. (1984). *Restoring our Competitive Edge: Competing Through Manufacturing*, Wiley, New York.

Jaikumar, R. (1984). Flexible manufacturing systems: A managerial perspective, Working Paper # 1-784-078, Harvard Business School, Cambridge, MA.

Kapur, J. N. (1986). Four families of measures of entropy, *Indian Journal of Pure and Applied Mathematics*, **17**(4), 429–449.

Kapur, J. N., Kumar, V., and Hawalashka, O. (1985). Maximum-entropy principle in flexible manufacturing systems, *Journal of Mathematical and Physical Sciences*, **35**(1), 11–18.

Kernighan, B. W., and Lin, S.(1970). An efficient heuristic procedure for partitioning graphs, *AT & T Bell Labs. Technical Journal*, **49**(3), 291–307.

Kumar, V. (1986). "On Measurement of Flexibility in Flexible Manufacturing Systems," in *Proceedings*, Second Special Interest Conference on FMS, K. E. Stecke and R. Suri (Eds.), Ann Arbor, MI, pp. 131–143.

Kumar, V. (1987). Entropic measures of manufacturing flexibility, *International Journal of Production Research*, **25**(7), 957–966.

Martello, S., and Toth, P. (1981). "An Algorithm for the Generalized Assignment Problem," in *Operational Research*, Vol. 81, J. P Barnes (Ed.), North Holland, Amsterdam, pp. 589–603.

Selim, H. M., Vakharia, A. J., and Askin, R. G. (1995). Flexibility in cellular manufacturing: A framework and measures, Working Paper 95-108, DIS Department, University of Florida, Gainesville, FL.

Sethi, A. K., and Sethi, S. P. (1990). Flexibility in manufacturing: A survey, *International Journal of Flexible Manufacturing Systems*, **2**(2), 289–329.

Shannon, C. E. (1948). A mathematical theory of communication, *The Bell System Technical Journal*, **27**(3), 379–656.

Vakharia, A. J. (1986). Cell formation in group technology: A framework for evaluation, *Journal of Operations Management*, **6**(3), 257–272.

Wheelwright, S. C. (1981). Japan—Where operations really are strategic, *Harvard Business Review*, **59**(4), 67–74.

Yao, D. D. (1985). Material and information flows in flexible manufacturing systems, *Material Flow*, **2**, 143–149.

11

QUALITY CONTROL IN CELLULAR MANUFACTURING

Tapas K. Das and W. A. Miller

Department of Industrial and Management Systems Engineering
University of South Florida
Tampa, Florida 33620

1. INTRODUCTION

Quality of a product is usually measured by how well the product conforms to its design specifications and/or how well it performs its intended functions. The quality measure of a product depends on: (1) how well the product has been designed and (2) how well the product has been manufactured, i.e., to what extent the design specifications of the product have been conformed to during the manufacturing process. Quality of a product is intimately related to the quality of its manufacturing processes. A Cellular Manufacturing (CM) system has many inherent advantages over a traditional (e.g., job shop, flow shop) manufacturing systems, and as a result, CM systems provide a much higher quality manufacturing process. Notable reasons for achieving higher process quality in CM processes are: (1) generally higher inherent capability of the processes, (2) ease of material and information flow control, and (3) reduced quantities of scrap and rework per unit product. The term *quality* encompasses both product quality and process quality. A well-designed product (namely, a product designed for manufacturability) enhances the process quality, and a high-quality process leads to a high-quality product.

Handbook of Cellular Manufacturing Systems, edited by Shahrukh A. Irani
ISBN 0-471-12139-8 © 1999 John Wiley & Sons, Inc.

1.1. Basic Issues

To maintain or improve quality in a cellular-type discrete-part manufacturing environment, the following basic questions need to be addressed: (1) Are the product and process in a stage of design and development? (2) Is it an existing process for which quality needs to be improved? (3) Is it an existing product that needs to be manufactured to a higher standard of conformance?

If a process and the product are in the state of design and development, the complete range of quality-related actions can be implemented to ensure higher quality. Such actions are as follows:

1. Design for ease of manufacture
2. Design for robust product parameters
3. Design for robust process parameters
4. Increase process capability
5. Monitor and control process during production
6. Use of acceptance sampling of finished products
7. Maintain quality system as per ANSI/ASQC Q9000/ISO 9000.

If a process and product are already in the stage of production, then only steps 4 through 7 may apply.

1.2. Advantages of Cellular Manufacturing

Of the many advantages of CM, improved product quality is paramount. This, however, is not well recognized because of the latent means by which the cellular mode of manufacturing improves product quality. The closeness of the locations of the various activities of a manufacturing process in a cellular environment provides the ability to exercise better control on how the work is being done and what, if anything, is going wrong. The effect of poor quality of an operation becomes evident quickly, giving the opportunity to take corrective actions faster.

Often the successive elements of a production process are carried out by the same individual or several individuals working in close proximity. This results in high levels of ownership, responsibility, and easy flow of formal and informal information through the production process. The basic approaches and tools of quality control remain the same, but the effectiveness with which such tools are employed improves significantly. For example, a production cell manufacturing elements of a gear box may detect the quality problems related to shaft and gear misfits much faster than a noncellular production arrangement. This may result in quicker readjustment of the process parameters and less rework and scrap. In regards to a better use of an established procedure, a cell operator studying the out-of-control trends of the shaft turning process may benefit from the firsthand knowledge of the problems associated with the shaft and gear misfit.

1.3. Logistics of Quality Control in Cellular Manufacturing

Simplified logistics of CM is a major contributor to quality. Maintaining quality systems requirements (as stipulated in standards such as ANSI/ASQC Q9000/ISO 9000) is much easier for a CM process than in a noncellular setup. For example, the problems such as missing identification of a bin full of in-process materials, improper handling and accounting of the rejects, and the like are less likely to occur in a cellular environment.

To derive full advantage of CM, it is essential that quality monitoring and control functions are done in the cell by the cell personnel. This, while minimizing inspection delays on the material movements, increases the need for operator training on quality monitoring techniques as well as quality awareness. Quality monitoring within the cell by the operating personnel also requires better access to monitoring tools. Decentralized inspection facilities should be encouraged in CM.

The rest of this chapter is organized as follows. We first introduce as set of management planning (MP) tools that are very useful in logically identifying the areas that need attention in order to improve product quality. We then outline a few basic steps (using the MP tools) that are required in seeking higher quality for the products. Some statistical tools, such as optimal product/process design, control charts, and acceptance sampling (uses of which are often recommended by the planning method) are introduced next. Finally, we introduce the ISO 9000 series of standards that state the requirement for a quality system.

2. SOME USEFUL QUALITY MANAGEMENT PLANNING TOOLS

In this section we will introduce simple and effective quality MP tools (Brassard, 1989) that were developed to fill the void left by the seven basic Quality Control (QC) tools (flowchart, Pareto chart, cause-and-effect diagram, etc.) in helping managers plan. Development of the MP tools were motivated by the following reasons:

- The graphical techniques, such as flowchart, trendchart, Pareto chart, and cause-and-effect diagram seem too basic to be truely valuable to managers.
- The statistical tools, such as histogram, scatter diagram, and control charts are often viewed as too technical.
- Majority of the seven basic QC tools and the statistical tools (e.g., control charts) are primarily good for numerical data, they cannot handle qualitative data.

The seven MP tools, presented next, are claimed to be simple yet sophisticated enough to warrant attention.

- *Affinity Diagram.* This diagram contains a large number of ideas, opinions, and issues put into groups based on the natural relationship between the items. The process of building an affinity diagram, which is largely a creative process rather than logical, allows a selected group of people to bring out as many ideas and thought patterns and organize them into logical groups for further elaboration. Affinity diagrams are particularly useful when a breakthrough in traditional concepts is needed.

- *Interrelationship Diagraph (ID)*. This diagraph displays the logical as well as casual relationships among the variables (or factors) of a problem. ID allows the patterns of relationships among the variables to emerge. ID is used when an issue is sufficiently complex that the interrelationship between and among ideas is difficult to determine.

- *Tree Diagram*. This diagram maps out all the paths and tasks that need to be accomplished in order to achieve a primary goal and other related subgoals. The tree diagram provides a tool for the team that must make sure that all the bases are covered and that the logic is sound. Tree diagrams are useful for tasks that are sufficiently complex and for which there are strong consequences for omission.

- *Prioritization Matrices*. These matrices present the tasks or actions in the order of priority decided based on known weighted criteria. These matrices are useful when resources (e.g., time, money, and manpower) are limited.

- *Matrix Diagram*. This tool shows the correlation between each idea in one or more groups of items. At each intersecting point between a vertical set of items and horizontal set of items of a matrix, a relationship is indicated as being present or absent. A common use of this tool is in showing the relationship of a task with people, functions, and other tasks.

- *Process Decision Diagram Chart (PDPC)*. This tool maps out all possible actions and contingencies that can occur when moving from a problem statement to the possible solutions. This is used to plan the chain of events that must take place, especially when the problem or goal is an unfamilier one.

- *Activity Network Diagram*. This tool is used to plan the most appropriate schedule for any complex task and all of its related subtasks. It is used in projecting the likely completion time and also for monitoring all subtasks for adherence to the necessary schedule.

2.1. Interrelationships among the Tools

All seven MP tools can be used alone effectively. However, the maximum power is achieved when the tools are used together to move from a chaotic situation to an implementable action plan for improvement. A typical diagram showing the flow of action among the tools is shown in Figure 1 (Brassard, 1989).

The seven tools described are quite general purpose and can be used very effectively for design and quality control in a CM system. Multitudes of ideas that can lead to higher quality can be channelized through these tools to arrive at implementable plans.

2.2. Guidelines for Implementing Management Planning Tools for Quality Control in Cellular Manufacturing Systems

In a cellular a manufacturing system, the seven management planning tools can be used in determining what the main quality control issues are, what actions will address

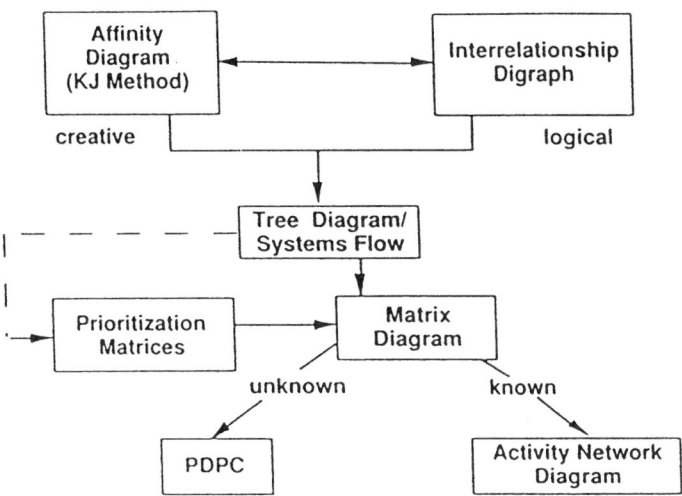

Figure 1. Interrelationship among the MP tools.

these issues, and how the action plans can be implemented. A brief outline of a possible sequence of steps follows:

- *Step 1.* Quite often quality control in manufacturing shops is done on an ad hoc basis. A thorough and systematic analysis of the activities (of both machine and shop personnel) and their effect on the product quality is often ignored. A team consisting of four to six people with good knowledge of the business processes and the management and analysis tools should be put together to brainstorm the issues relevant to increased product quality and develop an *affinity diagram*. A sample affinity diagram is shown in Figure 2.

- *Step 2.* Develop an *interrelationship diagraph* from the affinity diagram to show the logical links among the items. Figure 3 shows a sample ID developed from the affinity diagram shown in Figure 2.

- *Step 3.* Though an ID may look chaotic, it helps in forcing the key issues to rise to the surface. The *tree diagram* takes these issues and explodes them down to the lowest practical level of detail. Hence, a tree diagram develops the sequence of tasks that need to be completed in order to fully address the main issues of quality in CM. Some observations that can be made from the tree diagram are: difficulty of the implementation actions, interrelations among the actions, and possible priorities among the actions.

- *Step 4.* The remaining steps of the seven MP tools further develop the activities of the tree diagram, such as prioritizing the tasks, developing a relationship matrix for the activities of a task, and so forth. Some or all of these remaining activities can be implemented as necessary.

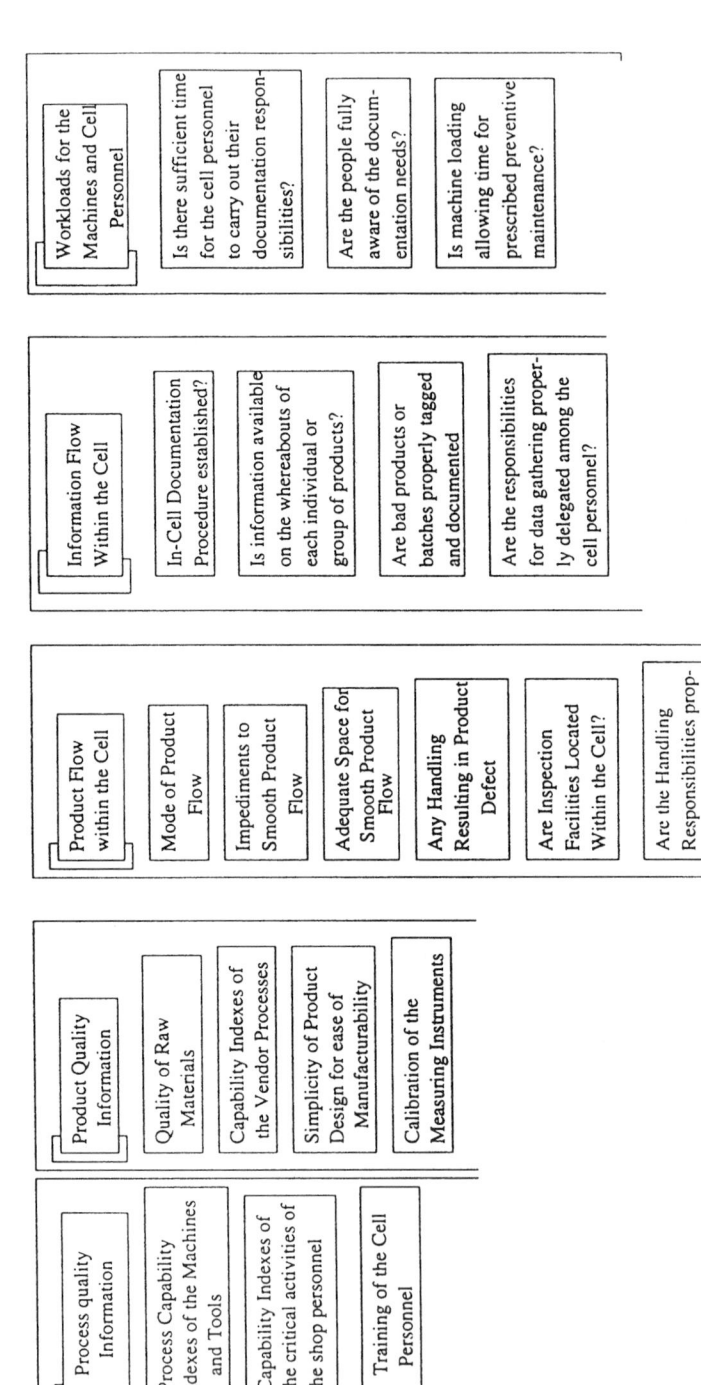

Figure 2. Affinity diagram for quality improvement in cellular manufacturing.

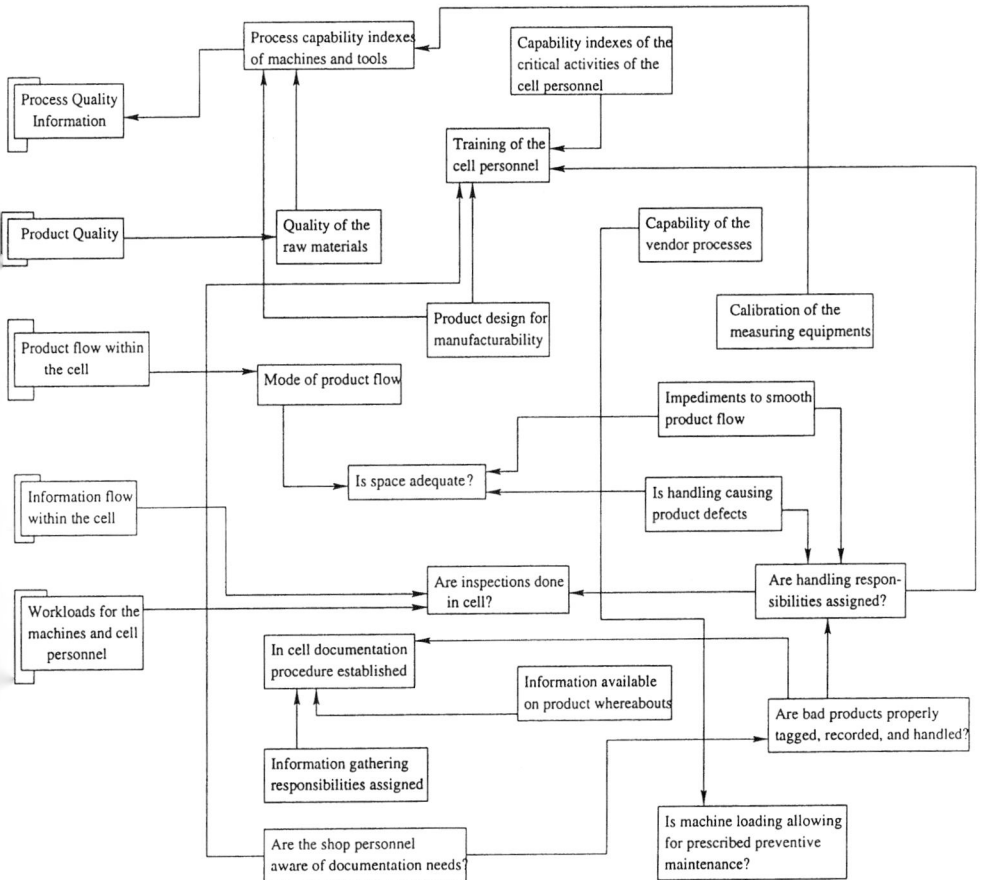

Figure 3. A First-round interrelationship diagraph for quality improvement in cellular manufacturing.

- *Step 5.* Some of the detailed action plans resulting from the use of the seven MP tools are very likely to be design one or more process/product parameters optimally, use variable or attribute-type control charts, and use acceptance sampling plans.

In the remaining part of this chapter, statistical tools needed to address these action plans are introduced. Numerical examples are provided in order to help in their implementation.

3. STATISTICAL TOOLS OF QUALITY CONTROL

The statistical QC tools that are used in a discrete parts manufacturing environment can be classified into three groups:

1. Quality design
2. Quality monitoring and control
3. Acceptance sampling

Manufacturing cells are excellent environments for applying the three groups of statistical QC tools listed and that will be discussed in detail later in this section. Because of the controlled manufacturing environment of the cell, it is ideally suited for applying these groups. Cells are devoted to one part type, or a family of part types, plus the cell's machines and operators are carefully selected for the cell. Capability of the cell's processing equipment is known and acceptable, or it would not be selected for the cell. Likewise, the cell workers are typically chosen because of their work ethics and their skills and knowledge of the cell's processes. With the proper equipment and personnel in a "closed, tight setting" of a cell, conditions are ideal to apply statistical QC techniques.

Quality engineers can develop well-designed experiments to establish required processing procedures as the cell is being initially put together. Initial studies can determine factors affecting product quality. This is discussed in Section 3.1. After the proper equipment and methods are established, it becomes an easy place to implement quality monitoring and control techniques. The cell workers are familiar with processes and can almost instinctively identify the problem when a control chart indicates a problem. This group of statistical QC techniques monitor the on-going processes to detect process variations such as changes in process mean and process variation.

If a manufacturing cell is properly setup, operated, and monitored, acceptance sampling could hopefully be avoided. But, if specifications or company directives require acceptance sampling, the situation is well suited. Cell workers, with their limited types of output products can easily become familiar with proper sampling plans and procedures.

The statistical quality control techniques that are discussed next lend themselves well to CM. Good process design, monitoring tools, and final product inspection all developed by appropriate engineering functions with cooperation of cell workers are some of the reasons cells are so successful.

3.1. Optimal Product and Process Design

The aspect of product and process design addressed here is how to effectively make the critical choices of various parameter levels of a particular design alternative (of a process or a product) to attain the best performance level for the design. For example, suppose the designer has selected a manufacturing process for manufacturing a rack and pinion arrangement. There is, however, a wide range of alternatives of process parameters, such as speed, feed, depth of cut, and tool material from which a designer has to make judicious choices for the process to perform effectively. A process performing effectively means that it can meet the desired product specifications with minimum quality loss (which is a function of the variance from the target quality level). Genechi Taguchi first introduced the concept of *cost of variation* from the target value (quality loss cost) for parts that meet specifications (Taguchi et al., 1989). In a CM system where the inherent variance is lower than traditional manufacturing processes, savings obtained from reduced quality loss cost could be significant and should be assessed carefully. The concept of quality loss cost is introduced next, and following that the technique of determining critical process parameters through variance analysis is presented.

Cost of Variation: Quality Loss Cost   Traditionally, the upper and lower product specification limits of a product have always guided the manufacturing community. It was thought that as long as the product dimensions/performances lie between their respective specification limits, the product should be considered to be of high quality. [Such a philosophy is referred to as *goalpost philosophy* by Ross (1988)]. For the most part, until Taguchi introduced the quality loss cost, no attempts were made to identify the so-called high-quality products, those that were closer to the target value than the others. Though, in reality, a product that is closer to the designed target is likely to perform better or provide higher satisfaction to the customer than those that are near the specification limits.

Taguchi's loss function philosophy recognizes the fact that as the products move away from the targeted value toward the specification limit(s), the loss to the society (caused by increased repair need, malfunction, etc. of the product) increases. Taguchi proposed a quadratic loss function as

$$L = k(y - m)^2,$$

where L = loss associated with a particular parameter (e.g., dimension) at a level y
m = nominal or targeted value of the parameter
k = constant depending on the cost at the specification limits and the width of the specification

The loss function is depicted in Figure 4, which shows that as the parameter moves away from the target or nominal value, the loss is obtained by a constant (k) times the square of the deviation $(y - m)^2$. The use of the loss function is demonstrated below with an example. Consider a typical machined part with a outer diameter tolerance of ± 0.10 mm. If the part dimension exceeds the Upper or the Lower Specification Limits (USL and LSL), the part is scrapped, which costs the manufacturer $50.00. It is also considered that as the part dimension reaches the specification limits, the finished product that it goes into is less likely to perform to the satisfaction of the customers. When a finished product fails to perform satisfactorily, it results in a loss to the society. This monetary loss can be obtained from the loss function as follows. Referring to Figure 5, we have that

$$\$50.00 = k(\text{LSL} - m)^2,$$

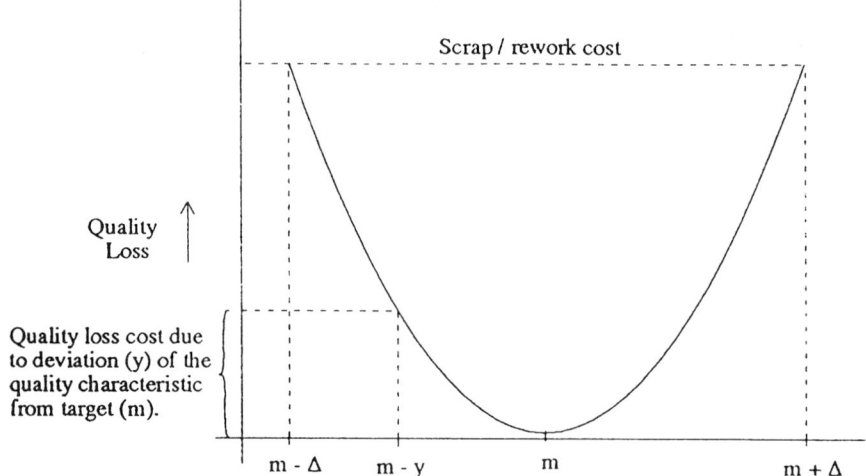

Figure 4. Quadratic quality loss function.

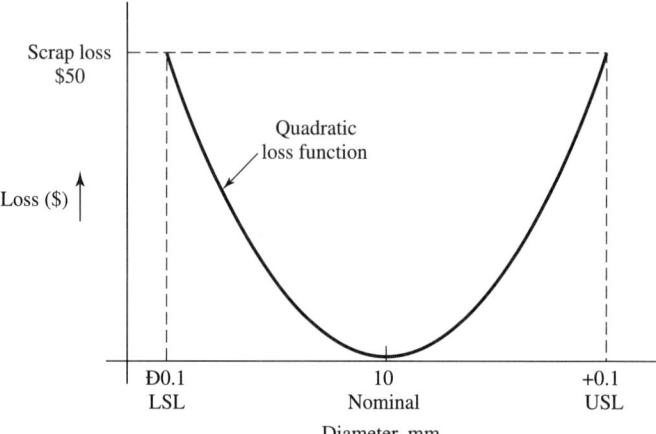

Figure 5. Loss function parameters.

which shows that when the dimension (y) is at either LSL or USL, then the loss is $50.00. Now k can be solved, for $m = 10.0$ mm (nominal value), as

$$k = \frac{\$50.00}{(9.9 - 10.0)^2},$$

$$k = \$5000 \text{ per mm}^2.$$

Therefore, the loss function is

$$L = 5000(y - 10.0)^2.$$

For a part with diameter $+10.05$ mm, the loss is

$$L = 5000(10.05 - 10.0)^2 = \$12.50.$$

That is, every part shipped to the customer with a diameter of 10.05 mm incurs a societal loss of \$12.50. The loss function approach can be used in determining (1) average loss per part of a production process and (2) the cost effectiveness of a quality improvement (variance reduction) process.

Analysis of Variance   Whenever a dimension/performance of a product deviates from the target, though within specification limits, it results in some loss to the society. In this section some of the basic techniques of identifying the significant sources (parameters affecting the outcome of a product or process) that contribute to the variation in process performance are addressed. Such sources of variations could then be studied individually and modified to reduce their contributions to process variance. The well-known statistical technique of ANalyis Of VAriance (ANOVA) serves as the basic tool for identifying the sources of variance. Variance analysis is accomplished through an experimental design method (Montgomery, 1991), which is explained in the next section.

Experimental Design   A designed experiment is a test or series of tests in which purposeful changes are made to the parameters (input variables) of a product/process so that we may observe and identify corresponding changes in the output response. Consider, for example, a process as shown in Figure 6. It can be visualized as some combination of machines, methods, and people that transforms an input material into an output product. This output product has one or more observable quality characteristics or responses. Some of the process variables $x_1, x_2, \ldots, x_p$ are controllable, while others $z_1, z_2, \ldots, z_q$ are uncontrollable (often referred to as *noise* factors).
The objectives of an experiment may include the following:

1. Determining which variables are most influential on the response, y
2. Determining where to set the influential x's so that the mean y is as near the target (nominal) as possible with least possible variance
3. Determining where to set the influential x's so that the effects of the uncontrollable variables z's are minimized

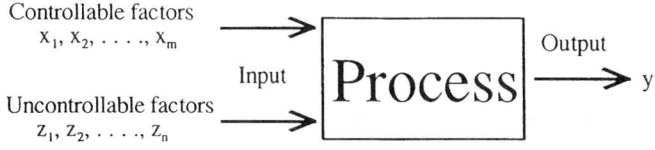

Controllable factors
$x_1, x_2, \ldots, x_m$

Input

Process

Output

y

Uncontrollable factors
$z_1, z_2, \ldots, z_n$

Figure 6. Process model.

In the context of product design, the applications of experimental design include the following:

1. Evaluation and comparison of basic design configurations
2. Evaluation of material alternatives
3. Determination of key product design parameters that impact performance

Use of experimental design in these areas can result in significantly improving the inherent quality of the product.

Single-Factor (One-Way) ANOVA   In this section, one of the simplest experimental design techniques that considers only one factor is discussed in detail. In the presentation, the general solution technique is supplemented with a numerical example.

Example: Tool Selection   An engineer with responsibility for the process design of a stainless steel turning operation is interested in testing a hypothesis that four different carbide tool inserts made by four different manufacturers are not significantly different with respect to their tool lives when used for the turning operation. After selecting the criteria that determine tool life, the engineer decided to make six test runs for each tool variety. Tool life data were collected from all 24 test runs (6 tests for each of 4 tool types) in a properly randomized order. The experimental data are shown in Table 1.

Suppose that there are a different levels of a single factor that are to be compared. The observed response at each of the factor levels is a random variable. The data would appear as in Table 2, where an entry, say y_{ij}, represents the jth observation taken under factor level i. Considered here is the case where there are an equal number of observations, n, at each factor level.

Let $y_{i.}$ represent the total of the observations for the ith factor level and $\bar{y}_{i.}$ represent the average of the observations under the ith factor level. Similarly, let $y_{..}$ represent the grand total of all observations and $\bar{y}_{..}$ represent the grand average of all observations. Expressed symbolically,

$$y_{i.} = \sum_{j=1}^{n} y_{ij}, \quad \bar{y}_{i.} = \frac{y_{i.}}{n}, \quad i = 1, 2, \ldots, a,$$

$$y_{..} = \sum_{i=1}^{a} \sum_{j=1}^{n} y_{ij}, \quad \bar{y}_{..} = \frac{y_{..}}{N},$$

where $N = an$ is the total number of observations. Thus, the "dot" subscript notation implies summation over the subscript that it replaces.

Let SS_T be the total sum of squares, SS_{Factor} is called the sum of squares due to the factor, and SS_E is called the sum of squares due to error. If the SS_{Factor} is large, it is due to differences among the responses at the different factor levels. Thus, by comparing

Table 1. Tool life data

Manufacturer of	Observations						Totals	Average
Tool Insects	1	2	3	4	5	6		
a	9	10	8	9	12	10	58	9.67
b	10	11	7	9	11	11	59	9.83
c	12	12	10	11	12	11	68	11.33
d	9	8	8	10	8	9	52	8.67

Table 2. Typical data for single-factor experiment

Factor Level	Observation				Totals	Averages
1	y_{11}	y_{12}	$\cdots$	y_{1n}	$y_{1\cdot}$	$\bar{y}_{1\cdot}$
2	y_{21}	y_{22}	$\cdots$	y_{2n}	$y_{2\cdot}$	$\bar{y}_{2\cdot}$
$\vdots$	$\vdots$	$\vdots$	$\cdots$	$\vdots$	$\vdots$	$\vdots$
a	y_{a1}	y_{a2}	$\cdots$	y_{an}	$y_{1\cdot}$	$\bar{y}_{a\cdot}$
					$y_{\cdot\cdot}$	$\bar{y}_{\cdot\cdot}$

From Montgomery (1991).

the magnitude of SS_{Factor} to SS_E one can see how much variability is due to changing factor levels and how much is due to error. This comparison is facilitated if these sums of squares are scaled by dividing them by their degrees of freedom. There are $an = N$ total observations; thus, SS_T has $N - 1$ degree of freedom. There are a levels of the factor, so SS_{Factor} has $a - 1$ degrees of freedom. Finally, within any factor level there are n replicates providing $n - 1$ degrees of freedom with which to estimate the experimental error. Since there are a factor levels, we have $a(n-1) = an - a = N - a$ degree of freedom for error. The ratio of a sum of squares to its number of degrees of freedom is called a Mean Square (MS).

For the example of carbide insert tool selection for the stainless steel turning process, the one-way ANOVA technique can be used to test the hypothesis that the tool lives do not vary significantly among the four different manufacturers of tool inserts. The sums of squares required for the ANOVA are computed as follows:

$$SS_T = \sum_{i=1}^{4} \sum_{j=1}^{6} y_{ij}^2 - \frac{y_{\cdot\cdot}^2}{an} = (9)^2 + (10)^2 + \cdots + (9)^2 - \frac{(237)^2}{4 \times 6} = 50.625.$$

$$SS_{\text{Factor}} = \sum_{i=1}^{4} \frac{y_{i\cdot}^2}{n} - \frac{y_{\cdot\cdot}^2}{an} \frac{(58)^2 + (59)^2 + (68)^2 + (52)^2}{6} - \frac{(237)^2}{24} = 21.79.$$

$$SS_E = SS_T - SS_{\text{Factor}} = 50.62 - 21.79 = 28.83.$$

The above computations are usually done using a computer software that supports ANOVA. The results of analysis of variance for the tool selection example are summarized in Table 3. The elements other than the sums of the squares of Table 3 can be explained as follows. The total Degrees of Freedom (DF) for this experiment is 23 [total number of data points (24) in Table 1 minus one]. The DF for insert type is total number of different inserts considered in the experiment minus one. Hence, the DF for the error component is $24 - 4 = 20$. The mean squares are obtained by dividing the sum of the squares with the respective DF. Since $F_{0.01,3,20} = 4.94$ and $F_0 = 5.02 > F_{0.01,3,20} = 4.94$, it can be concluded that the tool lives obtained from the different manufacturers are significantly different at the 99% level of significance.

However, most experiments for troubleshooting and improvement in industrial processes involve several variables. Often, one-way ANOVA is applied for several factors one at a time. The main limitation of this approach is that no interaction among the factors studied can be observed. If several factors are varied at the same time, this makes separation of any of the main factor effects impossible, let alone any interaction effects. Proper consideration of several variables can be achieved through *factorial experimental design*. In a factorial experiment, in each trial or replicate of the experiment, all possible combinations of the levels of the factors are investigated. For an excellent treatment on the factorial and fractional factorial experiments, readers are referred to Montgomery (1984).

In recent years, Taguchi has made a version of the fractional factorial design, called *Orthogonal Arrays* (OA), very popular in industry. For example, for the problem with 7 variables, each at two levels, a possible experiment is an eight-trial orthogonal array as depicted in the $L8$ OA matrix given in Figure 7. Notice that with only eight test combinations, $L8$ OA is one-sixteenth of a seven-factor full-factorial experiment.

In the remaining part of this section, an outline of the procedure for applying Taguchi's OA is given.

Table 3. ANOVA for tool selection example

Source of Variation	Sum of Squares	Degrees of Freedom	Mean Square	F_0
Manufacturer of Tool Inserts	21.79	3	7.26	5.04
Error	28.83	20	1.44	
Total	50.62	23		

	Column number						
Trial number	1	2	3	4	5	6	7
1	1	1	1	1	1	1	1
2	1	1	1	2	2	2	2
3	1	2	2	1	1	2	2
4	1	2	2	2	2	1	1
5	2	1	2	1	2	1	2
6	2	1	2	2	1	2	1
7	2	2	1	1	2	2	1
8	2	2	1	2	1	1	2

Figure 7. Matrix for $L8$ orthogonal array.

Steps in Designing, Conducting, and Analyzing an Experiment The major initial steps are as follows: Steps 1 through 4 concern the actual design of experiment

1. Selection of factors and/or interactions to be evaluated
2. Selection of number of levels for the factors
3. Selection of the appropriate OA
4. Assignment of factors and/or interactions to columns
5. Conduct tests
6. Analyze results
7. Confirmation experiment

Selection of Factors and/or Interactions to Evaluate The determination of which factors to investigate hinges upon the product or process performance characteristic(s) or response(s) of interest. Several methods are useful for determining which factors to include in initial experiments. These are:

1. Brainstorming
2. Flowcharting (especially for processes)
3. Cause-and-effect diagrams

Brainstorming This activity involves bringing together a group of people associated with the particular problems and soliciting their advice concerning what factors to investigate. It is appropriate to bring in product or process experts and statistically oriented people to discuss the factors and the structure of the experiment.

Flowcharting In the case of a process, flowcharts are particularly useful in the determination of factors affecting the process results. The flowchart adds some structure to the thought process and thus may avoid the omission of important factors.

An example of a process flowchart for a machining process can be given as follows:

Process Sequence	Factors
Select machine	Type of machine
	Machine capacity, cutting force, torque, etc.
	Jigs and fixturing arrangement
Select cutting tool	Tool material
	Tool geometry
	Type of insert
Select coolant	Type of coolant
	Viscosity
	Flow rate
Select machining parameters	Cutting speed
	Feed
	Depth of cut

Suppose that the problem observed at the machining process was that a percentage of the machined components did not meet its stringent tolerance specifications. The factors to include in an experiment should be the ones thought relevant to the problem of tolerance needs. All possible factors that are thought to influence the achieved tolerance of the machined part should be included in the initial round of experimentation. The purpose of the initial round of experimentation may be to eliminate some factors from contention and find those important few factors that do contribute to a product problem or to product quality improvement.

Cause-and-Effect Diagram Selection of factors and/or interactions to evaluate can be greatly helped by a Cause-and-Effect (CE) analysis developed by Japanese engineers. The structure for a CE analysis begins with the basic effect that is produced and progresses to what causes there may be for this effect. Primary, secondary, and perhaps tertiary causes are branched off the main trunk of the effect tree. The CE diagram for the machining problem (failure to meet the tolerance) would be similar to the diagram shown in Figure 8.

The steps in a CE analysis are:

1. Define the problem.
2. Select the method of analysis. Often the method of analysis involves brain-storming with a team of representatives from production, engineering, inspection, and any others potentially involved with the problem in question.
3. Draw the problem box and prime (center) arrow.

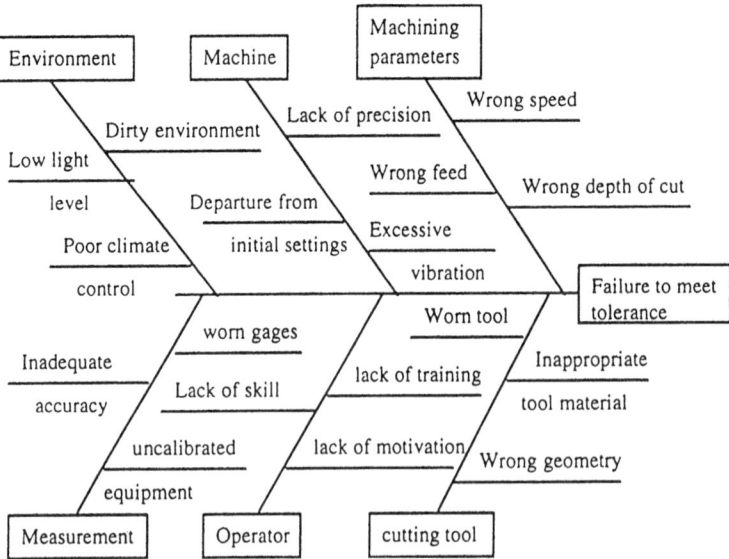

Figure 8. Cause-and-effect diagram of a machining process that does not meet a specified tolerance.

4. Specify the major categories of possible sources contributing to the problem.
5. Identify the possible causes of the problem.
6. Analyze the causes and take corrective action.

Selection of Number of Levels   Initial rounds of experimentation should involve many factors at few levels; two levels are recommended to minimize the size of the beginning experiment. The initial round of experimentation will eliminate many factors from contention and the few remaining can then be investigated with multiple levels without causing an undue inflation in the size of the experiment, which increases cost and/or time. Two kinds of parameters exist that may influence a product response, continuous and discrete parameters. Continuous parameters may be measured on a scale from a very low value to a very high value and may assume any value in between. Examples are speed, feed, and depth of cut. Discrete parameters may only assume particular values, such as on or off, material A, B, or C. If continuous parameters are being used, then the initial experiment should be at two levels only; interpolations and extrapolation may be used to predict other levels. If discrete factors are used, then interpolation or extrapolation may be meaningless. For instance, the use of three different materials is possible; there is no way to interpolate or extrapolate in order to predict results of a fourth possible material. Hence, if discrete parameters are studied, then more than two levels may be required in initial experiments.

Selection of the Orthogonal Array   The selection of which OA to use depends on these items:

1. The number of factors and interactions of interest
2. The number of levels for the factors of interest

These two items determine the total degrees of freedom required for the entire experiment. The degrees of freedom required for each factor is the number of levels (k) minus one:

$$v_a = k - 1.$$

The degrees of freedom for an interaction is the product of the interacting factor's degrees of freedom:

$$v_{A \times B} = (v_A)(v_B).$$

The minimum required degrees of freedom in the experiment is the sum of all the factor and interaction degrees of freedom. Examples of two basic kinds of OA are:

1. Two-level arrays: $L4$, $L8$, $L12$, $L16$, $L32$,
2. Three-level arrays: $L9$, $L18$, $L27$.

The number in the array designation indicates the number of trials in the array; an $L27$ has 27 trials, for example. The total degrees of freedom available in an OA is equal to the number of trials minus one.

$$\nu_{LN} = N - 1.$$

When a particular OA is selected for an experiment, the following inequality must be satisfied:

$$\nu_{LN} \geq \nu_{\text{required for factors and interactions}}.$$

Once the appropriate OA has been selected, the factors and interactions can be assigned to the various columns.

Assignment of Factors and Interactions   Taguchi has provided two tools to aid in the assignment of factors and interactions to arrays: linear graphs and triangular tables.

Each OA has a particular set of linear graphs and a triangular table associated with it. The linear graphs indicate various columns to which factors may be assigned and the columns those subsequently evaluate the interaction of the factors.

Conducting the Experiment   Once the factors are assigned to a particular column of the selected OA, the test strategy has been set and physical preparation for performing the test can begin. Some decisions need to be made concerning the order of testing the various trials.

Randomization   The order of performing the tests of the various trials should include some form of randomization. The randomized trial order protects the experiment from any unknown and uncontrolled factors that may vary during the entire experiment and which may influence the results. Through randomization, any uncontrolled factor effects can be evenly spread over the entire experiment. This will prevent bias of the factors and interactions assigned to the columns.

Selection of Sample Size   From a very practical viewpoint, a minimum of one test result for each trial is required to maintain the sample size balance (orthogonality) of the experiment (unbalanced experiments require special analysis technique). More than one test result per trial increases the sensitivity of the experiment to detect small changes in averages of populations. However, the gains in sensitivity diminish with increasing number of tests per trial. Often using higher size OA (e.g., using an $L8$ instead of two tests per trial of an $L4$) provides better results.

Analysis of Experimental Results   The ANOVA for an OA is conducted by calculating the Sums of Squares (SS) for each column. The sum of squares for factor A is

$$SS_A = \frac{(A_1 - A_2)^2}{N},$$

where A_1 and A_2 are the sums of the data associated with the first and second levels of factor A, respectively. Referring to Figure 7, if factor A is assigned to column 1, then A_1 is the sum of the data corresponding to trials 1 through 4.

Note that the total of the sums of squares for the unassigned columns is equal to the error sums of squares. The unassigned columns in an OA represent an estimate of error variation. Hence, the difference of the particular array selected for the experiment changes the analysis approach slightly.

Confirmation Experiment   The confirmation experiment is the final step in verifying the conclusions from the previous round of experimentation. The optimum conditions are set for the significant factors and levels and several tests are made under constant conditions. The average of the confirmation experiment results is compared to the anticipated average based on the factors and levels tested. The confirmation experiment is a crucial step and should not be omitted.

3.2. Process Capability Study

Before a production process can begin to produce end item(s), it is critical to establish the capability of the process to meet the product specifications. Of course, the word *capability* is meaningful only to the processes that are in statistical control. Hence, once a state of control is established on a process (which may require application of control charts), attention turns to the question: Is the process output meeting the design specifications and, if not, can the process be adjusted to a point where it will? Two important measures that characterize the process capability are (as shown in Fig. 9): (1) natural spread of the process variable and (2) the specification tolerance range (i.e., the difference between the USL and the LSL). Figure 9 and the subsequent discussion on process capability is based on the assumption that the process variable is a normally distributed random variable. Hence, $\pm 3\sigma$ limits on the mean of the normal distribution (with standard deviation σ) indicate the natural process spread.

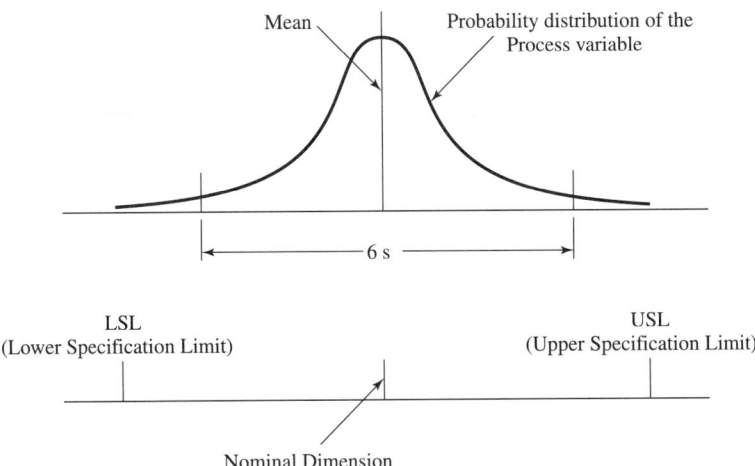

Figure 9. Natural spread and tolerance range: important measures that characterize process capability.

Steps in Determination of Process Capability   The steps outlined below may apply in their entirety or in part for a process depending on the situation. Before the study begins, decisions must be made regarding sampling procedures, timing, and collection and recording of data. Data must be recorded in such a manner that there will be no confusion as to how, where, or when they were taken.

1. *Bring the Process in Control.* The capability of a process is determined (as will be seen later in this section) by using the estimated values of the process mean and variance, which are obtained from the data collected from the process. Hence, it is crucial to ensure that the process is in control during data collection, that is, the process mean and its variance do not vary. Sources of extraneous variation for mean and variance should be identified and eliminated as much as possible before data collection.

2. *Analyze Data.* From the data collected, estimate the mean and variance. Assumptions are often made that the process variable is normally distributed. At the very least, a histogram from the data should be plotted to check the validity of the normality assumption. For critical applications, more rigorous statistical analysis should be done to assess the probability distribution of the process variable.

3. *Calculation of the Process Capability Indexes.* Though there are many capability indexes that are used in industry, the basic idea of those indexes are the same, namely the ratio of the specification range to the natural process spread. When a process is centered exactly at the middle of the specification range, the index used is C_p, which is given by

$$C_p = \frac{U - L}{6\sigma},$$

where U and L are the upper and lower specification limits, respectively, and σ is the estimated process standard deviation. It may be noted that the normality assumption for the process variable distribution is in place here, and hence 6σ represents the natural process spread around the mean (i.e., the process variable falls 99.765% of the time within the $\pm 3\sigma$ or 6σ limits). With any value of C_p at or higher than 1, the process will potentially meet the specification for a product characteristic (which is also referred to as process variable in the discussion above). Note here that, if there are more than one product characteristics critical enough to warrant process capability study, the above steps should be repeated for each characteristic, and separate C_p values should be calculated. A C_p value of 1.3 or higher is usually recommended.

Three other indexes used to assess process capability are

$$C_{pL} = \frac{\mu - L}{3\sigma},$$

$$C_{pU} = \frac{U - \mu}{3\sigma},$$

and

$$C_{pk} = \min(C_{pL}, C_{pU}).$$

The minimum target value again for each of C_{pL} and C_{pU} is 1. The C_{pk} provides the minimum of the two values. When the process mean (μ) does not coincide with the center of the specification range, the index C_{pk} should be used instead of C_p. Most manufacturing companies have adopted process capability index for their own as well as their suppliers operations.

Example: Process Capability Index Consider a gear hobbing process where the pitch diameter of the gear is a critical variable. In order to determine the process capability index, after eliminating the assignable causes of variations for pitch diameter from the production process, data has been collected as follows:

Observation	Diameter in mm
1	25.8
2	25.6
3	25.2
4	25.3
5	25.6
6	25.5
7	25.7
8	25.6
9	25.4
10	25.7

The mean and standard deviation of the process estimated from the above data are

$$\mu = \frac{X_1 + X_2 + \cdots + X_{10}}{9} = \frac{25.8 + 25.6 + \cdots + 25.7}{9} = 25.54 \text{ mm}$$

and

$$\sigma = \sqrt{\frac{(X_1 - \mu)^2 + (X_2 - \mu)^2 + \cdots + (X_{10} - \mu)^2}{n - 1}} = 0.1897 \text{ mm}.$$

The specification for the pitch diameter is given as 25.45 ± 0.6 mm, which gives $U = 26.05$ mm and $L = 24.85$ mm. Then the capability indexes are obtained, assuming that the data distribution is normal, as follows:

$$C_p = \frac{26.05 - 24.85}{6 \times 0.1897} = 1.054,$$

$$C_{pL} = \frac{25.54 - 24.85}{3 \times 0.1897} = 1.212,$$

$$C_{pU} = \frac{26.05 - 25.54}{3 \times 0.1897} = 0.896,$$

and then

$$C_{pk} = \min\{C_{pL}, C_{pU}\} = 0.896.$$

It may be noted from the above example that the natural spread of the process ($6\sigma = 1.138$) is less than the specification range ($U - L = 1.2$). But, since the process is not properly centered at the middle of the specification range, the process is deemed incapable ($C_{pk} < 1$). However, as $C_p = 1.054$ indicates, it is possible to ensure that almost 100% of the gear pitch diameters meet the specification just by recentering the process at 25.45 mm. Also note that this capability only refers to the pitch diameter. Similarly, there may be other critical parameters, such as height of gear tooth, various angles of the tooth geometry, and the like for which process capabilities may have to be established.

3.3. Control Charts

Once the product and the process designs are complete and the process capability is established, the process can go into production. At this time, appropriate control charts are placed into operation with an objective of monitoring the process. In fact, control charts are placed into operation to ensure that the process is in control before data is collected for process capability determination. During process operation, the purpose of control charts is to signal if and when the process goes out of control (deviates from its original operating parameter specifications). Upon obtaining a signal, QC personnel examine the process to identify the assignable causes, if there are any, of variation.

There are a variety of control charts that can be used in a CM environment. The major classification of the control charts are: (1) control charts for *variables* and (2) control charts for *attributes*. When the element of interest is a measured characteristic, such as a dimension expressed in millimeters, the quality is said to be expressed as variables. When the element of interest is the number of articles conforming or nonconforming to a specific requirement, quality of such an element is expressed by attributes.

All manufactured products must meet certain requirements, either clearly stated or implied. Many of these requirements may be stated as variables. Examples are dimensions, hardness in Rockwell units, and tensile strength in pounds per square inch. Most specifications of variables give both upper and lower limits for the measured value. Some may only have an upper or lower limit.

Many requirements are necessarily stated in terms of attributes rather than variables. For example, the number of defective spots per unit length of a conveyor belt, the horse power generated by an engine exceeding the desired level or not.

Control Charts for Variables: $\bar{X}$ and R Charts $\bar{X}$ (read as X-bar) charts are used for monitoring the mean of a variable (Fig. 10). $\bar{X}$ charts have a centerline and two control limits called Upper and Lower Control Limits (UCL and LCL). Mean values of the samples drawn from a process to be monitored are plotted on the $\bar{X}$

control chart. Generally speaking, as long as the points are within the control limits, the process is said to be in control with regard to the specific variable.

R charts (Fig. 10) are used for monitoring the range of the sample mean values, which in effect, monitors the variance of the processes, since variance is a function of the range. Like $\bar{X}$ charts, R charts also have a centerline and two control limits. The range of each sample (given by maximum value minus the minimum value) is plotted on the R chart. A point falling outside indicates the possibility of the process variance being out of control.

Calculation of the Control Limits   Suppose a quality characteristic has a mean μ and a standard deviation σ, where both μ and σ are known. The average ($\bar{X}$) of a sample of size n $(x_1, x_2, \ldots, x_n)$

$$\bar{X} = \frac{x_1 + x_2 + \cdots + x_n}{n}$$

is plotted on the $\bar{X}$-chart.

The control limits on the $\bar{X}$-chart are given as

$$\mathrm{UCL}_{\bar{X}} = \mu + 3\frac{\sigma}{\sqrt{n}},$$

and

$$\mathrm{LCL}_{\bar{X}} = \mu - 3\frac{\sigma}{\sqrt{n}}.$$

In practice, the mean and standard deviation (μ and σ) of the population are usually not known and are estimated from the samples drawn from the process. These estimates should be based on at least 20–25 samples. Suppose that m samples are available, each containing n observations on the quality characteristic. Typically, n will be of the order of 4, 5, or 6. Let $\bar{x}_1, \bar{x}_2, \ldots, \bar{x}_m$ be the averages of the m samples. Then an estimate of μ, the process average, is the grand average $\bar{\bar{X}}$ given by

$$\bar{\bar{X}} = \frac{\bar{x}_1 + \bar{x}_2 + \cdots + \bar{x}_m}{m}.$$

Thus, $\bar{\bar{X}}$ would be used as the centerline of the $\bar{X}$ chart.

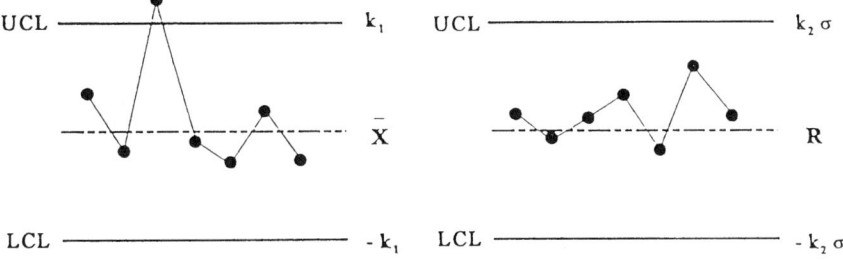

Figure 10. Typical $\bar{X}$ and R charts.

To construct the control limits, an estimate of the standard deviation σ is needed. Such an estimate can be obtained from the ranges of the m samples, as described below. If $x_1, x_2, \ldots, x_n$ is a sample size n, then the range (R) of the sample is the difference between the largest and smallest observations; that is,

$$R = x_{max} - x_{min}.$$

Let $R_1, R_2, \ldots, R_m$ be the ranges of the m samples. The average range is

$$\bar{R} = \frac{R_1 + R_2 + \cdots + R_m}{m}.$$

Then the control limits of the $\bar{X}$ chart are given as

$$\text{UCL} = \bar{\bar{x}} + A_2 \bar{R}$$

and

$$\text{LCL} = \bar{\bar{x}} - A_2 \bar{R}.$$

The constant A_2 is tabulated for various sample sizes (n) in Table 4.
The control limits for the R chart is given as

$$\text{UCL} = \bar{R} D_4$$

$$\text{LCL} = \bar{R} D_3.$$

The constants D_3 and D_4 are tabulated for various values of n in Table 4.
In what follows, an example [adopted from Montgomery (1991)] illustrating the development of $\bar{X}$ and R charts is presented.

Example: Piston Ring Forging   Piston rings for an automotive engine are produced by a forging process. It is intended to establish statistical control of the inside diameter of the rings manufactured by this process using $\bar{X}$ and R charts. Twenty-five samples, each of size five, have been taken while the process was supposedly in control. The inside diameter measurement data is presented in Table 5.
When setting up $\bar{X}$ and R control charts, it is best to set up the R chart first. Because the control limits on the $\bar{X}$ chart depend on the process variability, and unless the process variability is under control, these limits will not have much meaning. Using the data, the centerline is obtained as

$$\bar{R} = \frac{\sum_{i=1}^{25} R_i}{25} = \frac{0.581}{25} = 0.023.$$

For $n = 5$, from Table 4 $D_3 = 0$ and $D_4 = 2.15$. Then

$$\text{UCL} = \bar{R} D_4 = 0.023(2.115) = 0.049$$

$$\text{LCL} = \bar{R} D_3 = 0.023(0) = 0.$$

Table 4. Factors for constructing variable control charts

Observations in Sample, n	Chart for Averages			Chart for Standard Deviations						Chart for Ranges						
	Factors for Control Limits			Factors for Centerline		Factors for Control Limits				Factors for Centerline			Factors for Control Limits			
	A	A_2	A_3	c_4	$1/c_4$	B_3	B_4	B_5	B_6	d_2	$1/d_2$	d_3	D_1	D_2	D_3	D_4
2	2.121	1.880	2.659	0.7979	1.2533	0	3.267	0	2.606	1.128	0.8865	0.853	0	3.686	0	3.267
3	1.732	1.023	1.954	0.8862	1.1284	0	2.568	0	2.276	1.693	0.5907	0.888	0	4.358	0	2.575
4	1.500	0.729	1.628	0.9213	1.0854	0	2.266	0	2.088	2.059	0.4857	0.880	0	4.698	0	2.282
5	1.342	0.577	1.427	0.9400	1.0638	0	2.089	0	1.964	2.326	0.4299	0.864	0	4.918	0	2.115
6	1.225	0.483	1.287	0.9515	1.0510	0.030	1.970	0.029	1.874	2.534	0.3946	0.848	0	5.078	0	2.004
7	1.134	0.419	1.182	0.9594	1.0423	0.118	1.882	0.113	1.806	2.704	0.3698	0.833	0.204	5.204	0.076	1.924
8	1.061	0.373	1.099	0.9650	1.0363	0.185	1.815	0.179	1.751	2.847	0.3512	0.820	0.388	5.306	0.136	1.864
9	1.000	0.337	1.032	0.9693	1.0317	0.239	1.761	0.232	1.707	2.970	0.3367	0.808	0.547	5.393	0.184	1.816
10	0.949	0.308	0.975	0.9727	1.0281	0.284	1.716	0.276	1.669	3.078	0.3249	0.797	0.687	5.469	0.223	1.777
11	0.905	0.285	0.927	0.9754	1.0252	0.321	1.679	0.313	1.637	3.173	0.3152	0.787	0.811	5.535	0.256	1.744
12	0.866	0.266	0.886	0.9776	1.0229	0.354	1.646	0.346	1.610	3.258	0.3069	0.778	0.922	5.594	0.283	1.717
13	0.832	0.249	0.850	0.9794	1.0210	0.382	1.618	0.374	1.585	3.336	0.2998	0.770	1.025	5.647	0.307	1.693
14	0.802	0.235	0.817	0.9810	1.0194	0.406	1.594	0.399	1.563	3.407	0.2935	0.763	1.118	5.696	0.328	1.672
15	0.775	0.223	0.789	0.9823	1.0180	0.428	1.572	0.421	1.544	3.472	0.2880	0.756	1.203	5.741	0.347	1.653
16	0.750	0.212	0.763	0.9835	1.0168	0.448	1.552	0.440	1.526	3.532	0.2831	0.750	1.282	5.782	0.363	1.637
17	0.728	0.203	0.739	0.9845	1.0157	0.466	1.534	0.458	1.511	3.588	0.2787	0.744	1.356	5.820	0.378	1.622
18	0.707	0.194	0.718	0.9854	1.0148	0.482	1.518	0.475	1.496	3.640	0.2747	0.739	1.424	5.856	0.391	1.608
19	0.688	0.187	0.698	0.9862	1.0140	0.497	1.503	0.490	1.483	3.689	0.2711	0.734	1.487	5.891	0.403	1.597
20	0.671	0.180	0.680	0.9869	1.0133	0.510	1.490	0.504	1.470	3.735	0.2677	0.729	1.549	5.921	0.415	1.585
21	0.655	0.173	0.663	0.9876	1.0126	0.523	1.477	0.516	1.459	3.778	0.2647	0.724	1.605	5.951	0.425	1.575
22	0.640	0.167	0.647	0.9882	1.0119	0.534	1.466	0.528	1.448	3.819	0.2618	0.720	1.659	5.979	0.434	1.566
23	0.626	0.162	0.633	0.9887	1.0114	0.545	1.455	0.539	1.438	3.858	0.2592	0.716	1.710	6.006	0.443	1.557
24	0.612	0.157	0.619	0.9892	1.0109	0.555	1.445	0.549	1.429	3.895	0.2567	0.712	1.759	6.031	0.451	1.548
25	0.600	0.153	0.606	0.9896	1.0105	0.565	1.435	0.559	1.420	3.931	0.2544	0.708	1.806	6.056	0.459	1.541

For $4 > 25$

$$A = \frac{3}{\sqrt{n}}, \quad A_3 = \frac{3}{c_4\sqrt{n}}, \quad c_4 \simeq \frac{4(n-1)}{4n-3}, \quad B_3 = 1 - \frac{3}{c_4\sqrt{2(n-1)}}, \quad B_4 = 1 + \frac{3}{c_4\sqrt{2(n-1)}}, \quad B_5 = c_4 - \frac{3}{\sqrt{2(n-1)}}, \quad B_6 = c_4 + \frac{3}{\sqrt{2(n-1)}}.$$

From Montgomery (1991).

Table 5. Inside diameter measurements (mm) on forged piston rings

Sample Number	Observations					$\bar{x}_i$	R_i
1	74.030	74.002	74.019	73.992	74.008	74.010	0.038
2	73.995	73.992	74.001	74.011	74.004	74.001	0.019
3	73.988	74.024	74.021	74.005	74.002	74.008	0.036
4	74.002	73.996	73.993	74.015	74.009	74.003	0.022
5	73.992	74.007	74.015	73.989	74.014	74.003	0.026
6	74.009	73.994	73.997	73.985	73.993	73.996	0.024
7	73.995	74.006	73.994	74.000	74.005	74.000	0.012
8	73.985	74.003	73.993	74.015	73.988	73.997	0.030
9	74.008	73.995	74.009	74.005	74.004	74.004	0.014
10	73.998	74.000	73.990	74.007	73.995	73.998	0.017
11	73.994	73.998	73.994	73.995	73.990	73.994	0.008
12	74.004	74.000	74.007	74.000	73.996	74.001	0.011
13	73.983	74.002	73.998	73.997	74.012	73.998	0.029
14	74.006	73.967	73.994	74.000	73.984	73.990	0.039
15	74.012	74.014	73.998	73.999	74.007	74.006	0.016
16	74.000	73.984	74.005	73.998	73.996	73.997	0.021
17	73.994	74.012	73.986	74.005	74.007	74.001	0.026
18	74.006	74.010	74.018	74.003	74.000	74.007	0.018
19	73.984	74.002	74.003	74.005	73.997	73.998	0.021
20	74.000	74.010	74.013	74.020	74.003	74.009	0.020
21	73.988	74.001	74.009	74.005	73.996	74.000	0.033
22	74.004	73.999	73.990	74.006	74.009	74.002	0.019
23	74.010	73.989	73.990	74.009	74.014	74.002	0.025
24	74.015	74.008	73.993	74.000	74.010	74.005	0.022
25	73.982	73.984	73.995	74.017	74.013	73.998	0.035
					$\Sigma = 1850.028$		0.581
					$\bar{\bar{x}} = 74.001$		$\bar{R} = 0.023$

From Montgomery (1991).

The R chart is shown in Figure 11. When 25 sample ranges are plotted on this chart, there is no indication of an out of control condition.

Since the R chart indicates that process variability is in control, $\bar{X}$ chart may now be constructed. The centerline is

$$\bar{\bar{X}} = \frac{\sum_{i=1}^{25} \bar{X}_i}{25} = \frac{1850.028}{25} = 74.001.$$

For $n = 5$, from Table 4 $A_2 = 0.577$. Then the control limits are given as

$$\text{UCL} = \bar{\bar{X}} + A_2\bar{R} = 74.001 + (0.577)(0.023) = 74.014$$

and

$$\text{LCL} = \bar{\bar{X}} - A_2\bar{R} = 74.001 - (0.577)(0.023) = 73.988.$$

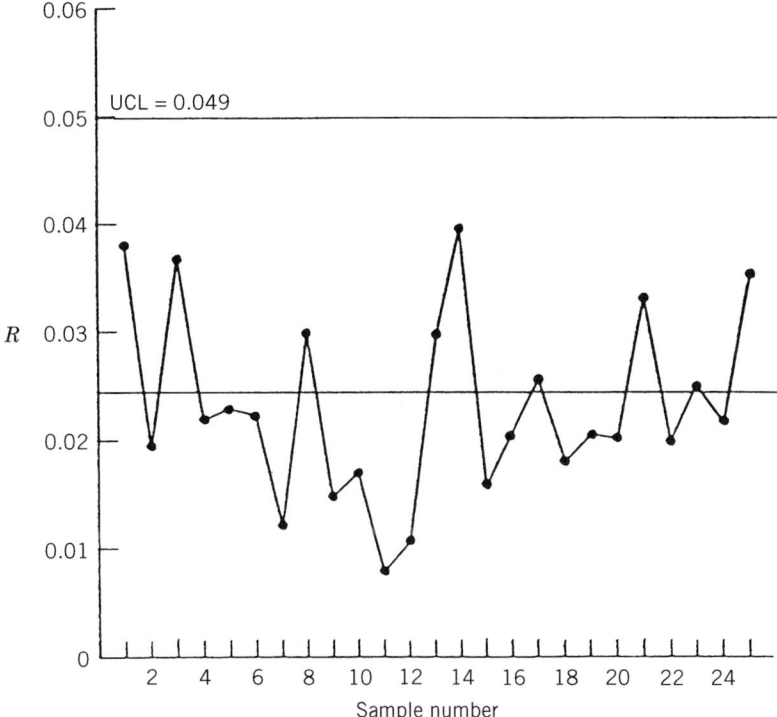

Figure 11. R chart for piston ring forging problem (Montgomery, 1991).

The $\bar{X}$ chart is shown in Figure 12. When the data, from which the chart parameters are calculated, are plotted on the chart, no indication of an out-of-control condition is observed. Therefore, since both charts exhibit control, it is concluded that the process is in control and the charts are passed on to production for use in on-line statistical process control.

Often some of the data points, from which limits are calculated, when plotted on the chart fall outside the control limits. This is an indication of the process being out of control during the data collection phase. Usual practice is to eliminate those out-of-control data points from the set and recalculate the control chart parameters. The purged data is then plotted on the new chart and the process is repeated, if necessary. During this process, exercising careful control of the process and additional data collection may be necessary.

Revision of Control Limits and Centerline   The initial control limits calculated should always be considered as *trial limits*, subject to subsequent revision. Generally, the effective use of any control chart will require periodic revision of the control limits and the centerlines. Some practitioners establish regular periods for review and

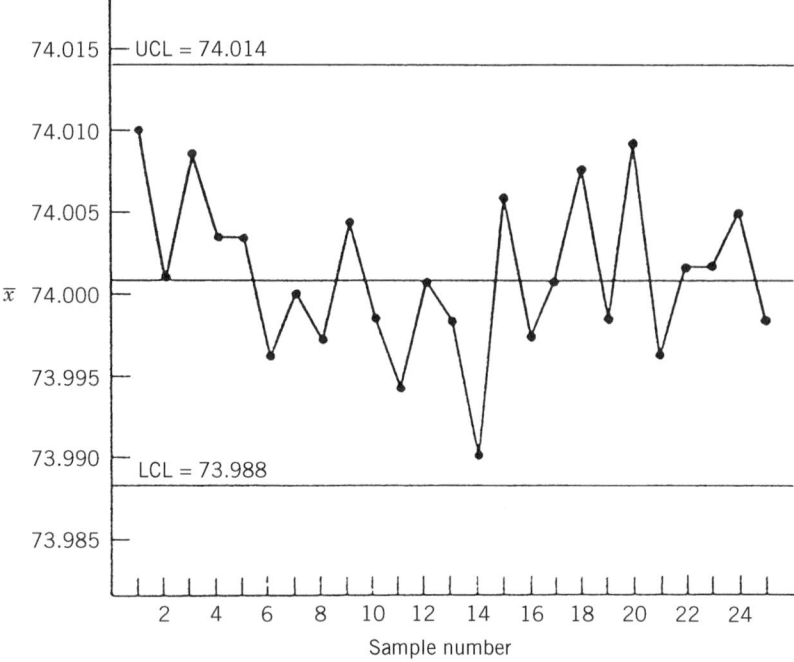

Figure 12. $\bar{X}$ chart for piston ring forging example (Montgomery, 1991).

revision of control chart limits, such as every week, every month, or every 25, 50, 100 samples.

Control Charts for Attributes In spite of the fact that $\bar{X}$ and R charts provide efficient means of identifying variations in a process, it is evident that their use is limited to only a small fraction of the quality characteristics specified for products and services, namely those that can be measured. Many quality characteristics that cannot be measured numerically are often observable as *attributes*, for example, conforming or nonconforming, defective or nondefective.

There are several different types of control charts that may be used in monitoring processes through product attributes:

1. p chart, the chart for fraction rejected as nonconforming to specifications
2. np chart, the control chart for number of nonconforming items
3. c chart, the control chart for number of nonconformities
4. u chart, the control chart for number of nonconformities per unit

The cost of collecting data for attributes charts is usually lower than the cost of collecting data for variable control charts because the attributes chart generally uses data already collected for other purposes. In addition to the cost advantages that may be gained, the use of attribute control charts provides management with useful records of quality history. For example, p charts give numbers related to fraction of products rejected which may be a lot more understanding to management than the mean value of a sample of a critical dimension in relation to the control limits.

Control Chart for Fraction Rejected: p Chart   The most widely used attribute chart is the p chart. This chart is used to monitor quality of a process through accounting for the fraction rejected as nonconforming, which is defined as the ratio of the number of nonconforming items in a population to the total number of items in the population. The items may have several quality characteristics that are examined simultaneously by the inspector. If the items do not conform to standard on one or more of these characteristics, the item is classified as nonconforming.

Suppose that the true fraction nonconforming p in a production process is known or is a standard value specified by management. Then the centerline and the control limits of a p chart would be

$$\text{UCL}_p = p + 3\sqrt{\frac{p(1 - p)}{n}},$$

$$\text{Centerline} = p,$$

$$\text{LCL}_p = p - 3\sqrt{p(1 - p)n}.$$

The operation of a p chart consists of taking subsequent samples of n units, computing the sample fraction nonconforming $\hat{p}$, and plotting the statistic $\hat{p}$ on the chart. As long as $\hat{p}$ remains within the control limits and the sequence of plotted points does not exhibit any systematic nonrandom pattern, the process is assumed to be in control at the level of fraction nonconforming p. If a point falls outside or a nonrandom pattern is observed, fraction nonconforming is assumed to have shifted to a new level (either high or low).

When p Is Not Known in Advance   Then it must be estimated from observed data. The usual procedure is to select m preliminary samples, each of size n. As a general rule m should be 20 or 25. Then if there are D_i nonconforming units in sample i, the fraction nonconforming can be computed as

$$\hat{p}_i = \frac{D_i}{n}, \quad i = 1, 2, \ldots, m,$$

and the average of these individual sample fractions nonconforming is

$$\bar{p} = \frac{\sum_{i=1}^{m} D_i}{mn} = \frac{\sum_{i=1}^{m} \hat{p}_i}{m}.$$

The statistic $\bar{p}$ estimates the unknown fraction nonconforming p. The centerline and control limits of the control chart for fraction nonconforming are computed as

$$\text{UCL} = \bar{p} + 3\sqrt{\frac{\bar{p}(1 - \bar{p})}{n}},$$

$$\text{Centerline} = \bar{p},$$

$$\text{LCL} = \bar{p} - 3\sqrt{\frac{\bar{p}(1 - \bar{p})}{n}}.$$

Alternative to p Chart: np Chart   Since p is fraction defective and n is sample size, np gives the total number defective in a sample. When subgroup size is variable, the control chart must show the fraction rejected rather than the actual number rejected. However, if the sample size is constant, a chart showing the actual numbers can be used. Often, such a chart gives more meaningful information to management.

Let D be the number of defective items in a sample of size n with an average of $p\%$ defective. Then D is a binomial distributed random variable with mean np and standard deviation $\sqrt{np(1 - p)}$. Then the parameters of a np chart will be

$$\text{UCL} = np + 3\sqrt{np(1 - p)},$$

$$\text{Centerline} = np,$$

$$\text{LCL} = np - 3\sqrt{np(1 - p)}.$$

A simple example in Table 6 illustrates the application of an np chart. Notice that both centerline and the control limits are functions of n, and hence sample sizes must be fixed if np charts are to be used.

Control Chart for Nonconformities: c Chart   A *nonconforming* product is one that does not satisfy one or more of the specifications for that product. Each instance of discrepancy between a product and its specifications is called *nonconformity*. Depending on the severity of the nonconformities, a product may not be classified as nonconforming with one or more nonconformities. For example, a conveyor belt may have one or more spots per meter length that are undesirable (i.e., nonconformities) and may still not be classified as nonconforming. The defective spots may be such that, even though they are undesirable, they do not affect the belts performance. However, if there are too many of those nonconformities (beyond an acceptable level), then the product is classified as nonconforming. It is possible to develop a control chart to monitor the total number of nonconformities per unit or the average number of nonconformities per unit.

The control chart for nonconformities, known as c chart, has a very restricted field of application, compared to $\bar{X}$, R, and p or np charts. However, c chart applications are unique and cannot be substituted by other charts. Some other examples of c chart applications (Grant and Leavenworth, 1996) are

- The number of nonconforming rivets in an aircraft wing or fuselage
- The number of breakdowns at weak spots in insulation in a given length of insulated wire subject to a specified test voltage
- The number of surface imperfections observed in a galvanized sheet or a painted, plated, or enameled surface of a given area
- The number of seeds (small air pockets) observed in a glass bottle

Table 6. np **chart application**

	Inspection of Reels of Electrical Components	
Reel Number	Number of Connectors for Reel n (lot size)	Number of Rejects r
1	1000	5
2	1000	8
3	1000	3
4	1000	6
5	1000	9
6	1000	4
7	1000	10
8	1000	5
9	1000	7
10	1000	4
Totals	10000	5

$$p = 61/10000 = 0.0061$$

$$np = 1000(0.0061)$$

$$\sigma = \sqrt{np(1-p)} = \sqrt{1000(0.0061)(1-0.0061)} = 2.462$$

$$\text{UCL} = np + 3\sigma = 6.1 + 3(2.462) = 13.486$$

$$\text{LCL} = np - 3\sigma = 6.1 - 3(2.462) = -1.286 = 0$$

Parameters of the c *Chart.* The c chart parameters are given as

$$\text{UCL} = c + 3\sqrt{c},$$

$$\text{Centerline} = c,$$

$$\text{LCL} = c - 3\sqrt{c}.$$

Example: c *Chart* The example considered here is adapted from Grant and Leavenworth (1996). Table 7 gives the number of errors of alignment observed at final inspection of a certain model of airplane. Figure 13 gives the control chart for these 50 observations. The alignment errors observed on each airplane constitute one subgroup for this chart.

The total number of alignment errors in the first 25 planes was 200. The average $\bar{c}$ is $200/25 = 8.0$. Trial control limits computed from this average are as follows:

$$\text{UCL} = \bar{c} + 3\sqrt{\bar{c}} = 8 + 3\sqrt{8} \approx 16.5,$$

$$\text{LCL} = \bar{c} - 3\sqrt{\bar{c}} = 8 - 3\sqrt{8} = \text{negative, therefore no LCL.}$$

Whenever calculations give a negative value of the lower control limit of a control chart for attributes, no lower control limit is used. In effect, the chart will exhibit an upper control limit only.

As none of the first 25 points on this chart is outside the trial control limits based on these points, the standard number of defects may be taken as c and the control chart can be continued for the following period with a central line of 8.0 and upper control limit of 16.5.

One point (airplane no. 236) out of the next 25 is above the upper control limit. The average during this period was $236/25 = 9.44$. (Even omitting the out-of-control value, the average is 9.08.) Of the final 16 points corresponding to airplanes 235–250, 12 are above the standard c, 3 are exactly at the standard, and only 1 is below. It seems evident that there has been a slight but definite deterioration in quality during this period.

Control Chart for Nonconformities with Choice of Sample Size: u Chart   In the above example, the sample size is restricted to one unit. However, there may be a need for varying the sample size. For example, it may be necessary to increase the area of opportunity for the occurrence of nonconformities, to obtain a positive lower

Table 7. Aircraft alignment errors observed at final inspection

Airplane Number	Number of Alignment Errors	Airplane Number	Number of Alignment Errors
201	7	226	7
202	6	227	13
203	6	228	4
204	7	229	5
205	4	230	9
206	7	231	3
207	8	232	4
208	12	233	6
209	9	234	7
210	9	235	14
211	8	236	18
212	5	237	11
213	5	238	11
214	9	239	11
215	8	240	8
216	15	241	10
217	6	242	8
218	4	243	7
219	13	244	16
220	7	245	13
221	8	246	12
222	15	247	9
223	6	248	11
224	6	249	11
225	10	250	8
Total	200	Total	236

From Grant and Leavenworth (1996).

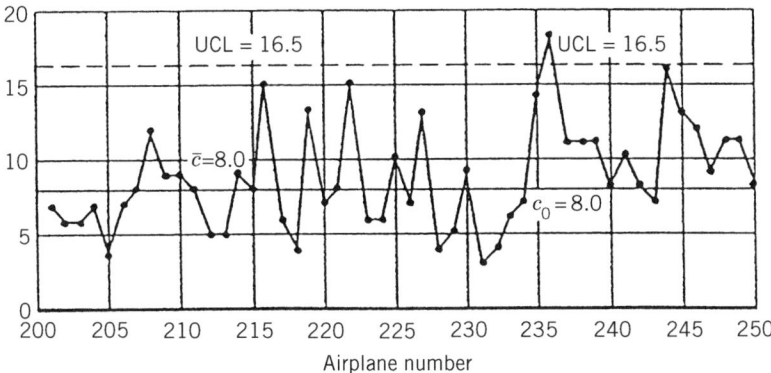

Figure 13. Control chart for nonconformities, c (Grant and Leavenworth, 1996).

limit, or to obtain a particular probability of detecting a shift. Moreover, economic considerations may enter into sample size determination.

Let the sample size be n inspection units. If c is the total number of nonconformities in the sample, then the average number of nonconformities per inspection unit is

$$u = \frac{c}{n}.$$

Hence the parameters of the u chart can be given as

$$UCL = \bar{u} + 3\sqrt{\frac{\bar{u}}{n}},$$

$$Centerline = \bar{u},$$

$$LCL = \bar{u} - 3\sqrt{\frac{\bar{u}}{n}},$$

where $\bar{u}$ represents the observed average number of nonconformities per unit in a preliminary set of data.

3.4. Acceptance Sampling

Acceptance sampling is a statistical method of accepting or rejecting a lot based on inspection of a small sample taken from the lot. In many cases, sampling is economically advantageous because inspection of the entire lot would be too costly, and impractical when destructive tests are necessary. A sampling inspection is likely to be more effective than 100% inspection because of the errors that are introduced due to the inspector's fatigue during 100% inspection. Also, when the vendor has an excellent quality history and their process capability index is high, there is no need for a 100% inspection. Acceptance sampling is commonly applied to incoming raw material and outgoing finished products. It is also common practice to do acceptance sampling between various stages of a production process.

The real challenge in most acceptance sampling is to design or select a satisfactory sampling plan. Because there are risks of accepting *bad* and rejecting *good* items. There are a number of different ways to classify acceptance sampling plans. One major classification is by attributes and variables. *Variables* are quality characteristics that are measured on a numerical scale. *Attributes* are quality characteristics that are expressed on an *accept* or *reject* basis.

Guidelines for Acceptance Sampling The major types of acceptance sampling procedures and their applications are shown in Table 8. In general, the selection of an acceptance sampling procedure depends both on the objective of the sampling organization and the history of the organization whose product is sampled. Furthermore, the application of sampling methodology is not static. That is, there is a natural evolution from one level of sampling effort to another. For example, for a vendor with excellent quality history, transition can be made from sampling each lot to skip-lot sampling. Also after extensive positive experience with a vendor who has a high process capability, acceptance sampling may become unnecessary.

Notation The following notation is used in the discussion of the sampling plans.

- N = number of items in a given lot
- n = number of items in a sample
- D = number of nonconforming items in the lot of size N
- r = number of nonconforming items in the sample of size n
- c = acceptance number, the maximum allowable number of nonconforming items in a sample of size n

Table 8. Acceptance sampling procedures

Objective	Attribute Procedure	Variables Procedure
Assure quality levels for consumer/producer	Select plan for specific OC curve	Select plan for specific OC curve
Maintain quality at a target	AQL system: MIL STD 105D, ANSI/ASQC Z1.4	AQL system: MIL STD 414, ANSI/ASQC Z1.9
Assure average outgoing quality level	AOQL system: Dodge–Roming plans	AOQL system
Reduce inspection with small sample sizes, good quality history	Chain sampling	Narrow-limit gaging
Reduce inspection after good-quality history	Skip-lot sampling; double sampling	Skip-lot sampling; double sampling
Assure quality no worse than target	LTPD plan; Dodge–Roming plans	LTPD plan; hypothesis testing

From Montgomery (1991).

- p = fraction nonconforming; D/N (for a lot); r/n (for a sample)
- p = true process average fraction nonconforming
- $\bar{p}$ = average fraction nonconforming in observed samples
- P_a = probability of acceptance
- α = producer's risk, type I error
- β = consumer's risk, type II error
- AQL = acceptable quality level (the maximum percent defective that, for the purpose of sampling inspection, can be considered satisfactory as a process average).
- AOQ = average outgoing quality
- AOQL = average outgoing quality limit

Single Sampling Plans for Attributes A common procedure in acceptance sampling is to consider each submitted lot of product separately and to base the decision on acceptance or rejection of the lot on the evidence of one or more samples chosen at random from the lot. When the decision is always made on the evidence of only one sample, the acceptance plan is described as a *single sampling plan*.

A single sampling plan requires that three numbers be specified, such as N, n, and c. The plan goes as follows: From a lot of size N a sample of size n is drawn and inspected for their conforming/nonconforming attribute. If the number of items in the nonconforming category exceeds c, then the lot is rejected.

Operating Characteristic Curve An important measure of the performance of an acceptance sampling plan is the Operating Characteristic (OC) curve. This curve plots the probability of accepting a lot versus the lot fraction nonconforming. Thus, the OC curve displays the discriminatory power of a sampling plan. That is, it shows the probability of accepting or rejecting a lot for a lot with given percentage of nonconforming items. A sample OC curve is shown in Figure 14 for the sampling plan N = very large (theoretically infinite), $n = 89$, $c = 2$.

The figure shows that for a lot having $p = 0.03$, the probability of accepting the lot, according to the plan, is 0.4985. That is, if it is okay to have a lot with 3% nonconforming, the plan will accept the lot only 49.85% of the time. On the other hand, if a lot with 3% or more defective is not acceptable, this plan will reject the lot only 50.15% of the time.

Rectifying Inspection When a lot is rejected, it requires corrective action generally in the form of 100% inspection. Such inspection identifies the nonconforming items for rework or repairs. Such sampling programs are called *rectifying inspection programs* since the inspection policy affects the final quality of the outgoing products. Suppose that the incoming lot to the inspection activity has fraction defective p_0. Some of these lots will be accepted (and will continue to have fraction defective p_0), and some will be rejected, which after 100% inspection will have 0 fraction defective. The average fraction defective of the outgoing lots (between 0 and p_0) is called it

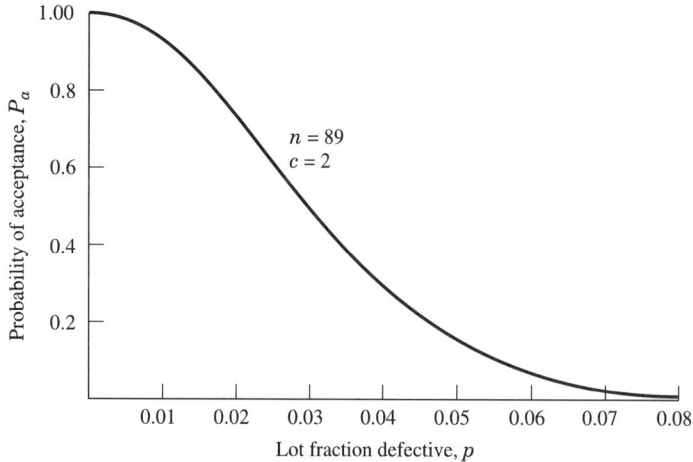

Figure 14. OC curve of the single sampling plan, $n = 89$, $c = 2$ (Montgomery, 1991).

Average Outgoing Quality (AOQ), which is widely used as a measure for evaluating sampling plan. The formula for calculation of AOQ can be developed as follows:

Let N be the lot size with fraction defective p, and n be the sample size. Defective items are replaced with good ones. Then n items after inspection have no defective items. If the lot is rejected, then the remaining $N - n$ items are also 100% inspected and hence the outgoing lot has no defective item. If the lot is accepted based on the sample, then $N - n$ remaining items still have $p(N - n)$ defective items. If P_a is the probability of the lot being accepted, based on the sample, then the AOQ is given as

$$\text{AOQ} = \frac{P_a p(N - n)}{N}.$$

When the lot size N is large compared to n, then

$$\text{AOQ} \approx P_a p.$$

Another important measure of rectifying inspection is the total amount of inspection required by the sampling program. If a lot is rejected, then only the lot is 100% inspected, otherwise only n items of the lot get inspected. Recall that $(1 - P_a)$ is the probability of a lot being rejected. Hence, the *average total inspected* (ATI) is

$$\text{ATI} = n + (1 - P_a)(N - n).$$

3.5. Military Standard 105D

Standard military procedures for accepting lots through sampling by attributes have been in place for over 40 years. Current standard designated MIL-STD-105D was issued by the U.S. Department of Defense in 1963 (MIL-STD-105D, 1963). The standard is essentially a set of sampling plans indexed with respect to AQL. The selection of a specific plan from MIL-STD-105D depends upon the lot size, the AQL that has been chosen, and the inspection level that has been chosen. A derivative of the MIL standard is the corresponding civilian standard ANSI/ASQC Z1.4.

The standard has three types of sampling plans: single sampling, double sampling, and multiple sampling. Corresponding to each sampling type, there are three alternatives, normal inspection (level II), tightened inspection (level III), and reduced inspection (level I). Sampling usually begins with normal category and later changes to tightened or reduced category if quality history of the vendor deteriorates or improves, respectively. There are also four special inspection levels, $S1$, $S2$, $S3$, and $S4$. The special inspection levels use very small samples and should only be used when small sample sizes are necessary and when large sampling risks can or must be tolerated.

3.6. Dodge–Romig Sampling Plans

H. F. Dodge and H. G. Romig developed a set of sampling inspection tables for lot-by-lot inspection of products by attributes in 1959. These plans are used extensively in manufacturing. Two types of sampling plans are presented in the Dodge-Romig tables: (1) plans for lot tolerance percent defective (LTPD) protection and (2) plans that provide a specified AOQ. For each of these sampling approaches, there are tables for single and double sampling.

Sampling plans that emphasize LTPD protection (e.g., Dodge–Romig plans) are often preferred to AQL-based sampling plans such as those in MIL-STD-105D, especially for critical components and parts. As can be seen below, even a small AQL implies a large number of defective parts that cannot be tolerated in today's competitive production environment (Montgomery, 1991).

AQL %	Defective Parts per Million
10	100,000
1.0	10,000
0.1	1,000
0.01	100
0.001	10
0.0001	1

3.7. Acceptance Sampling by Variables

In situations where a quality determining characteristic of a product is continuous and has a known distribution, it is sometimes economically advantageous to employ an acceptance sampling plan that utilizes the actual measurements of the characteristic rather than attributes of conforming and nonconforming. The advantage of acceptance sampling by variables is based upon the principle that a sample of n measurements contains more information than a sample of n counts. Hence, a plan based on measurements should require a smaller sample size than one based on attributes having comparable properties. Principal disadvantages of acceptance sampling by variables are the cost of variable inspection compared to attribute inspection, and potential for serious inaccuracies in the actual risks of accepting or rejecting lots when the distributions are not normal. However, increased cost of inspection may be offset by the reduction of the sample size.

When acceptable quality levels are very small, the sample sizes required by attribute sampling plans are very large. Under these circumstances, there may be significant advantages in switching to variables measurement. Thus, with increasing emphasis on the performance measure of *number of defective parts per million*, variables sampling becomes a very attractive alternative. Readers interested in implementing acceptance sampling plans by variables are referred to Montgomery (1991).

4. QUALITY SYSTEMS: ANSI/ASQC Q9000, ISO 9000

Products supplied by an organization are intended to satisfy customer's needs. With increased global competition, customer expectations of quality have increased tremendously. To maintain the level of quality required for today's market, organizations must employ and maintain increasingly effective and efficient production systems. Customer requirements in the form of specifications are not enough to guarantee that the requirements will be met consistently, especially when there may be deficiencies in the organizational system that produces and provides support for the product. These concerns have led to the development of the quality system standards and guidelines that complement the relevant product specifications. The international standards in the ISO 9000 family are intended to provide a generic core of quality system standards applicable to a broad range of industry and economic sectors. The American National Standard ANSI/ASQC Q9000 provides a roadmap for use of the ISO 9000 standards family.

According to the ANSI/ASQC Q9000, the key objectives and responsibilities of an organization are as follows. An organization should

1. Achieve, maintain, and seek to improve continuously the quality of its products in relationship to the requirements for quality.
2. Improve the quality of its own operations, so as to meet continually all customers' and stakeholders' (which include employees, owners, subsuppliers, and society) stated and implied needs.
3. Provide confidence to its internal management and other employees that the requirements for quality are being fulfilled and maintained, and that quality improvement is taking place.
4. Provide confidence to the customers and other relevant parties that the requirements for quality are being, or will be, achieved in the delivered products.
5. Provide confidence that quality system requirements are fulfilled.

Four facets that are key contributors to product quality are identified by the standards as follows:

1. Quality due to definition of needs for the products: The first facet is quality due to defining and updating the product, to meet marketplace requirements and opportunities.

2. Quality due to product design: The second facet is quality due to designing into the product the characteristics that enable it to meet marketplace requirements and opportunities and to provide value to customers. More precisely, quality due to product design is the product design features that influence intended performance within a given grade, plus product design features that influence the robustness of product performance under variable conditions of production and use.

3. Quality due to conformance to product design: The third facet is quality due to maintaining day-to-day consistency in conforming to product design and in providing the designed characteristics and values for customers and stakeholders.

4. Quality due to product support: The fourth facet is quality due to furnishing support throughout the product life-cycle, as needed, to provide the designed characteristics and values for customers and stakeholders.

4.1. What Is a Quality System?

The aim of a quality system is to ensure that the facility's product or service (generically referred to as *output*) meets the customer's quality requirements. A quality system incorporates both Quality Assurance (QA) and Quality Control (QC), where QA refers to planned, formalized activities intended to provide customer confidence on quality, and QC refers to the in-process activities and techniques intended to maintain certain quality characteristics to their specified levels.

A quality system is a management-driven, facility-wide, and process-wide program of plans, activities, resources, and events. An effective quality system (according to Johnson, 1997) is the philosophical and procedural glue that unites all elements of the facility including employees, plant, equipment, procedures, suppliers, and customers.

Facilities that operate quality systems tend to exhibit the following attributes:

- A philosophy of prevention rather than detection
- Continuous review of critical process plans, corrective actions, and outcomes
- Free and clear communication among the parties involved
- Thorough recordkeeping and efficient control of documents
- Quality awareness among the employees
- High level of management support

4.2. Evaluating Quality Systems

When evaluating quality systems, there are three essential questions that have to be asked in relation to every process being evaluated. They are as follows:

1. Are the processes defined and their procedures appropriately documented?
2. Are the processes fully deployed and implemented as documented?
3. Are the processes effective in providing the expected results?

The collective answers to these questions relating, respectively, to the approach, deployment, and results will determine the outcome of the evaluation. An evaluation of a quality system may vary in scope and encompass a wide range of activities, such as management review and quality system audits.

4.3. Documentation of a Quality System

A primary element of the quality system documentation is the *quality manual*. Of all the elements of the ISO 9000 quality system, none is as critical and central as the quality manual. It serves several purposes:

- It describes the objectives of the quality system.
- It details the procedures and actions needed to maintain a quality system.
- It demonstrates managements commitment to the quality system.
- It serves as a cross reference between the quality system and the quality standard to which the facility is certified.
- It serves as a reference document for the outside entities (such as customers, auditing authorities, and investors).

Sample Quality Manual for a Cellular Manufacturing Facility Typical elements of a quality manual prepared for a CM facility are as follows:

- Quality policy statement
- Company background
- Amendment record
- Controlled circulation list
- Management responsibility
- Quality system
- Contract review
- Design control
- Document and data control
- Purchasing
- Control of vendor-supplied products or services
- Product identification and traceability
- Process control
- Inspection and testing
- Inspection, measuring, and test equipment
- Inspection and test status

- Control of nonconforming products
- Corrective and preventive action
- Handling, storage, packaging, preservation, and delivery
- Control of quality records
- Internal quality audits
- Training
- Servicing
- Statistical techniques

Only a few of the above elements of the quality manual are discussed below.

Product Identification and Traceability   Clearly stated procedures exist for identification of all products (conforming and nonconforming) to allow easy and correct traceability. The activities controlled by the procedures include the following:

- Materials (raw materials, components, and subassemblies) entering the cell are carried in boxes with proper identification tags having date of entring the cell, source, batch number, process plan number, processing completion due date, and so forth. Details of the materials entering the cell are also entered in the computer "product trace" database. Computer data entry responsibilities are clearly stated in the job descriptions of the cell personnel.
- As the products move through different parts of the cell (according to the process plan), box labels are marked and the database is updated. Every nonconforming product produced in any part of the processing sequence are tagged individually with a "defective product report" generated through the computer (which links the report to the product database) and carried in a separate box with adequate label.
- The physical flow and the computer record of the products are maintained such that a product history is traceable starting from either end (the physical product or the computer identification number).

Process Control   Documented controls, plans, and procedures exist to govern the methods, practices, and responsibility allocations necessary to complete cell activities and processes. The activities governed by these controls include the following:

- Machine parameter settings and setup instructions
- Detailed process plans
- Detailed routing plans
- Tool change and other process maintenance procedures
- Quality control procedures
- Safety procedures
- Records keeping requirements

Inspection and Testing (I&T)   Detailed procedures exist for controlling all phases of receiving, in-process, and finished goods inspection and testing. Some of activities controlled by these procedures are as follows:

- Incoming materials are not allowed to enter the process until specified inspection and testing procedures are completed. The extent of I&T depends on the extent of vendor process capabilities and the audit results.
- Intracell movement of the products from one machining center to another requires that all documented I&T are completed.
- Finished products are not dispatched until required I&T are completed.
- All I&T records are maintained in computer database.

Control of Inspection and Test Equipment   Documented procedures exist for control of the calibration, accuracy, and repeatability of the I&T equipment. Some of the activities that are controlled are the following:

- Periodic calibrating and accuracy checking of all of the I&T equipment
- Availability of detailed technical specification for acceptable calibration and accuracy results
- Training and retraining of the personnel with the latest equipment and testing software technology
- Procedures to be followed when I&T equipment are found to be out of calibration
- Maintaining suitable environment for carrying out the I&T requirements
- Methods to protect I&T equipment from unauthorized calibration
- Maintaining records of all actions

For an excellent sample quality manual, readers are referred to Johnson (1997).

4.4. Selection and Use of Q9000 Standards Family

ANSI/ASQC Q9000-1, *Quality Management and Quality Assurance Standards—Guidelines for Selection and Use*: Reference to this standard should be made by any organization that is contemplating the development and implementation of a quality system. This standard clarifies the principal quality-related concepts and provides guidence for the selection and use of the ISO 9000 family.

ANSI/ASQC Q9001, *Quality Systems—Model for Quality Assurance in Design, Development, Production, Installation, and Servicing*: This standard should be selected and used when the need is to demonstrate the supplier's capability to control the processes for design as well as production of conforming product. The requirements specified are aimed primarily at achieving customer satisfaction by preventing nonconformity at all stages from design through to servicing.

ANSI/ASQC Q9002, *Quality Systems—Model for Quality Assurance in Production, Installation, and Servicing*: This standard should be selected and used when the need is to demonstrate the supplier's capability to control the processes for production of conforming product.

ANSI/ASQC Q9003, *Quality Systems—Model for Quality Assurance in Final Inspection and Test*: This standard should be selected and used when conformance to specified requirements is to be assured by the supplier solely at final inspection and test.

ANSI/ASQC Q9004, *Quality Management and Quality System Elements—Guidelines*: Reference should be made to this standard by any organization intending to develop and implement a quality system.

REFERENCES

Brassard, M. (1989). *The Memory Jogger Plus+*, Goal/QPC, Methuen, MA.

Dodge, H. F., and Romig, H. G. (1959). *Sampling Inspection Tables: Single and Double Sampling*, 2nd ed., Wiley, New York.

Grant, E. L., and Leavenworth, R. S. (1996). *Statistical Quality Control*, 7th ed., McGraw-Hill, New York.

Johnson, P. L. (1997). *ISO 9000: Meeting the International Standards*, 2nd ed., McGraw-Hill, New York.

MIL-STD-105D (1963). *Sampling Procedures and Tables for Inspection by Attributes*, U.S. Department of Defense, U.S. Government Printing Office, Washington, DC.

MIL-STD-414 (1957). *Sampling Procedures and Tables for Inspection by Variables for Percent Defective*, U.S. Department of Defense, U.S. Government Printing Office, Washington, DC.

Montgomery, D. C. (1991). *Introduction to Statistical Quality Control*, 2nd ed., Wiley, New York.

Montgomery, D. C. (1984). *Design and Analysis of Experiments*, 2nd ed., Wiley, New York.

Ross, P. J. (1988). *Taguchi Technique for Quality Engineering*, McGraw-Hill, New York.

Taguchi, G., Elsayed, E. A., and Hsiang, T. (1989). *Quality Engineering in Production Systems*, McGraw-Hill, New York.

12

ORGANIZATIONAL AND HUMAN ISSUES IN CELLULAR MANUFACTURING

Al Miller

Manufacturing Consultant
14871 45th Avenue North
Plymouth, Minnesota 55446

1. GENERAL OVERVIEW

The material in this chapter is the result of actual experiences in implementing change within several different companies. These companies have different cultures and management styles. These differences make implementing change a challenging, yet controllable, task. It is presented to identify the various considerations necessary to make a successful program for implementing change. It is easier to succeed by being aware of the adversities in the workplace and being prepared for them.

The experiences address implementation of several modern management philosophies that require the development of a "team" concept. This normally implies some form of employee empowerment. However, the definition, or in more precise terms, the perception of employee empowerment is clearly different from one example to the next. Any company may have a variation on these perceptions as its own culture dictates the way people in the organization are respected and valued.

Although many companies try to improve the value relationship of the employee, continuous cutbacks in the workforce, reductions in annual salary reviews, limited

Handbook of Cellular Manufacturing Systems, edited by Shahrukh A. Irani
ISBN 0-471-12139-8 © 1999 John Wiley & Sons, Inc.

growth potential, and added responsibilities without added benefits cause employee perception of what is really happening to be different from what is printed in the company mission statements.

The main theme of motivation in the corporate culture is provided to improve the reader's understanding. Such an understanding will assist the future manager or engineer be at the same level as employees in realizing how the organization functions and the variables in effect when decisions are made. Everyone should understand the basis for motivation and its power. Everyone will agree that communication is very important. However, motivation is the most critical and most important area to address. No matter how much communication is provided, it is the motivation that makes the individual perform on the job. If the motivation is down, their performance goes down as well. The manager must have a means to motivate the employee. Employees must be able to communicate what motivates them personally.

Communication is only one tool to help motivation. Far to often it is seen as the major problem in a company. In any meeting you can get people to agree with you by saying "We don't communicate well here." Very few, if any, companies feel their communication systems are perfect. If they were, would it matter to an employee who is not motivated? Experience says it would not matter if the employee is not motivated to act on the communications being received.

Information that needs to be communicated to improve motivation is the information necessary for employees to understand the job they have to perform. They have to understand the expected results. They have to understand the boundaries (cost, time, equipment, etc.) in which they are to do the job. They have to understand the support available to them to complete the job. They have to understand why they are important to do the job. The last reason for communication is most important for motivation. Unfortunately, it is the one usually left out because it is assumed that employees understand their importance to the organization.

Consider the situation of an employee who is working in a company that is "downsizing," "restructuring," "re-engineering," or any other 1990s corporate term used by the top level of management in that specific organization. The department this individual is in had 20 employees and is now down to 6. Employees were released as they completed their projects. The employee is asked to expedite his or her efforts to complete a project a month sooner. Will the employee perform to the best of his or her ability to achieve this goal? Maybe. Maybe not.

The situation is not clear where they are always the same. For example, a large company decided to restructure into strategic business units in an effort to streamline the operation and reduce cost of manufacture. The business area was eroding rapidly, and the company had to redirect toward other business areas to gain a competitive edge and use its resources for the best possible benefit. Several small operations, which were vertically integrated into the organization, were determined to be cost burdens on the overall operations and of no benefit to the new business areas. It was determined that these operations should be shutdown and sold off.

The plan was to inform the management of each operation and explain the situation to them. They were asked to support the overall company effort and to sell off their operations and reduce all liabilities to the company. They would be rewarded with a new assignment at the same level within the restructured organization of the "mother" company.

Each manager was allowed to implement the plan in his or her own manner. Each had clear communication. Each was provided with a date and the necessary support to accomplish the task. Each agreed the task was attainable. This is where the similarity ended.

The issue of trust played an important role. Not all of the managers trusted their superiors. Some of the managers trusted their superiors as if the superior were God. The results varied greatly.

One operations manager told his staff that the company had turned against them and were planning to sell the entire operation. Morale dropped. Jobs became late and costs ran over budget. Important people in this operation left the company on their own before a replacement could be found or other employees could be trained to take over the assignments. Contracts were canceled. Even some lawsuits developed. However, the end result was achieved. The operation was closed down and sold off to an old-time competitor. The manager was placed in the new organization as promised, but he soon left the company as well, leaving to work for a competitor.

On the other side of the scale, a manager called all of the employees together. They were told the company had to downsize to remain competitive. Their operation was selected to be eliminated, but they had several months to achieve this end. The dates were discussed and teams were developed to attain various parts of the overall plan. The entire operation was part of the process to determine how to achieve the goal and how the employees would be treated. The Human Resource Department was brought into the process, and the operation was provided with job opportunities throughout the larger company. As employees were accepted for new jobs in the remaining company operations, others were trained to finish the work they had left. Downsizing occurred in the steps as the workload was transferred to outside parties. All employees had faith that they would find a position within the original company organization.

Some employees did not feel comfortable with the newly restructured company. They decided to go to a competitor. Yet, these people were still supported by the operation to achieve their new job outside the company and they, in return, worked with the downsizing operation to assure their job was handed over to another employee with adequate expertise to complete the job on time and on budget.

It was almost a happy ending. Approximately 70% of the staff were reassigned into the remaining operations of the company. The operation was shutdown and sold 2 months ahead of schedule. No contracts were left unsatisfied. No lawsuits prevailed. Customers were sorry they could not continue business with the smaller operation. The manager of this operation was released the day the small operation was sold. No position was available at that time in the large company for the manager.

We do not do a good job a rewarding and valuing our employees. This is observed and discussed at water coolers, coffee machines, and lunches. It would be very difficult to go through this exercise again with the people who transferred back into the large company. They knew what happened to their manager. They would expect the same treatment toward themselves. Until the manager was released, they felt the company was the manager. Now they feel the company is something else, with no regard for loyalty or valuing the employee. It took several years to get the worker performance back to where it was before the restructuring.

The experiences addressed in this section pertain to implementing Total Quality Management (TQM), Cellular Manufacturing (CM), continuous improvement programs, and the relatively new ISO 9000 philosophy in American companies. This chapter will focus on ways to address an organization's culture, structure, and the manner in which to approach selling or motivating the organization to want to implement change.

2. ORGANIZATIONAL CULTURE, STRUCTURE, AND MOTIVATION

2.1. Organization Culture

Organizational culture is not easy to identify. There are usually hints when you interview for a position in a company as to how other people within the organization are referred to or the way they are treated. Even the way you are treated when being interviewed gives you some idea of the culture. The part that makes reading an organizational culture so difficult is individuals within the organization that have their own culture and apply them directly to a situation, masking the overall company culture.

Culture is very important when trying to implement change within an organization. The best way to change an organization is to use the culture that exists and apply it to motivate others in changing to better satisfy the culture in place. If you have to oppose the culture of an organization, you had better own the company or, at least, be the chief executive officer (CEO), president, or other top official with power and secure position in the organization. If not, you will require an astute power of persuasion.

Organizations have a culture in the same way a person does. The culture is varied and complex with many facets to it. However, a main theme is usually recognized as you deal with the organization over a period of time. It is this main theme that you want to address when implementing change into that organization.

As an example, people may be very nice to one another in a company and work well together. During a crisis situation, the culture becomes obvious. The crisis may be as minute as a simple phone call.

Everyone is chatting and working well together in the office area. Bob goes into Mary's office and asks if she has time to review a project with him to assure the plan is appropriate for what is to be the expected outcome. She says yes and the two start to go over the plan. They are sitting at a small round table in Mary's office. Bob is talking about the reasoning of his plan when the phone rings.

Mary picks up the phone on the first ring and answers it, cutting off Bob in midsentence. It was an informal call to make lunch plans with another manager. Yet, Mary completes the call and then hangs up. Mary sits back down at the table and says, "Go on," to Bob as if Bob had stopped the discussion.

If everyone reacts in this manner throughout the organization when a phone rings, a culture may exist that implies a crisis must be dealt with immediately. The phone call may or not be a crisis. Since Mary doesn't know if it is or isn't, she has to treat it as one in order to assure she complies with the culture. Bob is an insignificant being at this moment. The value of Bob to the company, or at least to Mary, is downgraded from "important enough to take the time and review his project" to "insignificant in comparison to an impromptu lunch appointment" by Mary's action. The crisis is most important. Hence, if you deal with crises very well, you have a good opportunity to succeed in this culture. You may even succeed very well in this company if you create crises.

Many of the modern and forward-looking companies want to have culture that addresses quality and customer needs. Customers are always considered to be both internal and external, until a crisis hits. One Fortune 500 company is very quality oriented and it shows in everything it does. Its products are well received; employees are treated with respect and treat others with respect. The company's costs are under control and profits are very good. It has quality as a culture. This also means the Finance Department has a quality system and produces quality results. Financial controls are very adequate to address issues and respond effectively. Products are manufactured as designed, and the design is such that productivity and manufacturability issues are always addressed and well thought out. Contracts are fair to protect the customer and the company. Human Resources provides a quality service in selecting the best candidates for a employment opportunity within the company. Everything is done by the best possible means. For quality to be a culture, it is more than having a reliable product or service. Everyone must be working with the same concept in mind. If not, it is not the culture of the company, only an individual or departmental culture.

One company said its culture was one of customer service. Customer service was preached from every corner. Memos and mission statements were published addressing customer service. Awards from customers were hung in the front lobby of the main office areas to demonstrate how well the company performed according to the customers' own standards. Yet each month the factory was pressed to meet its production quota. Engineering support went from development and process improvements to direct factory floor support the last week of each month. Overtime went up the last week of each month to ensure quotas were met. Quality issues were overruled or just ignored to meet month end production rates. In some cases the Quality Department staff helped in building product and stopped their routine quality evaluations to ensure Statistical Process Control (SPC) and other quality standards were satisfied.

If customers complained about a product built in the last week of the month, it was immediately replaced with a new product and the repairs made to the product and restocked in the warehouse. The Service Department was the largest of any of its competitors.

The real culture of this company was more likely to be schedule attainment rather than customer service. More resources were devoted to meeting schedules. Rapid response to immediate needs of a customer do not mean you are customer oriented. Your product should be such that it meets all the customer requirements the first time the customer receives the product. Having an excellent repair department does not warrant a company focus on customer service. It is merely a means to survival since the real culture could cause loss of customers if this *correction* to the system was not added.

The motto of a successful company in Minnesota tells the story of an antelope and a lion. It says the antelope gets up every morning and runs to avoid being eaten by the lion. The lion gets up every morning running to catch an antelope to eat. The moral of the story is you have to get up running to survive. This company's culture is based on speed. Speed implies survival of the fittest.

The company that operates under this type of culture does everything around dates and time. Everything would be measured in terms of speed. If you can do something faster in this company, it will be more readily accepted than if you can do it better over a longer period of time. Rewards for success are high and given in a timely fashion. Bonuses and special checks for immediate results are common. So are terminations. Planning is not a practice that is highly rewarded in this culture. Many people take unnecessary risk to get things done faster and are terminated just as fast for not succeeding.

Employee value is not a concept in this type of culture. In order to incorporate change in this type of company is easy if it has immediate paybacks. Culture change is timely and would not be acceptable under this type of company practice. Trying to implement a CM approach would be difficult unless the cell could demonstrate an immediate benefit to the organization. The problem is the perception of *immediate* in this culture. It does not mean the CM approach is not beneficial nor does it mean the implementation cannot be timely. It does mean the perception of timely under normal situations may not be acceptable under this type of culture.

Control is a major culture seen in many companies. It does not restrict itself to small companies that are privately owned. However, it is usually associated with the smaller companies. Small companies have to do more functions with less resources to survive. This means few people are involved and micromanagement becomes a way of life to ensure survival of the operation. Many larger companies have developed a similar culture through promoting from within without proper training to transition the individual from an engineer, as an example, to a manager.

If a company has a control culture, the implementation for teams becomes difficult. Team leaders become very important. Facilitators become important. These considerations have to be taken into account in implementing change to a team type of organization in one that has a strong culture based on control.

The 1990s have developed a new culture ethic. One that existed in the early 1930s is back in full force. Survival. This is a philosophy of culture based on the ancient belief of survival of the fittest. Many companies are in a survival mode but do not adopt the struggle to survive as a part of their culture. They may still fight to survive using strong quality issues, as does TRW with its "mistake proofing" culture incorporated in its Vehicle Safety Systems Division located at the Washington, Michigan, facility. The motor vehicle industry is very competitive and understands survival in real terms. Yet TRW bases its survival on being better at what it does to distance its competition.

When survival is accepted as the culture. Anyone subordinate to, or in competition with, another employee may be beaten to death, in a business sense, to allow the other person to survive. Career murder can go on to ensure the career survival of another. Firings, layoffs, and huge cutbacks take place to ensure budgets are met, regardless of the productive outcome of the department or the future capabilities needed. Survival means now. Tomorrow is another day.

Small and middle-size companies that have been bought and sold several times in the last few years tend to develop toward this type of culture as top management gets cut in the cycle of changeovers in ownership. The people remaining become survivalists and do what they are told without question or providing corporate logic. Corporate logic is assumed to be wrong, since it didn't prevent the last change in ownership. Therefore no one wants to stick his or her neck out by supporting a logical approach. They are motivated by the culture to survive.

Scott Heimes wrote in an article for American Society of Quality Control (ASQC) *Total Quality Newsletter* addressing employee motivation and the changing environment of the 1990s the following statement: "For managers, nothing about the future is certain except, perhaps, that failing to change might be lethal to their careers." This varies with the specific company, of course, but the feeling is felt throughout the industrial world that the old way of doing things is wrong. This panacea is taken to extremes in some organizations and change is made for change's sake alone.

If no change is made, than nothing has been done to try to improve the organization. Therefore, the idea of change dictates the motivation to survive.

To reduce operating costs, one company cutback to such a degree that the entire manufacturing engineering support staff was eliminated. The company proudly states it now has hundreds of manufacturing engineers in its factory workers ability and knowledge of the job. However, the development of new processes, production control, and future determination of manpower needs are falling in credibility to the point that the finance system is meaningless to control the operation's costs.

Only the year-end actual budget against the annual budget is the measure of performance for the operation—too late for any timely corrective action. Also, changes come about in the year that causes deviations in the original plan. Without clear cost controls, comparison to an annual budget is meaningless. Development and implementation of new products to compete are hindered, if not stopped. By trying to become more competitive they actually become less competitive in this sense.

Fortunately, many companies do not utilize this practice of survival as a culture to manage the operation. This is a unique example of what an extreme situation may create. However, many companies exist and many different people are involved in their management. Personalities of companies vary as much as the personality of the leadership in them. Culture must be accounted for in implementing change. Culture is very difficult to change in itself, so you must consider the culture to use it in support of what you want to accomplish. It is easier to go with the flow to incorporate change than to swim upstream. The salmon, swimming upstream, may accomplish its task. But many die in the attempt.

2.2 Organization Structure

The organizational structure identifies responsibilities, concerns, priorities, and power. Through the organizational structure, you can determine who has the control over areas and what they are concerned with, by whom they report. The way the organization is structured also identifies if the teams are acceptable or is the structure indicative of an authoritarian operation with one person at the top of a steep pyramid. The organizational chart is a graphic representation of the organizational power structure and responsibilities. However, it is difficult to determine from the organization chart who really controls the operation other than by who is at the top. Even that can be misleading.

Many small to medium-sized companies have an organization structure where the originating owner or founding father is at the top. Yet, as the organization grew, the wise owner assigned functional responsibility to a trained management staff in those specific areas and gave the power to operate the company to them. The founding father became a marketeer and negotiator for original customer contracts. He maintained relationships and represented the company in social events between the various organizations.

In some companies the organizations are such that the CEO is hired by the Board of Directors only to position the company for sale or a friendly takeover. The actual operation of the organization is performed by the operations executive, whatever the title may be. In some organizations the chief financial officer dictates what is to be done based on strict financial control. Other companies are run by the manufacturing area to ensure the productivity and manufacturability aspects are addressed to maintain a competitive position in the industry. Marketing and Sales can also be the primary departments to direct the operation of a company.

The organization charts of these companies may all be identical, except for the individual's names assigned to the tasks. This does not mean the organizational structure is the same. The structure, as an analogy, is the way the "bones" are clearly linked in a body. When one bone moves, another bone may have to move with it. Sometimes several bones are repositioned by the change in one bone. There are usually several bones in a organizations structure that cause several other bones to react. Organization structure is usually something intangible and not found to be documented. It is part of the organization's structure that cause several other bones to react. Organization structure is usually something intangible and not found to be documented. It is part of the organization's culture and the manner in which decisions are made. One department may say: "We cannot make that change until Department X approves it." If this is truly the case, even when Department X has no functional responsibility for the change taking place, than Department X is the primary function operating the organization's structure. It is the bone that moves the other bones so the body of the organization will start to move forward.

An example is when Marketing comes to Engineering and informs members of a customer concern. Engineering states it can make the change, but Manufacturing has to approve it. In this case Manufacturing controls the company actions. The company is not customer based since the feedback from Marketing did not cause an immediate response by Engineering. It is not design controlled since the design will not be changed without prior approval from a nondesign area. It is not a quality-controlled operation since no mention of quality was even considered. Hence, in this hypothetical situation, Manufacturing controls the organization's actions. After you work with a company for a while, you can easily see which functional group controls the company, regardless of what is printed, spoken, or advertised.

The way an organization is structured or the way it lacks structure has an effect on implementing change. There must be a structure that focuses on the project, one whose primary function is to implement the change. If there is no focus in the organizational structure, the likelihood of the change being successful is null.

Consider the situation where an organization is structured such that Finance controls the operation. The Marketing Department is trying to implement ISO 9000 to improve sales in Europe. If the Marketing Department tries to implement this change without addressing the financial aspects of ISO 9000, the project will most likely become a temporary activity to be dropped at a later time. It will not get the support because the organization is not structured to make the change work. The structure is such to minimize and control costs. Implementing ISO 9000 requires significant up front activity and cost for certifications and external auditors. Without the organizational structure of the Finance Department behind the project, it will fail.

This does not mean the Marketing Department cannot present and sell the concept of ISO 9000 to the financial group and gain its support. Then the structure would be in place to make the implementation a success. The ideas do not have to be generated by the structure that controls the manner in which an organization works. The ideas only have to come through the structure in the correct direction.

Using the analogy of a city drainage system, ideas in an organization have to flow in the correct way. As water runs through a city drainage system it is directed down toward the river or holding area until it can be dealt with later. The same is true for changes in an organization. They must be reviewed and approved at each juncture in the organization's structure, so the ideas will flow with the support of the necessary power behind them to make them continue in the direction the organization wants. As water cannot flow up the drainage system without a lot of support from below, the same is true for a change in an organization.

In order to initiate a change, the organization structure must be established to support the change. Support must be from the top down with people of authority to assist in the implementation. If this is not done, the project will most likely fail. If the organization is not willing to structure itself to support the project, it is most likely not willing to actually support the project at all.

The graphics in Figure 1 show a generic type of organizational structure that is the typical type necessary for implementing change within an organization. It involves all levels of the organization and the direction is top down. Regardless of how good a change agent you may be, you will fail 99.9999999% of the time if not supported by the top level of the organization in which you are trying to implement the change.

The executive level and management focus team deals with strategic issues. These teams are concerned with the overall organization's focus and direction. They will assure the focus is maintained and coordinate activities and resources to steer the outcome in this direction. They are focused on the long-term direction and activities of the organization.

The departmental teams are the implementors of the strategic plan. They have the tactical issues at hand. Departmental teams determine how the task will actually be accomplished and provide the resources and direction to support the implementation based on its own tactical agenda. These teams are usually focused on a relative period

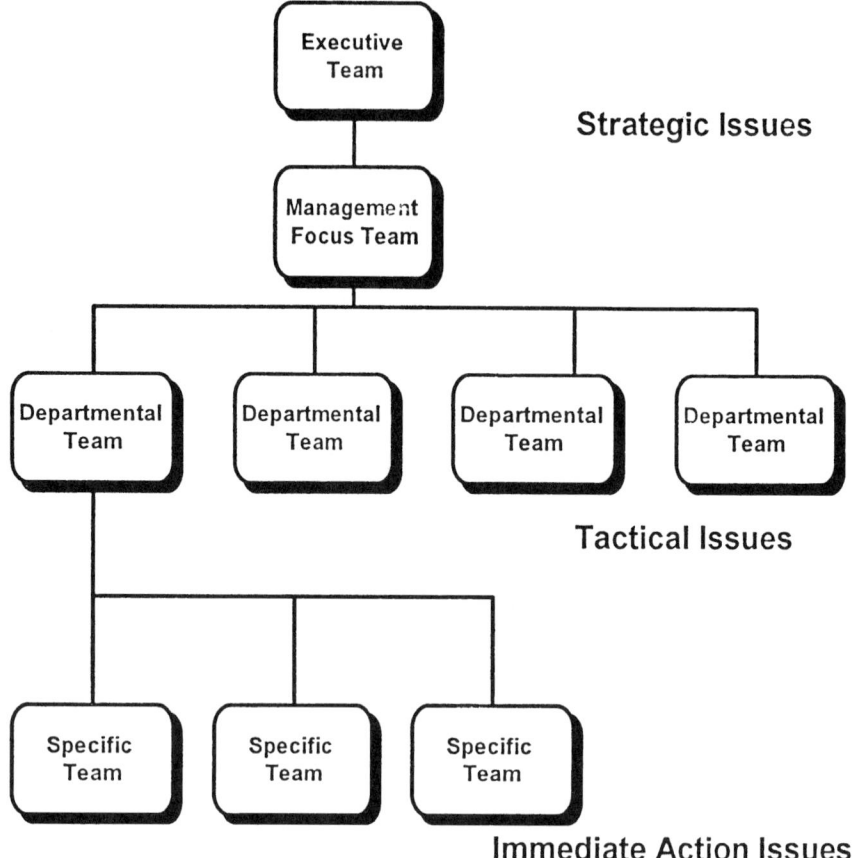

Figure 1. Organizational structure for change.

of time, such as a month, quarter, or year. Their focus is usually determined by budget concerns or other corporate controls which are in place.

Specific teams are the teams of hands-on people. The doers of the physical action. They will implement the ideas in a specific cell, department, or operation. They will set up the areas, implement the procedures, and direct or carry out the assignments. They do not care about long-term activities. These teams are focused on the immediate day-to-day activities. They are rarely, if ever, associated with long-term planning.

2.3 Three Types of Teams

Management Team The organization must be focused on the project. It has to assign resources and leadership that will ensure success. The leadership has to have authority to change the current system and processes. The structure must be such that it addresses all levels of responsibility in the organization to deal with matters that will arise. Usually this authority is controlled and never granted carte blanche. The manner in which this is best performed is to develop three levels of teams to address the issues at hand.

The management focus team is the highest level team other than the executive staff. This team is usually developed by the executive staff and will contain certain key people. These key people will be assigned to the team as development for them, as well as assuring success for the project itself. In some cases, the success of the team is a direct measure of the future success of the people on the team. If this is the case, these people will work very effectively to try and complete the task at hand with the best possible outcome.

The management team is the group that directs all activities associated with the project. Most often called the steering committee, this team is either comprised of executive-level management or has direct access to the executive management. They also have the authority to make the changes anticipated to implement the project.

Executive-level interface is crucial for this team. If it is not the executive level itself, the management team has to have interface with this group to assure the project is synchronized with the corporate strategies. This capability to remain synchronized with the corporate strategies is very important. Without this insight the project could easily veer off course and possibly cause it to be canceled.

Most major projects, such as CM, are part of a strategic plan to improve the competitiveness of the corporation. As the market and the competition changes and reacts to the industry and the environment the industry is in, the direction of the project may change as well. It is this team's responsibility to focus on the goals of the change. The steering committee has to keep the strategic issues of the organization in mind and continue to steer the course of the project as to benefit these strategic issues. The goals of the project must remain in harmony with the strategy. Hence, any change in strategy can immediately cause a change in the goals or the direction of the project.

A medium-sized company was undertaking a significant productivity improvement program. Part of this project was to implement several CM areas to reduce cost and start a transition to this philosophy. The company basically produces two product types. One was a commercial line and the other an industrial line. During the year, this medium-sized company took over a smaller company that had a significant sales advantage with the industrial products market.

Prior to this acquisition, the effort was focused on the commercial line since this was the forte of the original company, even though the company had a much higher margin of profit with the industrial products. After the acquisition, the strategy changed to divert resources to the industrial line and hold back on the commercial products. The steering committee changed priorities as well and implemented the next CM area in the industrial products.

This change in focus allowed for more savings and profit potential. Sales increased with the industrial products as the profit margins on the industrial products grew even greater. The same goal was attained, but the results had a greater benefit to the overall operation since the steering committee refocused the activity to remain in synchronization with the overall strategy of the corporation.

The management team also has to focus on resource utilization. No matter what the company is, there are only so many resources available to the organization to do things in a certain period of time.

This team level also has the project "live or die" authority: They decide whether or not to continue and devote resources to the project. Projects may last for years or days. It is always possible to have a project die even though the effort has gone on for several months or years.

You have to remember that management does not count money spent, only money to be spent. It is the concept of finance known as sunk funds. Once spent, they are gone. This does not mean you need to spend more money because you've already spent a large sum. This concept causes many projects to die even though they are close to completion.

I was directly responsible for a project that was 80% ready for implementation at a corporate level. Several divisions were trained and organization structure was in place to implement the project. The company changed leadership at the highest ranks. These new executives had different strategies for the company. They had a different understanding of the projects outcome. They killed the project and all funding in one week.

Unless the executive level and management teams continue to support your project, it can be stopped in an instant. Always keep them informed of its status and keep in touch with their feelings on it.

Sometimes these projects are not expected to be a success. In these cases, the team is selected by the executives to fail. If this is the case, the project has been given "lip" service and will not succeed no matter how hard the team works. One example of this was when a union suggested the implementation of CM to improve productivity. It was the union's belief that the individuals in the cell could better address the problems than the management that was in place.

The executive level decided to "go along" with the request to avoid conflict with the union. At the same time management put into place a team that was designed to cause failure. The team consisted of people who vehemently went by the rules and were very opposed to change or creativity. The project became so frustrating that the union support finally disappeared. As expected the project failed.

It is difficult to understand why some decisions are made in organizations. One such example was the paring of two very good managers who disliked each other on both a professional and a personal basis. The other members of the team became very frustrated with the constant bickering and squabbles seen in the meetings. After many long and tedious hours, the team finally did succeed. However, the results never reached the heights it could have succeeded, in the opinion of some of the other team members.

It appears executive management thought very highly of the two managers even though they disliked each other so apparently. Management also felt confident the project would succeed because the two managers were extremely competitive and succeeded in every assignment they were given. Executive management, however, did not feel the project was necessary to the survival of the corporation. They needed a project for the two managers where they could work together and develop a compromise. This would set the stage for future projects of more importance where the managers would have to compromise again. Hence, it was a training and development exercise for these two highly regarded managers. Everyone else on the team were mere spectators. In reality, the project itself was not important.

In developing personnel for executive positions, various techniques are applied. Many decisions, which appear to be simple in nature, are actually very complex. It is not necessary to understand all of the politics that may go on at the executive levels of an organization. It is only necessary to be aware they exist at times and not to become overly concerned when it is not readily obvious why a decision is made or how it is made. Support of change in an organization comes from many different directions. The important thing is that the support does come.

Some people may become paranoid with this type of action taken by higher management. Don't be. The project was still supported and you were able to get a foothold for future activity. Always see the sunnyside of the results. Your personal attitude is a very contributing factor to the way the organization will support your ideas and your career. If you always feel that negative things will happen . . . they will. If you feel positive things will happen . . . then they will. An individual's personal perception will often color results.

It must be clear that the executive and management teams deal with strategic issues and use various techniques to achieve these long-term goals. The team developed for the purpose of the project at hand may have a totally different reason than what the project is intended. Unless the project itself is vital to the strategic planning and long-term goals of the operation, it is merely a means to an end to reach or address another strategic issue of management. This is a very confusing and frustrating part of teams in many corporations. The "hidden agenda" creates many rifts in an organization and limits the trust that employees at all levels can give to the executive team.

Departmental Team   The next team to be established, formally or informally, will be the departmental team. This is the team within each department that will meet and discuss the actions and the results of the actions caused by the management team. This departmental team will be protecting its "turf," mostly in a positive way to assure stability and maintain the sacred cows developed over the years. This is a team that needs to be involved and receive excellent communication to avoid rumor and innuendo, which may cause unwarranted resistance later.

For the best results, identify this team and have members' input to be able to address their issues and support their needs. If you can support their needs, they will be more willing to support yours.

An example of not communicating with a departmental issue was when a company on the verge of initiating an ISO 9000 campaign developed a team from Marketing, Manufacturing, and Quality to investigate what was needed to become certified. Engineering was invited, but declined since members felt they were not responsible for quality issues. As the team members progressed toward ISO 9000, they realized the Engineering function was a major part of meeting the requirements. However, since Engineering declined to participate, the team decided to continue without them.

After several months of identifying key personnel for the project and developing a plan for implementation, a presentation was made to all functional areas. Engineering campaigned bitterly against the plan. It was interpreted that the design methodology would be severely restricted with all of the perceived procedures and requirements. This would hamper creativity and cause new designs to be less radical and less competitive. Since the company was always perceived to be the technological leader in the industry, this was a very strong point. Engineering won the battle and the ISO 9000 project sat idle for over a year.

When major customers started to request the company become certified to ISO 9000 or they would take their business elsewhere, the activity started again. This time the project was assigned to Engineering to implement. Engineering was given an ultimatum. Members were motivated; they were challenged; they were successful; they also had complete ownership to ensure they could protect their turf.

Make certain that all departments are involved formally, or informally, in your project. If they do not plan to support it, it is better to know this up front. Find out why they are opposed so you can address this issue. Never be surprised.

It is said an excellent trial lawyer is one who never asks a question to which he or she does not already know the answer. The same is also true for excellent project leaders. They know the opinions of all the people who will be involved or be impacted by a project. Knowing other individuals' opinions will help determine the best way to implement change. It always takes more than one person to implement change.

The focus of a departmental team is usually inclusive of the following items:

Functional responsibility

Focus on functional goals

Focus on functional resources

Focus on functional priorities

Focus on functional plan

Departments are very critical with functional responsibility, especially in large corporations. The streamlining, or downsizing, or whatever term you give it, tries to address this issue as large corporations make an effort to eliminate duplicity of effort in departments and reduce the staff and resources accordingly. This makes departments even more critical of their livelihood and assigned actions. They will fight, for survival, to maintain responsibilities of tasks.

The focus of functional goals is an area where departments see themselves as a company within a company. Their goals are, hopefully, developed to support a larger corporate goal. This larger goal in turn supports a strategic issue of the corporation. Hence, this departmental goal is perceived to be highly important to the management of the department.

Departmental goals may be ambiguous: Maintain a quality product that meets customer expectations. Others can be very specific: Reduce direct material cost per product line by $325,000 in the 1995 fiscal year from the 1994 actual material cost.

Focusing on the departmental functional resource deals with several issues. It could be the size of the departments labor force or the financial budget for the department. Some functional resource objectives of departmental teams are

Budget control

Labor resources

Capital expenditures

Revenue (especially where flexible budgets are in place)

Performance or efficiency

Utilization

Continuous improvement

Product service quality

Find out what departmental measurements are reported monthly to management. These measures will tell you what the departments interests are. These are the areas you will need to address to assure the department's support for your project.

Functional priorities are another issue. This point is often overlooked. Management always has "pet" projects. These projects will always find resources at the fate of other projects. If you know which projects are the pet projects of a department, try to align your project with it. If the two projects are seen as supporting each other, the likelihood of success for both projects will be greater. Hence, you will get support from the department at a much greater level.

An example of a department project is the new product. The Design Engineering Department was being streamlined to the point that only projects approved by executive management were supported by labor, money, and other company resources. There was one product that Engineering strongly felt was needed and was not being supported by the company. Budgets were slightly inflated. Engineers worked only on overtime basis on the project so as not to be seen during the regular workday. Many were asked to take this work home.

After 6 months the new product was demonstrated as a prototype to management during a strategic meeting to address Sales and Marketing issues. The product was accepted by Sales and Marketing as a great achievement. The department manager was seen as a shining star by the corporation. However, the executive management had a different perception.

Many departments have secret priorities. Many never are exposed to the level that this example was. However, they exist and it is beneficial to be aware of them if possible.

Functional plan is another item that is assumed to be in line with the strategic issues of a company. The fact is that this is true only 50% of the time. As the year starts to pass, the budgets of departments are reviewed and changes are made. Sometimes officially through channels and sometimes through departmental meetings.

The original plan was to purchase a new laser cutter for producing metal components. The budget was approved and the money allocated to the department for the expenditure. However, the machine was not purchased. Procurement found a new supplier that would reduce costs 15% over the current supplier. This meant the benefit of installing the new laser cutter was substantially reduced. The department could look at other areas to use the capital resources. The department plan changed from improving metal fabrication to expanding into new plastic manufacturing processes.

When the requisition was issued for the new injection molding equipment, it referenced the funding was available due to the reduced need for the laser cutter, whose funding was approved. It is a common practice. It demonstrates how the departmental plan can be different from the corporate plan.

The term "hidden agenda" also applies to departmental teams. You need to address these issues as well. Hidden agendas are not easy to identify or address. If you plan to implement CM into the Manufacturing Department, you need to address the supervisory issue.

The supervisory force is usually reduced as CM is implemented. Where do these people go? Will they be released from employment? Will they be reassigned to other areas in need of their services? This is definitely an issue that will become a hidden agenda for the supervisors. They may try to sabotage the project to protect their jobs. They may try to place blame on others so others lose their jobs in an attempt for them to keep their own. Many possibilities exist.

Other departments may feel sympathetic toward the manufacturing management and support their position. Other times the departments may oppose manufacturing and work strongly to support the project only to "get even" with them. Hidden agendas need to be addressed.

The best way to address the hidden agenda issue is to understand the culture of the operation. How do the various departments interface? What does one department feel about the other departments? How do the departments feel about the specific project . . . as a group and as individuals?

It is not easy to get the answers to these questions. It is important that you know there is a hidden agenda and that you try to address it as best possible.

Specific Team   When it comes right down to implementing a change, it is the specific team that has to do this. These are the people in the trenches. They are the ones that have to write procedures and direct people and train people and answer the questions that come up and keep a positive attitude to assure success. This is an important team, yet it has the least authority and has minimal input to the direction the task will take.

However, if you are implementing CM, this team becomes the authority. But, only after the change has been implemented completely. Members of this team will be setting up the cell and changing the procedures. They will be the backbone of the giant who will change the way the company operates. They will need authority and support from upper management, but they must be the ones to do the task. It is at this level that determines whether the CM approach will work in your organization or not.

The specific teams are usually the group abused the most. If they have a positive attitude toward the change, they will be pushed to succeed faster. If they oppose the change, they will be forced to change. If they want to change and the executive team does not want the change, they will be forced to stop the change. These change implementations are sometimes controlled by negative motivation techniques, such as threat of termination.

One energetic division of a top Fortune 500 company wanted to lead the way in implementing CM within the corporation. Division members actually had parties and made banners celebrating their expected success. Prizes were given to the people who would be in the CM areas complimenting them on the great work and effort expected from them. Everyone was cheered as a winner and a successful team player. There was no way this great team could fail.

Management praised and supported the efforts of the groups and then asked them how they planned to succeed at this task. As the departments and work groups identified what they needed and what they wanted to do, the praise and support continued. Publications of each area's successes were distributed with photos of the cellular teams as they met various goals.

This division succeeded and had the best performance record for production and quality of any division in the corporation. It was a top down–bottom up implementation. Top management developed a plan to get the bottom of the organization to support it. In turn top management supported the bottom part of the organization in that effort. From the outside, it was difficult to see who was making the decisions. Management controlled the budgets and capital expenditures while significant changes all through the operation took place. It was a joy to be part of this operation's success. No matter what level of the organization you were in, you felt as if you played an important role in changing it.

These specific teams focus on daily activity. What needs to be built? What needs to be shipped? What account needs to be adjusted? What customer needs service?

The next 24 hours is their long-term goal period. Nothing matters a month from now. They concentrate on the daily actions and do not have time to plan or even set priorities. They do whatever is the next item on the "list," which someone else had made for them.

Specific teams are the worker bees. They may or may not have goals. Mostly they do not have any goals. They work from day to day from minute to minute. Cellular manufacturing can have a significant impact on their daily activities. It may be very difficult to change these teams because of the cultural mind-set they have established.

Specific teams have extremely limited resources. All they have is within an arm's reach. They do not have budgets or determine how many people work with them on a project.

If a specific team has any priorities, they are assigned priorities or hidden agendas. Usually it is the dictated priorities that direct a team's future. The members of these specific teams have very limited responsibility. They have direct assignments and usually the responsibility is temporary.

Specific teams usually have limited abilities as well as limited resources. This requires them to have support from other areas. Without this support, they could not change anything. The support may be from their management to change their priorities. It may be from Engineering to change a product or process. It may be from Purchasing to change a part or just to communicate to a supplier about a concern.

The greatest challenge in implementing CM in a factory work area is to overcome the dependence these specific manufacturing teams have on the other parts of the organization.

2.4. Organization Motivation

No matter what your official job is within a company, you are in sales. You sell your ideas, your purpose, and your abilities every day to several different people. Regardless of what others may tell you, if you do not do this correctly, at some extent, you will not have a job. We are continuously selling ourselves and our ideas. We group with people to whom we can sell ourselves easier. At the same time we are buyers of what others sell.

The only difference in successful salespersons is the customers they develop. If you can convince executive management of your ideas and actions, you will do much better in business than those who cannot. When you convince executive management to support your ideas, you have a much better possibility of success on the project.

It is not a guaranteed success. People change their minds. Especially if the product fails them. You need to continue to sell the project by maintaining a relationship and communication link with the executive management that originally supported it. Be honest and inform executives of concerns as well as successes. Always have an action plan to address the concerns to assure that you are "protecting their interests." After all, you convinced them to buy your idea. No one wants to believe they bought a bad product.

Strategic issues need to be addressed when selling your project to management. How does this fit in with the long-term goals of the organization? Does it meet a plan already in place that had a weak scheme for implementation?

When addressing a strategic issue, several areas need to be satisfied. The benefit to the bottom line for this project needs to be evaluated and supported. The long-term impact the project will have on the organization and its potential marketing impact, if any, should be stated. How will this project relate to other long-term plans? Will it be in conflict or harmony? If it is in conflict, why should it be supported? Resource allocations are limited. What will it take to complete the project? Are there competitive advantages to this project? Can the competition implement similar tactics faster now that we have shown the way? How long will we maintain this advantage? What do our customers perceive of this project? Is it seen as a benefit or a loss to them?

Projects are best sold to management when they have short-term goals that can be monitored. This provides management with the controls to cancel the project at any point. This control is necessary since business is becoming more competitive and dynamic. Actions may be necessary that were not planned for earlier. If the project can be divided into small pilot projects, which can provide many successes, it is more likely the project will be accepted by top management.

If top management does not support the project, it is doomed to fail. Only an extreme small amount of projects developed from the bottom up succeed. Even these are usually directed to win over the top management to ensure their long-term success.

The best example of this is the Post-it notes at 3M. It is a well told story of how the inventor developed an adhesive that would stick but be easily removed. He originally used the product to mark his place in choir books as he went from song to song.

The 3M management did not feel this was a large enough market to develop a product. So the inventor supplied the secretaries of these 3M executives with the Post-it notes and just kept up the supply. When they became popular with the secretaries, he stopped. Secretaries complained to the executives. The executives called each other. Finally they realized the vast application of the product. They supported it and the Post-it notes were marketed. They provided free samples to all the executives and their secretaries at other major companies. The product is one of 3M's major profit makers.

This was an isolated instance where a project did succeed from the bottom up. The majority of ideas need the approval and support of executive or top management to become truly successful projects.

Middle management will need to be sold on the idea as well. In order to sell middle management, you need to address the tactical issues. Functional benefits of the project to the department and the managers themselves are issues that need to be addressed. Most middle managers are the problem solvers of an organization. This becomes another good selling point for the project. What is the problem reduction impact the project will provide? How will the middle manager's life be made easier by implementing your ideas?

Most middle managers have the responsibility of managing the budgets in order to remain competitive. They also design and manufacture parts to maintain the competitive position of the company. If they support your project, what are the competitive benefits?

The tactical part of the organization is usually conducted by middle management. This makes the methodology a concern of middle managers. They will want to know how you plan to implement this project and what effect the implementation will have on their current methods to meet their departmental or functional goals.

Priority becomes an issue with managers. They always have several things to do at once and never the time to do them. The joke that depicts a person sitting at his desk with someone standing over him and the quotation, "Do you want me to rush the rush I'm rushing or rush the new rush you just rushed over to give me?" is common among managers.

Ownership is the most critical of selling points. When you truly sell an idea, it becomes the idea or project of the other person. You can never do everything yourself. If you do not sell the idea in such a way that the manager does not feel ownership toward it, you have not succeeded. The idea remains yours and the process to implement remains yours.

You have to receive a commitment from the managers. Have the managers provide the manner in which they will implement the project and support it. If you tell them the ways for how, when, and who, they will add it to the list of things to do. They will not accept it as theirs and provide additional attention to details to assure its success. You should sell the idea based on why. The managers need to provide the how, when, and who. This forces them to participate and become part of the implementation team. They have assumed ownership for a part of the project, ownership you will need to achieve success.

Now that you have received the support of top management, and transitioned the ownership of the project to middle management, you need to sell the supervisory level. This really depends on the culture of the company. Regardless of the way we would want to think about how a company is managed, many companies still have an authoritarian type of management style at the supervisory level of the organization. In turn, supervisors operate the same way. If their manager tells them to do something, it gets done.

Depending on the culture of your company, this step is optional. If the company is very authoritative in its management style, you only need to sell your idea at the management level and these managers will dictate what is to be done at the supervisory level. The organizations that truly use or lean toward the participative management style will need additional effort. In these companies, you will have to sell the supervisory staff as well.

The areas that you will need to address for supervisory issues are the immediate area benefits. Consider specifically what will happen to the work area, the process, the people, the inventory, the material handling, and so on.

Another key issue is the personal benefits. What are the plans to assure employment for the supervisor. If CM is to take place, it is a known fact that the supervisory workforce is reduced and the role is changed significantly. How will this be conducted? What will become the new manner in measuring the performance of the supervisor to determine who stays, who goes, and so on?

This issue is a very important one and should not be avoided. Whether it is a primary concern in the beginning of the project or not, it will become a concern. It is better to deal with it up front and identify the rules with which to "play the game" of survival.

Productivity advantages will be a concern for the day-to-day fight at the supervisory level. Each supervisor will feel he or she is doing a very good job and there is little room for improvement from what he or she has already accomplished. This is a touchy subject. It can become a personal attack on the supervisor if not handled carefully. Discuss the overall benefits and reduction of handling. Discuss the improved communications and clearer responsibilities, areas which the supervisor had no control over and could not be directly associated with the area's specific performance.

Problem solving advantages for the day-to-day routine is another selling point. Again, this is most likely taking a responsibility away from the supervisor and giving it to the team. It could be seen as a threat to the supervisor rather than a benefit.

Supervisors usually get assignments and orders from everyone in the organization. Unless they know you personally, they will question your authority in making a change. Provide the identification of the project's "company backing." Explain who, and at what level, is providing the backing and support for the project. This is another reason to make certain management has ownership of the project.

The support provided or expected is another issue to address. The supervisory level has little control. Supervisors will require support and resources from other areas to do many of the things for CM implementation. Identify this support and the plan to fully implement the project.

The teams may stop at the supervisory or specific team level, but the selling activity needs to go one step further. The workforce, in CM, needs to be sold. Union issues and various work issues will need to be addressed. This should be done with the aid of the labor relation person at our company. The person who works with the union negotiations to define the contract and interpret the clauses needs to become involved. Do not try to discuss these issues with the union until you have first held a lengthy meeting with the labor relations person and agreed on a method together to address the union.

This is a very important step in selling your project. Unions do not operate as other parts of the company. The union is an elected body to represent the workers of the company. Union members do not always have the same interests in mind as the nonunion employees of the organization, not that they are not interested in the success of the corporation, since it provides the jobs for the workers. However, the perspective by which they judge the change is different.

Is the worker going to be compensated for the additional responsibility? How do we plan to retrain the material handlers to do other jobs as their jobs are eliminated? Is there a plan to lay off people? If so, how many and when? How many of the nonunion people will be laid off? When?

It is important to realize the union representatives are elected. They must look out for the benefits to the workers and the manner in which the workers will be treated. If they do not, they may be voted out and lose their jobs. What may be a very good change for the company may not be seen as a good change for the union workforce.

Be aware of this issue and consult with experienced people who deal with unions as a regular activity in their day-to-day work. Do not try to do this part of the sale alone. Even if the company is a nonunion operation, the same advice remains. It can severely impact a company as well as your career specifically.

2.5. Summary

The organizational culture is a difficult area to understand and cannot be assumed to be what is written in company literature. It takes several observations of actions to determine how a company actually makes decisions. The style and culture of the company dictate how the decisions and what type of motivation is applied to implement decisions.

Although communication is a very important activity that a company can improve and toward which it can provide resources, the manner in which a company provides motivation is more critical. The motivation is what gets the effort accomplished.

The organizational structure is varied from company to company, but all have three levels in common: the executive or top management level, the department level, and the operating of specific team level.

The executive level concentrates on the strategic plans and long-term goals of the operation. This level also controls the resources and the direction the operation will take to attain these long-term goals.

The department level is the tactical arm of the organization. It carries out the plans and uses the resources to change the direction of the operation from where it was to where the executive level wants it to be.

The specific team level is that part of the organization that actually does the work necessary to make the change. Whether the work is done by hand, machine, or computer, team members directly control these tools to make the project a success or failure.

The organization's motivation is different at the different levels. In selling an idea as a project for implementation, the motivation of each level of the organization needs to be addressed. Each should be addressed at an individual level.

The motivation for the executive level is to address the strategic issues and goals. The departmental levels need to understand why an idea should be implemented from a tactical sense. Yet departmental management needs to determine the how, when, and who for the project. This assures ownership and responsibility for the project in the respective areas.

Supervisory areas and the workforce itself should be motivated and sold on the idea of CM. However, the type of organization plays a role in how this is done. Regardless of the management style or culture, the union should never be consulted without the prior assistance from the labor relations function of the organization.

Unions deal through political agendas and do not operate in the same manner as companies. These differences need to be realized and appropriately managed.

All projects are to be supported by the executive management if they are most likely to succeed. Although some have succeeded from a bottom up or grass roots effort, the majority have always been fully supported by the executive levels of the company and been in complete accord with the strategic plans of the organization and its long-term goals.

3. TRAINING, DEVELOPMENT, AND IMPLEMENTATION

3.1. Training and Development

The philosophy of training is obvious to most people. If you know what you are to do, you do it better. However, in business this is not the important point. The important point is whether or not the training will improve the performance to a level that more than compensates the company for the resources it expends to acquire the training.

The training must be warranted and have a positive return on investment. For example, I am asked to train every employee in water safety. This will cost me $1000 per employee. I have 300 employees. My company is situated in the desert of Arizona. My product is the manufacture of small plastic parts used in telephones. What will be the return on investment to my company? Not much.

This example identifies two clear issues: The subject matter of the training must be relevant to the operation, and the cost must be offset by a benefit somewhere else. Clearly, there is no benefit to having a workforce, in my make-believe company, that knows the rules of water safety. Even though it may be very wise to understand the principles of water safety, it is not my company that is responsible to provided it. Nor is there a reason to spent $300,000 on the training.

Many times the idea of training is presented as a solution to problems. Yet the training does not address the real issue. Unlike my obvious example of the water safety in the desert, training outcomes are usually not as clear.

Quality issues are a common misuse of training expenditures. Companies with quality concerns hire consultants or use "canned programs" to teach quality to the employee. The reality is the employee does not mean to build a bad part in the first place. The employee is usually forced to produce a bad part because of other priorities within the company itself.

Company A receives a shipment of poor-quality parts from company B. The parts are slightly out of tolerance and, if used, cause the final assembly to stick and not turn properly. The executives of company A are strong believers in meeting shipping dates. The operators on the line tells the foreperson they cannot produce the required quantity of parts to meet the shipment because some parts are too big to fit into the socket. The foreperson forces the bad part into the socket, demonstrating it can be done and implying the workers are lazy. The foreperson walks away and the operators build the quantity for shipment with the bad parts.

Next week the customer complains that the parts don't work properly. No change was made by Engineering, so the design is correct. No supplier was changed so Procurement doesn't see an issue. The foreperson knows the operators are lazy so the only corrective action to make the customer happy is to train the operators in quality workmanship. This training falls on deaf ears as the operators know what the real world is like in company A.

In order to make a significant change in culture, such as is necessary for CM, the organization will have to undergo some training. The real question is what type of training and how much. This will vary from company to company and even from department to department.

Many companies still believe that a general training is best. Everyone hears the same thing. Everyone does the same thing. Everyone knows the same thing. Some companies even go to the extent of training the executives—then management, the supervisors, and the workforce take the exact same course and material. Does this really make sense? What activities do an executive and the hourly worker do in common? Not much. It is only for show that these things are done in this manner. The hourly person knows the jobs are different. It is embarrassing to think the American worker is this ignorant.

The visible application of this approach is in companies where the management dresses down in blue jeans and sweatshirts to "walk the walk" in the factory. You don't see the dirt and grim on the manager's sweatshirt or jeans as you do on the workers. The worker knows very well that if this person in the clean jeans asks them to do something, they better do it. Dressing in this way does not make a team. It is patronizing. If you want to have a truly effective operation, be honest with each other. Don't cover it up with training programs or dressing down. Everyone knows who the boss is and will react to that point alone.

Training issues need to be specific. If you plan to change the culture of an organization, what does management need to know about the new culture to which you plan to transition? What skills are needed for the workforce if they will be buying the materials directly in each cell? What training is required of the future supervisors to facilitate the cells rather than direct them?

Before you can answer these questions, you need to know where you are and where you plan to be. The training is a means to fill in the spots where you have pitfalls or weaknesses. You need to use the training as a road grader to level out the path so the organization can easily maneuver from point A, where you are, to point B, where you want to be.

Organizational Culture   Organizational culture is the first area that needs to be addressed. If the company has no intentions of changing the culture, then the training should not try to do so. Many managers feel they are great leaders of a world movement in changing business. They become frustrated or go to another company where the culture is already the way they want it to be. Unless the powers that be, the executive level of the organization, truly want to change the culture, DO NOT attempt it.

The training should always be in line with the management style and culture to which the executive level is adhering. If the executive management team is very authoritative and you are training the workforce to become empowered, there will be failure.

Some companies have CM at a level where the cells do what they are told to do and control specific items associated with manufacture. The management remains totally in control and only the supervisory levels of the company have been eliminated to provide payroll savings. These companies will not benefit from the full impact of the CM philosophy.

The other side of the pendulum is where a company totally gives all responsibility to the cell in a CM vein and eliminates all support functions to reduce costs. Supervisors are gone. Procurement buyers are gone. Manufacturing engineers are gone. The only thing that remains is the workers in the cells and the managers with the whips. This is CM in the prison sense of the work cell. No teams exist. They merely work and make change to survive. This works in industries were the people are either paid a high wage and cannot afford to go elsewhere and receive less or in areas where there is no place else to go. This may be perceived or real.

Regardless, the training must support the culture and the style of the expected new company format. If you want the team to be responsible for developing processes, you need to provide training in how to develop processes and tooling and ergonomics. If you plan to have the team become the buyers, you need to teach team members how to buy things. What is the procedure? How do they get authorization? Who do they contact to buy the material? You don't want a manufacturing cell in the Ford Plant in St. Paul, Minnesota, to buy tires for the F150 Series pickup trucks from the corner gas station. GMC may develop a competitive edge over you if you did that. Especially if GMC provided the tires to the local gas station on the corner from your Ford plant.

Strength and Weakness   Every company has its own strengths and weaknesses. Before the training program is developed, the strengths and weaknesses of the operation need to be evaluated. Many "canned programs" deal with general training, which may waste time and money for your company. Especially if you are already strong in the areas the "canned program" will emphasize.

Some companies already have a participative workplace where an open and honest trust relationship exists between the workers and management. Problem-solving skills are very keen and the company has high morale. Departments communicate well and everyone is respected as an individual. In this environment, the training would be minimum to prepare workers for a transition toward CM.

What are the strengths of the company? In order to determine this, list all of the strengths necessary to be the perfect company. This needs to be modified for the specific project you are working on. Then grade your company against the list— using 1 for poor, 2 for average, and 3 for good. It is not necessary to go into more detail than this. You are dealing with a subjective area and trying to apply objective data measurement to it. It only wastes time and effort to become more adapt at how good or poor your company is at something.

All the areas where the average response is below average requires you to develop training to improve it. Those where the responses indicate your company is good, you do not need to include in the training. From this basis, a simple and effective training program can be established.

Training should be relative to the department or specific team. If you want the cellular area to develop buyer skills, you do not need to train the Procurement Department in the same skills. Nor do you need to train the Engineering Department in those skills. You only need to train in the skills necessary for the individual areas to operate. The purpose is to improve their ability to operate effectively, not to develop a massive training operation.

The most effective training programs are those that train in the skills necessary at the time they are needed. If you train someone how to program a computer but never give her a computer to use for 3 years, it is most likely she will forget how to operate the computer when she does get one. Design the training to be pertinent and timely for the tasks at hand.

Resource Availability   Resource availability is another way to approach the cost-effectiveness issue of training. Especially in the example that dealt with training the cell to buy material, the understanding of resource availability can be very cost effective. The company already has buyers who are well trained in this area. These same people can become the trainers for the cells. As their jobs transition to some different level, they can be utilized to develop the CM teams to assume the responsibilities they are giving up.

Training instructors is only one area where resource availability has an impact. Time of the training is another area. To maintain production rates, training may be necessary on off hours or weekend shifts. Lunch hours could be an option with meals provided to deter the individual's loss of personal time.

The training can also be a very large cost area if not handled properly. Lengthy training courses cost the company salaries of the people going to the training, the lose of production time, and the cost of the trainers. By training only the necessary skills to the necessary people, this cost is reduced. Also, the cost of materials is reduced substantially by doing this. Training materials can be developed by managers in other departments who have skills in the area to be trained. These materials can be reproduced at the company's site to keep costs down and provide a specific company approach to the material.

By utilizing the available resources of the company and not having to go outside or receive approvals for unplanned expenses, the project is more palatable to executive management. Cost is always an issue. Effective use of the companies resources to achieve an end is another. By utilizing one to satisfy the other, you have the best of both worlds.

Time Constraints   Time constraints are usually self-inflicted. Companies determine that something has to be done by a set date. Why? Because they determined they need to be at market by a set time for a set reason. Few dates are determined by outside forces. It is mostly perception of the world around the company that develops the time constraints we learn to accept as everyday life.

The same is true for training. You will need to develop a time line for the training as to who will be trained when and in what skill. This will become the measure and be cast in concrete the moment it is accepted by management.

The plan has to include the training and the timing so people and material will be coordinated properly. However, do not make the time constraints a rigid thing. Most projects will allow for a pilot program. Use this pilot to develop the training requirements and the time constraints. Always allow for spaces in the plan to modify the training or activity. This flexibility will aid in the success of the project.

If constraints are placed on you, challenge them in a positive manner. Understand why they are necessary and if there are options you can take. Never agree to a time constraint when it is first presented. Always think about it. If you are rushed into the project, you are more likely to overlook something and more likely to fail. Planning and preparation are two strong ways to succeed in implementation of a major change.

Immediate and Future Needs   Usually training is developed to handle a future need. You are taking this course to meet a future need. Some of you may have an immediate need, but for the most part you train to meet a future need.

In industry you train to satisfy an immediate need. Most companies do not use the philosophy to training their employees for the future. This is something "preached" by the best of the best, but reality causes companies to hold back on training unless it is necessary. For the most part, training is performed as a corrective action to a situation.

It is under this type of situation that you can use training to improve the process a great deal. Design the training material to meet the current or immediate problem and also deal with some things that will be of benefit in the future. In this way you will be able to apply the training immediately to the current situation. Also, the more assertive people in the organization, at all levels, will try to apply some of the future ideas you placed in their heads.

This seed of training will cause the next phase of development to go easier and be an easier sell with management as well. You have to take care of the immediate need. This is why your project has been supported to begin with. Train the basics and help the areas to improve. This improvement will be recognized and increase future support.

Only deal with the future needs after the immediate goal has been accomplished. This seems simple, but many people try to attack windmills when they receive management approval. Refer back to why and for what you received approval. Always satisfy the immediate need first. If the immediate need is too great, then avoid any future visions in the training at this point.

Experience with a large company trying to improve its quality leads me to this example. The company approved a plan to train employees in the understanding of quality: the basics of no rework, no scrap, no rejects, no wasted material, and so on. The person who received the approvals from management to deal with this issue was a very energetic individual who truly believed in Total Quality Management (TQM) and the Malcolm Baldrige National Quality Award. He put together a very good training program on TQM and then explained the steps for applying to the Malcolm Baldrige National Quality Award. Even the ground rules for the award and the areas that would be measured were covered in the training.

Needless to say the first group to be trained were totally confused. The training was stopped in total. Instead the company assigned a type of punishment system if you created rework or scrap to improve quality in the factory.

This example may seem an exaggeration to the logical mind. However, it actually occurred. This person never did get to present his philosophy to anyone again. He left the company shortly after that and went on to another company to try and implement significant changes in quality. I do not know if he ever succeeded with his other efforts.

He could have changed the company to his approach if he kept in mind the original task at hand. What is needed immediately? No one tries to teach calculus to a child in first grade. Why? Because they do not have the basics yet. The same is true for any training course.

Teach the basics to meet the immediate need and, if you have time, lightly develop a vision for future improvement. Companies in the United States, especially, have to deal with the immediate. We are set up to meet quarterly returns and improvements. Stock markets and the Dow Jones control our time table for success or failure.

Always have a vision and try to persuade people to support it. But when you have an assignment to develop a training program, make certain the basics are well understood before taking on your vision. Immediate needs control the project's success. Future needs control the project's destiny.

Change Agents   When you perform the training exercise you will notice change agents. These may be people in the cell who will be very supportive and work wholeheartedly to make the changes. You need to nurture these individuals. Pamper them with support and information as they request it. Do not push them or try to become overly friendly with them. People see through this. You need to remain professional. However, a little extra attention to the change agents will go a long way.

They will help you in ways you could not imagine. The coffee machine is an excellent place to change culture and opinion. If you have several people commenting on how well the project is going and the support they get, the culture and support will go in your favor.

Change agents are also the powerful people in the organization who can make things happen on a large scale. Try to get these people to assist in the training projects and recognize them with the project. These people will get you support through association. Assuming you do your job well, their support will remain and benefit the project. In most cases, this support will benefit you personally in the organization.

Sometimes change agents are not people. They may be things or events that cause your project to gain support. One example is the competition. Let's say you are working on implementing CM as a way to reduce cost of goods. The competitor has just cut prices on a major product line to the point where your company's cost makes it prohibitive to have any profit if the prices were matched. This event could cause your project to take on a higher priority in the operation. It will most likely be rushed into completion as well. There is always the double-edged sword in industry with getting too much support.

Try to be aware of the various change agents in the organization. Use them in a way that will support your effort. Always remain honest and professional. You cannot fool people. It is a good practice not to try. The concerned person, who is trying to do the best job for the company, will be recognized.

Training Course Details   The actual course, whether it is an hour program or several days, needs to address the following in detail. You should have developed the topics and received support in the ways mentioned. Now you have to develop the material and present it.

The material and the presentations *need to be specific for the task.* This cannot be overemphasized. Stay with the main reason you were given the support by management to initiate the project. Cover all the topics necessary to make the initial project a success.

The saying "Success breeds success" is the basis for my concern. If you do what is expected and do it well, you get other chances to repeat it. Stay with the basics and meet the needs of the specific task at hand.

The second item is to design the material and your presentation so it meets the *needs of the unique company's culture.* If the company is authoritarian in style, it is not wise to train in participative management. The training has to be designed to agree with the culture that is in place. The only deviation from this would be if you are specifically assigned to change the culture. Even then, you should go in phases.

Never try to upset the apple cart. You need to get all the apples shined and polished and stacked very neatly. The person with the best looking apple cart is appreciated. Help others to understand what is needed. Associate to the current systems and practices. It also makes it easier for those you are trying to teach if they can associate with the practice in some way.

Most times training material simply directs a person and provides information. If you are trying to implement change in an organization you need to develop ownership. The people in the training have to accept the responsibility for making the change. You cannot do it all. The third thing is the training *needs to emphasize ownership.*

I try to associate this with the comparison of grade school and college. In grade school the teacher takes on most of the responsibility to train you. They make you practice reciting the ABCs over and over. You get to read aloud in class and everyone takes turns. The teacher takes a very active role in your training.

In college you are assigned vast reading responsibilities. Reference materials are suggested to assist you in your training. Many classes are lectures with little time allotted to questions. The responsibility to learn is yours. You have ownership in the grade you will receive.

In industry the training has to develop ownership on the part of the participants. The "students" are not just responsible for learning the material, but also for implementing the change implied by learning the material. The training has to make this clear. Phases of implementation and timeframes may be needed to follow up on the progress after the training. Training for the purpose of implementing change needs to initiate specific assignments that will be monitored. The results of the training have to become part of the immediate job assignment.

This does not mean you will not provide support and additional effort where it is needed. You will always have to assist in the implementation. Do not take on the responsibility for the implementation for a specific area. Only if it is your specific area do you take on the responsibility for it.

If you do not transfer ownership right from the beginning, you will need to start to raise *bananas*. Bananas is a term used when you take on a problem, or a monkey. You get too many monkeys on your back and soon they become pests. They get hungry and you have to raise bananas to feed them and calm them down. The best way to avoid raising bananas is never to take on the monkey in the first place.

The fourth and final item necessary for a good training program is the *need to show the end of the tunnel*. When you implement change you create stress. Most people do not like change, regardless of what they say. Change means something different. An unknown to any person causes stress.

If you can show people a vision where things will be better, it will help. They will have a goal to strive for and feel less anguish to work toward the change. Showing them an improved life for them helps even more. Include in the training the vision of the change after it is fully implemented. This should be in the beginning of the training. It will help with them accepting ownership.

Emphasize this by asking questions about what they see with this change. Address all the negatives to assure them these things will not happen. Provide support to those with positive comments and reinforce their statements.

I am a firm believer in motivation. Training is a form of motivation. It shows you care to take the time and improve the individual. The point of clearly understanding the "light at the end of the tunnel" is important to motivation. It shows them where you are going. It shows them a future. Make it as bright as possible, but always be honest with them. You will be amazed at how well their memories are of times you weren't honest with them.

Now that we know what is necessary for a training program, we need to see how it fits into the rest of the implementation process.

3.2. Implementation Process

The chart in Figure 2 is a simple yet complete visual display of the steps needed to implement change. Before you can change anything, you need to evaluate it.

The important thing to evaluate is the culture. The culture of a company is what makes it different from any other company. Just as people are individuals because of their personality, companies vary by their culture. The culture is the personality of the company. In order to get to know the company the first step to change has to be evaluating the **company's culture.**

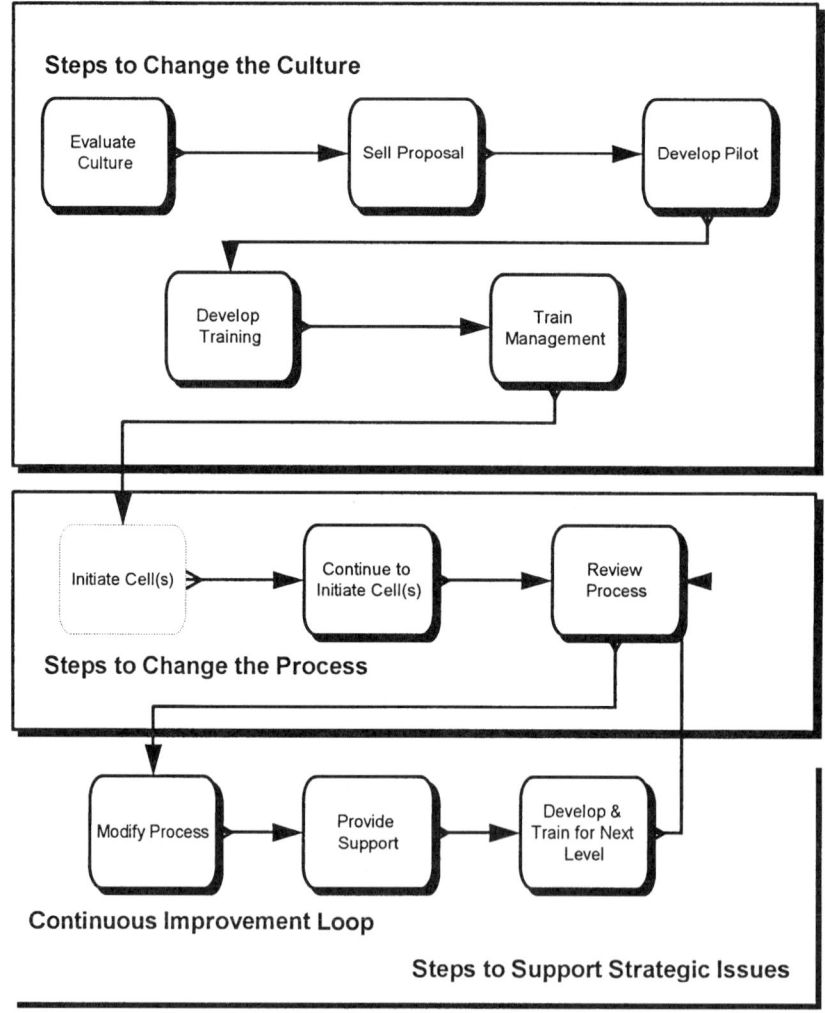

Figure 2. Process to implement change.

Company Culture Several things make up the culture of a company. Most important is the management style. Is it autocratic or participative or somewhere in between?

What is the company's real feelings about quality? Is it given lip service or clearly enforced? Is the quality a matter of pride or something done to avoid returns?

Can you tell what the company's focus is on? Does the company concern itself with bottom line issues 10% of the time?

Does the company concentrate on the market in which it participates or does it concentrate on its own product? Which is more important? What are the feelings about the customer? Is the customer involved with product design or improvements? How?

How does the company set goals? What are the goals and what issues do they address?

Is the company proactive or reactive? Are the long-term plans really strategic issues or tactical issues to keep up with the competition?

What is the employee value? How is it demonstrated? Are people this company's most important asset?

These are some questions to start understanding the culture of the company. They are even good questions for interviews. You will learn a lot about a company just by the way these questions are answered.

Sell the Project The next step is to sell the project. Until someone gives you permission to do a project, it is very unlikely you will get much support. The evaluation you did on the company should have provided you with information as to where the company is positioned today. It should also provide you with the way it considers projects and what values it places on various aspects. You need to use this in selling the project.

The project has to be designed to meet the needs of the company in a way that will be appreciated. It also has to be presented in stages. This step-by-step or stage approach gains you information, support, and additional selling points.

The first step is to make a presentation to management. This should not be the executive management at this point. Present it to the managers who will most likely be effected by the change. They will provide both positive and negative feedback.

Using this feedback, develop supporters and find reasons to overcome the negatives. Also use this as a barometer to determine how the executives feel about things. Many managers will know the opinions of their superiors. If not, they will have a strong opinion of what their superiors will think.

Review the data and revise your presentation for executive management. Based on the information you've received, determine who will be your supporters at the executive level. Arrange to see these people one-on-one. Review the presentation material with them and get their feedback. Develop more points to support your issues and relate them to projects or strategies being implemented.

If the feedback is positive, ask the executive to support your project when it is presented. Be certain the executive will actually speak in favor of it. Ask if you can address some of the issues the executive mentioned to you as reasons for support of the project.

Repeat this process with those you feel will strongly oppose your project. Do not tell them of the reasons other support you. Ask them why they feel the project is weak and how they would change it to make it acceptable to them. If possible, modify the project and the presentation to satisfy these requests.

Everyone wants to jump to the top of the stairs at a single bound with an exciting project or change. The best way to get to the top of the stairs is to take one step at a time. You are then in a much better position to take the next step.

After you have received your feedback from the management and the executives, one-on-one, you need to review your presentation. Revise the presentation to address all of the favorable aspects up front. Use names and quotations of supporting executives if possible.

Then go into the questionable areas. Present the options and how they would improve the overall impact of the project. Relate it to cultural issues, strategies, and other projects. Use the concept of synergy if it applies.

Once you are ready, make your presentation to the executive level. This is where you request the support and authority to initiate your project. Include sample plan or pilot project to prove your concept. This is optional depending on the size of the project and the size of the company. If you feel the project approval is in jeopardy, offer a temporary approval until after the results of the pilot project can be completed.

The idea of tentative approval is easier for management to approve. Executives can always change their minds at a later date. It may also give them some options with people development. It definitely is a way to soften the impact of a culture change on an organization.

If you have done all of these steps to a completely satisfactory level, the probability of success is in your favor. You should be granted the authority to do a pilot project. You have sold the concept and have been granted the opportunity to take the first step up the long flight of stairs to success.

Develop a Pilot   The pilot needs to be a success. Experience tells me to review what happened to get me to the first step. It is easier to look over things you are familiar with than to go blindly into a dark alley and not know what awaits. This means the concept of the culture you have developed should be tested.

You test the culture by working in the way you feel the culture would support. If the culture is one where you can make decisions and not have to ask for permission to do so, I would suggest you make small decisions and see what reaction comes from this. Assuming the culture you anticipated is correct, start to develop the pilot in tune with the culture to minimize the abrasiveness of making change in an organization.

If you did not interpret the culture correctly, you will definitely need to redefine and modify your plan. Never try to implement a change that goes directly across cultural behaviors unless you have absolute power. If you do not have absolute power in making the change, you may become the major change to the organization.

Evaluate your supporters. Why did they support you? It should be clear that if a superior supports you, you need to return the favor. You need to take steps to ensure the supporter is seen favorably by the organization. The philosophy of "you scratch my back and I'll scratch your back" is alive and well in corporate America. This is the way friendships and positive relationships develop in the business world. Everyone wants to succeed and be perceived as wise. Assist in this effort and you will benefit as well.

Do not create enemies. Those who opposed your idea have a different belief in how to accomplish things. Learn from them and their experiences. Thank them for the information and give them credit for the sound advice they provide. Try to take their input and apply it to the project to create a win-win approach to change.

After you have reviewed the lessons you've learned, apply them to the training. Determine what the basic training needs will be for the pilot project. Develop handouts for future reference. Use visual aids. If possible, have examples you can touch and feel. It is amazing how much people remember about something if they can see it and touch it.

Try to minimize lectures in your training. Lecturing is the least effective way to teach someone something. Unless the person takes notes to keep in tune with the material being presented, the mind will wander and the memory will fade. The more hands-on demonstrations and sample projects the people can work on themselves, the more they will retain from your training.

You can also hold their attention with humor. Do not become a comedian on stage. Humor is helpful to keep people alert during a presentation. It should only be in good taste. Being politically correct and funny is a very difficult thing. The better you stay on the material the better you will teach your audience.

You should have presented an implementation schedule to the executives to attain their approval. Evaluate this implementation schedule and make it the pilot project team's schedule. This is where you need to develop ownership for the project. If the team does not accept the challenge, you will have a difficult job in meeting the schedule. You have to sell the project team on the change at this point.

Part of the ownership is identifying the team's level of empowerment. What can team members do? What can't they do? When do they need to ask for additional authority? Who do they ask? You will need to address these issues in the training for the pilot project. Confirm your authority levels with management before you present them to the pilot project team. Seek a mentor from the executive level to provide you with guidance.

Motivation is always an important factor. During the training, employee values should be addressed. This should be addressed for two reasons. First, you want to assure the employees of their value to the company and the project. This is done to make the employees feel good about themselves. People do a better job when they feel good about themselves.

The second reason is to develop a feeling of team. Cellular Manufacturing requires people to work together as a unit. It is no longer the "not my job" syndrome. The team is unified by letting everyone on the team know he or she is important to the team. Teams are only as strong as the weakest person on the team. Team members have to understand that they need to support the weakest member on the team until they can satisfy the needs of the team by themselves. It should be understood that each member of the team may need support. It should be understood that each member of the team will be required to provide support.

After the training the pilot project team should go directly into action. The best of all scenarios is to have the training while the pilot project is underway. This allows the training to demonstrate immediate improvement over frustrations. However, if the frustrations become overwhelming, the training may never be applied. It is the nature of people to use what they know to survive. If they feel threatened, they will do what they know and not try new methods. Evaluate the situation before determining which approach you will take.

After the pilot has been completed, you should have learned a lot. By now you should have a clear understanding of culture and how things get accomplished in the company. You should also have experience in working with a new cell. You can identify with their needs and levels of skill. You can also identify with the support the cell will receive from other departments and groups within the company. This is the point where you develop the training program for the rest of the operation.

Develop Training The training must be developed based on your experiences and abilities. The training should be developed for each level of the organization. Prepare the training for the top management first. They will need to know what to expect as the CM process goes into place.

Base the training on what they have today and then transition them to the expected outcomes of the future. Identify strengths and weaknesses in the current system. Identify strengths and weaknesses realized in the pilot project.

Review what was learned from the pilot. Always tell top management what to expect, as well as what NOT to expect.

Prepare the training for the next level of management until you finally have the training planned for the next cell(s) to be implemented. Cover the sample topics, but relate them to the respective area or level.

Compare what you experienced with that of the management in the pilot project. If possible, have them assist in the training. At a minimum, have them review the training material and subjects to be covered. This management input is a way to maintain the ownership of the project with the managers that were involved. Whenever possible, involve the management and the members of the pilot project. These are now your champions. Give them credit and recognition whenever possible.

Some people are not good at presentations or conducting training sessions. If you have discovered this about yourself during the pilot project, find someone out of the champions who is a good trainer or presenter. Identify these training resources and utilize them to your advantage. The good manager knows where the good resources are to help them complete the project at high quality, on time, and under budget. This is a good training ground. Take advantage of the situation.

Normally the project will be reviewed by management at this point in time. Assuming you have successfully implemented the pilot project, you will be required to provide cost estimates for completing the entire program. These estimates were probably done earlier, but now they will become set in the financial concrete of the organization. Use your experience with the pilot to establish training time and cost. Based on the time it took to implement the pilot cell, establish a training schedule.

Review the needs of the pilot project team. Establish potential future phases of training based on actual experiences. This is another area where you can use the support of the managers associated with the pilot to, again, maintain their ownership in the project. It also is an opportunity to provide recognition while having them present this part of the reasoning to upper management.

Since this is the point in the project where finances will become a concern, address strategic issues. These concerns should be included in the training for upper management. These issues will develop the support and provide direction to the management team as you start to implement the program throughout the company.

Modify each training to the appropriate level you will be addressing as the training progresses down the corporation. Certain topics will be omitted and others added. Executives will not need to understand the specific way a team will be expected to solve problems. They will need to understand the specifics of how they will have to support a team in solving a problem.

Once the training is finalized and meets the approvals of all parties to be involved with the training, you need to start the training. Who do you train first? Management.

Train Management   When you start to train management you will be changing the culture of the company. It is important to know this. Often it is assumed the management is totally behind the project at this point because of the success of the pilot.

Reality is this is where the real change to culture takes place. Until now all parties maintained their own operation in the manner they did before. They may have assisted in the support of the pilot, but it was a mere test. Nothing in concrete. When you start to train the management you are initiating a new way of doing things that will cause change throughout the organization on a permanent basis. This is where the most resistance will come from. Prior to this point the project could have failed and never effected any of them. Now it is real and on their doorstep.

Emphasize how these changes will benefit them. Honestly address their concerns about the problems that developed in the pilot. Hopefully, you will have a method to overcome those same concerns in future implementation. It is also a good position to reinforce the old culture. Reference the similarities in the process. If the change to cellular is so drastic that no similarities exist, emphasize the benefits of the significant changes. Make them a welcomed alternative to what was done in the past. Never condemn the past. Only praise the manner in which the operation progressed and demonstrate how this is another progression for the company.

This leads into the "what's in it for me" attitude. Regardless of who you are addressing in the organization, you need to identify the personal advantages. Stoke the egos by identifying how this change can improve their position. Emphasize using their professional abilities to achieve even greater advantages than was seen in the pilot. After all, you did learn something from the pilot. Given the experiences, each cell implementation to follow should be smoother. The process should become accepted and improve in support, reducing the time to implement future cells.

Provide reasoning to management as to what future challenges await. One possible strategic reason for change may be an ISO 9000 certification.

Executives will need to know how they should review their supervision. What will be the new requirements? What skills will their supervisors need? How will they achieve these skills?

The topic of the supervisor's new role will lead to how the manager's job will change. The productivity benefits will be an interesting point of discussion here. Most managers are involved with improving productivity. The implementation of CM opens new opportunities for productivity improvements that were closed under the original method of production and supervising the process. This may relate back to the "what's in it for me" attitude.

Whatever track you take, productivity improvement is an important issue. It is the major reason most companies are willing to risk the effects of a major culture change to implement CM.

Ownership, ownership, ownership. At no time give up the idea that this is to be team's project. You are the facilitator of the effort. It is team's effort. If you become the scapegoat for the effort, it will fail. People will not support it. In turn, you will fail. It has to be the effort of all. It must remain the responsibility of the management teams and the production teams they represent. If you teach nothing else, teach this.

The steps you have accomplished, at this point, are directed to *change the culture* of a company or organization. These steps deal with the opinions of the management team and the "sacred cows" of the organization. These sacred cows' developed over years of practice and the trials and tribulations of the organization. As problems arose, various actions were taken. These actions become the ground rules for the company. The process was developed from trial and error.

The idea of change has to demonstrate reason and support it with proof. The reason is provided in the manner in which the change is sold to management. The proof is the pilot project. If the process has been completed effectively, the change is accepted for full implementation. The culture has been changed, or at least it is being honestly and openly questioned.

Culture can change back. Do not become overly confident. The task of making a change in organizational culture is now in your favor, but not a carte blanche operation. The real world has many variables in it. You must be prepared to take the cyclical support of your project with an open mind. The best way to succeed is to be prepared.

Initiate Cells   Once management has completed the training *and* accepted ownership for the project, the company needs to immediately start implementing the new process. If possible, complete the training on Friday and start implementing the cells the next Monday.

The first team you are scheduled to work with will have very different opinions of the change planned. The pilot project was experimental. The team was supported in a special manner to prove out a concept. The team itself was excited about the opportunity to do something different. There was no pressure in the pilot project to succeed. If the project did not succeed, the process would merely go back to the way it was. Now the line has been drawn in the sand. This team will need to change as a first step in converting over to the new method. The team's attitude may be completely different toward the implementation process.

The team's different attitude toward the project should not bother you. You must be prepared for it and not be surprised by it. This first step may seem like trying to force-feed a person who just finished a large meal. Nothing wants to go down.

The first teams to be converted should be in favor of making the transition. If you can, select the areas where the management is in complete support of your approach. As the areas transition, communicate their successes. Make them heroes in the organization. Develop new champions for your cause.

The best way I've seen in practice is to utilize the people from the original pilot project to train and assist in the transition of the first few cells or work areas. These people are supportive of the concept and already have ideas to improve on it. The relationships are also improved if it is a peer relationship. You personally, may be viewed as "one of them." There is always a group known as "them" in an organization. "Them" is always thought of as an enemy, to some degree, because "they" are not one of "us."

The concept of "us" and "them" is another area of the culture that needs to be changed. This will take longer than setting up a CM area. Keep this as a long-term goal, but continuously work toward it when the opportunity arises. The immediate moment requires achievement. It is easier to overcome this "us" and "them" hurdle by letting their peers support the new teams.

Keep the transition on an "us" and "us" level at ALL levels. Managers should interface with the managers. Engineers should interface with the engineers. Cell members should interface with cell members. All parties and all levels should work together as the CM process continues. It is only during the initial stages of the transition where you want the explanations of the change and the reasoning to be encouraged at a peer level.

The next concept may be obvious to most people. This concept, however, is often overlooked. Managers who are extremely supportive of the change will request their areas be transformed first. This positive support makes the implementer want to take advantage of the situation and say yes. However, the area must be trained first.

Companies have set up in-depth training programs for the areas to transition. SPC and other common tools to support the cellular process are included in the training for the future teams to be. Then they implement the new cells in areas that have not been completely through the training or only some have experienced it. The untrained team members become frustrated and cause a delay in the process. Sometimes they can even cause it to fail. Avoid frustrations and failures by training teams together.

Train the entire team for the areas to be transitioned prior to initiating the transition from the current process to the process being supported by the change. If possible, follow the same rule as with management training. Implement the change immediately after the training has been completed.

Support functions are also important to the success of a project. Some individuals in support groups, such as engineering or procurement, will be very proactive toward changing the process. Others will not be as energetic. Try to select the areas where the support people are in tune with making the change a success. Train these support functions first so they can prepare for the transition. Then involve the individuals with the team training, reinforcing their support of the team.

The two main concerns you should have in implementing the change to the process are frustration and stress. Frustration causes people to become less motivated. In order to maintain the motivation, frustration has to be kept at a minimum. The only time it is good to have a frustrated team is when they are frustrated about a practice they can control and improve. This frustration will bring about change. If the team is frustrated about a topic they cannot change, the frustration builds until it becomes detrimental to the motivation of the project or the team members themselves.

Stress is something we all deal with every day. The concept of reducing stress merely means providing the support in a timely fashion to reduce undue stress. Stress can never be completely eliminated in the real world. The level of stress can certainly be affected. Understanding that stress will be inevitable when implementing change is important. The actions taken to minimize this level of stress will improve the attitudes of the team members and assist in motivating the team and the individuals to be successful.

What is success? It has many meanings to various people. The approach taken that provides the most frequent success ratios is simple. Define the expectations clearly. Establish the measurement for success that reinforces the clear expectations. Visual aids are always useful. Have charts or the infamous thermometer posted in the areas. Clearly explain what the goal is and how team members should try to approach it. Have the team perform the measurement and keep the visual aid accurate and up to date.

This method of implementation creates ownership as well as keeping the goal in front of the team at all times. As the team accepts the goal, the goal will be accomplished. It becomes a self-motivational tool as well. I cannot tell you how the goal will be met, but experience has proven that it will be met within the agreed-upon time if ownership by the team is accepted.

As was accomplished with the pilot project, make champions from first cells that become implemented. This spirit derived from recognition will flame others to achieve similar successes or even improve on past accomplishments. If an organizational communication system is available, use it to communicate the successes. If a newsletter does not exist, start your own for this project. Post it on bulletin boards, especially in those areas where future change will be taking place.

Clear and correct communication is important. The communication must be complete. Include names of the individuals involved with the successes. Always be certain to include *all* of the names of *all* of the team members in *all* of the areas associated with the change. The names left out become the "us" and "them" division. Avoid creating any divisions.

Some organizations feel that competition is a good thing. It is the American way to compete. Experience tells me competition is good IF you are competing against the proper competitor. I have mixed feelings regarding the internal competition of cells or teams. It creates philosophical walls between the groups that may become problems at a future point. What happens when the project changes and the two teams who were arch competitors now have to work together to achieve the end?

Many organizational cultures use the spirit of competition as a way of life. Understanding the culture is necessary in dealing with this issue. Do not attack the culture, but use it in your favor. Address rates or standards or other targets as the competitors. Use the old process as the competitor. Address business competitors as competitors. Develop *benchmarks* from other companies to become the competitors. These are excellent ways to utilize the spirit of competition to motivate successful change. I would *not* recommend the use of internal competition as a way to develop teams. This conflicts with the goal of depleting the "us" and "them" philosophy from the organization.

Review the Process   The process will be changing as the teams continue to develop. It is necessary to continually review the process. The reason for this is to assure the team remains on track. Also, a team may develop a method that is better than the original change. This needs to be communicated to the teams already in place so they may apply a similar approach.

It becomes necessary to remain in touch with all levels of the organization as the change to culture is implemented. Evaluate the response from *all* areas. If the response implies a concern is developing, it will be necessary for you to address this issue.

Projects have been developing to an assumed completion when suddenly they are stopped. People are reassigned and the project is completely rejected with no further discussion allowed. Experience has demonstrated this where management receives feedback that a goal of the company is in jeopardy due to a project. The project must be monitored and this situation avoided. If a project is stopped, it is difficult to get it restarted. Also, the team will question the purpose since it was challenged and even halted.

The culture and the lines of communications to the executive management may permit unexpected changes in support. The status of the project should be monitored so issues can be addressed in a timely fashion. If left unattended, the fear of the change may overwhelm the manager. Unquantifiable risks may develop. Even personal opinions based totally on hearsay can develop conclusions that have no merit, yet become gospel.

The only way to short circuit these rumors or address the real issues is to be aware of them. Weekly or monthly meetings will give you some information. The informal discussions over lunch or around the coffee machine usually provide better feelers toward attitude and potential rumors. Be careful not to go chasing windmills. Confirm the problem or concern is real. Take action on the real concerns.

The training that was applied should be reviewed. Discuss with the teams what they need. Understand the parts of the training that was most beneficial. Use this information to continuously revise and improve the training or to develop future training programs.

Monitoring of the project is very important, especially with the first few teams. Their accomplishments will provide you with important information for continuing the project, as well as implementing future projects. Evaluate the implementation schedule. The projects will usually be ahead or behind schedule. You need to evaluate why. Continue to learn and improve the process for implementing change from what you have learned.

Evaluate the impact the change had on employee value. This is a long-term goal that each of us should have. It is a moral goal. Many well-known business evaluators who have written scores of books on successful companies relate to the value a company places on their employees. Those companies that continue to place a high regard on the value of the employee tend to do it better. Yet this issue is easily forgotten at the executive levels of management in most companies.

If the bottom line has any hint of deterioration, the first thing to go is the employee. Improved marketing activities, material changes to reduce cost of goods, process improvements, or many of the other possible choices are thrown out the window. Most companies only give lip service to the employee. Security is becoming a real issue. Company loyalty is fading rapidly in the United Stated. Changes need to be made to restore this loyalty: Not so a company has a devoted employee, but so the company has a motivated employee.

Motivation is the main aspect that provides a person a positive view on what he or she is doing. Quality improves if the person is motivated to make a good product. The job, whatever it is, is performed better if the person is motivated. Motivation deals with pride and the issue of ones self-worth. If the company has a low opinion of employee value, the self-worth of the employee is also down, as is the motivation, as is the performance.

The next concern to evaluate is the results of the success measurements. Were the goals met? If the project is a success, the goals will be met. In fact they will be surpassed. It is very typical for goals to be surpassed. The trick to the analysis is to determine if the goals were surpassed by a great enough margin. Sometimes the goals are developed to be very easy, attainable goals. If this is the case, than a very motivated team will greatly surpass the goals. The concern is if this is the case and the goals were barely satisfied.

Review the accomplishments with the management and discuss actions, if necessary, to improve performance. This is the opportune time to discuss employee value and positive motivation techniques.

Most of the time the implementation of CM is a means to support strategic issues. This concern needs to be reviewed to determine whether the strategic issue was satisfied and if further actions need to be taken. This part of the process will, potentially, open doors for future projects. It is highly unlikely a company will take on a major change in culture and its manufacturing practices and not associate it with a strategic issue. Well-managed companies address the strategic issues on a regular basis. Management determines if the organization is moving in the direction it had planned for long-term success. Clear understanding of strategic issues and being a part of the accomplishment is beneficial to an individual's success. If you have the opportunity to participate in these efforts, do so.

When these steps, initiation of cells and review of the process, have been completed, the process has been changed. Even if the process goes back to the original philosophy, ideas have been effected. The people will relate to this experience. Future methods will consider these ideas. Regardless of the degree of success the project had, the process has been effected.

The steps that follow deal with continuous improvement. These steps take the lessons learned and incorporate them into future changes. Many projects have a firm completion to them. When this completion date or activity is reached, the project ends. Project leaders may be reassigned to their original tasks or assigned to new projects. If the process is to address continuous improvement, someone must be responsible for the next three steps. If not, the continuous improvement process is merely circumstantial. It may or may not happen.

Modify the Process   This is a simple yet often overlooked action. The new process is implemented and the outcomes are reviewed. Hopefully, the organization has gained something from the effort. Now the company addresses the next issue and moves on.

It is expected the managers and teams associated with the change will improve on their own. This is not a true belief. Production rates, inventory, and customer needs change and many other manufacturing problems develop. The CM teams go to work on meeting the day-to-day problems. Remember they have no long-term plans or strategic issues to deal with at that level. They are given tasks and now learn to deal with them as a team. Additional changes to the process to improve it may never occur.

One company's experience demonstrates this point. The company was extremely interested in reducing costs to remain in business. It was a matter of survival for the company, as well as the management of the company. The management team had a 12-month period to improve the productivity. The owners, a financial organization would change the management team and possibly sell off the assets to regain their investment. Cellular Manufacturing was the approach the management team decided to take. They trained every person in the facility with the same instructor in the same classes on the same materials. Then the cells were put in place. Supervisors were released and the supervisory crew was reduced by 70%.

Material orders were set up for 12-month periods (blanket orders) and deliveries were set up to be released by CM areas themselves. The Procurement staff was then reduced by 80%.

Since the manufacturing processes were established and no new products were expected, Manufacturing Engineering was reduced by 85%. Since the manufacturing engineers were eliminated, standards for labor performance were not maintained, so the process to monitor labor performance was eliminated as well.

Cost was reduced substantially. Cellular Manufacturing was implemented. The idea of continuous process improvement is currently impossible. This company is locked into the processes it has. There is no one to generate significant process improvements. Only very small improvements can be made because no one has the ability to make the significant changes. Also, no one has the time. Everyone is working only on meeting the day-to-day obligations. However, the initial goal of survival was accomplished, or at least "life" was prolonged.

Other ways to modify the process includes changes in authority. Changes in authority lead to the practice of empowerment. Normally, the newly created cells are given some authority, but this is minimal. It is a real culture shock for most managers to give up authority. After all, they worked their way up the corporate ladder to achieve it. Now you want them to give away what they worked so hard to attain. The best way to deal with this issue is to gradually transition more and more authority to the manufacturing cells through the continuous improvement cycle.

Training programs need to be revised again or totally new ones set up. This time the training should address the next step in the vision you originally created. Perhaps computer skills are necessary to be able to access the system and research various data to solve problems. Whatever the criteria may be, it will change as you continuously improve the process.

Goals may change as well. Additional goals may be added. More complex goals could be included as people become familiar with systems. Development of the manufacturing cell personnel allows the cell to address larger issues. The day-to-day activity includes scheduling as well as doing the production. Verifying quality levels is included. Prioritizing also becomes part of it. Process changes and process evaluation should become another item. Inventory control and labor control will even be added to the day-to-day tasks for the cell.

As these changes take place, the cell instinctively starts to take on ownership, if the motivation and support is present. The team members will not continue to accept responsibilities if they are not supported. This is what happened to the desperate company that had to meet the cost targets set by the financial owners. The teams actually slowed down. They became frustrated with all of the new changes and very little support. The morale dropped and performance dropped with it. The first year's targets were met. The next year's targets became difficult to attain.

Provide Support   When dealing with support issues, it is important to identify what support is needed. The support groups need to plan accordingly as their role changes. Identify clear lines of support needed. The support groups can also become frustrated as the cells demand more and more from them.

For example, you eliminated the Manufacturing Engineering group. Now you expect the design engineers to pick up the responsibilities of the missing group. The design engineers need to have clear expectations. If not, they may refuse to support the requests of the cells. Then the cells become frustrated. They complain and Engineering becomes frustrated. It is easier to prevent frustration than try to eliminate it after it has taken hold in an organization.

The support areas will have to revise their priorities based on the new expectations. These types of organizational changes have to take place to ensure a complete implementation of the new culture. This will also change as the continuous improvement activities make other improvements in various processes.

For continuous process improvement you need to communicate, communicate, communicate, communicate, and communicate. This assumes everything else is perfect. Motivation exists. Support is present and adequate. Empowerment is in place. Now everyone needs to now what is going on and what is important. This allows them to work on the right things at the right time.

Develop and Train for the Next Level   This step simply puts you back in the implementation stage only starting at the next step. You merely need to reevaluate the process and culture and train for the next level you wish to attain. This should be a fairly easier task than the first implementation cycle. You will need to address the next emphasis for training, whatever that is to be. This depends on how far the first implementation cycle took you and what you want to achieve with the next cycle.

Develop the cell's potential and review it with management in the next training session. Management should be more involved with the training the second time around. Managers are now part of the culture and have accepted the changes at this time. Now they need to demonstrate their ownership and participate in the training development and the presentations themselves.

Repeat what you did before. Communicate the new successes. Clearly identify additional cultural changes. Prepare to address the negative issues that developed. It should be a copy of the first five steps of the process, condensed to meet the next level of attainment in the cultural change planned for the organization.

As well as satisfying the continuous improvement cycle of the process, these last three steps deal with supporting strategic issues of the company. The strategic issues continue to change as the company grows and moves ahead or as it flounders and falls behind. Either way the passing of time causes changes in the strategic plan and the tactical actions necessary to satisfy the strategy. These steps, which review the situation and cause action to change it again, should be in synchronization with the long-term course the company plans to take. Hence, the strategic issues of the company should be considered when reviewing the process and culture for future improvement and changes.

3.3. Summary

Training is an important issue with implementing change. The training must address the specific needs of the organization as well as the needs of the group being trained. It does not make sense to teach advance differential equations to the people who will be doing material handling. However, if you have developed the material handler to formulate a process by which they optimize the travel time of the trips they make with product, the application of differential equations may be the next step.

When the training issue has been resolved, the actual implementation plan can begin. The first five steps of the plan deal with changing culture. This requires an evaluation of the present culture, developing a sales plan to receive authority to make the change, prove your concept with a pilot, formalize a training plan for the entire operation, and then training managers to gain their support. These items formulate the culture that has been approved for further implementation. The various levels of management are aware of the plan and the expected outcomes. The areas that will be affected know what to expect. The mind-set is in place.

Implementation of the actual changes becomes relatively simple but very time consuming. You simply implement the plan in one area at a time. You learn from your experiences and improve as you go along. You communicate back to the areas already implemented to improve their processes and maintain an equal level of performance of all cells as the implementation continues.

The last phase of implementing change is the continuous improvement phase. This is where you step back and review what was done. Consider the next step or plateau you want to get to and repeat the process to get there. This phase deals with strategic issues and long-term goals. You can never change culture in such a drastic way that every problem is resolved with the first pass. You need to retrain and perhaps regroup to take on the next step, and the next, and the next.

Always keep the value of the employee in mind as well as the reason the employee will take on the challenges you present. Motivation is the key word. Positive motivation is best for a long-term impact. However, negative motivation does work . . . short term. As you progress with the business issues, try to incorporate the moral issues of employee value. These issues make you feel good about yourself and others, rather than just about your accomplishments in improving the bottom line.

4. ACTUAL EXPERIENCES IN CELLULAR MANUFACTURING IMPLEMENTATION

This section is a series of observations with company-specific information to allow the reader an insight of what variables take place in the real world. Each scenario is for a specific company dealing with implementation of a team concept. The scenarios look at the perspectives from *culture* and *leadership support* of the company in implementing change. The scenarios are provided not to measure whether a culture is good or bad or if the decisions are right or wrong. The scenarios provide a sampling of the variety in which companies operated and the changes that may take place during an implementation process. This information should be considered when trying to understand why a company acts, or reacts, a certain way. It is useful in understanding actions a person should consider to improve the possibility of success in a project implementation.

Each company has a different *opinion* on what is or is not important. Each company has a different *reason* for that opinion. Each have different *management styles*. Each has different *ends* or strategies for considering implementation or changing from the present norm. Each company, because of these variables, experiences significantly different *results*.

This section will provide insight. The conclusions are left to the reader. Life has no simple solutions to resolve future challenges. It does provide a learning experience from which some of life's future mistakes may be avoided.

4.1. Able Company

Able Company is a successful and large corporation that deals in "state-of-the-art" products. As well as providing a challenging product line and maintaining a challenge to the employees to keep up with leading-edge technology, the management maintains a pulse on the leading-edge management methods to provide positive motivation and job enrichment. Various types of changes were tried to continuously motivate and improve employee performance at all levels of the organization. Quality circles were initiated. Teams were developed and empowerment was a term used frequently in the workplace.

Employees were valued at a very high level. Benefits were excellent. They were provided at company expense to demonstrate the appreciation of the company to the workers. Vacation time was also very generous. Vacation was up to 6 weeks after 10 years of service. The respect shown to employees created a loyalty to the organization and many employees had service lengths of 20 or more years. The employees were allowed to make decisions to improve their personal work assignments.

Work areas were all ergonomically designed. Chairs and tables were completely adjustable. The high-tech culture permeated throughout the organization's actions and decisions.

The high-tech environment also created some frustration. It caused engineers and scientists to be promoted throughout the organization. Even in areas that did not require the skills and training of an engineer or technical person. Even though the culture provided for much employee appreciation, the technical staff was not in tune with the knowledge of good behavioral science techniques. Many employees became project engineers for a group that did not need a project engineer.

The management applied changes to the culture in the same manner in which a design engineer would apply a new technology. Managers would do a prototype. If the prototype worked, they would fine tune it and modify it for a production release design. If it did not work, it was simply dropped. This technique works well with product designs and research and development activities. It does not work well with people.

The "prototype" of the quality circles did not work. The technical personnel were not very good at being facilitators. The issues became gripes from the various functional areas about the other functional areas with which they worked. Quality was only understood as being the quality of the product. The performance of the process to produce a product at the optimum efficiency was not an issue. The "prototype" was dropped.

Communication was not an issue. The engineers and scientists communicated very little, except for the design teams working together on a specific project. They were basically given a task and did it. When the quality circle prototype was ended, management did not feel it was necessary to explain it. They did not evaluate strengths and weaknesses to develop a complete approach to make the change. The quality circles were set up and expected to develop from the bottom up. This went against some strategies, and teams lost support.

Now they wanted to prototype CM. Except for the management, most employees felt it was just a fad and would be dropped with no explanation as the previous changes. However, the culture was use to change so it was easily obliged. Initially, the attitude was one of skepticism. This was another prototype project and nothing else.

As the process of empowering the cellular areas, called product teams, took effect, the teams developed quickly. The engineering practice of working on a project, with limited management intervention and in small groups, was easy for the technically minded supervisors and management to grasp. They worked well with the other members of the team and treated them personally as equals. The transition worked exceptionally well and took a strong foothold in the company.

The teams grew in responsibility and empowerment. Many managers were actually roaming the halls with little to do as the teams did more and more independently. It was a very successful transition. The participative management style used in the design team environment quickly supported the new approach to CM. The past quality circles concept of quality first let to many improvements in process and waste reduction as the teams empowerment grew and these issues were addressed. As with the design teams, continuous improvement in the design transitioned into the product teams. They continually tried to improve and become better. The culture worked for them and developed the cellular manufacturing philosophy almost automatically.

Executive management provided the leadership and insight to try new things and utilize team concepts to resolve issues. The continuous innovation of the leadership made the culture acceptable to change. Even though the employees did not appreciate changes being canceled, they were still motivated to try a new method. This leadership and culture greatly supported the change to CM.

Participative management style was another benefit to implementing CM. It provided the style to deal with teams and allow for empowerment. It was also beneficial in dealing with the growth of CM as the project took on additional responsibilities and assignments.

The *strong emphasis on employee value* provided a respect of the individuals and supported transition of tasks and problem-solving responsibilities. Basically, it supported the empowerment transition form direct supervision to a team approach.

The high-tech environment could be said to support the transition. However, I feel the product or market was not an issue. Management had made a *specific effort directed at culture change*. Change was something that was a common event within this organization. Although the communication methods were not the best, the employees expected change. They even welcomed it. They were frustrated when the change was not fully implemented, but they seemed willing to try new things and take risks. This was due to management's willingness to continually take an effort in making change and learning from it.

The company had a very *strong emphasis on quality*. Although it did not dictate quality aspects to the lowest levels of the operation as most quality experts would suggest, management emphasized pride in workmanship and strong rapport with the customers. This noticeable interest in trying to please the customer was a perceived effort to make a quality part. This continuous effort to please the customer also continued through the organization. Each department or functional area tried to please the other. As the team developed and the CM philosophy was put in place, this customer issue supported it. Quality even improved in the teams as they interfaced with the teams they supported and those that supported them.

Able Company definitely had a *strong emphasis on empowerment*. This was probably the result of the various design teams that were empowered to develop new products based on specific instruction from management prior to initiating the projects. This same practice was readily available to apply to the CM teams. In turn it was readily acceptable from prior experience in working on many design projects. Able Company practiced simultaneous engineering when developing products: Manufacturing, Quality, and Marketing were involved on the design teams. As a result the products were highly producible, as well as having very low warranty claims, after being released to production. This also made all functional areas of the organization directly involved with the projects development, another factor that aided implementation to CM.

Since Able Company was a technology company, it dealt a great deal with data and testing. The company had a *strong emphasis on data collection* to support its designs and provide the continuous improvement requested by the customers. Although some companies implement CM with little data collection, Able Company already had SPC tools in place. This provided the teams with information and tools to resolve problems and improve the process with minimal risk. Again the culture was in place to provide the support and tools to assist a successful CM transition.

The reason the executive management wanted to implement CM methods was clear and simple. The executives wanted to *continue improvement*. Change was part of the culture. The reason for change was to improve; to do better than anyone else had done before. It was a strategy of the company to improve the designs over the competition. The company believed in improving the way of life for the employees so they would be motivated to do more for the company. It was the opinion of the leadership to improve the company performance by trying new ways to do things as well as trying new things.

Risk was acceptable to Able. It was the way it operated. The risk was not the important thing. It was the challenge to beat the odds by constantly improving over what was or could be done before. This part of the company culture was a motivational tool to provide employees with the drive to make CM work and work effectively. The company not only implemented the change, it reviewed it, improved it, and changed it again until it was satisfied the change was effective. In essence, Able Company believed "If you are going to do something, do it right," perhaps not the first time, but through change after change after change. Improve it until it becomes what you want it to be and do not become side tracked by failures. Learn from them.

Able Company had products that changed frequently due to the nature of the high-technology industry. This caused Able to be very flexible. This demand to be flexible drove its systems as well. Able's *systems were very flexible*. The flexibility of the systems allowed Able Company to provide information and change systems to support the CM transition with little interference with the overall company operation. It also provided information to the product teams in a very timely manner. This gave an effective tool to the product teams in analyzing what they needed to do to improve and correct problems that became evident as they transitioned to CM. The empowerment and transfer of responsibility would have become extremely frustrating if the teams could not get data and information to deal with the concerns. System flexibility was a great help to Able's implementation of CM.

4.2. Able Company Results

Many companies try to implement CM and plan on a minimum of 2 years. Most take 3 years or more, even though they try to say they implemented it in less. For the most part, companies successfully implement CM in a few main product areas or via group technology cells and say they are fully implemented.

Able company took 18 months to implement CM in all product teams and even had the philosophy spread to all the support function areas as well. This implementation only involved 150 people at Able. This was a small division of a larger corporation. It was allowed to happen because the larger corporation also had a culture similar to the smaller division. It should be noted that all divisions of a large corporation may not have the same culture as the corporate office. With mergers and divisional management it is very possible only some parts of the culture are maintained at a divisional level from the corporate office.

All of the divisions of Able's corporate center were involved in high technology, and all managers had been rotated from one division to another to introduce new concepts and provide variety to their lives as well as stimulate their careers.

As expected, Able had a *successful implementation.* The CM philosophy spread throughout the entire corporation. Cellular Manufacturing spread from division to division as management rotated into and out of the division that implemented it.

There was only one downside to the success of implementing CM. The company, in total, could not handle the excess management availability through attrition alone. Most employees stayed with Able for long careers, so turnover was low. After a period of 2 years, following the implementation of CM, Able made significant cuts in the management team. Streamlining was the term used by the corporate fathers. The efficiency of the operation was such that almost 60% of the managers were released. Mostly middle managers were let go. Many were provided with early retirement to keep in practice with the valued employee concept. Others were supported in finding new positions in other companies, even with competitors. Some were retrained to transition into different and growing industries. The concept of the valued employee was still a part of Able's culture.

However, it was not perceived in this manner by the employees. Turnover became a concern. The once long-term employee was now seeking "greener grass." As some of the managers left, they took some of the people with them to support the challenges they faced in other industries and other companies. Some people left on their own, perceiving it is better to leave on your own than be asked to leave. Many went because of the unknown. They had no feeling of security with Able. Even though the company was doing well, people continued to be "excused from their responsibility." No matter how it is said, when people are put out of work, the remaining employees get nervous.

This fast retreat from the supposed sinking ship gave Able problems that lasted for over 5 years. New college graduates avoided going to interview with Able. Suppliers questioned the financial strength of Able. Customers questioned Able's ability to meet schedules and deliveries.

Perception is a powerful thing. Any company must try to maintain a positive perception, regardless of what is actually happening.

Quality was improved. Many of the issues that were not addressed previously in manufacturing were being addressed now. Prior to CM the management did not have the time to help resolve many of the issues. As the product teams took responsibility of these concerns, they were dealt with and corrected. If the correction did not work, they were immediately back on the action register to be resolved again. Returns went from 10 to 0.01%.

Remember that communication was poor with this organization. As the teams went into effect the *communication improved.* Teams communicated with teams. Departments communicated with departments. Divisions communicated with divisions. The entire company realized improved communications that were also effective communication. They were communications that resolved issues and brought people into the team to resolve and prevent them from returning. Communication even improved with suppliers and customers. The impact of CM on the entire corporation was a positive one.

Productivity increased. Many tasks were accomplished by fewer people. The reduction of quality concerns, such as scrap and rework, allowed for greater output of quality products with less material and less labor effort. Schedule times were met on a higher rate. Production cycle times were also reduced.

Initially *morale increased.* The teams developed new levels of trust between functional areas and various departments within the company. They were all working toward the same goals. The improved communications also helped with morale. As the employees were provided with more and more information, they felt more ownership to the company and its successes and failures. They were able to realize how their efforts helped the company and had a stronger tie to its end results.

It was not until the streamlining took effect that the morale of the corporation went sour. The trust level between employees and their immediate superior was very high. When the employees were released from their assignments in waves of streamlining activities, the trust level fell. Five years later the ex-employees meet in restaurants for breakfast to discuss "old times" and put down the once loved company.

Required support was reduced. As the teams picked up more responsibilities and control over the products, they also absorbed many of the support functions. The typical one is Procurement. The use of the "blanket order" was increased. This is an order for several months of parts or components or material used by the product team directly. The deliveries of the items on these blanket orders is released by the product team as needed. In this fashion, inventory is kept to a minimum. As this practice continues, the need for many buyers in the organization is reduced. The buyer no longer needs to check on deliveries after receiving a complaint from the production floor that workers are out of material. The buyer is not tied up with discussions between manufacturing and the supplier to try and determine who did what and when. All of these issues are eliminated or taken over by the product team. All the buyer needs to do is establish a negotiated blanket contract with the supplier to deliver in a manner the product team will request. After that, the product team takes over.

Tooling and process was cell's responsibility. Another support function whose activities were reduced was the manufacturing or process engineer. These engineers were promoted to do a higher level of engineering support. They now did the necessary research and development to significantly improve processes. The maintenance of the processes and small, but continuous, improvements are now the responsibility of the product team. Basically, the hand holding that most manufacturing and process engineering groups do through the normal course of production support, before CM, has been eliminated.

Material was procured in Just-In-Time (JIT) fashion. This reduced inventories and material carrying costs. The most visible impact was more room was available in the facility. Storage of Work-In-Process (WIP) inventory was reduced. The impact of this triggered other actions such as reduction of setup times for machining processes and changeovers for assembly lines to produce different products. As these setup and changeover cycles reduced, the production quantities reduced. It became more efficient to build in smaller lot sizes. This in turn reduced the WIP even further. The final effect was a relative 50% reduction in WIP material costs.

Inventory control shifted to the cell. Inventory control transitioned from a complex Materials Requirement Planning (MRP) system to a visual inspection of material on hand and a short list of needs for the next day's or week's run. In most companies a complex MRP system is installed to control inventory, production, and financial records. What develops is that finance becomes the main focus. More controls are added to the MRP system making it more complex and less "user friendly." By simplifying the system, finance receives more accurate information while production is not tasked with non–value-added administrative duties.

The product teams were provided with the customer delivery schedules. They worked directly with suppliers and supporting product teams to satisfy the immediate needs first. This visual control of inventory could only take place after the product team had transitioned all supplier orders to a blanket type of order system with the buyers. The team also needed to understand the process in detail to coordinate its actions in meeting customer orders. This was an activity that resulted in the end of the implementation to CM. Many things have to be learned and in place before this can result.

There was *no change in design engineering process.* Even though the company evolved to CM based on the design engineering practices and the culture in place as a result of those practices, the design activities did not change. The reason for this was simple. He who has the gold makes the rules. The company was a high-tech company and its business was derived from its excellent design capabilities in state-of-the-art technology and innovation. The executive management knew this and provided design with top priority on all matters. There are some things, often called paradigms, that a company is not willing to change, regardless of the benefit that may be proven by actual experience. The design engineering process and departmental structure was Able's paradigm, the sacred cow of the corporation. Most managers were promoted out of the design group. They were not about to change that part of the corporation.

Scheduling was cell's responsibility. As the product teams assumed the responsibility of inventory control, it was only a matter of time before the teams accepted scheduling as well. They were prioritizing the work and the assignments within the cell. They were interfacing with the production coordinators on a regular basis. The next obvious step was to be empowered with the scheduling responsibility. The production coordinator's role transitioned to supporting setting up new products as they were developed and transitioned into manufacturing. These coordinators were promoted to doing more challenging activities as the day-to-day coordination went to the cells.

One area where the company did not have any improvement related to the CM implementation was with the customer interface. There was *no customer involvement in the cells.* Even though the customer did become more involved with the design group, the design group did not really change. Communication improved, but direct customer involvement stayed the same. The customer's input was sought after. Customers' ideas were investigated and sometimes applied to new products. However, the involvement of the customer was kept at arms length.

4.3. Able Company Summary

As an observer, I felt Able Company was very successful in implementing CM. It had many things in its favor. The one concern I would address from this experience was the lack of long-term planning for a successful implementation or the effect of changes in the environment around the company. It was an after thought to address the issue of the excess employees associated with all the improvements in efficiency. Also, the high-tech markets with which Able dealt were primarily military in nature. Able believed the military would not reduce in size regardless of what was on the evening news every day. Its long-term planning still included strong growth in military sales.

If the bottom line dictates you have to reduce the labor force to survive, I can understand the reasoning. If the reasoning to implement CM is for improving and you succeed, the next step should not be to reduce headcount. It should be, in my opinion, to develop new markets and areas for growth. I feel a company has an obligation, in some manner, to support the careers of the employees, unless fate says otherwise.

Management always has options. It is too easy for executives to take the streamlining option and avoid developing a company to even higher plateaus in the industry. Companies, in my opinion, who say they are world-class manufacturers and practice this method of resolving operational effectiveness, or ineffectiveness, are involved with the third-world class of manufacturing, the third world implying the company is still developing in professional management techniques. They have not taken on the challenge to develop their businesses to the next level by utilizing the resources they have just made available to themselves. The same was true with the reduction in military business.

Companies such as Hughes and Raytheon transitioned their efforts to new commercial markets. Utilizing the software and internal processes developed to meet military compliance to improve commercial manufacturing processes. They sold the technology they developed to companies to improve and compete worldwide. They transitioned with the marketplace and maintained a level of employment to support the cutbacks seen in their military products. Most of the streamlining in these companies was resolved by normal attrition.

The bottom line was already improved at Able by the actions taken in implementing CM. It was really not a necessity to streamline. There was an option. The 5 years of pain received by the Able employees was mere penance.

The other observation is that paradigms are extremely difficult to change. Even in a company such as Able where change was a way of life and part of the cultural fabric draped over the entire corporation, change was limited. It did not touch the sacred items of the company. In this case it was the design group. It is other groups or products or practices for other companies. Until a significant change is made, these paradigms will continue to exist. It is possible the paradigm may even conquer everything else and success will change to failure.

4.4. Baker Company

If you have seen the symbols for yin and yang, you will understand the difference in the culture between Baker Company and Able Company. It is the age-old comparison of black and white, day and night. Baker was a privately owned company that had much success. The owner developed the company from nothing to a mid-sized operation with sales of over $100 million a year.

Although the owner was extremely personable and a rather charming individual, he was very much in control of everything that happened in the company. It was his company and everyone knew it. Including the customers and suppliers.

As expected, the company had a *dictatorial management style*. Each manager was treated with little regard for personal or family concerns. The job was the only thing that was to be considered. Do what you were told with no reason for rebuttal. If something went wrong on the production line, you stayed over until it was fixed. Every customer delivery was to be met or you would be looking for employment elsewhere. If the company did not win, the employee responsible for the loss was traded for someone willing to make the commitment of heart and soul. It was always someone's responsibility for failure. However, success was a company achievement.

Ironically everyone seemed happy there. There was a spirit of camaraderie. They were all in the same boat. The company was extremely competitive as was the owner. Softball teams and basketball teams and bowling teams and such were all sponsored by the company. The front lobby was filled with trophies. The teams were filled with employees. The morale remained high. The team spirit and company spirit was developed by the team competitions. Team spirit was not developed by the management.

This may seem to be a very negative environment. In fact it is not. Everyone knew what was clearly expected of them. The salaries were excellent and individuals were rewarded almost immediately for their successes. They were also condemned for their failures. The immediate feedback mentioned in Deming's many lectures to management was a reality at the Baker Company. It was not done in a team atmosphere though.

The *culture was control*. There were no teams but many committees. The committees were basically used by some of the managers to cushion the personal effect of a failure. It also gave them control over the actions to be taken by directing the committees. Committees were usually "bossed" by the manager in charge. A manager gave into this form of management style. Each manager knew he or she would have the same type of control for the committees he or she specifically ruled. It was a mutual understanding that the committee you ruled had complete authority to dictate what the committee did. However, if something was done by a committee, the ruler was protected. It was a unique management style.

This style was developed by the owner. He wanted control. Yet he was well read on teams and the benefits of these teams on business performance. In his mind he tried to satisfy both of these opinions and developed the committees. He also realized they were not working well. He was a realist in this sense.

Emphasis on bottom line was always the primary issue in this company. Even with all of the company-supported teams and the outside appearance of a high employee value level, the goal was to improve the bottom line. The sports teams were the manner in which the owner perceived development of loyalty to the company. Everyone loves a winner. If you had a lot of winning teams, people loved you. It wasn't this simple, but it was along these lines that the company operated.

The company practiced *limited empowerment*. When a committee was established, the manager of the committee was empowered to make the final decisions, except in those cases when the owner did not agree. In reality, there was no empowerment. You were given the feeling of being empowered if the result was completely what the owner wanted. Committees and teams were mere facades and were perceived differently by the owner and the employees.

The owner saw the teams as a way to develop the organization toward CM and TQM and the team-oriented process improvement techniques. The employees perceived the teams to be a waste of time and effort. They went through the motions just to appease the owner. After being on several teams I realized a pattern. The teams would use the first few meetings to develop great ideas and unique changes. The teams would arrange a meeting with the owner and review these ideas. The owner would identify which ideas he would accept. The team could go on to implement any of those ideas. If the other ideas were developed, the team members could find their careers threatened by the owner.

The company operated as several small businesses. There were several small companies involved in production of similar types of products. Each company was a separate purchase and operated independently with a separate general manager who reported to the owner. The Sales Department was centralized and work was distributed to the individual operations based on cost capabilities and availability of meeting schedule. Although the individual operations were supposed to be specialized, work went to whomever had the capacity at the time of the order. This created many problems. The owner still expected the general managers to work together and support each other. Hence, if you were the general manager for plant 3 and your product specialty was to be run at plant 1, you were to support it. It was your responsibility, as well as that of the general manager of the other operation, to make certain it was delivered on time and at cost.

The problem with this situation was twofold. The plants were not close to each other. They were usually over 20 miles apart. Also, they had different equipment, tailored to do the product in which they were specialized. Although the products were similar, they were also dissimilar. Assuming the products were all machined parts, plant 1 handled high-volume low-cost parts that had wide open tolerances. Plant 2 was a medium- to high-volume producer of only turned parts made from bar stock or rods. Plant 3 was a low-volume high-precision manufacturer of complex components used in the aerospace industry. Using this type of situation, you can see how it becomes difficult to mix product lines and provide the appropriate quality controls to meet customer needs.

The only reason to implement CM was *market driven*. The company had several major customers in the European and military markets. The drive for ISO 9000 and TQM practices was perceived by the owner as a reason to implement CM. It was also driven by the purchase of a new operation that was from the dying military business of a large corporation. The purchase of the operation was sold with the emphasis that the operation was operating at this capacity. We will call this plant 4.

Plant 4 was completely shutdown when purchased. The employees of this operation were mostly union workers and went back into the union at another location within the larger company. The managers that had provided the leadership to the operation had found new opportunities as well. Although the operation was previously managed and run as a CM operation, all that remained was the equipment and the building.

A new general manager was hired and directed to set up the operation as a CM facility. The owner assumed this was an easy task since the operation had been a CM operation previously. The general manager was allowed to hire staff and employees as needed to meet the sales forecast and profit targets established by the chief financial officer (CFO) for all four plants.

The general manager started by hiring people who had experience in CM. They were trained to be strong facilitators and team developers. The workers were hired who had experience in working in teams. After the first few people were hired, they started to take part in the interviews for other employees to be working by their side. The new general manager was not aware of the culture that existed with the larger company of which he was a part.

Since the employees were all new, they were selected on their experience. Training was minimal. There were weekly team meetings and monthly facility meetings. Everyone knew what the operation had to do and what the results were as well as what was to be expected from them. Communication was excellent. However, in this case, it only added to the frustrations. As they tried to improve and make changes, they were denied. The denials were communicated and discussed openly by the teams. These feelings were directed at the general manager. He had established the teams and provided the direction for them. As time went on, the direction was hindered.

System capability was limited and controlled by finance. The new general manager was becoming familiarized with the systems available to him. They were extremely inflexible and did only what was approved by the owner. The CFO, who had little understanding of CM, controlled the changes to the system with an iron hand. The inexperience of the new general manager in working in this type of organization caused him frustration. He had been told by the owner to develop a CM facility. Yet the CFO opposed it vehemently. There was no compromise.

Scheduling was inflexible and controlled by sales. As with the financial systems, scheduling was also under strict control of another group. The teams developed in Plant 4 were not allowed to deal directly with customers to identify the real needs. If the order said 20,000 widgets in 6 months, you had to provide all 20,000 on that specific day, 6 months hence. You were not allowed to develop a rapport with the customer to deliver 5000 parts every 6 weeks or any other combinations of deliveries which may help your production cycles and reduce WIP inventories.

Basically the teams and Plant 4 were unable to perform as a team. They were not empowered except within their own four walls. The potential of implementing the CM approach was limited by the restrictions placed on the operation.

The other plants were asked to implement CM by the owner. They were told to follow the practices of Plant 4. The general manager of Plant 4 was instructed by the owner to implement the CM technique in the other three plants working with the other general managers. Unfortunately, the general manager of Plant 4 did not understand that he had the power to direct the actions of the other managers. He used a team approach to address their individual issues. When the other managers saw the new manager empowering them, they took the authority and fought the new manager's attempts to change their plant's operations. It became another frustrating experience for the new general manager.

4.5. Baker Company Results

The implementation of CM at Baker Company took only 12 months for the new operation at Plant 4. This was primarily due to the fact that the first 70 people were already trained in CM before they ever started at Baker. Training was minimal and the people already knew what was to be expected from them.

The conversion to CM failed. Even though the entire workforce of plant 4 was trained and had prior knowledge, they were never empowered to make the process work. *Systems failed* to be changed to support Plant 4. They could not get the information they needed to make many of the improvements that should have been made. This caused frustration and demotivated the teams over time.

Even with the entire facility trained and familiar with CM, the motivation was not there. Initially it was there. Over time it faded through the practices of the owner and other management. The one manager who supported the change was seen as an enemy to the process. Since he was not allowing the changes, he was directly responsible for the limitations at Plant 4. Perception is a strong power. It can work with you or against you. The people at Plant 4 only saw the general manager. The owner, CFO, and the other general managers were isolated from the people at Plant 4. They were never perceived as being the problem.

Staff support was refused to Plant 4 by the majority of the centralized departments for Baker Company. Finance, Procurement, Sales, and Marketing all maintained the same operational techniques and forced the old systems on Plant 4. These roadblocks only added to the resulting failure of the process.

High turnover followed in Plant 4. As frustrations grew, the employees hired to implement change resigned from the company. Unfortunately, they were replaced by employees from the other plants who wanted to work closer to home. Their cultural background was in direct opposition with that of Plant 4. The general manager was forced to take these employees from the other locations by direct order from the owner. This loss of authority by the general manager and the managers at Plant 4 increased frustrations. The managers and the other team builders of Plant 4 resigned as well.

Initially, the performance of Plant 4 was very high. In fact it was doing better than the other plants, including Plant 1, which was the main plant and office of the owner. As the project started to fail, *performance dropped*. The teams were forced to do things they did not want to do. They opposed the changes back to 1950 manufacturing methods with assembly line mentality. They requested meeting high-quality levels, processing smaller lot sizes, reducing inventory, and team work. As the support areas refused to abide by these requests, the requests became demands. The demands became stronger as did the opposition and the frustration. As one increased the other dropped. Even the quality fell off as the teams broke apart.

The Quality Department had remained separate from Plant 4. It was a centralized department, but the quality director did try to support the general manager of Plant

4. The owner became involved as the quality levels of Plant 4 took a nose dive. The quality director was forced to participate in the operation of Plant 4 as a policemen. Quality *held this policeman's role* in Plant 4 as the operation continued on its downward spiral. The performance of Plant 4 went from the best to the worst in a matter of 6 months. Finally the general manager for Plant 4 resigned. This was the last straw on the camel's back for teams at Baker. The camel fell to the ground and was crushed by the culture of the company.

Many support groups remained unchanged. *Material procurement was unchanged.* It caused many problems with the team building process and the CM transition. Inventories were huge. The Procurement Department bought in quantities to receive large price breaks. However, inventory was stored in the aisles and most work had to be done around the pallets of inventory. Material was damaged by the multiple handling. Material was moved from place to place to try and get it out of harms way. The damage to the material and the poor working conditions did not support a quality emphasis. It demonstrated a culture where quality was not as important as being able to deliver on time to the customer requests.

Inventory remained high (batch system). Scheduling dictated large run sizes, calculated from financial data that produced economical lot sizes based on the process of Plant 1. These systems controlled the inventories in large batches. Even though the teams at Plant 4 developed short turn around time from one product to the next, the batches remained the same. Dictated by Plant 1 systems, inventories of WIP and finished goods exceeded customer requests. Even the attrition rates of the process were in excess and provided additional unwanted inventories. This is an example where an effective MRP system for finance is not suitable to support production.

Even though many good suggestions to improve the systems for the company were submitted, *systems remained unchanged.* The authoritarian management style caused the support groups to have selective hearing when suggestions were made. If they did not see any benefit to their specific groups, nothing was done to assist another group. Each group was measured on its own performance. There was not an overall direction to improve the sum of all efforts made by all the groups. These changes were left to the owner and only the owner. Hence, if the owner did not dictate a change, it did not occur.

Within 18 months, the team operation of Plant 4 was eliminated. The new management, promoted from Plants 1, 2, and 3, took over the reigns of Plant 4 as the original management resigned. Most of the original employees from Plant 4 left with the management. The facility incorporated the same culture and systems from Plants 1, 2, and 3. The team process to implement CM was dead.

4.6. Baker Company Summary

Two very important lessons should be learned from this example. First, you cannot change a culture in a short period of time. You need to use the culture to implement the changes.

In the case of Baker Company, the general manager could have made some specific strides if he took the reigns of the "team" of general managers to implement CM across all plants. He could have even, potentially, directed system changes. Instead he tried to use the real practice of teams to develop a mutually agreed upon plan for the change. In doing so, the mutually agreed upon plan was to *not* change.

One can never say the alternative approach I am suggesting would have worked. What we do know is the approach taken was not successful. The reason the approach taken was not successful many would say was due to the second lesson to be learned from this experience. Executive management has to support the change.

The management of Baker Company did not support CM. It was a idea of the owner, but directed at one manager. Although that manager could have successfully implemented CM within his plant, the support systems and interface with the rest of the organization limited his ability. When the approach changed from implementing CM in Plant 4 to implementing CM in the entire operation, the manager ran into a stone wall of opposition. There was no support or effort from the companies executive management to implement the plan. No one believed in it or supported the effort. You cannot implement a change of this type from the bottom up on most occasions. If you do, be prepared to be frustrated for a long time. It will not be an easy task. If you do succeed, you will have to use the culture that exists to support your efforts.

The naiveté and inexperience of the new general manager allowed for the process to fail. If the proper actions were taken by an experienced manager, the results *may* have been different for Baker Company.

4.7. Charlie Company

Charlie Company is the third company we will examine. This is a middle- to large-size company with approximately $1 billion in annual sales. The company is incorporated and the president has to report to a Board of Directors. Aside from this change in the organizational structure, the management style was very close to that of Baker Company. It was a firm that used the *autocratic management style*.

The culture was power and control. Each executive was praised on his or her own specific area and function within the company. Executives did not work together well unless it was ordered by the president. The systems did tie into each other so other departments could see what was being done by another department and analyze the results on their own areas. However, it was left to survival of the fittest to determine who would win and whether the change was implemented.

The company had one main theme or emphasis. It was *speed*. Everything was done fast. Speed was the most praised activity in the entire organization. It was almost to the point that if you did something totally wrong, but did it fast, you were honored for your accomplishment. It was interpreted as, "doing something is better than doing nothing at all." Planning was not well accepted. It took time and was perceived to be inefficient. If you could shoot from the hip in this organization, you went very far, and fast.

Changes were made in the company in a matter of minutes. If it was decided to change the reporting systems or any systems, they were done relatively over night. The next morning the new system would be explained to the employees. If objections were made, the system would be changed again to satisfy the objections. Because of this approach, the systems of Charlie Company were extremely flexible. There was a large Management Information System (MIS) group staffed with computer programers that constantly changed and improved the system flexibility.

Charlie Company worked through, what I call, *artificial empowerment.* This is where the people perceive they are empowered to make a change but are not. The teams they had were actually infiltrated with "spies." These spies reported back to management what the team was working on and why. Management made decisions on these reports and took action on them. If something was not in tune with the direction management wanted the team to go, the team was immediately disbanded and the personnel reassigned to other projects. The concern was addressed by a new team and new guidelines. The method that was not acceptable to management was specifically expressed as an area on which not to waste time.

The term "spies" is unique to this situation. Individuals who were being considered for promotions were offered "carrots" based on their loyalty to management. These individuals remained secret to the organization. Sometimes they would make confidential reports even on other managers.

Teams at Charlie Company worked very well as long as they developed the specific ideas executive management had already chosen to be implemented. This was an advantage to implementing CM. Executive management felt there would be strong benefits to implementing CM. However, the president and CEO were not certain as to how they could allow teams to become empowered in the real meaning of the word.

The management of the company selected a key officer to oversee projects. The CM project was *assigned an executive VP to implement.* This was definitely a top down approach and provided significant benefits to a successful implementation to the project. The executive manager immediately accepted the opportunity, whether he believed in it or not. It was the culture of Charlie Company to do so. It was not necessarily true the manager assigned favored the project. In this specific case, he did favor the project. Another reason for success.

Charlie Company had a down side in implementing CM throughout its processes. The reason to implement CM processes was for *company recognition.* The company was run by a Board of Directors who felt it would be in the best interests if Charlie Company won, or at least placed, in the Malcolm Baldrige National Quality Award. Executives perceived the manner to do this was to implement a CM process. From this they would develop a TQM philosophy within the organization. This approach would win them the coveted award and assist to increased sales. It was a 3-year plan. One of the longest plans known at that time in the organization.

The newly developed *Training* Department was recognized as the *key change agent.* People were hired with experience in team development and behavioral science backgrounds. Others in positive motivational skills were also hired to train management and supervision. The training took several months to train all managers in all departments. The executive levels were not trained. Since it was their idea, they assumed they understood the techniques and practices so they did not participate in any of the training nor the issues covered by the training. Only the executive VP assigned became involved. He alone, other than some "spies," reported the progress of CM.

The company developed a *steering committee* as suggested by the newly hired experts in team development and CM. The steering committee consisted of the executive management, the operations director, and the director of engineering. The director of operations was responsible for the areas where CM was to be implemented. The director of engineering was responsible to assure technical support to the teams and areas developed. In reality, for Charlie Company, these people were on the steering committee to resolve issues that went wrong and to take blame for these issues. Both directors understood the culture and were prepared to avoid any negative issues. This technique actually supported the successful implementation of CM by developing a team to correct systems and make the transition work. The directors understood this to be a way to enhance their own careers.

Charlie Company operated with four divisions that supported each other extremely well. The plan to implement CM was to implement it in one division and then transition to the other divisions. The culture dictated speed, so the decision was made to implement the CM process on the largest division and have that division drive the implementation over to the other three.

After the project was in effect for 14 months, the executive management was changed by the Board of Directors and the president. The new executives did not agree with the CM approach, however, they strongly supported the aproach of the president and CEO.

4.8. Charlie Company Results

The CM transition was in effect for approximately 14 months. Over 250 people were directly involved with CM teams. Training had been completed for all managers and employees except the executive levels. Hence, only the vice presidents and president were not part of any training. The culture at the bottom three fourths of the organization was changing to a more participative management style and departments were working better together to resolve interdepartmental issues.

Productivity in the cells had *increased* by almost 50% through elimination of redundant activities performed by several departments that use to work independently and were now working together.

The *systems were changed* to support the new teams. Information became readily available to many departments. The information was used to improve communications and avoid duplication of effort.

After the turnover in executive management, the project was canceled. It was not until several months had passed that is was realized the project was seen as a loss of authority for the president. The president addressed the issues as to why CM would not be implemented. *Champions became the bad guys.* Employees who were once dedicated supporters of the project were now outcasts. Managers were either demoted in status or authority or released from the company completely. Many left on their own.

For the factory employees *morale dropped.* Teams were broken up and assembly lines reestablished. The performance on the lines dropped with the morale. However, the quality remained unchanged. It appeared that a self-pride was developed by the short-term taste of CM. This pride continued and showed itself in the quality of the products produced. The workmanship was the only facet left for the employees to control. As with every culture change, some residue remains. You never go back to exactly where you started.

Management turnover increased. Even after the releases made by the new executive management team was completed, many managers left on their own. Their morale was down. Apparently the return to the authoritarian management style was now too autocratic for them. Even the managers that were not directly involved with the CM process, but interfaced with managers working in these areas, decided to leave. It is not practical to believe all departures were from the rejection of the CM teams. Turnover before the teams was approximately 8% for management. After the CM implementation was rejected turnover for management went to approximately 42% in the areas where CM was implemented. Turnaround had gone up to 12% in the support areas associated with the teams. However, other areas not associated with the CM project also increased their turnover to approximately 10%. Without specific interviews with the people who left, it is difficult to conclude why they did leave. The fact does remain the management turnover increased substantially in the areas directly associated with the CM application and then rejection.

Some areas of the company had no noticeable effect from the attempt to implement CM. These areas included *material purchasing*. The material purchasing process and the suppliers remained unchanged. This was primarily due to the high volumes of production being handled. Lot sizes were always maintained at extremely large volumes. Because of these large production and process volumes, there was little or no improvement in inventory. Inventory remained at a high level even though the inventory turns ratio was over 12.

This high-volume rate was another reason it was easy for executive management to disregard CM. Cellular manufacturing has its most significant impact on smaller production run companies that require several changeovers in equipment a week, if not daily. The operation at Charlie Company allowed for changeovers to take place twice a month at most.

The attempt to implement CM within Charlie Company improved the *customer interface* significantly. Scheduling was changed to better meet the customer's needs. After this system was established, it would have been deadly to stop it. The customers became very involved and appreciated the access to the scheduling information. This practice actually secured customer loyalty and increased sales. This change was implemented during the transition to CM and remained in place after the CM practice was rejected.

Another benefit to the attempted implementation of CM was the change in the quality function. Quality was a *centralized function*. After the implementation to CM the quality function became decentralized and mostly the responsibility of the team. When the CM process was withdrawn, the quality function remained decentralized. The assembly lines were reestablished and they assumed the quality function.

4.9. Charlie Company Summary

Charlie Company is the perfect example to prove a project can be canceled at anytime. Even though the CM project was successful in its implementation, it was rejected. There was no business reason known to the project team or the teams that were in place and demonstrating improved performances. These teams were ignorant of the fact that significant changes were to be made in the executive levels of management.

Changes in management or of ownership of a company can mean changes in philosophy. The philosophy of what will make a company a success or a failure depends on the attitude management has toward the techniques being applied. Many companies exist and they have a variety of methods and styles. Many are successful even though they do not practice CM or TQM or general manufacturing practices or a variety of other modern management practices.

Companies operate on the direction of the executive management and the motivation of the people within the company. If these two items are working toward a common goal, the success of the operation is very likely, even if the most prominent management practices are not applied. The modern management practices would, most likely, improve the situation. However, it does not mean total failure for the company if it does not implement the methods expounded in the auditoriums of universities across the country.

4.10. Davis Company

Davis Company is a middle sized company with sales of approximately $100 million. The culture was strongly autocratic. The company was initially privately owned and management did what it was told. The family who owned the company sold it to several financial groups. Most of the executives remained in their positions except for the president. The ex-owner retired and a new president was appointed by the financial groups to lead the company on to continued prosperity.

After a short period of time, the company profits went from black to red. The financial institutions became nervous and changed the president to someone more aggressive. The profits continued to decrease. Management was changed at several levels. People were released if the slightest hint of poor performance was noticed. The company finally assigned a person who was a devout supporter of CM. He was also given the authority to implement it.

Davis Company was *product driven*. The competition basically competed by developing new products in the industry with various features. You also had to have a complete line of products or the fickle customer would immediately jump to a competitor. Price and delivery were the two primary issues in addressing customers needs.

When Davis Company hired the person to finally change the way it did business, he focused on CM. Culture change was done with a club. Not the "good old boys" club, but a large wooden one that is held in one hand. The club was swung to clear a path for the CM approach. Anyone in the way was removed from the path easily and without any second thoughts. *Culture change was immediate.*

Teams were put into place and they were immediately given all of the material needed to produce their products. The actual floor space was then reduced by 25%. Material started to stock up and the teams complained. The teams were told to take care of it. Some teams changed the procurement orders and reduced the inventory.

Others worked faster and longer to use up inventory and move it on to other locations. After a few short months, inventory levels dropped 30% below prior months levels.

At this time the floor space of the cells was reduced again. Again the reduction was almost 25% in floor space. Again the cells worked on reducing inventory and transitioned orders from bulk deliveries to JIT or kanban systems. Inventory for WIP was reduced by 70%. Processes were streamlined and combined to save space and reduce process time.

Change was inevitable. Without the change, the company was threatened to be sold off for its assets. The threat was not only the threat to the company, but it was a personal threat to all of management and all of the employees. This threat was explained and the plan to convert to CM was explained. Training programs were established and everyone was assigned mandatory attendance in a 40-hour class on CM. Everyone was trained. Engineering, administration, management, and the manufacturing personnel were all trained in mixed groups of 20. Immediately following the training, each group was given an assignment and members worked together to implement change in the organization. Employees who did not work toward implementing the change, regardless of which department they belonged, were introduced to the club.

Change was swift and direct. Although the means of motivation was not a positive means, it was very effective. This method is not presented in management courses at universities and colleges in the United States. This was a method to address survival. This is real. The management was given 12 months to demonstrate significant changes in the bottom line. This method of implementing the change to CM caused drastic reductions in material and carrying costs. As material was reduced, the impact to the bottom line was immediate. As productivity increased and sales were met, causing revenues to go up, the impact to the bottom line was immediate.

With all of the changes put into place, *no quality changes were made.* The cells continued for a year, focusing on cost drivers such as inventory and performance issues. No concern was placed on quality improvements unless it was directly related to a cost issue. The primary issue was survival. Survival meant dealing with immediate cost reduction. If the quality did not cause a significant impact on cost or profit, it was not investigated for change. The goal was extremely clear and concise.

Empowerment was granted to all of the cells, but it was *controlled closely.* In order to assure the correct actions were taken, the teams met weekly and were reviewed by management. If items were being addressed that did not deal directly with reducing cost or improving sales, they were immediately stopped. Only activities that dealt with cost reduction and improving revenues were allowed for action.

The statement used by Davis Company to implement CM was as follows: "Either get on the train or get off the train. If you stand on the track you will be run over by the train." This was stated often in explaining what was being done and why rather than getting into detailed discussions of right and wrong methods and other time-consuming agendas. Once a decision was made on something, the choice was up to you to support it or not. If your choice was not to support it, you had two options. Leave or be terminated. This rule was upheld on all issues associated with the process to implement CM. This was the motivation. For the most part, many people do not want to change jobs even if other employment exists. This attitude is changing with the younger generations. In my opinion, the practice of work force reduction from the late 1980s caused a philosophical devaluation of an employee to many companies.

Systems were home grown. They were very limited as to what they could or could not do. Most of the information was available, but it was not in the format needed to make decisions. Rather than change the system, which would have been costly, decisions were made with the data. Decisions were made for today and today alone.

Only what was necessary to maintain the operation of the company was left in place. This practice applied to all support functions and miscellaneous departments. Reports were stopped if not used. The administrators who manufactured such reports were released. Products were designed and in production. The processes and the tooling was in place. The need for manufacturing engineers was reduced, as were the manufacturing engineers. Inspection and test functions that were duplicated at the supplier and at Davis Company were eliminated at Davis Company. The test staff was reduced as well. Managers who provided communication to the executive level or communicated to the factory were eliminated. Communication was direct from the executive level of management to the floor personnel in the CM teams. Meetings were held every 2 weeks for immediate notification of changes. Routine communications were made every month.

Procurement function was very centralized. During the implementation of CM, the procurement function was transitioned to develop blanket orders only. The delivery of parts and material became the responsibility of the CM teams. They contacted the supplier directly and requested the supplies needed and specified a delivery date. This activity allowed the Procurement Department to be reduced by over 50%. Procurement was completely decentralized and buyers were assigned to support several CM teams. The buyers report directly to the manufacturing manager. The need for Procurement management was eliminated, as were the Procurement managers.

4.11. Davis Company Results

Davis Company was a success by most measurements. The change to CM process was *implemented* throughout the *entire company in only 9 months*. Costs of manufacture dropped significantly. Inventory costs dropped by 70% for WIP and 40% in finished goods while improving on-time deliveries to customers. Support costs dropped as unnecessary effort and the workforce associated with the effort was eliminated.

The change to *CM was successful*. Productivity increased as employees pulled together to satisfy a common goal and protect their own jobs. As the management and technical support were reduced, so were the costs associated with them.

No quality system improvements were implemented. Unlike the other examples, quality was never an issue to be addressed. In fact quality was less important than prior to the implementation of CM.

The company changed from being *product driven* to being *customer driven*. The teams opened communications with the customers to understand what was necessary in the product and what was not. This impacted design changes to reduce cost of manufacture without reducing the functional aspects of the products the customers wanted. This practice continued and the customer-driven philosophy developed to improve sales and increase revenue.

The manner in which material was received by the company changed. Material was ordered directly by the cells. Quantities and deliveries were provided JIT. The material was delivered directly to the cell. This change *eliminated the receiving inspection* area for over 70% of the material received. Now the receiving inspection, if done at all, was the responsibility of the CM area.

Inventory control changed from annual and semiannual audits to visual inventory control processes. All materials were stocked directly in the cell and could be counted within seconds. Weekly cycle counts were supplied to the Finance Department each Friday after production was complete for the week.

The only area of the company that remained unchanged was the Design Engineering Department. It seemed that management was not familiar enough with technical aspects to change the process used by Design. It was apparent the technical people could easily baffle management and make managers fear the designs would fall apart and Davis would lose its entire customer base if any changes were made in the Design group. Key designers in the company were threatening to leave for competitors if changes were made. Even without changes in the Design area, some of the key people did leave anyway.

Manufacturing engineering function was eliminated. Six months after the complete implementation of CM, the remainder of the manufacturing engineering function was eliminated. Processes are developed completely by the CM areas directly. Design has been directed to develop designs that do not require tooling to assemble. Special tooling is definitely avoided by all new designs. Standards are developed for finance based on how many people are in the CM area and what the production rate is for the product. Hence, if 5 people are assigned to the cell and the cell builds 10 units in 60 hours, the standard is 30 hours per machine $[(5 \times 60)/10]$. The standard system has been reduced to this simple method to reduce administrative costs as well as the manufacturing engineering overhead.

The unique change made during the transition to CM was the process change for new products. Because of the elimination of the manufacturing engineering function, tooling was difficult to develop for new products. The manner in which Davis Company skirted this issue was to direct the Design Engineering group to develop products that were *process focused*. Specifically, the designs could *not require any capital improvements*. New designs were developed for parts to snap together to eliminate the need for any tools in assembly. More plastic components were developed to eliminate paint booths and color matching concerns in the plant. Also, value engineering was applied early in the designs to avoid excess components and reduce the cost of manufacture through *functional analysis*.

Each CM area was given a *quota system for production*. The cells had to produce a predetermined quantity of product each week, regardless of holidays, illness, or other concerns that normally caused schedules not to be met. Once the quota was met for the week, the cell could work on improvements, call suppliers, and change deliveries. They could also meet with Design Engineering to review potential changes. The cell could even allocate people out of the cell to gain "credits" from other cells to improve overall performance. The cells were measured by the quota system and the actual hours of labor used to meet the quota.

If people could be allocated out to other cells, the labor hours were charged to the other cell, and the original cell was credited with the same number of hours. This is how sudden increases in demands by customers were met. People shifted from one cell to another. The cells called these people floaters. They received a slightly higher wage because they were cross trained in the operations for several cells.

The Davis Company took several risks in implementing CM. One was *reducing quality controls*. Inspection was basically eliminated. It was left to the cells to determine if quality controls were necessary. If they were, the cell would take on the responsibility. Some cells set up very simple SPC charts. Others eliminated all inspection and relied totally on the product testing after it was completely assembled. No significant problems have developed from this risk.

Quality became the cell's responsibility. Auditing functions still exist to a minor degree. Even the auditing is only performed in the cells that request it or in areas where management may deem it necessary. Usually new product releases are audited the first month. Otherwise there is no set procedure. If products are returned, they go directly to the cell and the repairs are done within the quota period. Hence, it is not beneficial to have a lot of returns. It makes it much more difficult to meet the quota.

Customers are separated from cells. Design Engineering has a more direct line to the customer as the Davis Company transitioned to a customer-driven organization. The cells are not part of this interface. Customers are allowed to tour the facility and see the cells operate. However, direct interface with the customer and the cells is not permitted. The reason for this is not certain. The only concern I see from this interface would be requests from the cells to remove some functional aspects of the machines that are difficult to install. Since the communications are open with the cells and the suppliers, I would not feel uncomfortable in having the same relationships with the customers and the cells. As with Charlie Company, the customers could have a direct communication line with the cell and potentially improve deliveries, increasing customer satisfaction and sales.

4.12. Davis Company Summary

Davis Company did not implement CM in the normal manner described in the lectures of many professors or industrial consultant. However, theirs was a matter of survival. It appears Davis has survived and is on the path to continued growth.

Davis took some calculated risks in establishing its CM process. Elimination or transition of manufacturing and engineering to the cells was a major change from the norm. The focus of the designs on a *"build without tools" philosophy* was another unique and creative approach. This was supported by adding a value engineer to the design engineering staff to provide a *functional analysis* approach to each design prior to release for production.

Long term, Davis Company may have to make some additional changes and possibly go back to some of the normal practices applied in manufacturing. I do not see this happening in the next 2 years. The company is stabilizing and at a point where revenues are greater than before. Cost of goods has decreased substantially and the cash is flowing back into the checking and savings accounts. Davis Company will operate this way for some time, continuing to work on continuous improvement and cost reduction activities.

As stated, Davis Company did not implement CM according to the normal process. I believe an organization, in order to remain competitive, *must take risk and be creative* in its approach. Davis Company definitely did that. The management also maintained a working operation and *provided for continued employment* rather than taking the easier way out and selling off the company assets. It is for this reason I agree with the approach taken.

5. SUMMARY

Four companies took four different approaches to implementing CM. Two failed and two succeeded. The common thread between the failures was the lack of support provided by the executive management. The common thread behind the two successes was the support provided by the executive management.

Baker and Charlie Companies both failed to implement CM. However, we can learn from their experience. We learned that even without the complete success of CM implementation, both companies gained from the attempt. Maybe in the future they will try again and be successful. The management team is the major concern in the success or failure of implementing CM. As time goes on, the management team changes. The individuals change their philosophy changes, or both. Time will also provide for potential future success in implementing CM at Baker and Charlie Companies.

Even though the process was successful, the processes for implementation and the results varied. I personally feel Davis Company wanted CM more than Able Company. Both companies have CM processing. Both have seen performance improvements and even increase in sales from the process. However, the creative nature of Davis caused other benefits and more significant savings in support areas. These changes could have been realized by Able Company as well. Able was too conservative in its approach even though it was a high-tech firm with supposedly forward thinking. Able made small changes and then built on them. Davis had to make a big change and took a big risk. It depends on how "hungry" you are as to how big your success becomes.

6. MOST OVERLOOKED AREAS IN IMPLEMENTING CELLULAR MANUFACTURING

The most overlooked areas in developing an effective CM system fall into three categories. In order of importance these areas are: systems, training, and motivation. The reason I prioritize them in this manner is because of the way they affect other problems and the impact they have on the total implementation. The systems can cause other functional areas to have insurmountable problems. The frustrations caused by poor systems can and will affect the effort placed in trying to take the implementation further. If the individuals do not know how to deal with these systems weaknesses, they will become frustrated to the point of giving up the project. No matter what the job is, if people do not have the proper tools to accomplish the job, it will not get done. At least it will not get done to the level it could or should be. Systems changes to support the project must be addressed in the beginning of the project.

The training can also cause people to become frustrated if they do not know or understand how to deal with the changes taking place. Although many companies use on-the-job-training to solve most training issues, it is not a good method. People take more pride in their work when they are treated with respect. Providing proper training shows you care about the individual and you want that person to succeed. You are not making the job unnecessarily hard for them by not providing all the information they may require.

The motivation will also cause frustrations. It usually leads to a lack of support on the project. It may even lead to potential sabotage or hinder the project implementation. All too often it is assumed that people will be motivated automatically to make a change in their normal methods. All too often implementation teams assume the authority for changing things without the involvement or input of the areas being affected. Take the example of a simple thing such as painting a wall in a work area. How often is the decision to paint a wall given to the people who have to work by the wall or have to see it everyday? This does not mean the correct color may not be selected by a person not associated with the area. It does mean that the people in the area were not involved, which could mean they do not like the color. This could lead to people being less motivated because of the terrible color they now have to stare at. Even if you only have two colors to choose from, it will improve the motivation of the people if they are given the authority to choose between the two.

6.1. Systems

Many times companies assume that their current systems for finance and inventory control are adequate to handle the change in manufacturing techniques from the standard method to a cellular approach. The cellular approach requires a faster and more precise system to control inventory in each cell. You do not want the system to confuse people with large stock inventory numbers. The system should identify the specific inventory in a cell. If total inventory control is needed by finance or inventory control, the inventories should be combined to provide a complete company or plant inventory. Using an MRP system would develop economical order sizes based on supplier input for company orders. After the material is received, it can be dispersed to the appropriate cells in the appropriate quantities. Many times the inventory control system in place is a "home grown" system that does not allow for a detailed breakdown, economically. Many of the older MRP systems do not allow separate storage areas for parts. The inventory system then dictates the inventory balances, and the cell is not allowed to minimize inventory or control costs as it should. Off-line systems need to be developed to support activities in the cells. In order to maximize the benefits of the CM approach, the cell must be able to control inventory and be held accountable for the inventory levels it has. Inventory should be regarded as a negative cost impact, as is scrap and salvage. The cell must be provided with information necessary to deal with controlling and minimizing inventories. This will reduce the cost associated with the inventory for which the cell is responsible for.

The other concern with systems is the financial system. The cells must be accountable for cost. Most costing systems do not provide the cost at the cell level. It should not be burdened with overhead and fringe costs, instead only the costs for which the cell is accountable. This cost breakdown needs to be provided for cost control and continuous improvements. Without it, the cell is firing shots in the dark, hoping to hit something significant.

If the systems are not changed to support the cells, the morale in the cells is likely to diminish. Poor morale is an enemy of implementing change. If the cells perceive they cannot succeed due to systems not supporting their efforts, they will *not* succeed. The teams must have an option to change the systems of inventory control and finance. These systems must become supportive to the implementation. Cellular Manufacturing controls must be available to empower the cells to managing their own destiny.

6.2. Training

Training is the second most overlooked area in implementing change. Training is not just the formal teaching of new methodology but should include the informal follow-up by facilitators to assure a proper understanding of what was taught. In a classroom setting, many people who are not used to attending meetings will get bored and disinterested. They may even fall asleep. Classes I attended were conducted by consultants and even company managers where people have slept through the presentations. Yet the perception was that everyone who attended the class was trained. No follow-up was provided. No testing of any kind was performed to see if everyone understood the new concept.

Business, in general, does poorly in conducting training classes. If a business expects the training to be beneficial to the productivity of the company, why does the business not take the training seriously enough to ensure that the personnel are trained? It is ironic. Experience tells me of only one company that takes training seriously. In the 1970s Deering Milliken had a training program for all professional new hires. They were transported to different parts of the country and taught many aspects of the company by formal and informal methods. Consultants were hired to provide some of the training. After each training session the newly hired employees were tested. If they failed, they were terminated. After the class had completed the entire training, employees went back to the location in the company for which they were originally hired and were given an increase in salary for their efforts. This was successful training. Everyone who went through it learned what the company wanted them to learn. Everyone was rewarded for the efforts made. Every time a new subject for training was initiated, people paid attention to it and took it very seriously, trainers and the trainees. Everyone knew the training was only being accomplished because it was important to the company. The employees' responsibility was to learn what was being taught and apply it as best they could. Training was a priority and only performed when necessary. It was never taken lightly at Deering Milliken.

It is not being suggested that people should be fired if they do not learn from a training session. However, I do feel training needs to be taken seriously by management and by the people being trained. All too often training is perceived as a break from work. Many do not put the effort in, nor do they apply the training after it has been received. They continue in the same manner they did before the training was provided. This is a waste of resources for the company. If you plan to have a training session, you should plan to apply what will be trained. If not, why train at all? You will only be wasting resources.

If someone sleeps in a class, it should be important enough to wake them up or to tell them that they have to attend another class. Do something so they clearly understand the importance of the training and take it seriously. All activities should be "valued added" as far as the company is concerned. If the activity is not a value-added activity, why waste your resources?

Provide the specific training necessary to implement the change. Follow up on the training to ensure it is being applied and applied correctly. Have reference material available for personnel to easily review if questions develop after the training. Make certain both the trainer and the trainee take the subject matter seriously.

6.3. Motivation

Motivation is another great assumption and mystery when implementing change. Companies appear to be naive in the way they treat motivation. It seems most companies believe people are naturally motivated to learn new things and do new things. This is a big mistake, yet an easy one to make. This is because the people making the assumption are change agents. They want to do things differently and improve on what is being done. They seek challenges and are willing to take risk. They see a benefit from the change and want to be part of the implementation. Others may not have these same feelings of enthusiasm.

What motivates people varies from organization to organization and from individual to individual. For the most part, pride motivates most people. Each level of an organization has different goals. Some of the goals of these different levels match those of other levels. Pride is a common thread in most of the goals.

A company making a significant change in its processes or cultures needs to identify a reason, a common purpose for the change. It should direct itself toward improvement. It may be to avoid going out of business. It may be to win the largest market share position in the industry. It may be to simply react to a competitor in order to maintain market share. It could be a variety of things. These reasons need to be communicated in such a fashion as to bring the entire organization together. Develop a unified approach that will motivate the operation to succeed. This "pep rally" activity is usually missing from a significant change.

Either the management is too "professional" to take such an action or it feels there is no benefit to it. In either case, this view is wrong. The transition may very well succeed, but it will take longer and become more difficult. Many more employees, at all levels, will challenge the change because they will not feel it is necessary. They have not been motivated.

In companies where management took the time to explain the reasons for the change to CM and pull the organization together to support it, the transition was more rapid and had better success. People were working together. Rather than doing only what was necessary, they addressed problem issues and improved the current system in many aspects not previously considered. This is what motivation and teamwork can do. It also develops a team atmosphere, which is extremely beneficial when implementing CM.

6.4. Summary

Consideration to systems, training, and motivation need special attention during the implementation of CM. This special attention is necessary only because these items are taken for granted. It is usually assumed these three items will correct themselves. They don't. Systems, training, and motivation can impact change in a negative fashion, if not addressed as part of the change.

Systems are far reaching and affect many departments and functions. Systems need to be analyzed carefully before any changes are made. Systems must be supportive and not hinder the activities of the change.

Training is a means to affect culture and behavior. If training is not conducted in a managed fashion, the lack of training can create the environment where people rely on old practices to address new situations. This may be a point of confusion and frustration that could impede the implementation of change. Training that is conducted properly and in a timely fashion can improve the outcome of the associated changes. People can relate to the new methods and have a clear understanding of their expectations.

Motivation is why we do things. If there is no motivation, the activity falls to a lesser priority and may never get completed. If it is a low priority, and does get completed, it may very likely be completed in a poor fashion, losing most of the benefits for which it was intended. People need to be motivated to support a change. Change is hard work. It becomes only harder if it is an uphill battle. If people are motivated to make the changes, the changes will succeed faster and with increased benefits.

13

ECONOMIC JUSTIFICATION OF CELLULAR MANUFACTURING

Peter L. Primrose

Total Technology Department
UMIST
Manchester, United Kingdom

1. INTRODUCTION

Managers who are considering the introduction of Cellular Manufacturing (CM) in their companies not only have to identify an application and plan its implementation, they have to ensure that the use of CM will be financially viable. In doing this, they have to do more than show that CM can provide the minimum required rate of return; they have to show that investing in CM will provide a better return than could be obtained from concentrating financial and managerial resources on improving other aspects of the company's operation. This means that introducing CM has to be compared with alternatives such as developing new products or investing in Advanced Manufacturing Technology (AMT); although introducing CM may itself require investment in AMT. It also means that managers who are advocating CM will have to use their company's existing investment appraisal procedures to justify the expenditure involved. However, financial evaluation should not be used to justify a decision that has already been taken, it should be used from the outset as an integral part of the planning process, the aim being to help identify both the optimum areas for the

Handbook of Cellular Manufacturing Systems, edited by Shahrukh A. Irani
ISBN 0-471-12139-8 © 1999 John Wiley & Sons, Inc.

application of CM and the most appropriate technology to use for those applications. Before discussing evaluation techniques, it is necessary to consider the economics of traditional batch manufacture.

2. ECONOMICS OF BATCH MANUFACTURE

Authors tend to categorize production into types, for example, job, batch, and mass, often illustrating this with a diagram that relates the choice of process to order quantity. Their implication is that batch-manufacturing techniques are only applicable for medium-volume production. The danger of this view is that it can predetermine the choice of process to be used so that, although CM can be used for any volume of production, companies may not consider using CM unless they are producing medium volume batches (e.g., 10 off–1000 off). Most companies manufacture components in batches; the nature of batch manufacture is that the manufacturing processes used are normally shared by batches of different components. However, manufacturing in batches is not restricted by volume or process. For example, the author has worked with high-volume companies making food products and metal fasteners that were using batch-manufacturing techniques for products where quantities ranged from 100 to over 500,000. In both companies they had linked machines together into cells with automatic component transfer, each cell being able to produce a family of similar products with the setup being changed at the end of each batch. Even companies in process industries use batch manufacture. For example, another company the author has worked with, which makes vinyl floor covering, has a wide product range that it makes in batches; the process lines are run continuously for several days and then stripped down and reset for the next product batch. At the other extreme, Primrose and Leonard (1991) show that Flexible Manufacturing Systems (FMS) can be financially viable when producing bathes of 1 off.

Although managers in companies that use Functional Layout (FL) for batch manufacture may try to rearrange some of their machines into cells, because they think that batch manufacture, is inefficient, doing so may not be financially viable. It is very easy to be critical of batch-manufacture, which appears to be a highly inefficient way of producing components in comparison with flow-line production. Components produced in batches in an FL machine shop can spend a very small part of their total lead time actually having work done on them. For example, batches of components may spend over 90% of their lead-time in a queue waiting for a machine, an individual component may then spend over 90% of the batch machining time waiting for its turn to go onto the machine. Even when a component is on the machine, various studies in the past have shown that the actual cutting time may represent less than 50% of the operation time. As a result, although lead times may be long, components can spend less than 1% of that time being worked on. For example, companies with an FL machine shop often plan production on the basis of one operation/week; for a component with a batch quantity of 10 and operation time of 30 minutes each, the batch would spend 5 hours on the machine; if the machine only spends 15 out of the 30 minutes operation time cutting metal, the result is 15 minutes work in 40 hours.

The fact that this is uneconomical seems self-evident, and it is always assumed that batch manufacture must be replaced by some other technique, such as CM.

Controlling batch manufacture is a highly complex balancing act that has to trade-off a number of conflicting interests. If managers did not have to try and keep costs to a minimum, controlling batch manufacture would not be a problem. Because they are under constant pressure to reduce costs, managers attempt the following:

- Keep expensive manufacturing facilities, such as Computer Numerical Control (CNC) machines, operating at maximum utilization.
- Keep direct labor fully utilized.
- Produce components in the largest batch sizes possible in order to reduce the number of setups, which are regarded as a nonproductive cost.
- Operate with the minimum levels of Work-In-Progress (WIP) and finished components.

At the same time they have to operate with various constraints, such as

- The need to keep lead times to a minimum in order to be able to respond to changes in customer demand
- The need to cope with variable factors such as machine breakdowns, labor absenteeism, scrap, and rework

Although there is a penalty of long lead times and high levels of WIP, batch manufacture using FL provides considerable flexibility to help balance these constraints. Grouping machines into cells can reduce this flexibility.

3. ALTERNATIVES TO BATCH MANUFACTURE

Since the days of F. W. Taylor, at the beginning of the twentieth century, people have been trying to overcome the apparent inefficiency of batch manufacture by developing ways to replace or control it. For example, Group Technology (GT) in the 1960s and FMS in the 1980s were both advocated as replacements. Unfortunately, the justification for such techniques has normally been subjective and based on the assumption that batch manufacture is "a bad thing" that must be replaced. Alternatively, the justification has been the perceived need to reduce WIP by reducing component lead times. The belief that reducing WIP is important is reflected throughout the literature relating to Just-In-Time (JIT) where reduced WIP is normally quoted as the primary benefit of introducing JIT. However, Primrose (1992a) shows that the value of reducing WIP can be negligible in comparison with the other potential benefits of JIT. The failure to evaluate the introduction of techniques such as CM, GT, and JIT means that the success of these techniques has been measured on the basis of technical rather than financial criteria.

About the time that transfer machine lines were being developed in the 1920s, the concept emerged of identifying a small number of high-value components that had the greatest effect on product delivery and concentrating on reducing their lead time. This was done by setting up groups of machines where each group was dedicated to the production of a family of critical components. For example, a company manufacturing diesel engines would have dedicated groups of machines for components such as crankcases, cylinder heads, and connecting rods; a gas turbine manufacturer would have groups for turbine blades, rotor discs, shafts, and casings. The selection of such groups is ocular rather than analytical and, as such, has not lent itself to academic study. The technique used to be called Family Planning (FP) and is still widely used, although rarely referred to in the academic literature.

When GT started to be widely used in the late 1950s and early 1960s, the aim was to use coding and classification to transform the informal and intuitive approach of FP into a formal and scientific procedure that could be applied to all those components still being manufactured using FL. However, the aim of FP is to improve delivery performance, whereas the aim of GT is to improve operating efficiency. The advent of FMS means that rather than setting up an FP group using stand-alone machines, managers can invest in an FMS as a high-technology FP group. In fact the techniques described by Primrose and Leonard (1991) for the selection of technology for investment in FMS are the same as those that should be used for investment in FP.

As well as GT, FP, and FMS, other attempts to improve batch manufacture have concentrated on improving the control of the existing layout, such as with Manufacturing Resource Planning (MRPII). However, despite the attempts to develop new techniques, most small- and medium-volume engineering companies still use the approach developed in the 1920s whereby they manufacture the small number of components that affect product delivery using FP groups, selected on the basis of practical experience, while using FL for the majority of parts. The fact that the components with the longest lead times are normally the most complex and expensive helps to improve the economics of FP groups.

4. IDENTIFYING APPLICATIONS FOR CELLULAR MANUFACTURING

Managers who are convinced that their company should introduce CM normally start by trying to identify potential applications. They then produce detailed plans for implementation and finally try to justify the costs involved. By doing this, they will not only find it difficult to justify the cost but they are also likely to select nonoptimum applications that are designed to solve technical rather than financial problems. Primrose (1991) shows that for investments to be financially attractive they must be aimed at solving one of the company's major problems. This means that the starting point should be trying to identify the main problems that the company is faced with, and then considering whether CM could help to solve any of them.

Many investments in manufacturing technology and systems are aimed at solving manufacturing problems or improving operating efficiency. However, manufacturing is an integral part of a company, and the aim should not be to optimize functional efficiency but to optimize the achievement of corporate objectives. Such corporate objectives are normally aimed at increasing the company's ability to supply customers with the products they want, when they want them, and by so doing, increasing competitiveness and profitability. This means that the starting point in deciding investment plans is to try and identify customer needs. In manufacturing companies the main factors that determine customer satisfaction, and affect a company's ability to obtain orders, are

- Product function and features
- Delivery performance
- Product quality
- Selling price

Having established market needs, the next stage is to try and increase the company's ability to meet these needs. Although CM can influence all four of the above factors, its most important contribution will normally be to help increase sales by improving delivery performance. However, improving delivery performance may produce other benefits in addition to increasing sales volume. For example, it may allow the company to charge a premium price, or to reduce product costs by only supplying customers with the exact specification they require rather than a more expensive standard product. In identifying market needs, what matters to customers is not what has been promised, such as with delivery, but the company's success in meeting these promises. Therefore, in comparing customers' requirements with the current performance, to identify the changes that are needed, it is important to measure the performance that is actually being achieved rather than what has been promised to customers. If actual performance is worse than promised performance, the need for improvements is greater, as are the potential benefits.

In considering the economics of CM one has to differentiate between different types of application, the main ones being:

- Flow lines for high-volume production
- Family Planning groups for key components that affect product delivery, possibly using an FMS
- Group Technology cells used to replace FL

Flow lines or transfer lines are used where the volume of production is such that batch manufacture and the manual transfer of components between machines would be uneconomical. Although the use of flow lines may be the only viable alternative for very high volume production, there will be a wide range of potential applications where alternatives have to be considered. As a result, managers may have to evaluate two or three alternative layouts and choice of technology when planning a cell. Primrose and Leonard (1991) describe how the development of FMS technology is allowing companies to obtain the benefits of control normally associated with transfer lines in medium- and small-volume production. At the same time, the development of flexible transfer lines allows companies to gain the flexibility of batch manufacture in high-volume production.

5. USING CELLULAR MANUFACTURING TO IMPROVE DELIVERY PERFORMANCE

Although CM is often aimed at reducing component lead times, it really should be aimed at improving the company's ability to supply customers with the products they want, when they want them. Doing this requires a lot more than reducing component lead times, companies may have to

- Reduce product delivery times
- Improve the reliability of meeting delivery promises
- Increase their ability to cope with fluctuations in sales volume
- Increase their ability to change product specification to suit individual customers

Traditional batch manufacture provides the flexibility needed to cope with fluctuations in sales volume and changes in product mix by having long lead times and high levels of WIP. However, this may disguise the magnitude of the short-term fluctuations in sales, so that production managers, who may have been providing capacity to meet average demand, can be unaware of the nature of the fluctuations. To illustrate the problem, Figure 1 shows how the receipt of orders actually varied within a company.

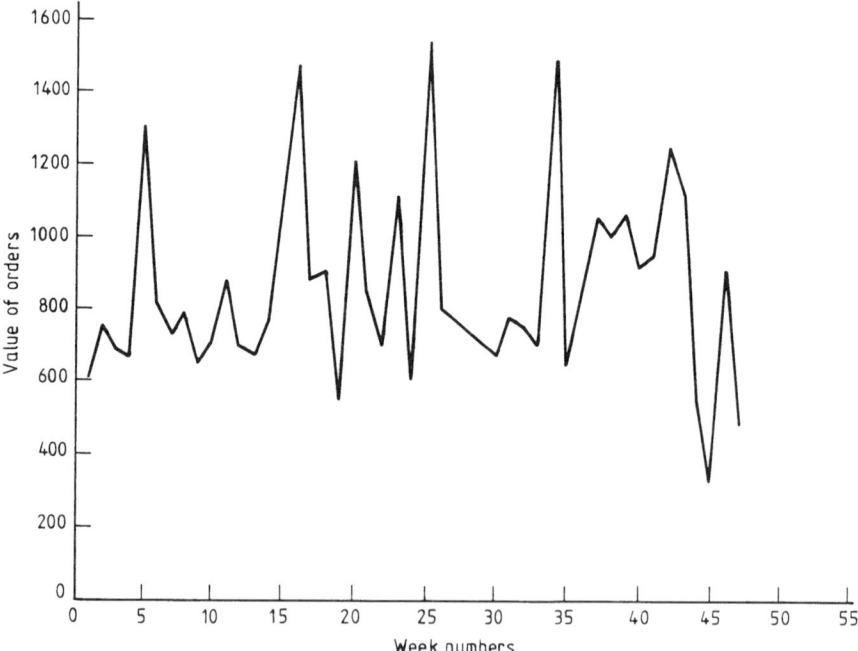

Figure 1. Fluctuations in order intake.

Fluctuations such as these do not matter very much when manufacturing lead times are long. However, as the company attempts to reduce both lead times and inventory levels, the factory output will increasingly have to be able to fluctuate in response to short-term changes in sales volume. While it may be sufficient to provide capacity for the average sales volume when lead times are long and output is insensitive to sales fluctuations, the closer the factory response becomes to sales orders, the greater the required capacity. For example, the company in Figure 1 may have been able to meet customer demands in the past with a capacity of 800 units but, as lead times reduce, may have to increase this to over 1000 units. At the same time there will be periods when the factory output will drop well below the 800-unit level. If a company is introducing CM in order to reduce delivery times, they may have to provide additional capacity in their cells to be able to respond to short-term fluctuations in demand.

Even when companies have a full order book, so that volume fluctuations are not a problem, it is likely that changes in product mix will result in load fluctuations. Figure 2 shows how, in a company that manufactures paper making machinery with a long delivery time, the required capacity on its machining centers varied during a year in which the company had a full order book. Figure 3 shows how the capacity required on the lathes varied at the same time. Plotting the load on all the other machine groups showed similar fluctuations, but with very little correlation between the timing of the peaks and troughs. This lack of correlation showed that the fluctuation in load was not being caused by seasonal factors or changes in sales volume but was being caused by random changes in the flow of work through the factory.

In trying to respond to changes in customer demands at the minimum cost, companies are likely to adopt a combination of approaches, such as

- Reducing manufacturing lead times so that they are less than the delivery time quoted to customers
- Carrying sufficient stocks of WIP and finished product to bridge the gap between customer needs and production capacity

To reduce manufacturing lead times, one has to start by identifying the components that affect product delivery, possibly using techniques such as Critical Path Analysis (CPA). In doing this there is likely to be a marked Pareto effect whereby only a small number of components have a significant effect on delivery. However, as the lead times of these are reduced, there will be an increasingly large number of other components whose lead time determines product delivery. Because the components that have the longest lead times will normally be the most complex and require the most operations, they will also tend to be the most expensive. This means that it may be economical to set up FP groups of dedicated machines to produce them rather than carry stocks of finished, or semifinished, components. This will be particularly the case where there is a need to customize products, and it is impractical to carry stock of a very large range of similar and expensive components.

For example, a manufacturer of large diesel engines sells 10 different engine configurations, which means having to produce 10 different crankcases; these represent

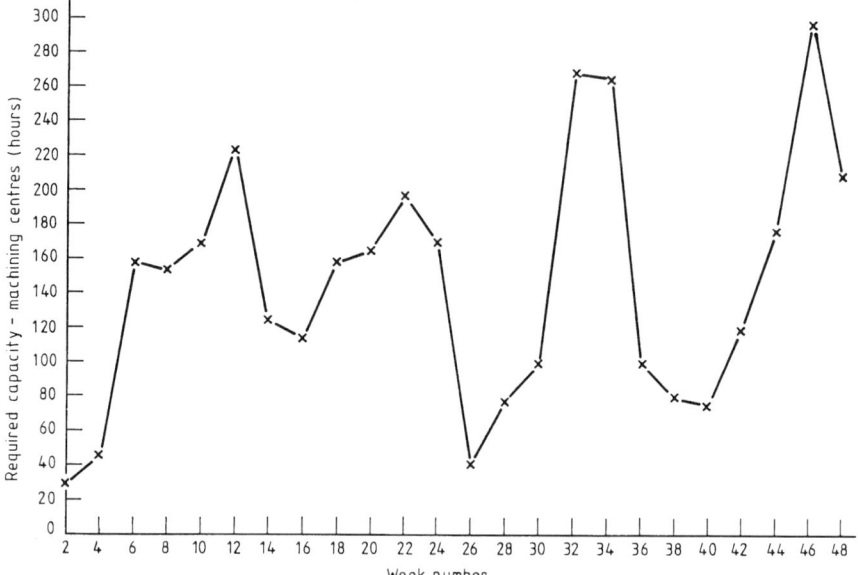

Figure 2. Fluctuations in machining center workload.

5% of total engine value, and their manufacture determines product delivery. The company sells 20–25 engines a year, but this does not represent 2 of each type of crankcase. The nature of orders is such that marketing cannot predict the specification of orders with sufficient accuracy for crankcase machining to be started before orders are received. At the same time, the lead time for producing crankcases in the general FL machine shop would be longer than the delivery that has to be offered to customers. Because it would be impractical for the company to carry a very large number of finished crankcases in stock, it has to set up a FP group of machines for crankcases.

However, as the product lead time is reduced and the number of components that have to be changed increases, the point will soon be reached where it is more economical to carry stocks of components than to set up additional FP groups. The importance of this is that, where setting up FP groups to improve product delivery is an alternative to carrying stocks of components, the financial benefit of setting up FP groups will not be the sales improvements that come from improved product delivery. The benefits come from the value of inventory saved by not carrying stocks of the components that would be produced on the group of machines. However, as suggested earlier, the one-off benefits that can be obtained from inventory reduction may be quite small. The dividing line on the Pareto curve of components, where setting up FP groups is no longer financially viable, will be unique for each company and is not determined by an analytical model. Managers who are introducing CM will start by considering the most obvious, most complex, most expensive and longest lead time component family. Having found that setting up a group of machines for that

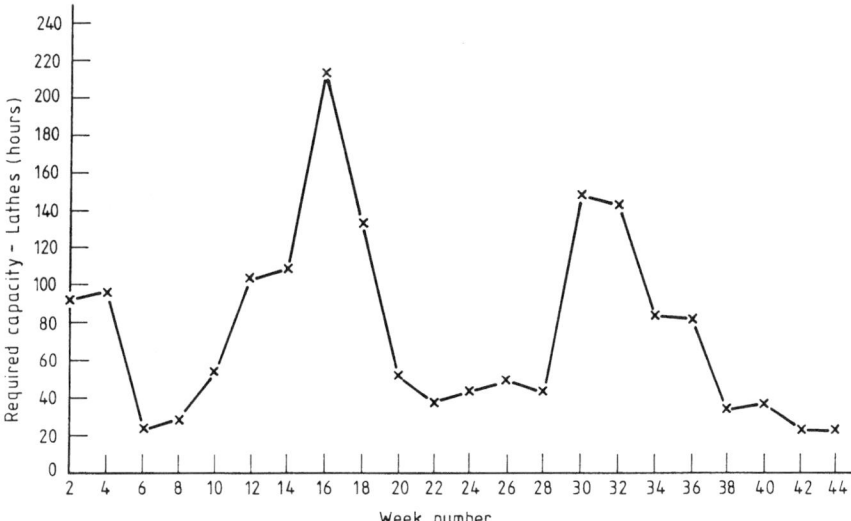

Figure 3. Fluctuations in lathe workload.

family is economical, they would go on to look at the next most obvious one. The process would be repeated, possibly for only three or four components, until the stage is reached where the next component considered does not justify setting up a group of machines.

The above discussion relates to the normal situation where CM is considered as an alternative to producing components in an FL shop, on machines that are used to produce other components. There is, however, an alternative situation where it may be possible to set up a cell that comprises machines that were previously only used for one family of components. For example, a diesel engine manufacturer may set up a cell to produce high tensile connecting rod and cylinder head studs. The cell would comprise an ending machine and centerless grinding and tread rolling machines. Because the cell can produce the components completely, and the machines have never been used for any other components, the only costs involved are likely to be those of moving the machines. The justification in this case is not improved delivery or reduced inventory but improved control and reduced material handling.

6. FINANCIAL EVALUATION TECHNIQUES

As well as providing benefits, introducing CM can be expensive. Therefore, whether it involves capital expenditure or not, it must be evaluated in the same way as any other investment to ensure that it will provide an adequate return on investment. Although companies set a minimum acceptable rate of return for investments, the finance available will be limited and the projects that are accepted will be expected to provide a much higher return than the minimum. One of the implications of this is that the techniques that are used to evaluate CM must be the same as are used for any other investment project so that the results are comparable.

There used to be a widespread belief that there was something wrong with established accountancy principles, and these could not cope with the complexities of modern manufacturing. One of the reasons for this was the belief that many of the benefits of new technology and management philosophies were intangible and could not be quantified in financial terms. However, the work done at UMIST (Primrose, 1991) has shown that there should be no such thing as an intangible benefit and that it is possible to state that "every benefit which can be identified can be redefined and quantified and included in an investment appraisal. No benefit should ever be excluded on the grounds that it is intangible."

One of the main reasons why the concept of intangible benefits came about was that when managers started to investigate the use of new technology and management philosophies they already believed that what they were considering (e.g., FMS, Robotics, MRPII, etc.) was essential. However, because the direct savings, such as labor, that they could quantify would not justify the high costs involved, they tried to identify additional benefits. They defined these in general terms, such as "increased flexibility of production," "better quality products," or "improved management control," but could not see how these elements could be quantified. As happened with manufacturing technology, people advocating the use of CM described the benefits in terms that made them difficult to quantify, such as "improved job satisfaction," "increased mobility of labor," and "increased management control." Although, as discussed later, all of these can be quantified, experience shows that their value may be limited, and in some cases they may represent a cost rather than a saving.

For any investment, whether it is evaluated or not, all the costs will be known and quantified because the bills have to be paid. Any unplanned or unforseen costs will be quantified when they occur, and costs that are originally underestimated will be correctly recorded. This means that, unlike benefits, there are no intangible costs, all the costs associated with a project will be known and can be attributed to the project and included in the company's costing system.

When GT was being introduced into companies in the 1960s, manufacturing industry was highly inefficient. As a result, comparisons were made between the previous poor performance using FL and the results obtained after concentrating considerable management effort into introducing GT. Some of the benefits that have been claimed for GT may have been the result of concentrating management effort on improving inefficient facilities and could have been achieved without GT. Although large savings were claimed for GT, it has been argued that the people involved in the well-publicized applications were of such a calibre that they could have achieved impressive results, no matter what system they had introduced. In some cases the benefits that had been obtained from introducing flow lines were quoted to justify the need for a widespread introduction of GT cells to replace FL.

A similar problem occurs in some FMS applications where the benefits that were claimed did not come from the use of the FMS, but from other changes, such as the redesign of components: In such cases most of the savings could have been obtained without the investment in FMS. Managers who are convinced that CM is essential often do not consider the alternative of making the existing facilities more efficient. It may be found that some of the benefits that have been claimed for CM could have been obtained without the costs of introducing CM. However, before excluding these from an evaluation, one has to find out why the required changes have not already been made, and question whether the changes will be made if CM is not used to force the improvements.

However, managers will not just reorganize manufacturing facilities into cells. While they are introducing CM, they are likely to spend a considerable amount of time and money on improving production methods. If the benefits of this are going to be included in an evaluation, so must any costs. One of the difficulties in identifying the benefits of GT from published case studies is trying to differentiate between the sources of the benefits. An extreme example of this was in the literature in the 1960s describing the Molin's System 24: Although major benefits were predicted for the system, examination of the details show that most of the savings were going to come from increased cutting speeds as a result of redesigning all steel and cast-iron components into aluminum. In fact almost all the savings predicted for System 24 could have been achieved by using proven technology.

Primrose (1991) shows that, although using payback to evaluate investment projects will invariably result in companies making incorrect decisions, using Discounted Cash Flow (DCF) overcomes the problems of payback. Although most of the accountancy literature dealing with DCF is concerned with the discounting process and the establishment of hurdle rates, the main problem that managers encounter in practice is the need to identify all the costs and benefits of a project in cash flow terms. Although managers can identify most of the costs and benefits of projects such as CM, they normally do so in non-cash-flow terms, such as the intangible benefits discussed earlier. With costs, the problems are mainly caused by the way that managers wrongly use standard cost information from their company's absorption costing system. For example, replacing FL with CM may increase the floor space needed; managers would quantify the cost of this by calculating the extra area required and value this using a standard cost rate of dollars per square foot. However, such a rate would be based on total cost of the factory building, including factors such as rates, heating, lighting, and maintenance, which may be unaffected by relatively minor changes in utilization. Although CM may occupy more floor area than FL, this may not involve additional cash expenditure, especially if in the past other activities had expanded to fill any space that was available and these activities can now be contracted.

7. BENEFITS OF CELLULAR MANUFACTURING

The benefits that are probably most frequently claimed for the introduction of CM are a reduction in WIP and a reduction in operating costs, such as from reduced setup times. Because inventory reduction is commonly regarded as the main benefit of JIT,

CM is often quoted as one of the techniques that can be used in the introduction of JIT. The danger of this is that, because CM is seen as an essential element of JIT, and because JIT is "fashionable," neither CM or JIT are evaluated. However, as described by Primrose (1992a), reducing WIP is not the biggest potential benefit of JIT, therefore applications whose main objective is to reduce WIP will be less profitable than applications designed to achieve alternative objectives. As with JIT, selecting CM applications designed to reduce WIP and operating costs are likely to be suboptimal.

Although reducing WIP has frequently been quoted as the main benefit of JIT and CM, Primrose (1992b) describes the complexity of evaluating inventory reduction and shows that the way this should be quantified has never been described correctly. Unlike all other savings, such as from reduced labor or increased sales, which can produce ongoing annual savings, inventory reduction provides a one-off saving. In addition, inventory reductions are not valued on the basis of the book value of the inventory; rather the valuation is complex and is based on the curtailment of production during the period when the level of inventory is being reduced.

In trying to show the relative importance of inventory reduction in the introduction of JIT, Primrose (1992a) suggests that managers should carry out an analysis to identify the areas of their company where resources should be concentrated. Carrying out such an analysis, and then listing the potential benefits in declining order of magnitude and feasibility, in most companies would provide the following list:

1. Increasing sales volume, or preventing sales being lost, by making products more competitive (e.g., delivery, price, quality, product specification)
2. Eliminating unprofitable orders (e.g., by improving quotation system)
3. Reduced cost of material content in products
4. Reduced operating costs (e.g., labor)
5. Reduced stocks of finished components and products
6. Reduced stocks of raw material and WIP

Primrose (1992a) suggests that for many companies the relative magnitude of benefits from investments would be similar to those shown in Figure 4. As with JIT, the main financial benefits of CM do not come from direct savings such as WIP or reduced operating costs, they come from increasing customer satisfaction and the effect this can have on sales volume.

The financial advantage of increasing sales volume is that, to produce the additional products needed for extra sales, only expenditure on the variable cost elements (e.g., material, direct labor, operating expenses) will be required. All the cost factors included in fixed overheads, including depreciation and indirect labor, will remain unchanged. In the case of large changes in sales volume, some fixed overheads will become variable, such as would be the case if extra machine tools have to be purchased to provide extra capacity. In order to illustrate this, an example is taken of a company with a turnover of $1,300,000, which comprises:

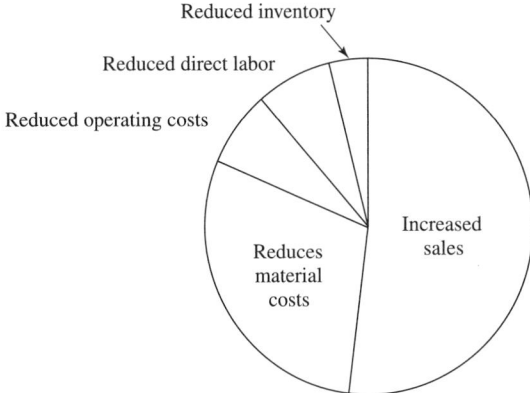

Figure 4. Relative magnitude of benefits.

Material cost	$400,000
Direct labor cost	$100,000
Operating expenses	$100,000
Manufacturing overheads	$400,000
Cost of manufacture	$1,000,000
Administrative overhead and profit	$300,000
Turnover	$1,300,000

Assuming that inventory represents 20% of annual turnover, the value of inventory will be $260,000.

Given that the value of overheads in the above turnover is $700,000, the value of a 10% increase in sales (from overhead contribution) will be $70,000. Although a 10% reduction in inventory would produce a saving of $26,000, this would represent a one-off financial saving compared with increased sales where the savings could be ongoing and annual. In the case of CM, the value of inventory reduction can be very small if CM is aimed at reducing WIP, which may represent a small percentage of total inventory value. However, despite the importance of benefits such as increased sales, because the relationship between the cause of the savings and their value is indirect, they were thought to be intangible and unquantifiable. As a result they have been rarely discussed, except in general terms.

Increased job satisfaction and increased management control have been claimed as benefits of CM. Because factors such as these may provide benefits in some applications, they should not be excluded on the grounds that they are intangible; however, such factors are only a benefit if they produce a cash flow saving. Taking the example of increased job satisfaction, if this is going to produce benefits, it presupposes that the workers were originally dissatisfied. If they were, their dissatisfaction would be reflected in symptoms such as high levels of absenteeism and labor turnover, poor-quality workmanship, low productivity, and the like, each of which would result in increased costs. Eliminating the causes of dissatisfaction would produce savings, however, one has to question whether it is necessary to introduce CM to do this.

The managers of a company visited by the author thought they were suffering from excessive absenteeism and a high level of staff turnover. They were looking at various ways of improving job satisfaction, including CM, but needed to justify the costs involved. They started by recording their existing levels of absenteeism and turnover, and then calculated the cost of absenteeism (i.e., a 10% average level of absenteeism implies that the labor force will be 10% larger than necessary to maintain production) and the cost of staff turnover (i.e., the cost of recruiting and training replacements for the number of operators who left in a year). However, when they compared their figures with other companies in the local area that had similar types of labor force, they found that the size of their problem was not untypical. Introducing changes, such as CM, in order to increase job satisfaction, but without first finding out the extent and causes of any dissatisfaction, would be unlikely to provide any savings.

Although it has been claimed as a benefit of CM, introducing CM in order to improve management control may well be counterproductive unless the reasons for the original problems are investigated and understood. Using CM to increase management control will only provide benefits if the original control is inadequate and cannot be improved by normal means, such as by improved training of staff or introducing improved control systems such as MRPII.

8. COSTS OF CELLULAR MANUFACTURING

The starting point in the introduction of CM, whether it involves GT, FP, or FMS, is to identify the group of components that are to be produced by the cell. Before doing a process planning for the components, one would normally examine the existing design and manufacturing processes to try and identify any potential improvements. It may be that, as a result of such improvements, there is no longer a need for CM. Having produced a process planning for the components, and ascertained their potential annual usage, one can define the types and numbers of machine tools required. Because some of the benefits of CM, such as improving delivery performance, may result in an increase in sales, this must be allowed for in the planned capacity. Allowance must also be made for the capacity that will be needed to cope with short-term fluctuations in demand. The following are the main stages of the planning process in which costs can be identified:

- Knowing the number and types of machine tools required, as well as the potential utilization of each one, it is possible to compare this with the existing machine shop facilities in order to identify the cost of any additional machines that have to be purchased.
- By taking machines out of the existing FL machine shop, it is likely that some of the remaining components may have to be produced on machines that are less accurate or productive than were previously used. For example, if a CNC machine is moved into a cell, but not all the components that were being produced

on it are moved, these may now have to be produced on slower manual machines. Any additional costs in the remaining machine shop resulting from such changes, such as labor or increased scrap and rework, need to be identified.

- In the same way that machine tool requirements can be estimated, the number of operators and their required skills can also be estimated. In the same way that the reduced utilization of machines in cells may require the purchase of additional machines, a reduction in operator utilization may require additional operators.
- Moving operators from an FL machine shop into a cell, with a relatively small number of machines, is likely to require an increase in operator flexibility. Any costs associated with this, such as operator training or additional payments for increased operator flexibility, need to be identified. Training costs should include any expected increase in scrap and rework during retraining.
- Even if changes are not made to designs or to process plannings, setting up cells is likely to require the purchase of additional tooling and equipment where this has to be duplicated. These costs will increase if improvements, such as using quick-change tooling to reduce setup times, are introduced.
- In addition to the cost of moving and reinstalling machines in the cells, costs are likely to be incurred in clearing the required area. There are also likely to be costs involved in reorganizing the remaining machines in the FL shop.
- If additional floor space is going to be needed, any costs of providing this have to be established.
- Moving machines into cells is likely to disrupt production. Any costs, such as additional shift or weekend working, or increased subcontract, need to be identified.

Because every application of CM will be different, it is not possible to list all the potential cost factors, however, the following are some of the costs that are frequently incurred in introducing and running CM:

- Any costs associated with reduced machine utilization, such as the need to purchase additional machines
- Any increase in manufacturing costs in the remaining FL shop if high production machines have been removed
- Any costs associated with reduced utilization of labor
- Operator training costs, including any scrap and rework during training
- Additional payments if increased operator flexibility is required
- Purchase of any shop equipment, such as tools and fixtures, that has to be duplicated
- Improving production methods, such as tooling and equipment costs for reducing setup times
- Plant movement costs
- Cost of additional floor space if this involves any cash flow expenditure
- Lost production during reorganization

Whereas some of these costs will be incurred at the time of the initial change, other costs will be ongoing. Although not all these costs may be relevant for a particular application, there may be additional factors, such as management or consultancy costs, that are not included in the above list. Although the list is not fully comprehensive, managers can use it to help identify some of the main areas of cost they will have to include in an evaluation.

9. LABOR AND MACHINE UTILIZATION

The need to keep labor fully utilized is one of the constraints of batch manufacture, and one of the advantages of having an FL machine shop is that high WIP and long lead times cushion the effect of short-term fluctuations in workload so that operators can be kept fully occupied. When they are taken out of FL groups and put into cells, the utilization of both machines and operators may be adversely affected. To minimize any reduction in the utilization of labor, operators will normally have to become more flexible and be prepared to work on more types of machine than they would in FL groups. As a result, companies may have to pay their operators more in order to obtain this increased flexibility. Another cost implication of this increased flexibility may be that skilled operators have to spend part of their time working on machines that would normally have a semiskilled operator. In extreme cases, operators may be given jobs that have been created just to keep them fully occupied. In most applications, increased mobility of labor is not one of the benefits of CM, as some authors have claimed, but is one of the requirements of setting up cells and, as such, represents a potential cost factor.

The machine tools in most FL shops will have been bought over many years so that the machines within any group are likely to have differing technical specifications and will not all work to the same quality standards. In FL this does not matter too much because supervisors will be able to schedule work on to the most suitable machine within a group. However, when machines are taken out of FL groups to be put into a cell, not only will the utilization of most of the machines in the cell be reduced, there may also be a problem with the capabilities of the machines that are left in the FL groups.

For example, if a company has bought a CNC lathe and put it into an FL group of manual lathes, the components that will be loaded on to it will be those where CNC shows the greatest savings. If managers wants to set up a cell that will produce some of the components that are planned for the CNC machine, they will have several alternative strategies available. For example:

- Move the CNC machine into the cell.
- Leave the CNC machine in the FL group.
- Buy an additional CNC machine.

In each case there will be a cost penalty. In the first two cases, some of the components that were previously produced on CNC will now have to be produced on less productive machines, increasing the cost of the components. If the CNC machine is left in the FL group, it may be possible to transfer additional components from the manual machines, but the productivity gains will not be as great as they were for the components that have been transferred to the cell and now have to be produced on manual machines. However, in addition to obvious cost penalties such as these, or the cost of buying additional machines, there are other penalties that may not be obvious because they are hidden in the company's costing system.

One of the reasons why the introduction of CM must be evaluated is to ensure that all the costs and benefits are correctly reflected in the company's costing system. In doing this, one of the factors which needs to be forecast is the expected level of machine utilization. In the past when GT was first being developed, capital costs were quite small and direct labor was the largest cost factor in most companies. Because of this, machine utilization was not seen as being very important and little attention was given to it. With the advent of AMT, capital costs have been increasing and the stage has been reached, such as with CNC and FMS, where depreciation is often a much larger element in a standard cost than labor. However, if the level of utilization has not been forecast correctly, the depreciation rate will be wrong. For example, if depreciation is calculated on the basis of 80 hours/week utilization, but the actual utilization is only 40 hours, the depreciation value included in the rate/hour will be only half its correct value.

The setting up of machine cells, whether FP or GT, will result in a reduction in machine utilization, both in the new cells and in the remaining FL groups. Reducing the utilization of a machine will increase the depreciation rate, and this will increase the standard cost of all the components planned for that machine. However, because standard costs are only revised once a year in many companies, any increase in standard costs caused by introducing CM may not be immediately obvious. Failing to evaluate the benefits of FP and GT to offset the effect of reduced utilization in the costing system may create the belief that their use is uneconomical. It is only if all the financial benefits are forecast and included in the new standard costs that these increases can be avoided. This is especially the case if any increases in standard costs are going to be reflected in increased selling prices, thereby reducing the company's competitive ability.

Unlike financial accounting, where systems have to comply with a rigidly defined set of rules, management accounting systems, such as absorption costing, are inexact and can vary from one company to another. Primrose (1992c) shows that many companies have problems because they have not updated their costing systems to reflect the way that new technology and management philosophies have changed manufacturing. Because there are no standard rules, and because of variations between companies, the effect on standard costs of changes in machine utilization has to be investigated for each application of CM. It is important that managers be aware of a potential problem and they discuss the implications with the company's management accountants.

An obvious way of overcoming the problem of machine utilization is to schedule additional components to a cell; however, doing this may be counterproductive. The aim of FP groups is to be able to respond directly to short-term fluctuations in sales, such as shown in Figure 1, or deal with fluctuations in machine load, such as shown in Figures 2 and 3. However, to do this the average utilization of machines must be low. Planning additional components for an FP group, in order to improve utilization, means that the ability to respond to fluctuations in demand can be reduced. A similar situation exists with FMS where managers think that, because of the high capital cost, they have to keep all the machines fully utilized. However, Primrose and Leonard (1991) suggest that this can be counterproductive because it can reduce the ability to respond to customer needs. Using simulation techniques would seem to be an obvious way of selecting additional components to increase the utilization of FP groups. However, Primrose and Leonard (1991) suggest that even with FMS, conventional method study is preferable to simulation for design and component selection. They support their argument with the fact that seven out of eight European FMS manufacturers were no longer using simulation on a regular basis.

10. CASE STUDY

The company makes hydraulic pumps and valves and provides customers with a large range of product options. This means that, although some standard components are made for stock on the basis of sales forecast, most components are made in batches to suit orders received, the batch sizes representing a month's requirements. The main castings are produced on conventional milling and drilling machines and some early first-generation NC machines. Some of these are single shifted and some double shifted. Because of the age and condition of the machines, reliability and declining quality standards is an increasing problem. The company has five alternatives to consider:

1. Do nothing: This was rejected because the consequence would be declining quality and delivery, with resultant long-term loss of market share.
2. Subcontract: This was rejected because it would have resulted in loss of control of the components critical for product delivery.
3. Replace most of the existing machines on a like-for-like basis.
4. Replace with CNC machines.
5. Replace with CM.

The last three alternatives were investigated in detail and, because replacing on a like-for-like basis was the cheapest and easiest option, it was considered first. Fourteen machines are used, almost all of them exclusively for casting production, and they are installed in an FL machine shop. Apart from reducing the cost of maintenance and eliminating much of the existing scrap and rework, there would be few productivity benefits from the investment, which would cost £300,000.

10.1. Investing in Computer Numerical Control Machines

The existing workload was studied in detail and considerable effort put into replanning the components for CNC, which in some cases involved design changes. On the basis of the new planned times and sales forecast, it was calculated that five CNC machining centers, double shifted, would be required at a cost of £700,000, which could be spread over 3 years. The CNC investment would achieve the same quality and maintenance savings as the like-for-like alternative, but would also produce considerable labor savings. Both investments would prevent the long-term loss of market share caused by declining quality and delivery, therefore the benefit of this could not be attributed solely to the CNC investment. Components would still have to be produced in monthly batches so there would be little saving in inventory or improvement in delivery performance.

Investing in CNC is an alternative to the investment in conventional machines; therefore, the additional cost of £400,000 has to be compared with the additional benefits. Increasing the basic wage to include on-costs (National Insurance, welfare costs, etc.) gives an annual cost per operator of £11,412 for days and £15,177 for nights. There were 14 operators on days and 4 on nights, compared with 5 on both days and nights that would be required with CNC. The labor savings would therefore be

$$(6 \times £11,412) = £68,472 \text{ in year 1,}$$

and

$$(9 \times £11,412) - (1 \times £15,177) = £87,531$$

in year 2 and onward.

In the evaluation the working life of the machines is assumed to be 8 years with a 10% residual value. Tax is calculated on the basis of a 30% rate; capital allowances are based on a 25% first year allowance, with the allowance in subsequent years being 25% of the remaining reducing balance. The cash flows are shown in Table 1. Assuming a 15% cost of capital, these figures give a Net Present Value (NPV) of −£6089 and an Internal Rate of Return (IRR) of 14.3%. The conclusion is that the additional cost of investing in CNC would provide a marginal return.

10.2. Investing in Cellular Manufacturing

Investment appraisal involves a comparison between two alternatives and, because investing in conventional machines was the minimum acceptable option, CM was compared against this. In addition to reducing labor costs, the aim of using CM would be to improve the delivery performance offered to customers by using a cell that was

Table 1. CNC investment

	Year 0	Year 1	Year 2	Year 3	Years 4–8	Year 9
Capital cost (£)	−200,000	−100,000	−100,000			
Capital tax (£)		15,000	18,750	21,562	Reducing	838
Labor saving (£)		68,472	87,531	87,531	87,531	
Revenue tax (£)			−20,542	−26,259	−26,259	−26,259
Net cash flow (£)	−200,000	−16,528	−14,261	82,834	Reducing	−25,421

based on FMS technology. In calculating the required capacity, it was assumed that machining times would be the same as for CNC, but there would be a considerable increase in both machine utilization and operating hours. In calculating the workload, an allowance was included for the increase in sales that was expected to result from the improved delivery performance. By operating the cell for 24 hours a day, 5 days a week, a three-machine system would have adequate capacity, leaving the weekends for maintenance or to cope with peaks in the workload. A coordinate measuring machine was also incorporated into the cell.

Unlike CNC, which could be run by setter/operators using relatively simple fixtures and standard tools, considerable expenditure would be required for the tools and fixtures used in the cell. All tools would need presetting, and so special presetting equipment was needed, along with a large number of special toolholders. More than one fixture had to be made for many of the components so that enough work could be loaded to run the cell between shifts, and also give it the flexibility needed to cope with fluctuations in workload. The aim was to plan the workload on a daily basis, which would considerably increase the fluctuations in workload compared with the previous monthly planning, thereby increasing the number of fixtures required. The total cost would be

Machine tools plus measuring machine	£825,000
Fixtures and tooling	£150,000
Programming, installation, etc.	£125,000
Total	£1,100,000

An 8-year working life was assumed, with zero resale value at the end of that time. Although there will be some expenditure on tooling and programming in year 2, for simplicity, when identifying the additional cash flows with CM, it is assumed that all expenditure takes place in years 0 and 1.

One potential problem with the cell was floor space because, unlike the other alternatives where machines could be fitted into existing spaces, a large area was needed in one location for the machines and peripherals (computer room and pre-setting area). Although this total area was not much different from that occupied by the old machines, the cell had to be installed and running before the old machines could be removed. Provision also had to be made in the layout for the possibility of future expansion of the cell's capacity. Fortunately, it was possible to find enough space, although some cost was involved in relayout.

Because the company had reached the stage where they had to spend at least £300,000 on conventional machines, great care was needed to avoid including benefits, such as savings in scrap and rework, that would have been achieved anyway. At the same time, because the use of FMS technology was perceived to have a considerable element of risk, the estimates of savings that were used were very conservative. The assumption was that if the project appeared viable using these figures, it would provide an attractive investment in practice. The cell would operate with 3 people on days and 2 on nights, thus saving 11 day-shift and 2 night-shift operators. By concentrating in the beginning on producing the critical components that affected delivery performance, many of the savings would be achieved by the end of the first year. However, it was assumed that the full level of savings would not be achieved until the end of the second year. The labor saving would therefore be

$$(5 \times £11,412) = £57,060$$

in year 1

$$(8 \times £11,412) + (1 \times £15,177) = £106,472$$

in year 2

$$(11 \times £11,412) + (2 \times £15,177) = £155,919$$

in year 3 and onward.

Because ordering would be done on a daily basis, the total lead time of all the components would be less than one week. Not only would this reduce shop floor work in progress, but the change in ordering would also reduce finished component stocks. The book value of the stock reduction would be £125,000 and, of this, the cash flow value would be £75,000 in year 2, the tax advantage of the stock reduction being £15,000 in year 3.

Because of the manufacturing lead time, and the consequent inflexibility of the ordering systems in the past, Marketing had been unable to quote for short delivery orders except for a limited range of stock models. The very short lead times provided by the cell would not only allow the company to quote for more enquiries, but as delivery would be better than its major competitors, a high proportion of these should result in orders. Marketing estimated that sales should increase by at least 5%, giving a contribution to overhead recovery of £157,500 a year from year 2. Although competitors would eventually improve their delivery performance, the assumption was made that the annual saving will still be valid because, without using CM, the company would have started to lose market share. Evaluating the investment in CM against the like-for-like alternative gives the cash flows shown in Table 2.

Using a 15% cost of capital, these figures give an NPV of £288,683 and an IRR of 26.5%, suggesting that the investment would be very profitable. Because the estimates

Table 2. Investment in Cellular Manufacturing

	Year 0	Year 1	Year 2	Year 3	Years 4–8	Year 9
Capital Cost £	−550,000	−250,000				
Capital tax £		41,250	49,687	37,266	Reducing	6,632
Labor saving £		57,060	106,472	155,919	155,919	
Stock reduction £			75,000			
Book value tax £				15,000		
Extra sales £			157,500	157,500	157,500	
Revenue tax £			−17,118	−49,192	−94,025	−94,025
Net cash flow £	−550,000	−151,690	371,541	286,493	Reducing	−87,393

of increased sales were conservative, and the estimates of cost were maximized, the actual return would probably be greater than this. If the value of increased sales is removed from the evaluation, the return becomes −£132,471 NPV and 8.7% IRR. This shows that, although the labor saving with CM are greater than with stand-alone CNC, they would not justify the increased cost.

11. CONCLUSIONS

The long lead times and high levels of WIP associated with it mean that batch manufacture is thought to be inefficient and must be replaced by something else, such as CM. However, investigating the economics of batch manufacture suggests that many of the assumptions that have been made about the benefits of potential replacement may have been wrong. For example, the financial penalties of high levels of WIP and long component lead times associated with batch manufacture may be small when compared with the benefits that result from having the flexibility they provide. The failure of managers in the past to evaluate the costs and benefits of CM means that many applications may be uneconomical. It also means that managers have not introduced some potentially profitable applications because they were unable to justify the costs involved. Unfortunately, managers have often introduced CM because they have been able to identify an application for a management philosophy that is currently "fashionable," rather than to solve a specific well-defined problem.

Despite innumerable attempts to find an alternative to batch manufacture, most companies still use the approach that was developed in the 1920s. They have a small number of dedicated FP machine groups that are used to produce the key A value components that affect product delivery and an FL machine shop to produce the remaining components. Although this approach is based on practical experience, rather than the result of any scientific analysis, it has stood the test of time. This is unlike the case of GT where very few of the companies that introduced it in the 1960s are still using it. Now that FMS technology has become established and understood by managers, companies that are investing in FMS are doing so as an alternative to using stand-alone machines for FP groups, rather than replacing FL machine shops using GT techniques. Managers are now able to plan and install FP groups and invest in FMS without using analytical techniques, such as simulation for cell design, or using coding and classification for component selection. As a result, there is probably more investment taking place in CM and FMS than there was when GT and FMS were "fashionable," although little is written about such investments.

REFERENCES

Primrose, P. L. (1991). *Investment in Manufacturing Technology,* Chapman & Hall, London.

Primrose, P. L. (1992a). Evaluating the introduction of JIT, *International Journal of Production Economics,* **27**, 9–22.

Primrose, P. L. (1992b). The value of inventory savings, *International Journal of Operations and Production Management,* **12**(5), 79–92.

Primrose. P. L. (1992c). Is anything really wrong with cost management? *Journal of Cost Management,* **6**(2), 48–57.

Primrose, P. L., and Leonard, R. (1991). Selecting technology for investment in flexible manufacturing, *International Journal of Flexible Manufacturing Systems,* **4**(1), 51–77.

14

PROJECT MANAGEMENT AND IMPLEMENTATION OF CELLULAR MANUFACTURING

Bopaya Bidanda, Rona Colosimo Warner,
Paul J. Warner, and Richard E. Billo

Department of Industrial Engineering
University of Pittsburgh
Pittsburgh, Pennsylvania, 15261

1. INTRODUCTION

Though the notion of manufacturing cells was first conceptualized in pre-Communist Russia in the mid part of this century (Burbidge, 1971, 1975; Mitrofanov, 1966), the success of the Japanese manufacturers over the past few decades has encouraged their recent popularity in the United States. With goals of reduced inventory, setup time, throughput time, and material handling, improved quality, and increased worker responsibility, cellular technology possesses some diverse manufacturing qualities unfamiliar to a large number of U.S. companies. In spite of these differences, many U.S. facilities have made a seamless changeover to cellular production while realizing high levels of percentage improvements along a variety of performance measures (Wemmerlov and Hyer, 1989). Many, however, have not been so fortunate, while some companies toiled for years until they achieved functioning cells, others simply abandoned the concept altogether. How did some organizations convert their operations with seemingly little effort, while for others a monumental struggle ensued?

Handbook of Cellular Manufacturing Systems, edited by Shahrukh A. Irani
ISBN 0-471-12139-8 © 1999 John Wiley & Sons, Inc.

More often than not, successful companies adequately prepare their organizations for the change to manufacturing cells.

Cellular manufacturing is relatively new to the United States, and the corresponding literature generally focuses on promoting this technology and the subsequent benefits. While the articles sufficiently describe how the ideal cellular factory functions, the steps a facility must take to get there are often incomplete and lacking in detail. Winning outcomes with manufacturing cells result from understanding the four basic components of cells: (1) *people* who utilize *(2) equipment* under (3) *operating rules* to transform (4) *material* into a salable product. Successful companies do not ignore the first component—people.

Who are these people? Many assume the only persons associated with manufacturing cells are production workers and managers. In actuality, cellular technology affects a wide range of people throughout the organization. These include design engineers, manufacturing engineers, industrial engineers, various levels of management, production and material control personnel, union representatives, scheduling coordinators, quality control engineers, marketing and sales persons, purchasing agents, cost accountants, trainers, warehouse personnel, facilities management, and maintenance technicians (Irani, 1997). In addition, the changeover to cellular technology will also impact persons from outside the organization such as customers, suppliers, subcontractors, and equipment vendors.

The successful companies include all key persons in a meaningful way in the transition from their present state of operation to manufacturing cells. The people who will eventually operate, manage, support, and maintain the cells are the people who must be involved in the design, development, construction, and testing of them. However, the people listed above represent a large cross section of skills, personalities, operating paradigms, and responsibilities. Simply gathering this group together and expecting them to produce manufacturing cells will only result in an unmanageable mob and wasted time, money, and effort. On the other hand, successful companies develop these people as a team and have them generate high performing manufacturing cells through the use of the tried and tested principles of project management.

1.1. Need for Project Management

Integrating manufacturing cells into an organization is similar to a track race. Every race has a beginning and an end. Every runner wants to be the best and reach the finish line in the fastest time. The key to winning a race is echoed each time at the start: "Get ready!" "Get Set!" "Go!" The runners who *get ready and set* have the best chance of winning. Runners who do not train, jump the gun, or do not understand the format of the race (the course, where it ends, etc.) usually end up in defeat. Companies that had a difficult time converting to cellular technology often paralleled the ill-prepared runners. Some started too soon toward manufacturing cells without fully understanding the technology or training and preparing their people. Many underestimated the magnitude of work required to transition themselves from conventional manufacturing to cellular technology. Unfortunately, defeat in the pursuit of manufacturing cells is often quite costly. Companies can find themselves saddled with equipment they cannot use and may never pay off, deteriorating performance, and a workforce that has developed a distrust of new technology. Incorporating cellular production into companies is more than just grouping parts, installing equipment, shuffling workers, and rescheduling production. An organization must first get itself, its people, and its facilities ready and set to *go* with this new philosophy. Project management allows companies to methodically integrate manufacturing cells into their factories and offers them an excellent chance of realizing their benefits.

Thus far, this handbook has detailed what manufacturing cells are, how they differ from conventional manufacturing, and the necessary changes required by a facility wanting to utilize them. This chapter provides a framework that prepares companies for the introduction, implementation, and execution of cellular technology based on project management. The chapter is organized as such. The next section provides the basics of project management in regard to implementing cells. Proven techniques are explained that companies can use to integrate organizational resources and put together a high performing team for the implementation of cells. This section also brings to focus some of the major human issues involved in cellular technology. After this, the steps in implementing successful manufacturing cells are detailed that involve utilizing a cross-functional team, cellular technology principles, and the techniques of project management.

2. PROJECT MANAGEMENT FOR CELLULAR MANUFACTURING

The changeover of a production facility from conventional manufacturing to manufacturing cells will entail a project or a series of projects. Cleland describes a project as "a combination of organizational resources pulled together to create something that did not previously exist that will provide a performance capability in the design and execution of organizational strategies" (Cleland, 1994, p. 4). During a cellular technology implementation project, a manufacturing cell (or cells) will be conceived and justified by a team of people, materialized through design, engineering, and construction tasks, and finally put to use, most likely by many of the original team of people. From feasibility studies to final installation and use, project management is a proven technique used by organizations to systematically plan, organize, schedule, manage, and control each activity such that the universal project objectives (Fig. 1) can be met (Cleland, 1994):

- Technical performance capabilities (e.g., reduced setup times, increased quality, etc.)
- Cost (e.g., implementation budget)
- Time frame (e.g., steady-state operation within 3 years)
- Synchronicity with the overall organizational direction (e.g., manufacturing cells are a natural extension of the corporate mission to reduce costs, increase manufacturing flexibility, satisfy customers, etc.)

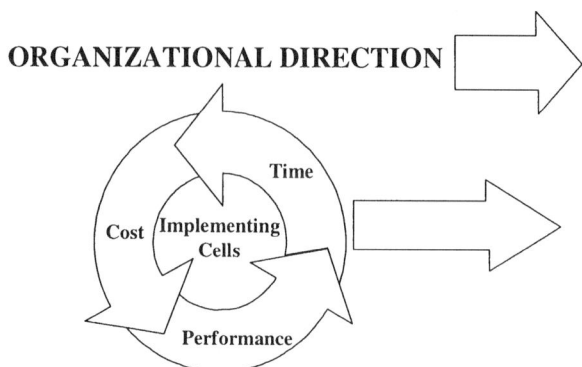

Figure 1. Factors to consider when implementing manufacturing cells.

Project management permits a flexible use of organizational resources and is one of the main forms for converting an organization from one state to another (Cleland, 1994). Thus, it is especially applicable when implementing manufacturing cells.

2.1. Human Issues in Implementing Manufacturing Cells

Companies implementing manufacturing cells must first understand why project management is necessary. The predominant causal issue in the failure of implementing new technologies is the lack of regard to human issues (Chung, 1996). This oversight can often be traced to the nature of conventional job shop operations. In a job shop, workers are isolated, decoupled from one another through buffers (work-in-process inventory), and other management practices. Surrounded by a pile of parts, workers continually cut, inspect, or assemble pieces while the company continually places more piles of parts in front of their stations to keep them productive. The focus of the worker is narrowed to only the requirements of the station. There is seemingly little need to coordinate the human issues involved in this type of environment.

Cellular Manufacturing (CM) strives for a more integrated facility, synchronizing workstations and placing responsibility on the workers to supervise the flow of parts throughout the entire plant. Under the use of cellular technology, everyone within the company must become aware of the larger operating environment. This will require an emphasis on previously neglected human issues whose relative importance soon becomes increasingly apparent under the new operating philosophy. For example, operators, whose prior extent of communication entailed occasional discussions of job schedules with their supervisors, will now need to arrive at workable schedules with their cell teammates, determine and verify improved methods with engineering and quality control, coordinate product flow with raw materials and finished goods warehousing, and interface on a continuous basis with many other support functions. Everyone who once focused only on his or her assigned task, will now focus on producing the company's products in the most efficient manner. This will include keeping an eye open for lights signaling a station in need of parts, stopping cell lines to assist a team member in need, recording and analyzing data for statistical process control, reviewing and discussing performance graphs and reports, and evaluating, as a team, proposals for new equipment (Irani, 1997).

To achieve more focus on the human aspect during the implementation of new technology, the following elements have been established as critical characteristics for the successful transfer of technology into organizations (adapted from Chung, 1996):

1. Establish clear project objectives.

2. Use of a *human-centered philosophy* that encompasses the concept of the computer-aided craftsman, who is both supported by and in control of the technology.

3. Early and significant *worker participation* in the planning and implementation of the technology.

4. Initial introduction of the technology by utilizing *pilot projects*.

5. The presence of a technology implementation *champion*.

6. *Employees are selected* and *trained* to be more capable in terms of knowledge, skills, and attitudes.

7. The organization specifically directs effort toward *overcoming resistance* to new technology.

8. Performance *evaluation and rewards systems* must be changed to meet the requirements of the new operating environment.

9. *Organizational design* is changed with technology implementation (e.g., cross-functional teams).

10. *Empowerment* by placing decision-making authority at the lowest level that has access to the necessary information.

The human skills necessary for successful cellular production—fluid communication, teamwork, decision-making ability, leadership, increased responsibility, and conflict resolution—are also important attributes of successful projects. Thus, a good way to achieve the requirements of successful cellular technology transfer is to develop these skills early and throughout the implementation process via the practices of project management.

2.2. Management of Workforce Change

It is often difficult for management and engineers to understand why some people are so adamant against change. What many do not realize is that the use of manufacturing cells will require employees to restructure how they do their jobs (Lewin, 1947) and, perhaps, the way their company rewards them for doing so. Change on a person's livelihood often has an unsettling effect (Lamarsh, 1994). A company cannot expect employees to eagerly participate in the implementation of manufacturing cells if they perceive that the end result will adversely effect them. Before implementing CM, a company would be wise to study the reasons that motivate people to resist improvements, even ones like manufacturing cells. Employees may interpret cellular technology as a threat in a variety of ways, such as more work for the same pay, loss of jobs, fear of failure, and the fear that CM will not work. Each of these are now detailed with solution strategies for addressing them.

More Work / Same Pay   Before the implementation of manufacturing cells, the company rewarded a worker in a certain way. This was centered around the skill set required to do the job and assessed through performance appraisals. Through the manufacturing cell philosophy, job assignments and requirements will change. For example, operations are typically engineered so that more value-added work is done. This may be accomplished by the operator assuming more decision-making responsibilities and becoming responsible for the operation of multiple machines. Another case is regarding maintenance, where the drive to maintain higher machine utilization will translate into more frequent maintenance schedules. These changes are viewed in two different ways. From management's point of view, people are still paid for 8 hours of work. From the workers' point of view, they must work more for the same amount of pay. The sentiment of doing more for the same pay could result in resistance to manufacturing cells from the people who will operate them.

The company needs to address this issue early in the implementation project. The team must be educated about cellular technology and the reasons why the company has chosen this mode of operation. This should include a discussion and explanation of the financial situation and future of the company in terms of job preservation, market share, competition, and even survival (Wellins et al., 1994). With a more global economy, employees need to be aware of the number of other companies who are competing for the same market and the advantages they have accomplished with CM. In addition, since compensation and rewards will have to be adjusted from purely individual efforts to team results, the company should be prepared to negotiate with the team to determine a representative family of performance measures (Gross and Safier, 1995; Thor, 1995). The company and the team need to arrive at savings targets, with the team able to share in the saved costs through profit-sharing and gain-sharing programs and small group incentives (Johnson, 1993). For instance, a savings target may be to decrease average Work-In-Process (WIP) inventory to $300,000 by the end of the year (Thor, 1995). The company could award members of the cell a percentage of this savings. It is very important for the team performance measures to be consistent with the rewards and recognition system, in addition to the management objectives of the organization.

Loss of Jobs   With cellular technology, there are numerous efficiencies that can be gained: production lines that took 5 machinists may require only three, supervisors that had 10 people reporting to them may have a number of cell *teams* reporting to them, and the responsibility of quality inspection by a few will now become the responsibility of quality management by all. These are all laudable goals to everyone except those who are part of the savings. If a company does not have the demand to fulfill the excess in personnel capacity, it will need to reduce headcount figures or displace people to other areas that can use them in to realize any labor savings.

The successful implementation of cells will require the input from a large number and wide range of organizational staff. However, it is difficult for people to be enthusiastic about putting themselves out of work. Any threat of job loss will not motivate employees to assure the success of manufacturing cells. The worst approach a company can make is to hide any downsizing or *right-sizing* intentions regarding its decisions to use manufacturing cells. These motives can be sensed immediately by company personnel and could lead to ill-will, sabotage, and bad publicity for the firm.

The company needs to address its down-sizing intentions early and honestly with the team and support its direction with financial facts. To offset the negative effects of job loss, the company should first explore attrition rates. This permits a graceful and nonimpacting reduction of headcount. If this is not feasible, buy-out options should be discussed and assistance offered in locating a new job. However, to maintain the motivation and commitment of people within the organization, the company should seek ways to avoid as much job loss as possible. This will include retraining employees so that other functions can use them (Collett and Spicer, 1995). The company must also not be blinded by its current business strategies. New markets should be investigated that could be attained profitably with the excess capacity (Goldratt and Cox, 1992). The company can draw upon the collective creativity of the team and allow the team members the opportunity to hold on to their jobs. The authors have found that in the large majority of cases, manufacturing cells help improve productivity, help manufacturing organizations boost capacity, and maintain current job levels.

Fear of Failure   Some companies practice nonmonetary rewards such as singling out the most productive employee. Even without formal recognition, each organization has implicit *people of excellence*. Manufacturing cells will depend on the collection of the efforts of individuals operating as a team. The company may change, remove, or combine operations, thus employees must become flexible. The new people of excellence will be able to perform many functions well, not just one function masterfully. Under cellular technology, employees may not understand their position within the shop structure and may even see themselves going backwards; no longer the MVP, just one of the workers. Many may not envision themselves as valuable under the new regime as they were under the old way, thus creating another potential source of resistance.

Fear of failure indicates that employees either do not know what will be their required duties (Lamarsh, 1994) or feel they do not have the necessary skills to perform them. The company can address this through education and training. First, people within the organization will need to know what CM is and how the company intends to use it. This can be attained through the use of seminars and meetings, developed in-house or through the use of outside consultants. The company must then allocate money and resources in the implementation project budget for training. This will demonstrate to employees how committed the company is to CM and how much it want its employees to succeed with the new technology.

Efficient and cost-effective selection and training will involve a number of steps (Warner et al., 1997). The first is to clearly *define* each of the required skills, both technical (e.g., machining, inspection) and nontechnical (e.g., communication, leadership), necessary for successful CM. This will provide a focus for the training and also a measurable target to evaluate the effectiveness of the training. Next, the company must *assess* the workers' current level for each of the skills. This will help the company determine which employees need what training and to prevent wasted efforts on employees completely overwhelmed by the training subject or well beyond it. Training modules will then need to be *selected and/or developed* to raise the workers' skill levels to that necessary for manufacturing cells. Finally, the company will need to *schedule* workers to participate in the training modules (Warner, 1998). The company should offer the training on a voluntary basis. However, to increase participation, the organization should communicate the shift from individual rewards to team rewards and rely on peer pressure to spur motivation to increase one's skill set.

Fear That Cellular Manufacturing Will Not Work Most employees are genuinely concerned about the performance of the organization and want the company to grow and be profitable. However, many of the tenets of CM are antithetical to the modes of operation or culture to which these people are accustomed. For example, in a job shop environment, employees have long associated good performance with large lot sizes to pay for setups, maximizing individual work effort to maximize overall company performance, and keeping a buffer of inventory to assure that production will continue in case of equipment failure. If a company just starts operating one day with one piece lot sizes and pacing production with the bottleneck operation, a panic will ensue throughout the company as "the rocks in the river" (Black, 1991) begin to surface. Without explanation and training, employees may view this new technology as another management fad in which poor results could seriously jeopardize the company's future, and in particular, their livelihood.

Management must realize how acutely different the cellular production philosophy may be to the current type of manufacturing. Again, this supports the thesis that management clearly communicate the reasons for the new cellular technology direction throughout the organization. To ease the transition, the company must demonstrate to its employees that CM will make the company stronger and more profitable. This can be accomplished through training seminars. Here, employees can compare cellular technology techniques to their current methods and realize themselves the advantages of manufacturing cells. During this time, they will also be introduced to some additional skill requirements they will have to attain (e.g., teamwork, decision making, leadership, etc.).

Another way to demonstrate the benefits of manufacturing cells is to arrange benchmarking tours of other companies utilizing cellular technology. This will allow employees to meet and interact with the people in these companies conducting similar functions and see firsthand what will be expected of them in this operating environment.

2.3. Project Management Fundamentals

The threat of the unknown simply has an unsettling effect on people. When changing to manufacturing cells, companies must discover and resolve the issues that trouble their employees; the earlier, the better. The success or failure of manufacturing cells depends on how soon companies address employee concerns and how effectively they manage these concerns along with the technological requirements during the implementation of the manufacturing cells. The human issues, coupled with the technological demands of manufacturing cellular technology, presents a tremendous challenge to organizations. Effective project management allows a company to successfully manage these issues throughout the transition while providing employees a means to contribute and become involved in the change (Fig.2).

Life Cycle of Manufacturing Cell Projects   A series of major steps is involved in the project of designing, developing, and implementing manufacturing cells. The total project of instituting manufacturing cells contains constantly changing levels of cost, time, and performance. Throughout each phase, the team acquires, utilizes, and then releases the numerous resources to perform a variety of tasks. The traditional hierarchical organization is just not designed to cope with managing this changing mix of resources (Cleland, 1994).

Activities or tasks that occur during a project fall under four major divisions of effort: conception, planning, execution, and termination (Adams and Barndt, 1988). During the conception phase, the company identifies and verifies the need for manufacturing cells, establishes clear objectives and goals and the resources necessary to achieve them, and then sells the approach to the organization to solicit support and commitment. In the planning phase, the project team is developed. During this time, the team, through the input from a wide variety of positions and responsibilities, puts together a detailed, manageable, and acceptable approach to implement manufacturing cells that address both the technical and human issues. Using the designed plan, the team performs the tasks and activities to construct and tests the cells in the next time period, the execution phase. Finally, the project is terminated in the final phase when the cells are in operation, documentation has been transferred to the cells, the area has been cleaned up, and the team has been reassigned. Traditionally, it is noted that the majority of effort occurs during the planning and execution phases (Adams and Barndt, 1988). Table 1 lists the major tasks involved in implementing a manufacturing cell.

Like the track runner who builds himself into a world-class champion, a tremendous amount of effort must occur in planning and building world-class manufacturing cells. The life-cycle model assists the project team when planning for manufacturing cells because it clearly defines the beginning, the end, and the necessary effort in between for each implementation project. At the conclusion of a project, the company will build something that did not previously exist. Thus, each team learns as the implementation progresses and when each cell is installed and in operation. The accumulation and documentation of information during the project is vital since the

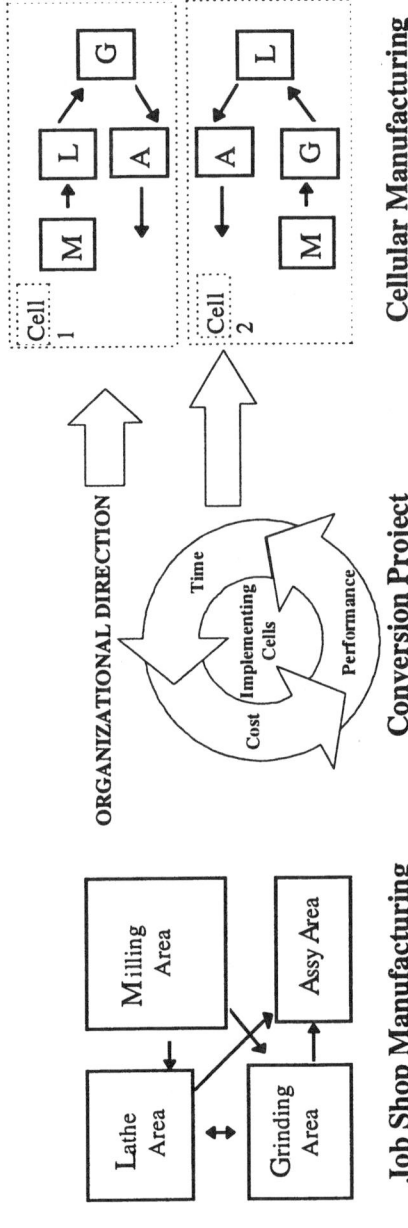

Figure 2. Using project management to implement cellular manufacturing.

Table 1. Major tasks involved in implementing manufacturing cells

Conception	Planning	Execution	Termination
• Company identifies need for manufacturing cells. • Educate employees about cellular technology via seminars, demonstrations, and through the aid of consultants. • A small cross-functional team prepares and presents a feasibility study, initial budgets, and schedule to upper management to convert operations to cells. • The team obtains approval and funding for the conversion. • Identify and select a cellular "champion" or leader along with employees within the organization for the implementation team.	• Develop the project team and team policies, rules, and procedures. • Prepare detailed schedule to build and phase in cells, assign work packages and responsibility, and develop control standards. • Conduct part population analysis, group parts, and design families of parts. • Design cells—determine processes, equipment, tooling, material handling, worker selection, communication devices, and operating procedures. • Develop cell information system and integrate with other production systems.	• Procure materials, equipment, and any contracted assistance for the cells. • Build cell—Clear floor space, install/relocate equipment, material handling, communication devices, etc. • Implement cell information system. • Integrate scheduling and parts into cells. • Train personnel and assign them to cells. • Collect data and test and verify cell performance. • Document support requirements, procedures, and policies.	• Transfer responsibility and documentation to cell "owners." • Clean up cell area. • Conduct post-project analysis. • Reassign project members.

team uses this information to update the life-cycle model during the implementation of each cell (Cleland, 1994).

Project Management Process   Management is typically defined as a continuous process consisting of five overlapping functions: planning, organizing, motivating, directing, and controlling (Cleland, 1994). The implementation team must understand that cost, time, and performance parameters are managed throughout the life of a project via the proven techniques of project management. Planning is performed in each of the phases to reset direction once more information has been gathered and control standards have been evaluated Organizing is continually performed as various employees are utilized at different times along the life cycle. The type of motivation and direction of the project leader must adapt to the changing personnel and fluctuating degrees of urgency. Finally, control must be exhibited throughout any project to ensure the team is constantly heading toward a successful end. The following are typical CM tasks in terms of these major functions.

Planning

- Develop cellular manufacturing objectives in accordance to the overall company direction.
- Define the major components of the manufacturing cell.
- Define the major tasks, precedence, and relationship of work effort required for the design and installation of cells along with the human, physical, and monetary resources necessary for the completion of each of the tasks.

Organizing

- Establish how the team will be organized (e.g., project leader, team structure) and the purpose for each member of the project team.
- Select team members and secure approval from each team member's functional unit.
- Clearly define and document the implementation project's policies, rules, and procedures.
- Assign each member's degree of responsibility and participation for the completion of tasks.
- Schedule each team member into the implementation project.

Motivation

- Address objections or lack of understanding people may have toward cellular production and provide appropriate training.
- Recognize that the variety of team members (e.g., engineers, production line employees, salesperson) are motivated in a variety of ways.
- Establish an appropriate rewards program for team member efforts and a conflict resolution procedure.

Directing

- Develop interpersonal skills and style of leadership used throughout the implementation project (e.g., open-door policy, level of command and control, participative).
- Establish limits of authority for purchasing, assigning team members, and other decisions.
- Develop methodology for team decisions (e.g., majority vote) and ensure that members will be committed to project in spite of differences of opinion.

Control

- Establish clear cost, schedule, and performance standards for the implementation and the mechanism and frequency of evaluation (e.g., weekly meetings, progress reports, variance analysis).

- Utilize an implementation management information system to record and report level of effort performed against task definitions, budgets, and time goals.

- Schedule time to adequately assess effort toward established milestones, define any corrective action, and rework task definitions, budgets, and time goals.

Planning for Manufacturing Cells Good planning does not guarantee the successful implementation of cells but eases the transition. However, as the familiar adage predicts, if a project team fails to plan, then it should plan to fail!

As the project life-cycle discussion demonstrated, all projects have a beginning and an end. Just as the finish line determines the end of a race, there must be a clear end to a cell implementation project; the point in time when the transition is complete and the company manufactures its products through the use of cells. Cleland (1994) provides a framework for project planning that takes the hierarchical path of defining the project's vision, then from this vision establishing the project's objectives (the future desired position in terms of cost, schedule, and performance), goals (milestones leading to the objectives), and strategy (plan of action and work packages). From these, the project's organizational structure and team roles, management and team culture, and human, physical, and monetary resources can be defined and charted. Systematic project planning lets an implementation team rationally determine how and when to initiate, sustain, and *terminate* the tasks necessary for the successful installation and operation of a manufacturing cell (Cleland, 1994).

Sound planning practices better ensure the manufacturing cell's fit within the overall company. While PERT and CPM are important components of planning and are familiar to many, implementing manufacturing cells involves a much wider scope of activities. Simply establishing a cross-functional team will not build and develop manufacturing cells. Effective project planning must determine a detailed list of activities and resources required to reach the venture's objectives to ensure that it is adequately executed. Thus, it is especially important for the team to understand the entire cell implementation life cycle and the organization in which cells will be developed and operated. Authority, responsibility, and accountability must be planned and *timed* so that members of the project team know what their specific roles are, when they are required, and how they relate to other activities of the project. An effective project leader must also know that different types of people require different types of motivation and to adjust the management style throughout the implementation to bring out the optimal effort of each team member.

When employees plan, build, and develop manufacturing cells, they obtain excellent training for the team that will be required to operate them. Involvement in the planning process provides employees with ownership and lets them address sources of resistance early and develop their own solutions to them. This typically anchors their commitment to the success of CM. The project management technique predominantly used to involve team members in the planning of the necessary implementation tasks is the *work breakdown structure* (WBS).

Work Breakdown Structure The information integration and communication between the team implementing cells is important to the success of this project. The key to completing a project on time, within budget, and meeting all technical parameters is the careful definition of the WBS. A WBS defines the tasks and establishes the resources necessary to fulfill project objectives. It helps ensure that no major step is overlooked and that all participants are fully aware of the total work required to completely and successfully implement manufacturing cells. A WBS provides a common communication channel by being the foundation upon which all estimates, schedules, and flow of project tasks are developed (Lavold, 1983).

A WBS contains various levels describing the work content required to fulfill tasks making up the implementation project. The project is defined at the highest level and broken into its major, natural work components at the next highest level. Then with each successively lower level, subtasks are further defined. The lowest level contains the information necessary to budget, schedule, perform, and control the tasks necessary in the implementation of cells. This information becomes the basis for project work packages. A work package is comprised of a clear task definition, standards to abide by, responsible parties, allocated budget, and due date.

During the implementation process, work effort is inputted by team members against their assigned work packages. Each progressively higher level can then become a management control point. Information at the lowest level can be compiled up through the WBS to determine the effort expended and the required effort remaining to determine adherence to schedule.

The following outline illustrates a comprehensive, general WBS for implementing manufacturing cells that can be adapted and customized for individual organizations. The WBS for each implementation project must be a collaborative team effort to gain insight into the specific needs of the organization for the implementation of cells.

1.0 **Implementing Manufacturing Cell(s)**
    1.1 **Conceptual Phase**
        1.1.1 Project Management Functions
            1.1.1.1 Ensure organizational vision is conducive to manufacturing cell(s)
            1.1.1.2 Gain organizational commitment to cellular technology
                1.1.1.2.1 Research cellular technology
                1.1.1.2.2 Conduct market study for sales forecast
                1.1.1.2.3 Benchmark manufacturing cells
                1.1.1.2.4 Conduct feasibility study
                1.1.1.2.5 Educate company personnel on cellular technology
                1.1.1.2.6 Educate company personnel on team concepts
            1.1.1.3 Solicit members for implementation team

1.2 **Planning Phase**
  1.2.1 Project Management Functions
     1.2.1.1 Form implementation team
         1.2.1.1.1   Select project leader
         1.2.1.1.2   Select team members
         1.2.1.1.3   Permission for release of members from functional areas
         1.2.1.1.4   Develop members to operate as a team
         1.2.1.1.5   Provide further education on how cell will operate
         1.2.1.1.6   Institute project policies and procedures
         1.2.1.1.7   Institute project control standards
     1.2.1.2 Develop project objectives
         1.2.1.2.1   Define technical performance objectives of the cell
         1.2.1.2.2   Establish budget
         1.2.1.2.3   Establish schedule and time lines
     1.2.1.3 Develop/utilize project information system
         1.2.1.3.1   Interface with payroll accounting
         1.2.1.3.2   Interface with purchasing
         1.2.1.3.3   Interface with time and attendance
         1.2.1.3.4   Educate team on use of system
     1.2.1.4 Develop work assignments
         1.2.1.4.1   Develop a thorough WBS to define work packages
         1.2.1.4.2   Assign work packages to members via a linear responsibility chart
         1.2.1.4.3   Schedule tasks via PERT and CPM
     1.2.1.5 Secure funding
         1.2.1.5.1   Estimate cost and time frame of installation
         1.2.1.5.2   Estimate cost of operation and schedule to steady-state
         1.2.1.5.3   Estimate operational savings and other benefits
         1.2.1.5.4   Develop final project plan and present to upper management
         1.2.1.5.5   Obtain project and funding approval
  1.2.2 Manufacturing Cell Implementation Functions
     1.2.2.1 Conduct part population analysis
         1.2.2.1.1   Determine which parts to group
         1.2.2.1.2   Select method of grouping parts
         1.2.2.1.3   Conduct necessary parts reengineering
         1.2.2.1.4   Define family of parts
         1.2.2.1.5   Determine number of cells
     1.2.2.2 Design manufacturing cell(s)
         1.2.2.2.1   Determine part process flow
         1.2.2.2.2   Select equipment
         1.2.2.2.3   Select tooling, jigs, and fixtures

1.2.2.2.4   Design/select material handling methods and equipment

1.2.2.2.5   Assess the capacity of the cells

1.2.2.2.6   Level, balance, and synchronize the cells

1.2.2.2.7   Design shop floor layout

1.2.2.2.8   Select communication signals (e.g., kanbans, report boards)

1.2.2.2.9   Determine quantity and type of support equipment

1.2.2.2.10 Select type and quantity of workers

> 1.2.2.2.10.1 Determine cell skill requirements
>
> 1.2.2.2.10.2 Assess skill levels of current workforce
>
> 1.2.2.2.10.3 Match workers to cell tasks
>
> 1.2.2.2.10.4 Develop rotation schedule
>
> 1.2.2.2.10.5 Develop training plans to attain cell skills
>
> 1.2.2.2.10.6 Develop performance measures and goals
>
> 1.2.2.2.10.7 Develop compensation and reward system

1.2.2.2.11 Develop cell operating rules

1.2.2.3 Quality issues

1.2.2.3.1   Develop inspection processes

1.2.2.3.2   Develop SPC requirements

1.2.2.3.3   Select inspection equipment (e.g., gauges, comparators)

1.2.2.4 Develop cell information and control system

1.2.2.4.1   Determine and define all data to collect

1.2.2.4.2   Determine means/frequency of data entry (e.g., bar codes)

1.2.2.4.3   Develop reporting requirements

> 1.2.2.4.3.1   Define types of reports and information to convey
>
> 1.2.2.4.3.2   Develop report layouts
>
> 1.2.2.4.3.3   Determine distribution for report
>
> 1.2.2.4.3.4   Define frequency of reporting (e.g., hourly, daily)
>
> 1.2.2.4.3.5   Means of distribution (e.g., computer screen)

1.2.2.4.4   Establish a control strategy

1.2.2.4.5   Interface cell with all necessary existing production systems

> 1.2.2.4.5.1   Product description database
>
> 1.2.2.4.5.2   Scheduling and MRP
>
> 1.2.2.4.5.3   Time and attendance

        1.2.2.4.5.4   Financial and costing

        1.2.2.4.5.5   Process routings and standards

        1.2.2.4.5.6   Production performance

        1.2.2.4.5.7   Purchasing

        1.2.2.4.5.8   Scrap and quality

        1.2.2.4.5.9   Asset management

    1.2.2.5 Determine cell implementation strategy

    1.2.2.6 Develop procedure to test performance of cell

1.3 **Construction of Cells**

    1.3.1 Project Management Functions

        1.3.1.1 Procure expense items (e.g., construction materials)

            1.3.1.1.1   Send out bids

            1.3.1.1.2   Send out purchase orders

            1.3.1.1.3   Receive and inspect items

        1.3.1.2 Procure capital items (e.g., machines)

            1.3.1.2.1   Send out bids

            1.3.1.2.2   Send out purchase orders

            1.3.1.2.3   Receive and inspect items

            1.3.1.2.4   Tag items

            1.3.1.2.5   Input into asset management system

        1.3.1.3 Procure contractors

        1.3.1.4 Conduct control meetings

    1.3.2 Manufacturing Cell Implementation Functions

        1.3.2.1 Build cells

            1.3.2.1.1   Clear floor space

            1.3.2.1.2   Relocate equipment

            1.3.2.1.3   Install new equipment

                1.3.2.1.3.1   Secure equipment

                1.3.2.1.3.2   Hookup electric

                1.3.2.1.3.3   Hookup coolant

                1.3.2.1.3.4   Hookup air

                1.3.2.1.3.5   Hookup waste removal

                1.3.2.1.3.6   Test and debug

                1.3.2.1.3.7   File all manuals and other relevant documentation

            1.3.2.1.4   Install material handling equipment

            1.3.2.1.5   Install communication devices

            1.3.2.1.6   Install work benches and other support equipment

            1.3.2.1.7   Implement quality processes

                1.3.2.1.7.1   Implement inspection processes

                1.3.2.1.7.2   Document SPC requirements

                1.3.2.1.7.3   Assign inspection equipment

            1.3.2.1.8   Install data recording devices

            1.3.2.1.9   Assign workers to cell

                1.3.2.1.9.1   Provide instructional training

       1.3.2.1.9.2   Provide instructional documentation

       1.3.2.1.9.3   Provide instructional documentation

   1.3.2.2 Install/test information system including interface with other systems

   1.3.2.3 Test cells according to test procedure

1.4 **Project Termination**

   1.4.1 Project Management Functions

      1.4.1.1 Clean up site

      1.4.1.2 Transfer final documentation

      1.4.1.3 Reassign project team members

      1.4.1.4 Conduct post-project analysis

      1.4.1.5 Report project findings for subsequent implementation projects

To be an effective communication device, the WBS must be understood by all of its users. This is accomplished via structuring and coding a manageable number of levels. In the previous WBS, the levels are defined as follows:

Level 1: Single-digit code—Manufacturing cell(s) to be implemented

Level 2: Single-digit code—Major implementation project phase (1, conceptual; 2, planning; 3, construction; 4, termination)

Level 3: Single-digit code—Classification of function for the phase (1, project management; 2, cell implementation)

Level 4: Double-digit code—Specific task for the function

Level 5: Double-digit code—Subtask

Level 6: Double-digit code—Additional subtask (if necessary)

The code for the WBS must be understood by every team member. They can uniquely identify each of the activities and work packages by concatenating the coded levels in a hierarchical order. Thus, for example, from the WBS above, 1.3.2.1.3.2 represents the electrical hookup required for the installation of new equipment. This activity is for the implementation of a manufacturing cell(s), is part of the construction phase, and is a cell implementation function, more specifically, for the building of the cell. Thus, if a manager wanted to know what work effort has been accomplished and the amount of money spent toward building the cell, the person would sum up all time sheets and purchase orders against work package codes that started with 1.3.2.1. If this component of the implementation project was over budget or past its due date, the manager could narrow the investigation to just the work elements under 1.3.2.1. For a more extensive review of WBS, please refer to Lavold (1983), Archibald (1992), and Warner (1997).

Organizing for the Implementation of Manufacturing Cells A successful project organization must be built around the purpose to be accomplished, analogous to the manufacturing cells the team will build (Cleland, 1994). In cells, various work-stations (lathe, mill, grinder, assembly) are put together in close proximity to produce a product. When implementing cells into a company, various personnel (engineer, designer, operator, marketer, financier) are pulled together in close proximity to build them.

The matrix organization is a structure that promotes the sharing of company physical resources, managerial skills, and expertise, and facilitates the integration of various organizational subsystems for use in projects (Badiru and Schlegel, 1994; Cleland, 1994). Figure 3 is a general representation of a matrix organization.

There are a number of advantages of the matrix organization including the sharing, definition, and pursuit of project objectives and resources by a number of functional units and that departments cooperate, thereby improving information flow (Badiru and Schlegel, 1994). However, the matrix organization does have some drawbacks. Under a matrix structure, an organization may be simultaneously conducting multiple projects in which various employees may belong to more than one project at a time. Since every project leader and functional manager have responsibilities and performance goals to meet, releasing *their* employees to serve elsewhere in the organization may interfere with their agendas. If an organization fails to adequately manage and control projects, conflicts and internal dissension could result. However, companies can overcome these disadvantages through effective education, planning, and use of the matrix organization (Badiru and Schlegel, 1994).

Linear Responsibility Chart   Within the WBS, work packages result from the intersection of the project requirements and the functional area(s) that will fulfill them. The team approach is used because these requirements will cross multiple functions. For example, manufacturing cells will necessitate the installation of new equipment. The various functional units involved in this activity include manufacturing engineering, equipment engineering, plant engineering, millwrights, electricians, pipefitters, and production workers, to name a few. Installing equipment is only one component of implementing a manufacturing cell. As each task gets broken down into its subtasks, it becomes evident that the implementation of manufacturing cells requires a significant amount of effort and many organizational resources. With such a large and diverse array of activities, members must be able to understand their roles and responsibilities. The *linear responsibility chart* (LRC) is a tool that communicates to team members the degree of responsibility and accountability they have concerning a given task. Table 2 is an example of an partial LRC for implementing manufacturing cells. An LRC includes project's tasks or work packages (along the left-hand side), the various team members (along the top), and the degree of involvement of each team member to complete the tasks. The definition of each team member's involvement is communicated through a key.

Since an LRC serves as a means of motivating the members and holding members accountable for their actions, the best way to develop one is to let the team form its own. With a WBS and an LRC, the team members understand how their work fits in with the overall objective of the project. They can then develop their own set of duties and degree of responsibility and accountability, which brings on a sense of commitment to the success of the implementation project. For a more in-depth discussion on linear responsibility charts, please refer to Cleland and King (1988).

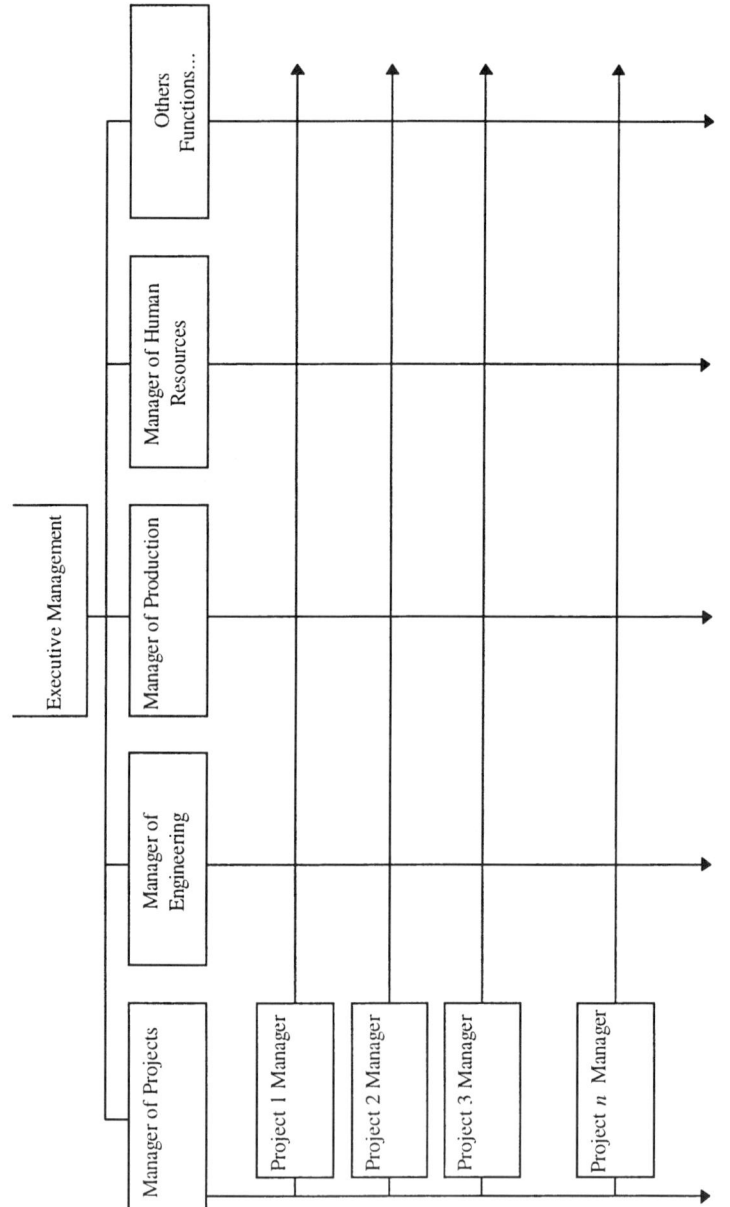

Figure 3. Matrix organization design.

Table 2. Partial linear responsibility chart for manufacturing cells

LRC	Industrial Engineering	Plant Engineering	Manufacturing Engineering	Production Supervision	Production Line Employees	...
Tasks						
Grouping parts	C		M	I	C	
Determining families	C		M	I	C	
Relocating equipment	I	M	C	C	C	
⋮						

Key: M, main responsible member; C, contributing member for completion; A, approving member; I, members to be informed.

3. IMPLEMENTING MANUFACTURING CELLS

With an awareness of the human issues involved in cellular technology and armed with the basics of project management, an organization can now begin to transition its operations to manufacturing cells. This section is organized around the WBS provided in the previous section and primarily covers the high-level work elements. Again, it is important that each organization recognizes its own unique requirements and, therefore, adapts the general WBS to include the steps critical to its cellular implementation project.

3.1. Conceptual Phase

The first step in any project is to gain organization support for the proposal. This phase concentrates on establishing the initial foundation for successful cellular implementation.

Organizational Vision and Commitment   Effective project management has two important prerequisites: organizational vision and commitment. For manufacturing cells to become a part of the overall corporate vision, organizations must recognize and understand the benefits of cellular technology. Company leaders must themselves be convinced of the value and benefit of manufacturing cells. They must then develop a vision of their production facilities operating under cellular concepts and a realistic, cost-effective means of transitioning to them. To make manufacturing cells a reality, these leaders must share their vision with the entire population of the organization.

Commitment to manufacturing cells is derived from the organization's mission, objectives, goals and strategies. If an organization's direction is not conducive to manufacturing cells, the implementation team will simply force something onto the organization that does not fit—the square peg in the round hole. An effective way to gain commitment is by providing to upper management ample supporting data. The company must thoroughly review CM sources (Burbidge, 1975; Joines et al., 1995; Wemmerlow and Hyer, 1989) and develop a definition of cellular technology that is easily understood by its employees. It must take time to understand the advantages and possible disadvantages of the technology and critically analyze forecasted sales volumes to ensure that cells will sufficiently benefit the organization (as opposed to a job shop setup or permanent production flow lines). Information, not normally shared with workers, must be presented to the entire organization to convey the new strategic direction. This may include such information as market position, trends in market share loss, and comparative studies of labor and cost structure with competitors (Wellins et al., 1994). It may also benefit the organization to benchmark other companies that have adopted CM. Seeing firsthand another company's progress and improvement through the use of manufacturing cells is an effective method to instill the vision of cellular technology. Finally, a feasibility study needs to be presented to upper management summarizing the supporting data. Besides a description of cellular technology and its advantages, this study should contain estimates on the costs to implement cells, the magnitude of enhanced performance they will bring to the organization, and a time frame to achieve them.

With cellular technology as the organizational future, top management will recognize that the successful management of implementation projects will translate into the organization accomplishing its mission in an effective and efficient manner.

Identifying Members of the Implementation Team—A Systems Perspective Once vision and organizational commitment are in place, a company can begin to transition itself to an organization that uses manufacturing cells. The next step will be the selection of a team to execute this objective. Taking an organizational systems perspective allows the company to locate the personnel vital to the success of the implementation (a discussion on the formation of the team and selection of a technological champion is reserved for later). The key to this endeavor is to consider what manufacturing cells are: a production philosophy. Production is simply the interaction of humans, machines, materials, and information. Therefore, the functional areas of the organization that are involved with these components of production are the areas requiring representation within the implementation team. Table 3 provides a summary of the primary activities and their respective functional areas for the implementation of manufacturing cells. This should be adjusted to the specific functions of each team's organization.

3.2. Planning and Construction Phases

The following section describes steps to follow and issues to consider when developing a complete plan of action for cellular implementation.

Form the Implementation Team Every team needs a leader, a champion who is willing to take the project to its fruition where it meets its cost, schedule, and performance parameters. Management must not haphazardly appoint a person to lead the implementation of cells. Management needs to recognize that the manufacturing cell(s) to be implemented will be a building block of the organization; its success is

Table 3. Activities and related functional areas for cellular manufacturing

Component of Production	Activities Related to Component of Production	Functional Areas Associated with the Component
Human	Employing personnel	Production
	Assessing workers	Union
	Development and enforcement of shop rules and responsibilities	Industrial engineering
		Human resources/personnel
	Rotating jobs	Safety and medical
	Compensating personnel	Payroll and benefits
	Scheduling workers	
	Measuring and controlling performance	
	Education and training	
	Keeping workers safe	
	Handling grievances	
	Participating in union activities	
Machines	Forming, machining, joining, heat treating, assembling, finishing, communicating, controlling, transporting, inspecting, etc.	Manufacturing engineering
		Industrial engineering
		Production
		Information support
	Setting up and changing over	Plant and equipment engineering
	Tooling and fixturing	Computer engineering
	Repairing and maintaining	Safety
	Rearranging machines	Machine maintenance
	Programming	Financial/accounting
	Complying to safety regulations	Purchasing
	Financing and purchasing equipment	Tool and fixture design
	Designing, fabricating, and purchasing tools, jigs, and fixtures	Quality control
	Managing and keeping track of	
	Measuring standard times	
Material	Routing throughout shop	Manufacturing engineering
	Operating upon	Industrial engineering
	Scheduling to workstations	Material and production control
	Handling methods: container, conveyors	Production
		Design engineering
	Quality responsibilities including sampling designs, testing equipment	Tool and fixture design
		Quality control
	Procuring materials	Purchasing
	Storing	Sales and marketing
	Selling and shipping of products	Shipping
	Costing of material	
Information	System, reporting capabilities, flexibility, availability, data requirements	All of the above

vital to the attainment of the organization's mission. The champion's enthusiasm, intellectual interest, and accountability will provide the base of support for the implementation of the manufacturing cells (Chung, 1996). This leader must be a person whose opinions and practices are respected by the workers and who also has enough power to make things happen when the team comes to decision points or roadblocks. In summary, this person must be at a decision-making position in the traditional hierarchy and a leadership position in the matrix organization.

The implementation team will consist of members from all functional areas within the organization. As noted in Table 3, this includes, but is not limited to, members from plant engineering, design engineering, manufacturing engineering, industrial engineering, information systems support, production and inventory control, human resources, marketing, and finance. When choosing team members, management should look for people who have worked in the company long enough to know the business but who are not set in their ways. These people will pave the way to success for the project and should therefore be drawn from informal or formal leaders within the organization.

Bear in mind that a project is temporary and the team implementing the cells is not permanently assigned to the project but is shared from some functional area within the organization. All members will have obligations to both the implementation project and to their home unit. The team leader or project manager must interface with all of the functional unit managers and arrive at a schedule for utilization of the team members' services that is agreeable to both parties. In some instances, it may be necessary to solicit the counsel of the organization's projects manager or a high-level organizational manager.

Develop and Educate the Team   The team must be trained in the concepts of project management, teamwork, CM, and the terminology that accompanies this technology. Training materials should be developed in accordance with the organizational mission and strategies, and through a collaborative effort including contributors from Human Resources, Manufacturing, Quality, Production, and Inventory Control, as well as other functional support areas. The training effort may also incorporate the use of outside trainers and consultants to effectively communicate this information to the organization's employees.

The first questions that surface are usually along the lines of: "Why teams?" "Why cellular manufacturing?" "Why does our organization want to change?" "Why am I going to be part of the change?" These questions can be easily answered if the company has adequately researched and prepared for cellular technology. If these questions cannot be answered at this time, the team should revisit the previous conceptual stage of the implementation project before proceeding with the education of its cellular implementation team.

Although the team is primarily comprised of people from within the organization, these team members typically come from focused functional areas. Therefore, when the team first comes together, it is important for all members to introduce themselves. They should explain their background and current duties within the company (i.e., accounting, scheduling), their understanding of the project to date, and feel free to request any information they feel they need to know regarding the implementation project. This interaction among the team will help remove some the communication *walls* between functions and develop a cohesive and operational implementation team.

The team should then receive basic training in some project management concepts such as developing a mission, working in teams, group interaction, handling conflict, training others, communicating with supervisors, and valuing differences (Cleland, 1996; Wellins et al., 1994; Wilson and Linscott, 1996). The team will also benefit from being taught the background, benefits, and terminology of CM along with how to develop policies, procedures, and performance measures and the basics of statistical process control and scheduling/production and inventory control. In addition, the team should be instructed on developing administrative procedures necessary to run a cell (i.e., production logs, procurement forms, shipping manifests). For example, a manufacturing organization that successfully used cells found the development of a cellular implementation manual that included all paperwork that the cell workers would need to complete, including a blank sample of each form, the purpose it served, steps for completion, who it should be sent to, and a completed example form to be most useful. The employees that will work in the cell(s) should eventually be provided appropriate technical cross training on the equipment within the cell. This idea is discussed more thoroughly in the section on selection and compensation for workers.

A trainer's primary responsibility throughout the training it to act as a facilitator. Accordingly, it is this person's duty to help the workers realize the benefits of implementing manufacturing cells through exercises. One such example (see Fig. 4) is to have the team compute the effects of a queue on cycle time and work in process (Billo, 1993). In order to do this, concepts such as cycle time, WIP inventory, operations, moves, queues, and storage should first be defined and communicated to the team. Then, beginning with part A, the team would trace each part as it traveled through a manufacturing process under a variety of scenarios. Figure 4 depicts the flow of a product through three machines with a processing time of 1 minute at each station. The first scenario represents traditional push production where parts move from machine to machine in batches of 50. The total cycle time to push part A through the entire system is 153 minutes and the total WIP at that time will be 150 parts. The 153 minutes is comprised of 1 minute processing plus 50 minutes waiting at each of the three stations. The second scenario represents a more flexible form of push production where the batch size has been reduced to 5. The total cycle time to push part A through the entire system is 18 minutes (1 minute processing plus 5 minutes waiting at each of the three stations), and the total WIP at that time will be 15 parts. The third scenario represents a near pull production where one part is pushed, or travels, to the next operation at a time. The total cycle time for part A to pass through the entire system is 6 minutes and the total WIP is 3 parts.

From this exercise, workers will learn the basis of both push manufacturing and pull manufacturing. It would then be beneficial for them to list the advantages (i.e., decreased inventory, shorter cycle times) and disadvantages (i.e., no buffer to offset breakdown or worker absenteeism) of CM to their current operating philosophy. This will provide a clearer picture of CM and plant the seeds for a new operating paradigm.

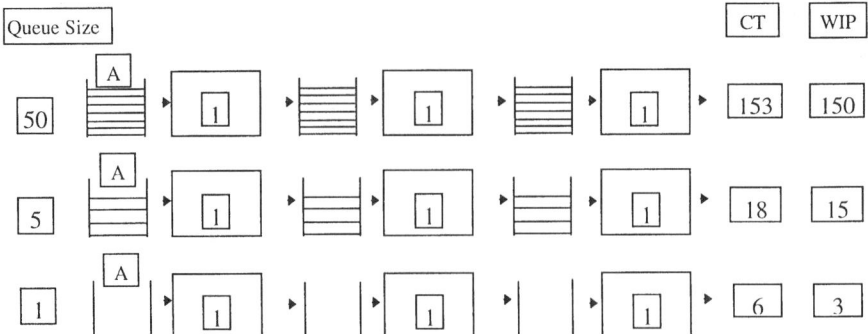

Figure 4. Example: Effects of queue time on Cycle Time (CT) and Work-in-Process (WIP). (From Billo, 1993)

Project Information System   Early in the implementation project, the team's estimates for task definitions, budgets, and schedule will contain much variance due to the lack of information at that time. As the implementation progresses, it is important to record and analyze the effort expended toward meeting the many tasks. Thus, the estimates can be reviewed and updated using the new information. To facilitate this control process, the implementation team should use a computerized project information management system.

The objective of an information system is "to provide the basis to plan, monitor, evaluate, and to illustrate the interrelationships among cost, schedule, and technical performance for the entire project and for the strategic direction of the organization" (Cleland, 1994, p. 268). Many off-the-shelf project management systems are natural extensions of the WBS and LRC. Through this system, information on the work packages (what, when, who, where, how, and how much) and the effort exerted toward completing them can be stored, retrieved, and reported in a timely and efficient manner. The team can use this information to assess their performance toward reaching the implementation's goals and milestones and make appropriate changes if necessary. To assist the team in other management functions, members should investigate interfacing the project's information management system with other systems, such as Payroll (to ensure team members are credited and paid for their work) and Purchasing (to ensure purchases are documented and conducted according to organizational policies).

Establish Objectives and Work Packages   Once the implementation team has been assembled and adequately trained, the planning process for implementation begins. Because this chapter only proposes a framework for implementing manufacturing cells, the team should conduct a brainstorming session, and perhaps employ the use of outside consultants, to arrive at the specific needs of its organization. The team must clearly define the desired cost, schedule, and performance measures of the cells and through an extensive WBS, chart out the required steps to meet them. Many of these high-level steps are discussed later in this section. Table 4 provides a list of sample measures that can be used to evaluate performance for the cellular implementation project and sample ranges that have been realized by U.S. manufacturing firms (Wemmerlov and Hyer, 1989). As described in the previous section, the team will then design an LRC to designate the amount of responsibility, authority, and accountability each team member has in regards to the various work package derived from the WBS.

The transition type is one of the most important factors to be considered at this time. It is generally not feasible for an established facility to transition completely over to manufacturing cells in the relatively short period of time that is needed to maintain production. Therefore, most companies begin with a pilot cell implementation before implementing the rest if the designed cells. This permits documentation of all problems encountered during the pilot cell implementation and provides for a smoother phase in of the other cells. (See discussion on constructing the cells for an enhanced explanation.)

Conduct a Part Population Analysis

Determine Which Parts to Group   When a company chooses to operate with manufacturing cells, the implementation team must decide where to begin grouping parts. According to observations made by Tompkins and White (1984), on average 85% of the production volume can most often be attributed to 15% of the products manufactured in a facility. They conclude that the facilities plan should consist of a cellular or flow production line for the 15% of high-volume items and a job shop arrangement for the remaining 85% of the product mix (Tompkins and White, 1984). These claims are substantiated by the fact that a cellular or flow line layout is most suitable for high-volume and low-variety products while a job shop layout is best for high-variety and low-volume products (Shaffer and Billo, 1994).

Based on the literature and our experience in implementing cells, companies are typically permitted to group only a portion of their parts due to constraints (i.e., machine justification, part similarities, etc.). We recommend that they group from 10 to 20%, or primarily their high-volume parts as determined by Pareto analysis. The benefits provided here are threefold: more balanced flows inside the cells, reduced setup hours per part, and most importantly a reasonable number of parts for the team to develop into families. For companies that choose to group all of their parts into

Table 4. Candidate performance measures for manufacturing cells

Performance Measure	Average Improvement %	Observed Range %
Decrease in throughput time	45.6	5–90
Decrease in WIP inventory	41.4	8–80
Decrease in materials handling	39.3	10–83
Decrease in setup time	32	2–95
Increase in quality	29.6	5–90

From Wemmerlov and Hyer (1989).

cells, approximately 10–20% will usually be assigned to "flow line cells" and 80–90% will be assigned to "job shop cells" (Shaffer and Billo, 1994).

Select Method for Grouping Parts   Once the implementation team determines which parts to group, it must choose a part grouping method. Current methods can be grouped into categories such as classification and coding, visual techniques, machine–component group analysis, similarity coefficients, mathematical programming, graphical approaches, artificial intelligence (genetic algorithms, neural networks, tabu search, simulated annealing, fuzzy logic), and heuristics (Joines et al., 1995; Singh, 1993). Both Singh (1993) and Joines et al. (1995) provide complete reviews of the cell formation literature.

To group parts or products into families, the appropriate information needs to be collected. This includes gathering data such as the machines the part runs across, the tooling and fixture it utilizes, the processing time it requires at each operation, and the setup time required. Some of this data does not directly apply to part grouping but will be used for later leveling, balancing, and scheduling activities and should therefore be gathered at this time.

Conduct Necessary Part Reengineering   Poor grouping of parts and the existence of exceptional parts are common problems encountered in forming manufacturing cells. In the first situation, the part grouping method forms an extremely high number of part families (Billo et al., 1994a). Often, the resulting number of part families cannot be efficiently transformed into manufacturing cells due to high equipment costs and low machine utilization. Billo et al. (1994a) researched this problem and did not attribute it to poor performance of the part family formation algorithm, but rather to inconsistent designs and process plans of the workpieces considered. Their research addresses part families produced in cells comprised primarily of Computer Numerical Control (CNC) machining centers and turning centers where both the machine tool and the tooling held in the tool magazine are considered. Five heuristic rules are presented for reengineering process plans in order to achieve better part family groupings. These rules are summarized in Table 5.

An exceptional part, however, cannot be reasonably assigned to an existing part family due to lack of similarities in machining requirements. This will result in the need to consider a make or buy decision. If the process plans for the part can be reengineered to permit acceptable entrance into an existing part family, then it may be best to produce the part in-house. If reengineering the process plans is not successful,

Table 5. Heuristics for reengineering process plans

1. Combine wholly contained subset families.
2. Combine families that share common tooling.
3. Standardize or reduce tooling where possible.
4. Employ pause machining and change a tool.
5. Create "children" part families.

From Billo et al. (1994a).

then the company should consider outsourcing the part to another manufacturer after a cost–benefit analysis. Once the reengineering considerations are complete, the process will result in the part families.

Design Manufacturing Cells

Determine Final Number of Cells   Due to constraints imposed on an organization, the final cell or configuration of cells may not exactly reflect what the grouping method and part reengineeing processes recommended as part families. For instance, a grouping algorithm may result in 12 cells that would require large capital expenditures and result in low machine utilization. The company may only have the means to support a 6-cell configuration, raising the question of which parts belong in the cells and which potentially belong back in the job shop portion of the operation.

Billo et al. (1994b, 1996) address the transition from a job shop to a cellular layout with the use of a groupability index as the basis of comparison for cellular solutions. Their model arrives at solutions via a genetic algorithm with a penalty function to deter the solution of a single part in a single cell where the groupability index value would be 1 (Billo et al., 1994b, 1996). Other researchers (Askin and Subramanian, 1987; Boucher and Muckstadt, 1985; Needy et al., 1998) propose methods that arrive at solutions based on an economic decision. Askin and Subramanian (1987) provide an extensive, data driven model for cost analysis. Needy et al. (1998) extend the work of Billo et al. (1994b, 1996) by examining each of the alternative configurations in terms of the costs to implement and subsequently operate the cellular system on the factory floor. Their model examines capital investment, material handling, and setup costs directly, and also considers the scheduling trends, cycle time, WIP, labor productivity, and quality costs (Needy et al., 1998).

Equipment Selection, Process Flow, and Cell Capacity   Each cell will consist of the machines required by the part family assigned to that cell. Regardless of the method the team selects to arrive at the number of cells, members must eventually consider the types, costs, and quantity of equipment required in their cell(s). This is driven by the flow of the parts through the cell and the machines and equipment necessary to produce them in a cellular fashion. The material transportation and communication signals are reserved for a later section.

The part process routings easily provides the team with a listing of the machines that are necessary. Research into the organization's machine database, along with vendor brochures, can provide the team relative costs for these machines. However, the question of whether each cell is capable of producing the demanded quantity of parts for its respective part family arises (Shaffer and Billo, 1994). This determines the quantity of equipment. If the capacity of the cell does not meet the demand for the parts, then the implementation team must replicate the cell in order to meet the demand. This analysis is based on the bottleneck operation, defined here as the operation with the highest cycle time, in each cell. In cell replication, the section of the original cell that produces the *overdemanded* parts is copied and the replicated subcells function independently (Shaffer and Billo, 1994). The operations comprising these cells may differ slightly from the original cell based on the process flow of the individual parts assigned to the cell.

Design/Select the Tooling, Jigs, and Fixtures   In a CM environment, an obvious goal is to reduce or eliminate non-value-added time involved in the production of goods. Setup time accounts for one of the largest proportions of non-value-added time in most production facilities (Black, 1991). Black (1991) defines total setup time as the time from the last good part from the previous setup to the first acceptable part from the new setup. Setup time most often includes changing the fixture and tooling on the machine tool, and then an iterative process of running sample parts, inspecting those parts, and making adjustments to the machine until satisfactory parts are obtained (Black, 1991).

A primary benefit of grouping parts into part families is that one initial setup is performed for the whole family, with small adjustments in between each new *batch* of different parts. This benefit may only be realized when a fixture is designed for the family that is flexible enough to accommodate each part within that family. Therefore, the implementation team should pay as much attention to designing a *family fixture* as it does to grouping the parts into a family. The team representatives from product design engineering and tool design engineering will shoulder the main responsibility for this task. They may also consider sequence-dependent setups where part families utilizing the most common number of tools are scheduled consecutively.

Determine Material Handling Method, Communication Signals   A pull system of material handling is a method that forms the basis for Just-In-Time (JIT) manufacturing. In a pull system the buffer sizes are controlled and minimized. In this type of production, "successive operations pull material from the buffer that exists between it and the previous operation" (Shaffer, 1992, p. 84). Also, a kanban square often exists between operations and acts as a signaling system. When the square is empty, it triggers the supply operation to start producing. In this case, the kanban square acts as both the buffer and the signal (Black, 1991). Alternative signaling methods includes the use of colored lights near each station and emergency lights for a line shut down.

In an ideal implementation, an operation would complete one part and then pass it on to the next operation. Often, small parts with extremely short cycle times are produced in a cell. In this case, there is a tradeoff between WIP and production rates, and a standard container size is often developed to hold a specified number of the parts (Shaffer, 1992). The implementation team should work directly with the production and inventory department of their organization to determine an optimum and constant container size to be used throughout the manufacturing facility. The reason for the constant container size is so that different containers do not have to be purchased for different parts.

Leveling, Balancing, and Synchronizing   Leveling is defined as "making the amounts of material on a production floor equal over time" (Black, 1991). Black (1991) refers to leveling as the process of planning and executing an even production schedule (see the section on scheduling). Balancing refers to the method of equalizing the work load at all the operations in a dependent process flow so as to minimize the idle time between operations (Black, 1991; Moodie, 1982). The basis of these two activities is to smooth demand spikes by regulating production output. Synchronization refers to the technique of "timing the flow of material between cells or other operations" (Black, 1991). Even after the amount of material and the times have been leveled and balanced, respectively, needless storage of WIP can occur between unsynchronized operations. Therefore, leveling, balancing, and synchronization must be carried out to reduce non-value-added times, WIP, and throughput times.

Develop the Shop Floor Layout   The machines are assigned to cells based on the part family formation method employed in the part population analysis. Now that the implementation team knows which machines are assigned to each cell, the process flow of the parts within the cell, and the material handling methods, members can begin designing the most efficient layout for the shop floor.

Floor space is not the only important factor to be considered in this planning. For instance, the structure of the building should be taken into account. In a recent implementation of manufacturing cells at a local company, a colleague facilitated a team design of the shop floor layout. The design was approved by the plant manager and the team implemented the layout. The only problem was that the building had previously been expanded and many of the precision cutting machines were placed over the expansion joints on the floor. This resulted in the inability of the machines to hold tolerances and necessitated a new shop floor layout. Other considerations should include the proximity of overhead lighting, power sources, and ventilation to the cells. With environmental issues gaining notoriety, the proliferation of hazardous materials such as plating tanks and painting booths must be addressed. In addition, an assembly cell should be placed as close to the packaging or shipping area as possible. The team must also address issues of placing assembly areas in proximity to machining stations (e.g., the effect of coolant mist on precision tools).

Selection and Compensation of Workers   Before implementing manufacturing cells, management must discuss and evaluate the impact of this technology on compensation plans and related activities such as performance appraisal and employee selection (Gerhart and Bretz, 1994). In the case of manufacturing cells, traditional time standards appear to have been pushed aside by other performance measurement methods (for a discussion on the concept of a family of performance measures see earlier discussed). The emphasis is no longer placed on the worker who can turn out the most parts in the shortest amount of time. It has shifted to team-oriented workers who can operate accordingly in a leveled, balanced, and synchronized environment. Within cells, workers must pay close attention to both upstream and downstream operations and also possess a breadth of both technical and nontechnical skills and knowledge. Therefore, specific considerations must be made by an organization when deciding on production standards for cells and on the compensation system for the workers. Also, government agencies, such as the Occupational Safety and Health Association (OSHA), have brought other work measurement considerations to the forefront when determining requirements for workers. These ergonomic issues include metabolic rates, neuromuscular functions, and psychological functions (Chaffin and Anderson, 1991).

The first step toward worker selection in a cellular system is the determination of skill requirements for each of the cells. There will be technical skills requirements such as the ability to read a blueprint or setup a part on a machine. There will also be nontechnical or human skill requirements such as communication with other cell team members or decision-making ability (Askin and Huang, 1997; Min and Shin, 1993; Warner, 1998). It is crucial that the organization take this time to clearly identify and define each of the skills necessary for the cells. Prominent skills definition sources include the Secretary's Committee on Achieving Necessary Skills (SCANS) Report (1992) which lists and defines 37 skills for world-class organizations and the National Tooling and Machining Association's Duties and Standards for Machining Skills (1995).

The next step requires skills assessments of the current workers. Traditionally, tests of cognitive ability were widely used to obtain a measure of a person's general intelligence. This provided a number with which to judge an employee. More recently, organizations have been turning toward job analysis, work samples, combinations of self, supervisor, and peer assessments with related forced distributions, and standard-ized trade exams (e.g., National Occupational Competency Testing Institute Precision Machinist Exam) to obtain more skill- and job-related measures rather than an overall assessment of general intelligence (Warner, 1998). Once the appropriate assessment information has been obtained, an organization can identify skill gaps, match workers to cells based on current skill levels, develop job rotation schedules, define realistic performance goals, and develop training materials and strategies to best meet the specific needs of their employees while achieving the biggest impact on the company's bottom line (Askin and Huang, 1997; Warner, 1998).

Researchers have argued that compensation decisions are of key strategic im-portance within an organization (Gerhart and Milkovich, 1992). Gerhart and Bretz (1994) cite two important attributes of compensation decisions. First, employee compensation and benefits typically account for a large portion of an organization's total operating costs. Next, compensation decisions can have important consequences for key outcomes such as "job satisfaction, retention, performance, flexibility, com-mitment, cooperation, skill acquisition, and ultimately organizational performance" (Gerhart and Bretz, 1994, p. 82).

A large amount of research has been carried out to investigate the best way to compensate workers in a CM environment. Gerhart and Bretz (1992) have concluded that no single pay program has the ability to achieve the entire set of complex and competing objectives to be met in this environment (Gerhart and Bretz, 1992). Therefore, in establishing individual differences in pay, some combination of plans (e.g., team awards, merit pay, skill-based pay, gain sharing and profit sharing) should be incorporated to attempt to meet the multiple objectives of the organization (Gerhart and Bretz, 1994). It is also important to continually evaluate whether the compensation plan supports the actions and behaviors that are required for a successful implementation of manufacturing cells.

Develop Cell Operating Rules   Whatever environment a team functions under, it is important that all members understand the rules by which to operate. Though the purpose of cellular technology is to be adaptive to various manufacturing scenarios, a core set of operating rules is necessary to ward off complete chaos. Irani (1997) lists a number of questions that teams must answer to operate a manufacturing cell that include the following:

- How are jobs assigned and instructions imparted to the operator?
- Are procedures such that an operator is ever without a job to do?
- Are materials, drawings, and tools stored orderly and secured? Who is in charge of replenishing them?
- What delays are incurred in issuing jobs, checking work, and recording data?
- Who sets up the machines and verifies first piece production?
- Who maintains the machines?
- How is responsibility transferred from shift to shift?

The purpose of these questions is to trigger the team to develop a set of operating procedures that will let members achieve the performance goals of the cell. With a team approach to this development, the experiences of many people can be captured to arrive at realistic and comprehensive rules. This will allow the team members to understand in advance the operating requirements of the cell to allow them to run it efficiently and effectively and remove some of the uncertainties that lead to scrap, rework, delays, and machine down time.

Quality Issues   Quality control is complementary to manufacturing cells. One of the tenets of CM is to instill quality into the production process and allow the cell team to manage and control its own quality. The workers control the quality within their cell and operate under the *make one–check one–pass one on* rule (Black, 1991). In this way, no defective parts get passed on to the next operation. Also, causes for defects are investigated as soon as a defect is discovered and then promptly corrected.

To meet this goal, the team must address key quality issues during the design of manufacturing cells. Team members will have to develop inspection processes to allow them to produce quality parts to the specifications required. This will involve instruction and training in the development and use of statistical process control techniques (e.g., $\bar{X}$ and $\bar{R}$ charts). Companies that still rely on acceptance sampling should understand that defects really are not acceptable at any level. However, if they continue to assess quality in this manner, the current product sampling plans should be taken into account. For instance, if a batch of 100 parts is currently produced and a random sample of 10 parts are inspected, then the sampling plan for a new container size of 10 parts will need to be revised. The inspection equipment may also require modification to accommodate the pull manufacturing philosophy.

To incorporate quality into the cell, the team will require inspection equipment (e.g., micrometers, calipers, surface plates, height gauges, comparators). Supplying each cell with these tools can quickly become expensive; thus, the team must incorporate the costs of inspection into its procedures. Also, members should be thorough trained on how to read and interpret all of the inspection tools.

It is often difficult to predict and quantify the quality benefits of implementing manufacturing cells. Shaffer (1992) documented the following benefits that occurred during one implementation:

1. Defective material no longer accumulates in queues. The cells find defective material as it is produced, and the problem is immediately resolved. Complete (100%) testing of *piles* of material has been eliminated.
2. Communication has substantially improved within the cells, and this results in a strong problem-solving environment.
3. All the workers in the cell gain a feeling of accomplishment when a quality product is produced.
4. Material damage that resulted from large amounts of material handling has been greatly reduced with the cells. The loading and unloading of material transports no longer occurs.

Manufacturing Cell's Information and Control System   The use of computers has exploded over the past 20 years. Computerized systems provide companies rapid consolidation of data and information that assists them in running, controlling, and growing their organizations. A cellular implementation team should plan to utilize a computerized information and control system for the manufacturing cell. Team members could ideally design a system in-house or contract the services of an outside computer software agency. A detailed plan to develop and implement an information system is beyond the realm of this chapter. However, regardless of who develops the system, the team should ensure that the three major capabilities are thoroughly investigated and designed: reporting information, inputting data, and linking to existing organizational systems.

The following are some high-level steps the team should take to develop an information and control system. First, the team must list all of the information members need to operate and control the manufacturing cell (e.g., job schedules, standard times, employee assignments, employee activities, tool databases). They should then assess why (e.g., to meet schedules on time, meet quality standards, continuous improvement) and how often they need this information (e.g., real-time, daily, weekly, monthly).

Once the team establishes the information requirements, members should investigate various means and responsibilities of inputting the data. Some data may be provided through support functions (e.g., engineering may describe products and machines). If data is to be collected in the cell, the team should investigate various means of performing this function. Members may want to install bar code printers and readers, touch computer screens, and other technology to facilitate fast and accurate data entry. The team should be prepared to justify the purchase and implementation of such technologies.

Finally, it is highly probable that other computerized systems already exist within the organization to which the cell's information system needs to interface. These include scheduling and an MRP system, time and attendance, financial and costing, process routings and standard times, production performance, purchasing, scrap and quality control, and asset management. The cell's information and control system cannot be isolated from the rest of the organization. It is vital that the links to other pertinent information systems be investigated and made.

Scheduling Parts and Workers Example   To illustrate the need for system integration, consider scheduling for parts and workers. With manufacturing cells, an organization has to successively sequence a family's parts in order to acquire the greatest advantage from setup similarity. Companies changing over from a job shop to cells will require new production planning and control techniques. If a company has grouped only about 10–20 % of its parts into families, an issue is raised regarding integrated scheduling. The company must employ a scheduling philosophy that will allow it to realize the advantages of cells while simultaneously maintaining its current production performance for the other 80% of parts.

The literature provides many group scheduling algorithms that focus on determining the sequence in which a company should process the jobs within a cell. These are generally based on optimizing some measure of effectiveness, such as machine utilization, product throughput, or operating costs. For example, in recent cell implementation, the authors' decided to download information from an existing MRP II system and upload into a scheduling module that was based on an existing algorithm.

However, scheduling is not just confined to machines, material, and tooling for a job order. Organizations must also consider scheduling their human resources. This task is complicated due to the breadth of skills and knowledge that is required for an operator within a cell. Because a worker is no longer classified under one job description (he or she must be capable of performing each operation with a cell), it is the duty of management to maintain current records on the capabilities of employees. This information will prove to be extremely useful in the instances of absenteeism because worker substitution can occur and production will not be halted.

Determine Cell Implementation Strategy   Once the plans for manufacturing cells are established, the team then needs to implement them. In establishing the objectives for cells, the team should have considered the phase-in strategy to be used. A company's conversion strategy may take the form of (adapted from Badiru and Schlegel, 1994):

Direct Conversion   This type of implementation falls under two conditions: (1) The old facility is removed totally and is replaced by the new manufacturing cells or (2) a totally new plant is being constructed with the use of cells.

Cellular

Conventional

Phased Conversion   Modules of new cells are progressively implemented one at a time using either direct conversion (of a particular area) or parallel conversion. Parallel conversion is when the company has the two operating methodologies, conventional and cellular, running concurrently until the company deems the cells satisfactory.

Cellular

Conventional

Pilot Conversion   A cell is fully implemented on a pilot basis in a selected area within the company. Upon full implementation of the pilot cell, the company may elect to continue implementing based on one of the above strategies.

Cell Pilot

Conventional

We do not recommend the direct conversion unless a company is building a new facility *and* is committed to manufacturing cellular technology. It is not often feasible to shut down production for the time it would take to convert the entire facility at once. Besides the obvious cost, a company loses the opportunity to learn as it implements cells. As a company transitions its existing facility to manufacturing cells, the implementation team will learn a great deal more about cellular technology— above what they have previously researched and planned. Each company has specific nuances that even the best formed plans may miss. To attain this chance to learn and grow, we recommend conducting a phased or pilot conversion when changing over an existing facility to manufacturing cells. By breaking the total plant implementation of cells into smaller projects through phased or pilot conversion, better control is gained over the transition, and the company can pass knowledge gained from each project onto the next area to be phased in.

Present Plan to Executive Management and Secure Funding   Before continuing any further, the team needs to present its detailed plan to implement the cell(s) to executive management to gain support and secure funding for the project. The project's cost, time, and performance objectives and plan of action to achieve them must be clear, concise, and realistic. Support documents (e.g., letters, memos) from functional unit managers for use of their employees and resources will be a plus for it demonstrates to upper management that cooperation exists and that potential sources for conflict have been addressed and resolved. After the presentation, the team leader must receive official approval for the implementation budget and use of company resources and have the project recorded into the official matrix organization.

Constructing the Cells   Once the team establishes a detailed cell design and implementation plan, the budget has been approved, and the overall direction of the team is supported by the organization, members can begin to construct the physical cells. Many of the project management functions at this stage center around ensuring that the resources, both physical and human, are available when they are needed. The team should develop as part of its administrative functions, procedures to procure all materials, expense, or capital items. During this time, contracts between outside vendors or consultants and the team member(s) authorized to sign them must be secured.

Building the cells requires first the clearing of space for the cells and installing all physical components such as machines, tooling, support equipment, computers, and data entry and communication devices. Once in place, all of the cell workers must be assigned to cells and trained in their new duties, procedures, and operating rules. Systems are then put into operation to schedule parts and job sequences.

It is important at this stage to conduct thorough testing to ensure that the cell is functioning appropriately. All equipment must be tested and debugged, worker performance tracked and measured, and control meetings scheduled to obtain feedback from cell operators. The test procedures should be developed BEFORE building the cells during the implementation planning phase. This allows the team to construct measurable targets to evaluate the cell's performance and know when the cell is successfully operating.

3.3. Closing Out the Implementation Project

A company must establish a definite end date to the transition project. This is when the project team takes all the appropriate actions to officially terminate the project. Team members will hand over control of the newly installed cells to those who will run them, along with the proper operating policies and procedures, control structure, and system maintenance requirements. All waste from building supplies must be removed from the area. The company must disband the team, providing them compensation for their work on the project and assigning them to their new positions (on another implementation team, back in their functional areas, or working in the new cell).

With projects as the building blocks for the organization, perhaps the most important function in closing out the implementation of cells is the post-project analysis. Often ignored, a post-project analysis provides the company an opportunity to learn, apply its new knowledge, and grow. The company can take appropriate actions based on this analysis to build on the strengths and resolve any weaknesses. For example, the team may have had an easy time in grouping parts because of readily available part information and sound algorithms. However, team members may have experienced difficulty in translating part groups into workable cells due to poor processes, standards, tooling, and information acquisition. This information needs to be documented and shared with the rest of the organization. At the very least from a post-project analysis, a leader may be recognized from the team who could be an ideal candidate to either facilitate the new operating cells or lead the next implementation team.

4. SUMMARY

Manufacturing cells can provide a company many benefits. Well-designed cells can yield a company exponential improvements in inventory, product throughput, worker satisfaction, and best of all profitability. To obtain these benefits, a company must make a successful transition from its current method of manufacturing to that of cellular production. Operating a facility under cellular technology is a tremendous challenge unto itself. A poor transition from the current mode of operation only adds to this challenge. The key to success is for the company to clearly identify three areas: where it is now, where it wants to be, and how it is going to get there from here.

We have proposed that companies utilize project management principles to lead them in this transition. Project management provides a framework for companies to conceive, plan, organize, build, and operate high-powered manufacturing cells. It does not "guarantee" success but is designed to let this challenge of implementing and operating under manufacturing cells be handled by the organization's greatest resource: its people. Project management involves a cross function of personnel along every step of the transition, personnel who will be impacted by the change to cells and personnel the company wants committed to its success. The structure of the project we have proposed provides companies proven techniques that better ensure them success while simultaneously providing them an opportunity to recognize and avoid the problems that have plagued the implementation efforts of others.

REFERENCES

Adams, J. R., and Barndt, S.E. (1988). "Behavioral Implications of the Project Life Cycle," in *Project Management Handbook*, 2nd ed., D. I. Cleland and W. R. King (Eds.), Van Nostrand Reinhold, New York.

Archibald, R. D. (1992). *Managing High-Technology Programs and Projects*, 2nd ed., Wiley, New York.

Askin, R., and Subramanian, S. (1987). A cost-based heuristic for group technology configuration, *International Journal of Production Research,* **25**(1), 101–113.

Askin, R., and Huang, Y. (1997). Employee training and assignment for facility reconfiguration, *Sixth Industrial Engineering Research Conference Proceedings*, Miami, FL, pp. 426–431.

Badiru, A. B., and Schlegel, R. E. (1994). "Project Management in Computer-Integrated Manufacturing Implementation," in *Organizations and Management of Advanced Manufacturing*, W. Karwowski and G. Salvendy (Eds.), Wiley, New York.

Billo, R. E. (1993). *Introduction to Manufacturing Systems*, Course Lecture Notes, University of Pittsburgh. Pittsburgh, PA.

Billo, R. E., Bidanda, B., and Kharbanda, P. (1994a). Re-engineering process plans for effective manufacturing cell formation, *International Journal of Manufacturing Systems Design*, **1**(3), 217–229.

Billo, R. E., Tate, D., and Bidanda, B. (1994b). "Comparison of a Genetic Algorithm and Cluster Analysis for the Cell Formation Problem," *Third Industrial Engineering Research Conference Proceedings*, Atlanta, pp. 543–548.

Billo, R. E., Bidanda, B., and Tate, D. (1996). A genetic cluster algorithm for the machine-component grouping problem, *Journal of Intelligent Manufacturing*, **7**, 229–241.

Black, J. T. (1991). *The Design of the Factory With a Future*, McGraw-Hill, New York.

Boucher, T. O., and Muckstadt, J. A. (1985). Cost estimating methods for evaluating the conversion from a functional manufacturing layout to group technology, *IIE Transactions*, **17**(3), 268–276.

Burbidge, J. L. (1971). Production flow analysis, *Production Engineer*, **50**(4–5), 139–152.

Burbidge, J. L. (1975). *The Introduction of Group Technology*, Halster Press and Wiley, New York.

Chaffin, D. B., and Anderson, G. B. J. (1991). *Occupational Biomechanics*, Wiley, New York.

Chung, C. (1996). Human issues influencing the successful implementation of advanced manufacturing technologies, *Journal of Engineering and Technology Management*, **13**, 283–299.

Cleland, D. I. (1994). *Project Management: Strategic Design and Implementation*, 2nd ed., McGraw-Hill, New York.

Cleland, D. I. (1996). *Strategic Management of Teams*, Wiley, New York.

Cleland, D. I., and King, W. R. (1988). "Linear Responsibility Charts in Project Management," in *Project Management Handbook*, 2nd ed., D.I. Cleland and W.R. King (Eds.), Van Nostrand Reinhold, New York.

Collett, S., and Spicer, R. J. (1995). Improving productivity through cellular manufacturing, *Production and Inventory Management Journal*, First Quarter, 71–75.

Gerhart, B., and Bretz, Jr., R. D. (1994). "Employee Compensation," in *Organizations and Management of Advanced Manufacturing*, W. Karwowski and G. Salvendy (Eds.), Wiley, New York.

Gerhart, B., and Milkovich, G. T. (1992). Organizational differences in managerial compensation and financial performance, *Academy of Management Journal*, **33**, 663–691.

Goldratt, E. M., and Cox, J. (1992). *The Goal: A Process of Ongoing Improvement*, 2nd ed. revised, North River Press, Great Barrington, MA.

Gross, S. E., and Safier, S. (1995). Unleash the power of teams with tailored pay, *Journal of Compensation and Benefits*, July/August, 27–31.

Irani, S. A. (1997). Cellular Manufacturing and Focused Factories: Design and Implementation Techniques, Workshop Lecture Notes, The Ohio State University.

Joines, J. A., King, R. E., and Culbreth, C. T. (1995). A comprehensive review of production-oriented manufacturing cell formation techniques, *International Journal of Flexible Automation and Integrated Manufacturing*, **3**(3–4), 225–264.

Johnson, S. T. (1993). Work teams: What's ahead in work design and rewards management, *Compensation and Benefits Review*, March/April, 35–41.

Lamarsh, J. (1994). "Managing the Change to Automated Manufacturing," in *Organizations and Management of Advanced Manufacturing*, W. Karwowski and G. Salvendy (Eds.), Wiley, New York.

Lavold, G. D. (1993). "Developing and Using the Work Breakdown Structure," in *Project Management Handbook,* 2nd ed., D. I. Cleland and W. R. King (Eds.), Van Nostrand Reinhold, New York.

Lewin, K. (1947). The ABCs of change management, *Training & Development Journal*, March, 5–41.

Min, H., and Shin, D. (1993). Simultaneous formation of machine and human cells in group technology: A multiple objective approach, *International Journal of Production Research*, **31**(10), 2307–2318.

Mitrofanov, S. P. (1966). *Scientific Principles of Group Technology*, National Lending Library for Science and Technology, London.

Moodie, C. L. (1982). "Assembly Line Balancing," in *Handbook of Industrial Engineering*, G. Salvendy (Ed.), Wiley, New York.

National Tooling and Machining Association and National Institute for Metalworking Skills (1995). *Duties and Standards for Machining Skills*, Fort Washington, MD.

Needy, K. L., Billo, R. E., and Warner, R. C. (1988). A cost model for the evaluation of alternative cellular manufacturing configurations, *International Journal of Computers and Industrial Engineering*, in press.

Shaffer, T. R. (1992). A Demand-Based Design Model Used for Cell Replication at Superior Valve Company, M.S. Thesis, University of Pittsburgh, Department of Industrial Engineering, Pittsburgh, PA.

Shaffer, T., and Billo, R. (1994). A demand-based method for manufacturing cell design and replication, *International Journal of Manufacturing Systems Design*, 1(2), 163–175.

Secretary's Commission on Achieving Necessary Skills: A SCANS Report for America 2000, (1992). U.S. Department of Labor, Washington, DC.

Singh, N. (1993). Design of cellular manufacturing systems: An invited review, *European Journal of Operational Research*, **69**, 284–291.

Tompkins, J. A., and White, J. A. (1984). *Facilities Planning*, Wiley, New York.

Thor, C. G. (1995). Using a family of measures to assess organizational performance, *National Productivity Review*, Summer, 111–131.

Warner, P. J. (1997). "How to Use the Work-Breakdown Structure," in *Field Guide to Project Management*, D. I. Cleland (Ed.), Van Nostrand Reinhold, New York.

Warner, R. C. (1998). A Systematic Approach to Worker Assignment in Manufacturing Cells, Doctoral Dissertation, University of Pittsburgh, Department of Industrial Engineering, Pittsburgh, PA.

Warner, R. C., Needy, K. L., and Bidanda, B. (1997). Worker Assignment in Implementing Manufacturing Cells, in *Sixth Industrial Engineering Research Conference Proceedings*, Miami, FL, pp. 240–245.

Wellins, R. S., Byham, W. C., and Dixon, G.R. (1994). *Inside Teams: How 20 World Class Organizations are Winning Through Teams*, Jossey-Bass Publishers, San Francisco.

Wemmerlov, U., and Hyer, N. L. (1989). Cellular manufacturing in the U.S. industry: A survey of users, *International Journal of Production Research*, **27**(9), 511–1530.

Wilson, R. G., and Linscott, J. D. (1996). *Top Shops: A Beginner's Guide to Team Building and Shop Management*, Hanser Gardner, Cincinnati, OH.

15

DESIGN AND IMPLEMENTATION OF LEAN MANUFACTURING SYSTEMS AND CELLS

J T. Black

Department of Industrial & Systems Engineering
Auburn University
Auburn, Alabama 36849

1. INTRODUCTION

The lean production company uses a system of linked manufacturing and assembly cells to produce low-cost products in a timely fashion. The manufacturing and assembly cells are designed for flexibility—they can handle changes in product design as well as external customer demand. This chapter presents a proven strategy to convert a job shop/flow shop into a lean manufacturing system. The manufacturing equipment in the cells is custom designed and custom built for the lean cells with the manufacturing system requirements in mind. This design of lean cells is discussed and an example is presented. Many companies have designed interim cells using machine tools designed for stand-alone applications in the job shop. The equipment and the tooling must be modified when it is utilized in cells.

2. LEAN PRODUCTION SYSTEMS DESIGN

The lean manufacturing system shown schematically in Figure 1 links manufacturing and assembly cells to final assembly by means of kanban inventory links. This kind

Handbook of Cellular Manufacturing Systems, edited by Shahrukh A. Irani
ISBN 0-471-12139-8 © 1999 John Wiley & Sons, Inc.

of manufacturing system produces superior quality products in a timely fashion at the lowest possible cost in a flexible way. The lean manufacturing system groups the processes into manufacturing and assembly cells. The processes can be changed over rapidly so products can be turned out in greater variety in an almost customized fashion with no cost penalty for small production runs (small lots). The manufacturing system is designed such that the throughput time (TPT) is as short as possible. This requires a continuous redesigning of the manufacturing system (which is just another way to say "continuous improvement") to shorten the TPT and increase the speed with which products move through the plant. Therefore, the Manufacturing System Design (MSD) must be as simple and flexible as possible.

The design of the manufacturing system is an important but neglected aspect of concurrent engineering that has focused on the relationships between product design and process design but has often ignored the system design. If lean MSD and implementation comes before concurrent engineering efforts (the latter involving

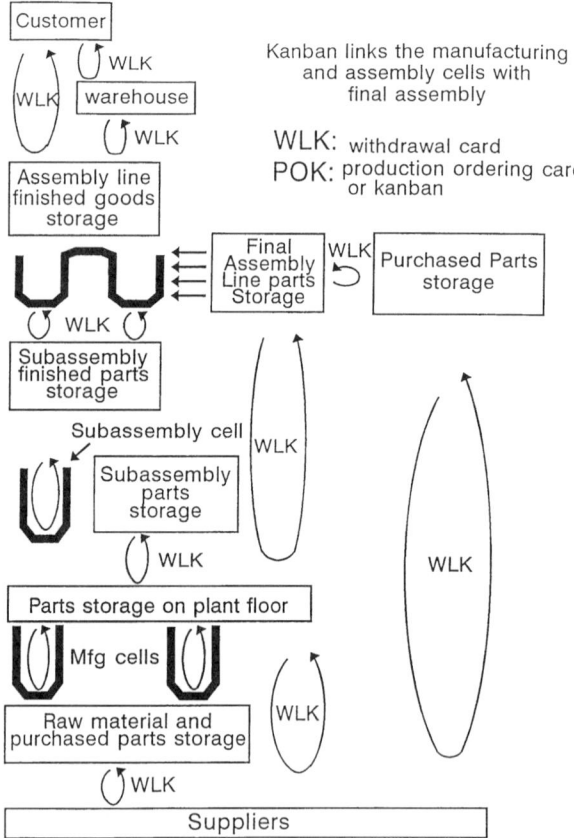

Figure 1. In the lean manufacturing system, the manufacturing and assembly cells are linked to the final assembly area by kanban inventory links or loops.

people from design engineering, manufacturing, logistics, marketing, and customer services for the development of new products), the ability of a company to compete in the global world of manufacturing will be enhanced. The manufacturing system is the beating heart of any manufacturing company.

Figure 2 provides a definition of a manufacturing system (Black and Schroer, 1988). The manufacturing system is defined as a complex arrangement of physical elements characterized by measurable parameters. The physical elements are machine tools, tooling (workholders), material handling equipment, and (most importantly) people. The people who work in the manufacturing system are the "internal customers" and the system must be designed to satisfy their needs. At the same time, the manufacturing system must produce products that satisfy the needs of the external customers. In terms of MSD, this is a key concept. That is, the manufacturing system is designed to satisfy the needs to both the internal and the external customers. The "complex arrangement" is the design of the manufacturing cells and of the manufacturing system that satisfies the needs of these two customers.

Different system designs will result in different levels of measurable parameters. Time is probably one of the most critical measurable parameters for evaluating manufacturing performance. By systematically reducing TPT and its variation, companies can achieve world-class status (Buzacott, 1995). In the lean manufacturing system, the manufacturing interval is reduced by the systematic and gradual removal of inventory. The inventory is captured in links (between the cells) in the system and controlled by the internal customers (the users of the inventory) using kanban (Black, 1991). The Linked-Cell Manufacturing System (L-CMS) shown in Figure 1 has the cells and assembly line linked to final assembly by inventory links. This is a surprise to many who think that the lean production system has no inventory. This is not correct. The inventory between the cells is in the links and is controlled and minimized by the internal customers.

Within the manufacturing cells, the material is referred to as stock on hand (SOH). This material is carried in the machines and the decouplers (Black and Schroer, 1988) and is strictly controlled. The addition of these decoupler elements in the cells and the inventory links between the cells results in a decoupled design (Suh, 1992).

The design of the factory and its physical elements must precede the design and manufacture of the product. Because the design includes people, ergonomic issues must be considered. A discussion of the application of design axioms (Suh, 1990, 1992) to system design can be found in Cochran (1994).

3. BRIEF HISTORY OF MANUFACTURING SYSTEM DESIGN

The L-CMS design is an outgrowth of previous manufacturing system designs. Currently most manufacturing systems are a combination of a job shop and a flow shop. The job shop as a manufacturing system design evolved during the 1800s. These early factories replaced craft or cottage manufacturing when it became necessary to have powered machines. A functional design evolved because of the *method* needed to drive or power the machines. That is, the first factories were built by rivers and the

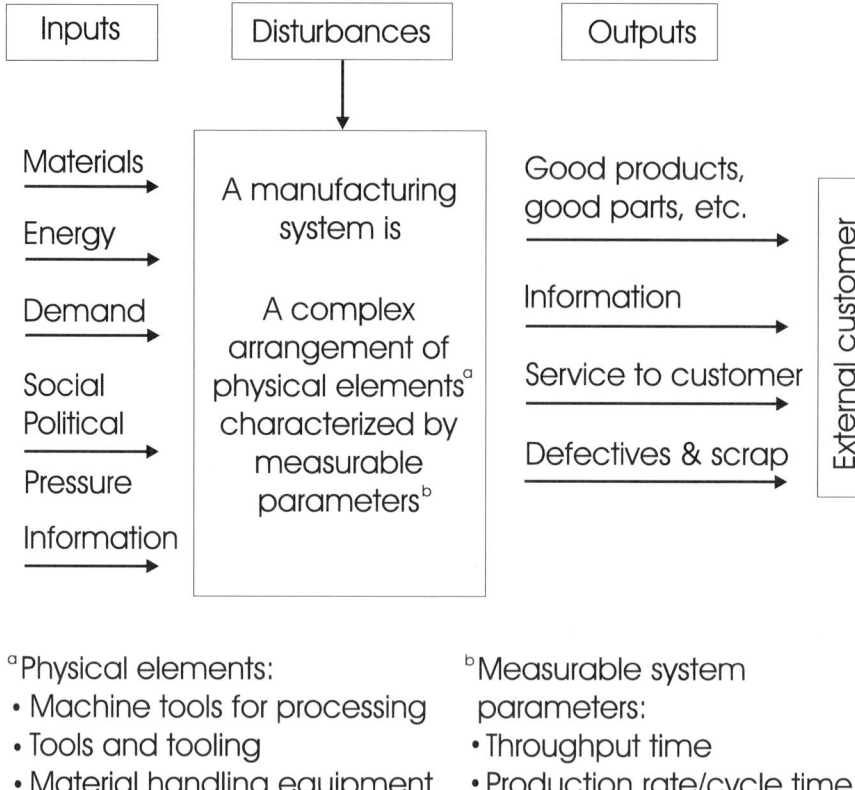

| Inputs | Disturbances | Outputs |

Materials →

A manufacturing system is

A complex arrangement of physical elements[a] characterized by measurable parameters[b]

Energy →

Demand →

Social Political Pressure →

Information →

Good products, good parts, etc. →

Information →

Service to customer →

Defectives & scrap →

External customer

[a]Physical elements:
• Machine tools for processing
• Tools and tooling
• Material handling equipment
• People (internal customers)

[b]Measurable system parameters:
• Throughput time
• Production rate/cycle time
• Work-In-Process inventory
• Percent on-time delivery
• Percent defective
• Daily/weekly/monthly production volumes
• Total cost or unit cost

Figure 2. Definition of a manufacturing system with its inputs and outputs (from Black, 1991).

machines were powered by waterwheels that drove shafts that ran into the factory. Machines of like type were set, all in a line, underneath the appropriate power shaft, that is, the shaft that turned at the speed needed to drive this kind of machine. So all the lathes were collected under their own power shaft, likewise with all the milling machines and all the presses. Belts were used to take power off the shaft to the machines. Later the waterwheel was replaced by a steam engine that allowed the factory to be built somewhere other than by a river. Eventually, large electric motors and then individual electric motors for each machine replaced the steam engine. Nevertheless, the job shop design was replicated and the functional design held. It became known to the historians as the American Armory System. The world came to see the American Armory System and it was duplicated around the industrial world.

In the early 1900s, the first vestiges of the flow shop began to emerge. Flow line manufacturing began for small items and culminated with the moving assembly line at the Ford Motor Company (Womack et al., 1991). This methodology was developed by Charles Sorenson (Ford, 1988). Just as in the 1800s, the world again came to see how this system worked, and this new design methodology was spread across the world, resulting in a hybrid system—a mixture of job shop and flow shop. This design permitted companies to manufacture large volumes of identical products at low unit costs, and the concept of mass production evolved. The job shop produced in large lot sizes according to the economic order quantity equation. Historians called this system mass production. These two eras, which resulted from the two classical MSDs, represent the first and second industrial revolutions, as summarized in Table 1.

4. LEAN MANUFACTURING SYSTEMS

We are now almost 40 years into the third industrial revolution. Again, this industrial revolution is not based on hardware or a particular process but once again on the design of the manufacturing system—the complex arrangement of physical elements

Table 1. Industrial Revolutions (IR) are spawned by new MSDs[a]

	First IR	Second IR	Third IR	Fourth IR
Time period (era)	1840–1910	1910–1970	1960–2010	2000–2040
Manufacturing system design	Job shop	Flow shop	Linked-cell	IM,C[b]
Layout	Functional layout	Product layout	One-piece flow via L-CMS	Linked assembly of large modules or subassemblies
Enabling technology	Power for machines Steel production Railroad for transportation	Moving final assembly line Standardization leading to true interchangeability	SMED; U-shaped cells; Kanban	Virtual reality/ simulation 3D design using low cost very high performance computers[c]
Historical name	American Armory System (Whitney, Colt, Root, Remington)	Mass production (Singer, Ford)	Lean production Toyota Motor Company[d], (HP, H-D, Omark)	Boeing, Lockheed, Chrysler
Economics	Economy of collected technology	Economy of scale	Economy of scope	TBN[e]

[a]MSD, Manufacturing System Design, including machine tools, material handling equipment, tooling, and people.
[b]IM,C, Integrated Manufacturing, Computerized.
[c]Single-source digital product definition and 3D design.
[d]System developed by Taiichi Ohno who called it the Toyota Production System (TPS).
[e]TBN, To be named.

characterized by measurable parameters. Again the world went to see the new design, but this time they went to Japan to try to understand how this tiny nation had become such a giant in the global manufacturing arena. Among the Japanese manufacturers, one company stood out as the best—The Toyota Motor Company. What made this company become the number one automobile manufacturer in the world?

I believe it was the development of a unique L-CMS [Black defined the new system as linked cell (Black, 1991)], known initially as the Toyota Production System (TPS) or the Just-In-Time/Total Quality Control (JIT/TQC) system or World Class Manufacturing (WCM) system. In 1990, it was finally given a name that would become universal, *lean production*. This term was coined by John Krafcik, an engineer working in the International Motor Vehicle program at MIT with Womack, Roos, and Jones (Womack et al., 1991). See Table 2. What was different about this system was the development of *manufacturing cells linked together* with a functionally integrated system for inventory and production control. The result is low-cost (high-efficiency), superior quality (no defects), and on-time delivery of unique products from a flexible system. That is, what was unique about the new manufacturing system was *its design*. The new design was able to operate in ways the old "mass production" system could not. The cells within the system could operate on a one-piece flow basis just like final and subassembly lines. The final assembly lines were designed (flexibly) to handle mixes of models so that demand for all the components pulled into the final product was leveled or smoothed (Monden, 1983).

5. STEPS IN DESIGN OF THE LINKED-CELL MANUFACTURING SYSTEM

The functional requirements of the new design are *flexibility*, *controllability*, *efficiency*, and *uniqueness*. The system is designed for flexibility—it can adapt to changes

Table 2. Names for the L-CMS

Lean production: MIT group of authors, researchers
Toyota Production System (TPS): Toyota Motor Company, Monden (1983), Ohno (1988), and Shingo (1989)
Ohno system: Taiichi Ohno, Inventor of TPS
Integrated pull manufacturing: AT&T (1990)
Minimum inventory production system: Westinghouse
Just-In-Time/Total Quality Control (JIT/TQC): Schonberger (1982)
Material As Needed (MAN): Harley Davidson
World-class manufacturing: Schonberger (1986)
Zero Inventory Production Systems (ZIPS): Omark; Hall (1983)
Quick response: Apparel Industry; Black and Schroer (1993)
Stockless production: Hewlett-Packard
Kanban system: Many companies in Japan
New production system: Suzaki (1987)
One-piece flow: Sekine (1990)
Focused Factory/Machining Process Design: Harmon and Peterson (1990)

in customer demand and changes in product designs. These new systems evoke a different strategy from that called concurrent engineering. The classical concurrent engineering strategy is turned around. The new strategy is functions as follows:

- Design the lean manufacturing system for existing products (to be flexible, unique, controllable, and efficient).
- Integrate the critical control functions (of quality, production, inventory, and process control).
- Design new products that can be made in this system, using concurrent engineering.

This design strategy is broken down into critical steps as outlined in Table 3. Notice that autonomation, the autonomous control of quality and quantity, comes very late in the process (step 9), after function integration has been achieved. That is, the system must be redesigned, simplified, and integrated before computers and automation are applied. Concurrent engineering becomes the last step as part of an effort to restructure the rest of the company, often called Business Process Reengineering (BPR).

This approach evolves a manufacturing system design that permits rapid deployment of new products and rapid production of existing products, as well as accommodating changes in the product demand, so it is the winner when it comes to time-based competition. Companies that employ this strategy become the factories with a future. The strategy outlined here simplifies the manufacturing system before technology is applied, avoids risks, and makes the implementation of automation easier.

The system is designed for controllability. That is, the design permits the critical control functions (of quality control, inventory control, production control, and machine tool reliability) to be directly integrated into the manufacturing system. This functional integration is quite different from getting the Computer-Aided Design (CAD) computer to talk to the Computer-Aided Manufacturing (CAM) computer. Because the new method is based on a different manufacturing system design rather than expensive technology, it requires less capital investment. The recent effort at the AT&T Denver Works (AT&T, 1990) typifies the revolution sweeping American factories. AT&T dubbed this conversion Integrated Pull Manufacturing (IPM), an excellent name.

One of the unique features of the integrated system is that information (about the materials in the system) is minimized and flows in the opposite direction to the materials movement. This required a decoupling of the material flow from the Material Information System, commonly called the MIS. In addition the amount of information will be minimized because the amount of material in the system is minimized.

The order of the design steps outlined in Table 3 is important. Many companies have implemented steps 6, 7, and 8 before steps 1–5. Such implementations often result in failure. Even worse, many companies have tried to implement Computer Integrated Manufacturing (CIM) strategies (Ayers, 1991), but now the realization that one must first implement lean production has become understood by even the most ardent CIM proponents (Ayers and Butcher, 1993).

Table 3. Ten steps in the design of an L-CMS

1. *Design or reengineer (reconfigure) the manufacturing system:* Design and implement manufacturing and assembly cells. The design of the manufacturing system must consider the design of the product and the needs of the internal and external customers (Black, 1988).
2. *Setup reduction, changing methods and designs to reduce setup time* (Shingo, 1985): Setup time is delay time. Affects lot size. Optimum lot size is one. Use SMED because it involves everyone on the factory floor (SMED = Single Minute Exchange of Dies).
3. *Integrate quality control into the manufacturing system* (Shingo, 1986): Does the manufacturing process satisfy the design specifications every time? Inspection to prevent the defect from occurring (pokayokes).
4. *Integrate preventive maintenance* (Nakajima, 1988): Do the machines and people behave reliably? Design equipment to be reliable. Design methods to check people and methods for people to check machines.
5. *Level and balance the manufacturing system, smoothing the material flow* (Monden, 1983): Leveling involves the development of mixed model final assembly. Smooth the demand on the cells. Balancing is getting the output from the cells to match the needs of final assembly.
6. *Integrate production control, link the cells, pull material to final assembly:* Control the where, when, and how much material. The design of the manufacturing system defines flow and the kanban operates within this structure. This is integrated production control or kanban (Black, 1991).
7. *Integrate inventory control:* Reduce the WIP in the links that connect the cells. This is control of the quantity of material in the links. Minimized and optimized and controlled by the internal customers, the users of the materials (Black, 1991).
8. *Integrate the vendors; make vendors JIT manufacturers just like you*: Vendors become remote cells. Vendors become partners. Relationship built on trust. This is how real technology transfer takes place (Black, 1991).
9. *Autonomation: autonomous control of quality and quantity within the manufacturing system:* Automate the integrated pull manufacturing system.
10. *Design new products concurrently with customers in mind* (Whitney, 1992).

6. TEN STEPS TO LEAN PRODUCTION

Integration of the production system functions into the manufacturing system requires commitment from top-level management and communication with everyone, particularly manufacturing. Total employee (and union) involvement is absolutely necessary, but it is not usually the union leadership or the production workers who raise barriers to lean production. It is those in middle management who have the most to lose in these systems-level changes as their job functions get integrated into the manufacturing system.

The preliminary steps are as follows:

1. All levels in the plant, from the production workers (the internal customer) to the president, must be educated in lean production philosophy and concepts.

2. Top management must be totally committed to this venture and provide the necessary leadership. Everyone must be involved in the change. The internal customer is empowered.

3. Everyone in the plant must understand that cost, not price, determines profits. Since the customers determine price and want superior quality with on-time delivery, customer satisfaction is the key.

4. Everyone must be committed to the elimination of waste. This is fundamental for getting lean.

5. The concept of standardization must be taught to everyone and applied to documentation, methods, processes, as well as measurements.

To summarize, *educate and communicate the entire plan to everyone*. This is not just turning a new leaf. This is planting a whole new tree.

Many companies have implemented lean production. The steps presented here convert a factory from a job shop/flow shop manufacturing system to a true lean production system.

6.1. Step 1: Form U-Shaped Cells—Restructure the Factory Floor

In L-CMS, cells replace the production job shop. The first task is to restructure and reorganize the basic manufacturing system into manufacturing cells that fabricate families of parts. This prepares the way for systematically creating a linked-cell system for one-piece movement of parts within cells and for small-lot movement between cells. Creating cells is the first step in designing a manufacturing system in which production control, inventory control, quality control, and machine tool maintenance are integrated.

Conversion of the functional system into a flexible, linked-cell system is a design task. Table 4 outlines some of the methodologies by which manufacturing cells can be formed. Most companies "design" their first cell by trial-and-error for expediency in gaining experience in cells. A family of products define the sequence of processes needed to produce the parts. The rack bar for a car is an example. A model of a car may have 10 variations of racks (right-hand drive versus left-hand drive, power steering versus mechanical steering, etc.). The operations in the cell include all sorts of metal cutting operations, heat treating, inspection, assembly, and even grinding and superfinishing all done as steps in one cell that makes all the racks for this model of car.

Sometimes a key machine is used to identify a family of parts. All the support equipment needs to be gathered about the key machine to form a cell.

In terms of more sophisticated approaches, *digital simulation* is gaining wider usage in designing and analyzing manufacturing systems with the advent of newer, more versatile languages.

This technology is starting to utilize virtual reality to simulate the workplace and the ergonomic involvement of the operators with the machines. This technique replaces another technique called *physical simulation*. This approach used small instructional-grade robots and scaled-down versions of machine tools (minimachines) to emulate

Table 4. How to form Cellular Manufacturing Systems

JIT manufacturing systems are based on a L-CMS design. Knowing how to design the cells to be flexible is the key to successful manufacturing.

1. Make tacit judgments based on axiomatic design principles.[a]
   a. Minimize function requirements (flexibility is chief criteria).
   b. Simplify the design of the system.
   c. Minimize the design information in the product.
   d. Decouple those elements that are functionally coupled.

2. Use GT methodology.
   a. Production flow analysis finds families and defines cells.
   b. Coding/classification is more complete and expensive.
   c. Other GT methods, including eyeball or tacit judgments.
      (1) Find the key machine, often a machining center, and declare all parts going to this machine a family. Move machines needed to complete all processes to key machine.
      (2) Build the cell around a common set of components such as gear, splines, spindles, rotors, rubs, shafts, etc.
      (3) Build the cell around a common set of processes, such as drill, bore, ream, keyset, chamfer holes.
      (4) Build the cell around a set of parts that eliminates the longest (most time-consuming) element in setups between parts being made in the cell.

3. Simulation
   a. Digital simulation of the system
   b. Physical simulation of the system
   c. Object-oriented and graphical simulation
   d. Virtual manufacturing simulation

4. Pick a product or products.
   Beginning with the final assembly line, converting final assembly to mixed model, and move backward through the subassembly to component parts and suppliers.

[a] Axiomatic design principles are presented in Suh (1990) and applied to manufacturing systems in Cochran (1994).

real-world systems. The small machines can employ essentially the same control computers and software as the full-scale systems. Virtual manufacturing can also do this. Thus, the development of the software needed to integrate the machines and design of the cell can be done prior to installation of the full-scale system on the shop floor. Unmanned cells and Flexible Manufacturing Systems (FMSs) can be simulated at quite reasonable cost.

Group Technology (GT) offers a systems solution to the reorganization of the functional system, restructuring the job shop into manufacturing cells. These conversions represent systems-level changes, which will create the potential for tremendous savings, but because of the magnitude of the changes, careful planning and full cooperation from everyone involved are absolutely required.

The application of this concept to a manufacturing facility results in the grouping of units or components into families wherein the components have similar design or manufacturing sequences. Machines are then collected into groups or cells (machine cells) to process the family. By grouping similar components into families of parts, a group or set of processes can be collected together to make a family. This is a cell. The arrangement of the machines in the cell is defined by the sequence of manufacturing processes.

Many companies begin with a pilot cell so that everyone can see how cells function. It will require time and effort to train the operators, and they will need time to adjust to standing and walking. Simply select a product or group of products that seems most logical. *The operators must be involved in designing the cell or they will not take ownership.* The pilot cell will show everyone how cells operate and how to reduce setup time on each machine. Machines will not be utilized 100%. Machine utilization rate usually improves but may not be what it was in the functional system where *overproduction* is allowed. *The objective in manned Cellular Manufacturing (CM) is to utilize the people fully*, enlarging and enriching jobs, allowing operators to become multifunctional. The operators learn to operate many different kinds of machines and perform tasks that include quality control, machine tool maintenance, setup reduction, and continuous improvement. In unmanned cells and systems, the utilization of the equipment is more important because the most flexible and smartest element in the cell, the operator, has been removed and has been "replaced" by a robot.

The cells are designed in a U-shape so that the workers can move from machine to machine, loading and unloading parts. Figure 3 shows an example of a manned cell. The cell has one worker who can make a walking loop around the cell in 110 seconds. See the breakdown of time in the table to the right. The machines in the cell are usually single-cycle automatics, so that they can complete the desired processing untended, turning themselves off when done with a machining cycle. The operator comes to a machine, unloads a part, checks the part, loads a new part into the machine, and starts the machining cycle by hitting a walkaway switch as he moves to the next machine. The cell usually includes all the processing needed for a complete part or subassembly. The table shows the typical average times for the operator time, HT, plus walking time. The times for the machining cycle are given as MTs. The cell is designed such that the machining time for any part in the family for any machine in the cell is less than the necessary cycle time, CT. That is, $MT_{ij} < CT$. Thus machining times are uncoupled from the cycle time.

However, note that the machining time for the third operation is greater than the necessary cycle time. That is, 180 seconds > 110 seconds. Therefore, this operation is duplicated, and the operator alternates between the two lathes, visiting each lathe every other trip he or she makes around the cell. This makes the average MT 180/2 = 90 seconds so that MT < CT holds.

Cells are typically manned, but unmanned cells are beginning to emerge with a robot replacing the worker. A *robotic cell* design is shown in Figure 4 with one robot and three Computer Numerical Control (CNC) machines. For the cell to operate autonomously, the machines may have adaptive control capability or be equipped with pokayoke devices to prevent defects from occurring. This kind of automation is only recommended after the manned cell has been developed and optimized within the L-CMS. The decouplers between the machines help the robot get parts in the

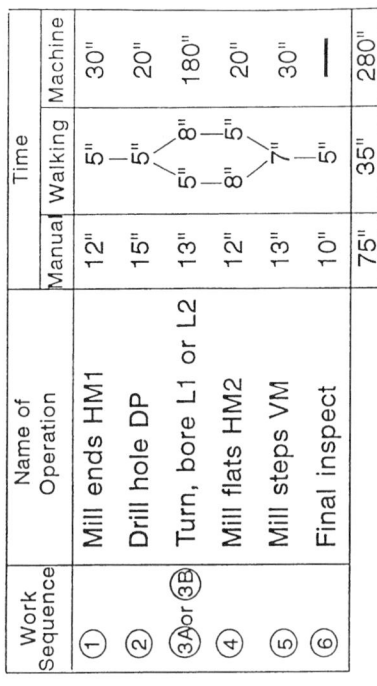

Work Sequence	Name of Operation	Time		
		Manual	Walking	Machine
①	Mill ends HM1	12"	5"	30"
②	Drill hole DP	15"	5"—5"	20"
③A or ③B	Turn, bore L1 or L2	13"	5"—8"—5" 5"—8"	180"
④	Mill flats HM2	12"	5"	20"
⑤	Mill steps VM	13"	7"	30"
⑥	Final inspect	10"	5"	—
		75"	35"	280"

Cycle Time = 75sec + 35 sec = 110 sec
Longest Machining Time = 180 sec
Total Machining Time = 280 sec

All machines in the cell are capable of processing unattended while the operator(s) are doing manual operations (unload, load, inspect, and deburr)and walking from machine to machine. The time to change tools and workholders (perform setup) not shown.

Key:
DP = Drill Press
L = Lathes
HM = Horizontal milling machine
VM = Vertical milling machine
→ Material flow
----- Operator's path
Ⓧ Operation sequence

Figure 3. Example of an interim manned cell part of a CMS. This cell operates with six machines and one multifunctional worker. The lathe operation, 3, is duplicated.

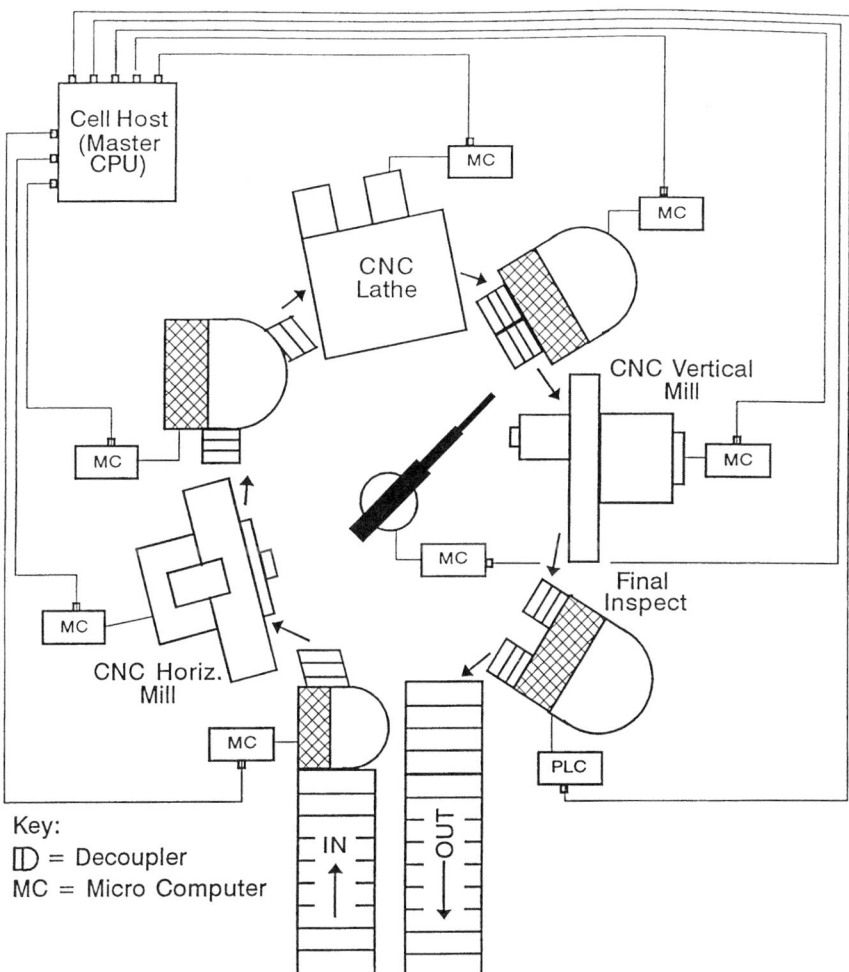

Figure 4. Unmanned robotic cells will have CNC machine tools, robot for material handling, and decouplers for flexibility and capability.

correct orientation and location to the next machine. The decouplers can also inspect the parts and perform other functions that were previously done by the operators.

At the same time that the manufacturing cells are being designed, the flow shop parts are reconfigured into U-shaped cells as well. See Figure 5. Using a conveyor, the flow shop can be redesigned to make these systems operate on a one-piece flow basis. This example comes from Sekine (1990) and shows the marked improvements in productivity that CM can achieve. As with the manufacturing cell, the long setup times typical in flow lines must be vigorously attacked and reduced so that the flow lines can be changed quickly from making one product to making another. This makes them flexible and compatible with the cells designed to make piece parts and with the other subassembly lines and final assembly lines. Again the cells are designed to manufacture specific groups or families of parts. In the U-shaped layout in Figure 5, workers 2, 3, and 4 cover multiple operations. Notice that workers 3 and 4 share operation 7. Workers 1 and 5 are covering operations 1, 11, and 12 using a rabbit chase (Suzaki, 1987). The need to line balance the flow line has been eliminated. This is accomplished by using standing, walking workers who are capable of performing multiple operations.

Cells have many features that make them unique and different from other manufacturing systems. Parts move from machine to machine *one at a time* within the cell. For material processing, the *machines are typically* capable of completing a machining cycle initiated by a worker. The U shape puts the start and finish points of the cell next to each other. Every time the operator completes a walking trip around the cell, a part is completed. The cell is designed so that this cycle time (CT) is equal to or less than the "necessary cycle time" called the takt time. The machining time (MT) for each machine needs only to be less than the time it takes for the operator to complete the walking trip around the cell. Thus, the machining time can be altered without changing the production schedule.

The cell is designed to make parts *as needed* by downstream processes and operations. There is no overproduction. Overproduction will result in the need to store parts, transport parts to storage, retrieve the parts when needed, keep track of the parts (paperwork), and so on. All this requires people and costs money but adds no value.

There is no need to balance the MTs for the machines. It is necessary only that no MT be greater than the required CT. The machining speeds and feeds can be relaxed to extend the tool life of the cutting tools and reduce the wear and tear on the machines so long as the MT does not equal or exceed the CT.

The fixtures in the machines are designed to hold the family of parts so rapid changeover from one part to another is possible. The fixtures are designed for easy load/unload, designed so well that parts cannot be loaded incorrectly and designed so that defective parts cannot be loaded. In some cells *decouplers* are placed between the processes, operations, or machines to provide flexibility, part transportation, inspection for defect prevention (pokayoka) and quality control, and process delay for the manufacturing cell. The decoupler inspects the part for a critical dimension and feeds back adjustments to the machine to prevent the machine from making oversize parts as the milling cutter wears. A process delay decoupler would delay the part movement to allow the part to cool down, heat up, cure, or whatever is necessary for a period of time greater than the cycle time for the cell. Decouplers and flexible fixtures are vital parts of both manned and unmanned cells.

6.2. Step 2: Rapid Exchange of Tooling and Dies

Everyone on the plant floor must be taught how to reduce setup time using SMED (Single-Minute Exchange of Die). A setup reduction team acts to facilitate the SMED process for the production workers and foremen and demonstrates the methodology

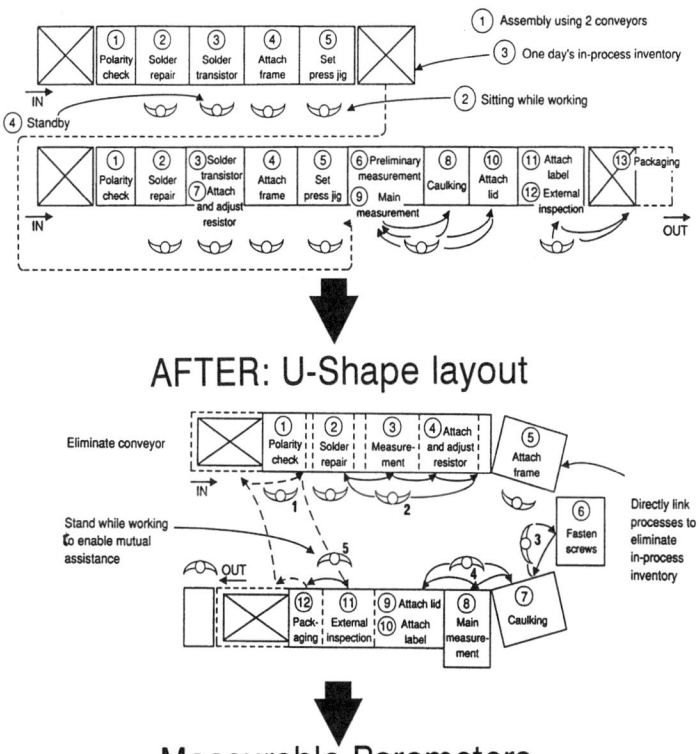

BEFORE: Layout with conveyor

AFTER: U-Shape layout

Measurable Parameters

	Before:	After:
Output	700	1056 units/shift
In-process inventory	750	8 units
Personnel	10	8 operators
Daily output per person	70	132 units
Cycle time	0.60 minute	0.43 minute

Figure 5. An assembly flow line, with conveyor, can be redesigned into a U-shaped cell. The cell uses walking operators (after Sekine, 1990).

on a project, usually the plant's worst setup problem. Reducing setup time is critical to reducing lot size. See Figure 6 for a summary of the SMED steps.

The lean production approach to manufacturing demands that small lots be run. This is impossible to do if machine setups take hours to accomplish. The *Economic Order Quantity* (EOQ) formula has been used in the United States to determine what quantity should run to cost-justify a long and costly setup time. The EOQ was a faulty

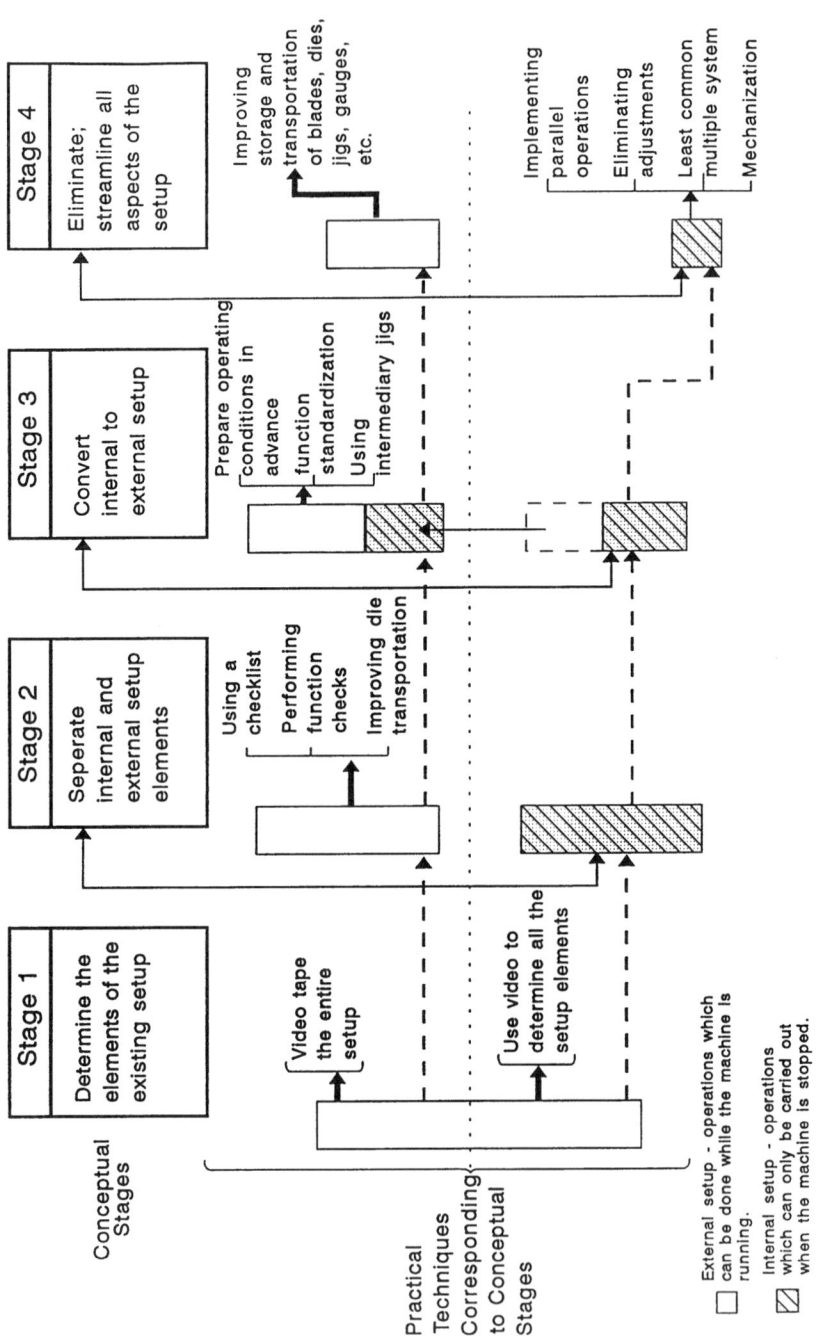

Figure 6. Conceptual stages of the SMED system for the rapid exchange of tooling and dies.

suboptimal approach that accepted long setup times as a given. Setup times can be reduced. This results in reduced lot sizes.

Successful setup reduction is easily achieved when approached from a methods engineering perspective. Much of the initial work in this area has been done by Shigeo Shingo and Taiichi Ohno. Large reductions can be achieved by applying time and motion studies and Shingo's SMED rules for rapid exchange of dies. Setup time reduction occurs in four stages; see Figure 6. The initial stage is to determine what currently is being done in the setup operation. The setup operation is usually videotaped and everyone concerned gets together and reviews the tape to determine the elemental steps in the setup. The next stage is to separate all setup activities into two categories, *internal* and *external*. Internal elements can be done only when the machine is not running; external elements can be done while the machine is running. This elemental division will usually shorten the lead time considerably. Stages 3 and 4 focus on reducing the internal time. The key here is for operators to learn how to reduce setup times, applying simple principles and techniques. If a company must wait for the setup reduction people to examine every process, a lean manufacturing system will never be achieved.

In the last stages of SMED, it may be necessary to invest capital to drive the setup times below 1 minute. Automatic positioning of workholders, intermediate jigs and fixtures, and duplicate workholders represent the typical kinds of hardware needed. The result is that long setup times can be reduced to under 15 seconds in relatively short order.

The similarity in shape and processes needed in the family of parts allows setup time to be reduced or even eliminated. Initially setup should be less than 10 minutes (SMED). As the cell matures, the setup times are continually reduced. The final goal is to get them down to around 10–15 seconds, what is commonly called One-Touch Exchange of Dies (OTED).

Reduce the setup time until it is equal or less than the cycle time (1–2 minutes) for the cell is usually quite easily accomplished. This will permit a significant initial reduction in lot size. After this, further setup time reductions will result in further reductions in the lot size. The next goal is to get the setup time down to less than the time needed to load, unload, inspect, deburr, and so on at a machine. As shown in Table 5, when numerous processes are involved in the CM of a family of parts, sequential setup changes are utilized. That is, the setups flow through the cell sequentially. After each setup, defect-free products should be made right from the start. The first part will be good. Ultimately, the ideal condition would be to eliminate setup between parts. This is called No Touch Exchange of Dies (NOTED).

Figure 7 shows the drilling machine from the cell described in Figure 3. This cell was designed for a family of four parts. To reduce the setup time to a few seconds, or "one touch," the drilling machine was modified to eliminate setup as far as possible. The total cost to modify the drilling machine was around $300. This is an example of one-touch exchange of fixtures.

Figure 8 shows a top view of a vertical milling machine from Figure 3, equipped with a digital readout device so that the four starting locations of the four machining operations were always the same. The fixtures for the four parts in the family are

Table 5. Example of changeover from part A to part B for four-process cell

Cycles Through the Cell	Four Processes in the Cell			
	Machine 1	Machine 2	Machine 3	Machine 4
1	A	A	A	A
2	A (last A part)	A	A	A
3	Setup change	A (last A part)	A	A
4	B (first B part)	Setup change	A (last A part)	A
5	B	B (first B part)	Setup change	A (last A part)
6	B	B	B (first B part)	Setup change
7	B	B	B	B (first B part)
8	B	B	B	B

never removed from the oversize table. The first part out of these fixtures was always a good part and there is no time spent on fixture exchange. This is an example of no touch exchange of fixtures.

Other setup reduction techniques include using group jigs (one jig to accommodate different parts, using adapters), training operators in rapid setup techniques, practicing rapid setups, and the intermediate jig concept. The *intermediate jig* is like the cassette for a VCR. The cassette is quickly loaded and locked in place. The tape inside every cassette can be different. To the machine (the tape player), every different fixture (tape) that is placed in the intermediate jig (cassette) looks the same. To the cutting tool (the tape head), every workholder looks different. Designing workholders so that they appear the same to the machine tool usually requires one to construct an intermediate jig or fixture plate to which the fixture itself is attached.

In summary, the savings in setup times are used to decrease the lot size and increase the frequency at which the lot is produced. The smaller the lot the lower will be the inventory, making shorter throughput time and improving quality.

6.3. Step 3: Integrate Quality Control

A *multiprocess* worker can run more than one kind of process. A *multifunctional* worker can do more than operate machines. Such a worker is also an inspector who understands process capability, quality control, and process improvement. In lean production, every worker has the responsibility and the authority to make the product right the first time and every time and the authority to stop the process when something is wrong. This integration of quality control into the manufacturing system markedly reduces defects while eliminating inspectors. Cells provide the natural environment for the integration of quality control. The fundamental idea is to inspect to prevent the defect from occurring.

When management and production workers trust each other, it is possible to implement an integrated quality control program. Japan was started on the road to superior quality with the visits of W. Edward Deming to Japan in the late 1940s and early 1950s. The Japanese were desperate to learn about quality, and statistical quality

Example of OTED

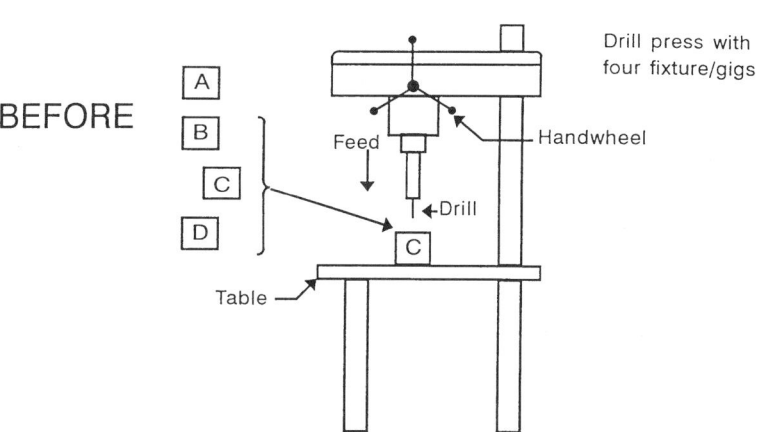

Typically, a machine that has four jobs, with four different fixtures or jigs, would need four different setups, each consisting of changing fixtures or jigs and alignment f each.

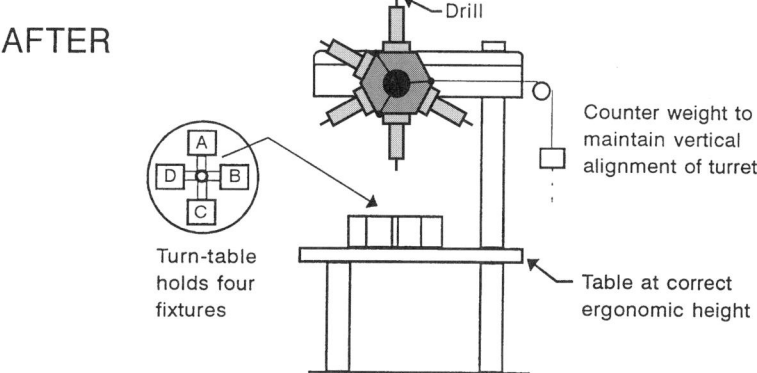

After redesign, the four fixtures are mounted on a turntable and are permanently aligned to the spindle when locked in position. Turret replaces single spindle. Automatic feed can replace handwheel.

Figure 7. Machines in the interim cell are modified to process a family of parts, reducing setup time, resulting in one-touch exchange of parts.

control techniques were readily accepted. They believed that everyone in America used statistical process control techniques. This of course was not the case. However, the Japanese readily accepted quality control concepts, and then they did something that even those American companies using Statistical Quality Control (SQC) did not do. They taught the techniques and concepts to everyone, including top management and the production workers, who even had a quality journal on the subject.

When every worker is responsible for quality and able to perform the seven basic tools of quality control, shown in Figure 9, the number of inspectors on the plant floor

Example of NOTED

Top view of milling machine with fixture one in place

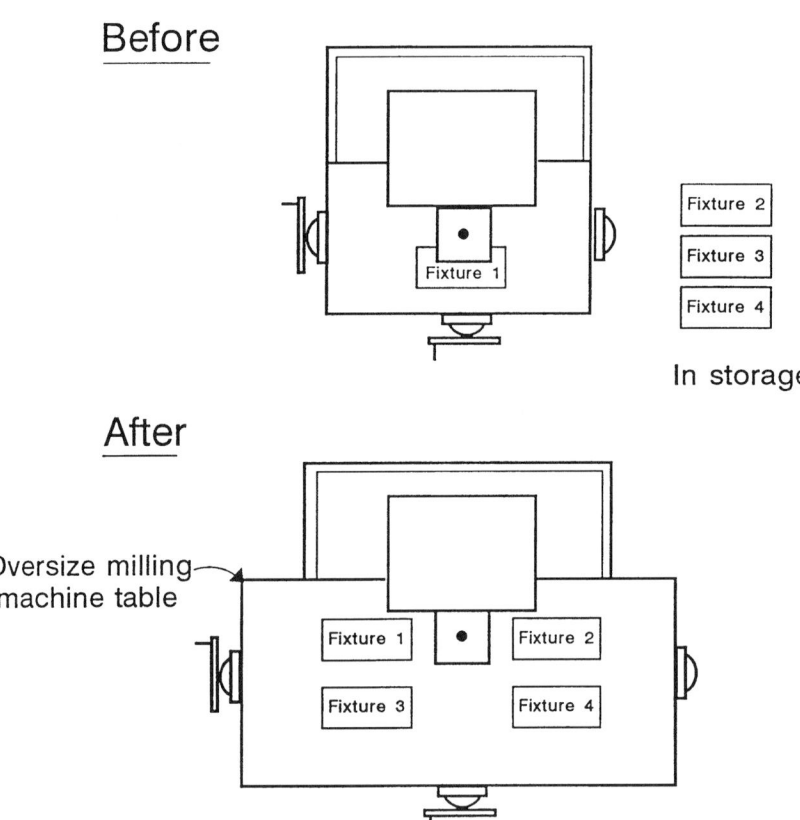

Before

Fixture 2

Fixture 3

Fixture 4

In storage

After

Oversize milling machine table

Figure 8. Top view of milling machine with four fixtures for four parts in the family—no setup required.

is markedly reduced. Products that fail to conform to specification are immediately uncovered because they are used immediately.

Under the leadership of Taiichi Ohno, a new idea took hold at Toyota, quite different in concept from our inspection philosophy. First, every worker was an inspector responsible for quality. And the workers on the assembly line could stop the line if they found something wrong. *Inspecting to prevent* the defect from occurring rather than to *inspecting to find* the defect after it has occurred became

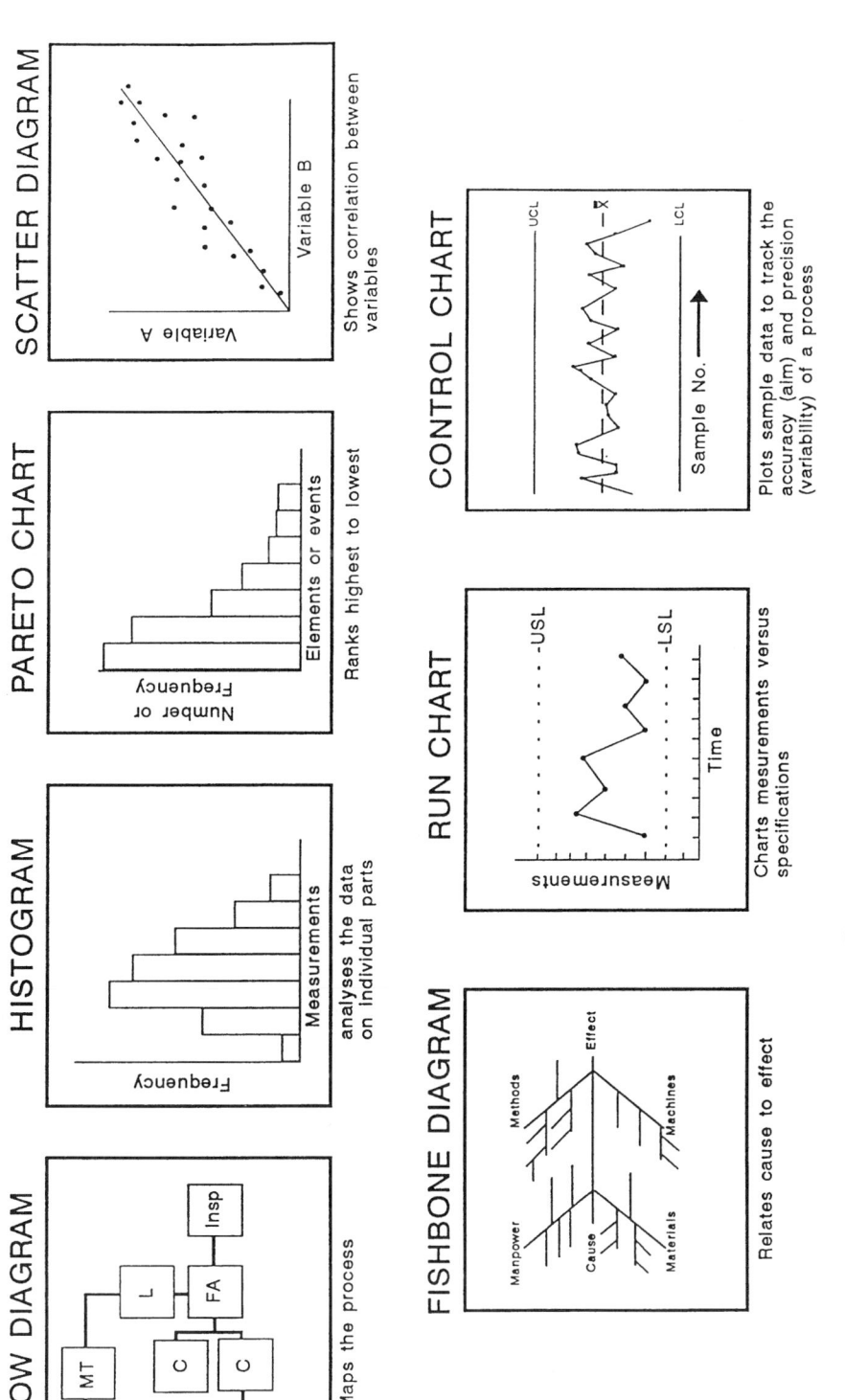

Figure 9. Seven tools of quality control.

473

the mode of operation in the cells. Ultimately, the concept of *autonomation* evolved (see step 9).

Cells produce parts one at a time, just like assembly lines. This is called one-piece flow (Sekine, 1990). Pull cords are installed on the assembly lines to stop the lines if anything goes wrong. If workers find defective parts, if they cannot keep up with production, if production is going too fast according to the quantity needed for the day, or if a safety hazard is found, they are obligated to stop the lines. The problem is fixed immediately. Meanwhile, the other workers maintain their equipment, change tools, sweep the floor, or practice setups; but the line does not move until the problem is solved.

For manual work on assembly lines, a system for tracking defective work is called *Andon*. Andon is actually an electric light board that hangs high above the conveyor assembly lines so that everyone can see it when everything is going okay, the lights are green. But, when a worker on the line needs help, he can turn on a yellow light. Nearby (multifunctional) workers who have finished their jobs within the allotted cycle time move to assist workers having problems (called mutual assistance). If the problem cannot be solved within the cycle time, a red light comes on and the line stops automatically until the problem is solved. Music usually plays to let everyone know there is a problem on the line.

In most cases the red lights go off within 10 seconds and the next cycle begins, a green light comes on, with all the processes beginning together. The name for this system is *Yo-I-don*, which literally means "ready, set, go." The stages on the assembly line are synchronized. Such systems are built on teamwork and a cooperative spirit among the workers, fostered by a management philosophy based on harmony and trust.

Contrast this with the way things often operate in the job shop. How long does it take to find a problem, to convince somebody that it is a problem, to get the problem solved, and to get the fix implemented? How many defective parts are produced in the meantime? Line shutdowns in lean factories are encouraged to protect quality. Management must have confidence in the individual worker.

6.4. Step 4: Integrate Preventive Maintenance

Making machines operate reliably begins with the installation of an integrated preventive maintenance program, giving workers the training and tools to maintain equipment properly. The excess processing capacity obtained by reducing setup time allows operators to reduce the equipment speeds or feeds and to run processes at less than full capacity. Reducing pressure on workers and processes to produce a given quantity fosters in workers a drive to produce perfect quality.

The multifunctional operators are trained to perform routine machine tool maintenance. Just adding lubricants (oiling the machine), checking for wear and tear, replacing damaged nuts and bolts, routinely changing and tightening belts and bolts, and listening for telltale whines and noises that signify impending failures can do wonders for machine tool reliability. The maintenance department must instruct the workers on how to do these things and help them prepare the routine check lists for machine maintenance. The workers are also responsible for keeping their areas of the plant clean and neat. Thus another function that is integrated into the manufacturing system is maintenance and housekeeping.

The following housekeeping rules are implemented:

1. A place for everything and everything in its place. Everything should be put away so it is ready to use the next time.
2. Each worker is responsible for cleanliness of workplace and equipment.

Naturally, the machines still need attention from the experts in the maintenance department, just as the airplane is taken out of service periodically for engine overhaul and maintenance. One alternative here is to switch to two 8-hour shifts separated by two 4-hour time blocks for machine maintenance, tooling changes, restocking, long setups, overtime, earlytime, and so on. This is called the 8–4–8–4 scheme. The main advantage that equipment has over people is that it can decrease variability, but it must be reliable and dependable. Smaller machines are simpler and easier to maintain and therefore are more reliable. Small machines in multiple copies add to the flexibility of the system as well. The linked-cell system permits certain machines in the cells to be slowed down and therefore, like the long-distance runner, to run farther and easier without breakdown. Many observers of the JIT manufacturing system come away with the feeling that the machines are "babied." In reality, they are being run at the pace needed to meet the demand.

True lean producers build and modify much of their manufacturing process technology. It is what makes them unique. In addition, they try to make equipment in multiple copies so that you have multiple sets of equipment making the similar products. Suppose you have a cell for making racks for a rack and pinion steering gear. The manufacturing cell for racks makes 6 different racks for the Accord. The cell for Toyota makes racks for Camry and Avalon and can make 10 different kinds of racks. But, the two cells are very similar in their design. In the event of a machine failure, a machine from another cell can be borrowed. Also, processing capacity is replicated in proven increments. Because the increment (the cell) has an optimal design, this is an economic choice as well as having the security of dealing with a proven manufacturing process technology. Modifying existing equipment shortens the time needed to bring new technology on stream. Manufacturing in multiple versions of small-capacity machines retains the expertise and permits the company to keep improving and mistake-proofing the process. In contrast to this approach is the typical job shop, where a new supermachine would be purchased and installed when product demand increases. That is, many companies try to increase capacity by buying new, untried manufacturing technology that may take months, even years, to debug and make reliable.

6.5. Step 5: Leveling, Balancing, and Synchronizing the Manufacturing System

The steps outlined here are the amalgamated experience of many companies that have Americanized and implemented some version of the Toyota Production System.

Machine layout follows the flow of processes wherein products having common or similar processes are grouped together and quick conveyance between the processes is provided, along with the means to reduce setup time. The basic premise of the system is to produce the kind of units needed in the quantities needed at the time needed. The lean production system depends upon *smoothing of the manufacturing system*. In order to eliminate variation or fluctuation in quantities in feeder processes, it is necessary to eliminate fluctuation in final assembly. This is also called *leveling* the final assembly. Here is a simple example to show the basic idea. First, we need to calculate cycle time.

$$DD = \text{Daily demand for parts} = \frac{\text{monthly demand (forecast plus customer orders)}}{\text{number of days in month}}$$

$$CT = \frac{1}{PR} \text{ where } PR = \frac{\text{daily demand (parts)}}{\text{available hours in day (hrs)}}.$$

This incredibly simple approach highlights the way in which JIT companies calculate cycle time, but life is simpler when the production job shop (PJS) has been eliminated and a linked-cell system has been installed.

Here is an example of how cycle time is determined as for a mix of cars at final assembly. Suppose that the forecast is for 240 cars per day and 480 production minutes are available (60 minutes × 8 hours/day). Thus, cycle time = 2 minutes. Cycle time for the final assembly is called takt time. Every 2 minutes a car rolls off the line. Suppose that the mix is as given in Table 6.

The subprocesses that feed the two-door fastback are controlled by the cycle time for this model. Every 4.8 minutes, the rear deck line will produce a rear hatch for the fastback version. Every 4.8 minutes, two doors for the fastback are made.

Every car, regardless of model type, has an engine. So engines are produced at a rate of one every 2 minutes. Each engine needs four pistons. Therefore every 2.0 minutes, four pistons are produced. Parts and assemblies are produced in their minimum lot sizes and delivered to the next process, under the control of kanban.

Balancing is making the output from the cells equal to necessary demand for the parts downstream. The parts or components are not made in sync with final assembly,

Table 6. Example of mixed model final assembly line that determines the cycle time for model

Q	Car Mix for Line Model	Cycle Time by Model (min)	Production Minutes by Model	Sequence (24 Cars)
50	Two-door coupe	9.6	100	TDC, TDF, TDF, FDS, FDW,
100	Two-door fastback	4.8	200	TDC, TDF, TDF, FDW, FDW,
25	Four-door sedan	18.2	50	TDC, TDF, TDF, FDS, FDW,
65	Four-door wagon	7.7	130	TDC, TDF, TDF, FDW, ...
	240 cars / 8 hr			480 min / 240 = 2 min per car

only the daily quantity is the same. In summary, small lot sizes, made possible by setup reduction within the cells, single-unit conveyance within the cells, and standardized cycle times are the keys to accomplishing a smoothed manufacturing system. One strives to make the cycle time the equal to the takt time for final assembly but at the outset, matching the daily demand is sufficient. Ultimately, every part, sequence of assembly operations, or subassembly has the same number of specified minutes as the final assembly line. For example, when the car body exits from painting, the order for the seats is issued to the seat supplier. The seats are made in the same order as the cars on the assembly line. They are made and delivered in the same amount of time as it takes the car to get from paint to the point on the line where seats are installed. That is, seat manufacturing is synchronized. The minimum number of workers needed to produce one unit of output in the necessary cycle time is used. This is called *leveling and balancing and synchronizing* the manufacturing system.

6.6. Step 6: Link Cells

Integration of production control is materialized by linking the cells, subassemblies, and final assembly elements, utilizing kanban. The layout of the manufacturing system now defines paths that parts can take through the plant. Begin by connecting the elements with kanban links. The need for route sheets is eliminated. The parts (i.e., the in-process inventory) flow within the structure. All the cells, processes, subassemblies, and final assemblies are connected by the kanban links, pulling material to final assembly. This is the integration of production control into the manufacturing system, forming a L-CMS.

The cells are linked to each other by the pull system of production/inventory control called *kanban*. Kanban is a visual control system that is only good for lean production with its linked cells and its namesakes; it is not good for the job shop. The cells are linked together by kanban, thus providing control over the route that the parts must take (while doing away with the route sheet), control over the amount of material flowing between any two points, and information about when the parts will be needed. To accomplish this, there are two kinds of kanban: *withdrawal* (or conveyance) *kanban* (WLK) and *production-ordering kanban* (POK). One can think of the kanban as a link connecting the output side of one cell with the input point of the next cell; see Figure 10. The link is filled with carts or containers that hold parts in specific numbers. Every cart has the same number of parts. Each cart has one WLK and one POK. If there are K carts, then the

Maximum inventory $= K$ (number of carts) $\times a$ (number of parts in cart).

The arrival of an empty cart at the manufacturing cell initiates the order to make more parts to fill the cart. The kanban cards tell the material handler where to take the parts.

The same kind of links connect the subassembly cells to final assembly. All the other cells in an L-CMS are similarly connected by the pull system for production control, as shown in Figure 11.

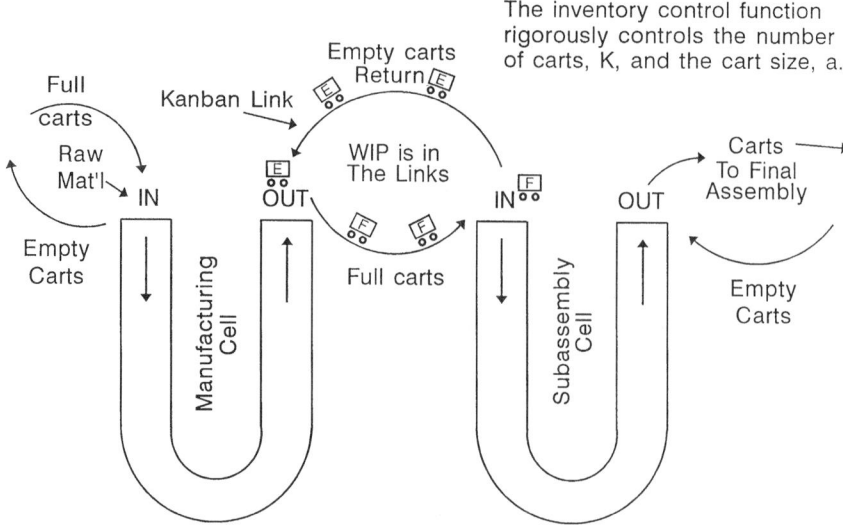

Figure 10. Cells are linked by a kanban system. The arrival of an empty container at the manufacturing cell is the signal to produce more parts.

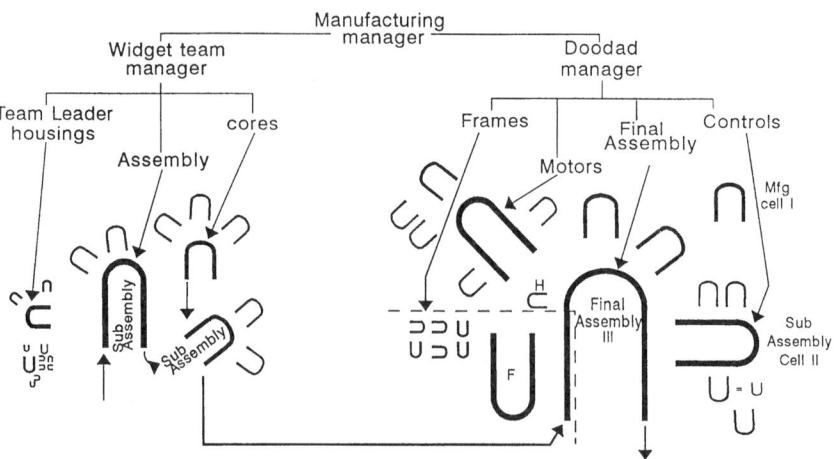

Figure 11. In the L-CMS, the cells are linked with controllable inventory buffers called kanban links or loops.

6.7. Step 7: Reducing the Work-In-Process

Step 7 involves the integration of the inventory control. The inventory in the system is held in the links and is called the *Work-In-Process* (WIP). The WIP inventory has been analogized to the water in a river, as shown in Figure 12. A high river level is equivalent to a high level of inventory in the system. The high river level covers the rocks in the riverbed. Rocks are equivalent to problems. Lower the level of the river (inventory) and the rocks (problems) are exposed. This analogy is quite accurate. The problems receive immediate attention when exposed. When all the rocks are removed, the river can run very smoothly with very little water. However, if there is no water, the river has dried up. The notion of zero inventory is incorrect. While zero defects is a proper objective, zero inventory is not possible. The idea is to minimize the necessary WIP between the cells. (Within the cell, parts are already handled one at a time, just as they are in assembly lines.)

The level of WIP between the stand-alone process, cells, subassembly, and assembly actually is controlled by the foremen in the various departments. The control is integrated and performed at the point of use. Here is how it works. Suppose that there are 10 carts in the link and that each cart holds 20 parts. The maximum inventory in this area is therefore 200 parts. The foreman goes to the stock area outside the cell and picks up the kanban cards (one WLK, one POK), which puts one full cart of parts out of commission. The (maximum) inventory level is now 9×20, or 180 parts. The foreman waits until a problem appears. When it appears, the foreman immediately restores the kanban, which restores the inventory to its previous level. The cause of the problem may or may not be identified by the restoration of the inventory, but the condition is relaxed until a solution can be enacted. Once the problem is solved, the foreman repeats this procedure. If no other problems occur, the foreman then tries to drop the inventory to $8 \times 20 = 160$ parts. This procedure is repeated daily in the links, all over the plant. After a few months, the foreman in the frame area may be down to 5 carts of 20 parts. Over the weekend the system will be restored to 10 carts between the two points, but this time each cart will hold only 10 parts. If everything works smoothly, with the reduced WIP lot size, the foreman will then remove a cart to see what happens. More than likely, some setup times will need to be reduced. In this way, the inventory in the linked-cell system is continually reduced, exposing problems. The problems are solved one by one. The teams work on solving the exposed problems. This is how continuous improvement works in the lean production factory.

The minimum level of inventory that can be achieved is a function of the quality level, the probability of a machine breakdown, the length of the setups, the variability in the manual operations, the number of workers in the cell, parts shortages, the transportation distance, and so on. It appears that the minimum number of carts is three, and, of course, the minimum lot size is one. The significant point here is that inventory becomes a controllable independent variable rather than an uncontrollable variable dependent on cravings of the users of the manufacturing system for more inventory.

6.8. Step 8: Integrate the Suppliers

Educate and encourage suppliers (vendors) to develop their own lean production system for superior quality, low cost, and rapid on-time delivery. They must be able to deliver parts to the customer when needed and where needed without incoming

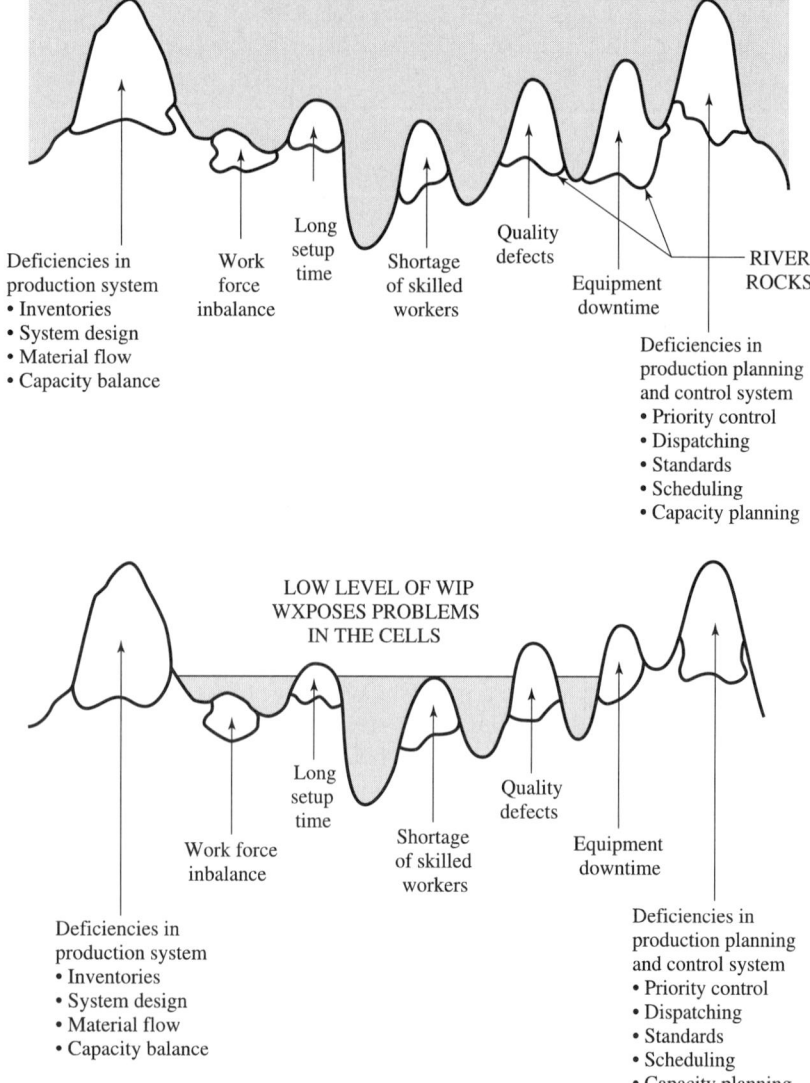

Figure 12. Rocks in the river analogy (Black, 1991).

inspection. The linked-cell network ultimately should include every supplier. Suppliers become a remote cell in the L-CMS.

In the traditional job shop environment, the purchasing department permits its vendors to make weekly/monthly/semiannual deliveries with long lead times—weeks/months are not uncommon. A large safety stock is kept just in case something goes wrong. Quantity variances are large and late and early deliveries are the norm. This situation leads to expediting.

As a hedge against vendor problems, multiple sources are developed. This may happen because one vendor cannot handle all the company's work. The purchasing department may claim that pitting one vendor against another gives the company a competitive advantage and lower cost parts.

The lean manufacturing system does it differently. Just-In-Time purchasing is a program of continual long-term improvement. The buyer and vendor work together to reduce lead times, lot sizes, and inventory levels. Both companies become more competitive in the world marketplace.

In this environment, longer-term (18–24 months) flexible contracts are drawn up with 3 or 4 weeks' lead time at the outset. The buyer supplies updated forecasts every month that are good for 12 months, commits to long-term quantity, and perhaps even promises to buy out any excess materials. Exact delivery is specified by midmonth for the next month. Frequent communication between the buyer and the vendor is typical. Kanban controls the material movement between the vendor and the buyer. The vendor is a remote cell. Long-range forecasting for 6 months to 1 year is utilized. As soon as the buyer sees a change, the vendor is informed; this knowledge gives the vendor better visibility instead of a limited lead time view. The vendor has *build schedule stability*, not "jerking" up and down of the build schedule.

The buyer moves toward fewer vendors, often going to local, sole sourcing. Frequent visits are made to the vendor by the buyer, who may supply engineering aid (quality, automation, setup reduction, packaging, and the like) to help the vendor become more knowledgeable on how to deliver, on time, the right quantity of parts that require no incoming inspection. This is truly technology transfer. The vendors learn from the customer. The buyer and the seller must be willing to work together to solve problems.

The advantages of single sourcing are that resources can be focused on selecting, developing, and monitoring one source instead of many. When tooling dollars are concentrated in one source, there is a savings in tooling dollars. The higher volume should lead to lower costs. The vendor is more inclined to do special things for the buyer. The buyer and the vendor learn to trust each other. The quality is more consistent and easier to control and monitor.

This chapter does not permit extensive discussion. Table 7 summarizes the key points in managing the lean production plant. This list was obtained from the plant manager at a first tier supplier to Toyota Camry.

6.9. Step 9: Autonomation

Autonomation means the autonomous control of quality and quantity. Stop everything immediately when something goes wrong; control the quality at the source instead of using inspectors to find the problem that someone else may have created. The workers in the lean factory inspect each other's work, called successive checking. Taiichi Ohno, former vice president of manufacturing for Toyota, was convinced that Toyota had to raise its quality to superior levels in order to penetrate the world automotive market. He wanted every worker to be personally responsible for the quality of the piece part or product that was produced (called source or self-checking).

Table 7. Managing the Lean Production System (LPS)

The Lean Production System is the basic philosophy and concepts used to guide production processes and environment. The LPS includes the linked-cell manufacturing system (cells linked by a kanban pull system), the five S's, standard operation, the seven tools of QC, and other key organizational elements.

Kanban pull system (see step 7): The production process that uses a card system, standard container sizes, and pull versus push production to accomplish just-in-time production.

Five S's (Seiri, Seiton, Seiketsu, Seisō, Shitsuke) (see step 4): The five S's are proper arrangement, orderliness, cleanliness, cleanup, and discipline.

Standard operation in manufacturing cells (see step 1): The production process used by technicians that combines people and process. The components of standard operation include cycle time, work sequence, and standard stock on hand in the cells.

Morning meeting: A daily meeting held for the purpose of sharing production and safety information, quite often by a quality circle.

Key points: Process sheets: The process sheets, which are visually posted at each workstation, detail the work sequence and most critical points for performing the tasks.

Tooling parts: Changeover and setup (see step 2): The machine setup that takes place when an assembly line changes products.

Seven tools of quality (see step 3): The seven tools to quality are Pareto's diagram, check sheets, histograms, causes-and-effect diagram, run charts for individuals, control charts for samples, and scatter diagrams.

Production behavior: Rules that include information on personal safety, safety equipment, clothing, restricted areas, vehicle safety, equipment safety, and housekeeping.

Visual management: Each line in the plant has a complete set of charts, graphs, or other devices, like Andons, for reporting the status and progress of the area.

The need for automation simply reflects the gradual transition of the factory from manual to automated functions. Some people think of this as Computer Integrated Manufacturing (CIM). Others recognize that people are the most important (and flexible) asset in the company and see the computer as just another tool in the process but not the heart of the system. These companies are moving toward Human Integrated Manufacturing (HIM) where a creative, motivated workforce is seen as the key to lean production.

Quite often, inspection devices are placed in the machines, source inspection, or in devices, called decouplers, between the machines, so the inspection is performed automatically. Again the idea is to prevent the defect from occurring rather than to inspect to find the defect after the part is made. Inspection by a machine instead of by a person can be faster, easier, and more repeatable. This is called *in-process control inspection.*

So autonomation means inspection becomes part of the production process and does not involve a separate location or person to perform it. Parts are 100% inspected by devices that either stop the process if a defect is found or correct the process before the defect can occur (required feedback to the controller). The machine may shut off automatically when a problem arises. This prevents mass production of defective parts. The machine may also shut off automatically when the necessary parts have been made to prevent overproduction. This is part of inventory control.

In *robotic cells*, the microcomputers of the CNC machine tools and a robot are networked together with a cell host computer. It is difficult, if not impossible, to conceive of this kind of arrangement without resorting to some method that collects the work into compatible families. All the machines in the cell are programmable, and therefore this kind of automation is very flexible. See Figure 4 for an example of a robotic cell.

While it is an easy task to draw the boxes and connect lines in Figure 4, it is quite a different matter to realize all of the mechanical, electrical, and computer engineering interfaces required to arrive at a fully unmanned, flexible, autonomous cell. In the real manufacturing environment, things are never as prescribed. The incoming work materials vary in geometry and specifications. The end effectors of the robot may not be able to accommodate such changes, or take note of random variations in the presentation of the components. Parts of the system will break, cutting tools will wear, and the quality control requirements will undoubtedly call for measuring more than one or two diameters.

6.10. Step 10: Restructure the Production System

Once the factory (the manufacturing system) has been restructured into a JIT manufacturing system and the critical control functions well integrated, the company will find it expedient to restructure the rest of the company. This will require removing the functionality of the various departments and forming teams, often along product lines. It will require the implementation of concurrent engineering teams to decrease the time needed to bring new products to market. This movement is gaining strength in many companies and is being called *business process reengineering*; it is basically restructuring the production system to be as waste free and efficient as the manufacturing system.

In shifting from one type of system to another, the change will affect product design, tool design and engineering, production planning (scheduling) and control, inventories and their control, purchasing, quality control and inspection, and, of course, the production worker, the foreman, the supervisors, the middle managers, and so on, right up to top management. Such a conversion cannot take place overnight and must be viewed as a *long-term transformation* from one type of *production system* to another. This kind of downsizing can be very traumatic for the business part of the company and usually has a negative impact on the morale of the company.

However, step 10 recognizes the need for the rest of the company to reorganize (get lean). This effort often begins with building *product realization teams* designed to bring new products to the marketplace faster. In the automotive industry, these are called *platform teams* and are an example of concurrent engineering. Platform teams are composed of people from design engineering, manufacturing, marketing, sales, finance, and so on. As the notion of team building spreads and the lean manufacturing system gets implemented, it is only natural that the production system will follow suit. Unfortunately, many companies are restructuring the business part of the company without having done the necessary steps 1 through 8 to get the manufacturing system lean and efficient. The reduction of the production system without simplifying and redesigning of the manufacturing system can lead to difficult times for the enterprise.

The design of the manufacturing process technology must be done early in the product development process. The manufacturing process technology within the manufacturing cells must be part of (i.e., elements within) a well-designed, integrated manufacturing system. Flexibility in the design of an integrated manufacturing system means it can readily accept new product designs. Flexibility in the process technology means the processes can readily adapt to product design changes and also be flexibly designed and engineered to accept new products.

The critical control functions of production control and inventory control are designed right into the manufacturing system. The critical control functions of quality control and process reliability are tied to process technology and are designed into the processes. Suppose a cell has six operations. The second machine may have a workholding device that checks to make sure that the first machine produced the correct part geometry before the second machine performs its processing steps. Sometimes the checking occurs in the part/holding transporting device between the first and second machines [called a decoupler by Black (1998; Black and Schroer, 1988)]. The decoupler device may simply check a dimension or it may provide feedback to the first machine to make process corrections. The worker is critical in this process as the operator is handling every part and checking every process.

In the manufacturing cells, the operators are considered to be the companies most important (fixed) asset, a point of view not traditionally held in U.S. factories where managers think of labor as an (unstable, costly) input to the manufacturing system. In L-CMS, operators must be multiprocess (can make different processes) and multifunctional (can make decisions about quality, maintenance, setups, process improvements, etc.). People are much more flexible than computers. Computers should be viewed as only a useful tool in the process.

7. CONTINUOUS IMPROVEMENT MEANS REDESIGNING THE SYSTEM

Technology, as in "manufacturing process technology," is what adds true value to the product for the external customer. *Methodology*, on the other hand, determines how the process technologies are sequenced together to produce the product. The methodology is captured or reflected by the manufacturing system design, which is defined as the complex arrangement of all the physical elements, including the people and processes, needed to make the product. *Thus, the methodology does not, in itself, add value, only additional cost.* Therefore, the most efficient methodology must be selected. This means that the most efficient design of the system must be evoked and that design must be continuously improved. *In other words, continuous improvement means continuously changing (redesigning) the manufacturing system.* Therefore, Manufacturing Systems Design (MSD) is emerging as a significant area of research and instruction and many universities are adding courses in this area.

8. LEAN MANUFACTURING CELL DESIGN

Product design (for manufacture and assembly) and MSD must go hand-in-hand, striving for customer satisfaction. The functional characteristics of a good MSD are flexibility, controllability, efficiency, and uniqueness. The factors needed to achieve satisfactory usage of the manufacturing system by the internal customers (the workers) are safety, equipment reliability, ergonomically sound equipment, and good service from engineering. Ergonomically sound equipment means that the processes are designed to be easy to operate, fail-safe, easy to maintain, and are not dirty, noisy, labor-intensive, or hazardous.

In the lean manufacturing system, manufacturing engineers are responsible for designing, building, testing, and implementing the manufacturing equipment that will be used by the internal customers in the manufacturing cells [machine tools and processes, tooling (workholders, cutting tools) and material handling devices (decouplers)]. Simple, reliable equipment that can be easily maintained should be specified. In general, flexible, dedicated equipment can be built in-house better than purchased and modified for the needs of the cell. Many companies understand that it is not good strategy simply to buy the manufacturing process technology from another company and then expect to make an exceptional product using the same technology as the competitor. When the process technology is purchased from outside vendors, any uniqueness aspects will be quickly lost. The company must perform Research and Development (R&D) on manufacturing technologies as well as manufacturing systems in order to produce effective and cost-efficient products. However, an effective, cost-efficient manufacturing system makes R&D in manufacturing process technology pay off.

What are the unique advantages of this custom or home-built equipment strategy?

1. *Flexibility* (process and tooling adaptable to many types of products): Flexibility requires rapid changeover of jigs, fixtures, and tooling for existing products and rapid modification for new designs. The processes have excess capacity—they can run faster if they need to, but they are designed for less-than-full capacity operation.

2. *Build exactly what you need:* There are three aspects to this. First, you are not paying for unused capability or options. Second, the machine can have unique capabilities that your competitors do not have and cannot get access to through equipment vendors. When you purchase equipment from vendors, you may be paying for capability your competition can get for free (from the vendor). Third, the equipment should allow the operator to stand. Equipment should be the appropriate height to allow the operators to easily perform tasks standing up and then move to the next machine in a step or two (i.e., have a narrow footprint).

3. *Maintainability/reliability/durability are all built-in features:* Equipment should be easy to maintain (oil, clean, changeover, replace worn parts, standardized screws). Many of the cells at the JIT suppliers are clones of each other.

The supplier company, being sole-source, has the volume and the expertise to get business from many companies making essentially the same components or subassemblies for many Original Equipment Manufacturers (OEMs). The vendors build a manufacturing cell for each OEM. The equipment can be interchanged from one cell to another in emergencies. The most skilled maintenance personnel must be given this task so that breakdowns in the L-CMS are eventually eliminated.

4. Design and build machines and material handling equipment (decouplers) and tooling to the needs of the cell and the system. Machines are typically single-cycle automatics but may have capacity for process delay. An example of process delay would be an Induction Heat Treatment (IHT) process that takes 4 minutes in a cell with a 1 minute CT. The IHT has the capacity for 4 units, each unit getting 4 minutes of treatment but outputs a unit every minute.

5. Equipment designed to prevent accidents (fail-safe).

6. Equipment designed to be easy to operate, load, and unload.

7. Equipment designed to process single units, not batches. Small footprint, low-cost equipment is the best. The MT (machining or processing time) should be modified so that it is less than the cycle time (the time in which one unit must be produced). Equipment processing speed should be set in view of the CT, such that MT < CT. The MT is related to the machine parameters selected. For example, for the turning lathe operation, suppose the part is of length L, then

$$\text{MT} = \frac{L + \text{allowance}}{\text{feed} \times N} \qquad \text{and} \qquad N = \frac{12V}{\pi D},$$

where N is the rpm of the spindle of the machine, V is the cutting speed, and D is the diameter of the shaft. Thus, increasing V decreases the tool life and usually requires more machine maintenance. The relationship here is (most simply)

$$V T^n = C,$$

where T is tool life, and n and C are empirical constants. Now suppose the MT is 30 seconds for the CT = 1 minute. The cutting speed can be reduced, thereby increasing the tool life and reducing downtime for tool changes. This approach also reduces equipment stoppages, lengthens the life of the equipment, and may improve quality.

8. Equipment can have inspection devices (such as sensors, pokayokes, counters) to promote autonomation. *Autonomation* is not the same thing as automation. Autonomation is the autonomous control of quantity (do not overproduce) and quality (no defects). Often the machine is equipped to count the number of items produced and the number of defects.

9. Equipment should be movable. Machines are equipped with casters or wheels, flexible pipes, and flexible wiring. There are no fixed conveyor lines.

10. Equipment should be self-cleaning. Equipment disposes of its own chips and trash.

11. Finally, equipment should be profitable at the production volume given to it. Equipment that needs millions of units to be profitable [Schonberger (1986) calls them supermachines] should be avoided because only once production volume even slightly exceeds the volume that the first supermachine can build, the purchase of another supermachine will be necessary, and the new supermachine will not be profitable until it approaches full utilization.

The equipment is designed and developed with priority on the internal customer factors even though the factors affecting the external customer are the highest priority of manufacturing engineering. Although many plants lack the expertise to build machines from scratch, most have the expertise to modify equipment to give it unique capabilities. This is the interim approach, using new machines instead of existing machines.

9. INTERIM CELLS

An interim cell uses existing equipment arranged in a U-shaped cell for one-piece flow. The equipment can be new equipment that was built for the job shop, stand-alone applications (see Fig. 13), or existing equipment in the plant. (Figure 3 was a good example.) The cell shown in Figure 13 replaced a transfer line using CNC equipment. This cell is designed for 10 different cylinder heads for small gas engines. The company brought together 2 machine tool companies and a workholding vendor as members of the design team. The part volumes ranged from 1000 to 600,000 parts per year. The machines were modified to allow the operators to quickly and easily load and unload the machines. Fixtures were designed for quick load and unload and rapid changeover from one part in the family to another. Some of the fixtures have part checking sensors (checking the operation from the previous machine) and pokayoke capability (defect prevention). Each machine can have a different fixture setup performed in less than one minute. The cell outputs about 2500 units per day and is now being cloned to add additional capacity (replacing other transfer lines). The design shown in Figure 13 is one of many that were evaluated using digital simulation and is not the actual layout being used by the company.

10. EXAMPLE OF A LEAN CELL DESIGN

The lean manufacturing system has some unique characteristics that are embodied in the manufacturing and assembly cells. The cells are manned by multifunctional workers. This means that they can perform tasks other than handling the material and operating the equipment. These tasks include quality control and inspection (to prevent defects from occurring), machine tool maintenance, setup reduction, and problem solving. The cells are usually U-shaped or rectangular. The cells are designed

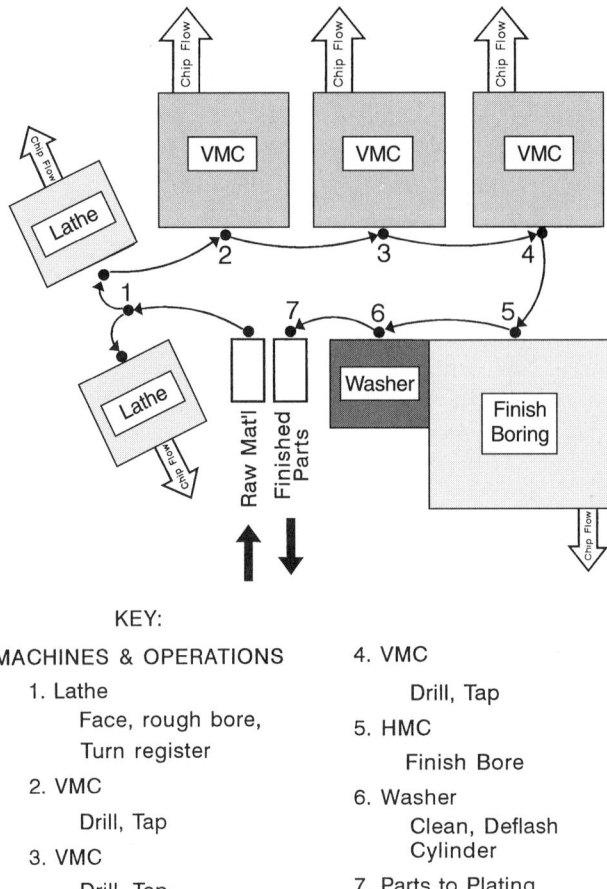

KEY:

MACHINES & OPERATIONS

1. Lathe
   Face, rough bore,
   Turn register
2. VMC
   Drill, Tap
3. VMC
   Drill, Tap

4. VMC
   Drill, Tap
5. HMC
   Finish Bore
6. Washer
   Clean, Deflash
   Cylinder
7. Parts to Plating

Figure 13. Interim cell for the manufacture of a family of cylinders, using six machines and one, two, or three operators. VCM = Vertical Milling Machine; HMC = Horizontal Boring Machine.

so the worker can step across the aisle and work on machines on the opposite side. The concept is called "separation of machine's work and man's work."

To implement this system requires hard/systematic work. The design of the cells and the system continues to evolve (change) and improve over time. This continuous improvement is forced through the gradual removal of inventory from the entire manufacturing system. See step 7 in Table 3. This gradual removal of inventory is from the inventory links between the cells. Within the cells, the amount of material on hand (called the stock on hand) is carefully controlled by the cell operators. Most of the equipment in the cell operates untended (completes the cycle initiated by the operator) and has devices built in to prevent defects from occurring (called pokayoke devices).

The cells operate on a one-piece flow strategy. Within the cell there is an exact number of parts that are either in the machines or in the decouplers between the machines. The material within the cell is called the stock on hand. When a part is finished and exits the cell, another part can be started into the cell. The true secret of lean production are these manufacturing cells that operate on the one-piece flow philosophy (Sekine, 1990). Very little has been written about lean cells but these manufacturing cells are the heart of the TPS.

Therefore, understanding the TPS is to understand how a manufacturing cell works. Manufacturing cells are the proprietary element in lean production. A typical lean manufacturing cell is shown in Figure 14. This is a fairly large cell capable of high throughput rates. The cell produces a part called the rack bar. The part requires heat treating, inspection, and mechanical straightening, in addition to numerous machining operations like drilling and tapping, gear teeth milling, deep hole drilling, grinding, and broaching. The point is that all the processing required to produce a finished bar, ready for subassembly, is in the cell.

This cell can make 10 different types of racks for the same model of product. The changeover at any individual machine occurs with "one touch" at the time the operator unloads the previous part for the machine (Shingo, 1985). Many machines are equipped with pokayoke devices that can prevent the machines or the operator from making mistakes.

This cell is designed a bit differently in that the work arrives and departs from the middle of the cell, but the cell still has a U shape. The operations at the start (the right end of the cell) are the same for all the bars in the family. Therefore, in this area, automatic transfer devices (small robots and mechanical arms and levers) can be used to move the part from machine to machine. For the transfer line portion of the cell, steps 2–10, the times for the machining processes (deep hole drilling) are longer than the cycle time for the cell. The cell puts out one finished rack bar per minute. Then one rack is started through the cell every minute and some of steps 2–10 are duplicated so that, on the average, the MT is less than one minute. In this area, the machines have automatic repeat cycle capability with automatic transfer devices moving the parts from step 2 to 10.

In total, the rack moves through 26 steps or operations, most of which are machining, performed by single-cycle automatic machines (that turn off after the machining process is complete). However, there are some manual operations in the cell other than loading/unloading. These include steps 13 and 20 to manually straighten the bar (which can warp) after heat treating or the assembly of parts onto the bar.

Figure 15 shows the standard work sheet for the manufacturing cell shown in Figure 14. Two operators are shown in the cell. These workers are standing, walking workers who move from machine to machine in loops as shown in the figure. The open circles indicate the positions at the machines when they perform some tasks. Each operator makes the loop in about one minute. Operator 1 addresses 10 stations and operator 2 addresses 11 stations. Mostly what the operators are doing is unloading a machine, loading another part into the machine, checking the part they have unloaded, and dropping the part into the decoupler elements between the machines. The Stock-On-Hand (SOH) in the decouplers and the machines help to maintain the smooth flow of the parts through the machines. The decouplers also can be designed to perform

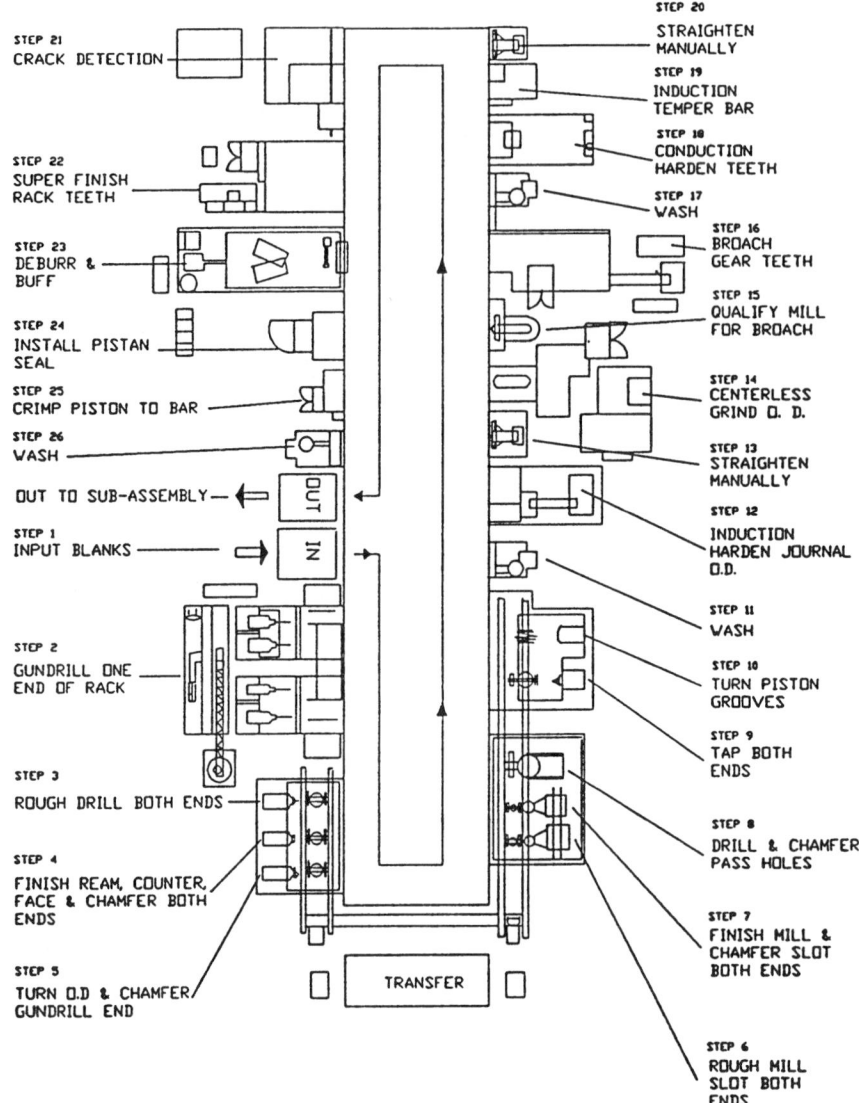

Figure 14. Typical lean manufacturing cell.

inspections for part quality or necessary process delays while the parts heat up, cool down, cure, and so forth. The SOH is kept as small as possible.

After they have completed all the tasks at a machine, the operators walk to the next machine, hitting a start switch for the machine as they leave (called a "walk away" switch).

Sometimes the decoupler elements perform the inspection or checking of the part, but mostly they serve to transport parts from one process to the next. Sometime

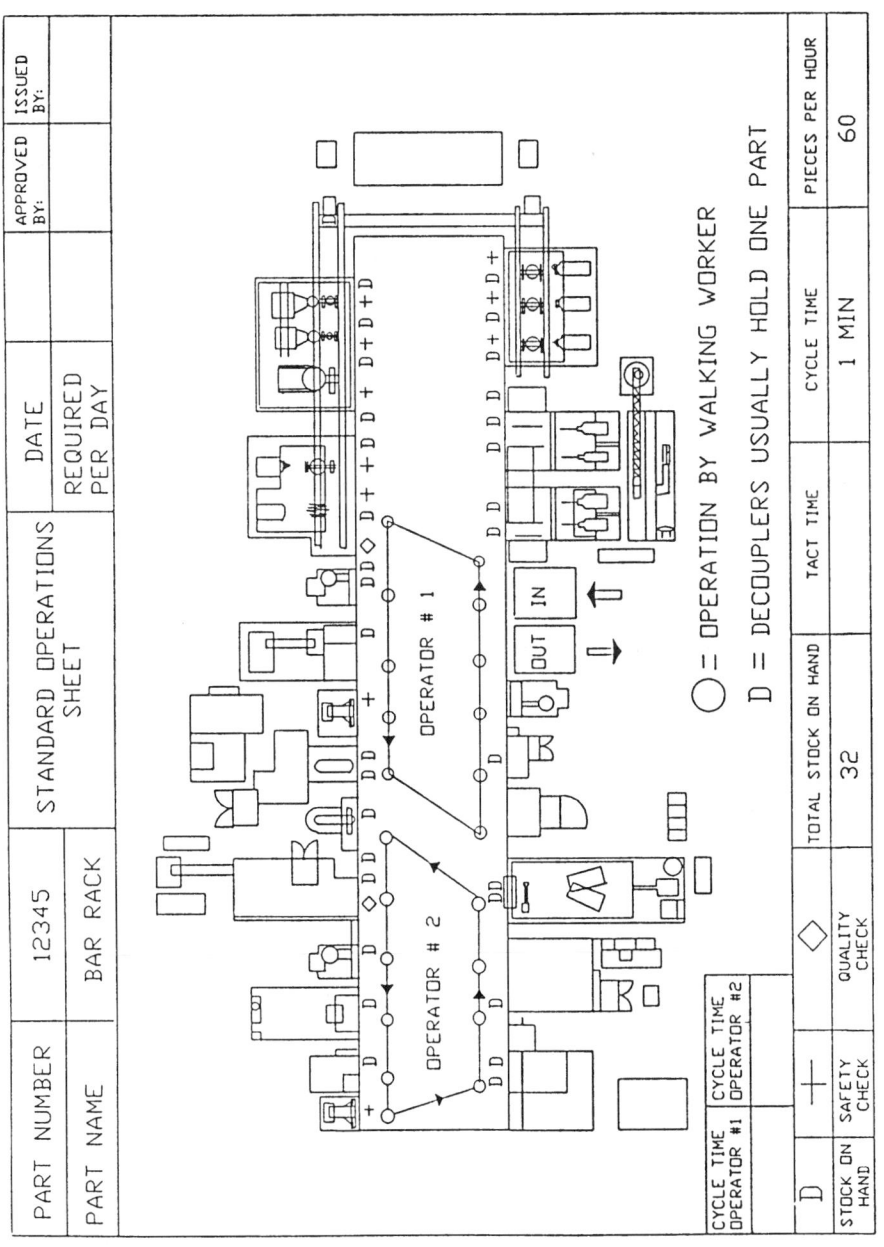

Figure 15. Standard work sheet for manufacturing cell.

the decoupler performs a secondary operation like deburring or degaussing the bar to remove residual magnetic fields. The bar is made from steel, which can become magnetized, causing small chips to adhere to the bar and perhaps cause the bar to be mislocated in the subsequent process.

Notice operator 1 controls both the input and output of the cell and this is by design. One operator always controls the volume of material going through the cell. This also keeps the SOH quantity constant and keeps the cell working in balance with the final or subassembly lines it is feeding. Operator 1 loads the centerless grinder then moves across the aisle to unload the operation called deburr and buff. Operator 2 unloads the centerless grinder and loads the part in the next process, moving in a counterclockwise loop in the cell from right to left.

At the interface between the two operators, either one can perform the necessary operations depending on when they arrive and when the processes in the machines are finished. That is, the region where the two operators meet is really not fixed but changes or shifts depending upon the way parts are moving about the cell. This is called the relay zone (Suzaki, 1987). This flexibility requires the workers be cross-trained on all the processes in the cell. The cell is designed so that it can be operated by one, two, three, or even four workers. Changing the number of workers changes the output rate. This is a key to flexibility.

11. RULES FOR LEAN MANUFACTURING CELL DESIGN

- Each machine or process or operation in the cell is designed for a standing/walking worker coming to the machine from the right or straight on, and leaving to the left.
- The material moves from right to left since most people are right handed, and it is easier to unload (with the left) than it is to load (with the right).
- The machines are designed to have walk-away switches that the worker hits when leaving the machine. The doors close and the machine begins the processing cycle untended. The machine completes the processing cycle in time MT. The machine is at least a single-cycle automatic.
- The machines are arranged in the sequence of operations needed to process the part.
- All the processes needed to make the part are in the cell and have an MT less than the CT or the average CT. This requires some rather unique process technology that cannot be described here for proprietary reasons.
- The width of the aisle in the cell is about 4 feet so that workers can pass each other in the cell but can also easily step across the aisle. Between each process is a decoupler that holds one part. The decouplers are designed to maintain the flow of the parts decoupling the dependency of one step or operation on the next. The decoupler may also transport the part, reorient the part, deburr the part, degauss the part, hold the part for heating or cooling or curing or drying or whatever (this

is called process delay). Decouplers are custom designed to hold all the parts in the family with equal facility.

- The cell is equipped with many pokayokes performing self-inspections or successive inspections. The devices can be in the decouplers or in the workholding of the next machine and should be simple.

12. ERGONOMICS OF LEAN MANUFACTURING CELLS

Ergonomics deals with the mental, physical, and social requirements of the job and how the job is designed (or modified) to accommodate the human limitation. For example, are the machines in the cell designed to a common height to minimize lifting of parts? Are transfer devices designed for slide on/slide off? Are automatic steps equipped with interrupt signaling to help the worker monitor the process? When the job is defined as primarily loading/unloading, ergonomic concerns with lifting and placing parts in machines and operating workholding devices must be addressed. In these systems human performance in detecting and correcting cell malfunctions will establish utilization and thus production efficiency. The design of machines for maintainability and diagnostics is critical.

The original designer of the cell must incorporate ergonomic issues initially rather than trying to come back later to implement fixes. We are working to develop the design rules for ergonomics in lean manufacturing cells.

12.1. Constraints to Conversion

Aside from the failure to recognize cells as a new form of manufacturing system, a major effort on the part of a business is required to undertake a conversion to lean manufacturing. The constraints on implementation are as follows:

1. The top management person (the real leader) does not buy in totally to the conversion.
2. *Systems changes are inherently difficult to implement.* Changing the *entire* manufacturing production system is a huge job.
3. *Companies spend freely for new manufacturing processes but not for new manufacturing systems.* It is easier to justify new hardware for the old manufacturing system than to rearrange the old hardware into a new manufacturing system (linked cells). However, anyone with capital can buy the newest equipment, often creating another island of automation.
4. *Fear of the unknown by top management.* Decision making is choosing among the alternatives in the face of uncertainty. The greater the uncertainty, the more likely that the "do-nothing" alternatives will be selected. Converting to linked cells will free up additional capacity (setup time saved) and capital (funds not tied up in inventory). Such conversions will require expenditure of funds for equipment modifications, employee training (in quality, maintenance, and setup

reduction), and so forth. *The long-term payback equals a highrisk situation in the minds of the decision makers.*

5. *Faulty criteria for decision making.* Decisions should be based on the ability of the company to compete (quality, reliability, unit cost, delivery time, flexibility for product change or volume change) rather than on price alone.

6. *Lack of blue-collar involvement in the decisionmaking process of the company.* Getting the production workers involved in the decisionmaking process is in itself a significant change. The managers of the manufacturing system have had problems adjusting to this situation.

7. The conversion to lean manufacturing represents a *real threat to staff and middle managers.* The functional tasks that they have been responsible for are being shifted and integrated into the manufacturing system. Also, the short-term life of the financially oriented middle managers is in conflict with the long-term nature of the program, which results in resistance to change in addition to the erosion of their functional empires.

13. SUMMARY: USE IM,C NOT CIM

There are many who believed that the only way in which manufacturing companies can compete is to automate. This was the CIM approach (Ayers, 1991). In a nutshell, the concept was to achieve integration through computerization and automation. But lean manufacturing cells is a different approach. I call it IM,C (or Integrate the Manufacturing system then Computerize and automate). The development of manufacturing and assembly cells are just the first step in the manufacturing systems design (Black, 1991). Experts on CIM now agree that lean manufacturing must proceed efforts to computerize the system (Ayers and Butcher, 1993). While costs of these systems are difficult to obtain, the lean cell approach is significantly less costly than the CIM approach.

We hear a lot of talk today about continuous improvement. Continuous improvement requires the continuous redesign of the manufacturing system. We hear a lot of talk about Technology Transfer (TT). Technology Transfer happens when the buyer shares their experience in L-CMS with their vendors on a one to one basis (step 8).

The lean factory is based on a different design for the manufacturing system in which the sources of variation in time are attacked and the delays in the system removed. In summary, the next generation of American factories (the factories with a future) will be designed with manufacturing and assembly cells linked together with a pull system for material and information control. In this L-CMS, downstream process will dictate upstream production rates. The L-CMS strategy simplifies the manufacturing system, integrates the critical control functions before applying technology (automation, robotization and computerization), avoids risks, and makes automation easier to do.

REFERENCES

AT&T Technical Journal, (July/Aug. 1990). Striving for manufacturing excellence, Vol. 69, No. 4.

Ayres, R. U. (1991). *Computer Integrated Manufacturing,* Chapman & Hall, London, UK.

Ayres, R. U., and Butcher, D. C. (1993). The flexible factory revisited, *American Scientist,* **81**, 448–459.

Black, J T. (1988). The design of manufacturing cells (step one to integrated manufacturing systems, *Manufacturing International '88,* Vol. III, p. 143.

Black, J T. (1991). *The Design of the Factory with a Future,* McGraw-Hill, New York.

Black, J T., and Schroer, B. J. (1988). Decouplers in integrated cellular manufacturing systems, *Journal of Engineering for Industry, Transactions ASME,* **110**, 77–85.

Black, J T., and Schroer, B. J. (1993). Simulation of an apparel assembly cell with walking workers and decouplers, *Journal of Manufacturing Systems,* **12**(2), 170–180.

Black, J T., Jiang, C. C., and Wiens, G. J. (1991). Design, analysis and control of manufacturing cells, *PED,* **53**, 000–000.

Buzacott, J. A. (1995). A perspective on new paradigms in manufacturing, *Journal of Manufacturing Systems,* **14**(2), 118.

Cochran, D. (1994). Manufacturing System Design and Control, Ph.D. Dissertation, Auburn University.

Ford, H. (1988). *Today and Tomorrow,* Productivity Press, Portland, OR.

Hall, R. W. (1983). *Zero Inventories,* Dow Jones-Irwin, Honewood, IL.

Harmon, R. L., and Peterson, L. D. (1990). *Reinventing The Factory II: Managing the World Class Factory Today,* Free Press, New York, NY.

Monden, Y. (1983). *Toyota Production System,* Industrial Engineering and Management. Press, Norcross, GA.

Nakajima, S. (1988). *TPM, Introduction to TPM: Total Productive Maintenance,* Productivity Press, Portland, OR.

Ohno, T. (1988). *Toyota Production System: Beyond Large-Scale Production,* Productivity Press, Portland, OR.

Schonberger, R. J. (1982). *Japanese Manufacturing Techniques: Nine Hidden Lessons in Simplicity,* Free Press, New York, NY.

Schonberger, R. J. (1986). *World Class Manufacturing,* page 78, Free Press, New York, NY.

Sekine, K. (1990). *One-Piece Flow: Cell Design for Transforming the Production Process,* Productivity Press, Portland, OR.

Shingo, S. (1985). *A Revolution in Manufacturing: The SMED System,* Productivity Press, Portland, OR.

Shingo, S. (1986). *Zero Quality Control: Source Inspection and the Poka-Yoke System,* Productivity Press, Portland, OR.

Shingo, S. (1989). *A Study of The Toyota Production System,* Productivity Press, Portland, OR.

Suh, N. P. (1990). *The Principles of Design,* Oxford University Press. New York, NY.

Suh, N. P. (1992). Design axioms and quality control, *Robotics and CIM*, **9**(4/A), Aug–Oct. 367.

Suzaki, K. (1987). *The New Manufacturing Challenge*, Free Press, New York, NY.

Whitney, D. (1992). Toyota (Plant visit), *Scientific Information Bulletin*, **17**(3).

Womack, J. P., Jones, D. T., Roos, D. (1991). *The Machine That Changed the World,* Harper Perennial, New York, NY.

16

INDUSTRIAL IMPLEMENTATION OF PRODUCTION FLOW ANALYSIS

Marc Barth and Roland De Guio

Laboratoire de Recherche en Productique de Strasbourg
Ecole Nationale Supérieure des Arts et Industries de Strasbourg
F-67084 Strasbourg Cedex, France

1. INTRODUCTION

Workshop designing is one of the many improvement methods followed by contemporary firms. Numerous factors such as the workshop's technical characteristics, its links with the information and control systems, its environment, its aim, and the firm's workforce make this designing a complex problem. It is advisable to entrust a work group with the designing task in order to integrate the various points of views. If it is to be efficient, the design group has to conform to a method. As early as the 1950s, Immer (1950) and then Murther stated that the general designing process of a production system is identical to any designing process in the engineering field, that is, definition of the objectives, search for ideas, valuation of the solutions, and final decision. Other authors such as Apple (1977), Nadler (1961), Reed (1961), Tompkins (1989), and White (Tumpkins, et al., 1996) have greatly enriched these methods.

It is worth noting that the general workshop designing process has not fundamentally evolved since the 1950s and that it does not include an element of any contemporary design process, that is, the functional analysis. The consequences of

Handbook of Cellular Manufacturing Systems, edited by Shahrukh A. Irani
ISBN 0-471-12139-8 © 1999 John Wiley & Sons, Inc.

this omission for the workshop thus designed are similar to the ones met whenever a product is conceived without making use of the functional analysis. On the one hand, the search for solutions is too often carried out on a technological basis, whereas it should be oriented toward the fulfilling of functions. Since the solutions' valuation step does not allow the technological solutions to be compared with the functions, the firm might be induced to select low-value options.

Furthermore, the functional approach gives the opportunity, via the concept of value, to design the workshop in the prospect of the factory's global performance optimization. It brings about a quantitative gain by cutting down the cost and the time required for the workshop design. Finally, by reducing the risk of forgetting major components or functions of the production system, it favors the quality of its design.

Independent of the patent stake raised by the introduction of functional analysis in the field of workshop designing, this chapter aims at defining, as precisely as possible, a method exploiting the new prospects that functional analysis brings about. In this chapter we present a workshop designing method based on the value engineering method that was tested on several industrial projects. We start by describing the complexity of the production system and showing the need for a systematic implementation approach. This leads us to define the draft specifications of the method. The operational specifications of the method are then given. Each step is illustrated with examples from industrial case studies.

2. THE NEED FOR A SYSTEMATIC IMPLEMENTATION METHOD

2.1. Complexity of the Production System

The complexity of designing a production system is due to various factors: the structure of the system, its environment, its aims, and participants.

Structure of the Production System   The production system consists of three interacting subsystems represented in Figure 1. An arrow between two subsystems means that the system the arrow is coming from acts on the system to which the arrow is pointing. The machines, tools, tooling, handling, and inventory control methods provide the physical transformation of the parts and the subassemblies into finished products and constitute the operating system. The information system generates the production reports. The manufacturing orders and the quality procedures are examples of elements of the information system. Lastly, the control system controls the operating system via the information system.

Because of these interactions, the selection of a solution for one of the subsystems will affect the solution chosen for the other two subsystems. Thus, choosing a control method such as the kanban (control system) imposes a minimum level of quality in the circulation of the information written down on the kanban labels (information system). The number of labels and their location must be carefully studied. The use of the kanban method also affects the operating system resources. The size of the containers and the number of containers to be handled are parameters affecting

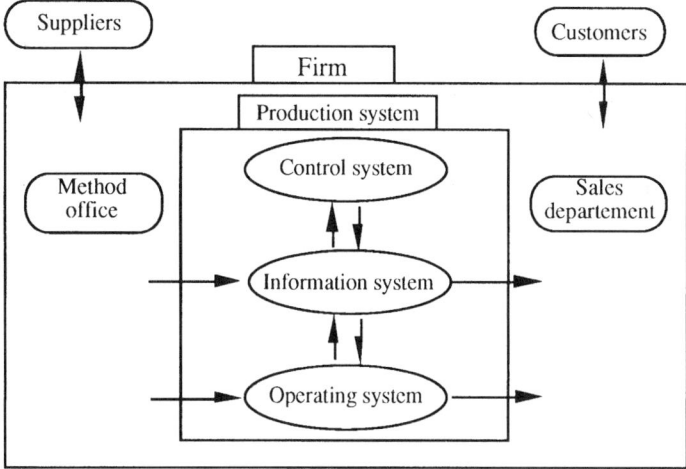

Figure 1. Production system and its environment.

the shop's physical system layout (operating system). Therefore, the design of the operating production system, which is the subject of this chapter, cannot be executed without referring to the information and control systems.

Environment of the Production System   The production system previously described is located in an environment within which two types of elements acting on the production system are commonly distinguished: the internal and external elements to the firm. The structure of the loading and unloading docks and the receiving and shipping points will partially depend on the external participants such as the customers or the suppliers. The participants internal to the firm, such as the Sales, the Research and Development, or the Maintenance Departments, have an effect on the shop's structure through their own objectives. Thus, the Sales Department, aiming at delivering on time, defines, with the agreement of the Production Department, a safety stock level for inventory control. The inventory level imposes area or volume constraints on the handling systems in use.

Purpose of the Production System   The development of the economic background has inclined firms to move from mass production to diversified production under strong market discipline. The constraints and objectives have been modified accordingly. To labor productivity and automation, which gave all its splendor to mass production, has been added delivery within time, lower production costs, higher quality standards, and product diversification. These constraints compel firms to structure their shops in such a way as to reach high productivity. The shop designer's ultimate aim does not any longer consist in locally optimizing this system. On the contrary, the designer must ensure that a global optimum of the system is searched for through the simultaneous integration of numerous constraints linked with the strategic objectives of the firm. Many skills mastered by various participants internal and external to the firm are then required.

Characteristics of the Participants Implementing the Factory Layout

The production operators often are the only members of the workforce to know every single detail about how the shop operates. They play a major role in becoming acquainted with the physical resources and in validating the proposed solutions. The production engineers are knowledgeable on issues such as the shop layout. The previous examples illustrate the fact that the design of a production system requires skills that are acquired at numerous hierarchical levels. The multiplicity of the objectives, due to the components, the environment, and the purposes of the system, requires skills available in numerous departments. Within these departments, differences in the initial training of the workforce induce different skills and expertise. They lead to different points of view about the production system.

2.2. Specifications of a Systematic Implementation Method

As just shown, the design of a production system is a complex issue. The number of points of view and their heterogeneity justifies the use of a systematic implementation method. This method must:

1. Guide the group's work at every hierarchical and functional level
2. Be understandable at every hierarchical level
3. Conceptually dissociate the search for functions to be performed from their solutions
4. Conceptually dissociate the search for solutions and their valuation
5. Promote the individual creativity within a group
6. Promote the contribution of each participant's knowledge at the opportune moment
7. Authorize the iterative nature of any design activity
8. Authorize the integration of the financial aspects
9. Authorize a global valuation of the solutions
10. Adjust to the extent of the breakdown desired by the designer

There exist some systematic implementation methods dedicated to design. They differ in complexity, accuracy, and in the type of product and services at which they are aimed. The scientific and industrial literature concerning systematic implementation methods lacks publications dedicated to the design of production systems. Working out such a method raises two problems. On the one hand, the stages and their contents must meet the set of subproblems raised during the design; on the other hand, the linking up of the stages must structure the design activity while complying to the 10 previous properties. The value engineering method, known in the field of product design for many years, offers a conceptual framework that addresses these concerns.

It comprises six main steps:

1. Orientation of the project
2. Search for information
3. Functional analysis
4. Search for ideas
5. Study and valuation of the solutions
6. Primary evaluation, submission of the solutions, and decision making

Classic examples of this literature illustrate these stages for the design of products. This chapter shows that the value engineering method is general enough to be applied in the specific framework of production system design. The method we introduce, which includes the six main steps of value engineering, is detailed and adjusted to the specific needs of production system design.

The use of this method is justified and original in many ways:

1. Value engineering is an amply tested method and is well known in industrial engineering.
2. The functional analysis stage specifies the various functions expected from the production system while not taking into account the technological solutions. This neutrality of the method makes it suitable for a great number of industries and allows it to cover a large range of solutions from the simplest to the most complex.
3. It directs the selection of solutions by means of the concept of value, which specifies the fulfilling of a need compared with the solution's cost. This concept of price/quality ratio prevents the selection of solutions often motivated by purely technological considerations.
4. The method facilitates the use of several cost and profitability measures. Each firm can make use of its own valuation technique.

However, no design method can guarantee the quality of a product or service. The method we recommend structures the activity of a design team. It drives the person in charge of the project and the designer to objectively ask questions. The use of this method minimizes the risks of a faulty design. By no means must the steps we recommend be considered as restraints. On the contrary, the whole individual and collective creativity can be expressed provided the project leader promotes and channels it.

2.3. The Method

First of all, brief comments are made on each step of the project described in Figure 2. The text in italic illustrates our remarks. Then, steps 1–6 specify all six stages of the method and make them operational. Although the steps are linearly described, the reader must keep in mind the iterative nature of the method.

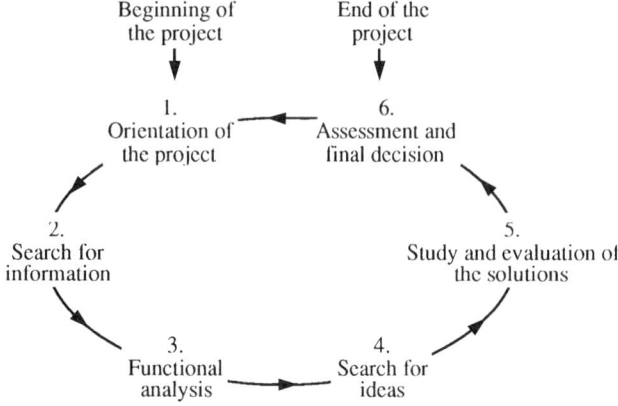

Figure 2. The method.

Step 1: Orientation of the Project   The orientation of the project puts into place the general framework of the project. The data, the objectives, the stakes, and the project management's structure are therein described.

Lower the production lead times from 3 to 1 week.
The term of the study is 4 months.
A multipurpose design team is created.

Step 2: Search for Information   During this stage, an inventory of the technical, industrial, economic, marketing, social, and regulatory information useful for subsequent steps is drawn up. Some pieces of information will prove to be necessary only at further stages of the project.

A plan of the shop on which the information and physical flows are mentioned is drawn up. The Sales Department provides the sales trend for the next 3 years. The method office inquires about the norms concerning the noise and the working conditions.

Step 3: Functional Analysis   It consists in defining, characterizing, and arranging the functions that the production system must carry out while avoiding reference to the technology in use in the concrete solutions.

Function: To transport the containers from a machine to another.
Criterion 1: Weight, from 1 to 13 kg.
Criterion 2: Transportation rate, 3600 containers per day.
Criterion 3: Ergonomics, to conform to the norms.

Step 4: Search for Ideas and Paths toward Solutions   The purpose of this stage is to suggest the maximum amount of solutions to the functions or set of functions. This step must be conceptually dissociated from any quantitative or qualitative valuation. It is at this very stage that must be expressed the creativity of the design team.

Function: To transport the containers from a machine to an other.

Solutions: Manual trolley, forklift truck, automated guided vehicle system, aerial conveyor.

Function: Visually clarify the flow.

Solutions: Layout into cell or Group Technology (GT) cell, linear layout.

Step 5: Study and Valuation of the Solutions   Full solutions are obtained from the partial solutions suggested during the previous step. These are appraised in terms of feasibility, costs, and various constraints. The selected solutions are detailed and then valued again. The process is continuing until a few good solutions are obtained.

Solution 1: Layout into cell with transport by manual trolley.

Solution 2: Layout into GT cell with transport by automated guided vehicle system.

Solution 3: Linear layout with transport by aerial conveyor.

Solution 4: Linear layout with transport by automated guided vehicle system.

The first selection turns solutions 2 and 4 down. Solutions 1 and 3 are studied further.

Step 6: Primary Evaluation, Submission of the Solutions, and Decision Making   The purpose of this stage is to draw up a primary evaluation of the selected solutions. The following justifications are necessary: motives of selection, level reached by each criterion, cost assessment, and terms of implementation.

Solution 1 is judged less efficient in terms of production lead times than solution 3, but it is less sensitive to variations in the quantities to be produced. The cost of both solutions is approximately the same. The decision raises a dilemma.

The main interest of the method we recommend rests in the use of the functional analysis. This stage leads to the definition of functions the solutions of which are found during step 4. Furthermore, all these solutions and their criteria allow the design team to judge the relevance of the solutions during step 5. As often as not in this kind of project, this stage is not considered. This omission can have harmful consequences. During step 4, the designer runs the risk of basing the design of the shop on solutions that are important merely according to a single point of view. Once the shop is laid out, some omissions are difficult to rectify. Likewise, the solutions imagined during step 4 must be judged in step 5. Without the comparison criteria expressed during the functional analysis, the comparison of the solutions in step 5 becomes difficult or even hazardous. An exhaustive description of the functions, constraints, and criteria prevents this methodological pitfall. Functional analysis often seems tedious. Step 3 describes a conceptual framework that substantially lightens the analysis.

The complete dissociation of the solutions from the functions is not easily assimilated and put into practice. The head of project must see to it. The approval of the method by the design team is a necessary condition to the success of such a project. It is essential to provide the training of the design team in the method. The training lasts between one and several days, depending on the participants' skills and the level of training to be reached.

3. OPERATIONAL SPECIFICATIONS OF THE METHOD

The aim of this section is to make the method operational. As a result of several layout experiences, general concepts have been revealed that are systematically dealt with at various steps. These concepts are described here. The reader can refer at any time to Appendix 1 in which the method is described by means of a synopsis of the steps of the method.

3.1. Orientation of the Project

The purpose of this step is to grasp the main components of the project in order to determine the resources and time necessary for its progression. The objectives and information required at this stage are roughly the same in most cases. We suggest a checklist that might trigger the design team's thinking.

Reasons for the Project   The origin of the project can be internal or external to the firm. The observation of competitors or the recommendations to reorganize from main trading partners forces companies to continuous improvement. If the project is enacted by the general management, its structure is easily put into place. If the project is initiated by a person who does not belong to the management, it is advisable to seek the approval of higher ranks in the hierarchy, otherwise the lack of resources will not allow the project to be executed in good working conditions for the design team.

Scope of the Study   The designer must set limits on the production system to be studied. In all events, a global plan of the factory must locate the area(s) to be laid out. A whole site, one or more shops, can be laid out.

Objectives   Nowadays, the objectives most commonly named by industrialists are the reduction of production lead times and costs and the improvement of the products' quality. These are often expressed indirectly. They are classified into the four categories defined as follows:

The objectives linked with the operating system:

- To adapt the required capacity
- To improve the working conditions
- To increase the volume of production

- To reduce the production risks
- To reduce waste in the production process
- To reduce the work-in-process inventory
- To integrate new machines
- To integrate quality control at production level
- To integrate new families of products into the existing system
- To modify the production processes
- To rationalize the means of production
- To reduce the production costs
- To simplify the physical production flows

The objectives linked with the information system:

- To improve the information system of the area to be laid out
- To improve the information flows
- To formalize and standardize the parameters of the process leading to the products' quality
- To master the customer/supplier relations within the firm
- To eliminate unnecessary information

The objectives linked with the decision system:

- To facilitate the communication between departments in order to improve the production management
- To decentralize the decision making
- To conceive a new production activity control system

The objectives linked with the production personnel:

- To reduce the number of industrial accidents
- To improve the working conditions
- To promote individual development
- To make the operators responsible for quality and on-time delivery

Data Requirements of the Problem   Synthetic data about the products, the machines, and their flows are necessary in order to measure the extent of the project. The following list is not exhaustive:

- Turnover
- Workforce
- Bill of material
- Volume of production

- Inventories and work-in-process report
- Main physical flows in the factory and shops
- Main production constraints (pollution, civil engineering, etc.)
- Type of production management
- Situation of the firm concerning other outstanding projects

Constraints   The constraints are limitations on the progress and execution of the project. These can be physical, such as the civil engineering constraints, or temporal, such as the date at which the project ends, or financial, such as the budget allocated to the study or to the layout. As for the constraints on personnel, the personnel's availability for this project as well as the need for external skills will be assessed.

Means   The means are expressed in terms of budget, man-hours per week, and skills necessary for the realization of the project.

Participants to the Project: Forming a Design Team   The method we recommend requires the forming of a design team. There exist numerous ways to work as a team. In some firms, the head of the project can easily gather and meet colleagues during informal sessions. In others, it is necessary to meet colleagues in especially formal sessions. It is thus very difficult to set a general rule. The head of the project must take into account a set of factors linked to the organization chart, to the type of management, and to the potential relations with colleagues. In all events, the head of the project plays a role of the utmost importance: to watch over the development of the project. The main difficulty is for the head of the project to remain impartial throughout the project.

As an example, we give the structure of a design team put into place in the case of a fairly complex project. The main factors of complexity of the project were the inexperience in teamwork, the number of parts and machines (3000 and 30, respectively), and the use of a new building. A formal structure, as illustrated in Figure 3, was necessary in order to clarify each participant's role.

The firms' general management controls the project. The entirety of the firm participates to the project, but the final choice will be endorsed by the *control group*. This group includes the general management, the production management, the administrative and financial management, and the research and development management. The members of the control group are in charge of specifying the quantitative and qualitative objectives of the project, approving the decision stages in the course of the project and selecting the final solution satisfying the objectives.

The success of such projects strongly depends on the *layout group*. This group is responsible for the project and must be conducted by a project leader capable of bringing together the participants of the project around the objectives of the control group, and hence around the general management. The layout group consists of the project leader, the heads of the Mechanic and Scheduling Departments, the head of the Assembly Department, and the head of the Methods Engineering office. The layout

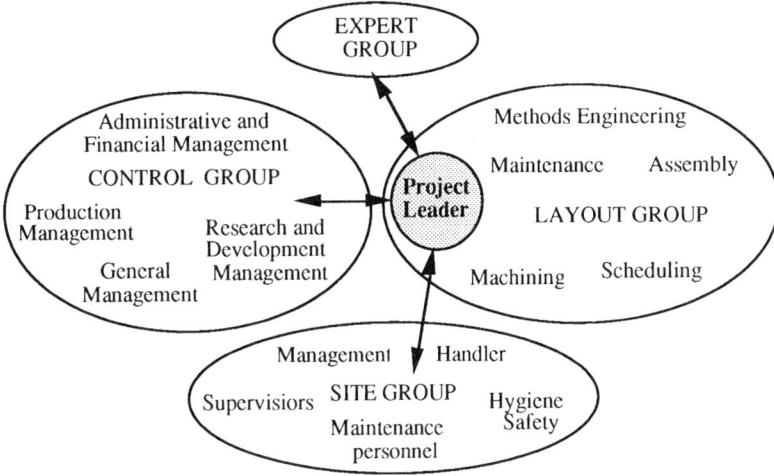

Figure 3. Formal structure of the design team.

group is responsible for the study and implementation of the project. It is in charge with gathering, analyzing, and then processing the data, and with conceiving layout scenarios as well as practical plans. It is also in charge of valuing each of the proposed solutions and with relating them to the objectives and constraints set out by the control group. In this firm, the head of project belongs to this group.

The *site group* is a multidisciplinary team representing the personnel directly concerned with the production site. It is made up of the shop management, the person in charge with the upkeeping and maintenance, the person responsible for the hygiene and safety, two shop supervisors, and a handler. This group participates mainly in the steps 2 and 5 of the method.

The usefulness of the production personnel participating to the project was unanimously acknowledged by all the firms we contacted. The foremen and operators must belong to site group because of their deep detailed knowledge of the following facts:

- Working and security conditions
- Communication between technical control data and fieldwork
- Over or underestimation of the routing times by the method office
- Surface necessary for the production and handling means
- Access of the machines for maintenance
- Access of the machines for adjusting
- Families of products
- Quality problems due to the thermal environment, the state of the floor, and the like
- Departments' difficulties in communicating with each other

The *expert group* gathers skills external to the firm. It sometimes includes a project leader. Their role is to assist the design team to perform Production Flow Analysis (PFA) with PFA softwares, to make use of simulations softwares and so forth.

Project Duration   The duration of the study must be estimated in advance. It can vary from a few weeks to a few months for a production site comprising several shops. To fix the duration of a project is a tricky task. The main variables of estimation of the duration are

- Surface area (the scope of the study)
- Number of products and machines studied (the data of the problem)
- Complexity of the analysis (the objectives)
- Number of man-hours per week spared to the project (the means)
- Firm's internal skills available for this project

Some investigations require external abilities, such as the use of simulation and routing analysis software, the civil engineering expertise for studying the opening or extending a shop floor, and the expertise in acoustics for ergonomic constraints. It is advisable to contact the corresponding experts in order to integrate the time necessary for the completion of their tasks to the duration of the project.

3.2. Search for Information

Dispersed information is gathered and noted down during this stage. By the end of this stage, the designers will have a *clear and global view* of their production system. The search for information is a resource consuming activity; between 40 and 100 man-hours must be allowed for a shop comprising of about 30 machines, 3000 items, and 20 operators. The scope of inquiry must hence be limited to the useful information, while keeping in mind that an inadequate gathering of information can be harmful to the understanding of the system and lead to faulty design. If, during the next stages of the project, the gathered information turns out to be inadequate, it is always possible to complete it. Note, though, that adding information a posteriori is more costly than if it is embedded in step 2.

In order to help the reader gather the relevant information, the rest of the section is divided into two parts. The first one describes the information that must be systematically gathered. The second one describes a set of information that is often used in this kind of project.

Minimal Information   There must be drawn up at least:

1. A plan of the existing shop with its flows
2. A diagram of the physical flows
3. A diagram of the information and decision flows
4. An ergonomic analysis of the workplace

1. The main flows of parts and handling means are drawn on a plan of the shop. This plan often reveals to the designers the complexity of their shop. It also gives them concrete ideas for potential improvements, which motivates the participants of the project straight away.

2. The diagram of physical flows, an example of which is represented on Figure 4(*a*), is a concise version of the plan 1. The semantics of the diagram is kept simple in order to be understood by all the participants of the project. The objects are represented by elementary shapes, such as triangles for the inventories and rectangles for the machines. The relationships between objects are represented by arrows. The diagram is complemented by a glossary [Figure 4(*b*)] that specifies the objects and approximately quantifies the flows. The diagram of physical flows, thanks to its simplicity, allows the design team to have a critical look at the existing solution, which would not be permitted by the often too detailed plan of the shop 1.

3. The *information and decision flows* are also represented with the help of the diagram in Figure 5(*a*). The information is represented by oval symbols. A glossary [Figure 5(*b*)] comments on the information and the relations. The decisions are represented by hexagonal symbols. A glossary specifies the type of activity, the inward and outward information, the support, and the person or machine executing the activity. Remarks regarding the difficulties linked with the information and activities complement the previous information.

4. A production system also comprises the workforce! The information about the working conditions is gathered by a direct survey of the shop. The questionnaire is made up of open questions, examples of which are given in the following list:

- What kind of parts do you carry? What is their weight?
- What are your working postures?
- How are the parts supplied to the workstation?
- How is the evacuation of the manufactured parts carried out?
- Do you feel pain due to your working conditions?

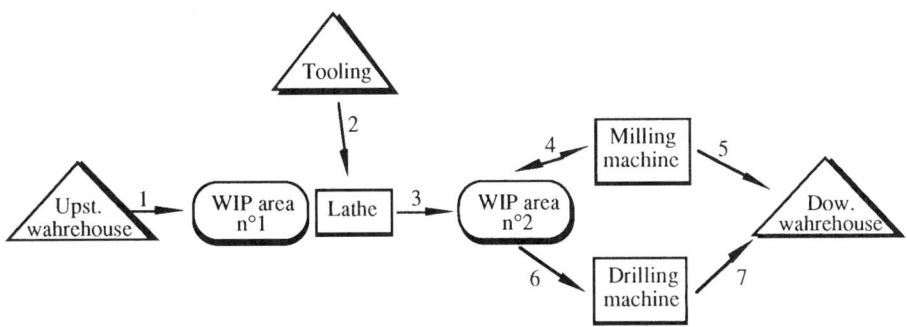

Figure 4(*a*). Example of physical flow diagram.

Symbol	Description	Remarks
Upstream warehouse	Raw materials store	
Work-in-process in area 1	Work-in-process storage area before the NC lathe	In case of a major increase in production, the containers are sometimes stored outside this area
Lathe	NC lathe	Workstation set up on strengthened foundations
Arrow 1	Handling of the raw materials toward the first work-in-process area. Approximately 300 items and 50 containers per day	
Arrow 2	Handling of the tooling necessary for the operations on the lathe; 40 types of tooling at the rate of one handling per day on average	Part of the tooling might come from the repair shop, which is not represented on the diagram (1–2% of cases)
Arrow 3	Handling of the parts manufactured in workstation toward the work-in-process area of the milling and drilling machines	Tricky handling of some fragile parts

Figure 4(b). Example of a glossary.

- Do you sometimes injure yourself?
- What tasks do you carry out?
- Are you working on a Numerically Controlled (NC) workstation?
- Are you alone in charge of your workstation?
- What are the problems that you would like to see solved?
- Where do your instructions come from?
- What information do you transmit and to whom?

These surveys, being extremely rich in sensitive information, must be carried out cautiously. Their aims must be clearly expounded by the management if they are not to be regarded as an invasion of privacy.

Complementary Information   The information essential for any layout project has just been described. Some more precise information can turn out to be useful. Let us cite a few examples:

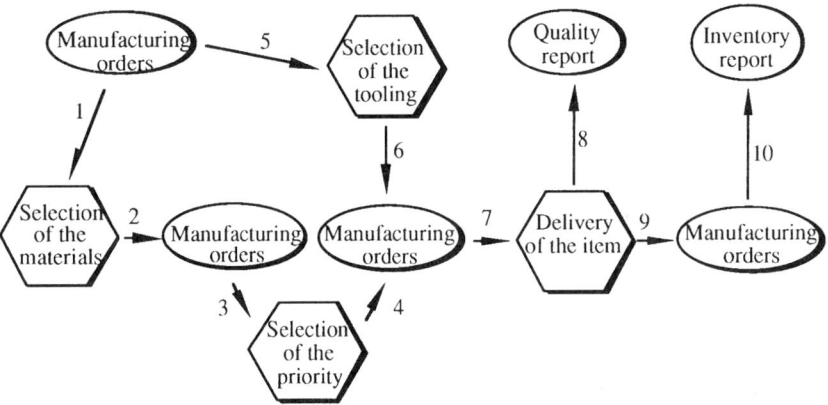

Figure 5(a). Example of diagram of the information and decision flows.

Information	Description	Remarks
Manufacturing orders	Orders for the current week	
Quality report	Report indicating the number of manufactured parts, the number of defects and their causes	One per week
Inventory report	List of the manufactured quantities of each part	

Activity	From	To	From whom	To whom	Medium	Remarks
Material selection	1	2	Planning	Shop foreman	Paper and computer	
Priority selection	3	4	Shop foreman	Operator	Paper	10 orders per day on average

Figure 5(b). Example of a glossary of information and decision flows.

- The diagram of the flows [Figure 4(a)] can sometimes be usefully complemented by a machine file comprising, for each machine, the weight, the volume, the type of energy, the working conditions, and the technical specifications.
- Step 4 of the method (searching for ideas and paths toward solutions, Figure 2) is generally based on a more developed knowledge of the physical flows than the

diagram in Figure 4. Mastering the flow analysis tools permits the anticipation of the demand, or even overlaps of some activities of the method. The useful information for the detailed flow analysis is given in the chapter on cell formation using PFA of this handbook.

- Some normative documents about the noise or the quality can be of some assistance.

The entirety of the gathered information must be approved by the design team. By the end of this stage, the head of project as well as the design team have a proper knowledge of the global functioning of the production system. The functional analysis can be triggered with full knowledge of the facts.

3.3. Functional Analysis

There exist a great number of functional analysis methods. Some make use of particular semantics and syntax, such as FAST (Function Analysis System Technique), SADT (Structured Analysis Design Technique), $IDEF_{0, 1, 2, 3}$, MERISE, and so forth.
    However, the following method can be used too:

- Study the product's life cycle (surveys, video films, etc.)
- User's behavior
- Study similar products
- Intuitive search
- Search for customer dissatisfaction with the existing products

The choice of the analysis method is not really a problem. Several of them can even be combined. On the other hand, methodological exactness is essential. The use of a structured method minimizes the risk of forgetting important functions of the system.
    The functional analysis method, inspired by the value engineering method, is defined in four steps:

- Search for the components of the system
- Exhaustive search for the functions of the system
- Search for constraints
- Ranking of the functions

Before describing this sequence of steps, the characteristics of the functions of a product or a service are briefly described in the value engineering sense:

- A function is an activity of the product or service being studied.
- A function acts between the components of the product and its environment or between the components themselves.

- A function is described without referring to the solutions that will be deployed.
- A function is expressed with a verb in the infinitive followed by several complements.

Search for the Components of System   Several generic components listed in Figure 5 emerge from the practice of the method. The definition of each component is given in Appendix 2. The list of the studied system's components is easily obtained from the generic list and from the minimal information of the information searching stage. In order to best exploit this list, the following pieces of advice are given:

- Rename the components of the generic list using the common vocabulary in use in the firm. For example, upstream warehouse should be called material warehouse.
- The components of the list in Figure 6 are generic names. Many components of the workshop may fit this definition and should be studied separately. For example, if it appears necessary to distinguish two upstream warehouse then create a row for each warehouse with their own ID labels.

The previous method is illustrated by an extract from a case study: a firm producing washbasins. The shop comprises a glazing kiln as well as cars on which the washbasins are kept after their glazing in the glazing station. Part of the physical flow diagram is given in Figure 7(*a*). The glazing station is run 16 hours per day, 5 days per week. The kiln is run 24 hours per day, 7 days per week. The glossary in Figure 7(*b*) explains the flow diagram.

Exhaustive Search for Functions   After having exhaustively defined the components of the production system, it is a question of searching the functions linking the components to each other. The search must be as exhaustive as possible in order not to forget functions, which could be prejudicial to the workshop design's quality.

1. Upstream warehouse	14. Tooling
2. Downstream warehouse	15. Quality control means
3. Means of transport	16. Maintenance means
4. Containing means	17. Environment
5. Item	18. Various personnel
6. Transfer path	19. Delivery area
7. Work-in-process area	20. Kitting area
8. Work in process	21. Loading means
9. Loading area	22. Unloading means
10. Unloading area	23. Information
11. Workstation	24. Energy
12. Production personnel	25. Civil engineering
13. Order	26. Other departments

Figure 6. List of the system's components.

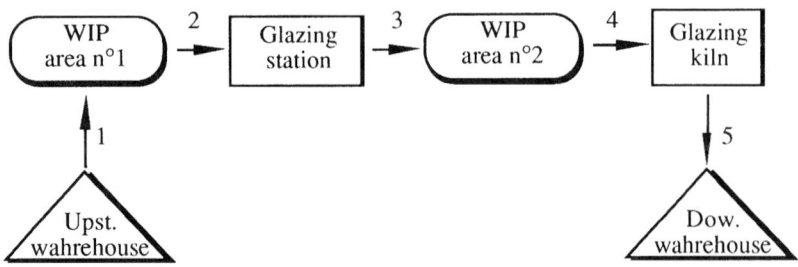

Figure 7(a). Physical flow diagram.

No.	Symbol	Description	Remarks
1	Upstream warehouse	Uncooked washbasins	Stored on shelves
7a	Work-in-process area 1	Work-in-process before glazing	
11a	Glazing station	Operators or automatons; 5 operators and 4 automatons	Station laid out in a area protected from dust
7b	Work-in-process area 2	Work-in-process before firing	
11b	Glazing kiln	Glazing kiln for all products	Running 24 hr/24 hr
2	Downstream warehouse	Glazed washbasins	
	Arrow 1	Handling of the unglazed washbasins toward work-in-process area 1	Tricky handling for the parts are fragile.
	Arrow 2	Handling of the washbasins from the containers toward the glazing stations	Manual handling

Figure 7(b). Glossary of the physical flow diagram.

An exhaustive search is obtained by systematically searching for every possible function linking all the components represented in Figure 6 and Appendix 2. A matrix showing the relations between the components is drawn up with that purpose [Figure 8(a)]. The matrix has 26 rows and columns corresponding to the 26 components. Since a function is not an oriented concept, a mere semimatrix is necessary for the search of functions. The cells of the matrix are initialize with a zero value. The matrix is filled up as follows: Is there a function linking component i to component j, for

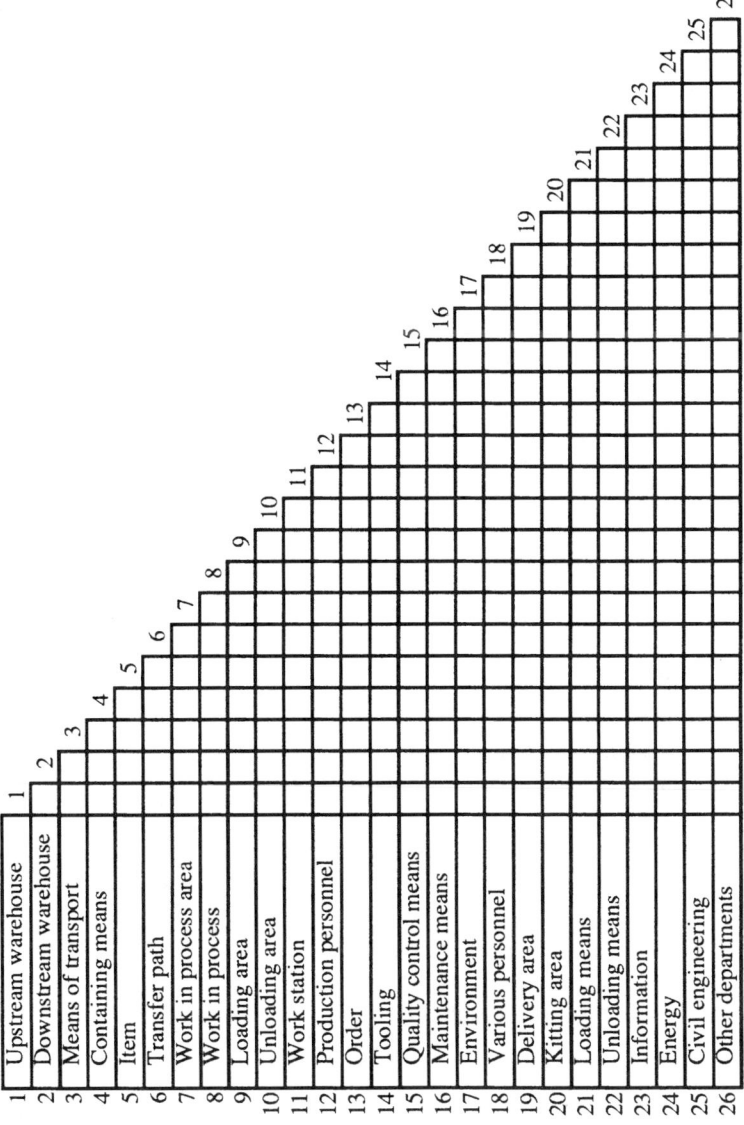

Figure 8(a). Function searching table.

1 Upstream warehouse
2 Downstream warehouse
3 Means of transport
4 Containing means
5 Item
6 Transfer path
7 Work in process area
8 Work in process
9 Loading area
10 Unloading area
11 Work station
12 Production personnel
13 Order
14 Tooling
15 Quality control means
16 Maintenance means
17 Environment
18 Various personnel
19 Delivery area
20 Kitting area
21 Loading means
22 Unloading means
23 Information
24 Energy
25 Civil engineering
26 Other departments

i and *j* varying from 1 to 26? For instance, is there a function linking the upstream warehouse and the downstream warehouse together. If yes, then the value 1 is added to the value at the crossing of the second row and the first column of the matrix represented in Figure 8(*a*). At the end of the process, the cell indicates the number of functions that tie these two components together. Altogether, the matrix comprises 325 cells [326 = (26 × 25)/2] and the systematic search becomes possible.

The functions are defined and characterized in terms of criteria, levels, acceptance limits, and a flexibility index. One or several criteria can be associated with a function. A criterion possesses a level, an acceptance index, an acceptance limit, and a flexibility class. The functions are gathered in a glossary.

The criterion is a qualitative or quantitative unit of measure of the function.

> *For a passenger transport function, the criteria can be the speed, the acceleration, the number of passengers, and so forth.*

The level is the magnitude on the scale measuring a criterion.

> *For the speed criterion of the previous function, the level is measured in kilometers per hour. Its magnitude is 160.*

The acceptance limit corresponds to the level under which or over which the function is considered as not being carried out by the system.

> *The acceptance limit of the previous level is, for instance, ±5 kilometers per hour (km/hr). If the maximum speed is less than 155 km/hr (155 = 160 − 5), the transport function is no longer considered acceptable.*

Lastly, the class of the flexibility indicates the compulsory nature of the fulfilling of a criterion. A semantic scale from 0 to 3 is usually used to grade the class. The classes F0, F1, F2, and F3, respectively, represent an imperative, not easily negotiable, negotiable, and very negotiable flexibility.

> *A flexibility of class 0 for the previous acceptance limit indicates that it is imperative. Any vehicle not reaching 155 km/hr at full speed will not be accepted.*

Examples of functions are represented in Figure 8(*b*), which is extracted from Figure 8(*a*) and from the example in Figures 7(*a*) and 7(*b*).

Function 1: To place the washbasins (5) in the containing means (4).

Criteria	Level	Acceptance Limit	Flexibility
Ergonomics	Defined according to the norm	Defined according to the norm	F0
Filling up	15 washbasins	± 3 washbasins	F1

Protection of the parts	Shock	A 0.5 cm maximum free fall	F0
Rate	3600 washbasins/hour	± 100 washbasins	F0
Weight of the parts	13 kg on average	3–25 kg	F0

Function 2: To carry out the weekend storing. Located at the intersection of column 7b (work-in-process area 2) and row 8 (work-in-process). Storage is necessary since the glazing kiln and the glazing stations do not produce at the same rate.

Criteria	Level	Acceptance Limit	Flexibility
Site	Minimum of m²	In the shop	F0
Site	4 m high	4 m maximum	F0
Quantity	50 hours of production	± 1 hour	F1

Function 3: To carry the containing means (4) in front of the glazing kiln (11b).

Criteria	Level	Acceptance Limit	Flexibility
Space	m²	In the shop	F0
Rate	Every 3 minutes	< 3 minutes	F1
Speed	5 km/hr	< 10 km/hr	F3

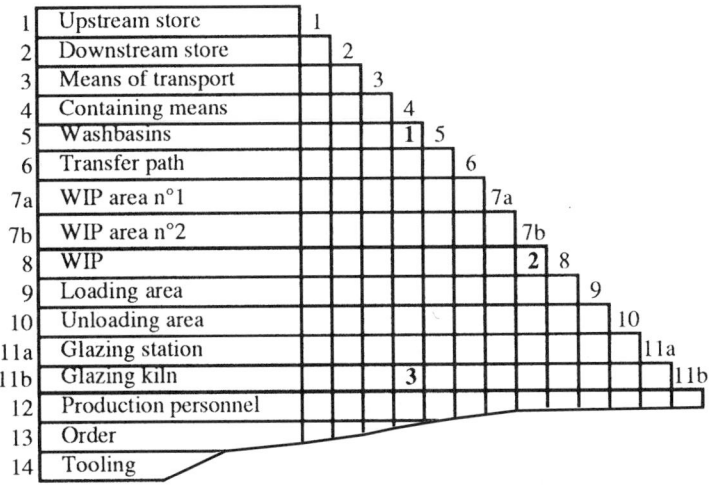

Figure 8(b). Example of function searching table.

Search for Constraints A constraint is a limit to the choice of solutions. Some materials or techniques are forbidden by the designer. These constraints evolve in time and must be strongly justified since they are a hindrance to the searching of ideas.

> *A firm competing in the wood sector aims at producing beach seats. The proposed solution will compulsory consist of wood. It is a constraint on the choice of materials.*

In the previous example, the constraint searching is based on the system's components list in Figure 6. An example of constraint: the environment.

Criteria	Level	Acceptance Limit	Flexibility
Dust	No mark on the washbasins	1/1000 washbasins	F1
Heat	25°C	$18°C < T < 25°C$	F1
Noise	80 dB	< 80 dB	F0

Other constraints than the components are possible.

An example of constraint: the compatibility of the layout works with the production.

Criteria	Level	Acceptance Limit	Flexibility
Length of the stops	20% of the capacity per month	< 20%	F0

Ranking of the Functions It is not unusual to find a great number of functions. The designer finds it difficult to simultaneously master all the functions. A way to solve the problem is the ranking of the functions. It is a question of understanding the relative importance of the functions. During the study and valuation of the solutions, the comparison of the function's technical realization cost and its mark lead to the introduction of the concept of value or price/quality ratio.

The function/function matrix represented in Figure 9 is used to rank and rate the functions. In each square is indicated the function considered as being the most important one, with or without weighting. The number of times the function appears in the matrix represents its relative importance and leads to the histogram in Figure 9.

If the solution's cost (obtained in step 4) corresponding to function F3 is high, its value is inadequate. The user will have spent a large amount of money on a function that is not worth it.

All the results stemming from the functional analysis must be validated by the design team. The quality of design depends on it. The quality of design is the capacity of a product or a service to fulfil the user's needs. This response is possible

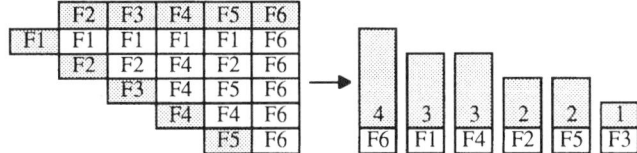

Figure 9. Classification matrix and the ranking of the functions.

provided that the functions expected by the user(s) are known. The functional analysis constitutes a proper assistance to this quest. The functions and constraints thus described can be studied in terms of solutions. It is the subject of step 4, the search for ideas and paths toward solutions.

3.4. Search for Ideas and Paths toward Solutions

The search for ideas is the creativity stage of the method. The result of this step is a set of concrete means fulfilling the criteria defined during the previous stage and leading to the execution of the functions. Each solution to a function or a group of functions is know as a local solution. Global solutions are established during step 5 through the combination of local solutions. For this step to be productive, the members of the design team must be careful not to limit themselves on the basis of the solution valuation criteria. The valuation of the solutions will be done in step 5. It is best to search the solutions by groups of functions. The classification of the functions must be done case by case. It is nevertheless advisable to group the functions related to the components of the system numbered 1, 2, 5, and 11 (respectively, upstream warehouse, downstream warehouse, item, and workstation) in Appendix 2, and to imagine solutions for this group independently of the other functions. This approach is justified by the results of a survey demonstrating the influence this type of component has on other functions such as the ones related to the workstations and to the containing means.

Search for Solutions to the Functions Related to Components 1, 2, 5, and 11 Analyzing the functions related to these components is equivalent to studying the flows independently of considerations involving the civil engineering, the shop's environment, and the personnel. The functions' criteria are both quantitative (the handling distances, the number of backtrack moves on a line, etc.) and qualitative (the proper visualization of the flows, the easiness of cell formation control, etc.). A systematic study of the four flow structures in Figure 10 can be considered. An independent cell is an autonomous control unit producing a part family. A dependent cell is a group of machines associated with a family of parts whose machine is gathered together in a shop. A line is a linear layout of workstations that minimizes the number of backtrack moves. Finally, the proximity layout minimizes the handling distances and the number of flow crossings. Solutions that consist of a compound of these flow patterns could also be considered.

It is at this stage that the methods for the search for cells and GT cells developed in this book are to be used. These methods make use of the data related to the items,

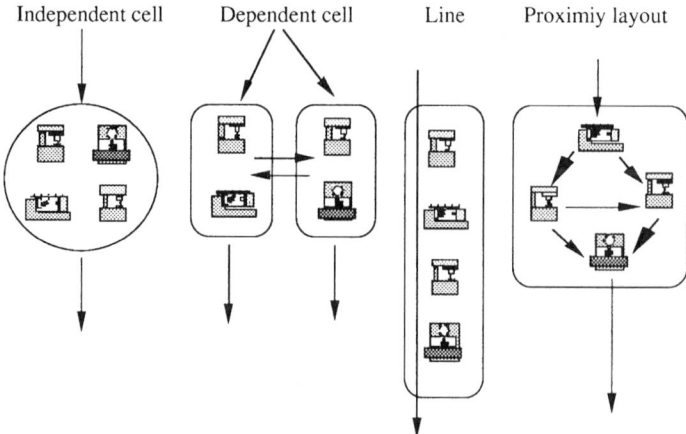

Figure 10. Flow patterns.

workcenters, and routings. Methods and tools dedicated to the identification of the linear flows and proximity layout solutions also exist. The results of this analysis is a flow as schematized in Figure 4(*a*).

Search for Solution to the Remaining Functions   The integration of other functions is based on the experts' knowledge and on the information gathered from consulting industrial equipment catalogs and from visiting shows as well as other firms. This stage, which might seem difficult, does actually not raise any problem if the leader of the design team promotes the expression of ideas. In any case, a plethora rather than scarcity of ideas is usually noted.

3.5. Study and Valuation of the Solutions

The purpose of this stage is to draw up a few comprehensive shop configurations, named global solutions and considered as interesting by the design team. These global solutions are combinations of local solutions elaborated during the previous stage. It would be difficult to evaluate all the possible layout combinations resulting from the local solutions. Indeed, if a functional analysis proposes 20 functions and if two local solutions are considered for each function, which is not many, the number of configurations to value is 2^{20} that is to say 1,048,576 global solutions. In practice, the designer limits the number of combinations thanks to the function ranking carried out during the functional analysis. In complex cases, between 10 and 20 succinct global solutions are first elaborated.

Only the most important functions are selected and roughly valued through the comparison of the functions with the solutions and through economic calculations. The best of these global solutions will be used as an outline of detailed global solutions embedding all the functions. These detailed solutions will also be valued by means of comparison of the functions with the solutions and economic calculations.

Tables similar to the ones represented in Figure 11 can be helpful for the comparison of the functions with the local solutions. The table in Figure 11(*a*) gives the definitions of the functions, criteria, and local solutions of the function F1: to transport the containers from workstations to workstations. The table in Figure 11(*b*) specifies the function F2: to visually clarify the flow. The suitability of the solution to the different criteria is given on a scale from 1 to 5.

The table represented in Figure 12 describes the global solutions as combinations of the partial solutions of the table in Figures 11(*a*) and 11(*b*). Six global solutions can theoretically be elaborated from this example ($3 \times 2 = 6$; 3 solutions for F1 and 2 solutions for F2). The design team has selected four solutions only, noted Sg1, Sg2, Sg3, and Sg4. Solution Sg4, for instance, is made up of the solutions Sl 1 for the function 1 and Sl 2 for the function 2. It is a global solution comprising manual trolleys and a linear layout.

The suitability marks in the tables are subjective when they are based on the design team's judgment; they are objective when they stem from a measurement or a calculation (surface, distance). Some measurements are calculated with the help of static and dynamic flow simulation. The selection of the best solutions is left to the initiative of the design team. A simple column summation or multicriteria analysis method could be used. The mere development of this table by the design team generally indicates the solutions best satisfying the functions expected from the system. The investment and operating costs corresponding to each solutions must be drawn up for the economic valuation.

The value of each solution is established from the results of the table and from the calculation of each solution's cost. The stage just described often is the longest in this

	Local Solutions		
	Sl 1	Sl 2	Sl 3
	Manual trolley	Aerial conveyor	Automated guided vehicle
C1 Weight of the containers	5	5	5
C2 Transport rate	3	5	3
C3 Load/capacity rate	5	1	2

Figure 11(*a*). Table of comparison of the function F1 with local solutions 1–3.

	Local Solutions	
	Sl 1	Sl 2
	GT Cell	Line
C1 Handling distances	2	4
C2 Backtrack moves	5	5
C3 Crossover flow	5	1

Figure 11(*b*). Table of comparison of the function F2 with the local solutions 1–2.

		Global Solutions			
		Sg 1	Sg 2	Sg 3	Sg 4
		Sl 1	Sl 2	Sl 3	Sl 1
F1	C1	5	5	5	5
	C2	3	5	3	3
	C3	5	1	2	5
		Sl 1	Sl 2	Sl 1	Sl 2
F2	C1	2	4	2	4
	C2	5	5	5	5
	C3	5	1	5	1

Figure 12. Table of comparison of the functions with the global solution.

type of project. It makes use of a considerable number of skills such as production management, and method engineering.

3.6. Primary Evaluation, Submission of the Solutions and Decision Making

The purpose of this step is to draw up a primary evaluation of the selected solutions as well as of the implementation conditions within the firm. The following justifications are essential: the criteria for selection, the rate of each criterion value, the costing, the implementation conditions.

The final decisions is taken by the firms' management following traditional investment approaches, which are split up into three steps:

- Analytic approach to investment that puts forward the necessary working capital and the differential expenses
- Economic valuation of the cash flow (discounted or not)
- Choice of the valuation criterion (net discounted value, recovery time)

The investment can only be globally considered according to its contribution to the firms' strategy. A multicriteria approach comprising the human, organizational, and technological aspects becomes indispensable.

Taking the working capital requirement into consideration can, for example, be quoted. Indeed, an increase in the working capital requirement follows most investments. Conversely, reorganizing a production shop must lead to a decrease in the working capital requirement through the lowering of the cycle times. In most accounting approaches, not taking into account the decrease in the working capital requirement tends to lengthen the return on investment time of the project and thus to reduce its economic impact.

The financial justification of a shop reorganization project can be done by two complementary approaches: static and dynamic. The static approach aims at reducing the volume of financial flow by cutting down fixed costs, working capital requirement, unsaleable hours, and hidden costs. As for the dynamic approach, it aims at quantifying the increase in financial flow speed. It is justified by cutting down the cycle times as well as by an improved operational and financial flexibility. This logic reinforces the diminution of the production cycle in the GT cells in Figure 13.

3.7. Remarks on the Ordering of the Method's Activities

The method has been presented as a series of sequential steps. Its entry point is the orientation of the project stage and its exit point is the primary evaluation and decision stage. Additional loops can be added. Thus, during step 2 for instance, step 1 can be reverted because a better understanding of the production system has been gained or because the analysis of the working conditions has led to the formulation of new targets with regard to the quality. Such loops can be added at each step of the system. The process ends when it is possible to cycle the steps of the method according to the diagram in Figure 2 without altering the results.

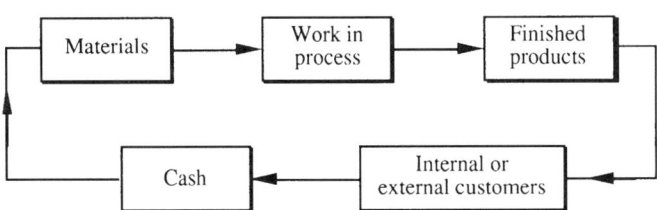

Figure 13. Diminution of the production cycle.

APPENDIX 1: FLOWCHART OF THE METHOD

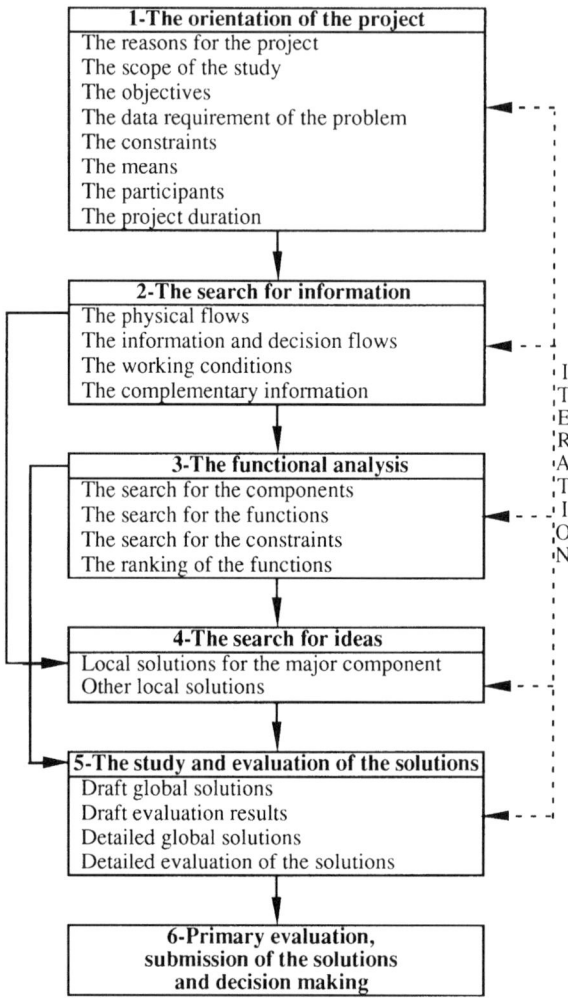

APPENDIX 2: COMPONENTS OF THE PRODUCTION SYSTEM

1. **Upstream warehouse:** place where the *items* are stocked in and taken out of to start the production of an *item* following an *order*
2. **Downstream warehouse:** place where the *items* are stocked at the end of the production of an *order*
3. **Means of transport:** resources utilized for the transport of the *items in process*, the *containing means*, the *tooling*, the production personnel
4. **Containing means:** resources containing the *items in process*, the *tooling*, and the *quality control means*
5. **Item:** raw material, elementary parts, subassemblies, or finished products produced within the production system and stocked in a *upstream* or *downstream store*
6. **Transfer path:** area where the *transport means*, the *production personnel*, and *various personnel* are moving
7. **Work-in-process area:** place where the *items* are stored awaiting to be transferred toward the *loading area* or the *delivery area*
8. **Work-in-process:** *item* being processed, taken out of the *upstream store* and not yet received in the *downstream store*
9. **Loading area:** place where the *production personnel* or the *loading means* supply the *workstation* with the *work-in-process*, the *tooling*, the *quality control means*
10. **Unloading area:** place where the *production personnel* or the *loading means* unload the *work-in-process*, the *tooling*, the *quality control means* of the *workstation*
11. **Workstation:** place where the products are elaborated
12. **Production personnel:** personnel assigned to the operation of the production system (operators, adjuster, maintenance personnel, shop foreman, planning and control personnel, handling personnel, warehouseman, etc.)
13. **Order:** information indicating to the production system which *item* to produce and any pieces of information related to it
14. **Tooling:** technical resources temporarily allocated to a *workstation* and being utilized for the production of an *item* (mould, tools etc.)
15. **Quality control means:** resources temporarily or permanently allocated to the control of the production quality (calliper square, measuring apparatuses, procedures etc.)
16. **Maintenance means:** resources temporarily or permanently allocated to the maintenance of the *workstations* (rag, oil, tooling etc.)
17. **Environment:** floor, air, temperature, noise, humidity etc.
18. **Various personnel:** people other than the *production personnel* moving around the production system (personnel from other departments, visitors)

19. **Delivery area:** place where the *work-in-process* is kept before the *downstream store*
20. **Kitting area:** place where the *work-in-process* is kept after the *upstream store*
21. **Loading means:** technical resources allocated to the supply of the *workstation* with *items*
22. **Unloading means:** technical resources allocated to the unloading of the *items* of the *working station*
23. **Information:** quality procedures, plans, routings, maintenance instructions etc.
24. **Energy:** electricity, gas, air, water, and any other supply
25. **Civil engineering:** building, basement, pole, wall etc.
26. **Other departments:** Design, Methods Department, Production Control Department, etc.

REFERENCES

Apple, J. M. (1977). *Plant Layout and Material Handling*, 3rd ed., Wiley, New York.

Immer, J. R. (1950). *Layout Planning Techniques*, McGraw Hill, New York.

Nadler, G. (1961). *Work Design: A Systems Concept*, Irwin, Homewood, IL.

Reed, R. (1961). *Plant Layout: Factors, Principles and Technique*, Irwin, Homewood, IL.

Tompkins, J. H. (1989). *Winning Manufacturing: The How To Book of Successful Manufacturing*, IEEE, Norcross, GA.

Tompkins, J. A., White, J. A., Bozer, Y. A., Frazelle, E. H., Tunchoco, J. M. A., and Trevino, J. *Facilities Planning*, 2nd ed., Wiley, New York.

17

DISHWASHING MACHINE ASSEMBLY: A CASE STUDY

Quarterman Lee

Strategos, Inc.
Kansas City, Missouri 64111

1. INTRODUCTION

American Dish Service (ADS) designs and manufactures low-temperature commercial dishwashers for restaurants, bars, and institutions. From the firm's beginnings in 1949 until the 1970s, ADS was a very small firm. Manufacturing was only one of several activities. The company produced one or two machines each week and offered only one model.

Some years ago, management focused the business on manufacturing and design. The product range increased to six basic models with variations on each. Volume rose to eight machines per day. ADS recognized the need for more efficient and flexible assembly. The work cell described here addressed this need.

This case study does not reflect current ADS operations, although its basic approach remains the same. Today, ADS is an international supplier of low-temperature dishwashers with a wide range of products and a solid reputation. Cellular Manufacturing (CM) has been a major contributor to the firm's success.

Handbook of Cellular Manufacturing Systems, edited by Shahrukh A. Irani
ISBN 0-471-12139-8 © 1999 John Wiley & Sons, Inc.

2. THE FACPLAN APPROACH

FacPlan is a structured, step-by-step methodology for facility planning. It uses the layout or space plan as the centerpiece for facility design. A space plan specifies amount, location, and type of space. Usually, a space plan takes the form of a drawing or layout, often in two dimensions. FacPlan uses five levels of space plan detail as Figure 1 shows. Figure 1 identifies each level and describes typical activities at that level. It shows the usual Space Planning Unit (SPU). An SPU is a block of space that the designer defines and arranges. The environment is the larger space that surrounds the space plan. The far right column of Figure 1 illustrates typical outputs for the planning process.

> *Site location* is the highest level of detail, as shown near the top of Figure 1. Site location decides the location, size, and other requirements for a site. We may also refer to this level as "level I" or "global."
>
> *Site planning* takes place at level II, the supra level. It specifies site features such as buildings, roads, and utilities.
>
> *Macro layout* locates departments or other large-scale features within a building. We sometimes call this a "block layout" or building layout.
>
> *Micro layout* locates equipment, furniture, and other items within each unit of the macro layout. This is the level at which most work cell design takes place.
>
> *Submicro layout* designs the space plan for individual workstations. It places tools, materials, and other small-scale features.

Ideally, the five levels translate into five sequential project phases leading to a complete space plan and facility design. In practice, however, we may enter the process at almost any level, depending on circumstances.

Work cell design (micro space planning) has the five major tasks of Figure 2. The 4 in the task identifier represents the fourth level of space planning detail—the micro space plan. The 01 represents the first task at this detail level. Task 04.01 examines the full range of possible products for work cells and selects a product or product family for each cell. Task 04.02 analyzes the current or proposed process. Task 04.02 examines alternate processes and process improvements. Task 04.03 designs the work cell's infrastructure. Task 04.04 produces space plans or layouts for each option.

While these tasks are sequential, they are also iterative. Each task gives rise to options that may force an iteration of the design. For example, the designers may recognize during task 04.04, design cell layouts, that a small change in process could allow a significantly different space plan. They might then return to the process definition, task 04.02, and generate a new design based on the revised process. The open arrows in Figure 2 illustrate this iteration.

When the designers arrive at the final task, task 04.05, they should have two–six significantly different designs. This allows managers, designers, and workers to select among a wide range of options. It forces consideration of new approaches that might otherwise suffer immediate rejection because of unfamiliarity. It generates richer solutions. It helps ensure a near-optimum work cell.

The Level Of Space Planning

Level	Activity	Typical SPU	Environment	Output
I Global	Site Location & Selection	Sites	World Or Country	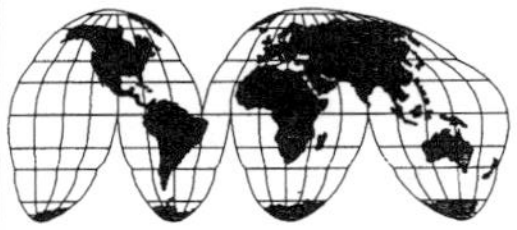
II Supra	Site Planning	Buildings Or Site Features	Site	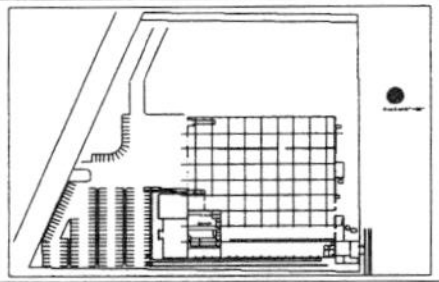
III Macro	Building Layout	Cells Or Departments	Building	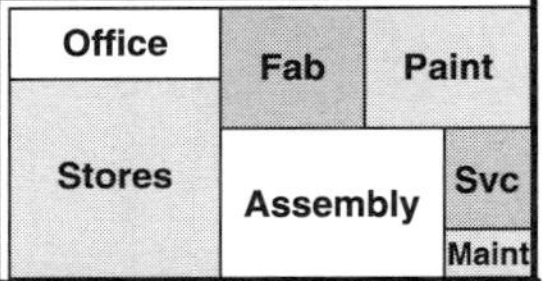
IV Micro	Department Or Cell Layout	Workstations Or Cell Features	Cells Or Departments	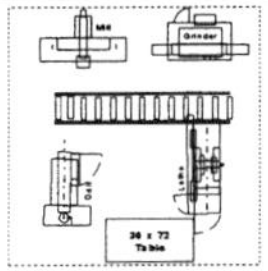
V Sub-Micro	Workstation Design	Tool Locations	Workstation	

Figure 1. Levels of space planning.

529

Level IV- Micro-Layout Model Project Plan

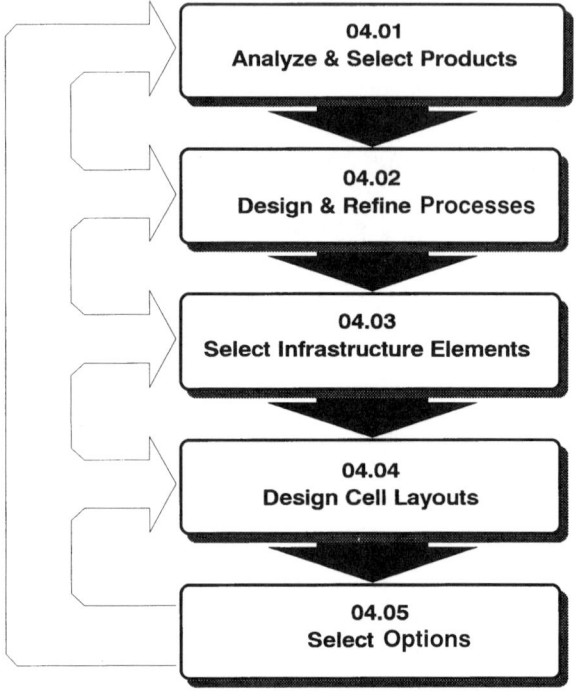

04.01
Analyze & Select Products

04.02
Design & Refine Processes

04.03
Select Infrastructure Elements

04.04
Design Cell Layouts

04.05
Select Options

Figure 2. Tasks for micro space planning.

3. TASK 04.01, ANALYZE AND SELECT PRODUCTS

Our first task is 04.01, product selection. Figure 3 is the FacPlan procedure. It shows, in detail, how to analyze and select the products for a work cell. Each block in the diagram represents a step, decision, input, or output. The small circles labeled C1, C2, . . . are connectors that link blocks in one task procedure to blocks in another task procedure.

In block 1 we gather product and sales information. If you are unfamiliar with the product, you will need basic information such as sales catalogs, operation manuals, and parts manuals. You will need a sales forecast and possibly sales history. Figure 4 shows the six primary ADS products. Each model is actually a product group. For example, model A may have any of three cycle times and several other options.

Task 04.01
Select Products
Level 04- Micro-Spaceplan

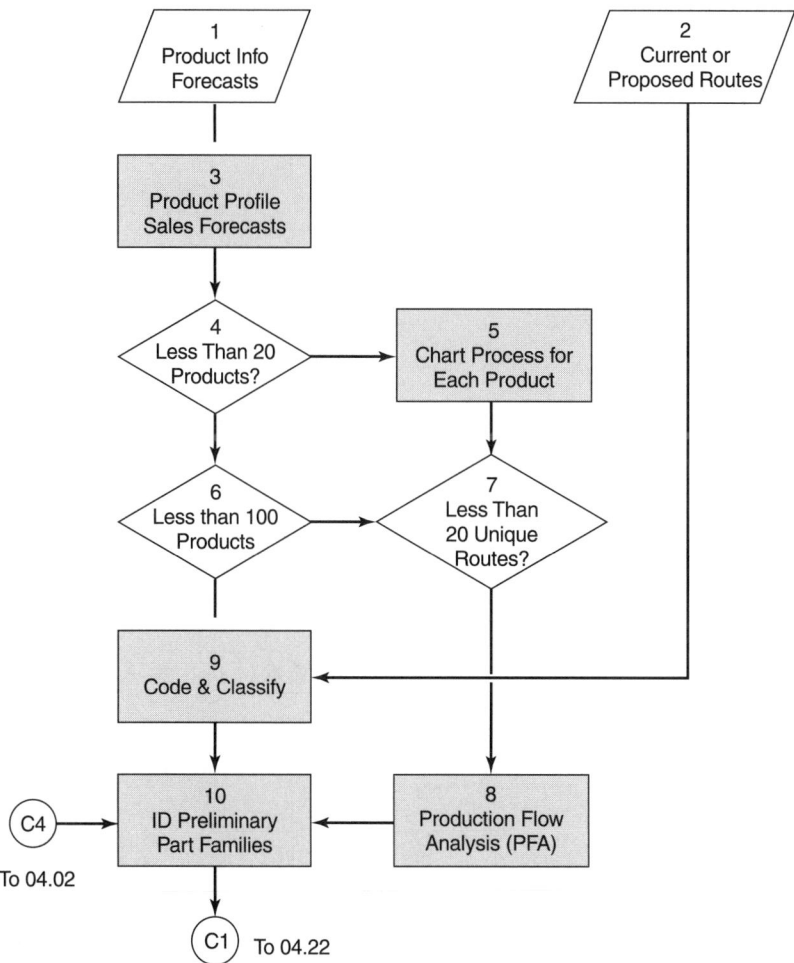

Figure 3. Procedure diagram—task 04.01.

A brief description of the machines and their operation will help to understand the manufacturing process. The superstructure of the machines is fabricated stainless steel sheet. Other parts are cast or fabricated from tube, rod, and other materials. The machines have a variety of purchased parts such as fittings and electrical components. The dish machines are installed in commercial kitchens with tables attached to two sides.

In operation, workers load soiled dishware in standard plastic racks approximately 30 × 30 in. They raise the doors on the machine with a lever mechanism and slide

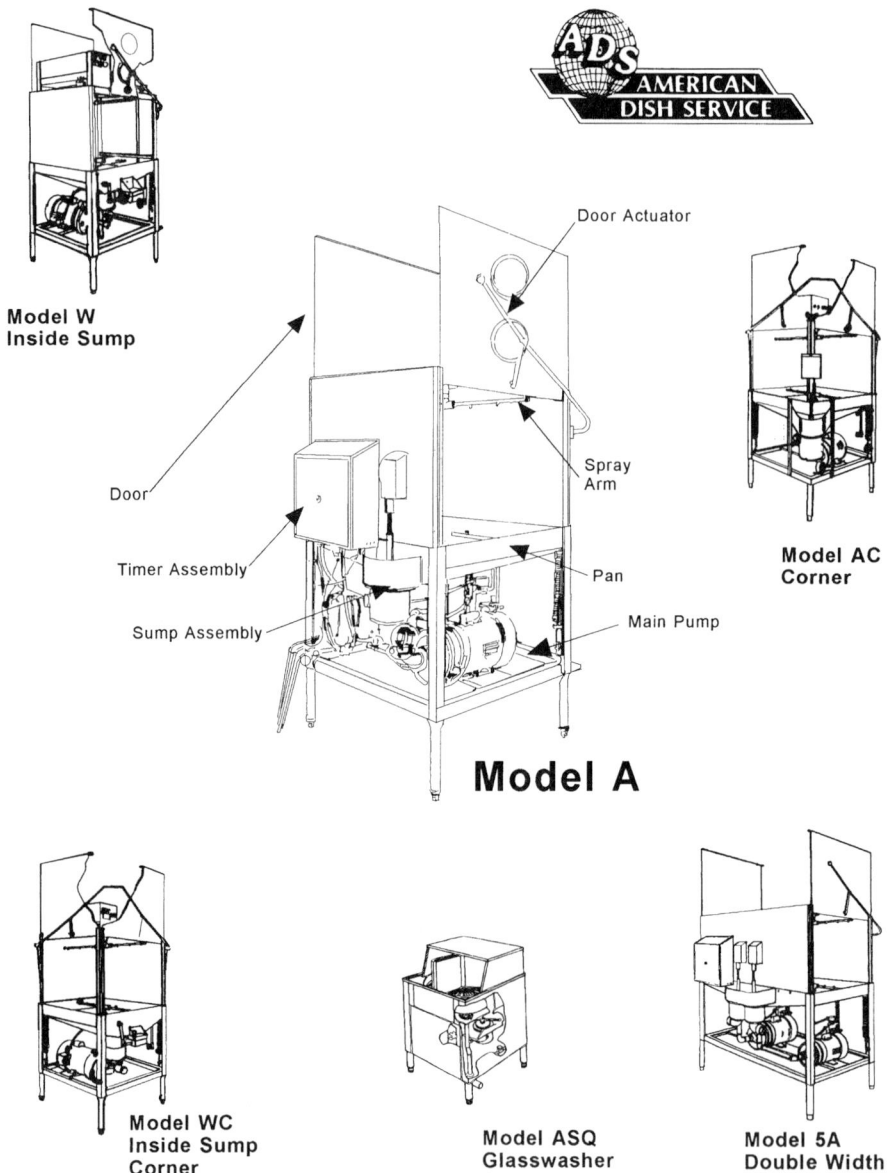

**Model W
Inside Sump**

Door Actuator

Spray
Arm

Door

Timer Assembly

Pan

Sump Assembly

Main Pump

**Model AC
Corner**

Model A

**Model WC
Inside Sump
Corner**

**Model ASQ
Glasswasher**

**Model 5A
Double Width**

Figure 4. ADS dishwashing machines.

a rack of dishes into the machine's cabinet. The operator then closes the doors and initiates the cycle. Water is held in the sump and pan of the machine. When the cycle starts, the main pump takes in water from the sump and pumps it through the spray arms over the dishware. Simultaneously, a small detergent pump meters a powerful liquid detergent into the sump. After a short time, which varies by model, an electromagnet raises the drain stopper in the sump. Soiled water flows out of the sump to a drain. Upon completion of draining, a solenoid valve admits fresh water through the fresh water plumbing assembly. Simultaneously metering pumps identical to the detergent pump introduce sanitizer and a rinse agent. The sanitizer is sodium hypochlorate, essentially concentrated liquid bleach. The rinse agent is a wetting agent that prevents spotting and aids drying. After rinsing the machine stops and the operator raises the doors. He or she slides the rack of clean dishes out the opposite side. Rinse water from each cycle remains in the machine to become wash water for the next load. The entire cycle takes 45–120 seconds, depending on the model.

The AC models function in the same way as model A. However, these machines are in corner locations with doors at 90 degrees. This contrasts with the straight-through construction of model A. Model AC is at the upper right of Figure 4.

Model W also functions like model A. In this model, the sump is inside the machine. The difference is largely aesthetic. The inside sump series also has a corner model, WC.

The 5AG is a double-rack machine. It is, essentially, two machines in a single cabinet. The 5AG washes two racks simultaneously and thereby doubles output.

The ASQ glasswasher uses the same operation principal but construction is quite different. It uses an integral rotary rack and fits under a bar counter.

Block 3 of task 04.01 constructs charts for past and future sales. This helps to establish the baseline design production rate. It may reveal uncertainties in the forecasts that dictate the necessity for high flexibility in the work cell design. Figure 5 shows the ADS production history in machines per week. It includes all models. From this history and information from marketing, we derived the sales forecast also shown in Figure 5. From this, we established and agreed on a baseline design volume of 12 machines per day. We also knew that the cell should have considerable volume flexibility.

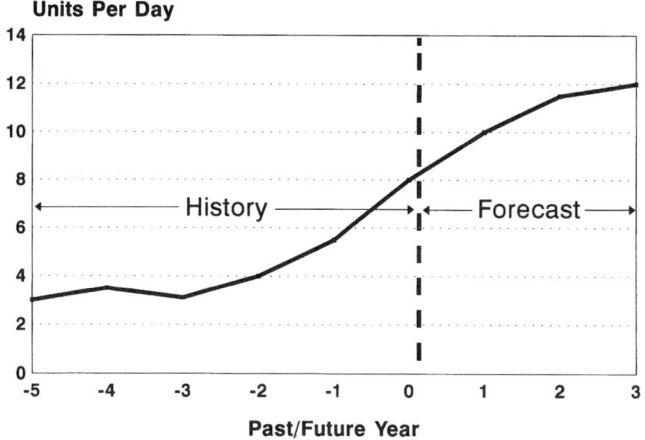

Figure 5. Sales history and forecast.

A product profile also comes from block 3. The product profile gives us insight into the product mix, both today's mix and tomorrow's mix. In situations where the product mix varies significantly in future years, you may want to construct a series of product profiles.

The ADS product profile is in Figure 6. It shows the relative volume of each model. For the ADS project, we assumed that the same relative volume would hold for future years.

Block 4 of Figure 3 is a decision or branch. It asks: "Are there less than 20 products?" We decided to define six products for ADS: A series, AC series, W series, WC series, and 5 series. Variations within each series were insignificant from a process viewpoint. This decision led us to block 5 where we charted the processes for each series.

Had there been more that 20 products, Figure 3 would have led us to a technique called Production Flow Analysis (PFA). Had there been more than 100 products, it would have led to Classification and Coding (C + C) as the most appropriate analysis technique.

Figure 7 shows the existing process chart for the model A series. Work cell designers have several process charting techniques from which to choose. The graphical chart using American National Standards Institute symbols is the most useful. This type of chart captures all "unofficial" elements such as movements, delays, and queues.

Many manufacturers do a good job of documenting value-added elements of a process but ignore non-value-adding elements. However, these non-value-adding elements usually contain the most opportunity for improvement.

With even a moderately complex process, the resulting chart can be quite large. If so, plot it on a large sheet of engineering drawing paper. This emphasizes the complexity of the process and helps in the subsequent simplification. Figure 7 is

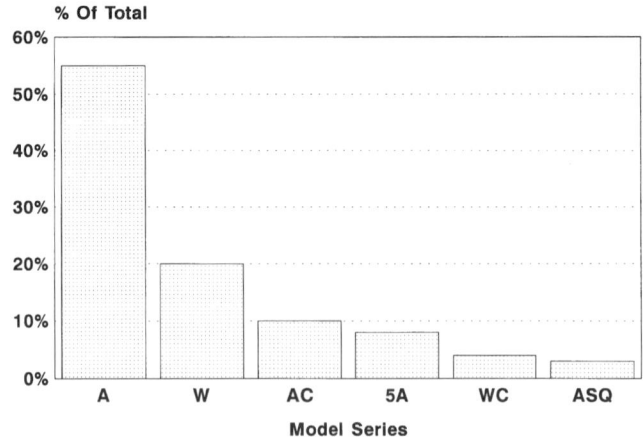

Figure 6. Product profile.

Dishmachine Assembly
Original Process

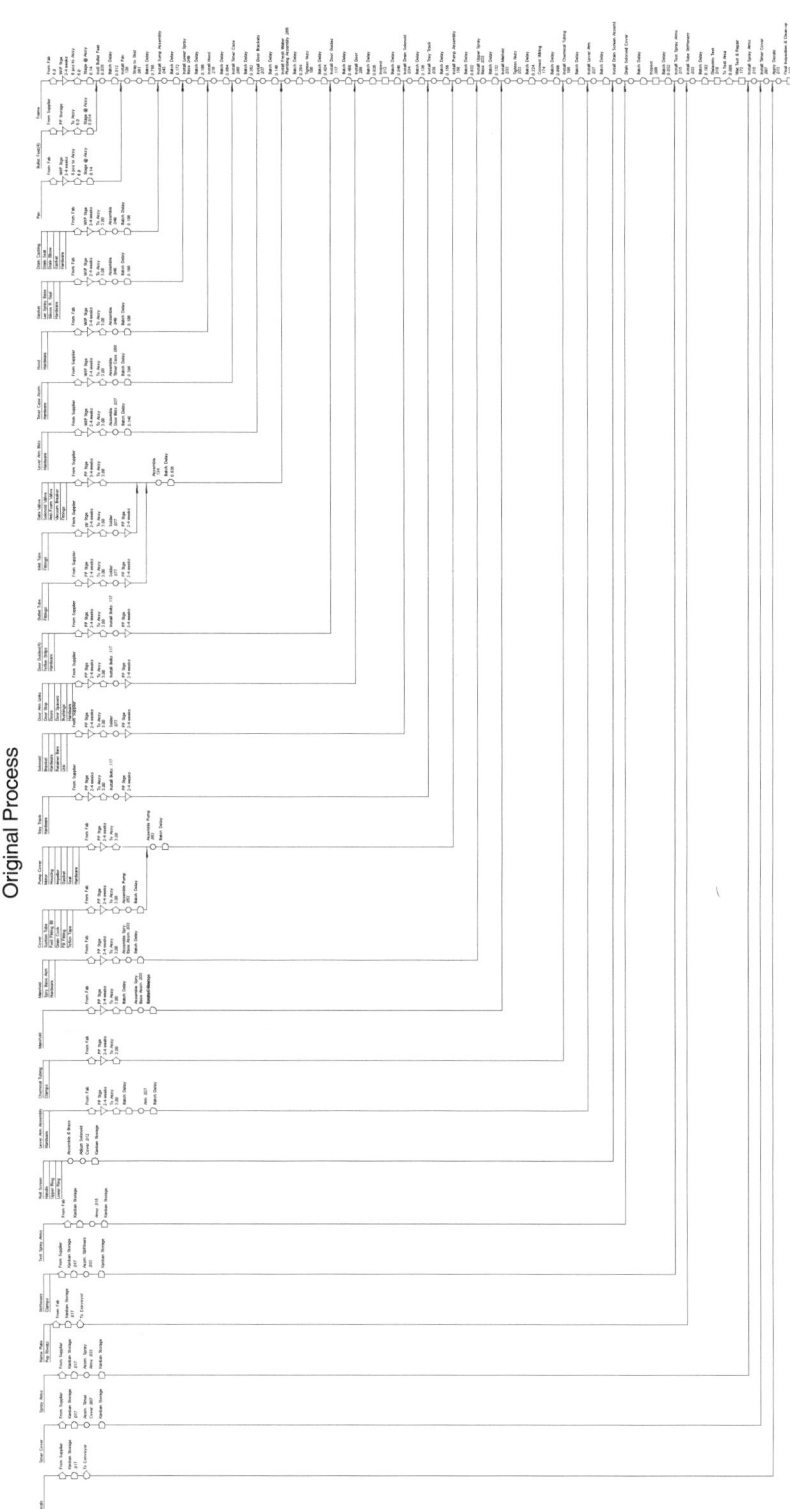

Figure 7. Existing process chart—model A assembly.

535

quite large. This size imparts useful information about the complexity of the process. Figure 8 enlarges a portion of the original process chart. This figure also shows a portion of the revised and simplified process for comparison.

From block 5 we proceeded directly to block 10. Here we make a preliminary selection of the products for each work cell. We have not shown the process charts

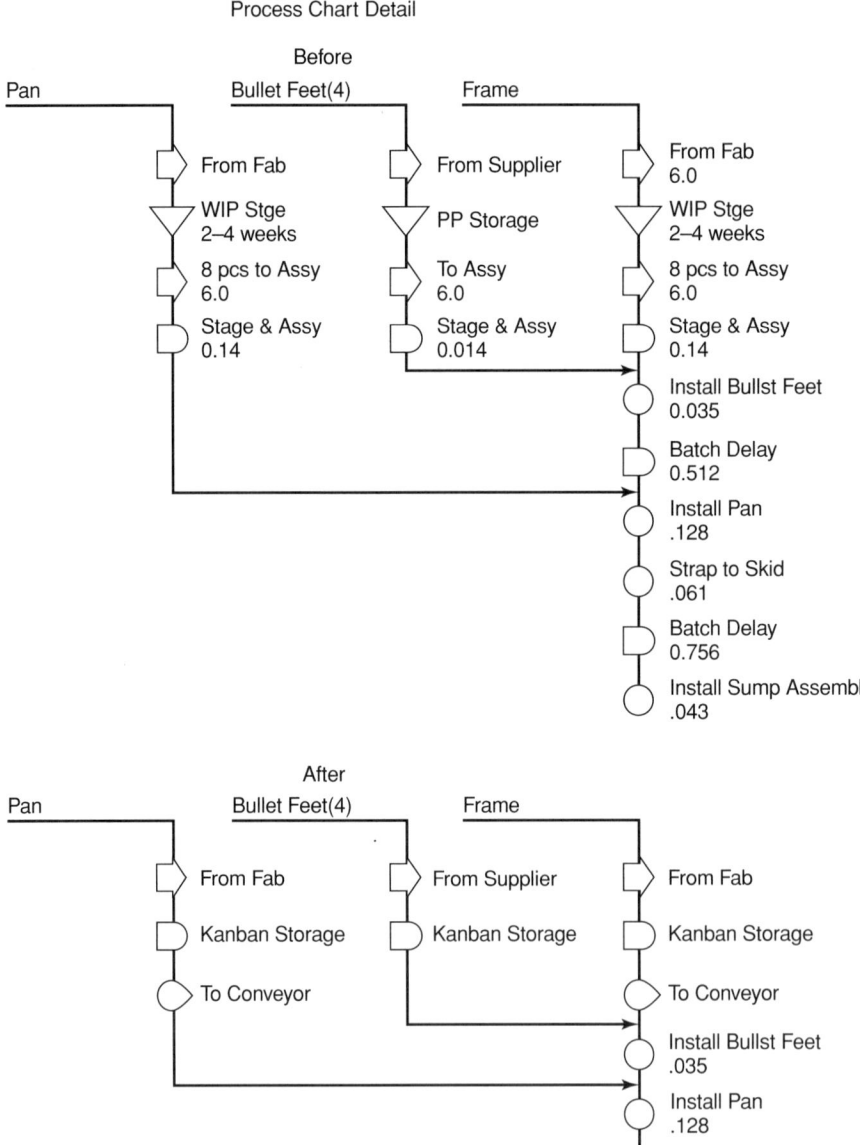

Figure 8. Process chart detail.

for all six models. However, an examination of the six charts showed that assembly operations for the dish machines were similar in type and sequence, although assembly time varied significantly. The dish machines also used many of the same parts. The Super Q glasswasher, because of its different design, had a very different process and much less part commonality. For this reason, we selected the five dish machine models for our assembly cell and excluded the glasswasher.

Our product family for this cell now consisted of five dish machine models but excluded the Super Q glasswasher. This concluded task 04.01 of the FacPlan procedure.

4. PROCESS DEFINITION

Henry Ford (1923), in his autobiography, *My Life and Work*, said "Every well-thought-out process is simple." Process is at the heart of every space plan. Simple, effective processes lead to simple and effective layouts. In task 4.02 we examine the existing or proposed process, improve it, and define a new process for the layout. There may be many processes or variations that can accomplish the required task. These differing processes may give rise to different layouts. Initially, designers should identify one–three fundamentally different processes. Other variations may evolve later as they iterate the design.

In blocks 11 and 12 of Figure 9, we started with the existing process charts for the five dish machines. The essential elements of assembly were sound, even at the higher design volume. However, we identified many improvements in tools, fixtures, and methods. We also found and eliminated many transports and delays.

Figure 8 shows the contrast between the original and revised processes. Figure 10 summarizes the elements by type. It shows that the total number of elements went from 202 down to 148, a 27% decrease. The figure also shows that the greatest reduction is in the non-value-added categories of storage and transport.

Block 13 asked us to calculate setup times for each operation. Since all operations were simple, manual assembly tasks, the setup times were insignificant.

Blocks 14 through 16 ask for equipment times, person times, and process times for each operation. The equipment time is the time for which a piece of process equipment is employed. The person time is the time an operation occupies a person. The process time is the time that a workpiece is occupied. For all of the ADS assembly processes, the person, equipment, and process times coincide. This simplified the time calculations.

In block 17, we calculated the number of people required for the target volume of 12 machines per day. This came to six persons. Equipment was, for the most part, hand tools and simple fixtures. Equipment calculations were largely unnecessary. The only exception was the functional test unit. This test required a water supply and drain. The calculations showed that one unit was sufficient.

Since setup times were minimal, the lot size was set at one piece. In other situations that require significant setup, you should perform an economic lot size calculation to help with this decision. This completed block 18 of task 04.02.

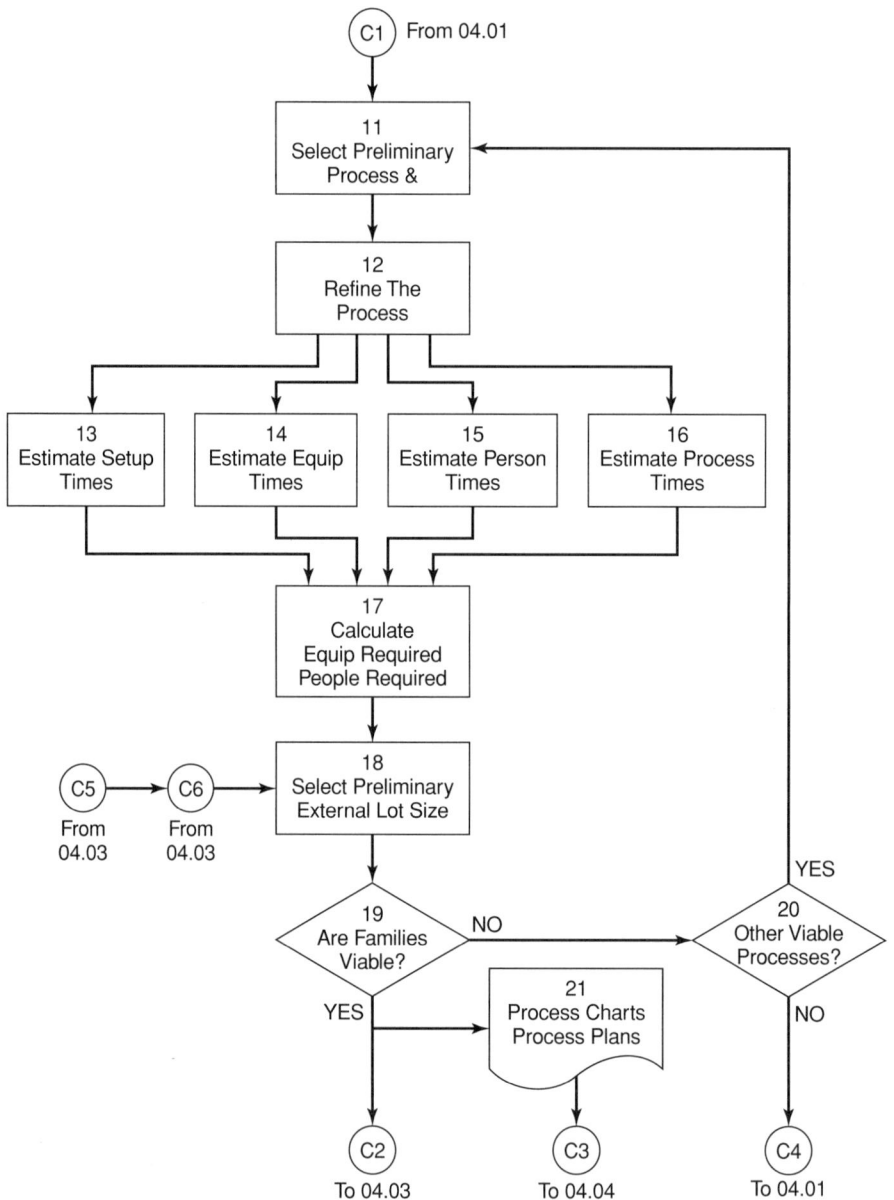

Figure 9. Design and refine process, task 4.02.

Lot size selection was one of the greatest process improvements. In the original process, machines were built in batches of 6 or 12 units. A large area in the center of the building was cleared and a batch of machines was built up. Each operation was completed on all machines before starting the next operation. This approach led to a lot of material and people movement. Operators went to a storage area for parts,

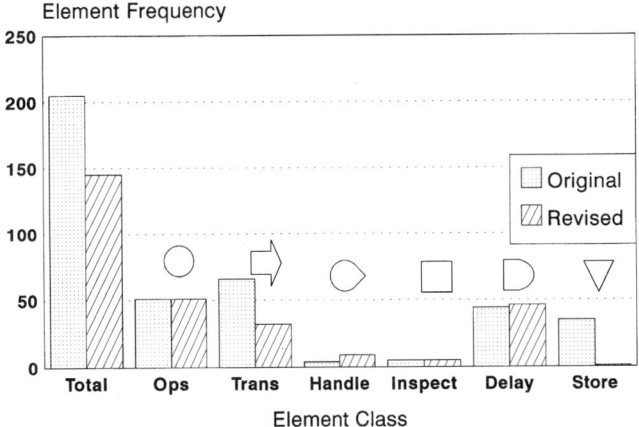

Figure 10. Process element summary.

carried the parts to the assembly floor, and distributed them near each machine. They also moved tools, test equipment, and fixtures to and from the assembly area and among the machines.

Perhaps the worst feature of the original process was the output characteristic. Twelve machines represented several days production. This system produced nothing for several days. Then, suddenly, 12 identical machines emerged. This output bore no relation to the customers' ordering pattern. Customers ordered a random mix of machines each day.

The consequence of this mismatch between production output and customer orders was a lot of finished goods inventory. In addition, machines in finished goods were often modified to meet specific customer requirements. For example, a 120-second cycle time machine might be converted to a 90-second machine. Or a 110-volt machine might be converted to a 220-volt machine. This added confusion, labor, and cost.

Quality was also an issue under the old system. Defective parts or an improper assembly method led to 12 identical units with 12 identical defects. This required 12 identical repairs. The result was high repair cost and disrupted schedules.

The work cell designers decided to build one machine at a time. This is the Toyota concept of mixed-model production. Looking ahead in the design process, they envisioned several possibilities. One possibility was a series of independent assembly stations that built complete units in one location. Another was a series of assembly stations with movement of the machines between stations as the assembly progressed. This sort of advance speculation is often necessary. However, it is important not to carry it too far. Design of the physical layout comes later.

In block 19 of Figure 9 we examined the proposed processes and families. The decisions made to that point looked viable and we proceeded to task 04.03. Had the family of products and corresponding processes not been viable, we would have moved to block 20. At block 20, the choice is to investigate other processes or revert back to task 04.01 and make a different product selection.

The output of task 04.02 was the process definition embodied in the revised process charts, element times, and lot size definition. These outputs show as block 21 on the task 04.02 procedure. They become important inputs for task 04.04, designing the cell layout.

5. INFRASTRUCTURE DESIGN

Task 04.03 selects infrastructure elements. The task procedure is in Figure 11. Infrastructure elements support the process but do not directly affect the transformation of the work product. Some infrastructure elements are tangible and some intangible. For an adequate design, you should define

- External containers
- Internal containers
- External material handling
- Internal material handling
- External production control
- Internal production control
- Internal lot size (transfer batch)
- Equipment balance method
- People balance method
- Quality assurance
- Supervision
- Compensation system
- Operator assignments and skills

Figure 12 is the work cell operations plan for the dish machine assembly cell. It captures and summarizes most infrastructural design decisions. It summarizes the day-to-day operations. The following paragraphs expand and explain the information in Figure 12.

FacPlan uses a size convention for containers and lot sizes. This convention uses the design production rate for the work cell. In the FacPlan system, a "large" container or lot has 4.0 hours or more of production material at the design rate. A "medium" container or lot has 1.5 hours of material but less than 4.0 hours. A "small" container or lot has less than 1.5 hours of material. "Single piece" is a container or lot with only one item. Single piece also refers to a single item without a container.

5.1. External Containers

The first block, block 22, in task 04.03 is selection of external containers. External containers carry material in and out of the work cell. The dish machine assembly cell uses single-piece containers for outbound material. This is the individual dish

Task 04.03
Identify Infrastructure
Level 04- Micro-Spaceplan

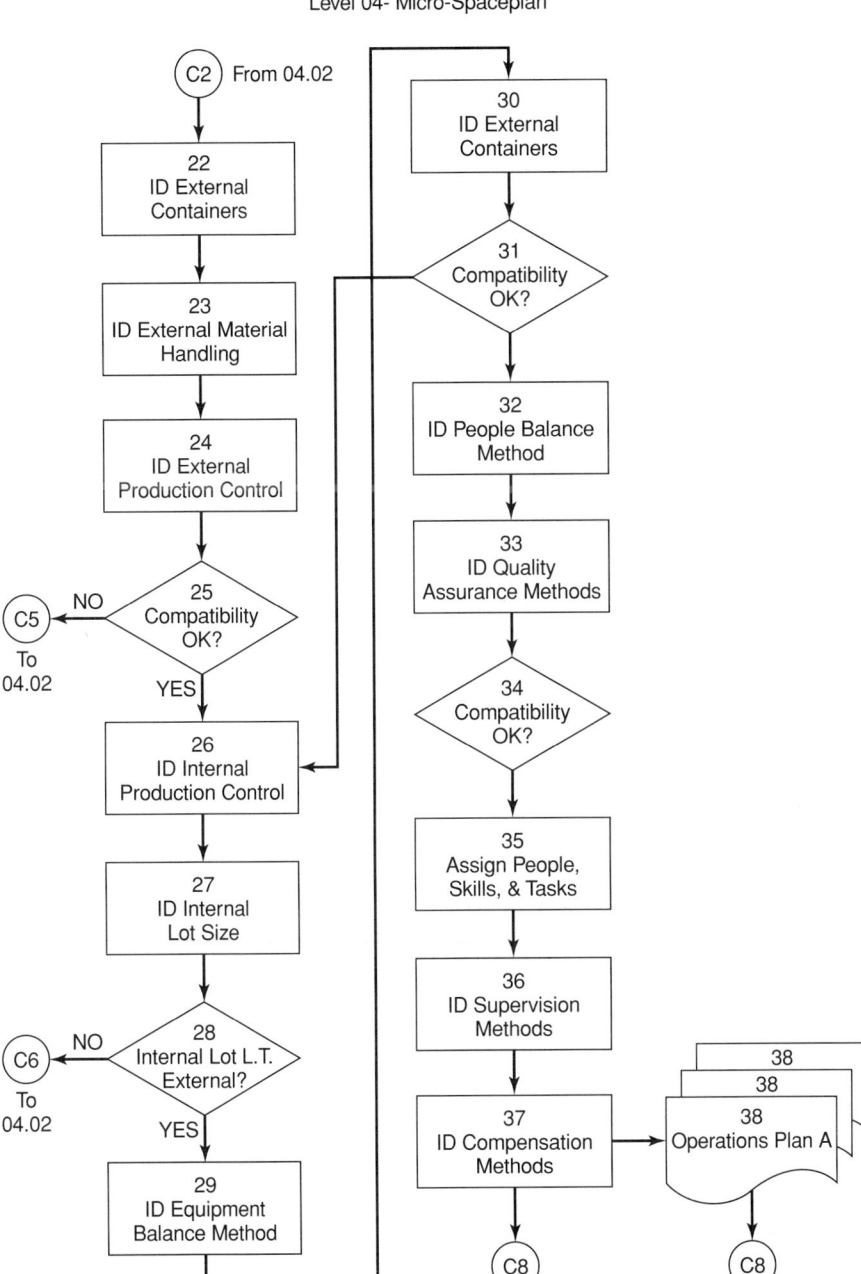

Figure 11. Select infrastructure elements, task 04.03.

machine sitting on skid and packaged in corrugated material. Inbound material arrives in large and medium containers. Sizable sheet-metal items arrive on semilive skids. Purchased items arrive on pallets or in cartons. In Figure 12, the area for "external containers" has checkmarks for "single piece" on outbound containers and "large" for inbound containers. A notation describes these containers briefly.

5.2. External Material Handling

For external handling, specify the load size using the convention discussed earlier. Also specify the route structure and type of equipment. You should coordinate these selections with the container selection discussed above. For a more complete discussion of route structures and handling systems, refer to Lee (1997). These selections occur in block 23 of Figure 11. Note the selections in the appropriate area of the work cell operations plan as Figure 12 shows.

For the dish machine cell, material coming in arrives in large loads, using a direct route structure. Sheet-metal parts are moved with a manual pallet jack. Other parts move on a walkie stacker truck. A few small parts are hand carried.

Outgoing material flows one-piece at a time using a walkie stacker truck. It is carried on a self-contained skid/carton.

5.3. External Production Control

Production control occurs both internally and externally. Block 24 of Figure 11 asks for a selection of the external production control system. The choices are physical link, broadcast, kanban, push scheduling, and reorder point. In addition, you must decide whether to make-to-order or make-to-stock. Lee (1997) describes these methods and discusses their application.

For the dish machine cell, outgoing units are scheduled by incoming orders, in the same sequence as orders arrive. Material coming into the cell is controlled using kanban to the fabrication areas and vendors.

Block 25 of Figure 11 asks for a compatibility check. At this time, we have made design decisions on products, processes, and several infrastructure elements. The compatibility check assures that these various decisions are consistent. For example, a "large" container is incompatible with "small" handling equipment. Compatibility is an important issue. It is the source of many difficulties when transitioning to work cells. Refer to Lee (1997) for additional discussion on the subject.

5.4. Internal Production Control

Internal to the work cell, operators or supervisors make decisions on what tasks to perform, what item to work on, and when to work. These decisions are often made minute to minute. For successful operation, the work cell needs a mechanism to help people decide quickly, consistently, and rationally. The choices are direct link, circulation, kanban, and push schedule. This is block 26 of Figure 11.

Strategos

Workcell Operations Plan

Project: DISHWASHER ASSEMBLY CELL	Prj #: 98XX
Company: AMERICAN DISH SERVICE	By: QL	Date: 02/02/95

Note:

External Lot Size

In / Out
- ☐ ☑ Single Piece
- ☐ ☐ Small <1.5 Hours
- ☐ ☐ Medium <4.0 Hours
- ☑ ☐ Large

Prod Unit: PIECE

External Containers

In / Out
- ☐ ☐ Small 1-PIECE
- ☑ ☐ Medium
- ☑ ☐ Large

Type: SKIDS/PALLETS

External Material Handling

In / Out
- ☐ ☑ Small 1-PIECE
- ☑ ☐ Medium
- ☑ ☐ Large

Route Structure:
- ☑ ☑ Direct
- ☐ ☐ Channel
- ☐ ☐ Terminal

Equipment:
- ☑ ☐ Hand Carry
- ☑ ☑ Other

Type: SEMI-LIVE SKIDS
WALKIE STACKER

External Production Control

In / Out
- ☐ ☐ Physical Link
- ☑ ☐ Broadcast
- ☑ ☐ Kanban
- ☐ ☐ Push Schedule
- ☐ ☐ Re-Order Point
- ☐ ☑ Make-To-Order

Internal Production Control

- ☐ Direct Link
- ☐ Circulation
- ☑ Kanban
- ☐ Push Schedule

Internal Lot Size

- ☑ Single Piece
- ☐ Small <1.5 Hours
- ☐ Medium <4.0 Hours
- ☐ Large

Equipment Balance

- ☐ Inherent Balance
- ☑ Queue
- ☑ Excess Capacity

Internal Containers

- ☐ Single Piece
- ☐ Small
- ☐ Medium
- ☐ Large

Type: SINGLE PIECE

People Balance

- ☐ Inherent Balance
- ☑ Queue
- ☐ Excess People
- ☑ Circulation
- ☑ Float

Quality Assurance

- ☑ Inspect & Reject
- ☐ SPC/TQM
- ☐ Inherent

Supervision

- ☑ Self Managing (Cybernetic)
- ☑ Kanban Signal
- ☑ Command & Control

Compensation

- ☑ Hourly/ Salary
- ☐ Individual Incentive
- ☐ Group Incentive

Operator Assignments & Skill Matrix

Operators	Skills						#1 ASSY	SUB ASSY	#2 ASSY	TEST	REPAIR	PACK	SHIP	Operations					
	MECH	ELECT	TEST	PACK	SHIP														
WILSON	●						●	○											
TAYLOR	●						●	○	○										
WASHINGTON	●							○	●										
DAVIS	●	●							○	●	●	○							
ROOSEVELT	○	○		●	●					○	●	●							
JOHNSON	●	●						●		○									

© Strategos, Inc.   CM0009.GED

Figure 12. ADS work cell operations plan.

For the ADS work cell, internal production control uses a kanban system. The kanban signal is the size of queues between operations.

5.5. Internal Lot Size

Block 27 of Figure 11 asks for the internal lot size. This refers to the number of production units that move between operations. In a work cell, the internal lot size is often much smaller than the external lot size. For example, a cell may produce a particular product in lots of 50 units. But, within the cell, units move in lots of five between operations.

For the ADS assembly cell, our internal lot size is one piece. This fits with our container selection and handling method.

The internal lot size should never be greater than the external. Block 28 of Figure 11 looks for this check. If you find an incompatibility between the internal and external lot sizes, Figure 11 directs you to review the lot size decisions.

5.6. Internal Containers

Within the cell, materials move between workstations. If the cell uses containers internally, they may be different than the external containers. In the dish machine cell, most items move without containers, one piece at a time. The location for "internal containers" in Figure 12 notes this. This design step is block 30 of Figure 11.

After choosing the internal containers, Figure 11 asks for another compatibility check. This is block 31. If compatibility is good, go to block 32. Otherwise, Figure 11 directs you to review certain infrastructural decisions.

5.7. Equipment Balance Method

Equipment balance refers to the method for balancing equipment workloads. The choices are inherent, queue, and excess capacity. Inherent balance requires that all equipment have the same cycle time for every item that moves through the cell. Queueing uses internal staging or queues to balance workloads. Excess capacity provides some equipment with less work and essentially circumvents the issue. Excess capacity is a common method and works well with low-cost equipment.

Since the ADS work cell requires no high-cost equipment, balance between workstation hardware is attained with excess capacity. In effect, no attempt is made to balance. Figure 12 notes this in the appropriate space.

5.8. People Balance Method

A successful work cell requires reasonably good balance between people. People are much more sensitive to imbalance than machines. Available techniques are inherent, queue, excess people, circulation, and float.

Inherent balance requires that each process step have an identical process time. This is often used with inherent equipment balancing.

Queueing uses internal queues to balance the workload. Queuing can help with short-term balance that results from differences in cycle times between various products. It cannot, however, correct long-term imbalances that result from differences in average cycle times over hours or days.

Excess people, like excess equipment, circumvents the issue. In a work cell, this is seldom satisfactory since the differences in workload will create conflict. This is an example of the socio-technical system effect.

The circulation method requires that people follow each part through the process. With multiple people and processes within a single cell, people follow each other around the cell. This method is very effective when products are small, the range of skills is not great and when the cell operates at less than about 70% capacity.

The ADS cell uses the floating technique. Here, some or all of the operators have primary and secondary task assignments. They use the kanban queues as guides to determine when to move from a primary to a secondary task. Identifying the people balance method occurs in block 32 of Figure 11.

5.9. Quality Assurance

Block 33 of Figure 11 identifies the methods for quality assurance. The ADS cell maintains quality with inspect and reject methods. Much of the inspection occurs at the final test. This is not generally the best approach. However, management did not feel they had the resources for a total quality approach at the time. Moreover, while quality was an expensive nuisance, other issues dominated.

Some processes have such good inherent quality that they require no additional process control or inspection. Do not overlook this possibility.

After selecting the quality assurance method, we perform a compatability check to ensure that our design decisions are are consistent. This is block 34.

5.10. Operator Assignments and Skills

In block 35, we assigned people, assigned tasks, and determined the required skills. The lower portion of Figure 12 shows the task and skill assignments for each operator. The six operators, Wilson, Taylor, Washington, Davis, Roosevelt, and Johnson, are in the left column. A list of skills is at the top left and a list of task assignments at the top right. The solid dots represent primary skill requirements and primary tasks. The open dots represent secondary skills and tasks.

Wilson and Taylor perform all two-person assembly tasks and enough single-person tasks to balance the workload. They also assist with subassembly work required in their area. From time to time, Taylor can shift to the No. 2 assembly station and assist Washington. Washington completes the remaining assembly tasks and assists with subassembly. Davis tests the machine. He may occasionally move to the No. 2 assembly station to assist Washington. He also assists Roosevelt with packaging from time to time. Roosevelt packages the machines, tracks the limited finished goods storage and coordinates shipping. Roosevelt can assist with repair when required.

Johnson performs the majority of subassembly work moving from station to station as required. He can also assist with repair.

All of the above movements and task shifts are based on kanban quantities in the various queues. When a particular queue gets low, operators shift to that product. When a queue fills, they cease work on that item and move to another location where the queue is lower.

5.11. Supervision

The ADS work cell uses several supervision methods. Kanban within the cell helps workers decide what to do and when to do it. Informal teamwork developed quickly because of the cell's configuration. Some elements of command and control remained.

5.12. Compensation System

Worker compensation used traditional hourly wages for the ADS work cell. ADS had never developed a highly structured job classification system. As in many smaller firms, people worked where they were needed and performed whatever jobs they were capable of. This fit well with the new work cell approach. Block 37 of Figure 11 made this decision.

6. THE LAYOUT

With the work cell operations plan complete, we went to task 04.04, design cell layouts. The details of this task are in Figure 13. This figure shows the steps that lead to a physical arrangement of the workcell. Inputs are

- Process definition from task 04.02
- Work cell operations plan from task 04.03
- Constraint summary created as part of this task

Every space plan (layout) has four fundamental elements: space planning units (SPUs), affinities, space, and constraints. The SPUs are the blocks of space that will be arranged. Affinities represent the level of proximity required between each SPU and other SPUs. Space is the area required for the SPU. Constraints are factors that limit the design. Constraints include building columns, walls, aisles, and regulatory requirements.

In block 39 of Figure 13, we define the space planning units. Figure 14(a) lists these SPUs in the affinity chart. The SPUs came from our analysis of the process, process times, and task assignments. The right portion of the chart has affinities for each combination of SPUs. These affinities are based on material flow, communications, and other considerations.

The affinity chart is like a mileage chart in an automobile atlas. Instead of actual distance between any two locations it shows the desired proximity. For example, the

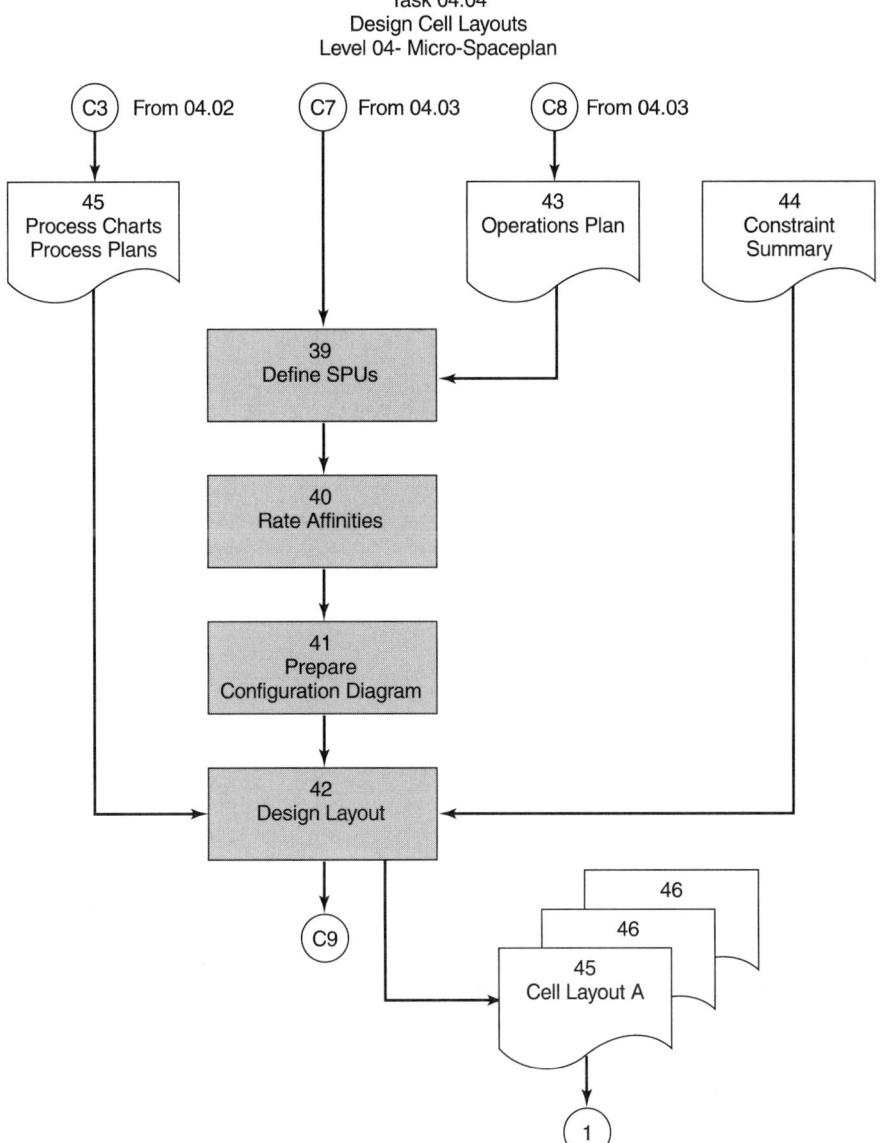

Figure 13. Task 04.04 design cell layouts.

desired proximity between No. 1 assembly and No. 2 assembly is rated as A. This
means that the need for proximity between these two locations is "absolute." Just
below the letter rating are two numbers. These numbers refer to the reason codes in
a box to the right of the figure. For the No. 1 and No. 2 assembly, the reasons for an
A rating are material flow and shared tasks. The diagram at the upper right in Figure
14 elaborates.

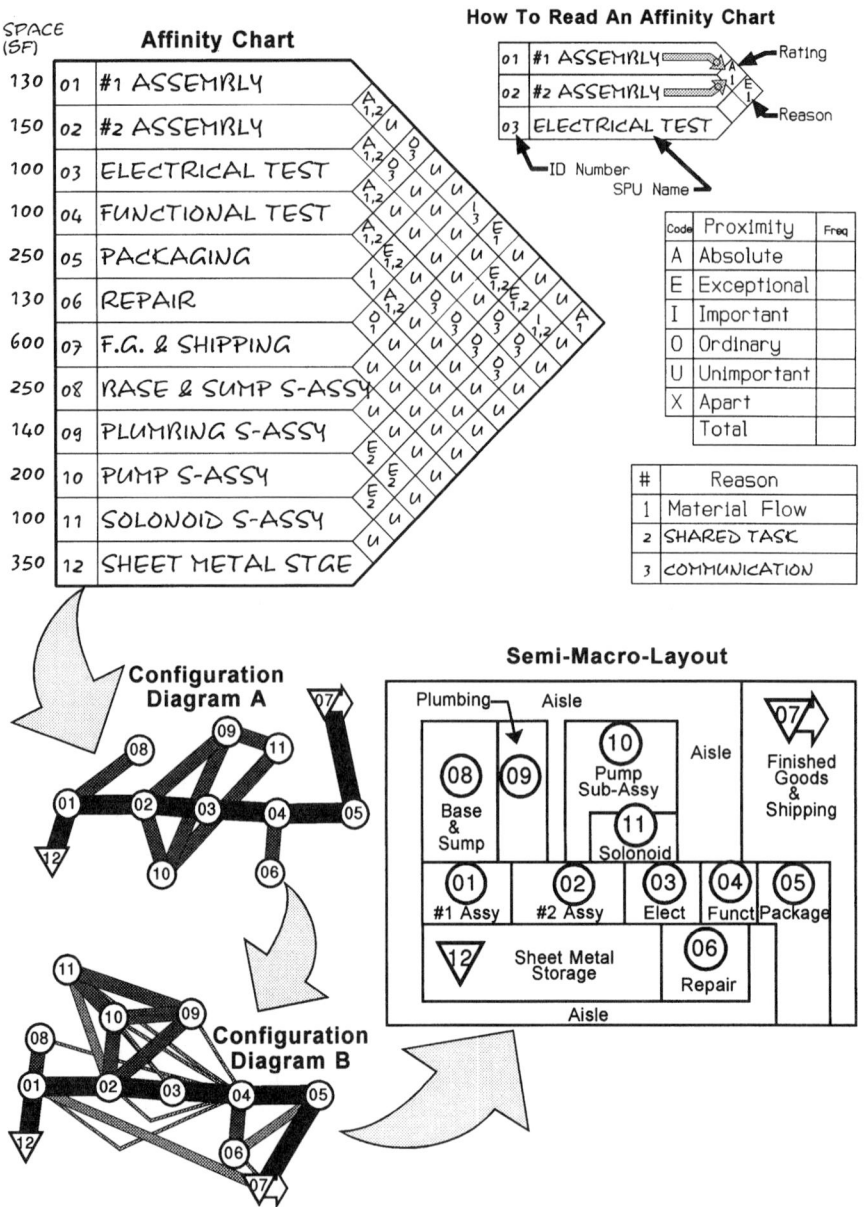

Figure 14. Layout development.

A box to the right of the affinity chart in Figure 14 lists the six possible affinity ratings. A is the highest level rating. E tells us that a proximity requirement is "exceptional." I tells us that a proximity requirement is "important." O tells us that a proximity requirement is "ordinary." U tells us that proximity is "unimportant" or neutral. An X rating indicates that a particular pair of SPUs should *not* have proximity. For a more detailed discussion of affinities and affinity charts, refer to Lee (1997). The estimated space for each SPU is at the left of the affinity chart.

Next, in block 40, we rate the affinities. For a work cell, you can usually do this through examination and consensus. More detailed quantitative analysis is seldom justified.

Block 41 creates a configuration diagram. Figure 14 shows this in the lower left. Diagram A is the first effort. It has only the A and E level affinities. Diagram B adds the lower level affinities and rearranges the diagram to a more optimum configuration.

The semimacro layout of Figure 14 shows how the configuration diagram translates into space. Figure 15 shows the final, detailed layout. In practice, we did not prepare a semimacro layout. We simply translated the configuration diagram into detailed equipment and feature locations.

In block 42, we prepared the detail layouts using a Computer-Aided Drafting (CAD) system. We started with a drawing of the building outline. We then added the equipment required for each SPU. We first placed major equipment in the areas suggested by the configuration diagram. With the knowledge of cell operations, we manipulated the equipment drawings into layouts. The final layout is in Figure 15.

Although we have not shown them here, task 04.04 resulted in several good layouts. During task 04.05, we examined the layouts and selected the best for implementation.

7. HOW IT WORKS

This work cell can operate as a large cell with four persons or even fewer. It can also operate as a main assembly cell in a line layout with subassembly cells feeding in. The east–west roller conveyor carries the main assembly as it progresses. Large sheet-metal components feed in from the kanban storage on the north side. Subassembly cells on the south side feed components directly to the final assembly conveyor.

Sheet-metal parts arrive on semilive skids from fabrication areas. The storage area on the north side of the line is an inbound kanban stockpoint. The skids have specific item assignments and serve as production orders for the fabrication area.

Machined components arrive on skids, pallets, and in cartons. They are staged in the subassembly areas or in the area north of the line. A kanban system also controls their delivery.

Purchased parts are primarily on a Reorder point system. A few items from local vendors use kanban.

At station 1 on the conveyor, two operators begin the assembly. Some parts of the machine are large and require two people. All two-person operations take place at station 1. Additional single-person operations approximately balance the workload with other cells. When their work on a machine is complete, the operators move the machine along the conveyor to a staging segment. Here the machine may sit for a short time until the operator at station 2 is free. This staging segment can hold two machines. It allows for variations in cycle time from one machine to the next. It allows

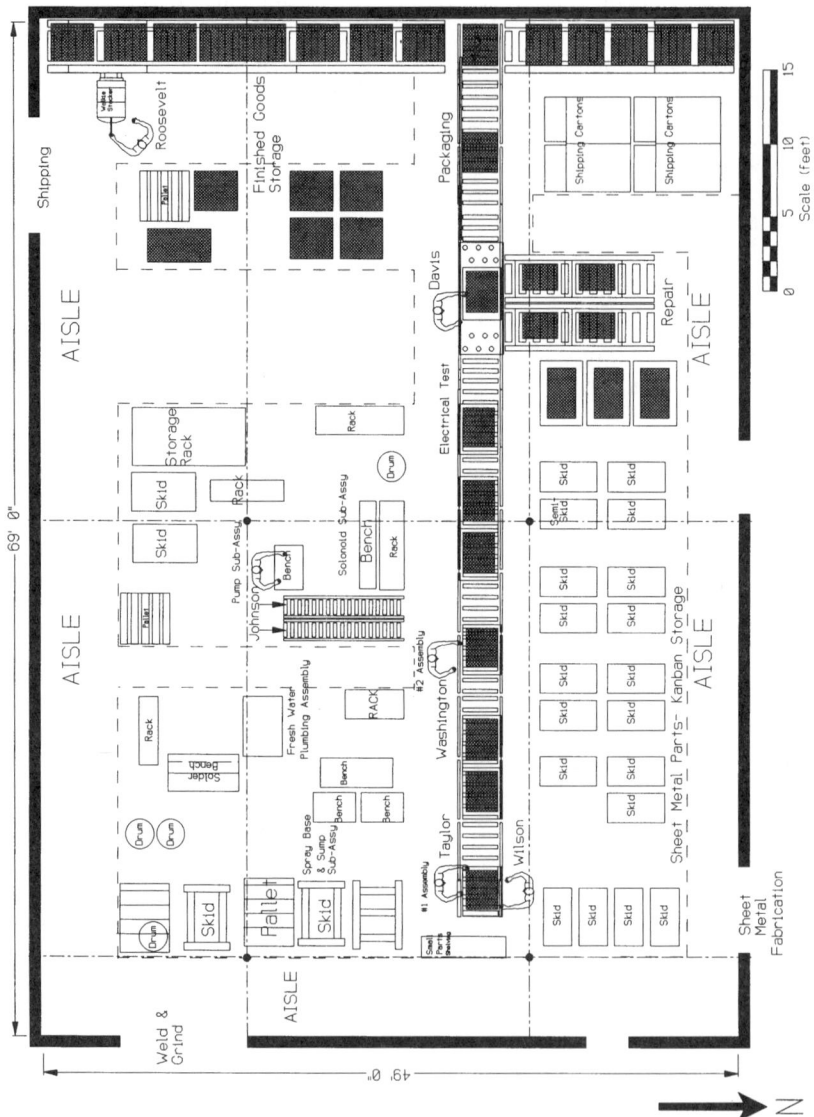

Figure 15. ADS assembly cell layout.

all five models to move through the process in customer order sequence. Station 2 completes assembly.

The next station accomplishes test and repair. This includes a dielectric test and functional tests using water. This station has a ball table to allow the machines to turn in any direction for access. Machines that require major repair move to the conveyors on the north side of the line. With repairs complete, they re-enter the test station.

The operator at the last station packages the machine and moves it to finished goods staging. This worker also ships finished machines and coordinates with the outbound trucks.

Figure 16 is a photograph of the main assembly line. This photo is taken from the functional test station looking east toward assembly station 2. You can partially see the test station operator in the foreground. Subassemblies from several subassembly stations that are in the background.

8. IMPLEMENTATION

Implementation of this cell was straightforward. Production people installed the conveyors, drains, shelving, and other equipment. They soon became accustomed to coordinating their efforts using the intermediate staging areas as signals.

The design of this work cell encourages teamwork. Limited Work-In-Process (WIP) storage areas demand close cooperation between workers. The level and location of WIP is immediately obvious. Every worker can see from the number and

Figure 16. Main assembly line.

location of machines which stations are lagging and which are ahead in the production process. Since most stations can accommodate one to three people, workers quickly move to assist their teammates.

The implementation plan did not include a self-managed work team. However, the inherent teamwork requirement quickly developed a high degree of self-management within the work group. This occurred without specific team training or development.

As with many work cell implementations, supervisors had the most difficult adjustment. They had to learn not to micromanage the cell. This destroys teamwork and incapacitates the operators' ability to move quickly between jobs and release bottlenecks.

Many seemingly minor problems in the production process quickly came to light. Workers and supervisors devised innovative, simple solutions such as racks, fixtures, and new tools. They addressed upstream quality problems because such problems quickly brought the entire assembly area to a halt. Many of these upstream quality problems had festered for years under the previous batch system.

As in most cellular implementations, the first few weeks and months were difficult, challenging, and frustrating. The rewards, however, were significant. For individuals, these rewards were intrinsic. They included a job well done, meaningful work, and comraderie. For ADS, the rewards were higher productivity, better quality, reduced inventory, and fast customer response.

9. CONCLUSION

Every work cell is different. Successful work cells result from an organic design that starts with customer requirements. An organic design incorporates the physical and psychological needs of workers. It accommodates the predominate culture of the organization. Most importantly, it follows and integrates with the underlying process.

In this chapter we have illustrated the FacPlan approach to work cell design. FacPlan uses a set of specific tasks and procedures to guide the designer. It identifies key parameters. It uses forms, charts, and other design aids that organize information and preclude omissions.

The FacPlan approach does not substitute for experience in work cell design. However, it supplements the experience of veteran designers. It accelerates learning for the novice. It hastens the conversion from functional operations to cellular. It increases the probability for success. It helps to achieve full realization of potential benefits.

REFERENCES

Ford, H. (1923). *My Life and Work*, Doubleday, New York.

Ford, H. (1926). *Today and Tomorrow*, Doubleday, New York.

Lee, Q. (1997), *Facilities and Workplace Design: An Illustrated Guide*, Engineering and Management Press, Norcross, GA.

Lee, Q. (1990). Manufacturing focus: A comprehensive view in *Manufacturing Strategy*, Christopher A. Voss (ed.), Chapman and Hall Scientific, London.

Muther, R. (1973). *Systematic Layout Planning*, 2nd ed., CBI Publishing, Boston.

Muther, R., and Haganas, K. (1969). *Systematic Handling Analysis*, Management and Industrial Research Publications, Kansas City.

18

CELL FORMATION IN A COIL FORGING SHOP: AN IMPLEMENTATION CASE

Marco Perona

Dipartimento d'Economia e Produzione
Politecnico di Milano
20133 Milano, Italy
and
Dipartimento di Ingegneria Meccanica
Universita di Brescia
25123 Brescia, Italy

This example of implementation of the Cellular Manufacturing (CM) approach in a real industrial context was carried out within a joint project of the Politecnico di Milano and an Italian company belonging to a large European group, leader in the market of house and industrial cooking, refrigerating, and washing appliances. The study was carried out in 1992–1993.

Due to the restrictedness of information, the company addressed prefers not to be mentioned with its real name. As a consequence, we will use the conventional name ALFA in order to address the company and BETA in order to indicate the group.

This report is organized as follows. In Section 1, we present a description of ALFA's scenario previous to the implementation of the project; Section 2 contains a general description of the project, its goals, and its organization. Sections 3 and 4 present the two informatic procedures that are at the heart of the project. Finally, in Section 5 the results of the project, as well as the problems encounterd, are discussed.

Handbook of Cellular Manufacturing Systems, edited by Shahrukh A. Irani
ISBN 0-471-12139-8 © 1999 John Wiley & Sons, Inc.

1. CONTEXT

The present section illustrates the status of the company at the beginning of the project. Many of the features discussed have greatly changed since the project implementation, as described in Sections 2, 3, and 4.

1.1. Company ALFA

ALFA is a mid-sized company, located in Northern Italy, that designs, manufactures, and sells kitchen appliances for industrial uses such as schools, hotels, hospitals, barracks, and communities. Its annual turnover, as of 1994, was around 300 billion lire or 180 million US$ at current change rates. Its workforce size is around 1000 units encompassing direct, indirect, and clerical employees.

ALFA belongs to BETA, a large European group, leader in Europe in refrigerating, washing, and cooking appliances for both house and industrial use. BETA has more than 20 production sites in 6 countries in Europe. The group policy has always been to search for the optimization of its operations, mainly through the standardization of products and components, and the consequent achievement of economies of scale. This target has been relatively easy to achieve in the house appliance market, while the industrial appliances sector allows for less standardized products to be put on the market due to the sharp differences that still characterize the cuisines of the different European countries.[1] As a consequence, the different companies of the group that operate in the market of industrial devices have been always left sufficiently free to decide the different aspects of their own operations management policies.

ALFA is structured in four small-sized production units located at a rather short distance from one each other, coordinated by an operative headquarter located across the street from the oldest production plant. The different production plants were progressively built following the development of the company's market and have ever since focused on specific product lines, as reported below:

- Air filtering and smoke conveying devices
- Ovens and cooking devices
- Dish-washing machines
- Refrigerators and freezers

The headquarters performs all the typical staff activities of the company. Among these, sales and order entry, since customers are more likely to buy complete kitchens rather than single modules. With relation to Figure 1, customer orders keep arriving to the central Sales Department in time (1); as soon as they arrive, they are entered to the Information System (IS) (2). Each order may consist of several order lines, each of which corresponds to one specific module of a kitchen (e.g., a dish-washing

---

1. For example, Italian customers typically require big vessels to boil their pasta, while Northern European customers such as the English, the Dutch, or the Scandinavians are more keen on deep-frying devices.

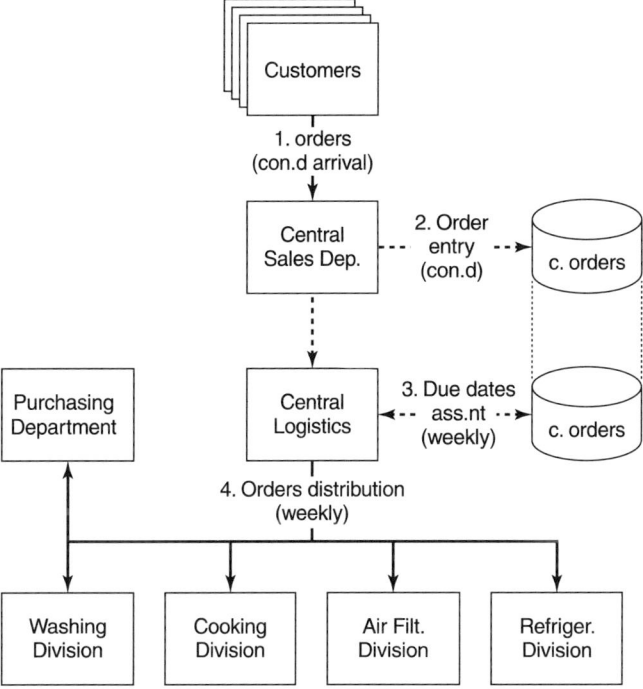

Figure 1. Scheme of the order entry process.

machine, an oven, or even an accessory). When entering orders to the IS, the Sales Department sets the order due date to the date required by the customer.

Once orders are completely stored in the customers orders table, they are elaborated by the Central Logistics Department, which assignes them a due date (3); this confirmed due date may respect or not the date required by the cutomer, depending on one or more of the following aspects:

- The standard average lead time of each division and each supplier involved in delivering the order
- The current workload of each division and the consequent estimation of its current lead time
- The possibility to expedite a specific order to a certain division, in a way that it can be completed sooner than the average lead time

Once the various order lines have been given a feasible due date, each of them is dispatched to the corresponding division of the company (4). Some complements and accessories may not be produced internally: therefore, the corresponding order lines are assigned to the central Purchase Department, which buys them from specialized suppliers. After the central planning activities are done, each processed order line is released to the corresponding division or department; apart from particular cases, this is done on a weekly basis.

1.2. Dish-Washing Machines Division[2]

The Dish-Washing Machines Division is, together with the Ovens Division, the biggest production unit inside ALFA. The production facility has an area of 2000 square meters and the overall number of employees is around 300 at the end of 1994. Its activity is to design and manufacture three product lines of dish-washing machines, differentiated by dimension.[3] Each product line is offered in different models and versions, while a rich variety of options is also available,[4] such that the different types of finished products offered are numbered in the hundreds.

With relation to Figure 2, the plant is traditionally organized by process and is therefore subdivided into the three following main departments, each of which has a specific organizational structure and manager.

1. *Assembly Department*, with three mixed model manual assembly lines to each of which is assigned one of the product lines.
2. *Metal Sheets Forging Department*, which provides the semifinished metal parts to be assembled into the dish-washing machine's structural frame. Inside the Metal Sheets Forging Department, the basic operations of drawing holes, shaping, and bending are performed on plain metal sheets.
3. *Coil Cutting Department*, which provides the Metal Sheets Forging Department with the required raw metal sheets. Inside the Metal Sheets Forging Department the basic operations of cutting and edging are performed, starting from standard metal coil, in order to obtain metal sheet in the required sizes.

After any of the manufacturing stages described has been carried out, the produced parts are stocked inside a dedicated warehouse, and remain there until they are picked to feed the following portion of the production process. All warehousing and material handling operations are performed by a crew of around 20 forklift carts that are organizationally part of the divisional Logistics Department. Each of the warehouses is controlled by a dedicated foreman who reports to the divisional logistic manager, as represented by the organizational structure presented in Figure 3.

All the components and subgroups that are not internally produced (such as electric motors, pumps, pipes, electric wires, electronic controllers, etc.) are externally purchased by the divisional Logistics Department; in many cases the supplier is another company of the BETA group.

1.3. Final Assembly Department

The Final Assembly Department consists of three manual assembly lines, each of which is dedicated to a specific dimensional class of dish-washing machines. Only

---

2. Data contained in this section refer to the commercial year 1994.

3. Roughly speaking they correspond with machines for bars (small size), restaurants and hotels (medium size), and hospitals, communities, and barracks (large size).

4. For instance, built-in drying systems or water purification devices.

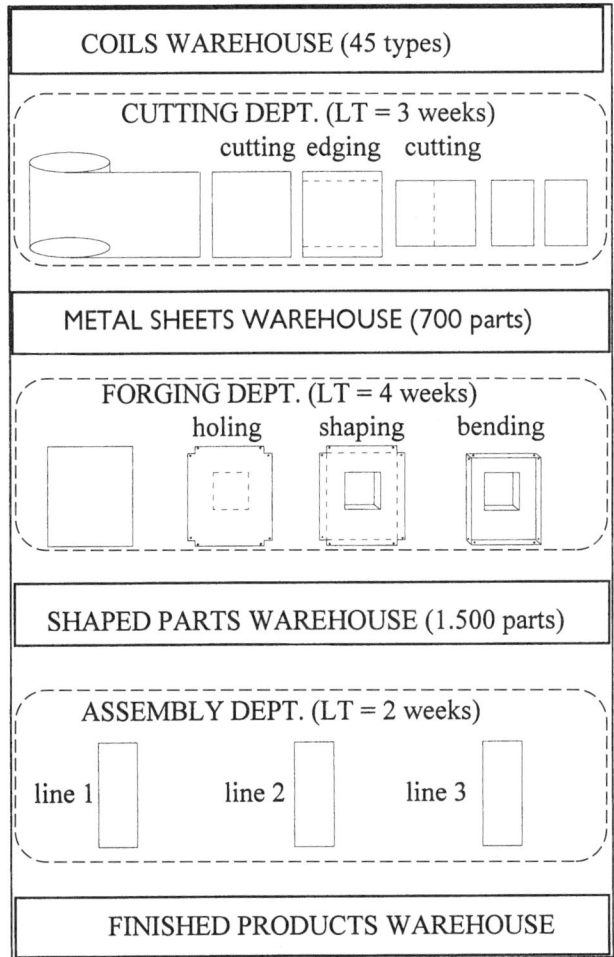

Figure 2. Layout of the production plant of the Dishwasher Machines Division.

simple manual or semiautomatic assembly devices are available at each line, so that changes in the product assembled can take place with virtually no setup time. For the same reason, it would be possible, in principle, to assemble any type of machine on any line; nevertheless, this would require the simultaneous availability at any assembly line of the components required by each finished product. This in turn would lead to an excessive space utilization: Therefore, a rigid assignment of products to lines, based on the dimensional criteria, was chosen.

Components used to assemble the washing machines are stored in buffers adjacent to the assembly station where they are used. When required by the line crew, a forklift cart performs a refill of the buffers, by picking the required materials from the corresponding warehouse. Bulk components, as an exception, are picked right before

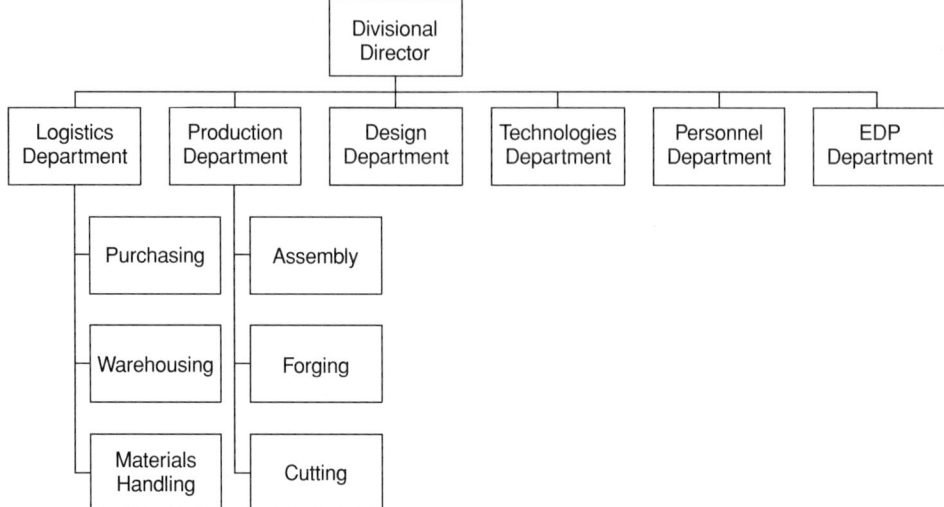

Figure 3. Organizational framework of ALFA's Washing Machines Division.

starting to assemble the lot of washing machines that requires them, in exactly the needed amount, so that the available space is not overcrowded.

Each line can be manned by a crew of 6–10 operators; each number of operators corresponds to a specific configuration of the line in terms of assembly operations assignment and workload balance, and as a consequence, to a well-determined standard production rate of the line. Each time the configuration is changed, one or two shifts are required to the line crew to adapt to the new configuration: During this transient time, the assembly efficiency is lower than standard. As a consequence, the assembly shop manager tries to modify the configuration of each line only when really required; furthermore, each configuration is maintained for at least one week.

A manager is in charge of the Assembly Department, while the work at each assembly line is coordinated by a foreman. Each line can output around one machine every 40 minutes, even if large variations can be expected depending on the type of products that are assembled, so that the overall assembly shop production rate is around 30–40 machines per shift. The normal working calendar of the lines consists of one daily shift; nevertheless, it is quite common that at least one of the lines is scheduled to work for two or even (albeit less frequently) three shifts per day. As a consequence, the normal weekly production rate is around 220–240 machines.

1.4. Metal Sheets Forging Department

The Metal Sheets Forging Department consists of around 40 machines, organized in a job shop and dedicated to the technologies of punching holes, shaping by press forging, and bending the plain metal sheets prepared in the Coil Cutting Department (see Section 1.5). The available machines can be grouped in the following categories, as represented in Figure 4.

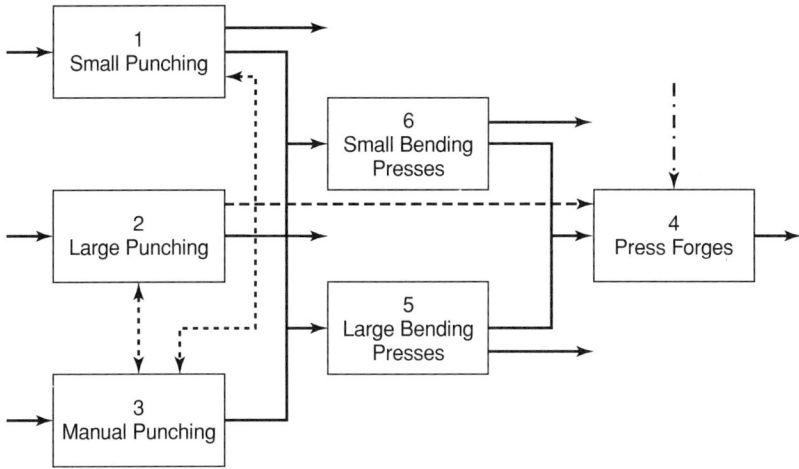

Figure 4. Scheme of the Metal Sheets Forging Department, with the main routings.

1. *Small Automatic Punching Machines.* Three identical numerically controlled machines that may draw holes of virtually any kind and shape on metal sheets with dimensions not exceeding 180 cm. Each machine has a tool magazine on which eight different tools can be fastened: Changing the tools as well as downloading the numerical control program is very fast, such that the setup operations only account for a small percentage of the utilization of this kind of machines. The tools are shared among the three machines.

2. *Large Automatic Punching Machines.* Five identical machines that differ from the previously described ones only by the dimension, since they can process metal sheets up to 360 cm long. The large punching machines can always be used as alternatives to the small ones, while the reverse is true only for small parts.

3. *Manual Punching Presses.* Two similar manual machines, which, equipped with piercing punches, can produce regularly shaped and positioned holes on metal sheets. The positioning and fixturing of the piercing punches require a large amount of time; nevertheless, once the tools have been blocked, the production rate is much higher than that of the punching machines, since all holes are simultaneously made by means of only one hit of the machine's working head. Therefore, these presses are preferred to the automated punching machines whenever the holes to be made are of regular shape and position and the batch of parts to be worked is substantially large. In any case, machines of type 1 and/or 2 may always be considered as alternatives to the manual punching presses.

4. *Press Forges.* This group consists of four different manual machines, dedicated to the shaping of a certain number of different items by means of molds. Due to the cost of molds, each press is dedicated to a certain number of item codes, since the molds are not interchangeable among presses and they are never duplicated due to their cost.

5. *Large Bending Presses*. Fourteen manually controlled machines 4 meters wide dedicated to the process of bending the large metal sheets. Each time a different type of bend is required, the operators must stop the machine and change the bending tools, which implies a time of around half an hour. Given that most lots can be processed in less than that, setup times are the biggest problem for these machines. Each machine has a vast set of tooling and fixturing, so that, with good approximation, any of the machines inside this class can perform any of the bending operations on large metal sheets. Some experiments have made clear that, if required, the bending presses can be equipped in order to substitute both the punching presses and the press forges: This result was obtained by means of special fixturing internally designed and built by the company.

6. *Small Bending Presses*. Ten manually controlled machines similar to the previously described ones, only smaller (2 meters wide). They can process any metal sheet not exceeding these dimensions, so that they cannot be an alternative to the large bending presses, while the reverse is almost always possible.

As much as 1500–2000 different part types are produced in the Metal Sheets Forging Department, with a number of manufacturing operations that may range from 1 to 9 (on average 4). The department has a normal working calendar of three shifts per day and 6 days per week; during each shift, a foreman is always around to ensure that the operators perform the scheduled jobs.

On a daily basis, the department manager decides at which machines to position the available operators[5] and what each machine should do the next day. Then, the foreman prepares the production orders relative to each operation to be performed at each machine and distributes these orders, together with the corresponding technical drawings to the appropriate operator. More or less 400–600 operations are completed in one shift inside the department.

Each time an operation has been completed by a machine, the operator calls a forklift cart to pick the processed batch of parts and store it to a departmental buffer, from where it will be retrieved for the next operation. At the same time, the operator asks the forklift cart to bring back the batch he wants to process next and gives to the driver the corresponding technical drawing in order to locate it in the departmental buffer. Since more or less 100–150 batches per shift are produced by the shop and the average lead time is around 4 weeks, an approximate evaluation of the Work-In-Process (WIP) level of the Metal Sheet Forging Department is around 7000 batches.[6] Since there are on average three batches per pallet, the average number of pallets circulating in the department is around 2500, which requires 8 forklift carts to be fully utilized.

---

5. Normally, the workforce of the department is smaller than that required to run all machines at the same time. Moreover, some machines (mainly the large ones) can be run either by one person or by a two-people crew, albeit with a different production rate.

6. 100 batches/shift $\times$ 3 shifts/day $\times$ 6 days/week $\times$ 4 weeks = 7200 batches.

1.5. Coil Cutting Department

The Coil Cutting Department consists of seven machines, organized in a classical job shop layout. The operations performed consist mainly in cutting the metal coil in sheets of the appropriate length and then dividing the obtained sheets in a way to obtain new sheets of the appropriate width, to further process them in the Forging Department (see Section 1.4).

The available machines can be grouped in the following categories, as shown in Figure 5:

1. *Coil Cutting Line.* One large Direct Numerical Control (DNC) machine provides for all of the coil cutting operations performed. One coil type at a time can be placed on board the machine. Cutting parameters such as the required sheet length and width are read from a part program; the operator can manually input from a keyboard the number of sheets of each measure that have to be cut. Large setup times are required in order to change the coil on board the machine and to change the cutting tools, due to different cutting parameters (e.g., different sheet width). As a consequence, the foreman tries to group the coil cutting jobs by coil and by width.

2. *Automated Cutting Machines.* Two large DNC machines provide for the cutting of large metal sheets that are directly purchased. No cutting tool changing is required, so that setups performed at the lot change are often negligible or small. These machines can be used as an alternative to the manual cutters (see below), normally used to process the smaller metal sheets internally produced by the coil cutting line.

3. *Small Cutters.* Three manual metal cutters of small dimension (2 meters wide) are used to process the majority of the metal sheets obtained by the coil cutting line that require further resizing. Of course, the constraint is that any processed part should not exceed the machine dimension. The cutting tool is always the same, so that no setup times at all are required. As an alternative, the automated metal sheet cutting machines can be always used, as well as the big manual metal sheet cutter (see below).

4. *Big Manual Cutter.* One big manual metal sheet cutter, almost identical to the small ones discussed above, apart from dimensions, is used to process the big metal

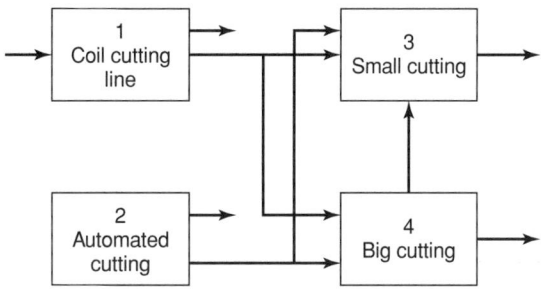

Figure 5. Scheme of the Coil Cutting Department, with the main routings.

sheets obtained by the coil cutting line that require further resizing, as well as the metal sheets purchased from the market. As a consequence, this machine can be used as an alternative for both the automated metal sheet cutting machines and the small manual metal sheet cutters.

Starting from around 20 different types of metal coil, around 500–700 differentiated metal sheets are produced using 1–4 manufacturing operations. The department has a normal working calendar of 2 shifts per day and 5 days per week; during each shift, a foreman is always around to ensure that the operators perform the scheduled jobs.

1.6. Production Planning and Control

After the weekly transmission of the customer order lines has been completed by the central Logistics Department (see Section 1.1), an MRP procedure takes place that considers all the customer order lines backlog as gross requirements and computes the net requirements profile in time by subtracting the amounts already available in stock and the production orders already in progress, while considering the static lead times connected to each department. The outcome of the MRP elaboration can be summarized as follows:

1. A weekly *Final Assembly Schedule* (FAS), matching the customer demand transmitted by the central Logistic Department. The schedule is not immediately released to the Assembly Department. Quite the contrary, a weekly meeting is held, attended by the managers of the Assembly, Metal Sheets Forging, and divisional Logistics Departments to ensure its feasibility. During the meeting, the following features are discussed:

- Availability of production capacity inside the assembly shop
- Availability of internally produced components
- Availability of externally purchased components

Once the FAS has been agreed upon, it is printed out and released to the Assembly Department. After that, it is the responsibility of the Assembly Department manager to decide the daily assembly schedule of each line.

2. A *weekly production plan* for the Forging Department, figured out by the requirements planning procedure in order to match the needs of the assembly lines, the planned stocks maintained at the forged parts warehouse, and the availability of semifinished metal sheets.

3. A *weekly production plan* for the Cutting Department, figured out by the requirements planning procedure in order to match the needs of the Forging Department and the raw metal sheets availability.

4. A weekly *purchase plan* for the divisional Logistics Department. All orders included in the first week have to be considered as operative ones, while orders inside

weeks 2–4 are just indications that are delivered to suppliers, in order to let them plan with some advance their production.

2. CELLULAR MANUFACTURING PROJECT

2.1. Genesis and Goals of the Project

The startup of the project is to be traced back to a bigger program, launched in 1992 by the BETA group, and aimed at achieving a global rationalization of the internal logistic system at a corporate level. This objective was a major issue of the group's top management that after the end of the 1980s had witnessed a large increase of the group's dimension, achieved mainly by means of acquisitions of small and medium companies throughout Europe.

Each of the companies belonging to the group was asked to take part to this project with proposals matching the peculiarities of its strategic position in the market. In 1992, ALFA had a clear leadership over competitors in the product range it could offer to customers; nevertheless, it felt aligned to competitors relating to costs, and inferior to some of them in response time and service. The larger response times could be partially explained with the bigger variety of products offered and therefore with the larger amount of assemblies and component types that had to be managed; nevertheless, ALFA reckoned that much of its time waste was due both to the internal organization of its production facilities and to the production planning and control strategies adopted (see Section 1). As a matter of fact, a considerable amount of customer orders ended up as late deliveries, and this fact was starting to undermine the company's image.

As a consequence, the management of ALFA decided to examine the possibilities to internally restructure its production means organization in a way to achieve shorter cicle times and, mainly, to establish more up-to-date and reliable production planning and control techniques. The Washing Machine Division of ALFA was first chosen as a test bed with the basic objectives to reduce the internal lead times and consequently to improve timeliness performances. The choice of the Washing Machine Division could be explained by the fact that it was at the same time the historical core business of the company and possibly the most tardy of all the divisions.

ALFA's top management submitted the whole division production system to a complete checkup, from the technical, managerial, and organizational point of view, starting from the standpoint that any type of modification of the system could take place, provided it could achieve the desired change while not harming the flexibility and agility of the production system. At the same time, given the company's struggle to slash costs, the direction set a rigid constraint on preserving (if not improving) the efficiency of the production system in both labor (direct and indirect) and fixed and running capital.

It was the intention of ALFA's management to apply any outcome of the project to other divisions of the company as soon as it could be possible.

2.2. General Approach

Given the above set of objectives and constraints, Group Technology (GT) and CM techniques were among the approaches chosen to achieve the desired improvements. It was very soon realized that the Metal Coil Cutting Department, as well as the Coil Forging Department redesign could be supported by means of this approach. Some classical techniques presented in the specialized literature were chosen as starting points of this analysis.

The theoretical operations research models commonly found in literature were not found satisfactory by the management of ALFA. Many of them, in fact, tend to misconsider the problem of production capacity. Often, the main objectives pursued in grouping the production means are the minimization of intercell movements or other such indicators of good grouping, computed only upon qualitative data. These approaches, in other words, tend to disregard the actual availability of machinery and tooling, as well as the real workload of each workcenter, as a relevant constraint. As a result, the manufacturing management of ALFA feared that by applying one of the above methods, the solution provided could prove to be inefficient due to the required duplication of production means and the consequent workload imbalance and scarce capacity utilization, ending up in increased costs.

Moreover, many of these methods fail to control the final number of obtained cells, in an attempt to maximize the qualitative goodness of the solution: Quite the contrary, it was a clear practical objective of this project to define as much as three cells, each of which should correspond to one of the already existing assembly lines.

Additionally, the operative management of the company disliked the way the classical GT models behave, that is, accepting a large amount of data in input, letting a complex algorithm process it, and finally producing an output with little intervention or decision to be made during the process. In many cases, it was reckoned, all the shop personnel, starting from the shop managers down to direct employees, can offer valuable knowledge and contribute to the desired final result.

As a consequence of these considerations, the problem was addressed by a mixed approach, both taking into consideration the research findings and complying to the specified requirements of the industrial context. Given the need for intensive human–computer interaction, it was chosen to develop an interactive software support in which the computer could provide its computing power and the humans their comprehensive judgmental ability.

The configuration of item codes, machines, and tooling described in Section 1 was the starting point to develop a solution method. The availability of production capacity and technological capability was initially taken as a constraint, as well as the production mix requirements in terms of both capacity and capability. This approach was based on the need to develop a quick, simple and cheap decision support system fitting the specific industrial environment described in Section 1.

ALFA required a software support that could run on a standard hardware and software platform, so that once developed and implemented it could be easily ported to other divisions. It was therefore decided to use Microsoft Excel, given its ease of use, large user base, and versatile functionalities.

2.3. Constraints and Objectives

The reasons discussed pushed strongly toward a pragmatic and simple approach, implementing a few simple decision support functionalities and working with well-defined and fixed constraints and limitations. Nevertheless, the problem addressed could be looked at in at least three perspectives, each of which connected with specific objectives and constraints.

    1.   In the *long-term* perspective,[7] it is possible to assume that almost any change of machinery is possible, provided it is justified by the capacity and capability requirements of the market that the company plans to serve. In this case, the objective pursued is the definition of the correct set of machines, tooling, and fixturing that should be available in order to cope with the planned market demand at the planned costs and quality levels. The only reasonable constraint to be taken into consideration at this level is the financial availability of the company.

    2.   In the *mid-term* perspective,[8] it is often impossible to plan and execute machinery acquisitions, unless they were already planned long before; quite the contrary, changes in machinery layout (redefinition of the set of machines that form any cell) and the corresponding manufacturing cycles routing can still take place at this level. As a consequence, at this decisional step, such parameters as the number of available machines of each type can be considered as a constraint, while the question to address is which is the best allocation of machines inside the manufacturing cells. The possibility to redefine the machine allocation to cells in time was considered a very important issue by the management of ALFA, since products and parts were fast changing both in technical features and demanded volumes.

    3.   In the *short term*[9] we can consider as fixed the decisions taken at level 2 about the configuration of the manufacturing cells: in fact, any layout change within too short a horizon could result in extra inefficiency, if not carefully planned in advance. Thus, the only decision variable remaining at this stage is the physical allocation of items to cells. In a context, like that of the Dish-Washing Machines Division of ALFA, where the production mix is in continuous evolution, the midterm solution is likely not to be optimal for each of the short periods inside the planning horizon: as a consequence, the possibility to change products allocation in the short term was considered by the management as a basic functionality of the system.

All of these perspectives are taken into consideration by the decision support system that was developed and described in the following sections.

2.4. Project Organization

The project was carried out jointly by ALFA and the Politecnico di Milano. Since the beginning, a steering committee was set to direct the project and to ensure that it would

---

7. We can assume in this scenario a planning horizon of at least one year.
8. Planning horizon of one month to one year.
9. One week to one month of planning horizon.

achieve the expected goals. The steering committee consisted of the manufacturing manager of ALFA, the director of the Washing Machines Division, and one professor from the Politecnico.

An operative team of roughly 20 people was set up by the steering committe in order to carry out the operative work connected to the project. The team mainly consisted of personnel from the Dish-Washing Machines Division, which was co-ordinated and supported by two reserachers from the Politecnico. People from the Dish-Washing Machines Division involved in the project pertained mainly to the divisional Logistics Department (the manager and those responsible for warehousing and materials handling where among them) and to Production (the managers of the three departments where among them, together with some foremen). People from the Design, Technologies, Personnel, and EDP Departments took part in the project, even if with a less active role.

The development of the project was divided into three main phases. The *design phase* encompassed such activities as the analysis of requirements and the specifica-tion of the possible solutions, which were examined from any possible point of view (technical, managerial, organizational, informative, etc.). This phase required more or less 6 months and was carried out through weekly meetings of the operative team, during which the team was from time to time split into task forces entitled to analyze and solve the specific problems that had arisen by the work previously done. At the end of the development phase, the work of the operative team was deeply examined and discussed by the steering commettee, which asked for some modifications and extentions to be made. The whole process required more or less 3 additional months.

The following *realization phase* was the longest and toughest one; it required the preparation and test of the software support procedures described in Sections 3 and 4, the revision and updating of the manufacturing data as discussed in Section 5.3, and the preparation of the shop floor personnel with training classes. It also required a great deal of dicussions inside the operative team and with the steering committee. Many of the decisions taken in the design phase were rediscussed and partially changed by the shop floor personnel and management when it became clear that the time was approaching to implement the plan. As a consequence, the realization phase required at least 1.5 years of work before it could be considered done.

The final *implementation phase* encompassed such activities as the startup of the new organization of the production means, as well as the connected renovation of the human resources organization, as decribed in Section 5. Moreover, the discovery of a wide set of problems (see Section 5.3) required the constant presence of one task force of the operative group at the shop floor for the first 2 months. At least 4 months were necessary to return the system to a steady state.

3. ANNUAL CELLS DEFINITION MODULE

The issues described in Section 2 and relating to the long and midterm decisional levels are addressed by one module called the Annual Cell Definition (ACD). The following sections are dedicated to the description of this module, focusing mainly on its objectives and costraints, input and output data, and logic.

3.1. Objectives and Constraints

The main objective of the ACD process can be stated as the definition of a set of manufacturing cells and the assignment of each manufactured part to one cell, in a way that the final solution minimizes both the intercell movements and the setup operations that must take place inside each cell. This objective has to be reached under certain constraints, summarized as follows:

1. The number of cells should equal the number of assembly lines, that is, three, in a way that each cell can serve one assembly line.
2. Each cell should have the availability of all the technological capability required by the parts processed and relating to the operations previously done in the Cutting and Forging Departments, in a way that little or no intercell movements are required.
3. The volumes assigned to each cell should not exceed the cell's capacity, in a way to distribute the workload evenly among different cells.

Nevertheless, the most important constraint is set on the available machinery and fixturing. Thus, the software module that was designed and used to support the process allows two well-defined funtioning modes (see Section 2.3):

- A *long-term mode* in which the available machines and fixturing can be modified by the user in order to match as well as possible the technological capabilities and the production capacity required by the processed parts mix.
- A *midterm mode* in which the technical context of the system is fixed and cells have to be worked out starting from the available machinery and fixturing appliances. This scenario implies, for instance, that any part requiring tooling and fixturing available in an unique set cannot be allocated to more than one cell at the same time.

3.2. Input Data

The input data required by the ACD procedure encompasses technical as well as managerial informtions. The following data tables are downloaded from the central Information System:

1. *Workcenters* table, containing a list of the available workcenters (see Sections 1.4 and 1.5). For each workcenter the main data gathered by the table are an identifier, a description, the number of machines available, the average availability and efficiency, and the technological type. A standard working calendar is also gathered for each workcenter, consisting in the specification of the standard working

hours per shift, shifts per day, and days per year. Machines belonging to the same workcenter are considered identical. The technological type that counterdistinguishes each workcenter defines the type of contained machines in a way that two workcenters sharing the same technological type may be interchangeable in the manufacture of some parts, even if with different production rates, setup times, or scrap percentages.

2.   *Items* table, containing the list of all components, subassemblies and finished products to be managed in the system. Each item is characterised by a code, a description, and a lot size.

3.   *Cycles routing headers* table, containing the list of all the cycles routing currently active for each of the referred items. Each header refers to one specific item, while more than one header may correspond to the same item, in the case of alternative cycles routing. Therefore each header is identified by one item code, a progressive version number and a flag, stating if the cycle routning is the main one for the considered item or a secondary one.

4.   *Cycles routing operations* table, containing the list of all the manufacturing operations required by each of the cycle routings contained in the cycles routing header table (see point 3). Each operation refers to one header and is further identified by a progressive number stating its relative position in the routing of the related item. Each operation is performed by one specific workcenter equipped with a specific tooling. Attributes encompass the preparation time (required to perfom the machine setup), the technological process time related to one unit, the average process scrap rate, and a flag specifiying if it is a main or alternative operation. Alternative operations are specified when it is technically possible to execute one specific operation at two or more altenative workcenters sharing the same type (see point 1).

5.   *Tools and fixtures* table, containing a list of all the tooling and fixturing devices that may be used inside the cutting and forging shops. Each tool is identified by a code, a description, and the number of available units.

6.   *Bills of materials* table, containing a simplified version of the bills of material, which puts in relation each item considered inside the item table to the list of its possible parent finished products and therefore with the corresponding assembly line.

7.   *Annual sales budget* table, expressed in terms of number of units per year for each considered finished product.

This data are organized in a relational database that is represented in the scheme presented in Figure 6.

3.3. Solution Building Procedure

Once input data has been downloaded, the user can start to work using any pre-existing solution, or interactively generating one if no solution is currently available. A spreadsheet interface supports this activity.

A solution consists of three cells; each workcenter has to be assigned to one or more of the three manufacturing cells in order to build a solution: if a multiple assignment is required, the rows of the workcenters table are split and two or three workcenters are

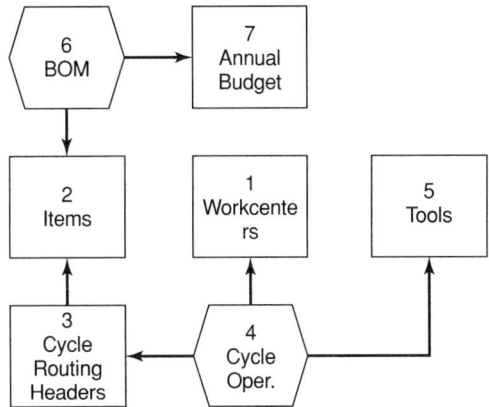

Figure 6. Scheme of the input data of the ACD procedure.

worked out from one, each of which is assigned to the appropriate cell. Accordingly, at the end of the process, each workcenter may be represented by one, two, or three rows inside the table, depending on how many cells it has been assigned to. Tools and fixtures are also assigned to cells in the same way. Finally, the assignment process regards manufactured parts as well. The procedure ends when all workcenters, tools, and parts have been assigned to one particular cell.

The user can create and modify the solution through one of the following actions.

1. *Part Assignment by Finished Product.* The software automatically assigns each part to the manufacturing cell associated to the assembly line that requires it. In cases in which a single part is required by two or more finished products assembled by different lines (not very frequent though), the part is assigned to all of the cells connected to the assembly lines resposible for the part's consumption, in the appropriate volumes. The assignment is achieved through a batch procedure that exploits the bills of materials to identify the finished products that may require a part and their cycle routings to determine the assembly line to which they are currently assigned. Each time an item is assigned to two or more cells, the corresponding row inside the manufacturing cycles table is doubled and a new cycle routing is created. This assignment strategy allows to assign manufactured items to each cell in an optimal way in terms of the intercell movements; nevertheless, it might prove infeasible from the point of view of technological capability or production capacity; moreover, it might prove inefficient to assign some items to more than one cell under the perspective of setup times.

2. *Part Assignment by Part Type.* This second batch assigment procedure works exactly like the previous one, but in cases of conflict assigns the whole amount of a part's requirement to the manufacturing cell that supplies the assembly line requiring the biggest percentage of the part's consumption. This second assignment strategy searches for a more efficient solution of the setup problem, but might incur

greater parts movements and still runs the risk of an uneven workload distribution among cells.

3. *Workcenters Balancing*. This batch procedure looks for an optimal workload balance by taking into consideration all the workcenters that are currently overloaded at each manufacturing cell. For any item currently assigned to be processed by any overloaded workcenter, any eventual cycle routing alternative operation requiring another workcenter currently inserted inside the same cell is identified. All operations allowing it are switched to an alternative workcenter inside the same cell. This action is automatically repeated until the currently processed workcenter overload has been solved or the addressed alternative workcenters have become overloaded on their turn.

4. *Interactive Parts Assignment*. The user can directly decide the parts assignment to cells by manually updating the annual budget table. If it is required to assign the same item to two differrent cells, the corresponding row is doubled.

5. *Interactive Workcenters Assignment*. The user can also decide to change the assignment of one workcenter to a manufacturing cell. Each workcenter can be as well doubled in order to assign it to more than one cell, provided it consists of more than one machine. This is achieved by doubling the corresponding row of the workcenters table.

6. *Interactive Tools Assignment*. By treating tools exactly as workcenters (see action 5), tools can be reassigned to cells. If one tool is available in more than one unit, it can be assigned to more than one cell by simply doubling the corresponding row of the table.

7. *Lot Size Modification*. In order to modify the workload profile of one or more workcenters inside one or more manufacturing cells, lot sizes can be modified by the user. This has an influence on the overall amount of setup operations that will be performed during production and on the amount of stock. Any reduction of the lot size will achieve a proportional reduction in stock, but an increase of workloads due to the larger number of setup operations required. The opposite effect is obtained through an increase in the lot size.

8. *Working Calendar Modifications*. In order to modify the capacity profile of one workcenter, the user can manually modify its working calendar, by updating the parameters: hours per shift, shifts per day, and working days per year.

9. *Workcenters Objectives Setting*. By modifying certain technical or managerial parameters of the workcenters, such as the scrap and rework percentages or the efficiency rate, the user can simulate the setting of corresponding performance objectives for the workcenter and the related cell. It is obvious how these parameters can influence the fulfillment of the problem's constraint by a specific solution.

10. *New Investments Simulation*. In order to use the ACD module in the long-term perspective described in Section 2.3., it is possible to modify the number of resources[10] assigned to each cell, so that eventual infeasibilities can be worked out.

---

10. Encompassing both machines and tools.

Each time a transaction is made with the above-described batch or interactive procedures, a set of performance indices that allow to roughly evaluate the current configuration is computed (see Section 3.4). Moreover, if any of the system constraints is violated (see Section 3.1), an appropriate warning message is displayed and the corresponding operation is canceled.

The solution is stored inside the workcenters table (machine–cells assignment) and the cycle routing headers and operation tables (parts–machines assignment). In order to confirm it, these data are transmitted to the central technical database.

3.4. Output Data

Data output by the ACD module provides an evaluation of the current solution goodness. A set of indices can be computed by the module after each change has taken place, so that the user can easily use ACD as a rough-cut what-if decision support system. The computed indices are of physical and monetary nature and are always computed by cell. The temporal base of these indices is always the year, while the monetary unit is the Italian lire (Lit). The computed indices are listed as follows:

- The *total setup time* per cell (TST_c), expressed in hours per year, accounts for the total estimated setup time of the current solution. The index is computed as follows:

$$\text{TST}_c = \sum_{i \in I_c} \sum_{o \in \text{CR}_{ic}} \text{ST}_o \, \text{IntSup} \left\{ \frac{(1 + \text{DR}_o) Q_{ic}}{\text{LS}_{ic}} \right\} \tag{1}$$

  where c = cell identifier
  i = item identifier
  o = cycle routing operation identifier
  I_c = set of items currently assigned to the manufacturing cell c
  CR_{ic} = cycle routing of item i at cell c
  ST_o = setup time of cycle routing operation o (hours)
  Q_{ic} = quantity of item i currently assigned to cell c (units/year)
  DR_o = defects rate of cycle routing operation o (%)
  LS_{ic} = lot size of item i at cell c (units)

- The *total man-hours* required at cell c (TMH_c), expressed in hours per year, that accounts for the total expected labor requirements at the considered cell in the current configuration. The index is computed as follows:

$$\text{TMH}_c = \text{TST}_c + \sum_{w \in W_c} \text{OM}_w \sum_{i \in I_c} \sum_{o \in \text{CR}_{ic}} (1 + \text{DR}_o) \frac{Q_{ic}}{E_w (\text{PR}_{iw})} \tag{2}$$

  where w = workcenter identifier
  W_c = set of workcentres assigned to cell c
  E_w = efficiency rate of workcenter w (%)
  PR_{iw} = standard production rate of item i at workcenter w (units/hour)
  OM_w = standard number of operators required per machine at w (op./machine)

- The *total utilization rate* of cell c (TUR$_c$), expressed as a percentage, that accounts for the expected ratio of the currently assigned workload against the workcenter's theoretical capacity. The index is computed as follows:

$$\text{TUR}_c = \frac{\text{TST}_c + \sum_{w \in W_c} \sum_{i \in I_c} \sum_{o \in \text{CR}_{ic}} (1 + \text{DR}_o) \left[\dfrac{Q_{ic}}{E_w(\text{PR}_{iw})} \right]}{\sum_{w \in W_c} \text{NM}_{wc} \text{HS}_{wc} \text{SD}_{wc} \text{DY}_{wc}} \tag{3}$$

where NM_{wc} = number of available machines at workcenter w inside cell c
HS_{wc} = standard number of hours per shift at workcenter w inside cell c (hours/shift)
SD_{wc} = standard number of shifts per day at workcenter w inside cell c (shifts/day)
DY_{wc} = standard number of working days per year at workcenter w inside cell c (days/year)

- The *total number of intercell moves per cell* (TIM$_c$), expressed in lots moved per year, that accounts for the expected runs of the forklift cart required to move from one cell to another all production lots that require so. The index is computed as follows:

$$\text{TIM}_c = \sum_{i \in I_c} \sum_{o \in \text{CR}_{ic}} M_{io} \times \text{IntSup} \left\{ \frac{(1 + \text{DR}_o) Q_{ic}}{\text{LS}_i} \right\} \tag{4}$$

where $M_{io} = 1$ if operation $o + 1$ is performed in a cell other than c or if the semifinished part is supplied to an assembly line different from that associated to c; 0 otherwise.

- The *average stock cost* of manufactured parts at cell c (ASC$_c$), expressed in monetary units per year, that accounts for the expected average level of stocks for the manufactured parts, due to the lot sizes. The index is computed as follows:

$$\text{ASC}_c = \sum_{i \in I_c} \frac{\text{VC}_i C_m \text{LS}_{ic}}{2} \tag{5}$$

where VC_i = variable cost of item i ($/unit)
C_m = holding cost discount rate (%/year)

- The *total production variable cost* at cell c (TPC$_c$), expressed in monetary units per year, that accounts for the expected total amount of production variable costs determined by the number of active machines at each workcenter of the cell and by the working calendar established for each of them. The index is computed as follows:

$$\text{TPC}_c = \sum_{w \in W_c} \text{HC}_{wc} \text{NM}_{wc} \text{HS}_{wc} \text{SD}_{wc} \text{DY}_{wc} \tag{6}$$

where HC_{wc} = hourly variable cost per resource of workcenter w in cell c (monetary units/machine $\times$ hour), basically accounting for energy and direct labor costs

- The *total fixed costs* at cell c (TFC$_c$), expressed in monetary units per year, that approximates the expected yearly portion of the discounted total cost of machinery and tooling. The index is computed as follows:

$$\text{TFC}_c = \frac{\sum_{w \in W_c} \text{NM}_{wc} \text{VM}_w}{\sum_{k=1,N_w} \frac{1}{(1+i)^k}} + \frac{\sum_{t \in T_c} \text{NT}_{tc} \text{VT}_t}{\sum_{k=1,N_t} \frac{1}{(1+i)^k}} \tag{7}$$

where k = identifier of the generic year

$\quad$ VM$_w$ = value of one machine of workcenter w, in monteray units

$\quad$ N$_w$ = number of expected years of utilization of machines inside work-center w

$\quad$ i = discount rate

$\quad$ T_c = set of tools assigned to cell c

$\quad$ NT$_{tc}$ = number of tools of type t currently assigned to cell c

$\quad$ VT$_t$ = value of one tool of type t, in monteray units

$\quad$ N_t = number of expected years of utilization of tools of type t

$\quad$ t = identifier of the generic tool

4. WEEKLY ITEMS ASSIGNMENT

The solution computed by the ACD module could be considered as satisfactory even in the short run, if only the production orders released to the shop floor each week by the MRP were roughly proportional to the average mix taken into consideration by the annual sales budget. Unfortunately, the production mix at the Washing Machines Department was particularly unstable. As a consequence, a solution that might be satisfactory on average, could often prove to be bad, if not infeasible, during a specific week.

The solution to this problem was achieved with the second module, to be executed weekly, immediately after the release of the new requirements plan, to define a new solution, starting from that one computed by ACD. This second solution is achieved by modifying the annual solution in a way to match the specific workload profile of the current period, while following partially different objectives and constraints. This procedure was called the Weekly Items Assignment (WIA).

4.1. Objectives and Constraints

The basic objective of the WIA procedure is to redefine the assignment of manu-factured parts to cells in order to guarantee the feasibility of the weekly plan, while minimizing the number of changes from the solution computed by means of the ACD module.

The main constraint that has to be respected in this process is that the physical structure of the cells should be preserved equal to the one decided through the ACD procedure, to prevent the shop from repeated layout changes.

As a consequence, the main way through which to change the solution is to redefine the assignment of manufactured parts to cells. The possibility to switch the cell assignment of one item is guaranteed provided the machine types and tooling required by the considered item is available in the new cell too. Yet, at this stage it is no more possible to switch workcenters from one cell to another one, since this would correspond to machine movements, which can hardly be planned and executed each week. Conversely, it is still possible, even if not recommended, to switch the assignment of tooling and fixturing to cells.

A second possibility still available is to change the allocation of man-hours to cells, provided that the overall amount of direct man-hours available is not increased. Finally, a lot size modification is also possible, provided that quantities required to fulfill customer orders are actually produced.

In any case, a good solution is achieved only when the number and entity of modifications is minimal. The basic reason for limiting the perturbation of the annual solution is that each change may have a considerable cost, mainly due to

- The lack of familiarity of direct employees in rotating jobs, which leads to a loss of efficiency in direct labor if too many changes are done
- The necessity to move to the appropriate cell the specific tooling and fixturing
- The necessity to remind the new routings of moved parts to the Logistics Department employees in charge of moving the materials in the shop

4.2. Input Data

The input data required to work out the WIA solution is the following, further described in Figure 7:

1. The *ACD solution* in terms of machines–cells and parts–cells assignment, coded inside the *workcenters*, *tools*, and *parts* tables
2. The weekly *MRP schedule*, expressed in terms of number of units required for each part number in the considered week
3. Technical data contained inside the *cycle routing header* and *operations* tables

4.3. Interactive Procedure

The user can modify the annual solution through the following set of actions; some of them are a subset of those defined inside the ACD procedure described in Section 3.3. Others are completely different, given the different set of constraint and objectives specified in Section 4.1.

1. *Workcenters Balancing*. Given the different quantities per part that may be scheduled by each weekly MRP run, the procedure can be reiterated in exactly the same way described in Section 3.3.

2. *Interactive Parts Assignment*. Given the different quantities per part that may be scheduled by each weekly MRP run, the procedure can be performed by the user in exactly the same way described in Section 3.3.

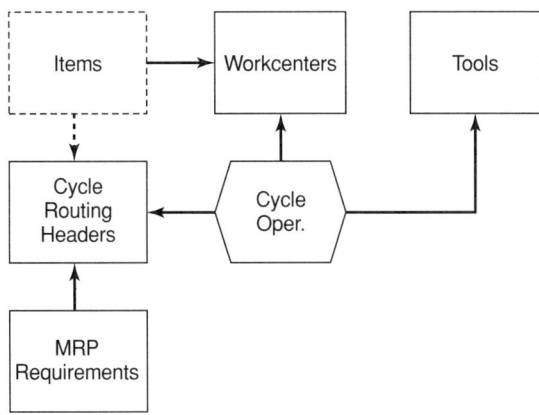

Figure 7. Input data of the WIA procedure.

3.   *Working Calendar Modifications.* To modify the capacity profile of one work-center inside one manufacturing cell, the user can manually update its working calendar, by modifying the parameters: hours per shift, shifts per day, and working days per week. If the working calendar is increased (e.g., a third shift is added) at a workcenter, this change must be counterbalanced by an equal decrease of the working calendar of another workcenter, in a way that the total amount of direct labor scheduled for that week is respected.

4.   *Lot Size Reductions.* The total lot size can be split into three well-defined components:

- *Needed quantity*, which corresponds to the actual quantity required by customer orders: This (and only this one) is the quantity that is really necessary to produce within a specified due date, because a late delivery of this quantity would surely lead to customer dissatisfaction.
- *Safety stock quantity*, which may be required only to reintegrate the item's safety stock in case it has been previously utilized. Of course, it is necessary to reintegrate as soon as possible the safety stock; nevertheless, a late delivery of this quantity would harm the customer service only in case a further unforeseen order should arrive: On the contrary, if demand is equal to or less than the forcast, no problem occurs.
- *Lot quantity*, which is the amount ordered only to round up the ordered quantity to the batch size (e.g., an economic order quantity). In other words, this quantity is ordered only for internal reasons, such as setup time and cost. As a consequence, a late delivery of this portion of the lot would cause no service problem at all.

With relation to the components discussed of the lots orderd by the MRP, a batch procedure can be activated in correspondence with any overloaded workcenter that scans all ordered lots that require that workcenter, sorted by decreasing processing time, and progressively cuts the lots quantity; the process ends when all lots have been examined or when the overload has been resolved. If at the end of the procedure, the overload has not been resolved yet, safety stocks are cut in a similar way.

Of course, it is necessary to resort to this action as little as possible because any reduction of the lot size would invariably cause a decrease of the workload in the current week, but a corresponding increase in the next ones, due to the increased number of setups.

Each time a transaction is made with the described procedures, the set of performance indices described in Section 3.4 is computed, with the only difference that the time basis here is the week. Moreover, if any of the system constraints is violated (see Section 3.1), an appropriate warning message is displayed and the corresponding operation is canceled.

5. IMPLEMENTATION ISSUES AND RESULTS

The implementation of the project was definitely not easy, since it encompassed technological, informational, and organizational issues, and it demanded the people involved to radically change their way of thinking and working. Due to these points, in many occasions the team that analyzed, designed, and implemented the project was forced to change its ideas, to rework parts of the general design of the project, or even to completely change them. Several people in the Washing Machines Division, mainly among those that mostly supported the project, probably thought, at least once, to resign. Nevertheless, mainly thanks to the strength and belief of these people, results were obtained that more than justified the phsycological and financial effort involved in this project.

The following sections of this report present an overview of the main difficulties connected with the implementation of the project, as compared with the most interesting results achieved.

5.1. Performance Results

The introduction of the ACD and WIA procedures did support the introduction of a new product-oriented organization of the shop floor, completely different from the previous, process-oriented one. An example of the new factory configuration gathered by the system implemented is presented in Figure 8.

The achievement of a cellular flow-based layout helped achieve in time considerable improvements in the internal and external performances of the shop floor. Achievements were not immediate, owing to the progressive steps with which the new organizational asset was actually put in practice (see Sections 5.2 and 5.3). One of the first performances to obtain large benefits thanks to the cellular manufacturing project was the flow time of the manufacturing process, which decreased from around 35 days (see Section 1.2) to less than 20, as shown in Figure 9.

Consistently, WIP decreased sharply, as shown in Figure 10, which refers to the first 8 weeks of the projects implementation. Figure 10 also shows that the improvements

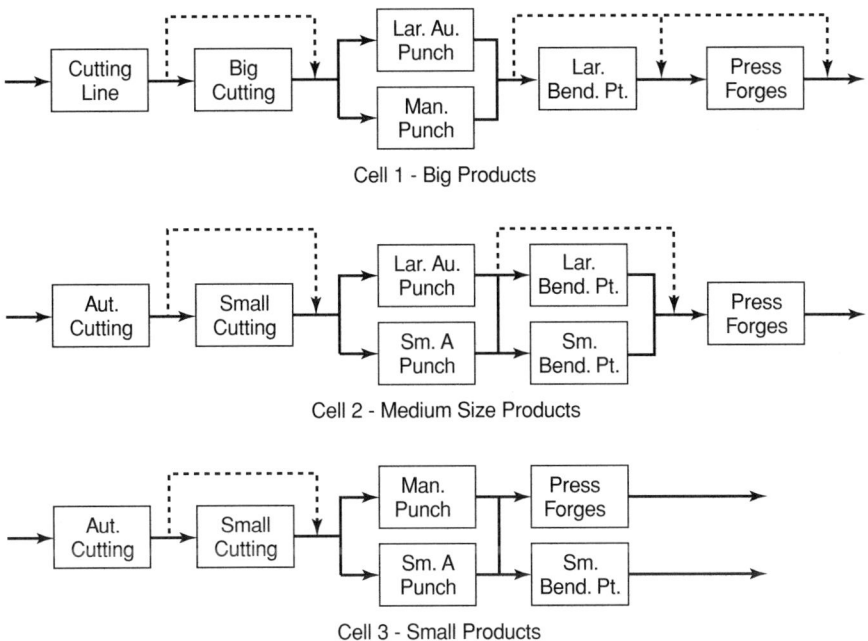

Cell 1 - Big Products

Cell 2 - Medium Size Products

Cell 3 - Small Products

Figure 8. New organization of production means.

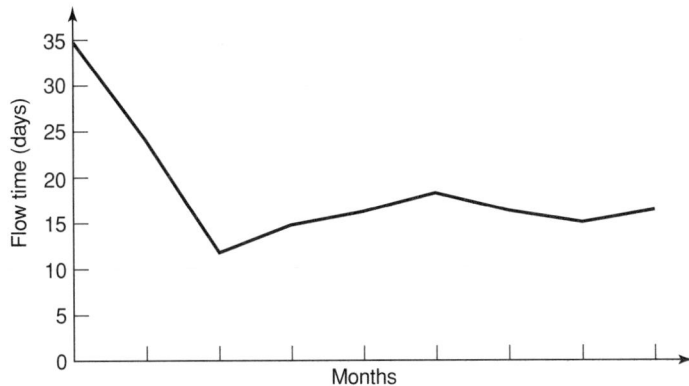

Figure 9. Manufacturing flow time evolution during months following the implementation of the project.

only started to show up after the third week, while in the first 2 or 3 weeks the WIP value was more or less constant.

The reduction of both WIP and flow time was the base for the gradual achievement of an improved effectiveness in serving the market. This result was particularly important since it matched one of the basic goals of the project. Nevertheless, despite the fact that both WIP and flow times had begun to be reduced almost immediately,

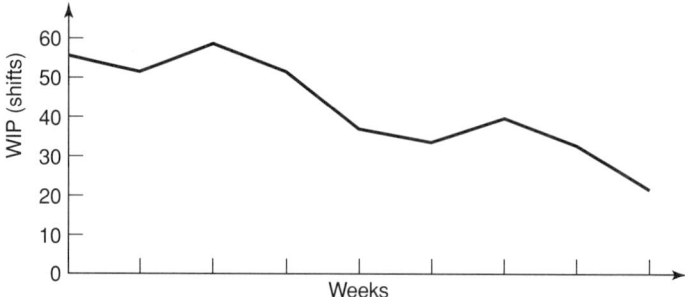

Figure 10. WIP evolution during weeks following the implementation of the project.

these achievements were capitalized in an improved response to the market only after 2 full months, as clearly shown by Figure 11. This late effect can perhaps be traced back to the organizational turmoil that characterized the shop floor when the organization by process started to be changed into one by product (see Section 5.2).

At the same time, the average lateness of late supplies started to decrease as well, as represented in Figure 12, which relates to the first 3 months after the full implemenntation of the project.

Despite the decrease of WIP, the new configuration of the system produced benefical effects also on the global output produced, even if not immediately. Figure 13 shows the values of average weekly output during the few months following the implementation of the project. The outup is here measured in terms of machines per average week, adjusted by considering the total amount of worked shifts, depending on the working calendars adopted and on the number of working days of the month. Thus, in the 2 months following the implementation, the output per shift decreased, mainly owing to the variuos problems encountered (see Section 5.3). In the following 2 months, it stayed constant, while only starting from month 5 did the system start to improve.

5.2. Organizational Results

Despite the interesting results reached in terms of exogenous and endogenous performances of the Dish-Washing Machines Division, perhaps the most brilliant achievements were those pertaining to the organization of the system. The introduction of a flow-dominant layout even in the manufacturing part of the production facility was accompanied by the setting up of a new product-oriented organization of the production means, completely different from the previous, process-oriented one.

The fact that the number of cells matched exactly the number of assembly lines, joined with the obvious consequence that parts assigned to each manufacturing cell were mainly those consumed by the corresponding assembly line, helped to build up a new concept of the shop floor organization, which matched the new configuration of the production means. Three production teams were set up, each of which was empowered to manage the complete production system consisting of

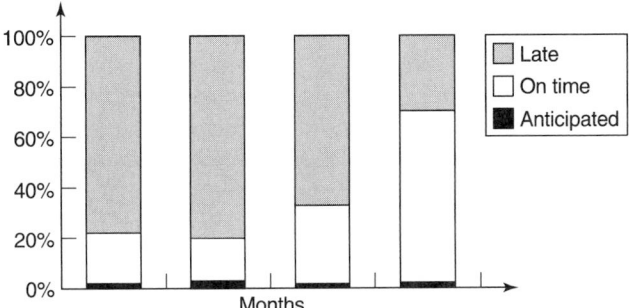

Figure 11. Evolution of order timeliness, during months following the implementation of the project.

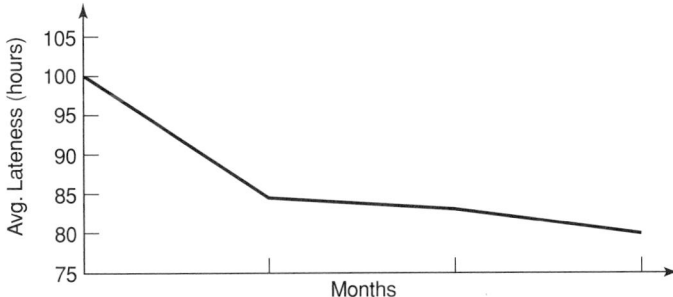

Figure 12. Evolution of order lateness during months following the implementation of the project.

one manufacturing cell and the corresponding assembly line. Given the different dimensional sizes of the products assigned to each assembly line, the teams were soon named the *big*, *medum*, and *small* products team.

The organizational change was extended not only to the Production Department, but also to the Logistics Department, which was involved in the transformation of the organizational configuration of the division. Activities regarding stock control and material handling were decentralized to each cell, as displayed by Figure 14, while only Purchasing remained centralized, given the big amount of common materials to be procured. Thus, each cell had its own manager and foremen, its own direct and indirect worforce (maintenance also was soon decentralised), and its own forklift carts and logistic personnel.

The managers of the three cells referred directly to the divisional director and were in charge of the operative aspects connected to production, maintenance, and internal logistics. The Logistics Department maintained control of purchasing as an operative activity, while the Production Department *de facto* disappeared and was put together with the Design and Technologies Department in a new Product & Process Design Department, which was entitled to completely develop the product and process design from scratch to production.

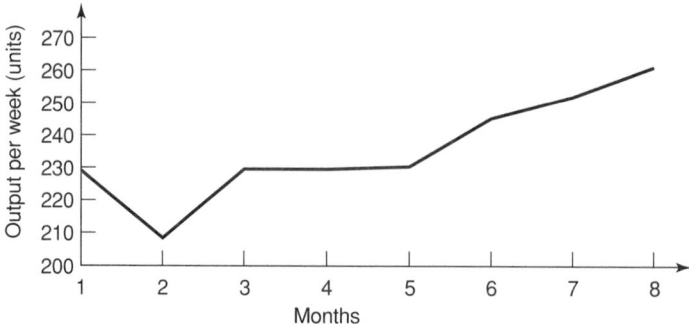

Figure 13. Evolution of average output per week during months following the implementation of the project.

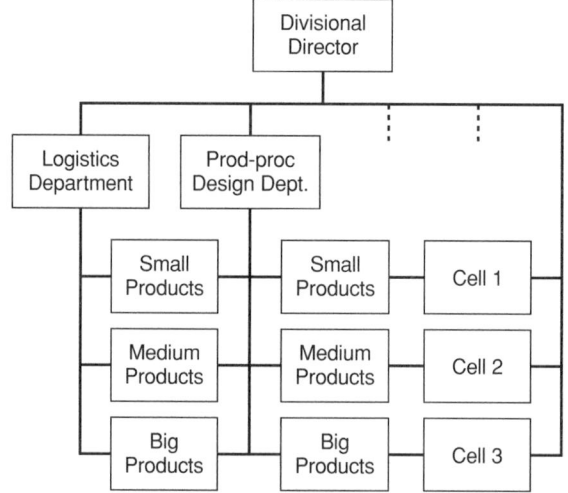

Figure 14. New organizational structure.

The new teams structure led to the possibility of assigning wider responsibilities to the shop floor. The general cells configuration was decided each year through a set of joint meetings of all the cells managers and their staffs and was personally confirmed by the director. In order to avoid any dispute, a fixed and roughly equal yearly budget was given to each cell manager, that he could invest how he wanted.[11]

Each week, a meeting was held among the managers of the three cells, together with the logistic manager, through which, thanks to the support of the WIA procedure, the weekly parts assignment was decided. From this point on, each cell was totally in charge of deciding its own way to meet the weekly objectives set by the central MRP system.

---

11. For instance, in hiring manpower or in buying new machine tools or again in improving the tooling and fixturing.

Consistently, the cells performance started to be measured in terms of such classical manufacturing indices as overall throughput per time unit, work efficiency, production means utilization rate, inventory, lead time, or percentage of respect of the MRP schedule, as well as scrap and rework rates. What is most interesting, these indices were not measured by single departments, as was done before. They were measured by cell, thus encompassing the complete production process. Each team was asked quarterly to establish a set of objectives in terms of the performance measures discussed and to give a weight to each of them. A reward was assigned to the team that could set the most ambitious objectives with respect to performances actually achieved in the previous quarter. In this way, each team was stimulated to avoid objectives too easy to fulfill. The EDP Department was put in charge of computing weekly the performance indices achieved by each cell and of displaying them on big charts located at focal points on the shop floor. In this process, evidence was set on any sharp success or failure in achieving the goals. At the end of each quarter, each team could sum up its performance achievements and check to which extent it had achieved its own objectives. A new reward was given to the most successful of the teams.

As soon as these new responsibilities were fully achieved, they very soon led the work teams to figure out different ways to improve their work. The small products cell, for instance, decided to avoid bringing materials back to a buffer after each manufacturing operation had taken place; this had been tried some years before, but had flooded the shop floor with a huge amount of pallets. Thanks to the reduced WIP inventory allowed by the linearized flow of parts, no particular inconvenience occurred, so that the experiment was soon extended to the other two cells, for parts of reasonably small size. This saved much of the work time of forklift carts, which in turn meant letting three to four employees free. These forklift drivers were put back to production.

Another cell found out that, by simply setting a punching machine and a bending press very close to each other, the corresponding operations could be synchronized and coupled, so that the two machines could be operated simultaneously by one operator: The set of machines was called an *island*. This sequence of operations was very common among the parts mix, so that very soon each cell had its islands. Further on, cells begun to refine their production scheduling techniques, and some of them even started to use a pull kanban flow control with the most regularly consumed parts.

Many innovations such as the reduction of forklift activity or a better scheduling were soon capitalized in increased system throughput, as well shown in Figure 13. Naturally, as made clear from the analysis of the performances discussed in Section 5.1, none of these achievements was reached within days from the initial implementation of the project. Quite the contrary, the first months were filled with problems and controversies, as discussed in the following section.

5.3. Implementation Problems

The problems encountered during and after the implementation of the project had been foreseen largely before they came to reality. They were no less urgent or easier to solve for this fact. Three main types of problems occurred: technological, informational, and organizational.

On the technological side, the main problems depended on the scarse product and process standardization that had been the norm in previous years, yielding an enormous amount of different raw materials, parts, subgroups, and final assemblies to be managed. For decades, information exchanges among the Design and the Production Departments had been delegated to oral communications, while the use of multifunctional design teams had been almost unknown. As an excuse, managers maintained that they had desired and planned to increase variety, in order to preserve the company's prominence over competitors in terms of finished products range and diversification.

Indeed, the result of the limited attention payed to standardization was that the large variety of finished products was easily translated into an even larger variety of groups and parts at almost any level of the bills of materials. For example, around 45 different types of metal coil were used, differing from one each other in terms of type of steel, thickness of the metal sheet, width of the metal sheet, and superficial treatment of the metal sheet.

In many cases, as an analysis made clear, a new type of metal coil was adopted by the Design Department just because it was much easier to design the technological process necessary to obtain the required part with the new coil than with an already used one. A second reason for choosing a new coil was the desire of designers to test its behavior; almost without exception, the tested coil was adopted for permanent usage, but no other coil was suppressed instead. Another analysis showed that there were more than 100 different types of stands for the bases of the dishwashers, with differences in strength, dimensions, and regulation mode.

It is clear that such a situation can make it very difficult to implement a process standardization project like the one described by this report. It was therefore necessary to undergo a long and difficult products standardization process, before, during, and after the implementation of the CM project. This took a long time, since the Design Department was busy (as it always happens) with the design of a new product line.

A second major problem was the scarse usage that had been made in previous times of the company's technical data (mainly the workcenters, fixtures, and tools tables and the cycle routings header and operations tables). Regardless how peculiar it might seem, technical data were used by ALFA mainly for accounting and costs control purposes. As a direct consequence of this, scarse attention was payed to the correctness of the technical aspects of the data contained in the central database.

Many times, machines that were not exactly the same from a technical point of view[12] were set in the same workcenter just because this was a possible and desirable approximation from a cost accounting point of view. More frequently, alternative cycle routings were simply not coded inside the technical database, since it was considered nonsense to process one part with machines other than the optimal one. This is surely

---

12. For example, they could not process exactly the same parts or could not process them with the same process rates.

correct when only production direct costs are taken into consideration; quite the contrary, it is well-known among manufacturing experts how much the flexibility of a production system can benefit from the exploitation of alternative routings. In other cases, the cycle routings operations were even assigned to the wrong workcenter, but nobody really cared because the production operations had been always scheduled manually by the different shop managers. Furthermore, the tools availability stored in the technical database was far from up to date.

Of course, the lack of correct and formalized information is a big drawback when developing a decision support system like the one here envisaged, since inconsistent data can undermine the system's effectiveness by generating incorrect results. It was therefore necessary to go through large quantities of the company's technical data in order to correct errors and integrate the missing information. Only part of the job was done in time: The majority of it was performed during the implementation of the project, and it gave a strong contribution in postponing the achievement of the desired results, as described in Section 5.1.

During the first weeks and months, it was common to spot on the shop floor a group of direct operators discussing whether a part could or could not be manufactured by this or that machine, how this should be done, and by whom (or by which cell). To put this system into practice was a powerful way of detecting the incorrect pieces of information; correction teams made up of production and design people were set up, in a way that each time a part was spotted being assigned to the wrong workcenter:

- The manufacture of that part would eventually be blocked.
- The corresponding technical data woluld be carefully examined while searching for the error.
- The error and any missing data would be definitively fixed.
- The manufacture of the part would be restarted again.

Of course, this procedure meant a significant loss of efficiency and effectiveness during the months in which errors were very frequent (one or two per day was normal).

Quite the contrary, it was more difficult to spot and integrate the missing information where it was not associated with wrong data. In fact, when scheduling is done manually, alternative cycle routings must not be stored inside the technical database, provided the foremen remember them. To this purpose, a multifunctional team consisting of people from the Design, Production, and the Central Technologies Departments was set up and started to examine the different parts in a decreasing production volumes order, trying to figure out if any cycle routing other that those already present inside the technical database was possible. Often, direct workers remembered they had processed one part in a particular way when a major breakdown had occurred in the main machinery required, but it was often hard to extract and formalize this base of kowledge.

One of the most serious problems encountered regarded the production planning of those parts that had to be manufactured by one cell and supplied to the assemby line of another one. Given the reward system described in Section 5.2, any cell would tend to be late mainly on these parts, in order not to harm its own performances. Countermeasures to this tendency were to include the supply tardiness of these parts among the monitored performances of the cells and, at the same time, to reduce, as much as possible, the assignment of parts to be supplied to other cells, even if this could require a penalty in terms of WIP or increased setups. In addition, interchanged parts were eventually assigned in equal quantities, so as to balance the risks described.

Finally, difficulties connected with the human processes were, as usual, among the hardest to cope with. The change from a functional organization, configured to manage production *by process*, to a new context in which the organizational pattern would be *by product* had a tough influence on almost everyone. Nevertheless, the main resistance came, at least in the first months, from managers and foremen, while the change was quite well accepted from the beginning by the direct and indirect workforce.

The reasons for this behavior can be perhaps traced back to the loss of identity perceived by foremen when their old manufacturing departments were about to disappear. Many of the low-level managers and foremen had once been machine workers and had a very strong technical experience in the precise manufacturing process of which they were responsible. As a consequence, they felt that their prominence was based on their technological expertise over *their* process. Thus, they feared that their power could have been eroded by the change. As a matter of fact, the new system in the beginning created some confusion, mainly when critical problems emerged, because it was not clear which manager to address in order to get them solved.

Quite the contrary, the positive attitude of the shop floor workers toward the new organization of the manufacturing shop was due to the fact that most of them kept on controlling their machine, so that nothing dramatically changed for them, at least at the beginning. Despite this initial confidence with the new organizational configuration, problems started to arise for direct workers as soon as they started to be fixed at higher levels. Following the continuous improvement process described in Section 5.2, it was soon demanded of workers to increase their flexibility in the number of hours worked per day and in the rotation of jobs. Some of them (almost without exception the younger and more motivated ones) accepted enthusiastically to perform different jobs in different workdays and with different timetables. Others, the majority, were too old or undermotivated to do so, even if rewards could be gained. As a solution, volunteers were oragnized in special teams, named k-teams, that could be employed, week by week, at any workcenter and in any shift, Saturday included. Each cell got its k-team, which was scheduled weekly to help the bottleneck workenters.

In order to cope with these organizational issues, training classes were organized in advance for production and logistics people to explain the philosophy behind the organizational changes. Two levels of classes were set up: the higher level, attended by the cell managers and the foremen, were taught by the academicians that had helped develop the project. Then, the attendees taught the operative personnel of the cells the second level classes. Topics taught were general issues about production management, from MRP to just-in-time and lean production. More specific discussions were dedicated to organizational issues such as the setting up of multifunctional work teams, process management, decisional decentralization, and job enrichment/enlargement/rotation.

Many of the topics taught were supported by examples collected on the field, during the project development. Despite this effort, it can be easily stated that the best teacher was time.

19

CASES IN CELLULAR MANUFACTURING

William Wrennall and Frank C. Kerns

The Leawood Group, Ltd.
Leawood, Kansas 66211

1. INTRODUCTION

The development of cells and, for that matter, the implementation of a Cellular Manufacturing (CM) system is more than arranging people and equipment in a pattern that fits into some available space. It is the adoption of a manufacturing strategy that will have both social and technical effects within and on an organization. What is the best way to observe this process? Case histories provide an excellent window to view this approach in operation.

The following case examples are from actual cellular installations and will demonstrate the techniques utilized in implementing CM. Each will provide the company's situation at the beginning of the project, the mechanics of the data gathering and analysis, and the outcome or result of the implementation.

Given the range of variables that must be considered, it can be difficult to identify, evaluate, and maintain the information necessary to design an appropriate and successful cellular system. This is where following a well-thought-out guide can be invaluable in emphasizing the appropriate direction or path to enhance a company's unique circumstance.

Handbook of Cellular Manufacturing Systems, edited by Shahrukh A. Irani
ISBN 0-471-12139-8 © 1999 John Wiley & Sons, Inc.

2. FOCUSED OPERATIONS

The key characteristics of the focused factory are process technologies, market demands, product volumes, quality levels, and manufacturing tools (Skinner, 1974). The focusing of operations results from the comparison of an organization's operations strategy and product profiling. Profile analysis is a way to ascertain the degree of fit between the choices of process that have been made or are proposed to be made and the order-winning criteria of the product(s) under review (Hill, 1985).

A benefit this activity offers is the identification of matches or mismatches between products and their process arrangements. Process arrangement, or modes, can be organized in the following ways:

- Project
- Cellular
- Integrated cellular or Toyota
- Line
- Process

The product–volume plot in Figure 1 illustrates how the cellular modes fill the void between project, or job shop, and line manufacturing. This classification style has repeatedly proven itself more useful, especially for layout purposes, than terms such as fixed location, batch, and mass production. The latter terms gave us the choice of batch manufacture or mass production. Cellular manufacture provides us with the benefits of line operations with reduced setups that are inherent in classical batch mode. The integration of cells within manufacturing fills the gap between line and batch modes without the disadvantages of both.

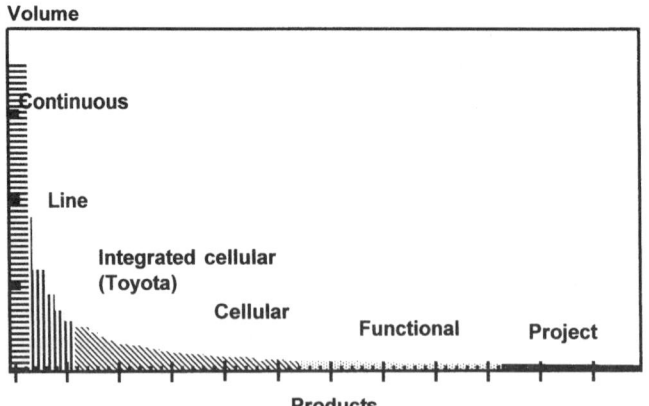

Figure 1. Production modes.

This mode continuum is the link between product mix, product velocity, processes, focus, and manufacturing strategy. In other words product–volume plots are an indicator when to consider cellular operations.

Product profiling (Figure 2) is a companion tool, developed by Hill (1985), that indicates where there is an inherent fit between product characteristics and a manufacturing mode. We have used the production modes described earlier in the example. Figure 3 shows the poor profile when all components in a lamp factory use one mode, and the smoothing effect of a product split is shown in Figure 4. In the processes described in the case histories, we have extended Skinner's (1974) focus factory concept from organization focus, through factory, to a factory within a factory, to cell, and workplace. A work cell is a focus choice at the mode level (see Figure 5).

3. ELEMENTS OF CELL DESIGN

As defined by Lee (1994, p. 205), "An optimum work cell requires engineering. As a science, work cell engineering demands a knowledge of principles. As an art, work cell engineering needs an eye for unique combinations and applications of basic principles."

A work cell has the fundamental elements of hardware and software. Not to be confused with computers or other forms of automation, these categories relate to the "hard" product manufacturing elements and the "soft" people and flow elements. The components that make up each are listed below:

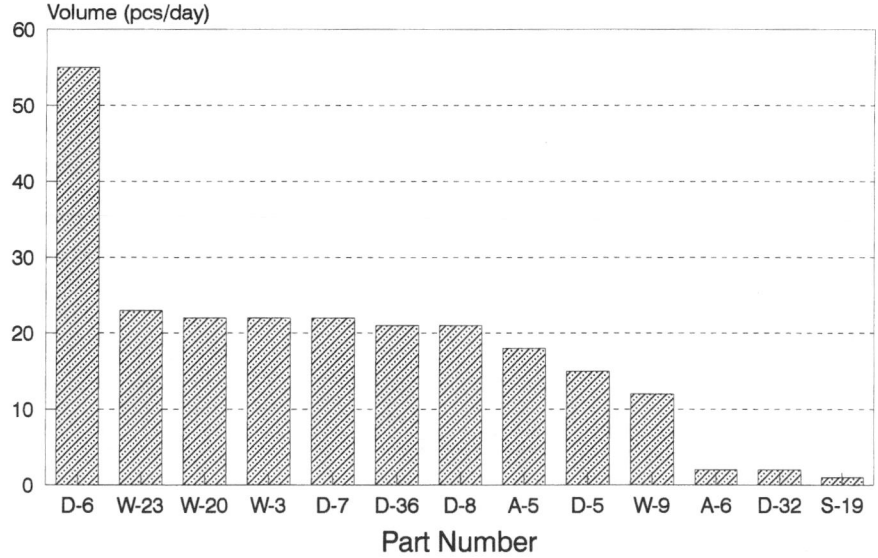

Figure 2. Product profile.

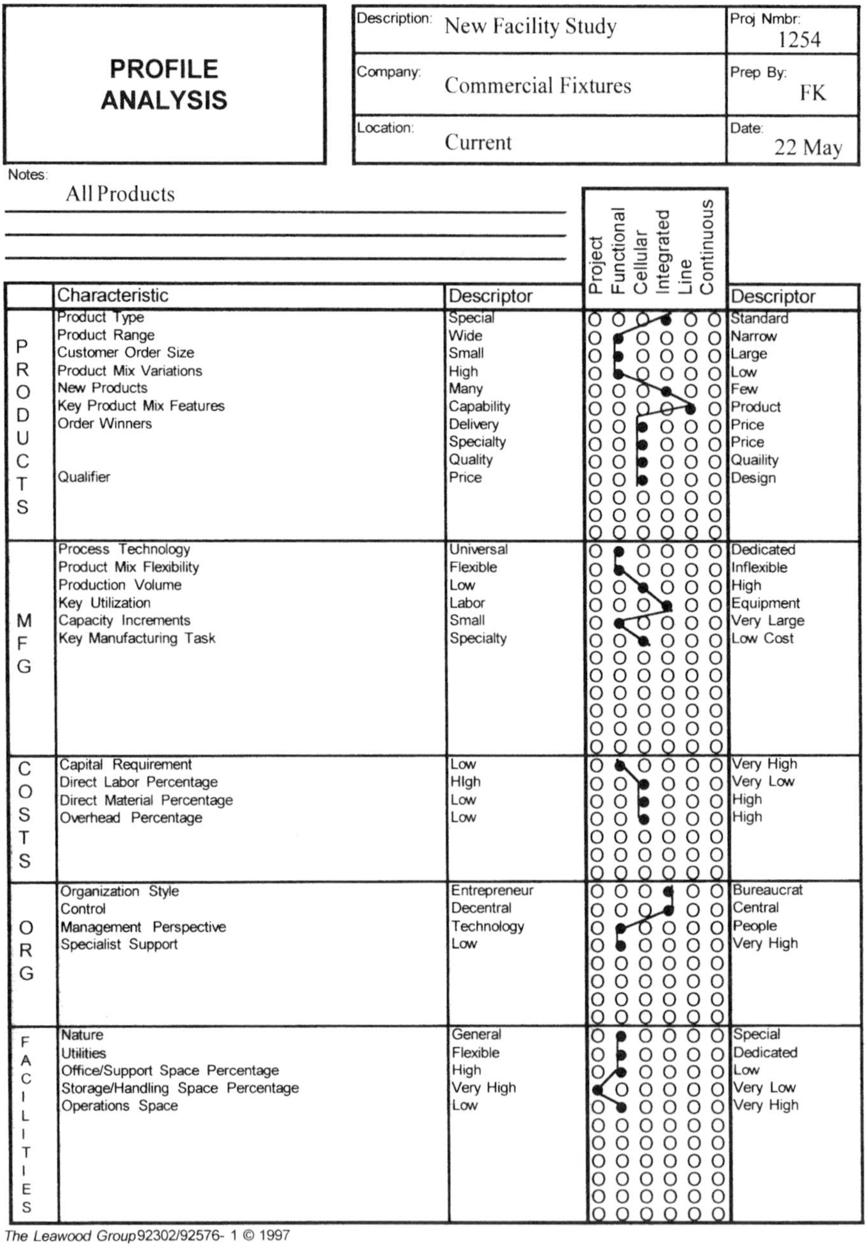

PROFILE ANALYSIS	Description: New Facility Study	Proj Nmbr: 1254
	Company: Commercial Fixtures	Prep By: FK
	Location: Current	Date: 22 May

Notes:

All Products

Characteristic	Descriptor	Project	Functional	Cellular	Integrated	Line	Continuous	Descriptor
PRODUCTS								
Product Type	Special			●				Standard
Product Range	Wide		●					Narrow
Customer Order Size	Small		●					Large
Product Mix Variations	High		●					Low
New Products	Many			●				Few
Key Product Mix Features	Capability				●			Product
Order Winners	Delivery			●				Price
	Specialty			●				Price
	Quality			●				Quality
Qualifier	Price			●				Design
MFG								
Process Technology	Universal		●					Dedicated
Product Mix Flexibility	Flexible		●					Inflexible
Production Volume	Low			●				High
Key Utilization	Labor			●				Equipment
Capacity Increments	Small		●					Very Large
Key Manufacturing Task	Specialty			●				Low Cost
COSTS								
Capital Requirement	Low		●					Very High
Direct Labor Percentage	High			●				Very Low
Direct Material Percentage	Low			●				High
Overhead Percentage	Low			●				High
ORG								
Organization Style	Entrepreneur				●			Bureaucrat
Control	Decentral				●			Central
Management Perspective	Technology		●					People
Specialist Support	Low		●					Very High
FACILITIES								
Nature	General		●					Special
Utilities	Flexible		●					Dedicated
Office/Support Space Percentage	High	●						Low
Storage/Handling Space Percentage	Very High		●					Very Low
Operations Space	Low		●					Very High

The Leawood Group 92302/92576- 1 © 1997

Figure 3. Poor product profile.

			Description: New Facility Study		Proj Nmbr: 1254

	Description:	New Facility Study	Proj Nmbr: 1254
PROFILE ANALYSIS	Company:	Commercial Fixtures	Prep By: FK
	Location:	New Facility	Date: 22 May

Notes:

Exit Signs Only - Proposed

	Characteristic	Descriptor	Project	Functional	Cellular	Integrated	Line	Continuous	Descriptor
P R O D U C T S	Product Type	Special	O	O	O	O	●	O	Standard
	Product Range	Wide	O	O	O	◐	O	O	Narrow
	Customer Order Size	Small	O	O	●	O	O	O	Large
	Product Mix Variations	High	O	O	O	O	O	O	Low
	New Products	Many	O	O	O	O	O	O	Few
	Key Product Mix Features	Capability	O	O	O	O	●	O	Product
	Order Winners	Delivery	O	O	O	O	●	O	Price
		Specialty	O	O	O	O	●	O	Price
		Quality	O	O	O	O	O	O	Quaility
	Qualifier	Price	O	O	O	O	●	O	Design
			O	O	O	O	O	O	
			O	O	O	O	O	O	
M F G	Process Technology	Universal	O	O	O	●	O	O	Dedicated
	Product Mix Flexibility	Flexible	O	O	O	◐	O	O	Inflexible
	Production Volume	Low	O	O	O	◑	O	O	High
	Key Utilization	Labor	O	O	O	●	O	O	Equipment
	Capacity Increments	Small	O	O	O	◑	O	O	Very Large
	Key Manufacturing Task	Specialty	O	O	O	◑	O	O	Low Cost
			O	O	O	O	O	O	
			O	O	O	O	O	O	
			O	O	O	O	O	O	
			O	O	O	O	O	O	
C O S T S	Capital Requirement	Low	O	O	O	●	O	O	Very High
	Direct Labor Percentage	HIgh	O	O	O	●	O	O	Very Low
	Direct Material Percentage	Low	O	O	O	●	O	O	High
	Overhead Percentage	Low	O	O	O	●	O	O	High
			O	O	O	O	O	O	
			O	O	O	O	O	O	
O R G	Organization Style	Entrepreneur	O	O	O	●	O	O	Bureaucrat
	Control	Decentral	O	O	O	◐	O	O	Central
	Management Perspective	Technology	O	O	O	◑	O	O	People
	Specialist Support	Low	O	O	O	●	O	O	Very High
			O	O	O	O	O	O	
			O	O	O	O	O	O	
			O	O	O	O	O	O	
F A C I L I T I E S	Nature	General	O	●	O	O	O	O	Special
	Utilities	Flexible	O	●	O	O	O	O	Dedicated
	Office/Support Space Percentage	High	O	O	O	O	O	O	Low
	Storage/Handling Space Percentage	Very High	O	O	O	O	O	O	Very Low
	Operations Space	Low	O	O	O	O	O	O	Very High
			O	O	O	O	O	O	
			O	O	O	O	O	O	
			O	O	O	O	O	O	

The Leawood Group 92302/92576- 2 © 1997

Figure 4. Adjusted product profile.

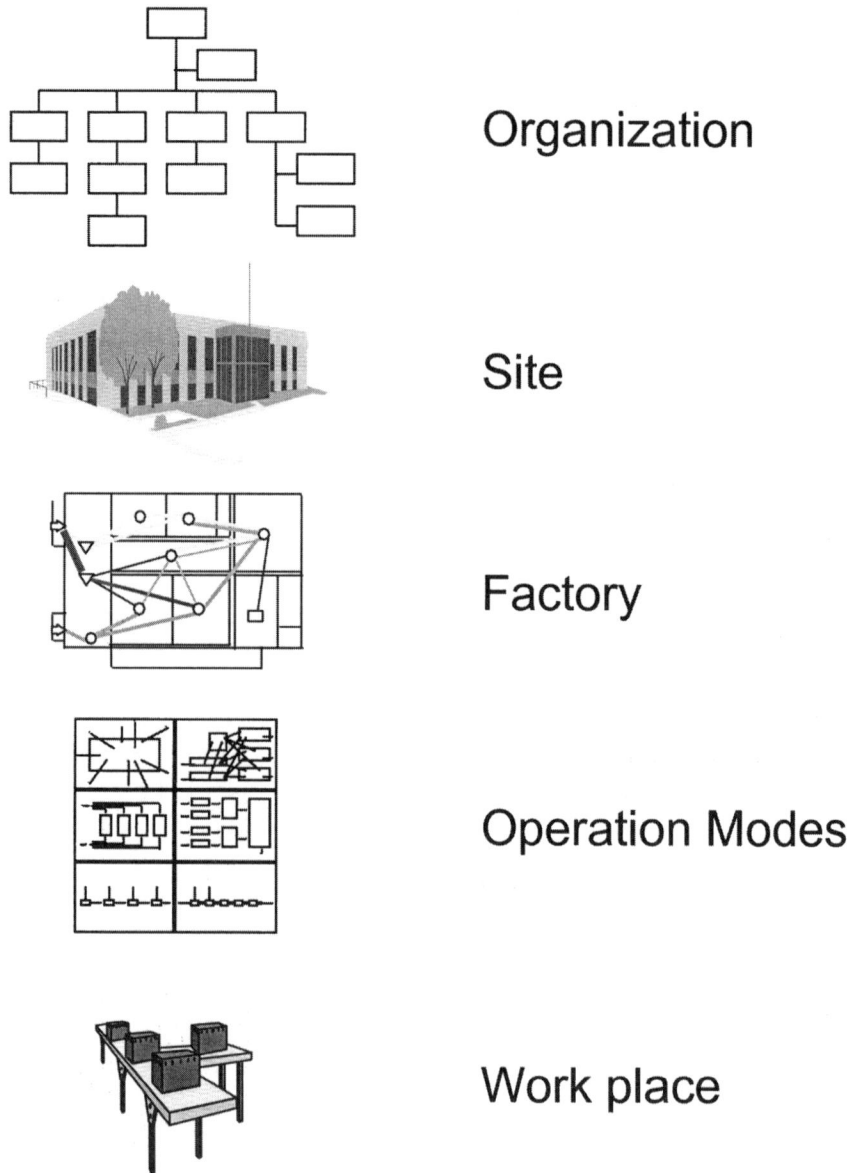

Organization

Site

Factory

Operation Modes

Work place

The Leawood Group 94411 © 1997

Figure 5. Focus spectrum.

Hardware consists of

- Product grouping
- Process selection
- Containers
- Material handling methods
- Layout

Software includes

- People selection
- Skills training
- Lot sizing
- Production control
- Capacity balance
- People balance
- Task assignment
- Quality assurance

4. CELL SYNTHESIS ALGORITHM

The Leawood Group's Cell Synthesis Algorithm (CSA) structures the decision process to enhance the design process. The CSA has the following elements:

- A logic flowchart
- A compatibility chart
- Decision and design aids (i.e., forms, checklists, guides).

The CSA design outputs are an equipment list, a cell layout, and a work cell operations plan. The algorithm and compatibility chart are given in Lee (1994). Several cell layouts are given in this chapter and an example of a Work Cell Operation Plan is provided as Figure 6. The CSA was used to complete the following case examples.

5. CASE HISTORIES

We have selected examples of cell designs from different industries. They are all from manufacturing and none of them are group technology parts' manufacturing cells.

Case 1. Product-Focused Cable Cell

The first case is a manufacturer of control cables who decided to rearrange the manufacturing operations from assembly lines to cells. The company is a marginal profit business with a small market share. Its customers needed to continuously reduce costs to maintain its market share.

Description:		Proj. No.
Prototype Mfg. Cell - Flexible Cable		*p1254*
Company:		Prep By:
Control Cable Company		*HT*
Location:	Page: of:	Date:
Main Plant	1 1	*12- Jul.*

Cell ID:
Brake Cable

External Lot Size

- [ ] Single Piece
- [ ] Small <.5 hours
- [ ] Medium < 4 hrs.
- [x] Large

Prod.Units: 2100 units/ship week

External Containers

- [ ] Small
- [x] Medium (70 each)
- [ ] Large

Type: 2100 / 70 = 30 boxes

External Material Handling

Scale:
- [x] Small
- [ ] Medium
- [ ] Large

Route Structure:
- [x] Direct
- [ ] Channel
- [ ] Terminal

Equipment:
- [x] Hand Carry
- [ ] Other

Kanban items into cell
Forklift boxes on pallets
away from cell

External Prod. Control Method

- [ ] Physical Link
- [x] Broadcast
- [x] Kanban
- [ ] Push Schedule
- [ ] Re-Order Point

Internal Prod. Control Method

- [x] Direct Link
- [ ] Circulation
- [ ] Kanban
- [ ] Push Schedule

Internal Lot Size

- [ ] Single Piece
- [x] Small
- [ ] Medium
- [ ] Large

Equipment Balance Method

- [ ] Queue
- [ ] Inherent Balance
- [x] Excess Capacity

Internal Containers

- [x] Small
- [ ] Medium
- [ ] Large

Type: carts

People Balance Method

- [ ] Queue
- [x] Circulation
- [ ] Float
- [ ] Inherent Balance
- [ ] Excess People

Quality Assurance

- [ ] Inspect & Reject
- [x] SPC/TQC
- [ ] Inherent

Supervision

- [x] Self-Determination
- [ ] Kanban Signal
- [ ] Command & Control

Compensation

- [x] Hourly / Salary
- [ ] Individual Incentive
- [ ] Group Incentive

Operator Assignments & Skill Matrix

Operators	Skills					Operations															
	Setup	QA Test	Die Cast	Assemble		Obtain Matl.	Deburr	Press & Adj.	String Cable	Press Tube	Position Adp.	Off Cast	Form Tube	Sprg & Cage	Die Cast	Clip & Assbl.	Insp.& Pack		Setup	QA Test	Runner	Make Boxes
Alan	X	X	X	X		X	X	X	X	X	X	X	X	X	X	X	X		X	X	X	X
Betty		X	X	X		X	X	X	X	X	X	X	X	X	X	X	X			X	X	X
Bonnie	X	X	X	X		X	X	X	X	X	X	X	X	X	X	X	X		X	X	X	X
Doug		X	X	X		X	X	X	X	X	X	X	X	X	X	X	X			X	X	X

The Leawood Group 92615 / 93133

Figure 6. Completed work cell operations plan.

The methods of identifying product candidates for CM are group technology, production flow analysis, and inspection. Since the variety of control cables being manufactured was small, the method used to identify those cables with similar manufacturing characteristics in this instance was by inspection.

The first product selected for cellular arrangement was for a high-visibility customer suffering from overseas competition. Through inspection we found that the family of cables produced for this customer, we will call HD products, employed about seven people for the required quantities. This is an excellent match with cell size criteria.

As is usually the case, an objective of this project was to improve productivity. It is necessary to determine the current productivity for use as a baseline, so quantitative evidence of the benefits can be demonstrated. Since plant relayout provides an opportunity for method improvement at the macro level, the operations from material receipt to product shipment were mapped. The reengineered process included the following method changes:

- Packaging material and purchased parts were removed from the main warehouse and located in kanban stock points near the point of use—the cell.
- The setup person was made part of the cell team.
- Products were inspected, packed, and staged at the cell for weekly transfer to the truck at the loading dock.
- The equipment capacity was calculated and additional items purchased to allow for a continuous, linear flow at a capacity greater than the forecast peak weekly demand.
- Actual production volume was varied to meet weekly shipments only.

Direct and indirect labor were consolidated in the cell team. Raw materials, purchased parts, operations, and packaging were co-located for greatest impact. The team members were cross trained in all of the workplace activities and their number was varied to match customer orders. There was no building to stock. The completed work cell operations plan is given in Figure 6.

Cell Layout   The layout of the cable cell was sequential by operation. We adopted the method and operation plan shown in Figure 7. The cell layout is shown in Figure 8. When the cell design was accepted by the company management, several activities took place in quick succession. Included among these were the team members being trained, the kanban stock points were built and stocked, the utility drops were provided, and the relayout was then implemented over a weekend.

The cell layout is U-shaped and the initial people balance was calculated to determine workstation allocations. An initial operation objective was to achieve a batch size of one, but it was soon found that the handling times were greater than the value adding times. The product was packed in quantities of eight. Using this batch size created a significant productivity gain. The operator who did the final inspection also packed the product and included in the pack a personalized inspection ticket. The operators assumed ownership of the product with pride in their work. Since the cell focused on customer and product characteristic, we identified the cell with a sign using the customer name. The next time representatives from the client visited the company they saw that their product was made by their team in their cell. It was learned later that they had requested this in the first place.

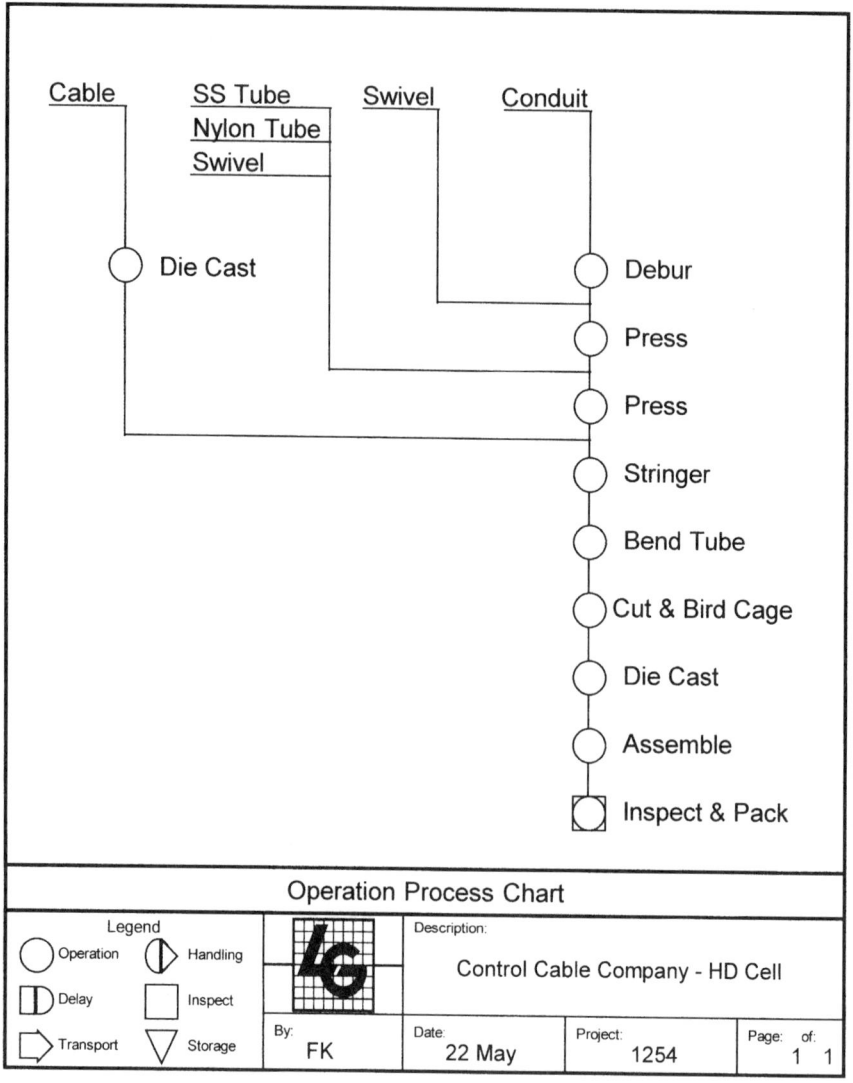

Figure 7. Method operation process chart.

The team size and work allocation was adjusted to satisfy product volume variation. Peak demand was met with operators working at each workstation. To support lower volumes, the transition was made from an operator staffing two or more workstations to each operator circulating through the cell and performing all operations.

One of the early difficulties was the relationship with production planning and cell operations. Production planning had previously driven operations, but its role was diminished in the customer pull method. Production planning demanded a building

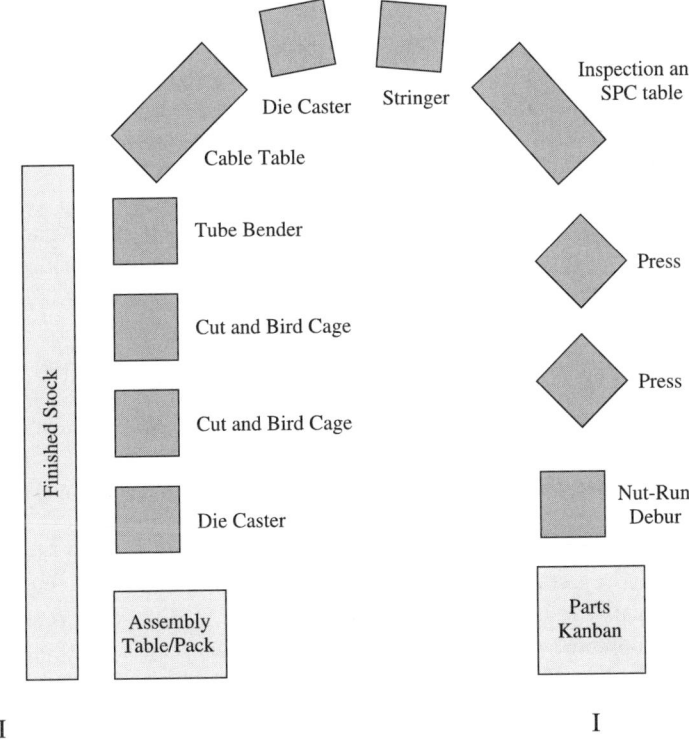

Figure 8. Control cable cell layout.

to stock operation and gave false customer order quantities to support its claims. The inherent flexibility in the cell capacity and the team's determination to succeed soon overcame these disruptive efforts.

By now, in spite of early disclaimers about needing productivity and cost-saving justifications for extension of the project, top management asked what was the benefit from the cell operations. Fortunately, we had worked with the cost accountant who was able to supply independently determined productivity gains of 25%.

The benefits were

- Reduction in raw materials, purchased parts, and finished product inventory
- Elimination of indirect labor cost
- Integration with the plant Total Quality Management (TQM) process
- Increased operator skills and morale
- Reject elimination
- Higher throughput velocity
- Improved order completion performance
- Operator productivity improvement

- Space savings
- Employee requests to be members of the next cell
- Customer satisfaction
- Increased orders and profitability
- Contribution to meeting ISO 9001 registration standards.

Cellular operations were extended to all cable making operations and integrated into an overall plant relayout. The company achieved its growth and profitability plans as well as price reduction targets for its customers.

Case 2. Saw Assembly Cell

Abrasives Inc., a producer of commercial abrasive products, decided to reengineer its operations. Profit was low and volume growth was high. Following a survey of the company activities, we recommended that its saw assembly operations be converted to CM.

Manufacturing produced blades as its only component of the finished product. The Assembly Department used standard purchased parts and locally fabricated and painted internally designed parts. It employed about 45 shift workers in assembly. They were supported on the day shift by 12 purchasing agents, 10 planners, and 3 material handlers. Purchased and subcontracted parts were unloaded at the dock, held until accepted by quality control, counted and transferred to the warehouse, and entered in the information system as inventory. Stock was replenished by purchase order. Order quantities were at purchasing agent discretion, often based on quantity discounts. Inventory of purchased parts had risen to $10 million on equipment sales of $27 million. Obsolescent inventory was transferred regularly to an offsite rented warehouse. Annually, the obsolete inventory was written off at management's discretion. Current items in excess of the warehouse capacity were also transferred to the offsite warehouse so the company was considering expanding the warehouse.

Finished goods were built to shop order issued by production planners. The product was built to shop order and transferred to the adjacent finished goods warehouse. Finished goods inventory of about $5 million was held against catalog items. Work-in-process inventory was about $530,000.

A total of 20 product lines were assembled with each product having 5–20 different configurations. Product types were assembled at workstations dedicated to a particular product group or family.

The work order was sent to the warehouse and the parts were picked and delivered to the appropriate workstation/area. If there were parts shortages, the incomplete pick was still delivered and the shortages recorded. The shortages occurred because of Materials Requirement Planning (MRP) driven accuracy counts. Even when real shortages were known, the available parts were picked and staged to allow for incomplete assembly work to proceed. The inventory problem was compounded.

Ironically, an inventory control group, consisting of an engineer, two people who made continuous inventory counts, and a computer operator who reconciled system

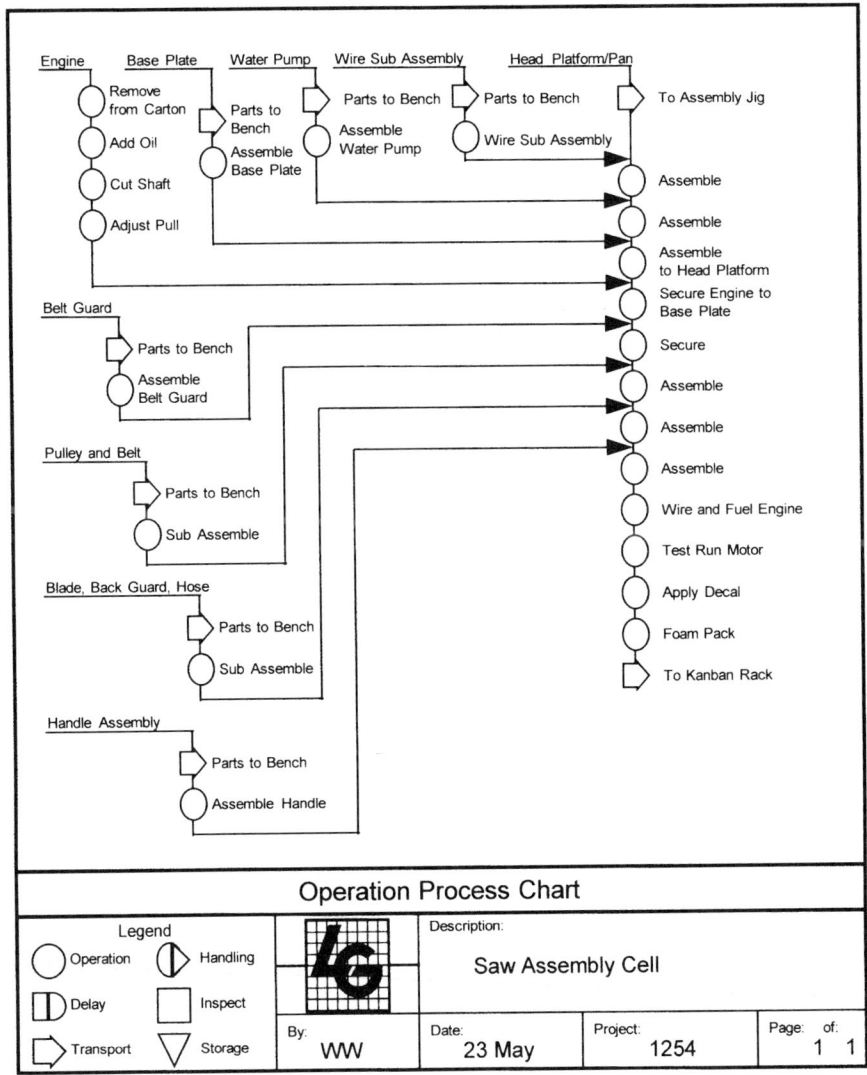

Figure 9. Saw assembly flow process chart.

and actual inventory in the MRP system, managed to maintain continuous shortages with growing inventory levels.

Cellular Approach   Subassembly and assembly work was performed on benches in batches of 10 (see Figure 9), and work in process accumulated due to parts shortages. The success of the sales force and the inefficiency of the production processes resulted in a flood of complaints from customers and top management.

It was agreed to make the following changes:

- Transform assembly to a series of cells
- Provide as many as possible product group dedicated cells
- Assemble other products in general-purpose cells
- Design and implement a kanban purchasing system
- Locate purchased parts at the point of use with a small warehouse reserve
- Contract with a fastener company to provide a continuous small fastener replenishment service
- Select a product group for the first cell
- Appoint and train a cell manager responsible for purchasing parts, planning and scheduling the work, and managing the cell

The product volume required two assembly workers who had previously assembled one product at a time. The shared work was performed on benches, tables, and the floor.

The MRP system was converted for use on long-range master purchase contracts with a visual kanban system managing the real-time control. The company's Information System staff objected strongly saying that control would be lost. However, this had been the case for some time. It was suggested that they continue with their "information system," but the "new way" would operate as a trial.

We rearranged the assembly process (see Figure 10) in the following way:

- Kanban stock points for purchased materials
- Kanban storage for finished product
- Two free standing assembly jigs
- Bench-top subassembly areas with point of use parts
- A foam packing area for the cell

Observations had indicated that there was a large performance difference between the two operators allocated to the cell that was restricting output. It was arranged for each operator to assemble an output unit of product. All steps of assembly were performed in batch sizes of one. Each operator subassembled at the bench, then completed the assembly on the assembly jig. The product was moved by hoist into a carton at the packing unit. The packed and labeled product was placed in the kanban product storage rack by a material handler who served all cells.

An analysis of past and forecast product demand was used to calculate preliminary kanban quantities by product configuration. Management's approval of a safety factor was an enticement to gain its support. The rack spaces were labeled with the product configuration identifier.

The buildup of inventory in the kanban product racks was based on order priority until back orders were eliminated. The cell operators then decided product assembly sequence, as spaces became vacant. Product was pulled and moved to shipping, by material handlers, to fulfill shipping orders.

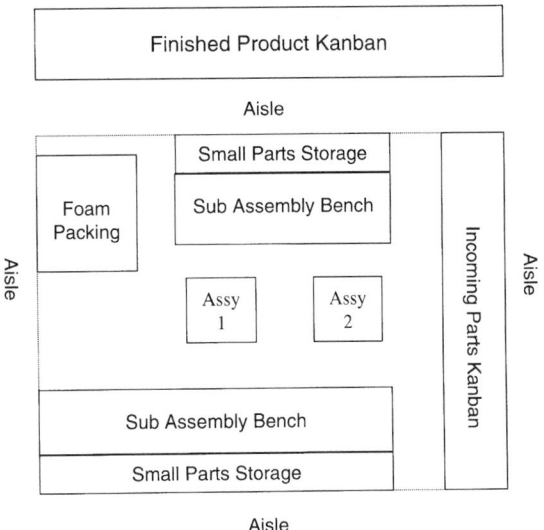

Figure 10. Saw assembly cell layout.

Purchased parts were controlled by kanban cards. As delivered parts were consumed, the release cards were posted. The manager and material handlers used these visual signals to replenish from supplier or warehouse reserve.

Continuous Improvement   The progressive disappearance of purchased parts and work in process, with faster output of completed products, with less intrusion from "support" departments met with some trepidation at several levels. With consultant protection, and local decision making within the cell team, continuous improvement occurred. Within 2 weeks, design flaws were corrected, parts shortages were eliminated, part shipments of this family of products and late deliveries were eliminated, and inventory turns doubled.

The cost standards and operators' performance levels were still disappointing. The work standards were recalculated for improved methods and new output targets based upon these sound standards were introduced. The finance group treated the reduced inventory levels, output quantities, and cost standards with suspicion. A performance and inventory audit, witnessed by the cost and financial accountants, satisfied them that the figures were correct. They retreated into their own cells.

Operator performance was another matter. With the support of a skilled operator, it was demonstrated that the production targets were fair and reasonable and job security was enhanced not threatened.

Results

Flexibility. Instead of one product configuration completed in a 10-hour period, 4 configurations can be assembled per assembly station or 8 per cell with 16 units available to ship every day.

Picking time by unit configuration was eliminated. Order pickers became material handlers replenishing cell stock.

Lower inventory costs. Work in process was reduced by 77%. For one product line, finished goods inventory was reduced from 180 to 80 units—a reduction of 56%. The goal is to produce to customer order and no more.

Reduced space requirements. With 25 work cells in operation and continuous inventory reduction space saving in each cell, warehouse space will become available for value-adding operations.

An interesting side benefit arises from lowering finished product inventories. The obsolete finished product becomes virtually nonexistent. Components are only used for products that are sold. This sounds obvious, but there were many times when a customer could not get an immediate shipment because the needed parts were on warehoused product that would not be required for several weeks.

Other benefits stemmed from this transition to CM:

- Reduction in direct labor costs
- Fewer computer and key punch transactions
- Better visibility
- Simplified process flow

This example was to meet the critical competitive advantages for the company. The demands for fast response and higher labor productivity provided an improved return on assets employed. This in turn created higher product sales and customer satisfaction.

Case 3. Erie Windows

Erie Windows (EW) designs and produces replacement vinyl windows. It was a profitable fast growing company that was running out of space. It had submitted a capital expenditure plan to the head office for the expansion of its building and production capacity. Management were told to engage the services of facility planning consultants to plan and design a building addition with a financial justification.

The consultants investigated the current situation and reported the following findings:

- The building was congested.
- The customer response time was in excess of 4 weeks.
- Orders were staged by customer and delivery route.
- Deliveries were delayed for reworked items.
- The shipping area was full of incomplete orders and growing.
- Production operated on day shift with regular overtime, but shipping staff were working until the early morning hours.

- Plant and equipment maintenance was slow and reactive.
- Customer order processing took 2 weeks in administration and 2 weeks in production and delivery.
- Raw material inventory was high—three turns per year.
- An offsite warehouse was rented for raw material and finished product overflow.
- The management style was autocratic.

The following actions were recommend:

- A strategic master site plan should be prepared.
- An operations strategy should be developed.
- A building plan, material handling, storage, and strategic master site plan and plant layout should be developed.

The recommendations were accepted and the following were to be incorporated:

- A new technology glass cutting and gas filling of insulated window units were to be purchased and installed as functional cells to serve all product lines.
- Window manufacturing was to be in product-focused cells.
- Team concepts were to be implemented.
- The facility plans should be sized for single-shift operations.
- Customer orders should be delivered complete within 2 weeks.
- Production should be scheduled by truck delivery load in reverse drop order.
- Product defects should be reprocessed immediately and replaced in the customer order lot.
- Completed product should be loaded on the trailer, spotted in the loading dock, as produced.
- The computer-generated bill of materials should be adapted for the new order process.
- Just-In-Time (JIT) raw material purchasing should be adopted.
- The new lean manufacturing cell layouts should be implemented as developed.
- The cell locations should be flexible but as far as possible conform to the expanded building layout plan.
- Provide for three new product cells.
- The need for an offsite warehouse should be eliminated.

Product-Focused Cellular Window Manufacturing   The introductory material in this case is an example of the strategic steps that lead to CM. The next step was which cell to design first. The criteria were

Maximize customer satisfaction
High probability of success

Ease of implementation

Double-hung windows were the product family that was selected. The concept and design were new, and the product was seen as providing a company competitive advantage in the marketplace. The core process to support the advantage was sash manufacture. Volume was high but more suitable for cellular than line production.

We were also influenced by the quality of the key people experienced in sash manufacturing. Their attitudes were favorable and the lead hand had a high work ethic with facilitating skills and was competent on all the production processes. The current location of the major equipment was acceptable in the new layout plans.

Vinyl Window Sash Cell Design   Window sash vinyl parts were currently produced in batches of 250. They queued in carts at each process station and were transferred on carts from process to process. The process sequence is mapped in Figure 11. The typical work in process was about 45 sashes.

The optimum batch size, to pull through the cell, was determined by trial and error. A batch of 13 minimized the process capacity loss and allowed for a good workload balance. Parts carriers were designed for each sash family and conveyors linked the saw, punch, drill, and weld processes. The input conveyor was designed to hold 13 sash sets only. This prevented regression to a push system. The cell layout is shown in Figure 12 and the operating plans in Figures 13 and 14.

Early Problems   Production was restricted in the early days because of the low standard of equipment maintenance. For instance, saw guides could not be adjusted and solenoids in the welders frequently stuck. These reasons were noted on the production record. The maintenance inadequacies had not been obvious to management until now, nor had the ways of dealing with them. The saw operators did not use the guides but judged where to make the cuts, hence quality problems. To free the solenoid valve when it became stuck, a hole was drilled in the weld casing, and the operator used a small screw driver to free up the valve on each occurrence.

The production manager did not readily accept the cellular concept with real-time production decisions made in the cell. The manager frequently gave new instructions to individual operators then departed. When the manager was engaged in offsite activities, output invariably increased. This did not improve the manager's behavior.

The team was informed of the daily production quantity and the work orders were provided each morning. Operators were allocated to workstations to match their experience. The lead hand also floated between workstations as a relief worker and to maintain balance. An hourly production record was posted in the cell for the operators to observe their output. When the daily production target was met, the team held meetings to discuss how to make improvements, worked on minor maintenance tasks, or cross trained their colleagues.

Gradually, the team performance improved, equipment reliability increased, rejects were eliminated, overtime became unnecessary, and cells were established throughout the plant.

The overall results were

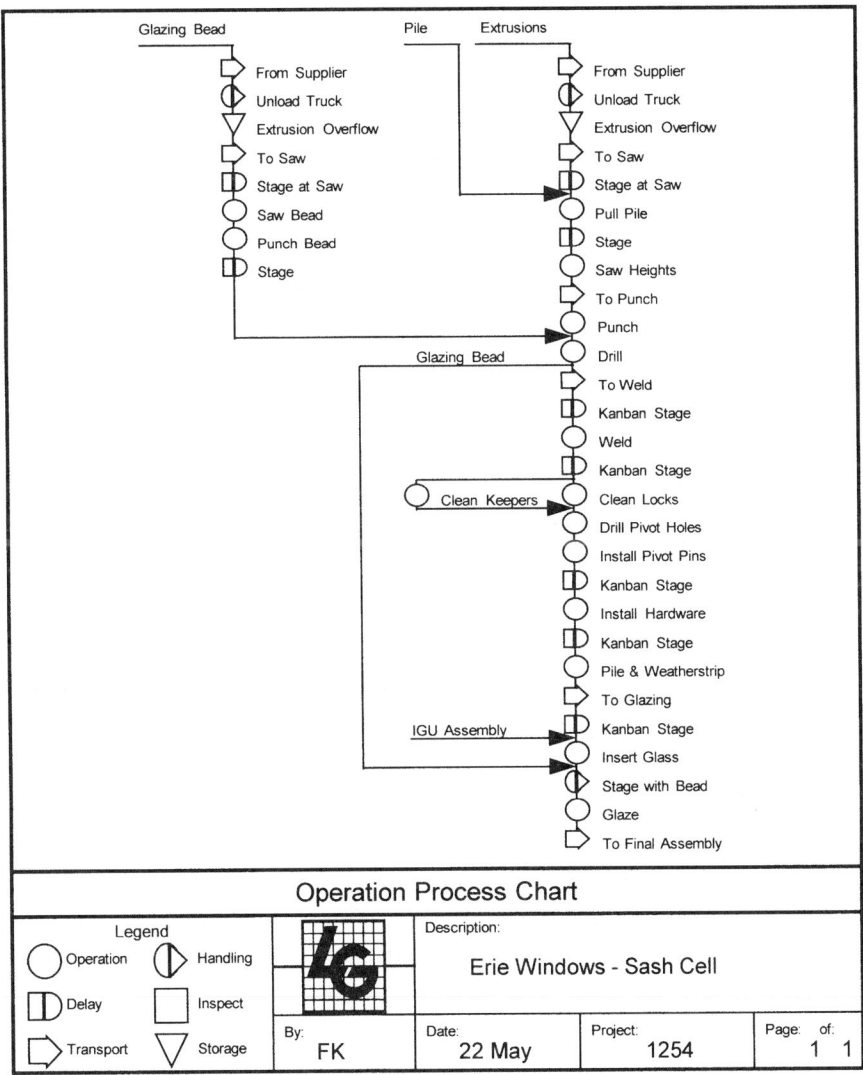

Figure 11. Erie Windows operation process chart.

- Sashes were produced for orders to be assembled into windows for immediate delivery to customers.
- Orders issued to the plant at 8:00 A.M. were loaded complete and on their way to the customer by noon the same day.
- Inactive work in process disappeared.

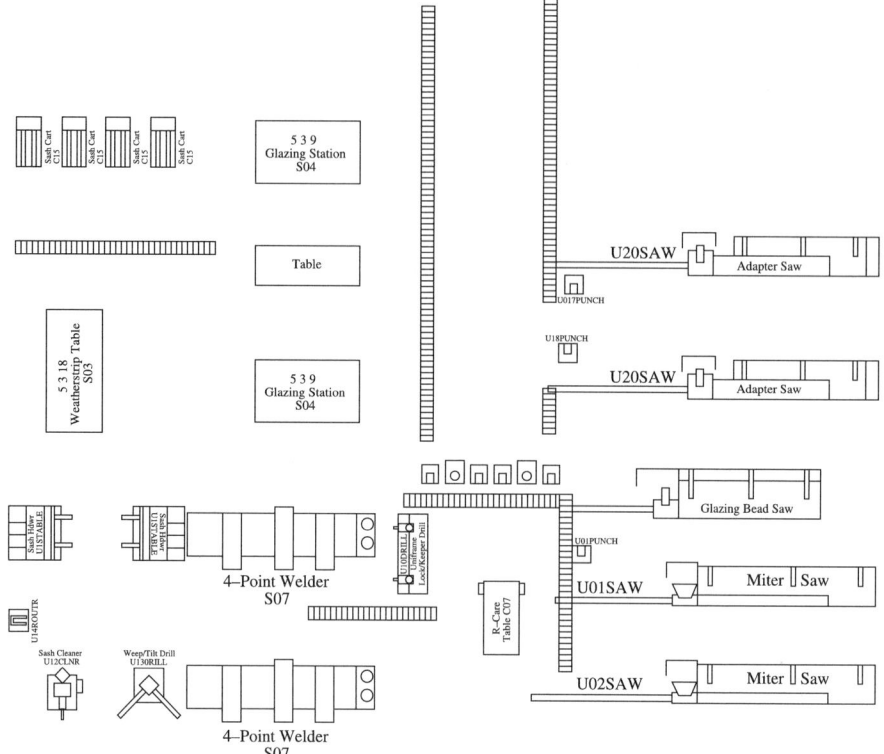

Figure 12. Erie Windows sash subcell layout.

- Overtime came under control.
- Late hour staging and shipping was now unnecessary.
- Order lead time exceeded expectations and gave an advantage over the competition.
- Productivity increased.
- Inventory reduction released funds and space for extra capacity.
- The building expansion was not so urgent.
- Record financial results recorded.
- The justification for an expansion of operations was readily accepted by the head office.

6. PITFALLS

The case examples given describe what happened in fact. Cellular operations grew out of a family parts approach to the machining of metal parts. The examples given in

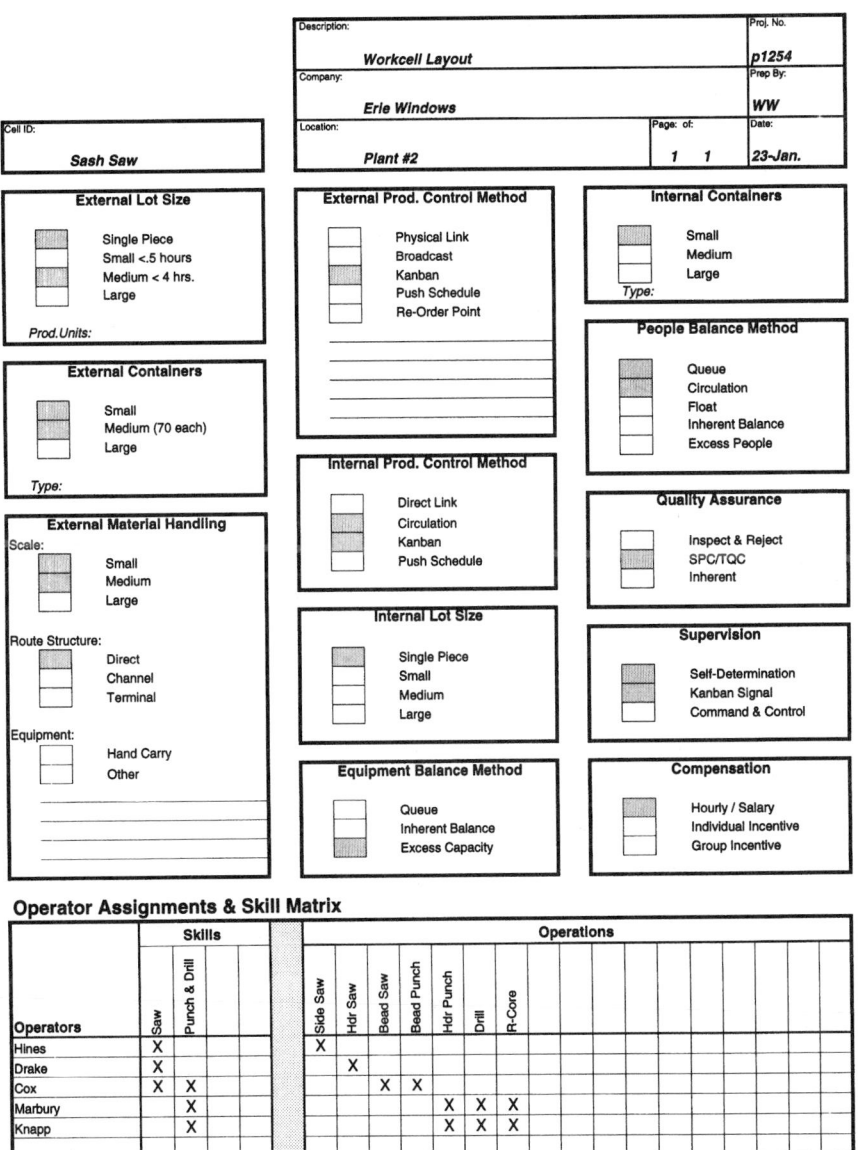

Figure 13. Work cell operations plan—sash saw.

this chapter are not machining operations. Indeed cellular concepts have applicability to many nonmanufacturing work such as office operations.

With hindsight, critics may say that the solutions were nothing new. In fact cell design is complex and sophisticated. This is not always apparent to the uninformed after-the-fact observer.

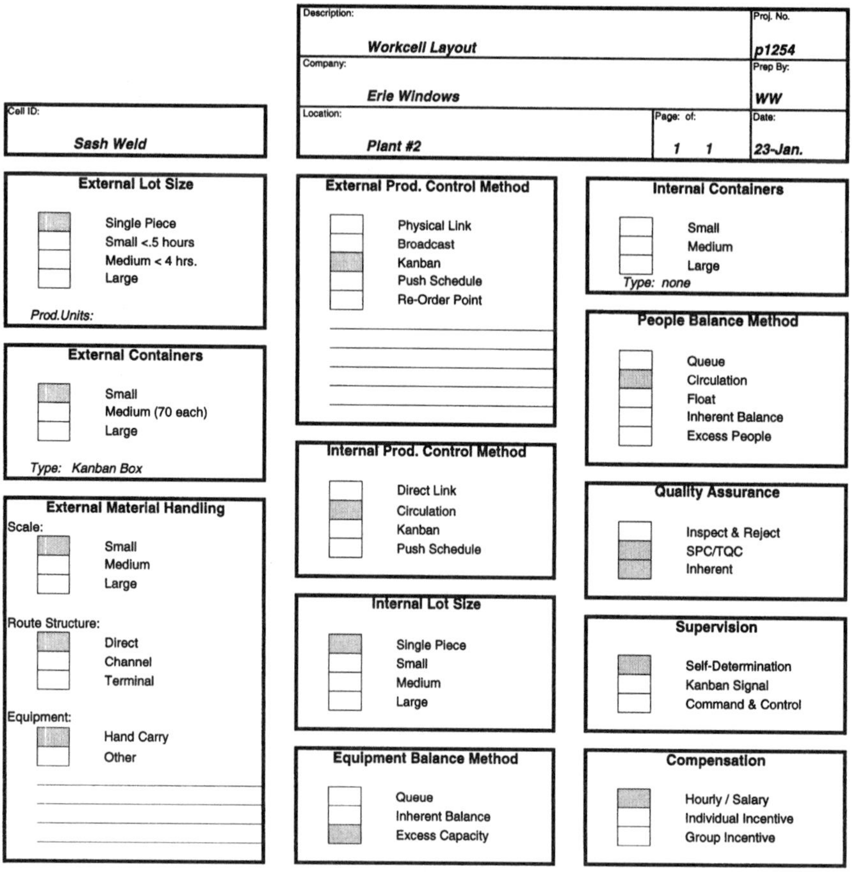

Description:			Proj. No.
Workcell Layout			**p1254**
Company:			Prep By:
Erle Windows			**WW**
Location:		Page: of:	Date:
Plant #2		**1 1**	**23-Jan.**

Cell ID:

Sash Weld

External Lot Size
- ▨ Single Piece
- ☐ Small <.5 hours
- ☐ Medium < 4 hrs.
- ☐ Large

Prod.Units:

External Containers
- ▨ Small
- ☐ Medium (70 each)
- ☐ Large

Type: Kanban Box

External Material Handling

Scale:
- ▨ Small
- ☐ Medium
- ☐ Large

Route Structure:
- ▨ Direct
- ☐ Channel
- ☐ Terminal

Equipment:
- ▨ Hand Carry
- ☐ Other

External Prod. Control Method
- ☐ Physical Link
- ☐ Broadcast
- ▨ Kanban
- ☐ Push Schedule
- ☐ Re-Order Point

Internal Prod. Control Method
- ☐ Direct Link
- ▨ Circulation
- ☐ Kanban
- ☐ Push Schedule

Internal Lot Size
- ▨ Single Piece
- ☐ Small
- ☐ Medium
- ☐ Large

Equipment Balance Method
- ☐ Queue
- ☐ Inherent Balance
- ▨ Excess Capacity

Internal Containers
- ☐ Small
- ☐ Medium
- ☐ Large

Type: none

People Balance Method
- ☐ Queue
- ▨ Circulation
- ☐ Float
- ☐ Inherent Balance
- ☐ Excess People

Quality Assurance
- ▨ Inspect & Reject
- ☐ SPC/TQC
- ☐ Inherent

Supervision
- ▨ Self-Determination
- ☐ Kanban Signal
- ☐ Command & Control

Compensation
- ▨ Hourly / Salary
- ☐ Individual Incentive
- ☐ Group Incentive

Operator Assignments & Skill Matrix

Operators	Skills				Operations										
	Weld	Clean	Assemble		Weld	Hand Clean	Drill	Router	Assemble						
Marcus	X				X										
Williams	X				X										
Gonzales		X	X			X	X	X	X						
Getz		X	X			X	X	X	X						
Dodanski		X	X			X	X	X	X						

Figure 14. Work cell operations plan—sash weld.

The design and implementation of CM are very dependent on the coupling of operations. This can lead to problems when there is an attempt to couple operations that, because of the processes involved, have dramatically different output quantities. Therefore, sometimes the best course may be the counterintuitive action of decoupling operations.

Another pitfall is forced single-piece flow; see case 1. Single-piece flow is a goal not a law to apply when it is not economically advantageous.

As Maxey (1994, p. 4) has stated: "It is interesting to note that operators accepted cellular manufacturing much easier than middle management. In fact, some hourly employees said they mentioned the need to have (locate) components in the work areas several years earlier—a move towards the cellular approach."

A natural outcome of cell design is the introduction of empowered manufacturing teams. This can cause considerable concern for middle management who see no role rather than an enhanced roll in a cellular environment. This is especially necessary when the team members need to make the real-time decisions that had been the prerogative of middle management. Other sections cover this material.

7. SUMMARY

The cell implementations in the cases provided describe what happened. The emphasis is on the critical elements or deliverables of each example. A layout is a single deliverable reflecting the implementation of many elements and other chapters lead you through the layout design process in more detail.

As stated by Wrennall (1992, p. 1): "Work cells appear simple. They are, in fact, sophisticated Socio/Bio/Technical systems. Proper functioning depends on subtle interactions of people and equipment. Each component must fit with the others in a smoothly functioning, self regulating and self-improving operation."

Cellular operations may make the difference between a company that is competitive and one that is not. The order process, from order receipt to payment for delivery cycle, can be substantially reduced when product-focused cells are introduced. This was the case in all the examples given. The emphasis is on material speed through the operations. Combined with a kanban, visual, real-time production control system leads the way to *lean management*.

There is always a request for a set of rules or model for all activities. They are for perpetration of mediocrity. Innovative solutions distinguish leaders from followers. The above case examples all had innovative content. They resulted from participation of innovative individuals in a creative atmosphere and the guidance of the CSA as a framework.

REFERENCES

Hill, T. (1985). "Choice of Process," in *Manufacturing Strategy*, Macmillian, London, pp. 89–95.

Lee, Q. (1994). "Workcell Design for World Class Manufacturing," in *Handbook of Commercial and Industrial Facilities Management*, W. Wrennall and Q. Lee (eds.), McGraw-Hill, New York, pp. 201–240.

Maxey, D. (1994). Cellular manufacturing, A case study, *The Leaword*, **12**, Summer, 1–5.

Skinner, W. (1974). The focused factory, *Harvard Business Review*, May–June, 182–183.

Wrennall, W., and Lee, Q. (1992). Manufacturing Workcell Design: Getting it Right—Step by Step, Unpublished monograph, Institute of Industrial Engineers 20th Systems Integration Conference, Chicago, IL. May.

Bibliography

Dhavale, D. (1995). Justifying manufacturing cells, *Manufacturing Engineering*, December.

Dhavale, D. G., (1996). *Management Accounting Issues in Cellular Manufacturing and Focused Factory Systems*, Institute of Management Accountants' Foundation for Applied Research, Inc. Montvale, N.J.

Nyman, L. R. Ed. (1992). *Making Manufacturing Cells Work*, Society of Manufacturing Engineers, Dearborn, MI.

20

CLASSROOM TUTORIAL ON THE DESIGN OF A CELLULAR MANUFACTURING SYSTEM

Lifang Yan
MVE, Inc.
New Prague, Minnesota 56071

Shahrukh A. Irani
Department of Industrial, Welding and Systems Engineering
Ohio State University
Columbus, Ohio 43210

1. INTRODUCTION

Cellular Manufacturing (CM) is the formation of independent groups of functionally dissimilar machines and workstation types, located together on the floor, dedicated to the manufacture of a family of similar parts or products. In other words, it involves the breaking down of a job shop type production environment into units called "cells." Cells are primarily self-sufficient production units that produce a family of geometrically similar parts or a certain product line. Like everything else in the world, CM has both its advantages and disadvantages. The advantages of CM are reduced material handling, reduced tooling, reduced setup time, reduced work-in-process (WIP), improved human relationship, improved operator expertise, and simplified production scheduling and control.

Handbook of Cellular Manufacturing Systems, edited by Shahrukh A. Irani
ISBN 0-471-12139-8 © 1999 John Wiley & Sons, Inc.

However, the disadvantages of CM are

- *Lower Machine Utilization.* This is primarily caused by the fact that current machines are now dedicated to a smaller family of parts, and the fact that machine duplication is usually needed to smooth the production flows between cells.
- *Less Flexibility.* With only a certain combination of machines producing a certain group of products, load imbalances may occur when the part mix changes over a period of time. In addition, a part with some new design features will not likely be able to be produced in a current cell without adding extra machines. Machine failures will have more damaging effects on cell output since this will require sending parts or products to other cells, causing production delays.

As the discussion above has illustrated, there are advantages and disadvantages associated with CM. Most of these disadvantages, however can be avoided or minimized by using a proper technique for initial cell design and configuration. Furthermore, CM alone cannot maximize productivity. Rather, its physical organization along with effective capacity planning, production control, and inventory control can lead to maximum productivity (Greene et al., 1984).

2. CELL FORMATION

Numerous techniques are available to configure or reconfigure a CM system. However, Production Flow Analysis (PFA), proposed by Burbidge (1989), is probably the most well-known and most widely accepted method. Unlike the Group Technology (GT) approach, PFA does not utilize part classification and coding as input data. Instead it relies upon the parts' route sheets for information, a complete list of machines and specification of machine capabilities. As mentioned in Irani and Ramakrishnan (1995), when applied to a single factory, PFA consists of four stages, each stage achieving material flow reduction for a progressively reducing portion of the factory. In Factory Flow Analysis (FFA), if parts are observed to backtrack between any of the shops such as the machine shop, forge, foundry, press, or assembly shop, these flows are eliminated by a minor redeployment of equipment. Factory Flow Analysis may often be redundant for a factory that essentially consists of a single machine or fabrication shop. In Group Analysis (GA), the flow in each of the shops identified by FFA is analyzed. Group Analysis analyzes the flow interactions among the facilities and operation sequences of the parts to identify manufacturing cells. Machine loads are calculated for each part family to obtain the machine requirements for each cell. Each cell usually contains all the equipment necessary to satisfy the complete manufacturing requirements of its part family. Due to equipment sharing problems, some intercell material flow may exist. In Line Analysis (LA), a layout is designed for the machines assigned to each cell. It considers the operation frequencies and sequences of the parts and develops a cell configuration that allows efficient transport within the cell. The cell layout must also encourage multimachine tending by some operators. In Tooling Analysis (TA), the principles of GA and LA are integrated with the design and manufacturing attributes of the parts such as shape, size, material, tooling, fixturing attributes of the parts. Tooling Analysis helps to schedule the cell by identifying subfamilies of parts with similar operation sequences, tooling, and setups. It seeks to sequence parts on each machine tool and to schedule all the machines in the cell to exploit setup similarities or dependencies on the machines in order to achieve short throughput times for the parts.

Despite its popularity, PFA is a rather qualitative procedure requiring considerable manual sorting and resorting of the route sheets. Due to this aspect, experience and knowledge concerning part mix and machine capabilities are often required for the application of PFA in practice. This chapter presents a case study used for instructional purposes to demonstrate a computer implementation of PFA based on data obtained from the literature.

3. CASE STUDY

This case study assumes that FFA has been completed. The essential data used in this study are the routing sheets (Table 1) and machine availability list (Table 2). The goal of this study is to evaluate different alternatives for cell formation and compare and evaluate them to select the system configuration which minimizes intercell flows.

Table 1. Routing sheet

Part Number	Sequence of Machines	Total Batch Machining Time (minutes per operation)	Batch Size
1	1,4,8,9	96-36-36-72	2
2	1,4,7,4,8,7	36-120-20-120-24-20	3
3	1,2,4,7,8,9	96-48-36-120-36-72	1
4	1,4,7,9	96-36-120-72	3
5	1,6,10,7,9	96-72-200-120-72	2
6	6,10,7,8,9	36-120-60-24-36	1
7	6,4,8,9	72-36-48-48	2
8	3,5,2,6,4,8,9	144-120-48-72-36-48-48	1
9	3,5,6,4,8,9	144-120-72-36-48-48	1
10	4,7,4,8	120-20-120-24	2
11	6	72	3
12	11,7,12	192-150-80	1
13	11,12	192-60	1
14	11,7,10	288-180-360	3
15	1,7,11,10,11,12	15-70-54-45-54-30	1
16	1,7,11,10,11,12	15-70-54-45-54-30	2
17	11,7,12	192-150-80	1
18	6,7,10	108-180-360	3
19	12	60	2

Table 2. Machine availability list

Machine Type	Number Available
1	2
2	1
3	1
4	2
5	1
6	2
7	4
8	1
9	2
10	3
11	3
12	1
Production horizon	480 minutes
Machine availability	80%

4. GROUP ANALYSIS

4.1 Machine Grouping and Part Family Formation

Given the above data, GA can be implemented by first converting the routing sheets into an initial machine–part matrix, and then calculating similarity coefficients.

Creation of Initial Machine–Part Matrix   Using the original information on machines and parts, we can create an initial machine–part matrix (Table 3). The initial machine–part matrix is a chart with parts on one axis and machines on the other. This chart is created by looking through the routings to determine which parts visit which machines, and placing a 1 at the intersection of each machine–part pair that defines an operation. The matrix shows which parts have to be processed on which machine. The 1 in each position represents a machine that is being used by a part. For instance, part 1 needs to visit machines 1, 4, 8, and 9. At this stage, each machine is assumed to be unique and sequencing information for the parts is neglected in the matrix.

Calculation of Similarity Coefficients   From this initial machine–part matrix, we can calculate similarity coefficients for machines and parts. This calculation gives us an idea about the relationships between each pair of machines and each pair of parts. The formulas for these calculations follow:

Similarity Coefficient for a Pair of Machines

$$S_{KL}^{M} = N_{KL}^{P} / \left(N_{KK}^{P} + N_{LL}^{P} - N_{KL}^{P} \right)$$

where S_{KL}^{M} = similarity coefficient for machines K and L
N_{KL}^{P} = number of parts processed by both machines K and L

Table 3. Initial machine–part matrix

Parts	Machines											
	1	2	3	4	5	6	7	8	9	10	11	12
1	1	0	0	1	0	0	0	1	1	0	0	0
2	1	0	0	1	0	0	1	1	0	0	0	0
3	1	1	0	1	0	0	1	1	1	0	0	0
4	1	0	0	1	0	0	1	0	1	0	0	0
5	1	0	0	0	0	1	1	0	1	1	0	0
6	0	0	0	0	0	1	1	1	1	1	0	0
7	0	0	0	1	0	1	0	1	1	0	0	0
8	0	1	1	1	1	1	0	1	1	0	0	0
9	0	0	1	1	1	1	0	1	1	0	0	0
10	0	0	0	1	0	0	1	1	0	0	0	0
11	0	0	0	0	0	1	0	0	0	0	0	0
12	0	0	0	0	0	0	1	0	0	0	1	1
13	0	0	0	0	0	0	0	0	0	0	1	1
14	0	0	0	0	0	0	1	0	0	1	1	0
15	1	0	0	0	0	0	1	0	0	1	1	1
16	1	0	0	0	0	0	1	0	0	1	1	1
17	0	0	0	0	0	0	1	0	0	0	1	1
18	0	0	0	0	0	1	1	0	0	1	0	0
19	0	0	0	0	0	0	0	0	0	0	0	1

N_{KK}^{P} = number of parts processed by machine K
N_{LL}^{P} = number of parts processed by machine L

Similarity Coefficient for a Pair of Parts

$$S_{KL}^{P} = N_{KL}^{M}/\left(N_{KK}^{M} + N_{LL}^{M} - N_{KL}^{M}\right)$$

where S_{KL}^{P} = similarity coefficient for parts K and L
N_{KL}^{M} = number of machines used by both parts K and L
N_{KK}^{M} = number of machines used by part K
N_{LL}^{M} = number of machines used by part L

The results of the calculation are shown in Tables 4 and 5.

4.2 Generate Machine and Part Permutations

The calculation of similarity coefficients is intended to capture the similarity between two machines or two parts numerically in order to generate a Block Diagonal Form (BDF) for the final machine–part matrix (Table 6). Manual generation of permutations can easily be done in problems with a small number of machines and parts. Problems with a large number of machines and parts, however, are at best solved by using a computer program such as STORM (Irani and Ramakrishnan, 1995).

As explained in Irani and Ramakrishnan (1995), the similarity coefficients matrix for machines and parts can be treated as travel charts in the Facility Layout Module of

Table 4. Similarity coefficients for machines

S^M	1	2	3	4	5	6	7	8	9	10	11	12
1	0.125	0.000	0.364	0.000	0.077	0.462	0.250	0.364	0.300	0.182	0.182	
2		0.333	0.250	0.333	0.125	0.077	0.250	0.250	0.000	0.000	0.000	
3			0.250	1.000	0.286	0.000	0.250	0.250	0.000	0.000	0.000	
4				0.250	0.250	0.250	0.778	0.600	0.000	0.000	0.000	
5					0.286	0.000	0.250	0.250	0.000	0.000	0.000	
6						0.188	0.364	0.500	0.300	0.000	0.000	
7							0.250	0.250	0.500	0.385	0.286	
8								0.600	0.077	0.000	0.000	
9									0.167	0.000	0.000	
10										0.333	0.200	
11											0.714	
12												

STORM. Using the STORM program, we use the matrices of similarity coefficients as input and generate near-optimal permutations of machines and parts. This is because the Facility Layout Module of STORM uses the steepest descent pairwise interchange heuristic. Generally, the steepest descent pairwise interchange heuristic yields desirable permutations. The best way to avoid local minima is to repeatedly change the initial solution (this can be done easily in STORM) for multiple runs. Figures 1 and 2 are STORM outputs for machine and part permutations. Since part 11 uses only machine 6, we can put this part anywhere within the part group that uses machine 6. In these permutations, we switch the position of part 11 and get a better part permutation (Figure 3).

4.3 Formation of the Final Machine–Part Matrix

The final machine–part matrix can be created (Tables 6 and 7) by combining the machine and part permutations obtained using STORM. Notice how the entries in the initial matrix have been clustered along the diagonal to yield a BDF. However, it is not always possible to partition the BDF into diagonal blocks. These off-diagonal 1's are called exception elements (Arvindh and Irani, 1994). Each exception element is an indication that a machine is required by a part outside its part family and the presence of this machine implies intercell flows that will prevent the formation of independent cells.

The final machine–part matrix can be used to suggest groups of parts that should be placed together into part families and groups of machines that can be placed into machine cells. As we can see from both of these machine–part matrices, it is not easy to form cells, and there is no definite way to form cells. In order to determine a good configuration of cells, we need to consider other factors, such as available capacity, intercell flows, intracell flows, and the like. Since the second matrix in Figure 3 has a better parts permutation, we selected this matrix for further analysis.

Table 5. Similarity coefficients for parts

S^P	1	2	3	4	5	6	7	8	9	10	11	12	13	14	15	16	17	18	19
1																			
2	0.600																		
3	0.667	0.667																	
4	0.600	0.600	0.667																
5	0.286	0.286	0.375	0.500															
6	0.286	0.286	0.375	0.286	0.667														
7	0.600	0.333	0.429	0.333	0.286	0.500													
8	0.375	0.222	0.444	0.222	0.200	0.333	0.571												
9	0.429	0.250	0.333	0.250	0.222	0.375	0.667	0.857											
10	0.400	0.750	0.500	0.400	0.143	0.333	0.400	0.250	0.286										
11	0.000	0.000	0.000	0.000	0.200	0.200	0.250	0.143	0.167	0.000									
12	0.000	0.167	0.125	0.167	0.143	0.143	0.000	0.000	0.000	0.200	0.000								
13	0.000	0.000	0.000	0.000	0.000	0.000	0.000	0.000	0.000	0.000	0.000	0.667							
14	0.000	0.167	0.125	0.167	0.333	0.333	0.000	0.000	0.000	0.200	0.000	0.500	0.250						
15	0.125	0.286	0.222	0.286	0.429	0.250	0.000	0.000	0.000	0.143	0.000	0.600	0.400	0.600					
16	0.125	0.286	0.222	0.286	0.429	0.250	0.000	0.000	0.000	0.143	0.000	0.600	0.400	0.600	1.000				
17	0.000	0.167	0.125	0.167	0.143	0.143	0.000	0.000	0.000	0.200	0.000	0.600	0.667	0.500	0.600	0.600			
18	0.000	0.167	0.125	0.167	0.600	0.600	0.167	0.000	0.000	0.200	0.333	1.000	0.000	0.500	0.333	0.333	0.200		
19	0.000	0.000	0.000	0.000	0.000	0.000	0.000	0.000	0.125	0.000	0.000	0.333	0.500	0.000	0.200	0.200	0.333	0.000	

Table 6. Final matrix 1 from Machine Permutation (Fig. 1) and Part Permutation (Fig. 2)

Parts							Machines					
	3	5	2	4	8	9	6	1	7	10	11	12
9	1	1		1	1	1	1					
8	1	1	1	1	1	1	1					
11							1					
7				1	1	1	1					
1				1	1	1		1				
3			1	1	1	1		1	1			
10				1	1				1			
2				1	1			1	1			
4				1				1	1			
6					1	1	1		1	1		
5						1	1	1	1	1		
18							1		1	1		
14									1	1	1	
15								1	1	1	1	1
16								1	1	1	1	1
17										1	1	1
12										1	1	1
13											1	1
19												1

```
            Col  1      Col  2      Col  3      Col  4      Col  5
        +-----------+-----------+-----------+-----------+-----------+
Row 1   |    ME3 |       ME5 |       ME2 |       ME4 |       ME8 |
        +-----------+-----------+-----------+-----------+-----------+

            Col  6      Col  7      Col  8      Col  9      Col  10
        +-----------+-----------+-----------+-----------+-----------+
Row 1   |    ME9 |       ME6 |       ME1 |       ME7 |       ME10 |
        +-----------+-----------+-----------+-----------+-----------+

            Col  11    Col  12
        +-----------+-----------+
Row 1   |    ME11|       ME12|
        +-----------+-----------+

        Total objective function value = 64.578
```

Figure 1. STORM output for machine permutation.

4.4. Cell Formation and Capacity Calculations for Each Cell

To determine how many cells are needed, capacity analysis is required. Detailed explanations can be found in Irani and Ramakrishnam (1995).

Five-Cell Solution   When we consider forming cells, we first want to put parts that use the same machines in the same cell. From this point of view, we can have a preliminary five-cell proposal (Figure 4).

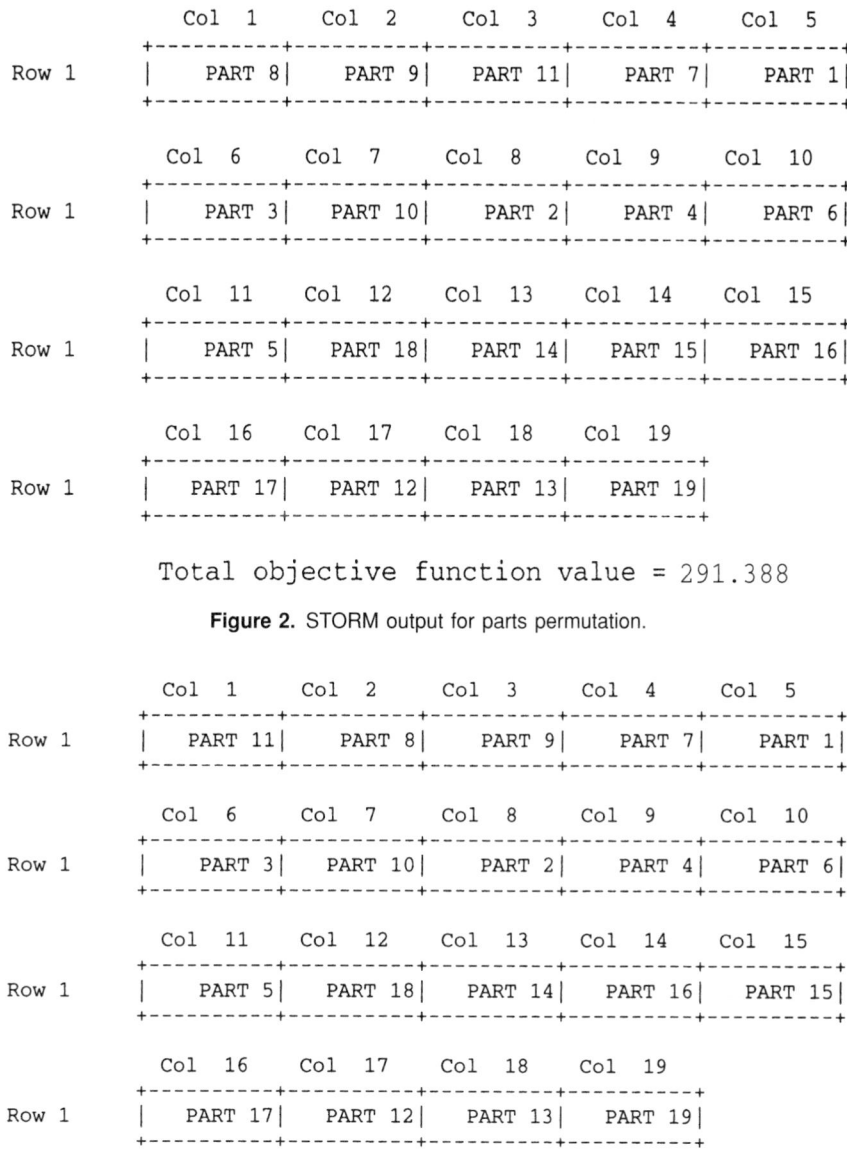

Total objective function value = 291.388

Figure 2. STORM output for parts permutation.

Total objective function value = 284.038

Figure 3. Revised STORM output for parts permutation.

Before checking machine capacity, we might, at a glance, notice that this cell partition gives us small part groups in each cell, but with more duplication of machines, that is, at least one more of machines 1, 2, 9, and 12 and two more of machine 8. This requires either buying six more machines or having more intercell flows. For example, if we decide to put the single machines 2 and 8 in cell 2, then part 8 needs

Table 7. Final matrix 2 with Machine Permutation (Fig. 1) and Part Permutation (Fig. 3)

Parts	Machines											
	3	5	2	4	8	9	6	1	7	10	11	12
11							1					
8	1	1	1	1	1	1	1					
9	1	1		1	1	1	1					
7				1	1	1	1					
1				1	1	1		1				
3		1		1	1	1		1	1			
10				1	1			1				
2				1	1			1	1			
4				1		1		1	1			
6					1	1	1		1	1		
5						1	1	1	1	1		
18						1			1	1		
14									1	1	1	
16								1	1	1	1	1
15								1	1	1	1	1
17									1		1	1
12									1		1	1
13											1	1
19												1

to be moved to cell 2 for machines 2 and 8; parts 7 and 9 have to be moved to cell 2 for machine 8; likewise, part 6 has to be moved from cell 3 to cell 2 for the operation on machine 8. A similar situation exists for machine 12. In this case, the five-cell solution might be feasible only when duplications of those machines is feasible. We get the same solution if we use more detailed machine capacity calculations.

The capacity calculations use the machine–part processing time matrix, which is obtained by replacing each 1 in the final machine–part matrix with its respective processing time obtained from the routing sheet. The number of each machine type needed for each cell shown in Table 1 is the ratio of total processing time required on that machine in that cell to available machining time for that machine. In this case study, available machining time for each machine is considered to be

Production horizon × Machine availability = 480 min/day × 80% = 384 min/day.

The result of capacity calculations for the five-cell solution is shown in Figure 5.

This calculation shows that in order to meet the needs of five cells, we will have to buy nine extra machines, or have significant intercell flows. Neither of these is acceptable, and we will not consider this cell formation option further.

Two-Cell Solution When we consider grouping parts without any duplication of machines, we obtain a different system configuration (Fig. 6).

In this two-cell solution, we have no machine duplication and only one intercell flow (part 6 has to be moved from cell 2 to cell 1 for machine 8. There will be 10

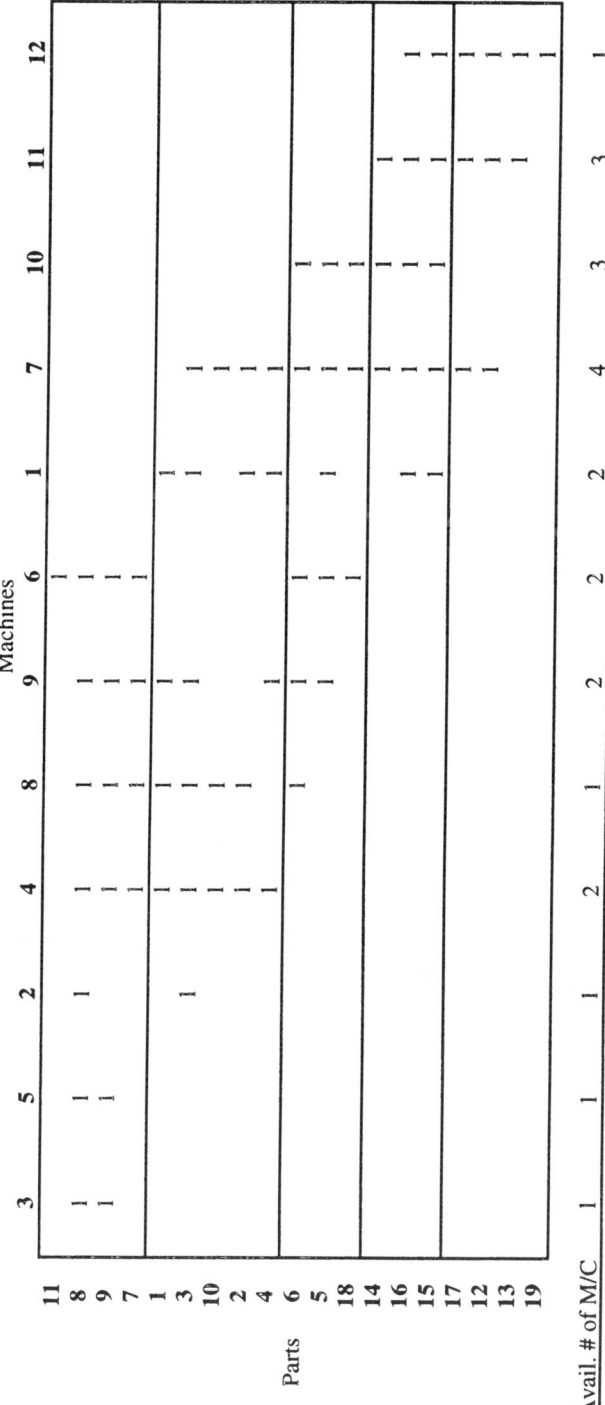

Figure 4. Five-cell solution.

Figure 5. Capacity calculations for five-cell solution.

Production Horizon (min)	480 (minutes)	Machine Availability	80%

Parts vs. Machines (minutes)

Parts	3	5	2	4	8	9	6	1	7	10	11	12
11	144			36	48	48	72					
8	144				48	48	72					
9		120		36	48		72					
7		120	48	36		48	72					
1			48	36	24	72		96	120			
3				36	24	72		96	120			
10				240	24			96	40			
2				240	24			36	20			
4				36	24	72						
6						108	108		180	120		
5					24		72	96	120	200		
18							36		60	360		
14								15	180	360	288	30
16								15	70	45	108	30
15									70	45	108	
17									150		192	80
12									150		192	80
13											192	60
19												60

# of M/C to:

	3	5	2	4	8	9	6	1	7	10	11	12
Cell 1	288	240	48	108	144	144	288	0	0	0	0	0
Cell 2	0	0	48	588	120	216	0	324	300	0	0	0
Cell 3	0	0	0	0	24	108	216	96	360	680	0	0
Cell 4	0	0	0	0	0	0	0	30	320	450	504	60
Cell 5	0	0	0	0	0	0	0	0	300	0	576	280

	3	5	2	4	8	9	6	1	7	10	11	12
Cell 1	0.750	0.625	0.125	0.281	0.375	0.375	0.750	0.000	0.000	0.000	0.000	0.000
Cell 2	0.000	0.000	0.125	1.531	0.313	0.563	0.000	0.844	0.781	0.000	0.000	0.000
Cell 3	0.000	0.000	0.000	0.000	0.063	0.281	0.563	0.250	0.938	1.771	0.000	0.000
Cell 4	0.000	0.000	0.000	0.000	0.000	0.000	0.000	0.078	0.833	1.172	1.313	0.156
Cell 5	0.000	0.000	0.000	0.000	0.000	0.000	0.000	0.000	0.781	0.000	1.500	0.729
Avail.# of M/C:	1	1	1	2	1	2	2	2	4	3	3	1

Required # of M/C

	3	5	2	4	8	9	6	1	7	10	11	12	Total
Cell 1	1	1	1	1	1	1	1	0	0	0	0	0	7
Cell 2	0	0	1	2	1	1	0	1	1	0	0	0	7
Cell 3	0	0	0	0	1	1	1	1	1	2	0	0	7
Cell 4	0	0	0	0	0	0	0	1	1	2	2	1	7
Cell 5	0	0	0	0	0	0	0	0	1	0	2	1	4
													32

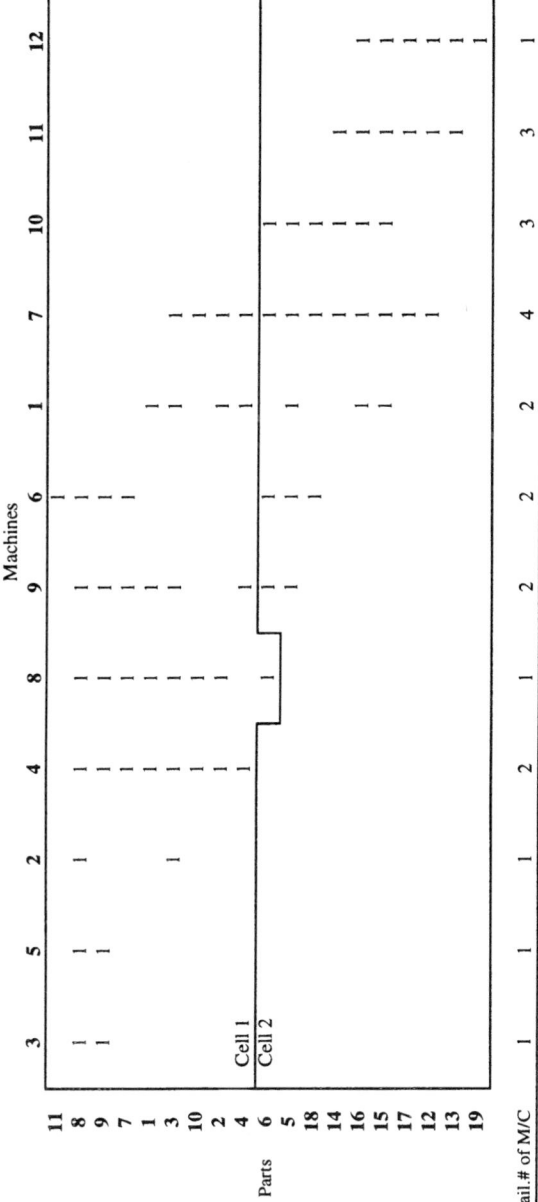

Figure 6. Two-cell solution.

Production Horizon (min)	480		Machine Availability	80%

Parts	3	5	2	4	8	9	6	1	7	10	11	12
11	144						72					
8	144	120	48	36	48	48	72					
9		120		36	48	48	72					
7				36	48	48	72					
1				36	36	72		96	120			
3			48	36	36	72		96	20			
10				240	24				40			
2				240	24			36	120			
4 — Cell 1				36	24	72		96				
6 — Cell 2						36	36	96	60	120		
5						72	72		120	200		
18							108		180	360	288	
14									180	360	108	
16									70	45	108	30
15								15	70	45	192	30
17								15	150		192	80
12									150		192	80
13												60
19												60

	3	5	2	4	8	9	6	1	7	10	11	12	Total
Cell 1	288	240	96	696	288	360	288	324	300	0	0	0	
Cell 2	0	0	0	0	0	108	216	126	980	1130	1080	340	
# of M/C to:													
Cell 1	0.750	0.625	0.250	1.813	0.750	0.938	0.750	0.844	0.781	0.000	0.000	0.000	
Cell 2	0.000	0.000	0.000	0.000	0.000	0.281	0.563	0.328	2.552	2.943	2.813	0.885	
Avail. # of M/C:	1	1	1	2	1	2	2	2	4	3	3	1	23
Required # of M/C													
Cell 1	1	1	1	2	1	1	1	1	1	0	0	0	10
Cell 2	0	0	0	0	0	1	1	1	3	3	3	1	13

Machines

Figure 7. Capacity calculations for two-cell solution.

machines with 9 parts in cell 1 and 13 machines with 10 parts in cell 2. The capacity calculation for this two-cell solution is shown in Figure 7.

This cell solution has no extra machine requirements with only one intercell flow of part 6 to cell 2 for the operation on machine 8. To avoid part 6 traveling to cell 1 for machine 8 and then traveling back to cell 2 for the last operation on machine 9, the following adjustment, shown in Figure 8, was made. This adjustment also includes the last operation of part 5 into cell 1 (on machine 9) to balance the utilization of machine type 9 among the cells (Figure 8).

Alternative Two-Cell Solution There are many different way of forming cells. For example, another two-cell solution could look like Figure 9 with capacity calculations shown in Figure 10.

This two-cell solution (Figures 9 and 10) will have a large group of machines and parts in cell 1, and will present difficulties during the layout and scheduling steps.

This two-cell solution has no intercell flows, but needs two extra machines. We can avoid duplication of machine 1 by shipping parts 15 and 16 to cell 1 for the operation on machine 1 as shown in Figure 11. The problem with this solution is that cell 1 is too big and will be difficult to schedule and control.

Three-Cell Solution When we consider the available numbers of each type of machine and the operation sequences of the parts, we have the following alternative three-cell solutions (shown in Figures 12–15).

This cell formation (Figures 12 & 13) will have most parts of cell 1 shipped to cell 2 for the latter portions of their operation sequences. Even though this eliminates bidirectional intercell flows, cell 2 is very big and will be difficult to schedule and control. Also, this three-cell solution will need one extra machine and create considerable intercell traffic.

Figure 15 shows a three-cell solution with more balanced machine utilization, moderate intercell flows, and requiring three extra machines. Part 3 will need to go to cell 2 for later operations after it finishes the machine operations in cell 1; parts 15 and 16 need to finish their first operation at cell 2 before they can start other operations in cell 3. This three-cell solution looks acceptable at this stage.

From the above analysis, we decided to choose the modified two-cell solution (Figure 8) and three-cell solution, proposal 2 (Figure 15) for the next step of PFA— Line Analysis. It should be cautioned that these solutions are subject to revision after subsequent analysis. Please see Arvindh and Irani (1994) for a detailed explanation on interactions among the different stages of analysis in PFA.

5. LINE ANALYSIS

In this step of PFA, we need to do intercell or intracell flow analyses to capture the interrelationships among machines in each cell and then design the layout for each cell.

To quantitatively represent the interrelationships among machines within a cell, the routing sheets are converted into a from–to chart (shown as Fig. 16). During the conversion, it is decided that all parts that go into a cell will start from one "input"

	Production Horizon (min)	480					Machine Availability		80%			

Machines

Parts	3	5	2	4	8	9	6	1	7	10	11	12
11	144	120	48	36	48	48	72					
8	144	120		36	48	48	72					
9				36	48	48	72					
7				36	36	72	72					
1			48	36	36	72		96	120			
3				240	24			96	20			
10				240	24				40			
2								36	120			
4 *(Cell 1)*				36	24	72		96	180			
6 *(Cell 2)*						36	36	96	60	120		
5						72	72		120	200		
18							108		180	360	288	
14									180	360	108	
16								15	70	45	108	30
15								15	70	45		30
17									150		192	80
12									150		192	80
13											192	60
19												60

# of M/C to:	3	5	2	4	8	9	6	1	7	10	11	12	Total
Cell 1	288	240	96	696	288	468	288	324	480	0	0	0	
Cell 2	0	0	0	0	0	0	216	126	800	1130	1080	340	
Cell 1	0.750	0.625	0.250	1.813	0.750	1.219	0.750	0.844	1.250	0.000	0.000	0.000	
Cell 2	0.000	0.000	0.000	0.000	0.000	0.000	0.563	0.328	2.083	2.943	2.813	0.885	
Avail. # of M/C:	1	1	1	2	1	2	2	2	4	3	3	1	23
		1					1	1	3	3	3	1	

Required # of M/C	3	5	2	4	8	9	6	1	7	10	11	12	Total
Cell 1	1	1	1	2	1	2	1	1	2	0	0	0	12
Cell 2	0	0	0	0	0	0	1	1	3	3	3	1	12

Figure 8. Modified two-cell solution.

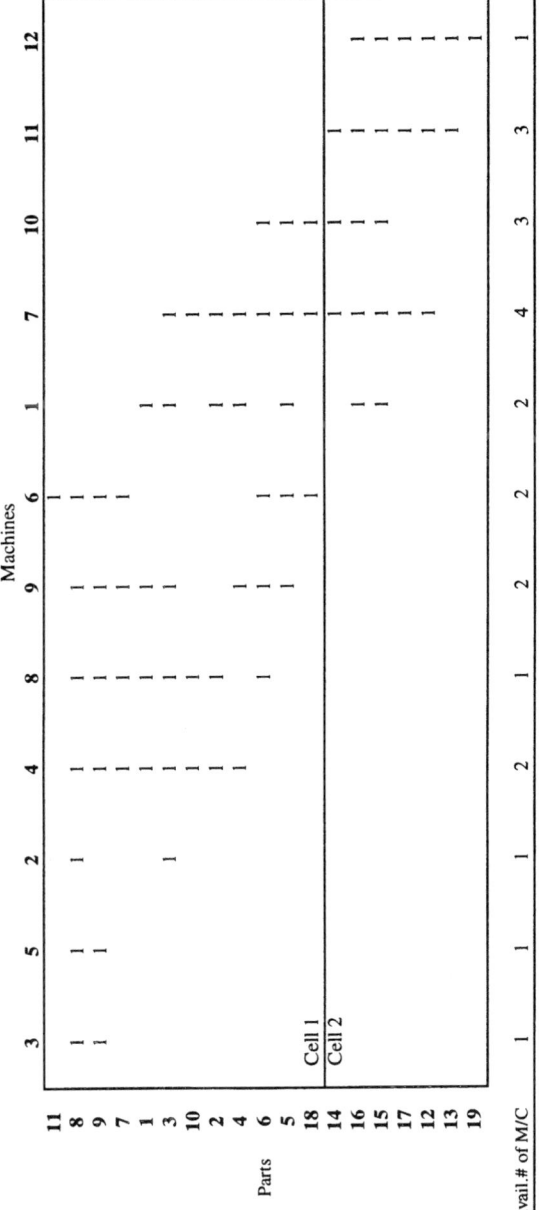

Figure 9. Alternative two-cell solution.

Figure 10. Capacity calculation for alternative two-cell solution.

Production Horizon (min)	480		Machine Availability	80%

						Machines							
Parts	**3**	**5**	**2**	**4**	**8**	**9**	**6**	**1**	**7**	**10**	**11**	**12**
11	144	120										
9	144	120	48									
8				36	48	48	72					
7				36	48	48	72					
1				36	48	48	72	96	120			
3			48	36	36	72	72	96	20			
10				240	36	36			40			
2				240	24			36	120			
4				36	24			96	60			
6				36	24	72	36		120	120		
5						72	72		180	200		
18						72	108	96		360		
Cell 1												
Cell 2												
14									180	360	288	30
16								15	70	45	108	30
15								15	70	45	108	80
17									150		192	80
12									150		192	60
13											192	60
19												

# of M/C to:	**3**	**5**	**2**	**4**	**8**	**9**	**6**	**1**	**7**	**10**	**11**	**12**	
Cell 1	288	240	96	696	288	468	504	420	660	680	0	0	
Cell 2	0	0	0	0	0	0	0	30	620	450	1080	340	
# of M/C to:													
Cell 1	0.750	0.625	0.250	1.813	0.750	1.219	1.313	1.094	1.719	1.771	0.000	0.000	
Cell 2	0.000	0.000	0.000	0.000	0.000	0.000	0.000	0.078	1.615	1.172	2.813	0.885	
Avail. # of M/C:	1	1	1	2	1	2	2	2	4	3	3	1	23

Required # of M/C	**3**	**5**	**2**	**4**	**8**	**9**	**6**	**1**	**7**	**10**	**11**	**12**	**Total**
Cell 1	1	1	1	2	1	2	2	2	2	2	0	0	16
Cell 2	0	0	0	0	0	0	0	1	2	2	3	1	9

Production Horizon (min) 480   **Machine Availability** 80%

Parts	3	5	2	4	8	9	6	1	7	10	11	12
11	144	120					72					
9	144	120					72					
8			48	36	48	48	72					
7				36	48	48	72	96				
1				36	48	48		96	120			
3			48	36	36	72		36	20			
10				36	36	72		96	40			
2				240	24	72			120			
4				240	24	36			60			
6				36	24	72	36	96	120			
5							72		180	120		
18							108		180	200		
14									70	360	288	30
16								15	70	360	108	30
15								15	150	45	108	80
17									150	45	192	80
12											192	
13											192	60
19												60

Cell 1 (parts 11, 9, 8, 7, 1, 3, 10, 2, 4, 6, 5, 18) — Cell 2 (parts 14, 16, 15, 17, 12, 13, 19)

	3	5	2	4	8	9	6	1	7	10	11	12
Cell 1	288	240	96	696	288	468	504	450	660	680	0	0
Cell 2	0	0	0	0	0	0	0	0	620	450	1080	340
# of M/C to: Cell 1	0.750	0.625	0.250	1.813	0.750	1.219	1.313	1.172	1.719	1.771	0.000	0.000
# of M/C to: Cell 2	0.000	0.000	0.000	0.000	0.000	0.000	0.000	0.000	1.615	1.172	2.813	0.885
Avail. # of M/C:	1	1	1	2	1	2	2	2	4	3	3	1

Avail. # of M/C Total: 23

Required # of M/C	3	5	2	4	8	9	6	1	7	10	11	12	Total
Cell 1	1	1	1	2	1	2	2	2	2	2	0	0	16
Cell 2	0	0	0	0	0	0	0	0	2	2	3	1	8

Figure 11. Revised alternative two-cell solution.

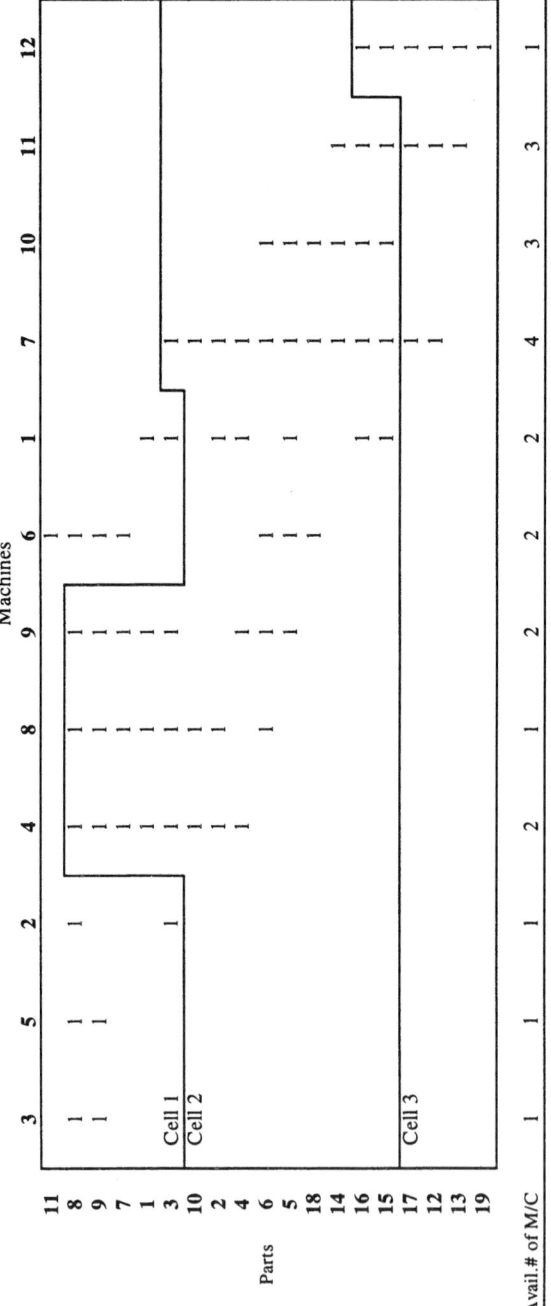

Figure 12. Three-cell solution—proposal 1.

Production Horizon (min)		Machine Availability		80%
480				

Machines

Parts	3	5	2	4	8	9	6	1	7	10	11	12
11												
8	144	120	48	36	48	48	72	96				
9	144	120		36	48	48	72	96				
7				36	48	48	72					
1	Cell 1			36	36	72	72	36	120			
3	Cell 2			36	36	72		96	20			
10			48	240	24	72			40	120		
2				240	24	36			120	200		
4				36	24	72				200		
6							36	96	60	360		
5							72		120	360		
18							108		180		288	
14									180			
16								15	70	45	108	30
15	Cell 2							15	70	45	108	30
17	Cell 3								150		192	80
12									150		192	80
13											192	60
19												60

# of M/C to:

	3	5	2	4	8	9	6	1	7	10	11	12
Cell 1	288	240	96	0	0	0	288	192	0	0	0	0
Cell 2	0	0	0	696	288	468	216	258	980	1130	504	0
Cell 3	0	0	0	0	0	0	0	0	300	0	576	340
Cell 1	0.750	0.625	0.250	0.000	0.000	0.000	0.750	0.500	0.000	0.000	0.000	0.000
Cell 2	0.000	0.000	0.000	1.813	0.750	1.219	0.563	0.672	2.552	2.943	1.313	0.000
Cell 3	0.000	0.000	0.000	0.000	0.000	0.000	0.000	0.000	0.781	0.000	1.500	0.885

	3	5	2	4	8	9	6	1	7	10	11	12	Total
Avail. M/C:	1	1	1	2	1	2	2	2	4	3	3	1	23

Required # of M/C

	3	5	2	4	8	9	6	1	7	10	11	12	Total
Cell 1	1	1	1	0	0	0	1	1	0	0	0	0	5
Cell 2	0	0	0	2	1	2	1	1	3	3	2	0	15
Cell 3	0	0	0	0	0	0	0	0	1	0	2	1	4

Figure 13. Capacity calculations for three-cell solution—proposal 1.

633

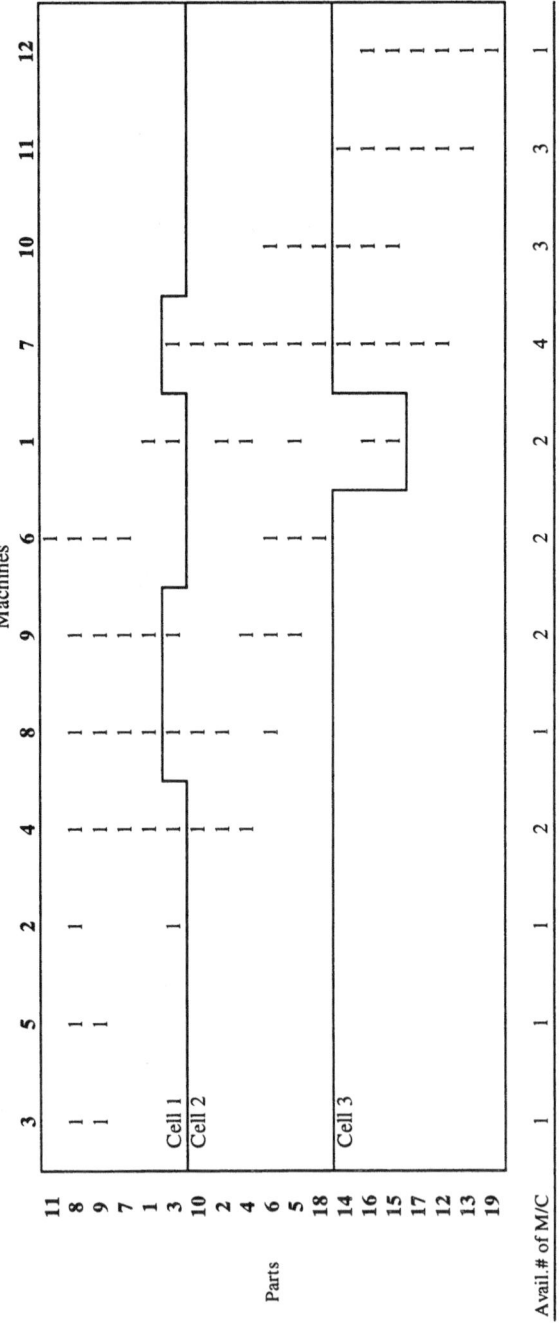

Figure 14. Three-cell solution—proposal 2.

Production Horizon (min)	480					Machine Availability	80%				

Figure 15. Capacity calculations for three-cell solution—proposal 2.

						Machine						
Parts	3	5	2	4	8	9	6	1	7	10	11	12
9	144	120		36	48	48	72					
8	144	120	48	36	48	48	72					
11							72	96				
7				36	48	48	72	96				
1				36	36	72		96	120			
3 (Cell 1)				36	36	72		36	20			
10 (Cell 2)				240	24				40	120		
2			48	240	24				120	200		
4				36					60	360		
6					24	72	36	96	120			
5						36	36		180			
18 (Cell 2)						72	72					
14 (Cell 3)							108		180	360	288	30
15								15	70	45	108	30
16								15	70	45	108	80
17									150		192	80
12											192	60
13											192	60
19									150			
# of M/C to:												
Cell 1	288	240	96	180	180	216	288	192	0	0	0	0
Cell 2	0	0	0	516	108	252	216	258	660	680	0	0
Cell 3	0	0	0	0	0	0	0	0	620	450	1080	340
Cell 1	0.750	0.625	0.250	0.469	0.469	0.563	0.750	0.500	0.000	0.000	0.000	0.000
Cell 2	0.000	0.000	0.000	1.344	0.281	0.656	0.563	0.672	1.719	1.771	0.000	0.000
Cell 3	0.000	0.000	0.000	0.000	0.000	0.000	0.000	0.000	1.615	1.172	2.813	0.885
Avail. M/C:	1	1	1	2	1	2	2	2	4	3	3	1	23
Required # of M/C													Total
Cell 1	1	1	1	1	1	1	1	1	0	0	0	0	8
Cell 2	0	0	0	2	1	1	1	1	2	2	0	0	10
Cell 3	0	0	0	0	0	0	0	0	2	2	3	1	8

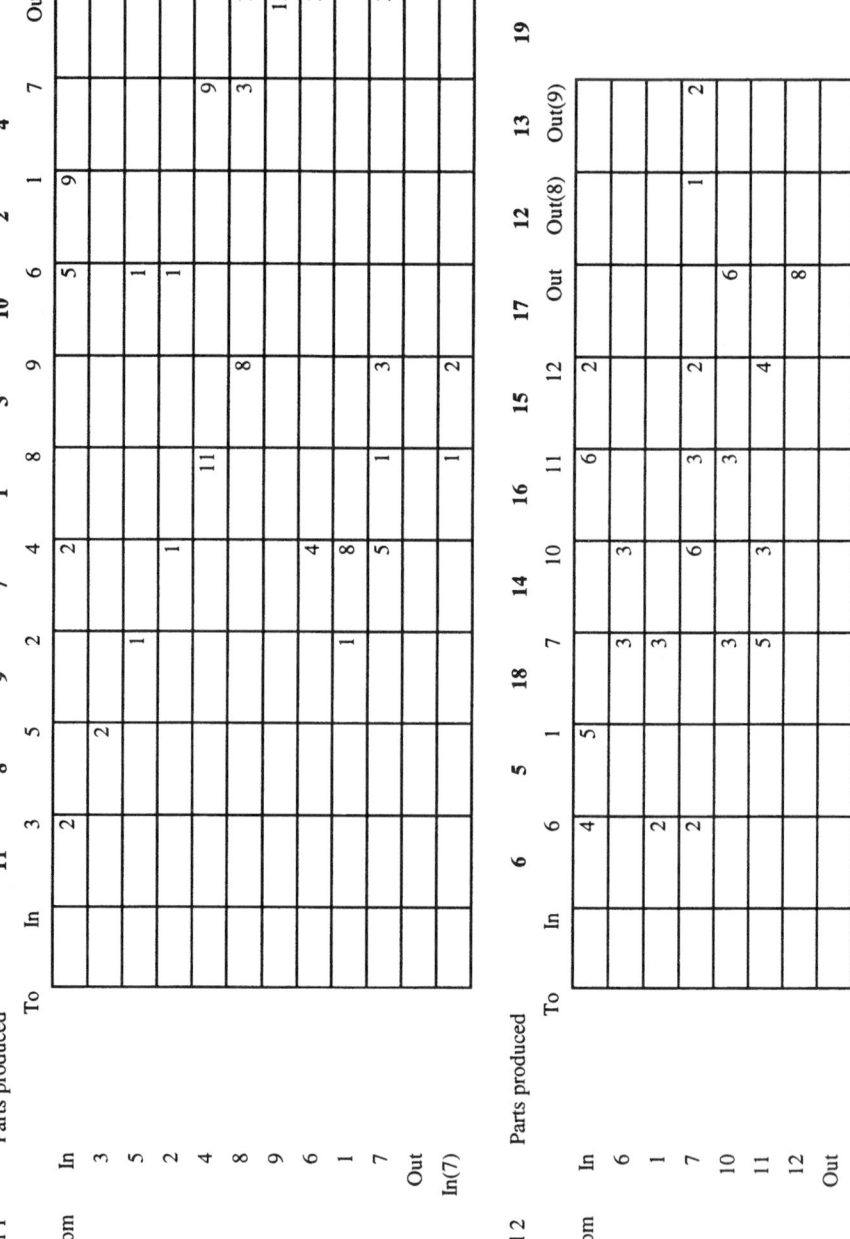

Figure 16. From–to chart for modified two-cell solution.

station and all the trips out of the cell will be from one "output" station. For now, the duplications of machines within the cells are not considered. Another important assumption that must be made here is that any machine can be placed adjacent to any other machines without any difficulties. This may not be the case in real life. Shapes and sizes of machines are also neglected in this analysis and the machines are assumed to be placed in any orientation.

5.1. Line Analysis for Modified Two-Cell Solution (Shown in Figure 8)

In further analysis, the depiction of the parts flow pattern between machines can be used to generate the initial layout. The Facility Layout Module of STORM can be used for this step. STORM can be used to generate a linear layout for the machines in each cell (Figure 17), which can then be modified into the desired layout shape. In generating the linear layout, it is necessary to fix the input and output station of each cell at both ends of the linear layout. As pointed out in the previous analysis,

```
Cell 1:      LAYOUT

          Col  1     Col  2     Col  3     Col  4     Col  5
          +----------+----------+----------+----------+----------+
Row 1     |       IN|       M1|       M3|       M5|       M6|
          +----------+----------+----------+----------+----------+

          Col  6     Col  7     Col  8     Col  9     Col  10
          +----------+----------+----------+----------+----------+
Row 1     |       M2|       M7|       M4|       M8|       M9|
          +----------+----------+----------+----------+----------+

          Col  11
          +----------+
Row 1     |      OUT|
          +----------+

       Total objective function value = 164


Cell 2:      LAYOUT

          Col  1     Col  2     Col  3     Col  4     Col  5
          +----------+----------+----------+----------+----------+
Row 1     |       IN|       M1|       M6|      M11|       M7|
          +----------+----------+----------+----------+----------+

          Col  6     Col  7     Col  8
          +----------+----------+----------+
Row 1     |      M10|      M12|      OUT|
          +----------+----------+----------+

       Total objective function value = 147
```

Figure 17. Layout generated using STORM.

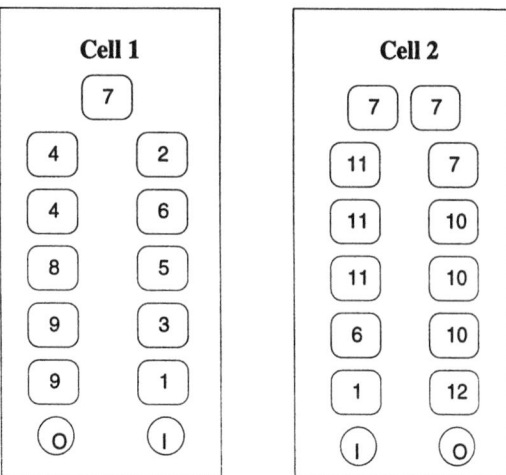

Figure 18. Tentative layout.

the results obtained by STORM may not be optimal. However, since a large number of layouts was generated, the results in Figure 18 represent a viable solution.

From this tentative layout, we can see that some parts backtrack in this cell, especially in cell 1 (Figure 19). To avoid these, the following adjustments were made:

1. Switch the position of machines 2 and 6 in cell 1 to avoid backtracking of part 8.
2. Switch machine 4 with 7 in cell 1 to suit the flows of parts 10 and 2.
3. Insert a machine 11 between machines 10 and 12, and a machine 7 between machines 6 and 11 to suit parts 15 and 16 in cell 2.

These adjustments give us the final two-cell layout (Figure 20).

There is an improvement in the part flows (Figures 21 and 22) even though backtracking is not eliminated completely.

5.2 Line Analysis for the Three-Cell Solution: Proposal 2

The analysis will follow the same procedures we used for the two-cell solution earlier. We first generate the from–to chart (Figure 23) from the routing sheets, run STORM, obtain the STORM output (Figure 24) to generate the tentative layout (Figure 25) for each cell.

From this tentative layout, the flow of each part is checked. Some parts with backtrack flows are shown in Figures 26–28. The following adjustments were made to get a final three-cell layout as in Figure 29 and improved part flows shown in Figures 30–32.

1. Part 3 will have operations on machines 1, 2, and 4 in cell 1 and operations on machines 7, 8, and 9 in cell 2.

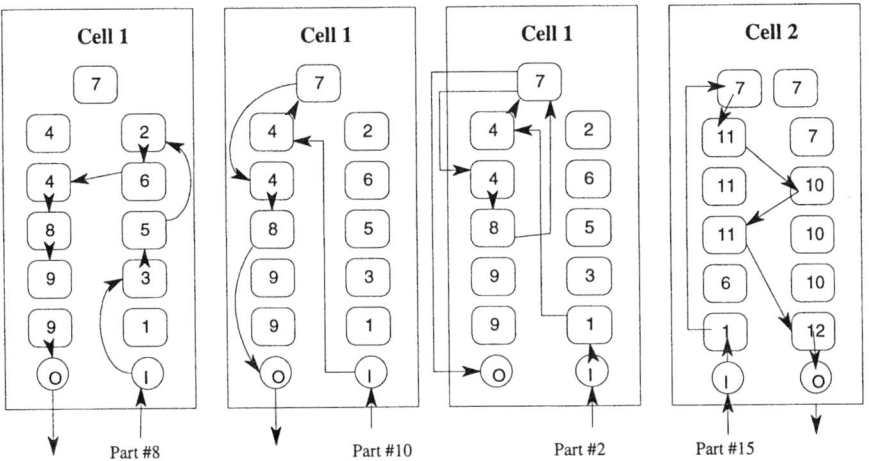

Figure 19. Flows of parts 8, 10, and 2 in cell 1 and part 15 in cell 2.

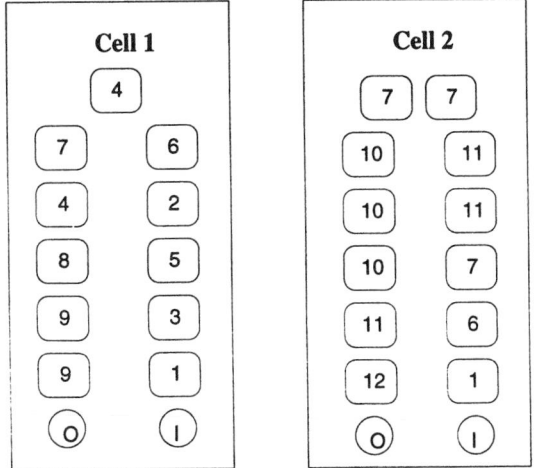

Figure 20. Final layout for two-cell solution.

2. Switch one machine 4 with machine 7 to suit parts 2 and 10, and insert one machine 7 between machines 9 and 8 to suit part 2 in cell 2.

3. Insert a machine 10 between machines 4 and 7 in cell 2 to suit parts 5 and 6.

4. Switch one machine 7 with 11 in cell 3 for part 14.

5. Insert one machine 10 between machines 7 and 11 for parts 15 and 16.

6. Because more parts have operations on machine 11 prior to the operation on machine 7, two copies of machine 11 were put in between machine 7.

7. Since parts 15 and 16 need to flow from cell 2 to cell 3, output point of cell 2 should be close to the input point of cell 3.

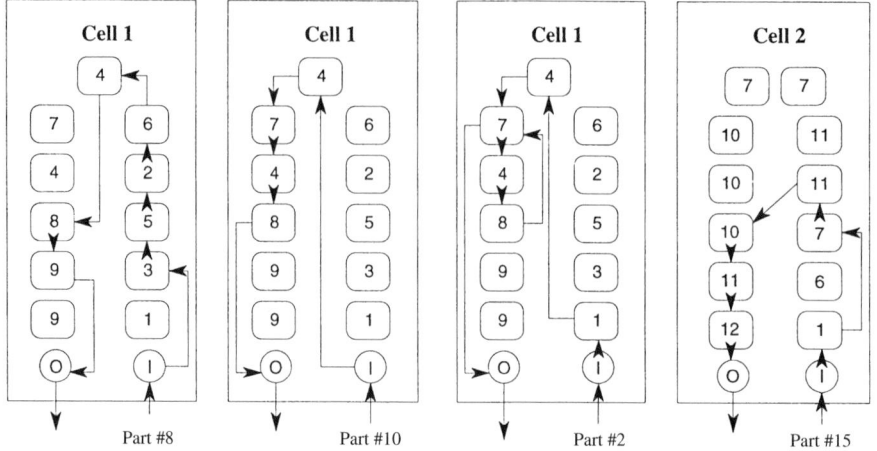

Figure 21. Part flows in revised layout.

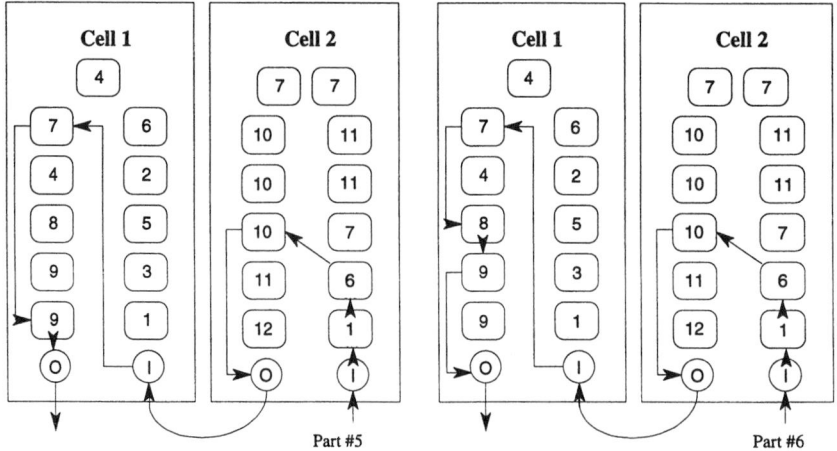

Figure 22. Flow of parts 5 and 6.

Based on the above calculations and analysis for both two-cell and three-cell solutions, it can be observed that each solution has its own advantages. The two-cell solution has less intercell flows with less machine requirements; the three-cell solution has less machines in each cell, which makes each cell easier to schedule and control.

6. SHOP LAYOUT ANALYSIS

Once LA has been completed, we need to analyze the overall shop layout. Ideally, one would like to develop the layout and then construct the building around the layout. Unfortunately, there are often constraints of building configuration and amount of

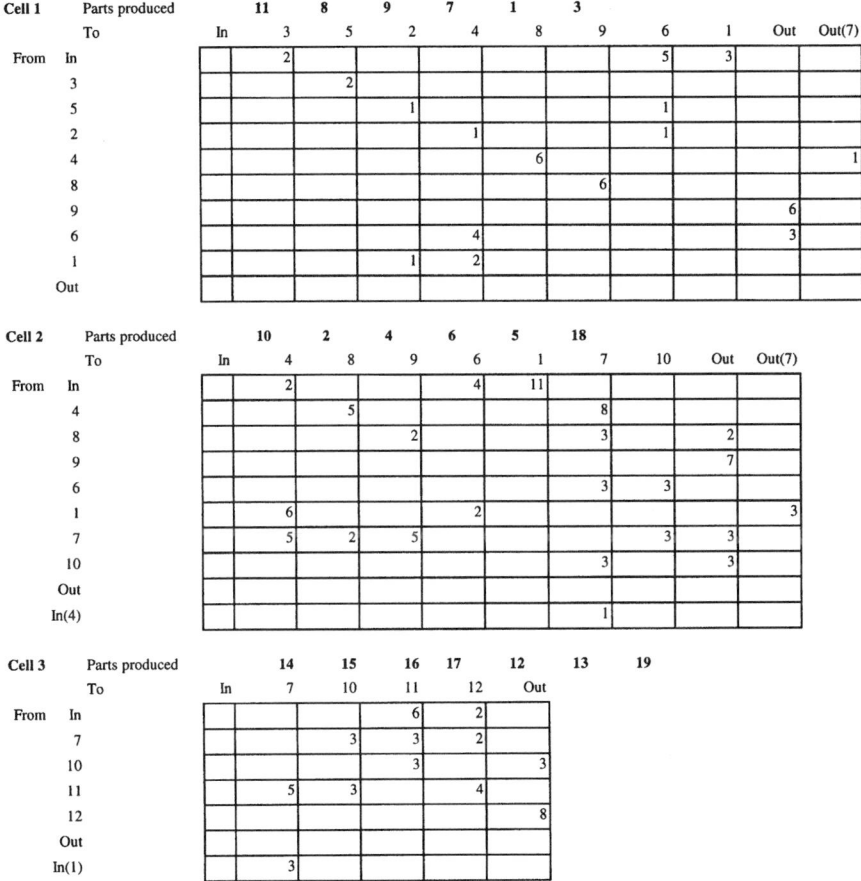

Figure 23. From–to chart for three-cell solution.

available space for raw material storage, in-process inventory storage, finished goods storage, material handling equipment storage, types of material handling (forklift truck, cranes, etc.), toolrooms and tool cribs, maintenance space, additional support services such as heat treatment, painting, warehouse, lavatories, packaging, and quality control. These constraints will affect one's approach to the design of a shop layout. For this case study, however, it is assumed that there are no constraints and the machines or the cells can be placed anywhere in any orientation. This analysis does not cover space requirements analysis.

As discussed in Irani and Ramakrishnan (1995), STORM can again be used for shop layout analysis. However, due to the simplicity of this particular case study, computer generation of the layout for a two-cell shop was unnecessary.

With no shape constraints, cells could have any shape subject to space availability. For instance, the two-cell solution can be organized into a layout shown in Figure 33. Special arrangements in this layout are bottleneck machine 7 in two cells are put close to each other, so that machine 7 in cell 2 can be reached in case machine 7 in cell 1

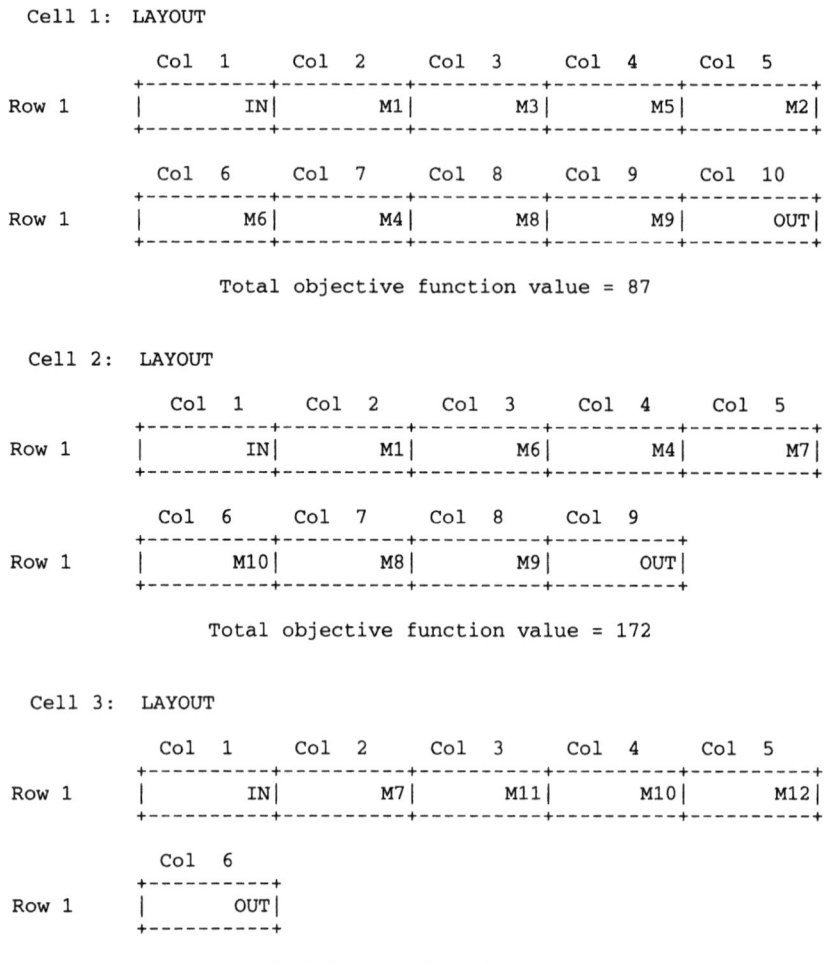

Figure 24. STORM output.

breaks down; put output point of cell 2 close to the input point of cell 1 for intercell flows; in cell 2, put machine 6 close to machine 10, and machine 10 close to output point to benefit flows of parts 5 and 6; in cell 1, put machine 7 close to input point for the intercell flow of parts 5 and 6. The shape of the individual cells in this layout is "semi-U" and "S." Generally, a U-shape cell occupies less space and promotes better communications inside the cell, and is therefore recommended for use in practice.

7. SCHEDULING CONSIDERATIONS

At this point, PFA is traditionally considered to be done and the overall shop layout has been accepted as a good enough layout. Based on these results, jobs can be scheduled

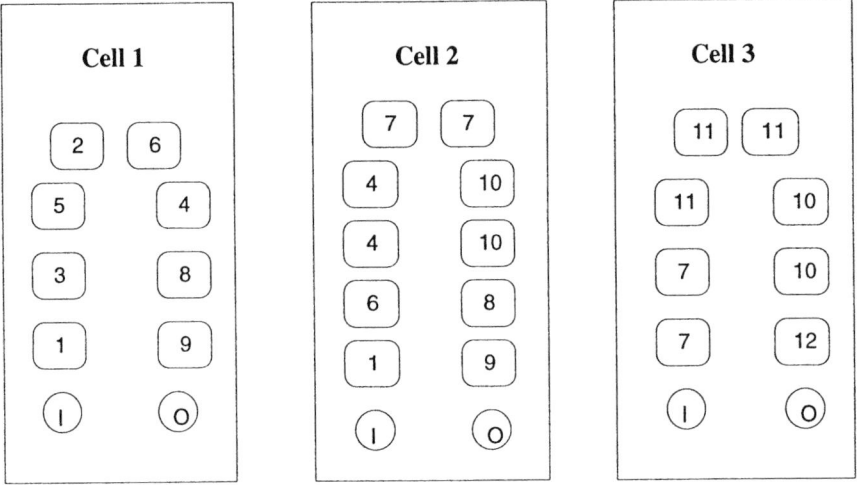

Figure 25. Tentative layout based on STORM output.

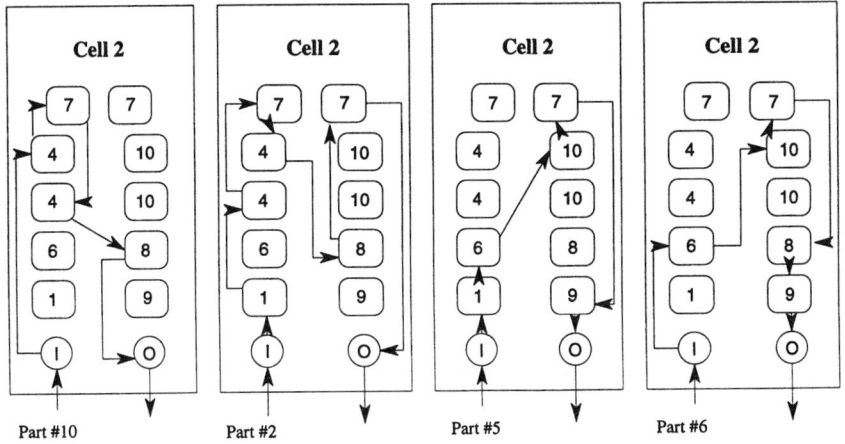

Figure 26. Part flows in cell 2 with tentative layout for three-cell solution.

within and between the cells. The cell scheduling problem is very complex and it is hard to reach an optimal schedule. In the case of dynamic arrival patterns, stochastic processing rates, routing flexibility, and sequencing flexibility, the scheduling problem is impossible to solve optimally. One approach to solving the scheduling problem is the use of heuristics and dispatching rules. The procedure used in the case study is to generate a schedule by scheduling one operation at a time. At any point in time each job is either finished or has one operation that can be scheduled. The procedure will simply move forward in time whenever one or more operations are ready to be

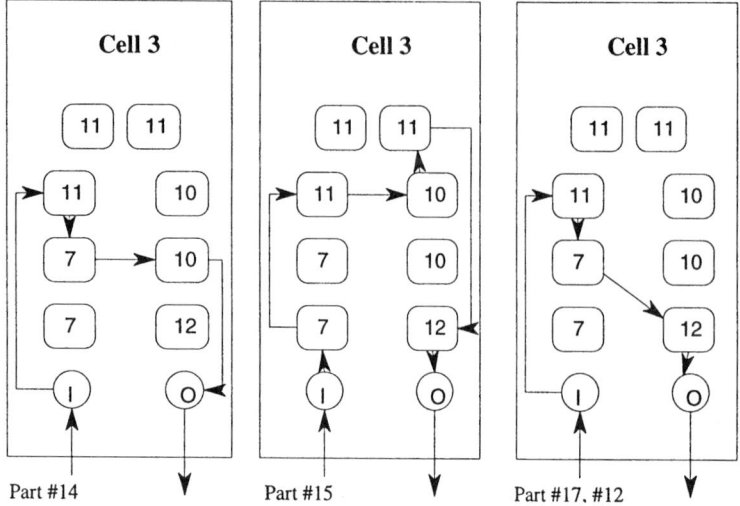

Figure 27. Part flows in cell 3 with tentative layout for three-cell solution.

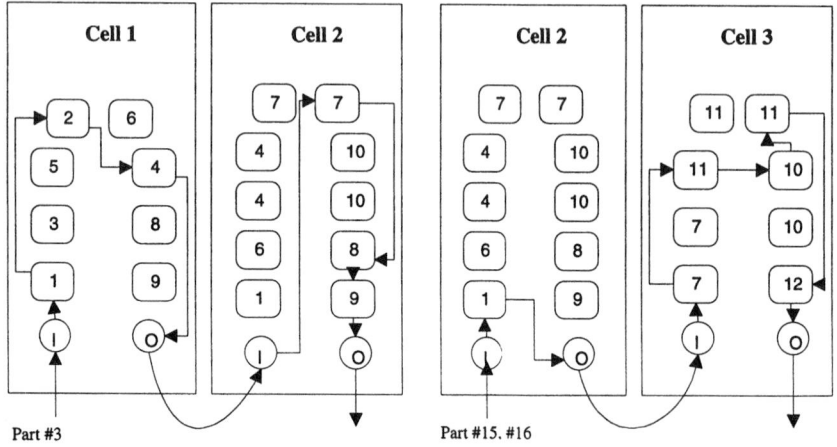

Figure 28. Intercell part flows in tentative layout for three-cell solution.

scheduled and their desired machines are available until all operations are assigned. *Since it is very tedious and time consuming to manually solve the scheduling problem, MS-Project was utilized.*

The previous analysis gave two scheduling problems: one for the two-cell solution and the other for the three-cell solution. *Each of these two solutions was scheduled using MS-Project and then these schedules were compared to see the differences between these two solutions.*

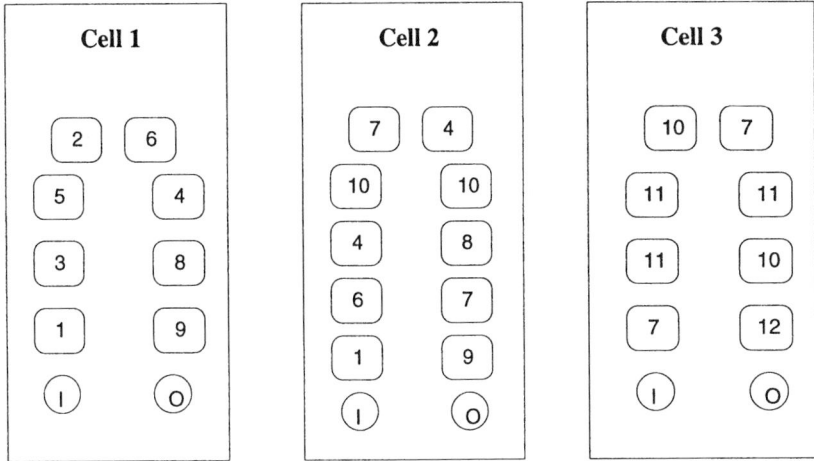

Figure 29. Final three-cell layout.

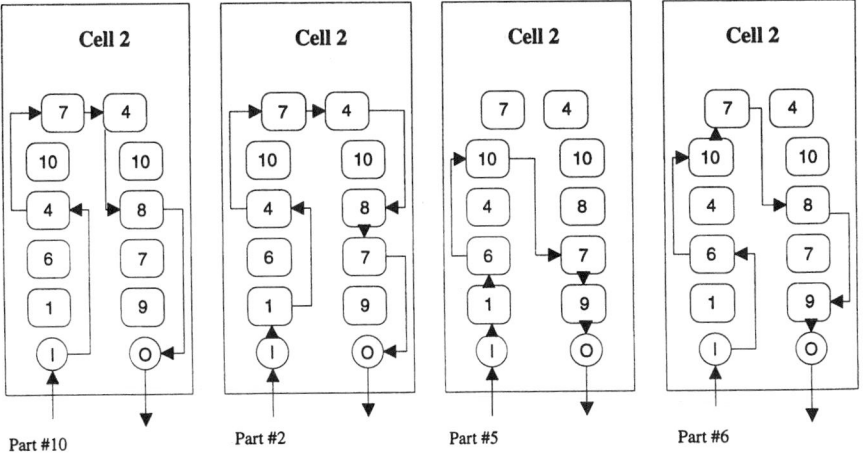

Figure 30. Part flows in cell 2 with final layout for three-cell solution.

7.1. Three-Cell Solution

This three-cell solution has a cell partition shown as in Figure 15, and machine allocations in each cell are shown in Table 8. Parts produced in each cell are as follows:

Cell 1 produces parts 1, 7, 8, 9, 11, and partial 3.

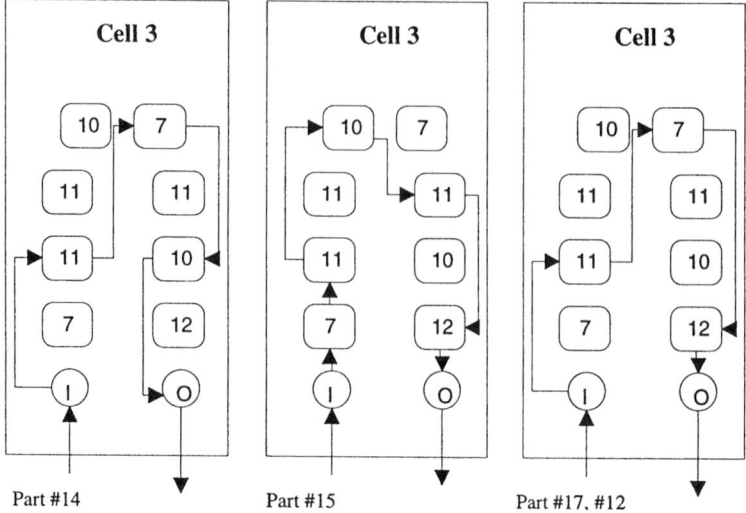

Figure 31. Part flows in cell 3 with final layout for three-cell solution.

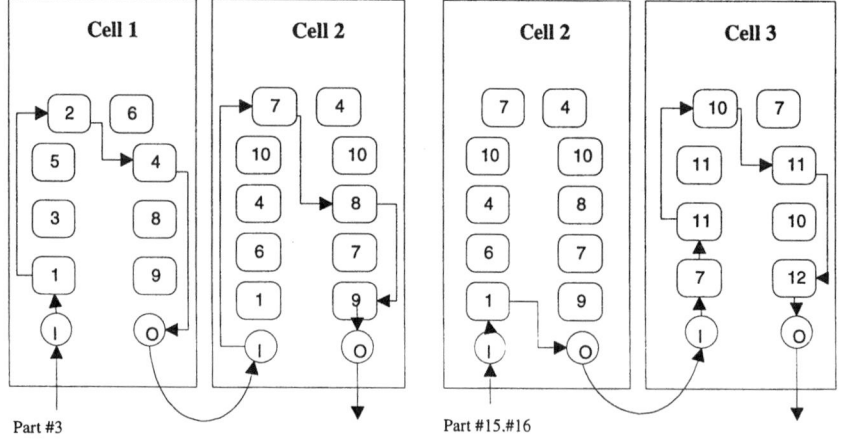

Figure 32. Intercell flows in final layout for three-cell solution.

Cell 2 produces parts 2, 4, 5, 6, 10, 18, partial 3, 15, and 16.

Cell 3 produces parts 12, 13, 14, 17, 19, partial 15, and 16.

These data were input into MS-Project for the scheduling of production for each cell. Like other computer programs, MS-Project will not generate an optimal solution automatically with only machine and part information. The user will need to check

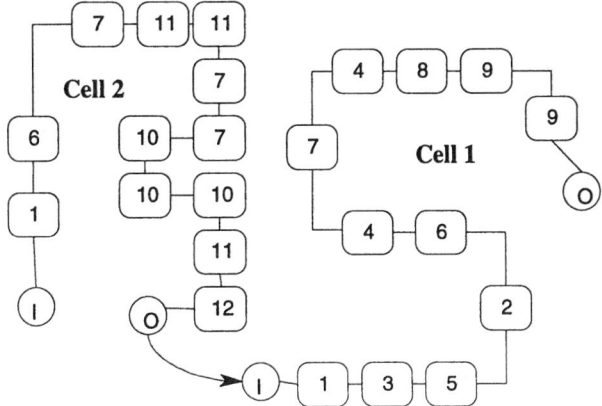

Figure 33. One alternative two-cell layout.

Table 8. Machine allocation in each cell

Required Machine	3	5	2	4	8	9	6	1	7	10	11	12	Total
Cell 1	1	1	1	1	1	1	1	1	0	0	0	0	8
Cell 2	0	0	0	2	1	1	1	1	2	2	0	0	10
Cell 3	0	0	0	0	0	0	0	0	2	2	3	1	8

the initial schedule generated by MS-Project and make necessary adjustments based on machine usage.

Initial schedule is shown as Appendix 1 (summarized in Excel spreadsheet). Here are some notations used:

Time unit:	Hour	
M_k:	Machine k	(M1 represents machine 1)
P_{i-j}:	Part i, Operation j	(P1-1 represents part 1 in operation 1)

The processes are described as follows:

1. Assume all parts start processing at same time. Assign machine to each part based on parts' routing and process time. This is the initial loading of each machine.
2. The initial loading will also result in initial schedule for parts on each machine. Check this initial schedule to see if there is any time conflicts on any machine.
3. If there is, rearrange them according to parts' routing and machine availability. At this time, we apply rules such as shortest processing time and longest remaining process time to determine the sequence of parts on an individual machine.

4. For parts that have intercell flows, once they finished processing in the previous cell, they will be transferred to the next cell and will be processed as soon as the required machine in the next cell is available. For instance, when part 3 finished processing in cell 1, it needs to be processed at M7 in cell 2. Since M7 in cell 2 is busy at this very moment, part 3 will have to wait until M7 is available. Once M7 is available, part 3 has the first priority for using M7.

5. For a detailed description of the adjustments on each machine, please refer to the initial schedule arranged by machines:

Cell 1

M1:   Because P3 needs to be processed in another cell after it is finished machining in this cell, process P3 before P1.

M2:   Works OK, no adjustment is needed.

M3:   P8 has MWKR (most work remaining), process P8 first.

M4:   On machine 4, only P1, P3, and P7 have conflicts. Since P1-2 arrives a little late than P7-2, there is no reason to hold the machine and wait to process P1-2 first. In this case, we will process P7-2 first.

M5:   Since we have made adjustment of P8-1 and P9-1 on M3, no adjustment is needed here.

M6:   P-7-1 and P11 have conflicts. Even though P11 has shorter process time, moving P7-1 behind P11 will affect the next three steps for P7. To the contrary, moving P11 behind P7-1 will only affect 1 operation, which has no effect on other machines. We chose to process P7-1 first.

M8:   With the adjustment made on M4, no further adjustment is needed.

M9:   Schedule works fine on this machine due to adjustment on previous machine.

Cell 2

M1:   P15 and P16 need to go to next cell; they should be processed first. Because P15-1 and P16-1 have same process time and remaining process time, we choose arbitrarily P15-1 to be the first one, and then P16-1. Next one should be P2-1. P4-1, P5-1 have same process time, but P5-1 has MWKR. We chose P5-1 to be the next, P4-1 to be the last.

M4:   There are two M4 machines in cell 2:

    M4-1:   P10-1 → P10-3 → P4-2

    M4-2:   P2-2 → P2-4

M6:   P6-1 has SPT, it is processed first. While we are waiting for P5-2, we start process P18-1. When P5-2 arrives, it will have to wait until P18-1 is finished.

M7:   There are two M7 in cell 2:

    M7-1:   P10-2 → P6-3 → P2-3 → P3-4 → P2-6 → P5-4

    M7-2:   P18-2 → P4-3

M8:   P6-4 → P10-4 → P2-5 → P3-5

M9:   P6-5 → P3-6 → P4-4 → P5-5

M10: There are two M10 machines in cell 2:

  M10-1: P6-2 → P5-3

  M10-2: P18-3

Cell 3

M7:   There are two M7 machines in cell 3:

  M7-1: P12-2 → P14-2

  M7-2: P15-2 → P16-2 → P17-2

M10: There are two M10 machines in cell 3:

It looks as if these two machines will have low utilization. One way to solve this problem is to have one machine in this cell and have some part done in another cell (cell 2). Here, we chose still to use two M10s.

M11: There are three M11s in cell 3:

  M11-1: P12-1 → P17-1

  M11-2: P13-1 → P15-3 → P15-5 → P16-5

  M11-3: P14-1 → P16-3

M12: P19-1 → P13-2 → P15-6 → P12-3 → P16-6 → P17-3

6. After rearranging parts' scheduling on each machine, check the sequence of each part again to see if they are ordered as per the required process routing. If not, rearrange them.

7. Final scheduling is shown in Appendix 2. They are listed by cell.

7.2. Two-Cell Solution

When considering scheduling, it is found that the two-cell partition needs some further adjustment (Table 9). In order to have smoother production flow of parts 5 and 6 and to reduce the intercell flow, one machine 7 is moved to cell 1 to process parts 5 and 6. The revised two-cell partition is shown as Figure 34; the revised layout is shown in Figure 35.

Parts produced in each cell are listed below:

Cell 1 produces parts 1, 2, 3, 4, 7, 8, 9, 10, 11, partial 5, and 6,

Cell 2 produces parts 12, 13, 14, 15, 16, 17, 18, 19, partial 5, and 6.

Table 9. Machine allocation in each cell

Required Machine	3	5	2	4	8	9	6	1	7	10	11	12	Total
Cell 1	1	1	1	2	1	2	1	1	2	0	0	0	12
Cell 2	0	0	0	0	0	0	1	1	2	3	3	1	11

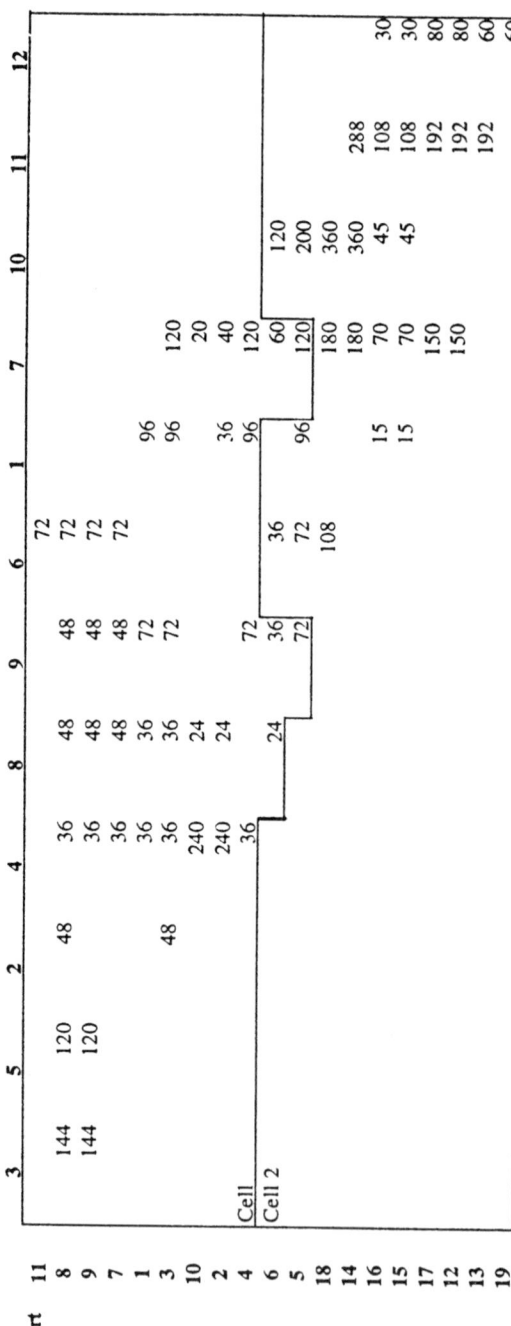

Figure 34. Revised two-cell partition.

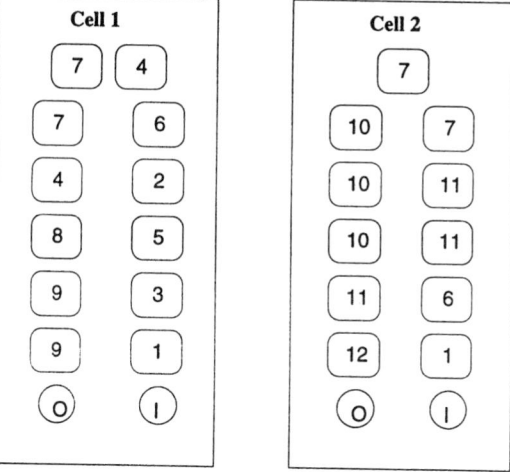

Figure 35. Machine layout in each cell.

Scheduling of production for each cell is as follows:

Initial Scheduling   The printout shown in Appendix 3 is the results of the initial schedule for the two-cell solution. They are arranged by cell.

1. Assume all parts start processing at same time. Assign part to each machine based on parts' routing and process time. This is the initial loading of each machine.

2. The initial loading will also result in initial schedule for parts on each machine. Check this initial schedule to see if there is any time conflicts on any machine.

3. If there is, rearrange them according to parts' routing and machine availability. At this time, we apply rules such as shortest processing time and longest remaining process time, to determine the sequence of parts on an individual machine.

4. For parts that have intercell flows, once they finished processing in the previous cell, they will be transferred to the next cell and will be processed as soon as the required machine in the next cell is available. For instance, when parts 5 and 6 finished processing in cell 2, it needs to be processed at M7 in cell 1. Once M7 is available, parts 5 and 6 have the first priority for using M7.

5. For a detailed description of the adjustments on each cell and each machine, please refer to the initial schedule arranged by machine:

Cell 1

M1:   P2-1(SPT) → P3-1 (MWKR) → P4-1 (MWKR) → P1-1

M2:   P3-2 → P8-3

M3:   P8-1 (MWKR) → P9-1

M4:   There are two M4 machines in cell 1:

M4-1: P10-1 (FIFO) → P7-2 (FIFO) → P10-3 (FIFO) → P4-2 → P1-2

M4-2: P2-2 (FIFO) → P3-3 (SPT) → P2-4 → P8-5 → P9-4

M5:   Fine after adjustment of M4

M6:   P7-1 (MWKR) → P11 → P8-4 → P9-3

M7:   There are two M7 machines:

M7-1: P2-3 → P3-4 → P2-6 → P5-4

M7-2: P10-2 → P6-3 → P4-3

M8:   P7-3 (FIFO) → P6-4 → P10-4 → P2-5 (SPT) → P1-3 (SPT) → P3-5(SPT) → P8-6 → P9-5

M9:   There are two M9 machines:

M9-1: P7-4 → P1-4 → P4-4 → P8-7

M9-2: P6-5 → P3-6 → P5-5 → P9-6

Cell 2

M1:   P5-1 (intercell flow) → P15-1 → P16-1

M6:   P6-1 → P18-1 (even though P5-2 is supposed to have priority, to avoid idling one hour time, process P18-1 right after P6-1 is finished) → P5-2

M7:   There are two M7 machines:

M7-1: P15-2 → P12-2 → P14-2

M7-2: P18-2 → P16-2 (to balance) → P17-2

M10: There are three M10 machines:

M10-1: P6-2 → P5-3 → P15-4 → P16-4

M10-2: P14-3

M10-3: P18-3

M11: There are three M11 machines:

M11-1: P12-1 → P13-1

M11-2: P14-1 (MWKR) → P16-3

M11-3: P17-1 (MWKR) → P15-3 → P15-5 → P16-5

M12: P19-1 → P15-6 (SPT) → P12-3 → P13-2 → P16-6 → P17-3

6. After rearranging parts' scheduling on each machine, check the sequence of each part again to see if they are ordered as per the required process routing. If not, rearrange them again.

Appendix 4 is result of final schedule of two-cell solution. They are listed by cell.

8. COMPARISON OF TWO SOLUTIONS

Tables 10 and 11 show that average process duration and time spent in system by the parts in the three-cell solution is shorter than the one for two-cell solution. This

Table 10. Process duration and time spent in system of each part

Part No.	Process Duration (hr)			Time Spent in System (hr)		
	Three-Cell	Two-Cell	Diff (2C–3C)	Three-Cell	Two-Cell	Diff (2C–3C)
P1	4.00	4.00	0.00	5.60	7.80	2.20
P2	6.13	7.52	1.39	6.67	7.52	0.85
P3	8.20	7.80	−0.40	8.20	8.40	0.20
P4	6.70	6.80	0.10	9.40	9.00	−0.40
P5	9.50	9.08	−0.42	10.60	9.60	−1.00
P6	4.60	4.60	0.00	4.60	4.60	0.00
P7	3.40	4.20	0.80	3.40	4.20	0.80
P8	8.60	9.80	1.20	8.60	9.80	1.20
P9	7.80	8.00	0.20	10.20	10.40	0.20
P10	4.72	5.00	0.28	4.72	5.00	0.28
P11	1.20	1.20	0.00	2.40	2.40	0.00
P12	7.57	7.57	0.00	7.57	7.57	0.00
P13	4.20	5.37	1.17	4.20	8.57	4.37
P14	14.70	14.70	0.00	15.20	15.20	0.00
P15	6.25	4.65	−1.60	6.25	6.30	0.05
P16	7.82	7.77	−0.05	8.07	9.60	1.53
P17	7.02	10.93	3.91	10.20	10.93	0.73
P18	10.80	10.80	0.00	11.90	11.90	0.00
P19	1.00	1.00	0.00	1.00	1.00	0.00
Average	6.54	6.88	0.35	7.30	7.88	0.58

might be caused by the fact that there are more machines in three-cell solutions than two-cell solutions. The three extra machines in three-cell solution give more flexibility for production scheduling and more capacity. More machines also cause lower average utilization, but more balanced machine usage and shorter average flow time in three-cell solutions.

Based on Tables 10 and 11 and previous analysis, we can summarize a few points of differences between two-cell solution and three-cell solution in Table 12.

We can conclude that there is no easy way to say which solution is better. Depending on what priority the manufacturing company has, different decisions can be made. For example, if a company is financially strong and requires shorter average processing times for parts, it may choose the three-cell solution. On the contrary, if a company has a tight budget and prefers higher machine utilization rate, the two-cell solution would be the better choice.

9. CONCLUSIONS

This chapter demonstrated the step-by-step execution of the different steps in PFA for formation and layout of manufacturing cells. Depending on the size of an industrial problem and the availability of data, several of these steps could be suitably modified.

Table 11. Machine usage, and utilization rate

	Three-Cell Solution			Two-Cell Solution	
Mach. No.	Usage (hr)	Utilization	Mach. No	Usage (hr)	Utilization
M1-1	3.20	50%	M1-1	5.40	84%
M1-2	4.30	67%	M1-2	2.10	33%
M2	1.60	25%	M2	1.60	25%
M3	4.80	75%	M3	4.80	75%
M4-1	3.00	47%	M4-1	5.80	91%
M4-2	4.00	63%	M4-2	5.80	91%
M4-3	4.60	72%			
M5	4.00	63%	M5	4.00	63%
M6-1	4.80	75%	M6-1	4.80	75%
M6-2	3.60	56%	M6-2	3.60	56%
M7-1	5.96	93%	M7-1	5.81	91%
M7-2	5.00	78%	M7-2	4.49	70%
M7-3	5.50	86%	M7-3	5.50	86%
M7-4	4.84	76%	M7-4	5.50	86%
M8-1	3.00	47%	M8-1	4.80	75%
M8-2	1.80	28%			
M9-1	3.60	56%	M9-1	4.00	63%
M9-2	4.20	66%	M9-2	3.80	59%
M10-1	4.00	63%	M10-1	6.00	94%
M10-2	6.00	94%	M10-2	6.00	94%
M10-3	6.00	94%	M10-3	5.50	86%
M10-4	1.50	23%			
M11-1	6.40	100%	M11-1	6.40	100%
M11-2	5.90	92%	M11-2	5.90	92%
M11-3	5.70	89%	M11-3	5.70	89%
M12	5.64	88%	M12	5.64	88%
Average	4.34	68%		4.91	77%

Table 12. Comparison of two solutions

Item	Two-Cell Solution	Three-Cell Solution
Intercell flow	Less	More
Intracell flow	More	Less
Machine usage	Higher	Lower
Machine utilization	Higher	Lower
Machine balance	Lower	Higher
Process duration of each part	Longer	Shorter
Time spent in system of each part	Longer	Shorter
Production scheduling	Harder	Easier
Number of machine needed	Less	More

APPENDIX 1: INITIAL SCHEDULE—THREE CELL

Time Unit: 1 2 3 4 5 6 7 8 9 0 1 2 3 4 5 6 7 8 9 0 1 2 3 4 5 6 7 8 9 0 1 2 3 4 5 6 7 8 9 0 1 2 3 4 5 6 7 8 9 0 1 2 3 4 5 6 7 8 9 0 1 2 3 4 5 6 7 8 9 0 1 2 3 4 5 6 7 8 9 0

Cell #1

P1	m1-1	m4-1	m8-1	m9-1
P3	m1-1		m2-1	m4-1
P7	m6-1	m4-1	m8-1	m9-1
P8	m3-1		m2-1	m5-1	m6-1	m4-1	m8-1	m9-1
P9	m3-1		m2-1	m5-1	m6-1	m4-1	m8-1	m9-1
P11	m6-1

Cell #2

P3	m1-2	m4-2	m7-2	m8-2	m9-2
P2	m1-2	m4-2	7-2	m4-2	8-2	7-2
P4	m1-2	m6-2	m7-2	m9-2
P5	m10-2	m10-2	m7-2	m9-2
P6	m6-2	m10-2	m7-2	8-2	m9-2
P10	m4-2	7-2	m4-2	8-2
P18	m6-2	m7-2	m10-2
P15	1
P16	1

Cell #3

P12	m11-3	m7-3	m12-3
P13	m11-3	m12-3	m7-3
P14	m11-3	m7-3	m10-3
P15	m7-3	m11-3	m10-3	m11-3	12-3
P16	m7-3	m11-3	m10-3	m11-3	12-3
P17	m11-3	m7-3	m12-3
P19	m12-3

Note: Each Time Unit represents 0.2 hours.

APPENDIX 2: FINAL SCHEDULE—THREE CELL

APPENDIX 3: INITIAL SCHEDULE—TWO CELL

Note: Each Time Unit represents 0.2 hours.

657

APPENDIX 4: FINAL SCHEDULE—TWO CELL

Note: Each Time Unit represents 0.2 hours.

658

REFERENCES

Arvindh, B., and Irani, S. A. (1994). Cell formation: The need for integrated solution of the subproblems, *International Journal of Production Research*, **32**(5), 1197–1218.

Burbidge, J. L. (1989). *Production Flow Analysis for Planning Group Technology*, Oxford University Press, Oxford, UK.

Irani, S. A. (1995). Lecture Notes and Class Handouts, IEOR 5990, Winter Quarter, Department of Mechanical Engineering, University of Minnesota.

Irani, S. A., and Ramakrishnan, R. (1995). "Production Flow Analysis Using STORM," in *Planning, Design and Analysis of Cellular Manufacturing Systems*, A. K. Kamrani, H. R. Parsaei, and D. H. Liles (eds.), Elsevier, Amsterdam, Netherlands, pp. 299–349.

21

SYSTEMATIC REDESIGN OF A MANUFACTURING CELL

Yosef S. Allam and Shahrukh A. Irani

Department of Industrial, Welding, and Systems Engineering
The Ohio State University
Columbus, Ohio 43210

1. COMPANY CONCERNS AND OBJECTIVES

The small wheels manufacturing cell in Company X needed to increase utilization on the mill turn machining centers, capacity on the conventional machines, and order throughput. The company desired to reduce lead times, work in process, queue and cycle times, and idle times. The objective was to redesign the cell layout, with assistance from company personnel and outside resources. The Appendix includes a questionnaire that was used to make an assessment of the existing cell layout and operations.

2. ORIGINAL LAYOUT, PART ROUTINGS, AND FLOW PATTERNS

The original layout, as seen in Figure 1, features the washing station in the upper left corner, labeled W, and to the right of the washing station, the horizontal and vertical machining centers, labeled H and V, respectively. Below these machines, located in the lower half of the layout, are the four mill turn machining centers, each labeled with an M. Scattered among these machining centers are various items such as drill presses, unused slotting machines, tool and fixture racks, tables, and the like. All mill turn machining centers are capable of running the same products, with the exception

Handbook of Cellular Manufacturing Systems, edited by Shahrukh A. Irani
ISBN 0-471-12139-8 © 1999 John Wiley & Sons, Inc.

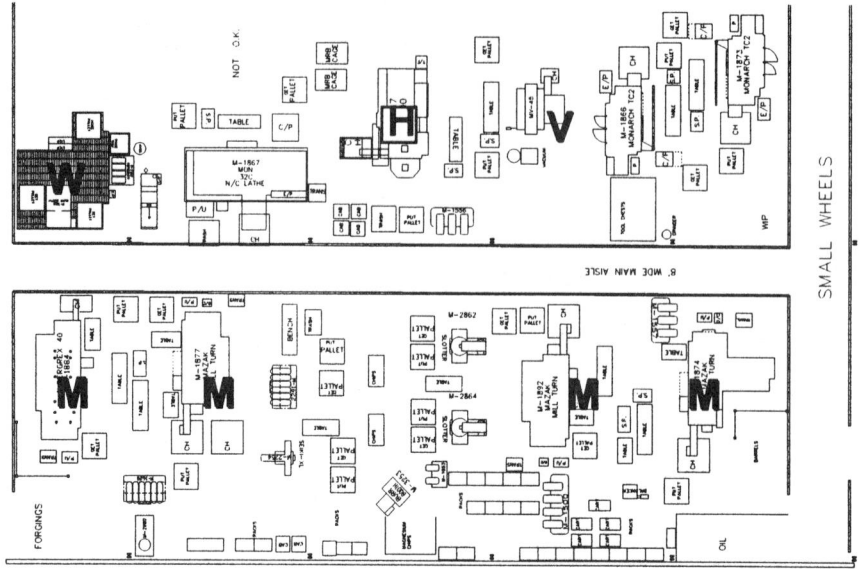

Figure 1. Current layout of the cell.

of M-1874, located in the far right bottom corner, which can run additional parts due to its extra fixtures. This mill turn machining center is also the slowest, relative to the other three mill turn machining centers. The middle two mill turn machining centers, M-1877 and M-1892, are identical in speed and capability. The newest mill turn machining center, M-1884, located on the far left, is the fastest of the four.

As can be seen in Table 1, there are three basic types of routings for the mill turn machining center parts. All parts begin with mill turn machining center operations and end with wash/deburring operations, followed by an exit from the cell, to the materials finishing area, located to the upper left of the washing station (not shown in figure). The third routing in the table actually refers to parts that incorporate a

Table 1. Routings for parts produced on mill turn machining centers

Routings	From	To
1	Mill turn	Wash/deburr
	Wash/deburr	Material finishing
2	Mill turn	Vertical machining center
	Vertical machining center	Wash/deburr
	Wash/deburr	Material finishing
3	Mill turn	New vertical
	New vertical	Wash/deburr
	Wash/deburr	Material finishing

number of different combinations of steps on the vertical and horizontal machining centers prior to and after the mill turn machining center operations. Because there are operations at other machining centers occurring between these vertical and horizontal machining center operations, considerable backtracking is caused. These parts have been combined into a modified routing (Routing 3 in Table 1) that reflects the changes that would occur to their original routings upon the acquisition of a new vertical machining center. This new center can combine the original vertical and horizontal machining operations into one operation, thus eliminating the backtracking.

An example of a commonly run part not yet converted to routing 3 in Table 1 is part 10-1515. As can be seen in Figure 2, this part flows from the horizontal machining center to the mill turn machining center, from the mill turn machining center to the vertical machining center, and from the vertical machining center to the horizontal machining center again, with much backtracking, before it finally stops at the washing station.

Part 10-1585-1 (Figure 3) is another unconverted part, although it requires fewer steps and experiences less backtracking in its flow path. Part 10-1384 (Figure 4) is an example of routing 1 in Table 1, requiring only mill turn machining center and washing station operations. This flow diagram best illustrates the awkward input point of the washing station for incoming parts. The input point is located in the upper left area of the washing station.

Conventional parts, on the other hand, are not as simple. The only rule-of-thumb to determine the conventional part routings is shown in Figure 5 (see steps 1–3, 5, and 8). These are generally the required lathe operations, but sometimes after steps 3 or 5, horizontal machining center or drill press operations may be required. This

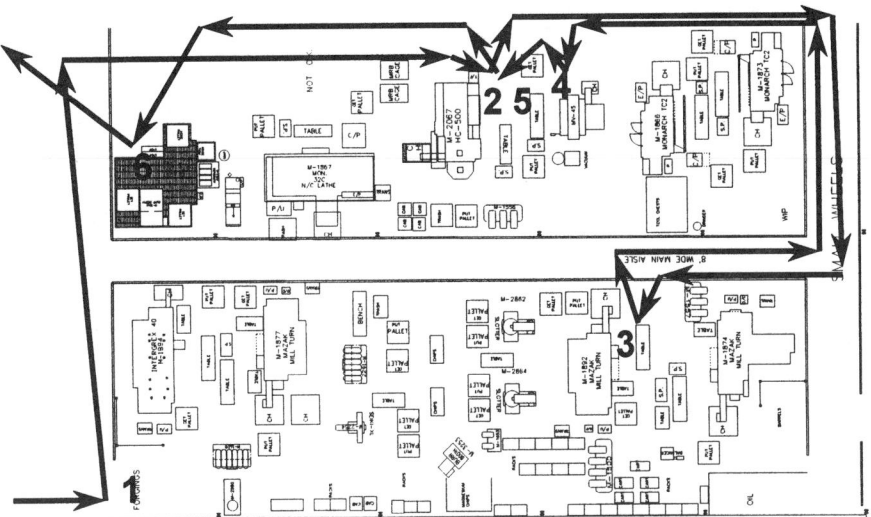

Figure 2. Original flow of part 10-1515.

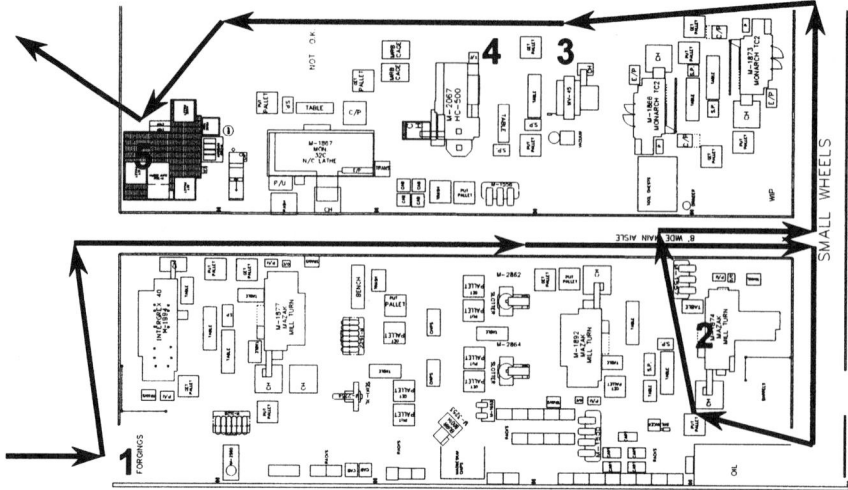

Figure 3. Original flow of part 10-1585-1.

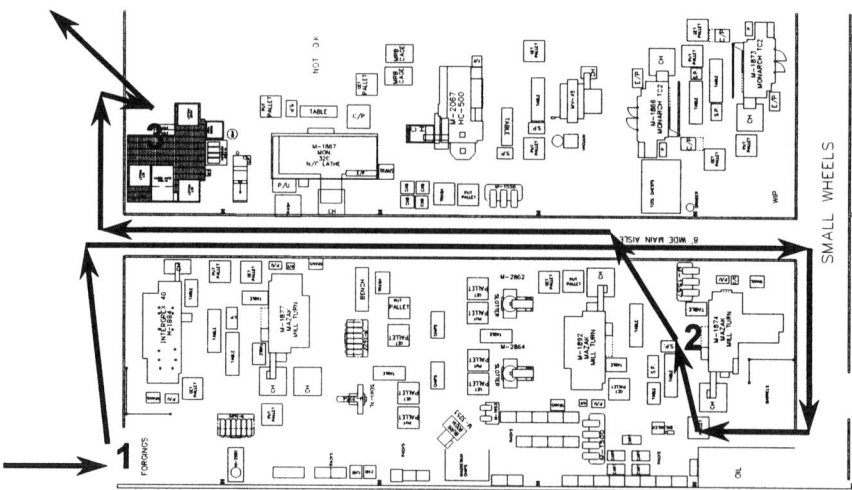

Figure 4. Original flow of part 10-1384.

varies widely across conventional parts and depends on the existence of windows or holes for tying down the wheel, as well as several other features. As can be seen from Figure 6, there is extensive backtracking in this example of the flow pattern for a conventional part. The situation seems even worse when considering the fact that this is an example of a conventional part with only one optional step, and there may be as many as three such steps.

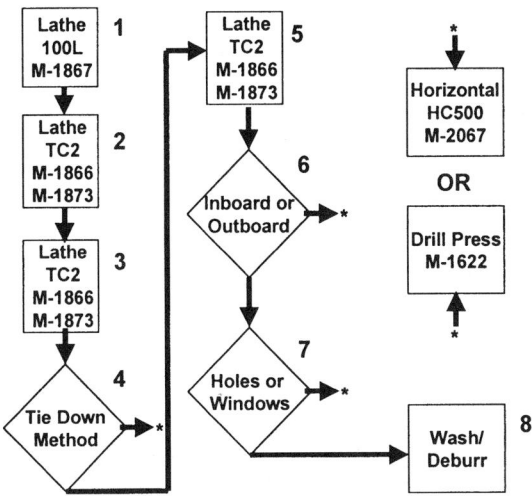

Figure 5. Material flow in conventional subcell.

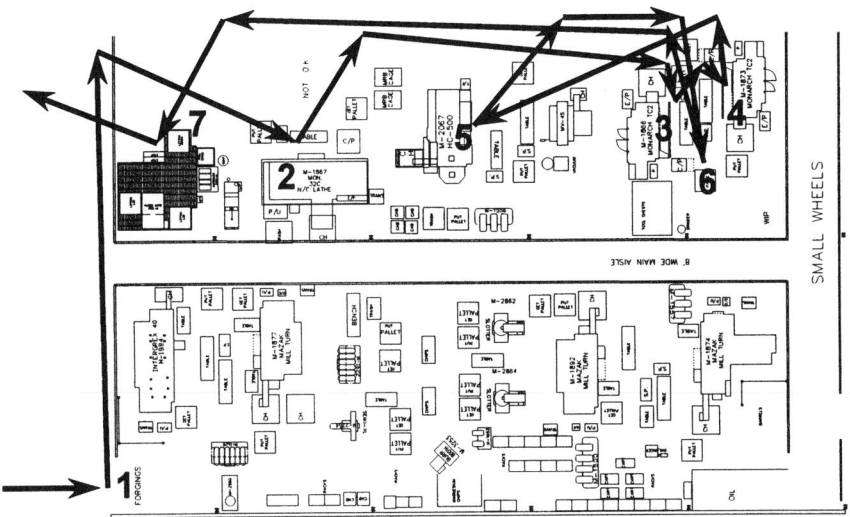

Figure 6. Current flow of conventional parts.

3. ALTERNATIVE LAYOUT AND NEW FLOW PATTERNS

Redesign of the manufacturing cell was approached with several specific guidelines in mind. First, there was a desire to maintain separate subcells based on typical operation sequences. Second, the concept of product ownership was incorporated in the final orientation of the machinery and other items. This reinforced the task of maintaining separate subcells. Next, multimachine tending by the operators was to be provided for, wherever possible. Loading flexibility, constrained by the above guidelines, pushed similar machines closer together. Finally, resource sharing, if feasible, was to be incorporated in the new layout.

The alternative layout, shown in Figure 7, was broken down into five distinct subcells, and the washing station, used by all parts, was left as a separate entity. The location of the washing station was not changed, but it was rotated so that its input point was closer to the other subcells. Also, space was reserved for the queuing of parts. In the case of increased workloads, the washing station could be run constantly for three shifts to complete the machined parts, thus reducing the queue that could accumulate during the first and second shifts, when metal removal machinery is typically running.

The five new subcells feature dedicated drill presses with locations in proximity to machining centers with operations directly preceding those of the drill presses. Also, similar subcells share tooling and fixture racks, as in the case of subcells 1, 2 and 3, 4. The mill turn machining center subcells also stress multimachine tending by operators. Subcells 1 and 2, which would both typically run type 1 routings (Table 1), feature deburr booths, which can increase operator utilization, as well as maintain the concept of one operator running two machines in the four mill turn machining center subcells.

Three mill turn machining centers adjacent to each other provided for emergency transfer of orders if maintenance issues arose. For the same reason, the vertical machining centers were also placed adjacent to each other.

The conventional subcell was reorganized so as to match the most typical rule-of-thumb conventional routings. It also featured a drill press (M-1622) and tool setting device (M-2980). These were located as close as possible to the conventional machining centers with operations directly preceding those of these two machines.

The flow path of part 10-1384, a part with a type 1 routing (Table 1), was greatly improved (see Figure 8). When compared with the original layout and flow, it can be seen that the distance traveled has been reduced. There is no lack of backtracking also. The same is true with parts flowing through subcells 3 (Figure 9) and 4 (Figure 10). Parts can now travel from the forgings area, along the new lower aisle to the appropriate mill turn machining center subcell. They then leave the subcell, travel along the dividing aisle (which runs across the center of the layout) to the washing station on the left.

The conventional subcell features an input buffer location opposite to the output buffer location, allowing parts to travel straight through from right to left, to the washing station input buffer (Figure 11). Until process improvements are made, however, some conventional parts will have to cross the dividing aisle to the vertical machining center and back. Some of these parts may also require operations on the drill press or horizontal machining center before they complete their lathe operations, thus causing backtracking. The advantages of the new layout are summarized in Table 2.

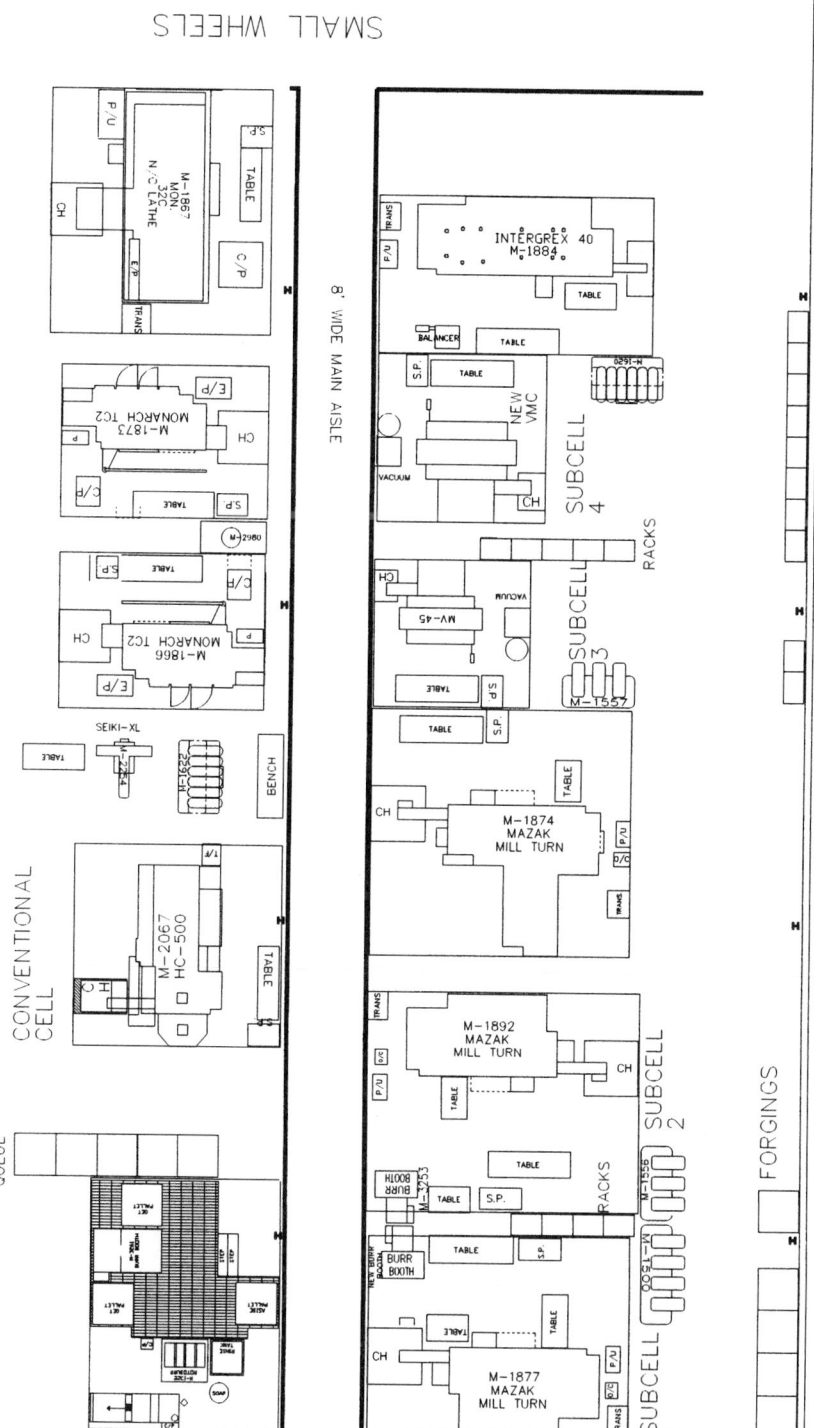

Figure 7. Alternative layout for the small wheels cell.

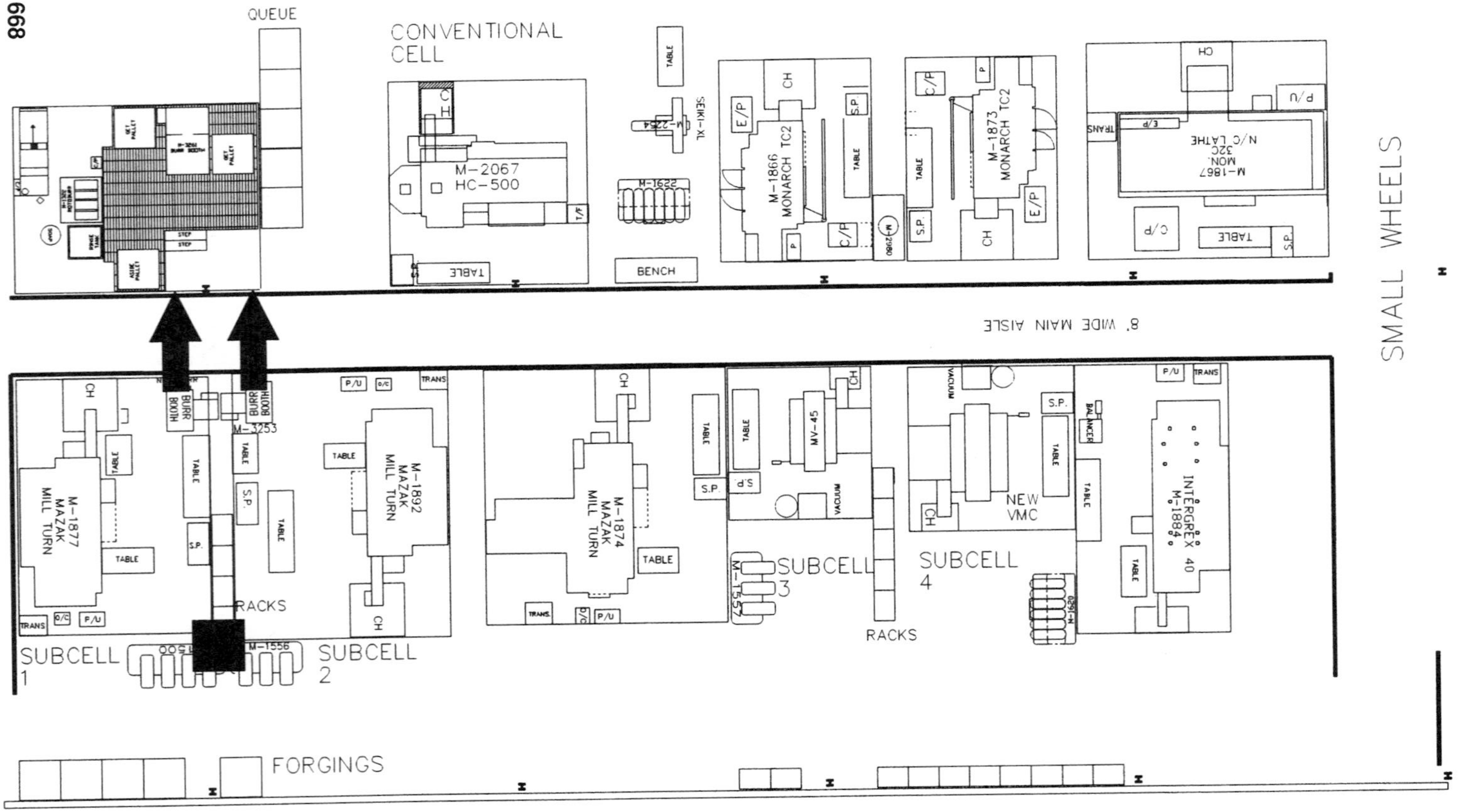

Figure 8. Material flow in subcells 1 and 2 for part 10-1384.

668

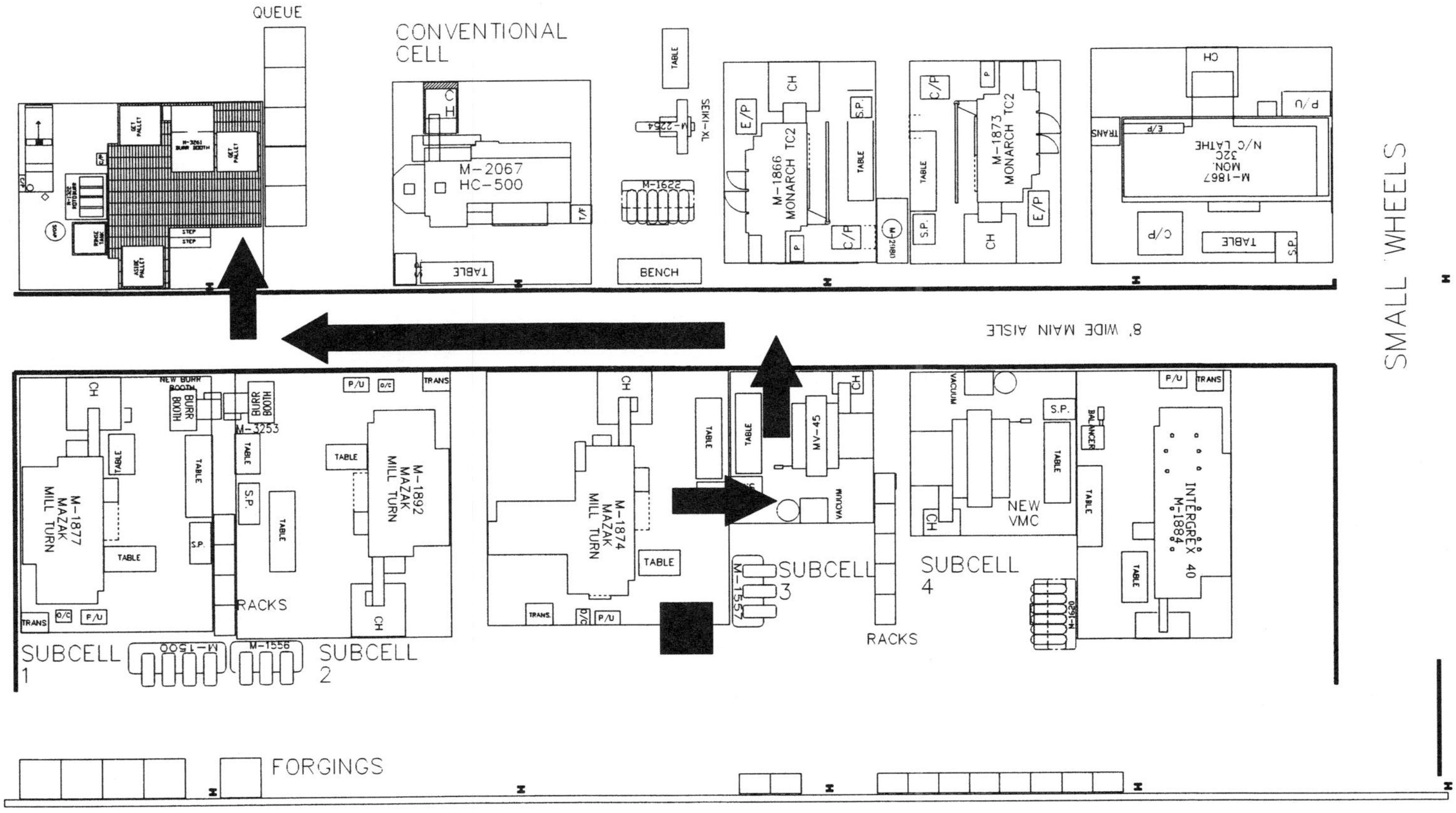

Figure 9. Material flow in subcell 3.

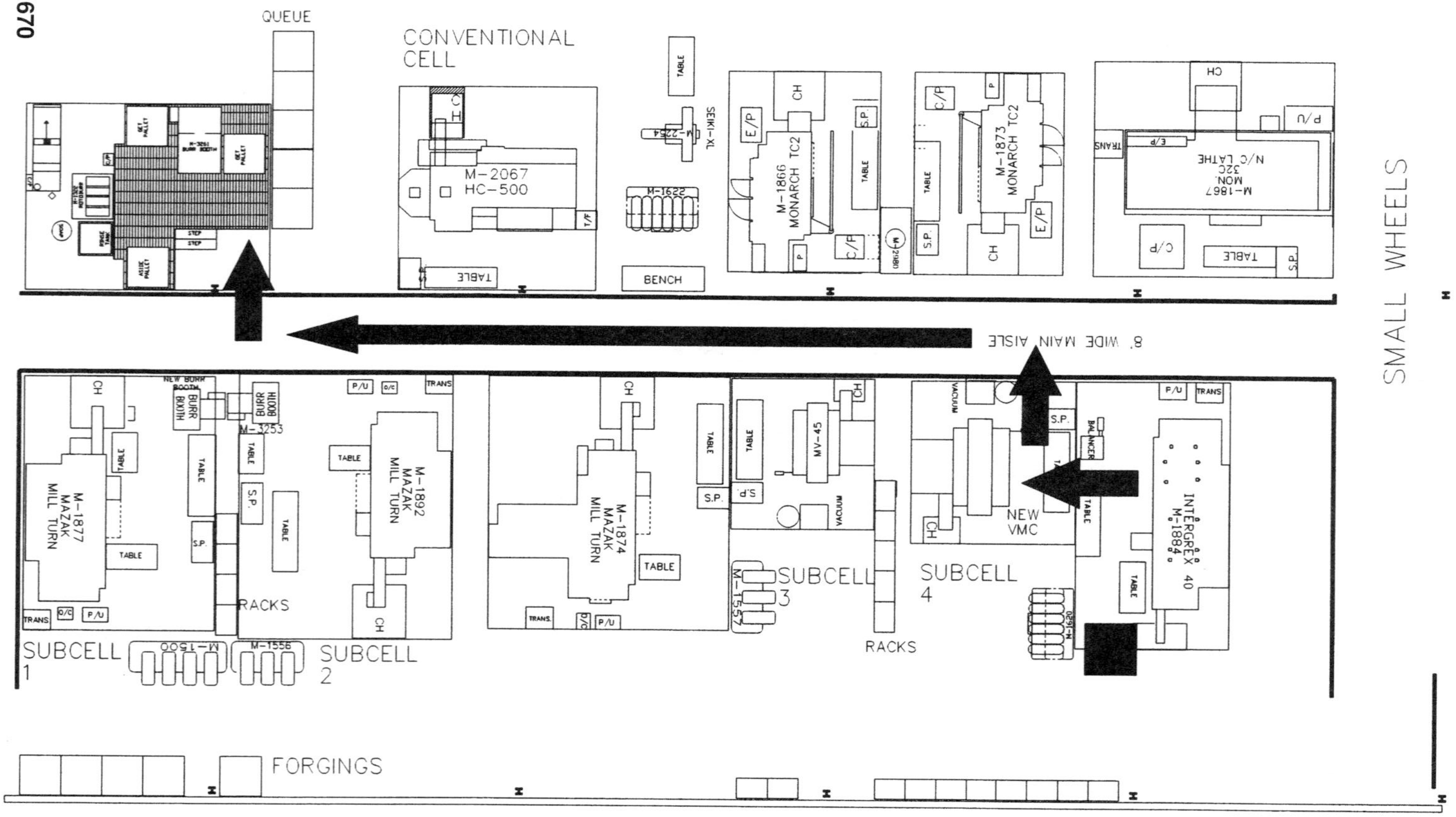

Figure 10. Material flow in subcell 4 for part 10-1515.

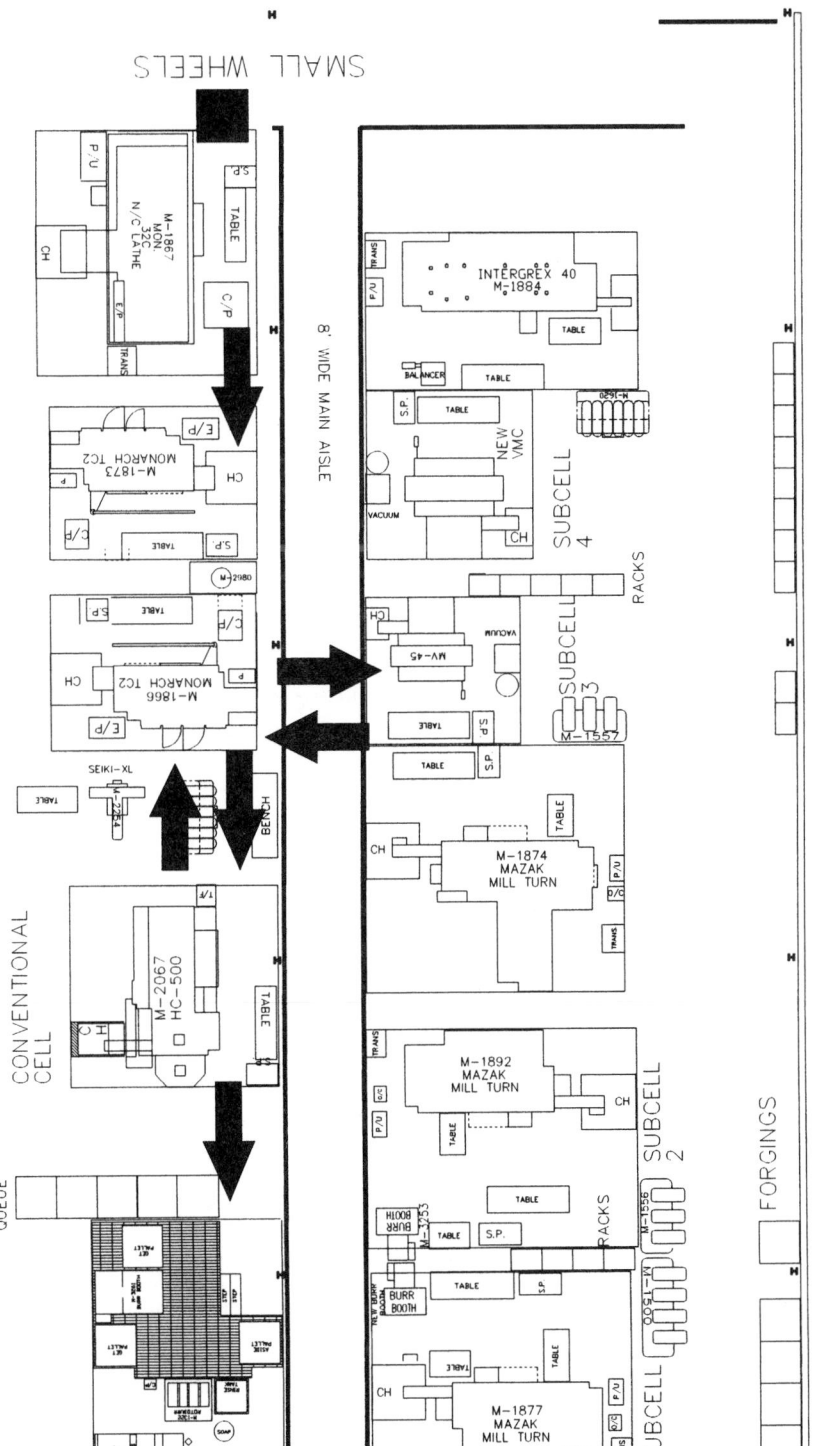

Figure 11. Material flow in conventional subcell.

Table 2. Advantages of new layout

Reduction in distances traveled by materials (46%)
No backtracking
Product ownership by cell operators
Multimachine tending by operators
Local fixture racks
Subcells visually defined and separated
Drill presses located within subcells
Stand-alone mill turn machining centers (subcells 1 and 2)
    feature individual deburring booths

4. CAPACITY PLANNING

The procedure used for capacity planning to determine machine requirements is displayed in Figure 12. Scrap calculations were not necessary because they were already included in the data collected. Also, setup times were not incorporated into the calculations because it was feared that they were inaccurate. In addition, the available capacity figures were based on the total of all four mill turn machining centers, and individual machining center hours were not available. Because the mill turn machining centers were not equivalent in their process capabilities, it was not possible to calculate the need for more or fewer machines when considering the future cumulative run hours based on incoming orders.

The graph in Figure 13 shows the fraction of mill turn machining center run hours as compared to total cell run hours. In all past and future plan data, the fraction of

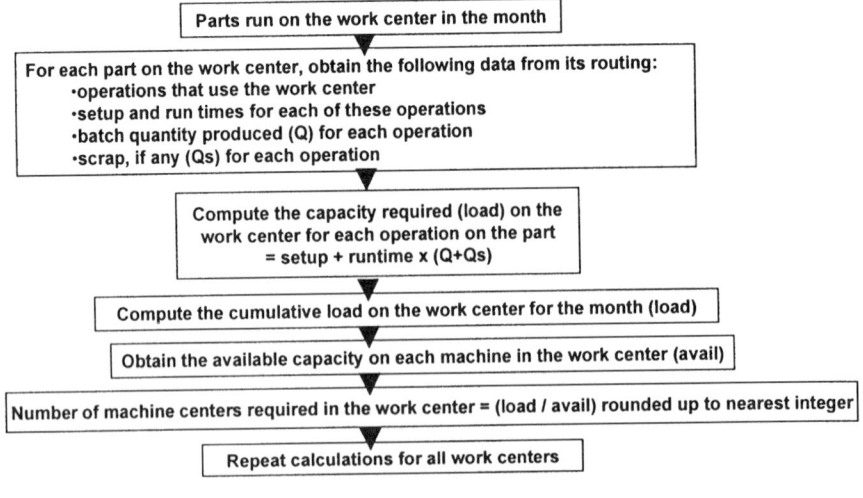

Figure 12. Capacity planning for machine requirements analysis.

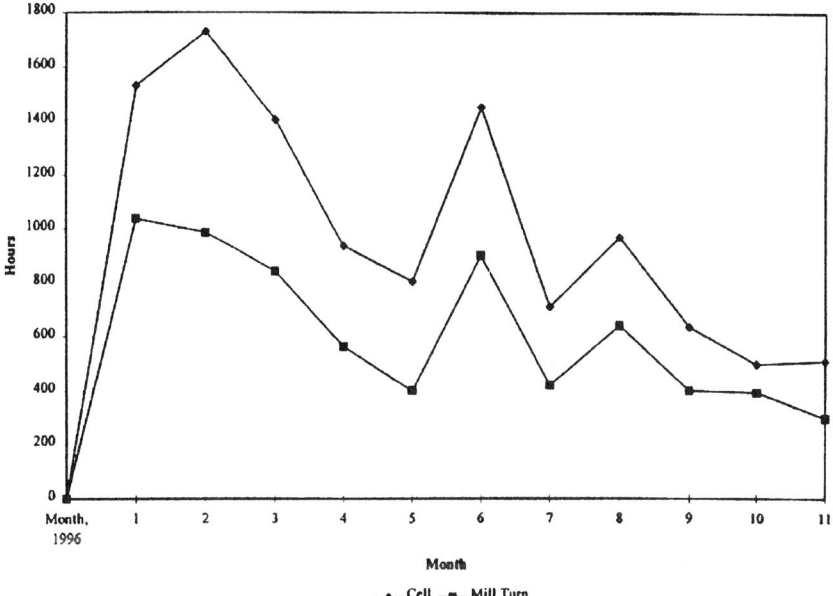

Figure 13. Mill turn machining centers load vs. total cell load.

mill turn machining center hours to the whole cell's hours ranged from 52% to 59%. This shows the importance of focusing on improving the mill turn machining center operations.

In Table 3, future run hours are almost evenly divided among mill turn machining centers in subcells 1, 2, and 4. Each mill turn machining center will run in the neighborhood of 1350 hours. The mill turn machining center in subcell 3, however, is planned to run for only 743 hours. This works out well because this mill turn machining center (M-1874) happens to be the slowest of the four mill turn machining centers. If it becomes underutilized, any surges in load can be placed on this mill turn machining center. Increasing available workcenter capacity can help deal with the increasing demand forecasted at Company X. Ways in which this can done are summarized in Table 4.

Table 3. **Available capacity in cell through January 1998**

Subcell	Sequence	Hours
1 and 2	Mill turn	2700
3	Mill turn, vertical	743
4	Mill turn, new vertical	1364
Conventional	(varies—see Fig. 5)	2676
Remaining machinery		1689
Total		9172

Table 4. Strategies for increasing available capacity in cell

Reduce setup and machining times for key operations
Reduce idle times on all machines
    Tool and fixture availability
    Material availability
    Parts delivery/removal
    Operator availability
    Production control communications

5. SUMMARY OF RESULTS

The redesign of the small wheels manufacturing cell will virtually eliminate back-tracking of parts through the cell. The redesign will also reduce the distance traveled by mill turn machining center parts by 46% according to calculations performed with FactoryFlow. The redesign will allow operators to own the product as it progresses through various metal removal processes. Subcell separation and multimachine tending are supported in the new layout. Positioning similar machines together allows for resource sharing and loading flexibility. The redesigned cell can act as a guide, or pilot, for the reorganization of other manufacturing cells in the plant. The activities involved in producing the new cell layout can be repeated as necessary and applied to other cells. Activities that were performed are listed in Table 5.

6. SCHEDULE FOR CELL RELAYOUT

A tentative relayout schedule (Table 6) was shown to the company. This schedule did not account for the time it would take during or between machine moves but showed a suggested order for moving the machines in the mill turn machining center subcells.

Table 5. Activities performed to complete cell relayout

Collection of background information
Update of current layout
Determination of machine footprints
Complete inventory of objects in cell
Part routing, machine sequencing, and part quantities resulting in cell traffic
Tracking of a sample of commonly run parts
From–to chart analysis
Generation of alternative layouts
Generation of layouts for conventional subcells
Estimation of machine loads in subcells

Table 6. Schedule for relayout of cell

Move out M-1884
Slide M-1887 left into position of M-1884
Move M-1622, M-2254, M-2862, and M-2864 into conventional area
Slide M-1892 next to M-1877
Move M-1884 next to M-1892
Move M-2063 next to M-1892
Slide M-1874 right (if necessary)

7. COSTS OF CELL RELAYOUT

Costs were budgeted for machine movement and reorientation, such as rental of moving machinery, new foundations, plumbing, electrical, leveling, and laser alignment. The cost of the new vertical machining center was included in the budget. Costs not included were the cost of a new deburring booth, six-spindle drill press, and the costs of moving smaller, less significant items such as tables and racks.

8. EQUIPMENT CHANGES IN THE CELL

The most likely changes to occur in the placement and orientation of the equipment located in the cell upon implementation of the relayout included removal of slotters M-2862 and M-2864 and the dual spindle drill press M-1653. Purchases included a six-spindle drill press, vertical machining center, and a deburring booth.

9. FUTURE ACTIVITIES

To facilitate improved operation of the cell, it was recommended that several improvements be made. They include the reduction of machine footprints, better cell scheduling, and more accurate cell loading. The individual footprints of machining centers can be reduced by optimizing the placement of objects at each machine, considering material handling options, considering parts storage options, revising the location and spacing of aisles, and revising the input/output points of the machining centers with relation to the rest of the factory.

Better cell scheduling can be accomplished by establishing order dispatching and sequencing rules. Performance measures can be devised to evaluate any schedule. Determining the set of orders and the percentage of each that can be produced in any week subject to material, labor, machine, and other resource capacity constraints will result in improved cell loading strategies.

Future cell redesign projects can be facilitated by providing individual machine capacity data, in addition to workcenter capacity data. Accurate setup time data, feedback from cell operators, and operator utilization data would also greatly aid the redesign process.

APPENDIX: QUESTIONNAIRE FOR ASSESSMENT OF EXISTING MANUFACTURING CELLS (WITH ANSWERS FROM COMPANY X, WHERE PROVIDED)

I. GENERAL QUESTIONS

1. What problems or reasons prompted a switch to cells?
   - Originally, it was a machine shop with a functional layout.
   - Changeover times and, later, quality control prompted the switch.

2. How or by whom was the idea for cells championed in Company X?
   - The former plant manager.
   - The shop committee was completely opposed.

3. How was the idea introduced or sold to the employees?
   - Cells were mandated, and there was mistrust at first.

4. What portion of the facility was converted to cells?
   - All of it.

5. What is the objective of the cells?
   - Safety, then quality, then production.
   - Production involves timeliness and utilization.

6. What are the demand volumes?
   - Direct labor hours were 1443 and 1087 pieces per month for 1996.

7. How many machines are in the cell?
   - There are 4 mill turns, 4 conventional lathes, 1 horizontal, 1 vertical.

8. How many operators run the cell?
   - 1st shift: 6, with 2 seniors
   - 2nd shift: 3, with 1 senior
   - 3rd shift: 4, with 1 senior

9. What operations are performed inside the cell?
   - All metal removal such as turning, drilling, milling, and deburring.

10. What operations are performed outside the cell?
    - Finishing and assembly.

11. What is the level of automation?
    - Between 80% and 90%, with manual operations involving deburring, drilling, and some fixture changes.

12. How long has the cell been in operation?
    - About 8 years.

13. What changes have been made to the cell over time?

- There was a switch from just a foreman to a team consisting of a Numerical Control (NC) programmer, production engineer, manufacturing engineer, and a foreman (a management team).

14. How has the manufacturing focus and content of the cell been determined?
    - The cell is organized as a product family, with all the machining operations in the cell.

15. What, if any, software was used for the analysis?
    - Manual study of a few areas, and the same logic was assumed for the remainder of the plant.

16. What problems arose during cell planning?
    - Nonstandardized processes.
    - Converting fixtures for new machines.
    - Lack of cell autonomy.
    - Converting NC programs.

17. What problems arose during cell implementation?
    - Operator job skills.
    - Quality issues with new processes.
    - Equipment capabilities.

18. What post-implementation problems were encountered?
    - Quality.
    - Safety later became a company-wide concern as safety standards were pushed to a higher level.

19. What benefits have been realized since the cells were introduced?
    - Throughput was improved from 120 days to 20–30 days.
    - There was an inventory decrease.
    - There was greater focus on quality.

20. From your experiences, what conditions at Company X were favorable for the introduction of the cells?
    - High production volumes.
    - A need for a quick response manufacturing system.

II. DETAILED QUESTIONS

1. What is the cell's layout?
2. What is the layout for the whole system of cells?
3. What intracellular material handling methods are employed?
    - Personnel carrying parts by hand.
    - Lift trucks.
4. What intercellular material handling methods are employed?
    - Lift trucks, operated by material movers, a.k.a. "handymen."
5. How are orders released and progressed?
    - Customer order.

- Material Requirements Planning (MRP) system.
- Bill of Materials.
- Start dates.
- Production foreman.
- Release status.

6. What strategies or rules are used for scheduling operators and machines in the cell and in the system of cells?
   - Assembly areas share operators.
   - Experience comes first.
   - Start dates and shipping dates are significant factors.
   - Generally unstructured.

7. Is there machine sharing by cells, in particular, the cell under study?
   - No.

8. How is part movement coordinated?

9. Is there operator sharing by the cells?
   - No, but more communication between cells is desired.

10. How is operator movement coordinated?

11. What quality control techniques are used?
    - Electronic Statistical Process Control (SPC).
    - Workmanship standards.
    - Pareto analyses are done on defects.

12. How do equipment breakdowns affect cell performance?
    - There is only a single one-of-a-kind piece of equipment, the Miles lug mill.

13. What preventative measures are taken to minimize equipment breakdowns?

14. Is there more management control or delegation of control to the operators?
    - There is more management control than desired, which is why Company X is trying to move more toward delegation to lower levels in the organizational hierarchy.

15. What is the level of operator cross-training?
    - Limited across cells.
    - But, an operator must be able to run all equipment inside a cell.

16. Which of the following supporting activities were done before or during the cell introduction?
    a. Setup time reduction?
       - No, but it is in the works.
    b. Equipment changes?
       - Yes, from manual processes to NC.
    c. Just-in-time control?
       - Yes, to the customer.

- No, from the vendor.
   d. Quality training?
- There is a certified operator program with a bonus.
   e. Group problem solving?
- Problem-solving techniques.
- Team building techniques.
   f. Teamwork and leadership training?
- An external consultant helped to teach team building techinques.
   g. Computer Numerical Control (CNC) programming/Computer Aided Design (CAD)?
- Yes. The geometry is pulled from NC prints.
   h. Cost estimation?
   i. Repair and preventative maintenance?
- More prominent in the last 3 years.
   j. Communication with vendors and subcontractors?
- Recent developments, but just scratching the surface.
   k. Direct electronic or phone link with customers?
- No. There are walls between departments preventing direct communication with customers.
   l. Bar coding, computerized shop floor control?
- Yes, in the warehouse only.
   m. Direct electronic or phone link with sales and marketing?
- Yes, CINCOM is our system.
18. Are there group payment methods or group incentive schemes?
19. How is cell performance evaluated?
- Safety [Occupational Safety and Health Administration (OSHA) recordable, lost time].
21. Is cell performance communicated to cell members, and how? If yes, how? If no, what is being done?
- The operators favor it today, although it was a mandated transition.
22. What were the economic justifications to justify the conversion to cells?
- Team building techniques. Customer pressure.
- Needed forced change.
23. What, to date, are the plans to build on cell experiences?
- More independent or empowered workforce.
- Single class of operators per cell in future.
- Apply Single Machine Exchange of Dies (SMED) principles.

22

PLANTWIDE CONVERSION TO CELLULAR MANUFACTURING

Quarterman Lee

Strategos, Inc.
Kansas City, Missouri 64111

1. INTRODUCTION

Cellular Manufacturing (CM) and other product-focused strategies offer tremendous benefits. Reductions in inventory of 50–90% are not uncommon. With this inventory reduction comes corresponding decreases in cycle time and the ability to respond quickly to customer needs. Material handling is often reduced by 70–90% with resulting reductions in cost, handling damage, and material handling accidents. Work flow in the factory is simplified and allows simpler scheduling methods. Teamwork is improved and employee satisfaction increased.

Work cells appear simple. In reality they are complex and delicate socio-technical-bio-systems. They require careful design that often proves difficult and demanding. At present, the literature provides little guidance on the design process. This chapter presents an approach to the design of work cells and the factory that leads planners through a design in a structured, step-by step way. This design procedure, referred to as FacPlan, tells the designers and planners what, when, and how to do each of the required tasks.

A previous chapter emphasized FacPlan's approach to work cell design. This chapter concentrates on the plantwide or macro design of a cellular facility. A macro layout shows the location and space for major work cells, functional departments, and

Handbook of Cellular Manufacturing Systems, edited by Shahrukh A. Irani
ISBN 0-471-12139-8 © 1999 John Wiley & Sons, Inc.

other significant features. It includes the building envelope and relevant site features such as loading docks.

The macro layout is designed prior to the overall conversion of a facility to CM. It guides the project with a manufacturing strategy. It determines the overall material flows and information flows. It establishes the type and degree of manufacturing focus. Sometimes an organization develops prototype work cells in advance of the macro layout to gain experience and test concepts. This is useful when the organization has had limited experience with cellular approaches.

Such a broad-based plan requires consensus from every part of the organization. Accordingly, it should be a team effort. Typically, a macro-layout project uses a core team of employees and perhaps consultants to plan the project, execute the work, and frame the issues. A larger team of managers advises, provides information, and makes important decisions. Informal participation comes from many others in the organization who participate in meetings from time to time or provide information and advice.

This chapter uses a project called Induct Manufacturing to show the FacPlan system and concepts at work. The Induct example was first developed for Turnkey Manufacturing Seminars and used in its popular programs on plant layout and work cell design. It derives from a series of successful projects that the author conducted with a Midwest firm.

2. LEVELS OF DESIGN

FacPlan is a structured, step-by-step methodology for facility planning. It uses the layout or space plan as the centerpiece for facility design. A space plan specifies amount, location, and type of space. Usually, a space plan takes the form of a drawing or layout in two dimensions.

FacPlan uses five levels of space plan detail as Figure 1 shows. Figure 1 identifies each level and describes typical activities at that level. It shows the usual Space Planning Unit (SPU). An SPU is a block of space, which the designer defines and arranges. The environment is the larger space, which surrounds the space plan. The far right column of Figure 1 illustrates typical outputs for the planning process. Each level can be a, largely, independent project. This breaks facility planning into manageable units or subprojects. This is important because even a medium-sized facility of, say, 100 persons requires huge amounts of data and thousands of design decisions.

2.1. Levels of Space Planning

Site location is the highest level of detail, as shown near the top of Figure 1. Site location decides the location, size, and other requirements for a site. We may also refer to this level as "level I" or "global."

Site planning takes place at level II, the supralevel. It specifies site features such as buildings, roads, and utilities.

Level	Activity	Typical SPU	Environment	Output
I **Global**	Site Location & Selection	Sites	World Or Country	
II **Supra**	Site Planning	Buildings Or Site Features	Site	
III **Macro**	Building Layout	Cells Or Departments	Building	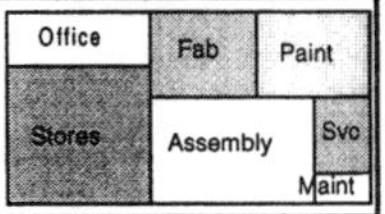
IV **Micro**	Department Or Cell Layout	Workstations Or Cell Features	Cells Or Departments	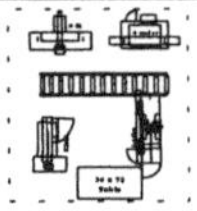
V **Sub-** **Micro**	Workstation Design	Tool Locations	Workstation	

SP0002 GED

Figure 1. Levels of space planning.

Macro layout locates departments or other large-scale features within a building. We sometimes call this a "block layout" or building layout.

Micro layout locates equipment, furniture, and other items within each unit of the macro layout. This is the level at which most work cell design takes place.

Submicro layout designs the space plan for individual workstations. It places tools, materials, and other small-scale features.

2.2. Time Phasing

Normally, facility designers should work from the highest to the most detailed level, as shown in Figure 2. This sequence prevents detail from clouding long-range, strategic decisions. It ensures that strategic decisions and policies guide the detail planning.

In practice, it rarely happens that we have the luxury of starting with site selection and moving down to workstation design in logical sequence. Usually, we are given a building or department and must rearrange it without significant change to the larger space plan. A company might also want to experiment with a work cell prototype before committing the entire plant to a rearrangement.

3. MACRO LAYOUT

Every layout has four fundamental elements: *SPUs, affinities, space, and constraints.* In the design process we generate the additional derived elements of configuration

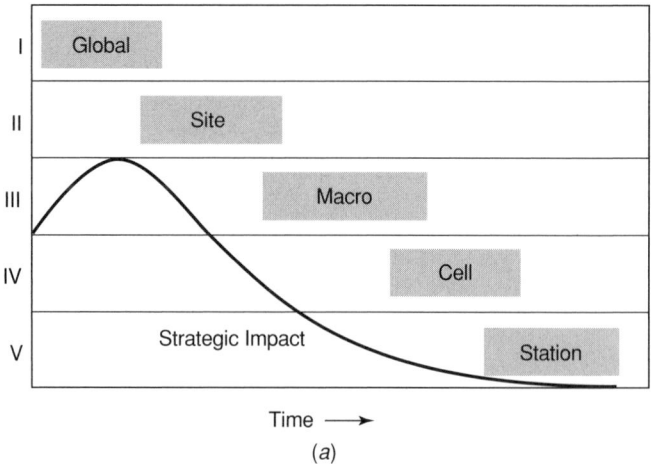

Time ⟶

(*a*)

Figure 2. Time phasing the planning levels.

diagrams and space plan primitives. Figure 3 shows how the fundamental and derived elements converge to form a layout.

Space Planning Units are the "chunks" of space that we arrange. Their definition is fundamental for the layout and should reflect a coherent and well-thought-out manufacturing strategy.

Affinities reflect the needs for proximity among the SPUs. Affinities may result from material flows, communication flows, or resource flows. We represent affinities graphically with lines of varied width, colors, and hues.

Space is the area required for each SPU. Designers may use several of the six methods for calculating this space.

Constraints limit the design in some way. They are varied and often depend on conditions. Examples are building envelopes, column spacing, fixed locations such as docks, and need for supporting facilities such as overhead cranes.

Figure 3 does not help very much in planning a project. The tasks required overlap several elements. The sequence of task execution does not correspond to the elements as shown in Figure 3. To plan a layout project, we need a different model based on specific, definable tasks.

3.1. Model Project Plan

The model project plan of Figure 4 shows the tasks necessary to develop a macro layout and their sequence. It is the basis for virtually any macro-layout project. The plan identifies tasks, deliverables, and sequence of the work. For a specific project, the project team would define the scope, depth, responsibility, and target dates for each task. These tasks fall into three broad groups: (1) information acquisition, (2) strategy development, and (3) layout design. Two exceptions are the project planning task at the beginning and the selection task at the very end. Induct used this model to plan its layout project. Below, we describe the tasks and results.

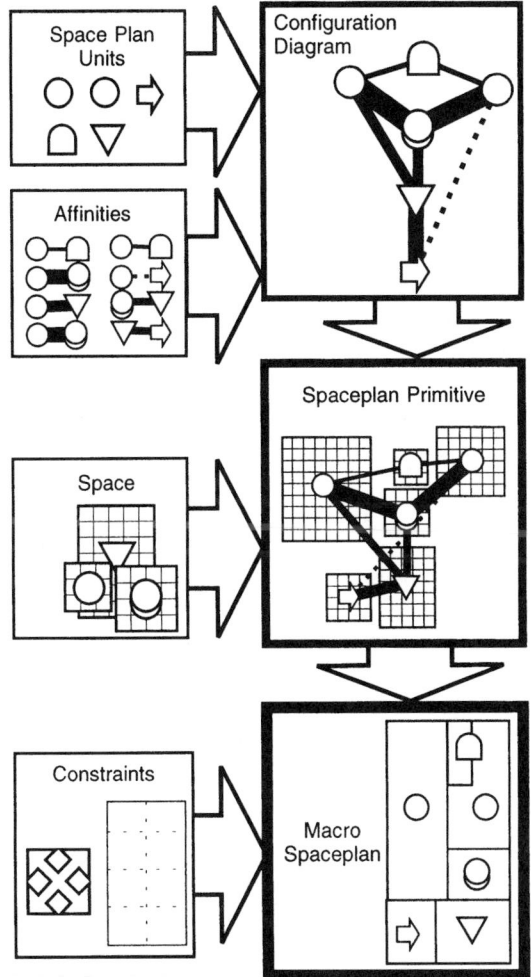

Figure 3. Elements of a space plan.

Facilities and Workplace Design—An Illustrated Guide (Lee, 1997), shows procedures for each task. This book has additional details, discussion, and examples, which can assist a project team in applying the concepts to a wide range of projects.

3.2. Information Acquisition

Task 3.02 Products and Volumes   In this task we gathered information about Induct's products and volumes. Induct manufactures prefabricated components for industrial ductwork of the type used in dust collection. Figure 5 illustrates the principle components. These components are fabricated from sheet metal, MIG welded, and

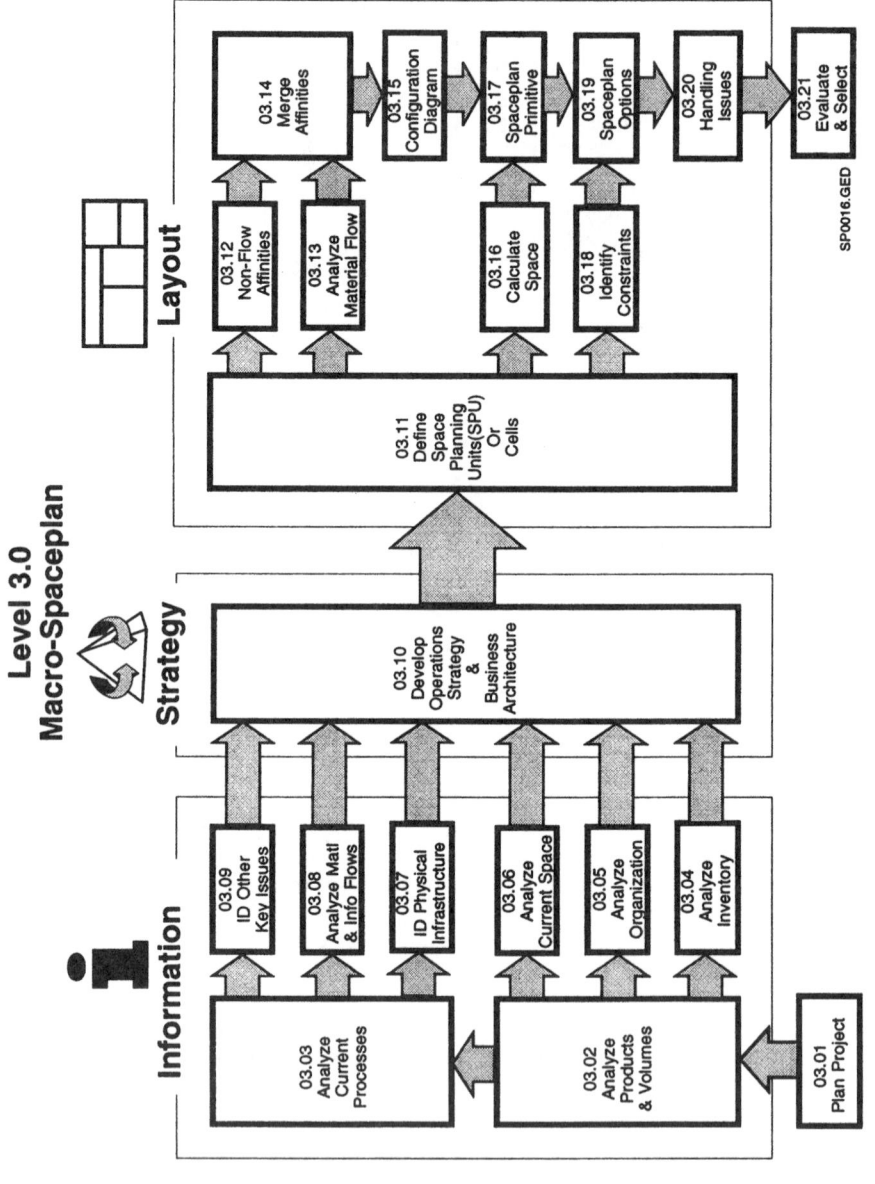

Level 3.0
Macro-Spaceplan

Information

Strategy

Layout

03.01 Plan Project

03.02 Analyze Products & Volumes

03.03 Analyze Current Processes

03.04 Analyze Inventory

03.05 Analyze Organization

03.06 Analyze Current Space

03.07 ID Physical Infrastructure

03.08 Analyze Matl & Info Flows

03.09 ID Other Key Issues

03.10 Develop Operations Strategy & Business Architecture

03.11 Define Space Planning Units(SPU) Or Cells

03.12 Non-Flow Affinities

03.13 Analyze Material Flow

03.14 Merge Affinities

03.15 Configuration Diagram

03.16 Calculate Space

03.17 Spaceplan Primitive

03.18 Identify Constraints

03.19 Spaceplan Options

03.20 Handling Issues

03.21 Evaluate & Select

SP0016.GED

686

then painted. Each component comes with one of two types of connection: plain or flanged. The plain duct has a slip fit similar to the pipe on a household vacuum cleaner. Flanged duct has a circular angle welded to one end that mates with a similar angle on the adjoining duct. At the opposite end, a circular angle floats to allow holes to align with the mating duct. The floating angle is held on the duct by turning a small flange on the end of the duct skin. The connection then bolts together.

The work for this task also included gathering information on overall sales volumes for the past 8 years and a forecast for 7 future years. This is shown in Figure 6. Product profiles show the distribution of total sales between product groups, (Figure 7), connection (Figure 8), and duct size (Figure 9)..

Each product comes in a wide range of diameters. The duct size distribution, shown in Figure 9, provides insight into the relative volumes of various diameters.

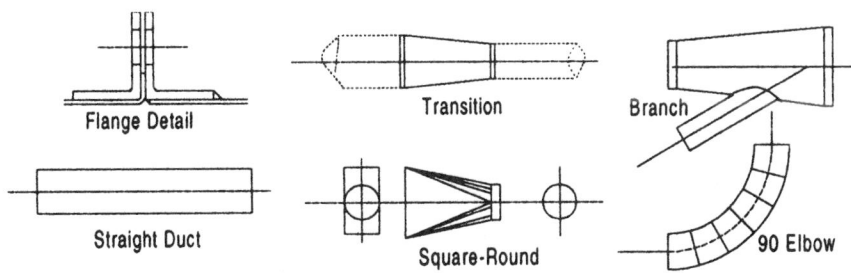

Figure 5. Induct product groups.

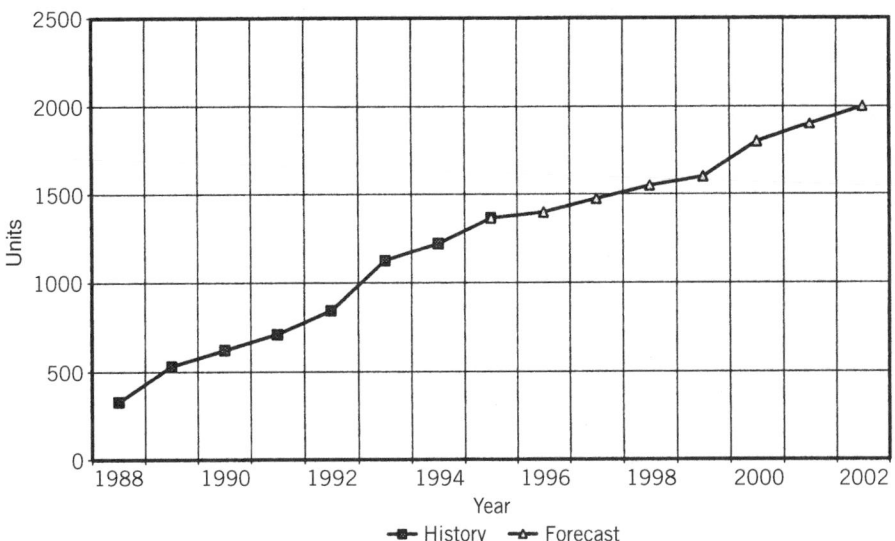

Figure 6. Sales history/forecast.

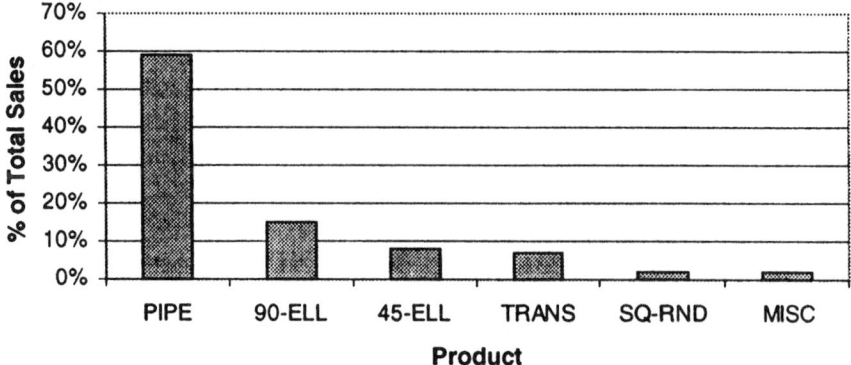

Figure 7. Product profile by group.

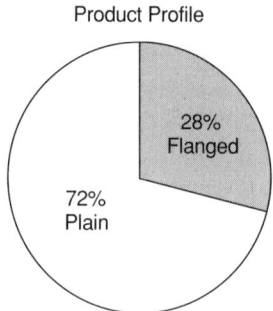

Figure 8. Product profile by connection.

Toward the end of this task, we had new insights concerning Induct's products, markets, and sales volumes for the past, present, and future. While much more needed to be learned before arriving at firm decisions, we nevertheless came to some preliminary conclusions:

- The new layout would be based initially on 1560 pieces per day. This should provide capacity for 2 years.
- The largest production volume is in straight pipe and will probably require high-volume, semiautomated processes.
- The smallest volume is in miscellaneous and special fittings, which will probably require a craft-type shop.
- Other fittings appear to be good candidates for work cells.

Task 3.03 Existing Process Figure 10 is typical of the process charts developed for each major product group. These charts were developed by a broad-based team that included workers and managers from each area. Note the large proportion of non-value-added events, primarily delays and transports.

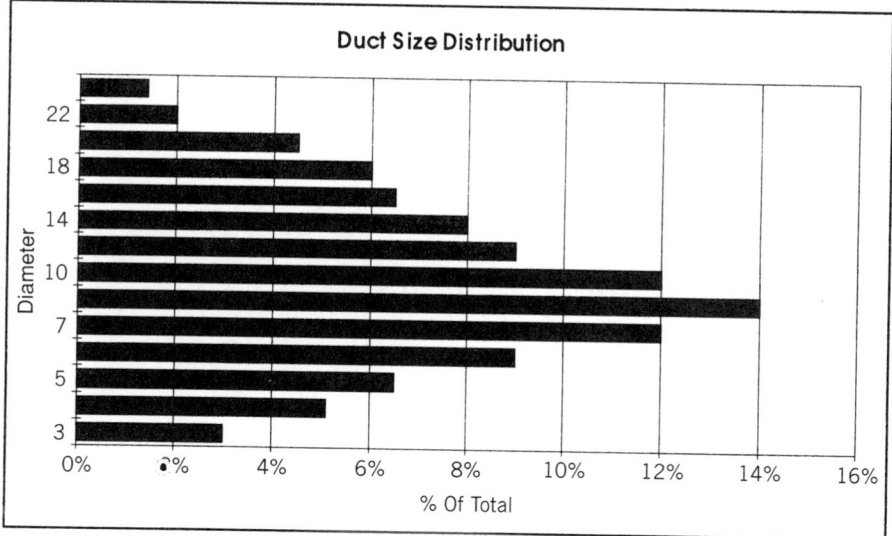

Figure 9. Product profile by duct size.

Task 3.04 Analyze Inventory Figure 11 shows the inventory history and a forecast for the next several years assuming no operational changes. Inventory turns are on a significant downtrend. If this trend continued, the new facility would need much more space than currently available. In addition, Induct would need to raise significant capital funds to pay for the inventory.

Figure 12 shows that most inventory is in finished goods. This is important because it illustrates the mismatch between customer requirements and manufacturing's ability to produce. Reducing this mismatch by making the process more flexible became an important goal.

Task 3.05 Analyze Organization A cellular implementation and the associated reengineering of processes usually leads to changes in organization and power structure. This is a delicate and potentially divisive issue. Figure 13 shows Induct's current organization chart. Note that the descriptions of position correspond roughly to the descriptions of space in Figure 14. With a new layout these position descriptions could drastically change. Helping people to recognize this possibility and prepare for it was an important benefit from this task.

We also evaluated the culture and organizational style at Induct. In the past, it reflected the personality of a long-time production manager who had only retired a few years before. The style tended to be confrontational and dictatorial. It became clear that significant changes would be required to make a cellular layout work well.

Task 3.06 Analyze Current Space Here we determine how space is currently used. This can take the form of a table or coded layout as in Figure 14. Here we see that a large part of the current space is used for inventory and material movement.

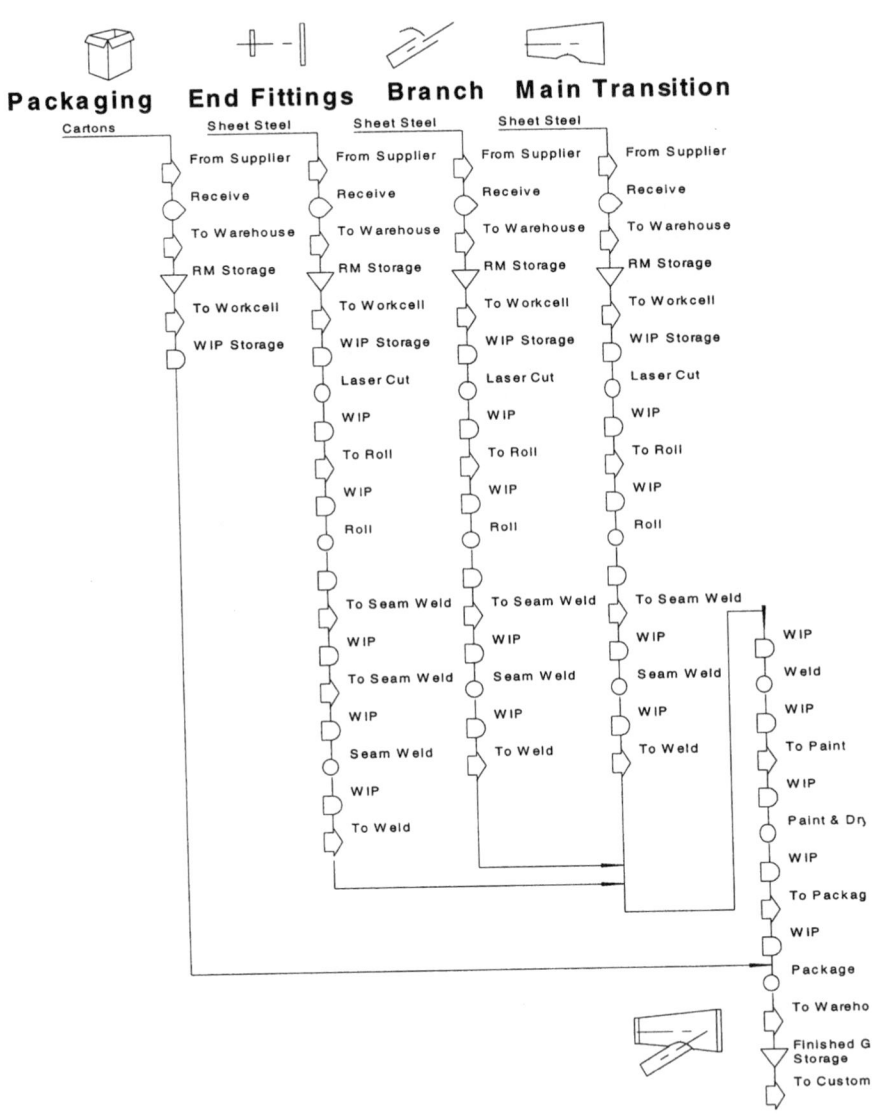

Packaging   End Fittings   Branch   Main Transition

Figure 10. Existing process—branch fittings.

At the beginning of the project, insufficient space was a major concern. Our analysis of current space usage, however, demonstrated that the real issue was how to utilize the space available. As the management team came to this recognition, the project emphasis shifted from expansion to reengineering.

Task 3.07 Physical Infrastructure Physical infrastructure are the spaces and facilities that support various processes. The project team made a list of items such

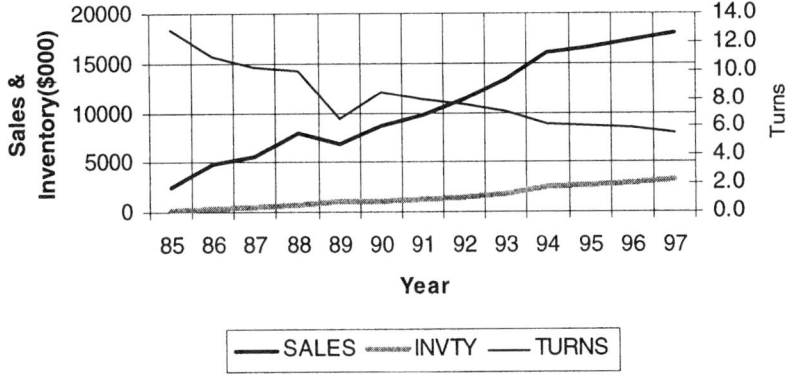

Figure 11. Inventory history/forecast.

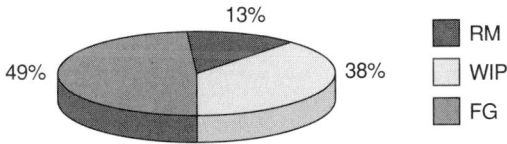

Figure 12. Inventory distribution by production stage.

as maintenance, employee services, electrical systems, and other items necessary for the new layout.

Task 3.08 Material and Information Flows   Figure 15 shows the material flow for one product. The flows of other products would be similar in nature. Note the long, complex flows and the large number of moves. If all the products were charted on this layout, the picture would be far worse. This analysis provides a baseline for comparing the new layout. It also helped Induct's management appreciate the nature and extent of the problem.

3.3 Manufacturing Strategy

Induct's management then went through the development of a manufacturing strategy. This is a difficult process for organizations to do well. With training and outside assistance, Induct prepared the Manufacturing Strategy Summary discussed next.

Manufacturing Strategy Summary

Site Mission Statement   Supply our customers with prefabricated duct components within 24 hours and quality within the top 20% in our industry. At Induct we expect to be a good neighbor and integral part of our community. Induct should be known as a satisfactory employer.

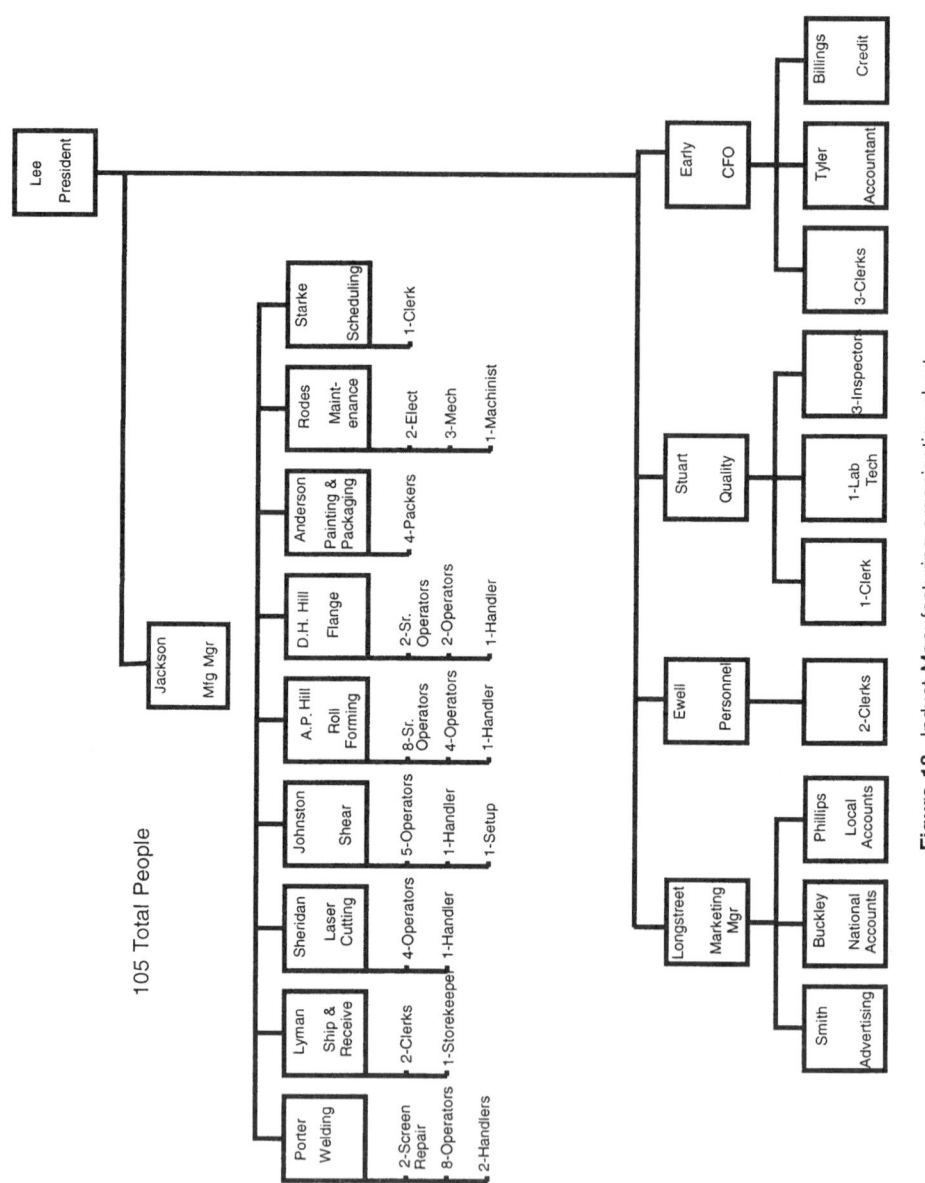

Figure 13. Induct Manufacturing organization chart.

692

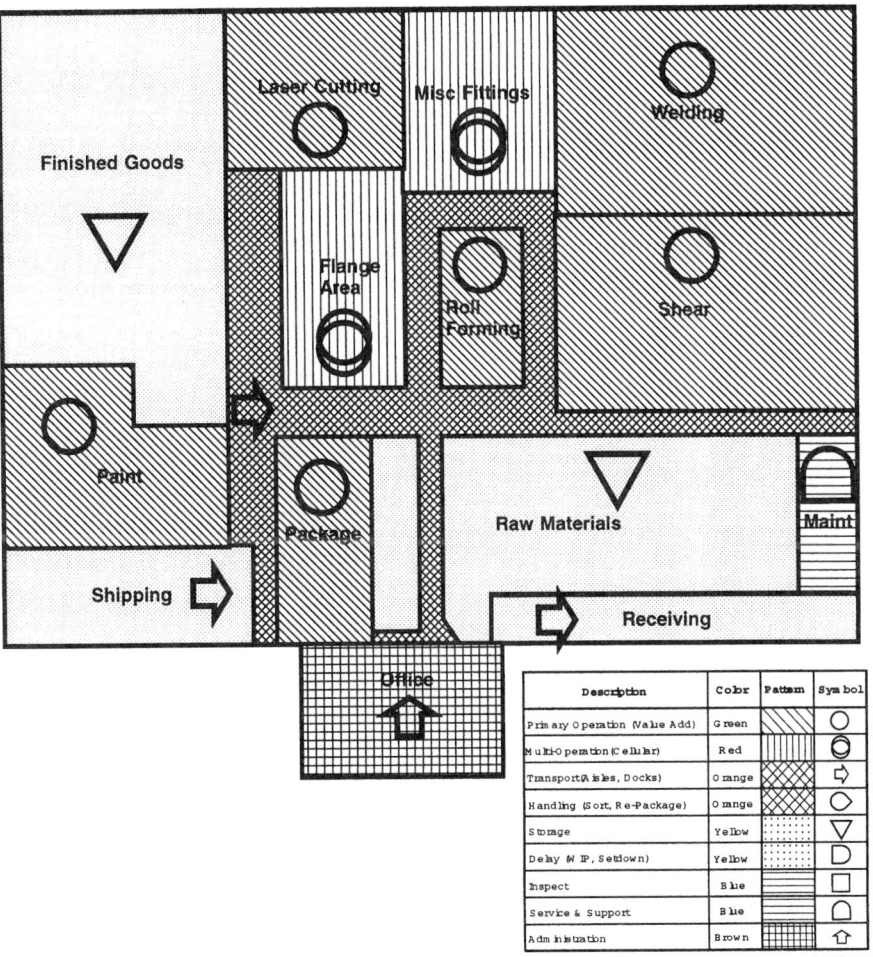

Description	Color	Pattern	Symbol
Primary Operation (Value Add)	Green		○
Multi-Operation (Cellular)	Red		⊚
Transport (Aisles, Docks)	Orange		⇨
Handling (Sort, Re-Package)	Orange		○
Storage	Yellow		▽
Delay (WIP, Setdown)	Yellow		�yD
Inspect	Blue		□
Service & Support	Blue		◠
Administration	Brown		⇧

SP0012GED

	TOTAL	OPER	TRANS	HANDLE	DELAY	INSPECT	STORE	SUPPT	OFFICE
Finished Goods	8672						8672		
Paint	3760	3760							
Shipping	2460		2460						
Laser Cut	2744	1150	200		1244	150			
Misc Fittings	2900	1173	345		1332	50			
Flange	2800	1536	267		997				
Packaging	1920	264	100		1556				
Raw Materials	6528						6528		
Shear	4200	2008	356		1447	155	234		
Weld	4480	2034	300		1865	137	144		
Office	2242								2242
Aisles	9172		9172						
	51878	11925	13200	0	8441	492	15578	0	2242
		23%	25%	0%	16%	1%	30%	0%	4%

Figure 14. Existing space analysis.

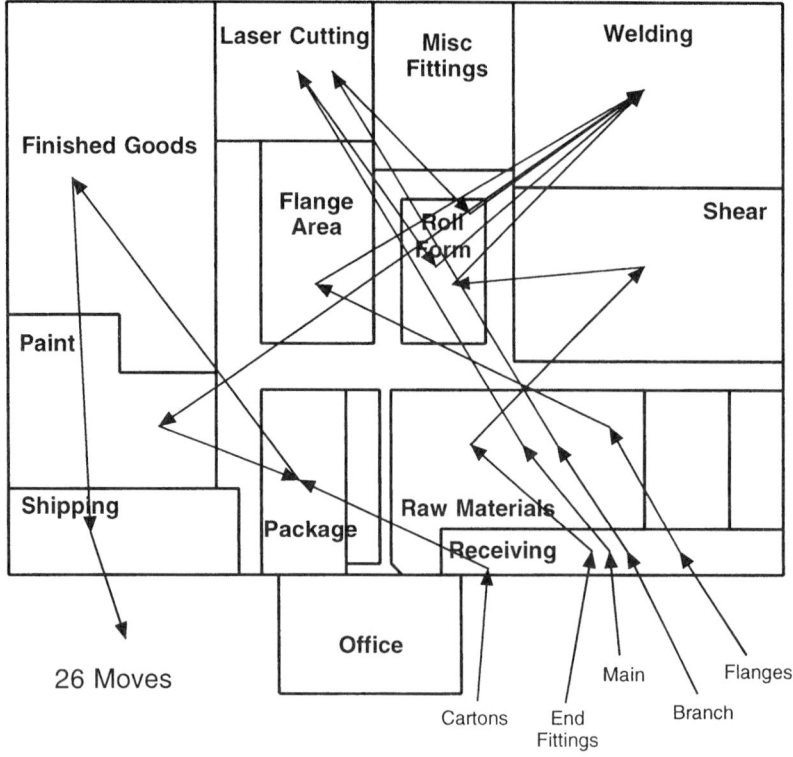

Figure 15. Induct material flow.

Process   Induct will strive for a product-focused factory. We will have a mix of large and small-scale equipment in Group Technology (GT) cells. Rapid setup is an important priority for equipment selection and operation. We will attempt an average equipment utilization between 60 and 85%. We will add process capacity 6–12 months ahead of demand. Processes should have a minimum capability index of 1.4. We will move gradually to higher technology processes provided they are consistent with our focus strategy, cost justified, and have adequate support.

Infrastructure   Induct will strive for a product-focused, shallow, consultative, and informal organization. We will gradually move toward a participative team-based organization over the next 5 years.

Our accounting system should accommodate activity-based costing using cost drivers for overhead allocation. We recognize the limitations of conventional accounting systems for management decisions.

Production control will use Materials Requirement Planning (MRP)-type scheduling for suppliers. We will use kanban systems for internal scheduling. We will use small finished goods stocks for the highest volume 20% of line items. The remaining

80% of low-volume products will be made to order. Unusually large orders for any product will have extended deliveries and be made to order.

At Induct we will strive for long-term relationships with reliable suppliers. We will select suppliers on the basis of quality, delivery reliability, and cost in that order.

Facilities Our facility will emphasize flexibility of layout. Bus duct will distribute electricity. Compressed air distribution will be designed for easy addition of new drops in any location.

3.4. Layout Design

Task 3.11 Define Space Planning Units Space Planning Units are the units of space, which we arrange on the layout. The definition of these units determines the organization of manufacturing. It determines large-scale material flow. It is, perhaps, the most important step in the layout process.

Working from the Manufacturing Strategy Summary and the information documents, we began to identify SPUs. There was broad agreement that some SPUs would be functional. These included shipping and receiving, raw materials storage, finished goods storage and, maintenance, employee services, and the office.

The manufacturing work cells required considerably more analysis. We used Production Flow Analysis (PFA). In this technique, each product is listed on the rows of a matrix and each process or workcenter on the columns. The initial matrix is shown in Figure 16.

By rearranging the rows and column, we can aggregate families of similar products with groups of machines or workcenters. This required some changes in the original process. It also required some additional equipment. Where a wide range of products use the same processes or workcenters, these were placed in the far right columns. Such processes must appear in most or all of the resulting workcells. The resulting matrix is shown in Figure 17.

Tasks 3.12–3.14 Affinity Development These three tasks result in a set of affinities. The material flow analysis was quite complex and we will not show the details here. To perform the flow analysis, we defined a standard material flow unit. From process charts and the SPU definition, we identified each interdepartmental move. The design volume and other data from the product–volume analysis determine the number of units moved per day.

Many affinities result from nonflow needs such as communication or shared resources. These affinities do not lend themselves to quantitative methods. Accordingly, these affinities were developed with a consensus meeting. Next we merged the flow and nonflow affinities using a weighted average. The results are shown in Figure 18.

Task 3.15 Configuration Diagram From the SPUs and affinities, the designers made an idealized diagram. Here we attempted to maintain close proximity for those SPUs with high valued affinities while allowing lower valued affinities greater distance. The diagram for Induct's layout is shown in Figure 19.

TYPE	STYLE	SIZE	ROLL SMALL	FLANGE	ANGLE SHEAR	SEAM WELD, SMALL	LASER CUT	PUNCH	SHEAR	BRAKE	SEAM WELD, LARGE	ANGLE ROLL	MANUAL WELD	ROLL, LARGE	SEAM WELD, LARGE	ROLL, LARGE	ROLL, SMALL	SEAM WELD, SMALL
0202-XXXXXX	90-ELBOW	FLANGED	SMALL	4		4								4				4	
0101-XXXXXX	DUCT	PLAIN	LARGE								4				4	4			
0501-XXXXXX	TRANSITION	FLANGED	SMALL	4		4								4				4	4
0401-XXXXXX	30-ELBOW	PLAIN	LARGE								4			4			4		
0102-XXXXXX	DUCT	FLANGED	SMALL	4	4		4			4				4					
0301-XXXXXX	45-ELBOW	PLAIN	SMALL								4			4				4	
0902-XXXXXX	45-BRANCH	FLANGED	SMALL	4		4								4				4	4
0302-XXXXXX	45-ELBOW	FLANGED	LARGE	4		4								4				4	
0101-XXXXXX	DUCT	PLAIN	SMALL	4			4			4									
0402-XXXXXX	30-ELBOW	FLANGED	SMALL	4		4								4				4	
0501-XXXXXX	TRANSITION	PLAIN	LARGE								4		4				4		
0601-XXXXXX	SQ-RND	PLAIN	LARGE								4		4	4			4		
0401-XXXXXX	30-ELBOW	PLAIN	SMALL								4			4				4	
0102-XXXXXX	DUCT	FLANGED	LARGE	4						4				4	4	4			
0202-XXXXXX	90-ELBOW	FLANGED	LARGE	4		4								4				4	
0501-XXXXXX	TRANSITION	FLANGED	LARGE	4		4							4	4			4		
0702-XXXXXX	FLANGE	FLANGED	LARGE			4			4			4							
0201-XXXXXX	90-ELBOW	PLAIN	LARGE								4			4			4		
0602-XXXXXX	SQ-RND	FLANGED	LARGE	4		4				4				4			4		
0902-XXXXXX	45-BRANCH	FLANGED	LARGE	4		4							4	4			4		
0602-XXXXXX	SQ-RND	FLANGED	SMALL	4		4				4				4				4	
0702-XXXXXX	FLANGE	FLANGED	SMALL			4			4			4							
0501-XXXXXX	TRANSITION	PLAIN	SMALL								4							4	4
0801-XXXXXX	30-BRANCH	PLAIN	SMALL								4							4	4
0302-XXXXXX	45-ELBOW	FLANGED	SMALL	4		4								4				4	
0301-XXXXXX	45-ELBOW	PLAIN	LARGE								4			4				4	
0201-XXXXXX	90-ELBOW	PLAIN	SMALL								4			4				4	
0901-XXXXXX	45-BRANCH	PLAIN	LARGE								4		4	4			4		
0802-XXXXXX	30-BRANCH	FLANGED	LARGE	4		4							4	4			4		
0802-XXXXXX	30-BRANCH	FLANGED	SMALL	4		4								4				4	4
0402-XXXXXX	30-ELBOW	FLANGED	LARGE	4		4								4				4	
0601-XXXXXX	SQ-RND	PLAIN	SMALL								4		4	4				4	
0901-XXXXXX	45-BRANCH	PLAIN	SMALL								4							4	4
0801-XXXXXX	30-BRANCH	PLAIN	LARGE								4			4	4		4		

Figure 16. Initial PFA matrix.

Tasks 3.16–3.18 Space and Constraints The project team calculated space requirements for the design volume using several of the six methods. This space was then superimposed on the configuration diagram to obtain a layout primitive. We identified constraints such as fixed locations, building envelope, material handling, and others.

	TYPE	STYLE	SIZE	ROLL SMALL	SEAM WELD, SMALL	SHEAR	ROLL LARGE	SEAM WELD, LARGE	ROLL LARGE	SEAM WELD, LARGE	LASER CUT	ROLL SMALL	SEAM WELD, SMALL	BRAKE	ANGLE SHEAR	ANGLE ROLL	PUNCH	MANUAL WELD	FLANGE
0101-XXXXXX	DUCT	PLAIN	SMALL	4	4	4													
0102-XXXXXX	DUCT	FLANGED	SMALL	4	4	4												4	4
0101-XXXXXX	DUCT	PLAIN	LARGE				4	4	4										
0102-XXXXXX	DUCT	FLANGED	LARGE				4	4	4									4	4
0501-XXXXXX	TRANSITION	PLAIN	LARGE						4	4	4							4	
0501-XXXXXX	TRANSITION	FLANGED	LARGE						4	4	4							4	4
0801-XXXXXX	30-BRANCH	PLAIN	LARGE						4	4	4							4	
0802-XXXXXX	30-BRANCH	FLANGED	LARGE						4	4	4							4	4
0901-XXXXXX	45-BRANCH	PLAIN	LARGE						4	4	4							4	
0902-XXXXXX	45-BRANCH	FLANGED	LARGE						4	4	4							4	4
0201-XXXXXX	90-ELBOW	PLAIN	LARGE								4	4						4	
0201-XXXXXX	90-ELBOW	PLAIN	SMALL								4	4						4	
0202-XXXXXX	90-ELBOW	FLANGED	LARGE								4	4						4	4
0202-XXXXXX	90-ELBOW	FLANGED	SMALL								4	4						4	4
0301-XXXXXX	45-ELBOW	PLAIN	LARGE								4	4						4	
0301-XXXXXX	45-ELBOW	PLAIN	SMALL								4	4						4	
0302-XXXXXX	45-ELBOW	FLANGED	LARGE								4	4						4	4
0302-XXXXXX	45-ELBOW	FLANGED	SMALL								4	4						4	4
0401-XXXXXX	30-ELBOW	PLAIN	LARGE								4	4						4	
0401-XXXXXX	30-ELBOW	PLAIN	SMALL								4	4						4	
0402-XXXXXX	30-ELBOW	FLANGED	LARGE								4	4						4	4
0402-XXXXXX	30-ELBOW	FLANGED	SMALL								4	4						4	4
0501-XXXXXX	TRANSITION	PLAIN	SMALL								4	4	4						
0501-XXXXXX	TRANSITION	FLANGED	SMALL								4	4	4					4	4
0801-XXXXXX	30-BRANCH	PLAIN	SMALL								4	4	4					4	
0802-XXXXXX	30-BRANCH	FLANGED	SMALL								4	4	4					4	4
0901-XXXXXX	45-BRANCH	PLAIN	SMALL								4	4	4					4	
0902-XXXXXX	45-BRANCH	FLANGED	SMALL								4	4	4					4	4
0601-XXXXXX	SQ-RND	PLAIN	LARGE								4	4		4				4	
0601-XXXXXX	SQ-RND	PLAIN	SMALL								4	4		4				4	
0602-XXXXXX	SQ-RND	FLANGED	LARGE								4	4		4				4	4
0602-XXXXXX	SQ-RND	FLANGED	SMALL								4	4		4				4	4
0702-XXXXXX	FLANGE	FLANGED	LARGE												4	4	4		
0702-XXXXXX	FLANGE	FLANGED	SMALL												4	4	4		

Figure 17. Final PFA matrix.

Task 3.19 Macro-Layout Options

Task 3.19 Macro-Layout Options   From constraints and the configuration diagram, we synthesized several viable layout options. The preferred option is shown in Figure 20.

This layout has material flows superimposed that correspond to the branch and transition flows in Figure 15. Note the 50% reduction in material moves and relative simplicity of flow. Induct's products are relatively simple compared to most

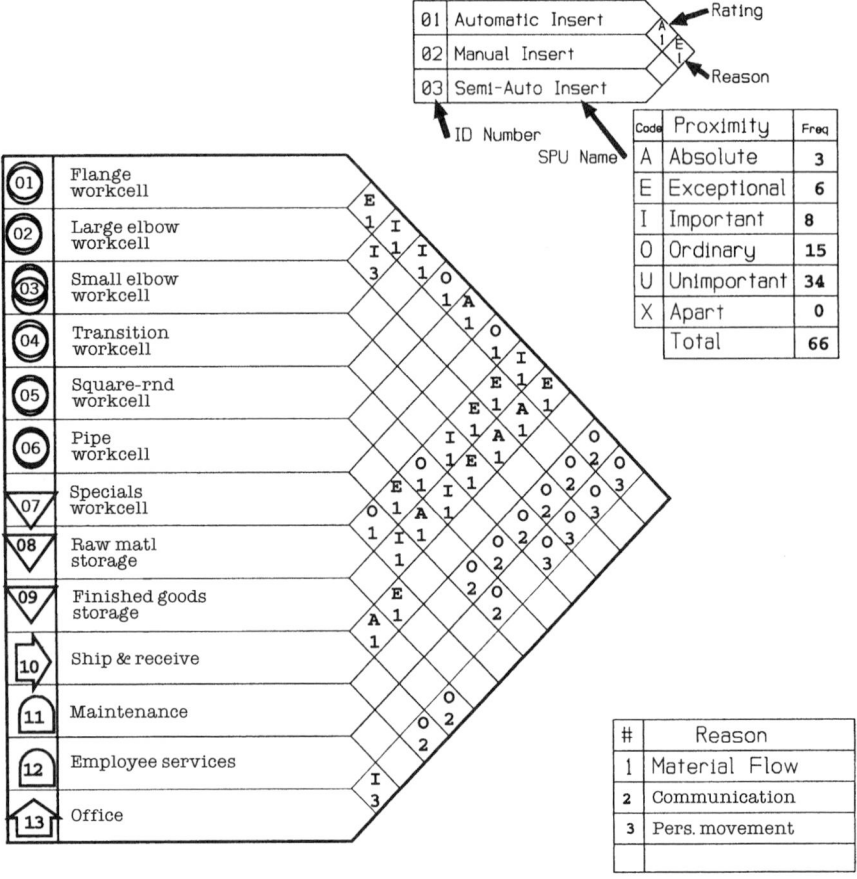

Figure 18. Total affinities for Induct Manufacturing.

manufacturers and the 50% reduction is actually quite modest. It is not unusual to see 80–90% reductions in the number of material moves with more complex products.

The branch and transition work cell is in a remote location compared to other work cells. This resulted from the affinities that in turn reflected material, communication, and resource movement. For this reason the resulting movements of material for this product remain relatively long. If the comparison were made for other products, their moves would be significantly shorter.

4. MICRO LAYOUT

4.1. Model Project Plan

Work cell design (micro space planning) has five major tasks. Task 4.01 examines the full range of possible products for work cells and selects a product or product

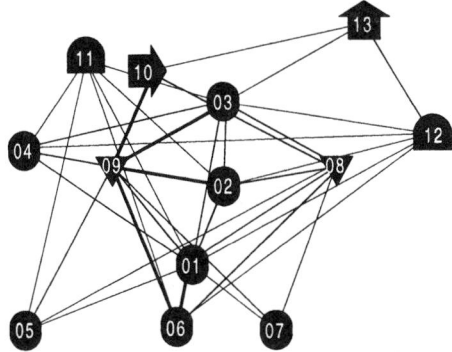

Figure 19. Induct configuration diagram.

family for each cell. Task 4.02 analyzes the current or proposed process. Task 4.02 examines alternate processes and process improvements. Task 4.03 designs the work cell's infrastructure. Task 4.04 produces space plans or layouts for each option.

4.2. Select the Products

Selecting products for a workcell is the most fundamental of the four tasks. It determines the number of people and the equipment for the cell. Remaining design decisions flow from this task.

The layout designer has three methods available for product selection: (1) intuitive grouping, (2) PFA, and (3) coding and classification. Intuitive grouping uses process charts and general knowledge to identify product families. Production Flow Analysis uses a matrix of existing products and processes. Coding and classification is a powerful but complex technique that can handle many thousands of products and processes.

Product selection for the work cells can be done at either the macro or micro level. For Induct, we chose to make the selection at the macro level. This task was described earlier.

4.3. Engineer the Process

The project team started with the process shown in Figure 21. A critical review concluded that the equipment and technology was appropriate for our work cell. However, the many transfers and delays that resulted from batching could probably be eliminated. This revised process is also shown in Figure 21.

4.4. Infrastructure—The Hidden Design

Some elements of a work cell are not part of the process, yet they are necessary to support the process and the operation of the cell. These are the infrastructural elements.

Option "A"

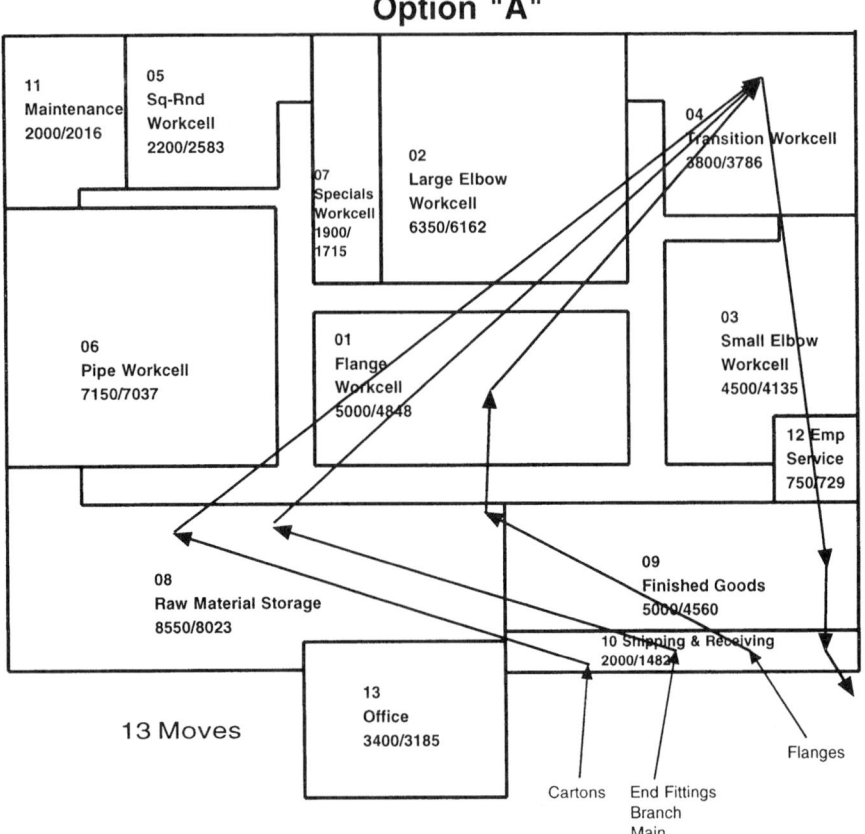

Figure 20. Induct macro space plan.

The decisions made in this task frequently involve intangibles such as scheduling approaches and compensation. Because they are intangible, work cell designers can easily overlook them. Here is a brief summary of infrastructure elements.

External Lot Size This is the quantity of product produced by the entire cell before changeover to a different product or order. Lot sizes should be as small as possible. If the setup of the cell requires significant time and expense, larger lot sizes amortize the cost over multiple items. The best work cells use processes that are quick to setup and therefore need small or one-piece lots.

External Containers External containers convey units of incoming and outbound materials for the cell. The outbound containers should always be smaller than the external lot size. When going from a functional layout to a cellular layout, you may need smaller containers and different material handling equipment.

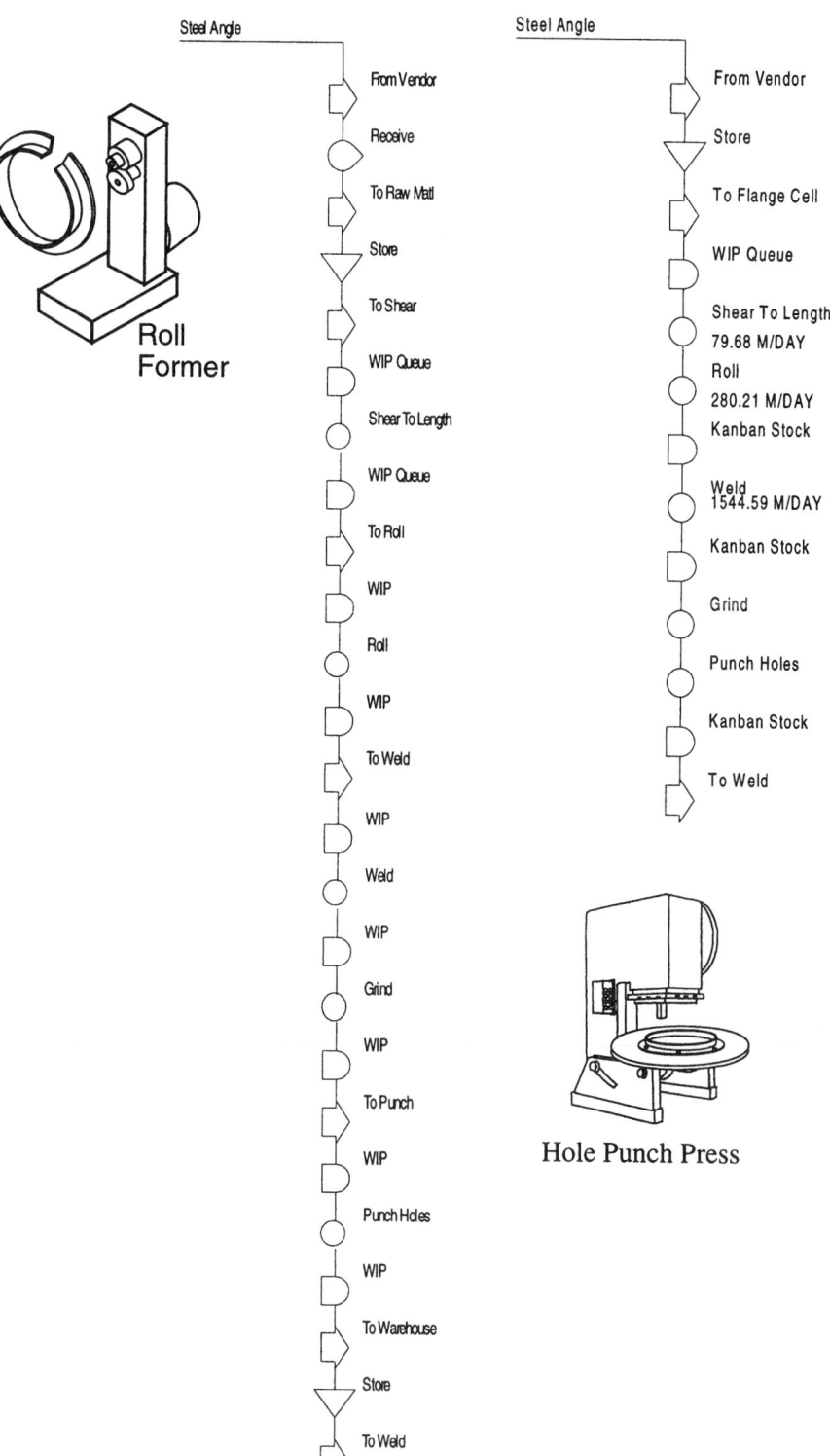

Steel Angle

From Vendor
Receive
To Raw Matl
Store
To Shear
WIP Queue
Shear To Length
WIP Queue
To Roll
WIP
Roll
WIP
To Weld
WIP
Weld
WIP
Grind
WIP
To Punch
WIP
Punch Holes
WIP
To Warehouse
Store
To Weld

Roll Former

Steel Angle

From Vendor
Store
To Flange Cell
WIP Queue
Shear To Length
79.68 M/DAY
Roll
280.21 M/DAY
Kanban Stock
Weld
1544.59 M/DAY
Kanban Stock
Grind
Punch Holes
Kanban Stock
To Weld

Hole Punch Press

Figure 21. Flange process—before/after.

External Material Handling Materials must come into the cell from outside. Here we specify the methods and equipment for this. Equipment should be appropriate for the container and external lot size. Also, specify the route structure. Direct route structures carry a single load of material from origin to destination. Fork trucks, for example, use direct route structures. Channel route structures are like bus routes. In channel routes, the conveying equipment follows a fixed route and picks up partial loads at various points. Terminal structures are like airline routes. The transported items move to a central terminal for consolidation. They then move to a destination terminal where the load is separated and carried to the final destination.

External Production Control The work cell must know what to build and when. External production control refers to the method for doing this. Physical linking ties the work cell to upstream and/or downstream operations, often with a conveyor. Broadcast systems use a single schedule that is sent or broadcast to all operations in the production stream. Kanban uses small stockpoints, which are replenished as material is withdrawn. Push schedules tell each production unit what to make during a given time period.

Internal Production Control Internal production control tells each operator and each workstation what to make and when. Direct link and broadcast systems are common means for controlling cell production internally. Kanban stockpoints between operations are also common. With the circulation method, each operator carries the product through all operations. Push scheduling rarely is appropriate within a cell.

Internal Lot Size The internal lot size is also known as the "transfer batch." It is the number of units moved from operation to operation within the cell. The ideal internal lot size is a single piece. Larger internal lot sizes may be appropriate when cycle times are very short or when equipment requires several units for each cycle.

Equipment Balance To balance the workload on equipment, we usually design the cell around the bottleneck operation and allow excess capacity for other operations. For short-term imbalances from changing product mix, you may use short queues within the cell. The queues can double as kanban stockpoints. Inherent balance occurs when all operations have precisely the same work times for every product. This is seldom feasible.

Internal Containers Internal containers are often different than those external to the cell. You may use the same container if the internal and external lot size is the same.

People Balance Balancing the work among work cell operators is more critical than balance among machines. Inherent balance is rarely achieved. Some means of distributing and sharing work is therefore necessary for balance. Circulation serves this purpose. An internal kanban system is also effective. The floating method, often used with kanban, allows operators to float between primary and secondary tasks. They can float over to assist another operator who is overloaded and float back as necessary.

Quality Assurance Work cells fit well with Total Quality (TQ) concepts such as statistical process control, team-based problem solving, and worker participation. Conventional quality control is acceptable if the process is stable.

Supervision In most well-designed work cells, fast coordination of all operations and people is essential. They must function like a basketball team. Command and control supervision rarely enhances team performance. Self-directed work teams and coaching style supervision are most appropriate.

Compensation Work cells do not function well with individual incentive systems. The individual incentive destroys teamwork and focuses the workers' attention on only their individual tasks. Group incentives work better since they reward the entire work cell team. Hourly and salary pay schemes function well in most instances.

The Work Cell Operations Plan provides a convenient format for specifying and summarizing infrastructural design decisions. The operations plan for Induct's flange work cell is shown in Figure 22. For each infrastructural element there is a box containing the full range of possible options. As decisions are made, the designers check them off in the appropriate box. At the bottom of this form, the people who work in the cell are listed along with skill requirements and task assignments. A solid bullet indicates primary task responsibility. The unfilled bullets indicate a secondary task responsibility.

4.5. Layout the Work Cell

If the tasks discussed are done well, the detailed layout of the work cell is easy and straightforward. The fundamental and derived elements of Figure 3 still apply. However, we can often skip some of the intermediate steps and move directly from a process chart to the layout. Figure 23 is the result of this process for the branch fitting work cell at Induct Manufacturing.

5. SUMMARY

In this chapter we have demonstrated a step-by-step, structured process for reengineering a plant and designing the layout for CM. FacPlan improves both the productivity of facility planners and the quality of their designs. It allows relatively inexperienced designers and design teams to produce effective work cells.

The FacPlan approach starts by segregating the facility planning process into distinct levels of detail that become, largely, independent processes. At the macro level, we design a layout, which establishes the basic organization and material flow for the plant. To do this we use three groups of tasks: information acquisition, strategy,

Strategos

Workcell Operations Plan

Project: FLANGE WORKCELL		Prj#: IND 201
Company: INDUCT MFG	By: Q L	Date 10/02/96
Note:		

External Lot Size
In Out
- • Single Piece
- • Small <1.5 Hours
- • Medium <4.0 Hours
- ✓ • Large

Prod Unit: _____

External Containers
In Out
- • Small
- • Medium
- • Large

Type: _____

External Material Handling
In Out
- ✓ Small
- • Medium
- ✓ • Large

Route Structure:
- ✓ ✓ Direct
- • • Channel
- • • Terminal

Equipment:
- • • Hand Carry
- ✓ ✓ Other

Type: IN: FORK TRUCK
OUT: CARTS

External Production Control
In Out
- • • Physical Link
- ✓ ✓ Broadcast
- • • Push Schedule
- • • Re-Order Point
- • • Make-To-Order

Internal Production Control
- ✓ Direct Link
- ✓ Circulation
- ✓ Kanban
- • Push Schedule

Internal Lot Size
- • Single Piece (Cutoff)
- • Small <1.5 Hours
- • Medium <4.0 Hours
- • Large

Equipment Balance
- • Inherent Balance
- • Queue
- • Excess Capacity

Internal Containers
- • Small
- • Medium
- • Large

Type: Single Piece

People Balance
- • Inherent Balance
- ✓ Queue
- ✓ Excess People
- ✓ Circulation
- • Float

Quality Assurance
- ✓ Inspect & Reject
- • SPC/TQM
- • Inherent

Supervision
- • Self Managing (Cybernetic)
- • Kanban Signal
- • Command & Control

Compensation
- ✓ Hourly/Salary
- • Individual Incentive
- • Group Incentive

Operator Assignment & Skillmatrix

Operators	CUTOFF	ROLL	WELD	GRND	PUNCH		CUTOFF	ROLL	WELD	GRND	PUNCH						
	Skills						Operations										
MACARTHUR	●	●	○	○	○		●	●	○	○	○						
BRADLEY		●						●									
MONTGOMERY		●						●									
KLUGE			●						●								
RUNDSTEDT				●						●							

© Strategos, Inc.   CM 0010.GED

Figure 22. Flange work cell operations plan.

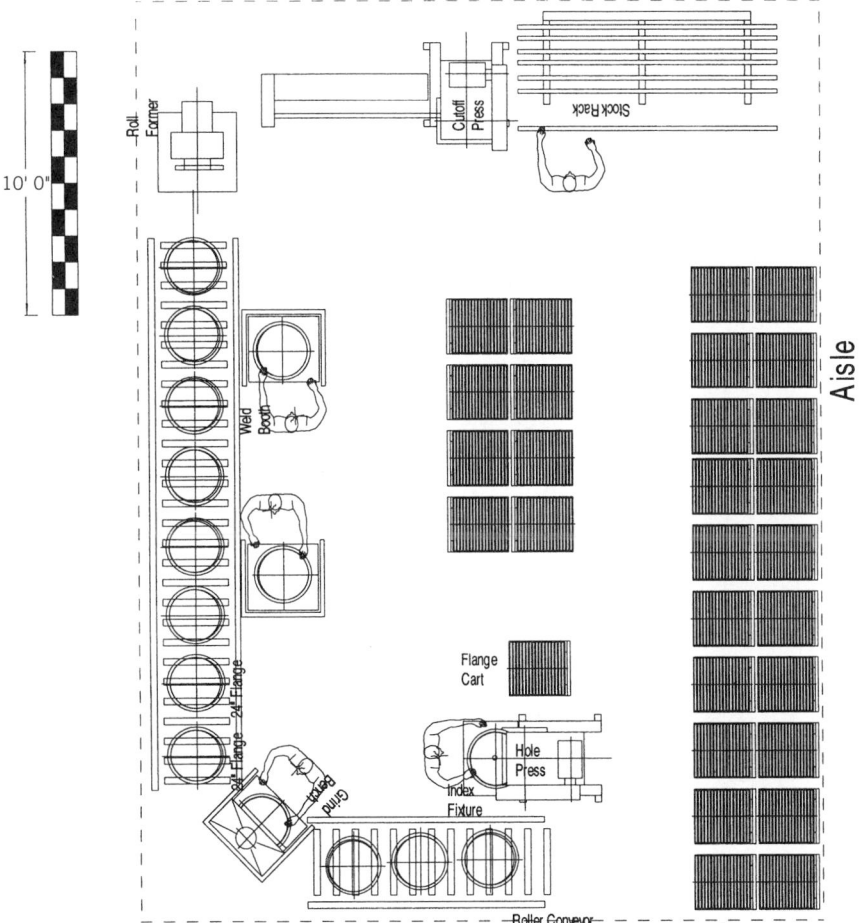

Figure 23. Flange work cell layout.

and layout. At the micro level, the design process is organized around four tasks: (1) selecting the products, (2) engineering the process, (3) selecting infrastructure, and (4) layout the work cell.

In January of 1940, Charles E. Sorensen planned the layout for Ford Motor Company's Willow Run bomber plant. By 1944 this remarkable factory was producing a four-engine heavy bomber every hour with half the labor content previously required. Willow Run was the crowning achievement of American industry during World War II. The principles and methods used at Willow Run are incorporated here in the FacPlan system. Sorensen made one error. He assumed that the product-focused, flow-line methods developed at Ford Motor Company applied only to "mass production." In fact, they apply across a broad range of manufacturing situations. Our challenge is to adapt them using imagination and ingenuity.

Bibliography

Lee, Q. (1990a). "Manufacturing Focus—A Comprehensive View," in *Manufacturing Strategy*, E. Voss (ed.), Chapman & Hall, Scientific, London.

Lee, Q. (1990b). Manufacturing Focus—A Comprehensive View, Operations Management Association (OMA) Conference Proceedings, Warwick, England, June.

Lee, Q. (1995a). *Cellular Manufacturing*, 2nd ed., Institute of Industrial Engineers, Nashville, TN.

Lee, Q. (1995b). *Kanban Scheduling—Effective, Simple, Demanding*, 2nd ed., Institute of Industrial Engineers, Nashville, TN.

Lee, Q. (1997). *Facilities and Workplace Design: An Illustrated Guide*, Engineering & Management Press, Atlanta, GA.

Lee, Q. (1996b). *Reengineering Through Facility Layout*, 2nd ed., Institute of Industrial Engineers, Minneapolis, MN.

Singh, N., and Rajamani, D. (1996). *Cellular Manufacturing Systems*, Chapman & Hall, London.

Wrennall, W., and Lee, Q. (1993). *Handbook of Commercial and Industrial Facilities Management*, McGraw Hill, New York.

23

CONVERSION TO CELLULAR MANUFACTURING AT SHEET METAL PRODUCTS

Danny J. Johnson

College of Business
Iowa State University
Ames, Iowa 50011

1. INTRODUCTION

1.1. Background

This chapter details the conversion to Cellular Manufacturing (CM) at Sheet Metal Products (SMP). (The plant's real name and the actual sheet metal products produced have been disguised for confidentiality reasons.) It is based on information obtained through on-site interviews with employees, through the examination of company documents, and by observing the manufacturing system in place at the plant. The study was conducted over a one-year period that ended in October of 1997. The classification and number of individuals interviewed at the plant is shown in Table 1.

1.2. Company Description

Sheet Metal Products manufactured sheet metal products of various dimensions and configurations and light- and medium-duty ball slides used by the drawers contained

Handbook of Cellular Manufacturing Systems, edited by Shahrukh A. Irani
ISBN 0-471-12139-8 © 1999 John Wiley & Sons, Inc.

Table 1. Classification and number of employees interviewed at SMP

Employee Classification	Number Interviewed
Plant manager	1
Engineering manager	1
Human resources manager	1
Materials manager	1
Accounting supervisor	1
Plant superintendent	1
Industrial engineer	2
Drawer slide cell supervisor (original)	1
Drawer cell supervisor (original)	1
Drawer cell supervisor (present)	1
Enclosure fabrication cell supervisor	1
Twin line welding supervisor	1
Drawer slide cell operator	2
Drawer cell operator	3
Enclosure fabrication cell operator	2
Spot welder	3
Brake operator	3

in some products. The plant was part of a multi-plant corporation and 15 of the larger products produced accounted for 75–80% of the plant's total sales. Ninety-five percent of all products produced were sold in the United States in both commercial and industrial markets.

Approximately 95% of the products were made to stock while the remaining 5% were made to order. The made-to-order percentage of the business was expected to grow in the future due to increased demand for customized products. Most of this customization was expected to occur at the assembly stage where customers could choose from a standard set of product configuration options and SMP would manufacture and assemble the products according to those specifications as the orders were received.

The total size of the plant was approximately 420,000 square feet. Seventy thousand square feet was office and engineering space, and the remaining 350,000 square feet was devoted to the manufacturing floor and the shipping area. The plant had 140 major pieces of equipment directly used for manufacturing. This number did not include forklifts, roller conveyors, and other material handling equipment; machines used for facility maintenance, tool sharpening, equipment repair, and fixture and die manufacturing; or equipment contained in the plant's painting system or the zinc plating line (used for plating the ball slides used in some products).

The plant had 10,000 active part numbers and operated one full and two partially staffed, 8-hour shifts per day, 5 days per week. Approximately 30% of all direct labor hours and 30% of all machine hours were spent in cells. The plant had plans to install more cells over the next 2 or 3 years.

Sheet Metal Products employed 400 individuals. Seventy-six of these employees were office workers (administration and clerical workers, engineers, supervisory personnel, managerial employees, etc.), 68 were indirect labor (maintenance people, tool and die makers, material handlers, facility maintenance, etc.), and the remaining 256 were direct labor (machine operators and assemblers). All direct and indirect labor employees were paid on an hourly basis and all office workers were salaried. The rate of employee turnover was very low for both salaried and hourly employees, and the average employee had been with the plant for approximately 20 years. The plant was unionized and was represented by the International Association of Machinists and Aerospace Workers.

2. PRECELL ENVIRONMENT

2.1. Historical Overview

Prior to 1986, SMP manufactured sheet metal products and assembled printed circuit boards and hose and gauge assemblies used in products manufactured at other plants. Twenty-five percent of the plant's floor space was devoted to the assembly of printed circuit boards and hose and gauge assemblies, and the remaining 75% was devoted to the fabrication of sheet metal products. In 1985, the corporation decided to focus each of its plants. As a result, SMP's plant became focused on sheet metal products and the assembly of printed circuit boards was moved to a different plant. The assembly of hose and gauge assemblies was moved to a different plant several years later.

As the focusing of the plants took place, the managers at SMP began to investigate the feasibility of changing the layout of equipment and flow of material on the plant floor to improve performance. In December of 1988, SMP implemented two high-volume fabrication lines called the Twin Lines. These lines marked the starting point of the conversion to CM, and the precell environment described in this section includes the production system prior to their implementation. In addition, since the plant was focused on sheet metal processes at the time of this study, the discussion concentrates on the manufacturing system for the sheet metal products.

2.2. Previous Organizational Structure

The organization structure of the plant prior to the implementation of the Twin Lines followed a departmental orientation with six individuals reporting directly to the plant manager: an accounting supervisor, a materials manager, an engineering manager, a human resources manager, a plant superintendent, and a quality control (or total quality systems) supervisor (see Figure 1).

2.3. Previous Production Processes

The exact sequence of operations and the number of times a particular operation had to be performed to build a sheet metal product varied from unit to unit. However, the basic steps required to produce the units were very similar and can be generalized as follows: (1) The different panels and sheet metal components that comprised the unit

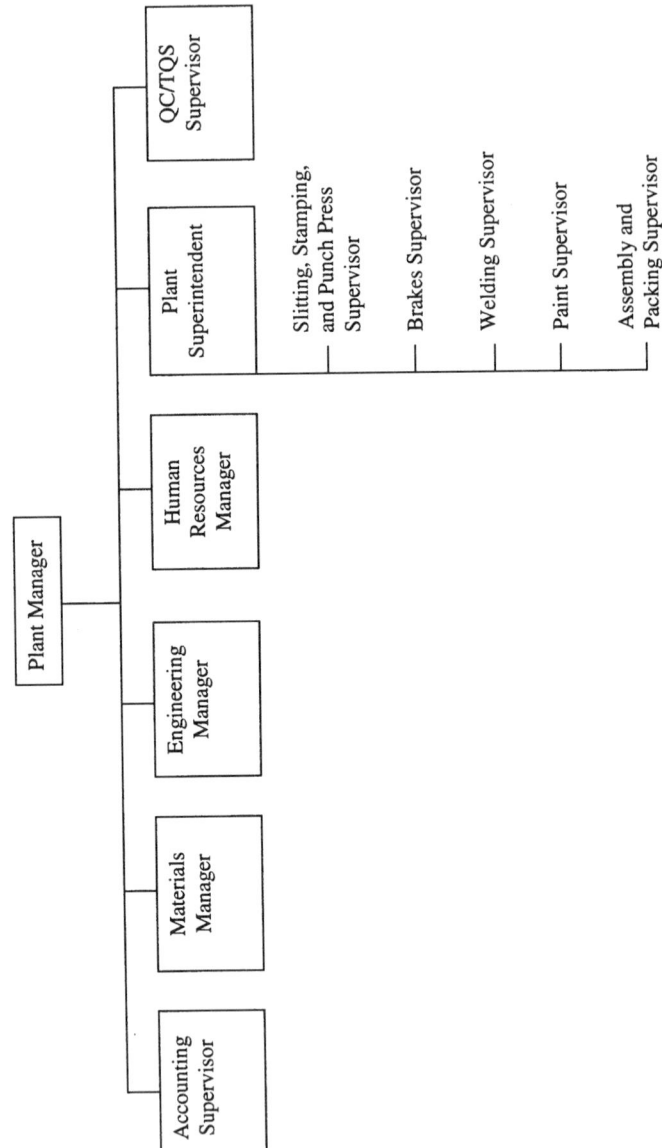

Figure 1. Sheet Metal Products organizational chart prior to cells.

were either stamped or punched out of sheet metal to form the flat shape of the parts needed (these flat parts were called blanks), (2) the blanks were sent to a press brake where one or more bends were made in the part, (3) the formed parts were spot welded together to form a semifinished unit, (4) the semifinished unit was painted, (5) the painted unit was assembled, and (6) the completed unit was packaged for shipment to a distribution center or to the final customer.

2.4. Previous Manufacturing System

Equipment   Prior to 1988, most of the equipment in the plant consisted of machines for slitting coiled steel to the correct width, stamping presses, turret punching presses, sheet metal brakes, rocker arm welders, gun welders, multispot welders, and an electropainting system. Some of the turret punching presses were computer numerical controlled (CNC) while the remaining equipment was either manually operated or had mechanically operated cycles (as in the case of the stamping presses and some turret presses).

Plant Layout   The plant was laid out in a departmentalized fashion. Machines performing similar operations were grouped, resulting in a Stamping Department (which included both punch and turret presses), a Press Brake Department, a Welding Department, a Painting Department, and a Final Assembly, Packaging, and Shipping Department. Each department had a supervisor and a group of workers that were assigned to that particular department. A representative block layout of the plant prior to cells is shown in Figure 2.

Production Flow   A diagram showing the production flow of the parts and products produced in the plant prior to cells is shown in Figure 3. As the figure shows, production began in the Stamping Department where coiled or sheet steel was fed into stamping or turret presses, respectively, to form the required blanks. Slitting was often required to get the coils of steel to the correct width for the dies used in the stamping presses, and this was done prior to the stamping operation. The blanks produced by the machines were placed into containers for transfer to a storage area or, if needed immediately, for transfer to the press brake area.

Containers of unformed parts were brought by forklift to the press brakes for forming. The worker would take the unformed part out of the container, make one or more bends to form the part into its required shape, and then place the formed part into a different container (a single brake would usually do all forming required on the part). Full containers of formed parts were moved by forklift to a storage area until the parts were needed in the Welding Department. While some parts could only be processed on certain brakes due to equipment capabilities, most of the press brakes were not dedicated to families of parts.

The Welding Department had a welding line for the large products (originally referred to as the Large Line), a welding line for the smaller products (referred to as the Small Line), four welding lines for drawers, and a number of single-operator welding

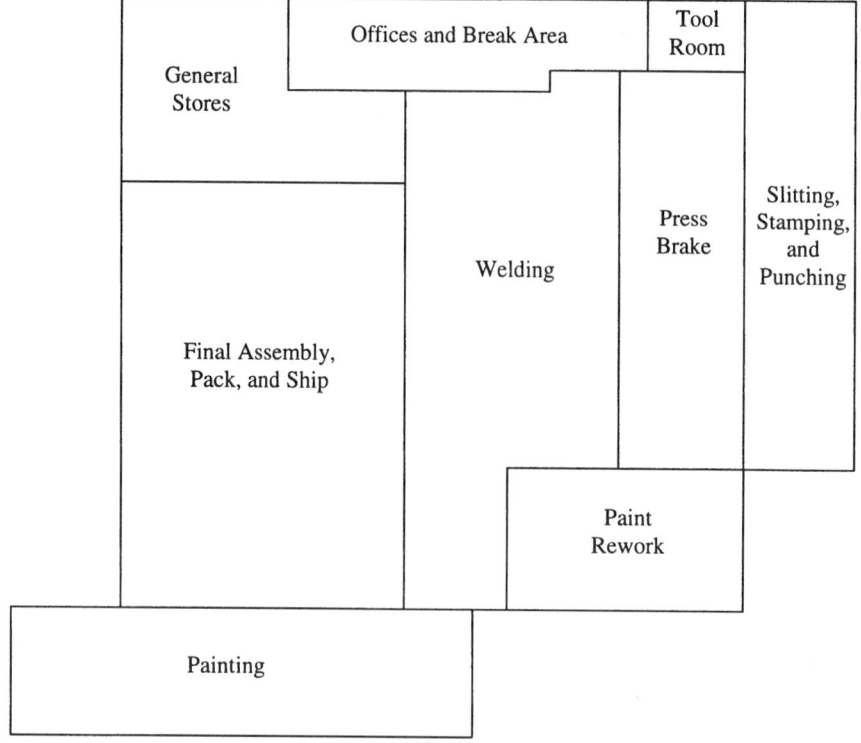

Figure 2. Departmental block layout.

stations. The large products were quite heavy, averaged 50 component parts per unit, required numerous welding steps to join all of the parts together, and had one long welding line devoted to them (the Large Line). In comparison, the smaller products were considerably lighter, had fewer component parts, required fewer welding steps to join the parts together, and had one short welding line devoted to them (the Small Line). The sequence of welding steps was virtually identical for all drawers, and they had four lines dedicated to them. The single operator welding stations were used for welding subassemblies required at one of the welding lines or at final assembly and most were not dedicated to any particular subassemblies or families of parts. At the single operator welding stations, the operator would remove the required components from the containers, place them on a rocker arm welder for the required spot welding operation, and then place the completed subassembly into a different container. Full containers of welded subassemblies were either moved by forklift to a storage area or to the station on the welding line needing the parts.

Each of the welding lines operated in an assembly line fashion and all functioned similarly. Containers of parts were brought to each station on the welding lines by forklift or hand cart. At the first station on the line, the worker would remove two or more different component parts from their respective containers and position them on

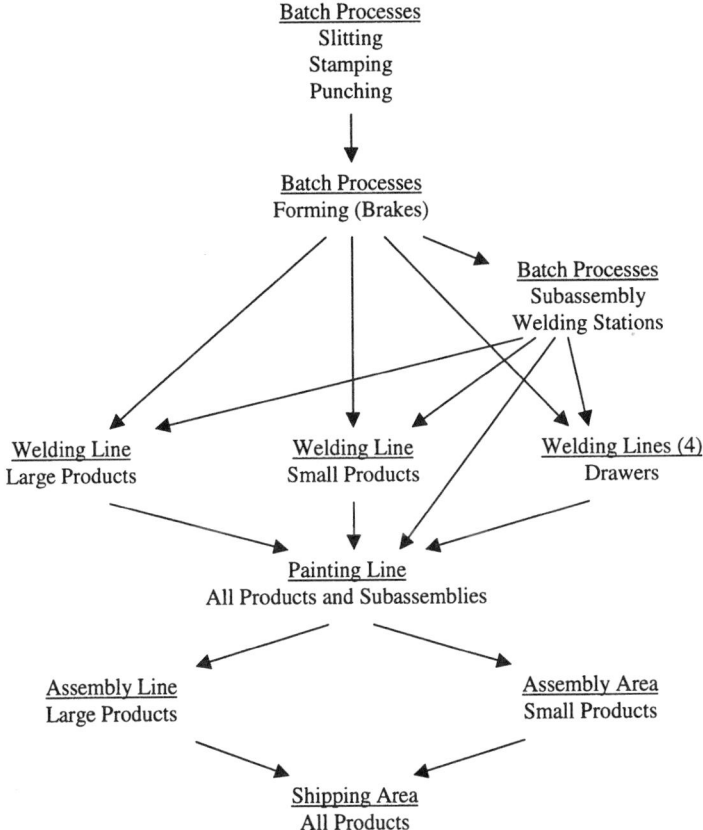

Figure 3. Part/product flow diagram for functional system.

a rocker arm or multispot welder for the required spot welding operation. Once the operation was finished, the unit was placed on a roller conveyor and manually pushed to the next station. At each succeeding station, a worker would remove the partially completed product from the conveyor and place it, along with the other component parts to be joined, on a rocker arm welder to perform the next spot welding operation required. Once the operation was completed, the product was again put on the roller conveyor and manually pushed to the next welding station. This process continued until the entire unit was welded together.

Over the years, a number of gun welders were purchased to replace some of the rocker arm welders. With the gun welders, the component parts were held in place at a particular station by a fixture and the gun welder was moved to the part for the spot welding operation. The gun welders were suspended from overhead tracks and balanced so that a minimal amount of effort was required to move and position the welders. The purchase of these gun welders significantly reduced the amount of effort required to weld the units together.

Once the large products, small products, or drawers were welded together, they were sent to the Painting Department where the unit was cleaned, sent through a dip tank where an electropainting process coated the unit with paint, rinsed to remove excess paint, and finally sent through a drying oven. Once the unit was dry, it was transferred to the Assembly Department for final assembly.

At the Assembly Department, the bodies of the large products were placed on the beginning of a powered roller conveyor that had multiple assembly stations. At each station, a different manual assembly operation was performed. Component parts were brought to their respective stations by forklift or a hand cart. When the assembly operations at all stations were completed, the line was started and the unit was moved to the next station. Once assembly of the unit was complete, it was packaged and placed into storage to await shipping. The Small Line products were assembled in the Assembly Department on either a small roller conveyor or an assembly bench (depending on the weight of the product). The assembly operations for the Small Line products were usually done by one or two individuals.

Job Design and Responsibilities   Historically, workers were assigned to operate a particular machine within a department and very little, if any, rotation occurred between machines. Rotation was eventually instituted within the departments to relieve boredom, but the extent and frequency of the rotation appeared to vary with the supervisor in charge. Workers did not rotate between different departments. If a worker wished to change departments, he or she was assigned to that department on a permanent basis.

A worker's responsibilities were usually limited to the operation of the machine. Some workers did set up their own equipment but this was usually performed by a lead setup person. While workers were responsible for the quality of the parts produced, an inspector would check the first parts coming off the machine after a setup had been performed and then occasionally thereafter to ensure that quality had been maintained.

Material Handling   The movement of containers in and out of storage areas was done by forklift, and forklift drivers were dedicated to particular areas of the plant. In the stamping and press brake areas, the individuals operating the forklifts also coordinated which containers needed to be moved next. Forklift operators followed a dispatch list that indicated the jobs to be processed on each machine and the number of containers in each job. Forklift drivers were responsible for transferring empty containers to machines needing them and for transferring full containers to and from the storage areas as needed.

Forklifts were also used to move material to the Welding, Final Assembly, and Packing Departments. The forklift drivers were aided in this task by material handlers who coordinated the flow of material to the stations to ensure that the lines were not stopped for lack of material and who helped position containers at the correct station on the line.

Batch Sizes   Batch sizes for the large products ranged from 1000 to 2000 units per order. If a unit required four or five drawers, all of the same size, this could have translated into an order of 10,000 units for drawer bodies. Several skids or pallets were needed to hold a batch of parts and usually the entire batch was processed before a new setup was incurred.

Production Control   Forecasted demand and firm customer orders for each end item were subtracted from the inventory in the distribution centers to determine when a particular product should be replenished. This information was used to form a master production schedule. These demands, in combination with orders for customized units, were input into a Material Requirements Planning (MRP) system as gross requirements for the end items. The MRP system was regenerated on a weekly basis and the typical gross-to-net explosion process produced the planned order releases for the individual component parts. A typical component would have an updated part number assigned to it after each processing step had been completed (i.e., stamping, forming, welding, painting, etc.). A typical large product had four levels in the bill of material, averaged 50 different components, and could require 100–120 order releases (i.e., an order release at stamping and forming for each component part, an order release at welding for each unit and each type of drawer, and an order release at assembly and packing). Each time a work order was completed, an entry was made into the MRP system to indicate the receipt of the order into a storage area and to record the amount of scrap produced. Thus, each batch of end units required a significant number of transactions to the MRP system before it was ready for shipment.

Order releases were given to dispatchers who scheduled the orders to be processed on each machine. The dispatcher had to take into account the current setups on the machines, the available capacity of the machines, and machine capabilities when performing this task. These schedules were used by the forklift drivers and material handlers when moving parts to and from the machines.

2.5. Problems with the Previous System

There were several problems with the functional system in place at SMP in 1988. First, Work-In-Process (WIP) inventories averaged $1.9 million and required a significant amount of storage space. The stamped blanks were usually flat and could be stored quite efficiently. However, the volume required for the storage of the parts increased drastically as the parts were formed to their required voluminous, three-dimensional shape. Storage space requirements increased even more as the parts were welded together.

Second, throughput time averaged 47 days for the large products. This throughput time included the time to perform all slitting, shearing, stamping, punching, forming, welding, painting, assembly, and packaging operations; the time spent waiting in queue at each of these operations; and the time spent being moved. Throughput times this long made it difficult to react quickly to changes in market demand or special orders from customers.

Third, despite the high level of inventories in the plant, the service level to the distribution centers (percent of the master scheduled orders that were not backordered) was only 88%, and $18–20 million worth of inventory was kept on hand in the distribution centers to provide a high service level to the final customers.

Finally, the plant's workmen compensation payments averaged $2 million a year. Most of the injuries were attributed to the constant bending, lifting, and twisting required to load and unload containers of parts at the forming and welding operations. Sheet Metal Products needed to reduce the number and severity of these injuries in order to safeguard the health of the employees and to reduce the associated costs of the injuries.

Sheet Metal Products wanted to address these problems in the most cost-effective manner. Cellular Manufacturing was chosen as the best way to solve these problems, and the first cell was implemented in December of 1988 (a complete analysis of the reasons for converting to cells is given in Section 5).

3. CELLULAR SYSTEM

3.1. Overview

The primary individuals leading the push toward manufacturing cells were the plant manager and the engineering manager. Most of the cells were designed by the engineering staff with input from the supervisors selected for the cells and from selected lead people in the Brake and Welding Departments. Cell operators were selected after the cells had been designed and most were not involved in the design process. However, management was willing to listen when workers had suggestions for improvement to the cell design or operation.

Workers were selected for the cells according to a job bidding procedure, with the most senior workers bidding on the cell given first priority. On all cells except for the Twin Lines, an external consultant was hired to provide 2 or 3 days of team building and teamwork training for the initial cell workers. In addition, a member of the management staff at SMP spent 1–2 days training the initial cell workers on how to lead and conduct meetings, and this training was repeated 6 months later. Training on team building, teamwork, and the like for any new individuals that entered the cell after this period was performed by the cell supervisors. On-the-job training was used to train the cell workers on how to operate the equipment in the cell. This was usually conducted by the workers in the cell who were most experienced with the equipment in question.

A diagram showing the production flow through the cellular system at the time this case study was written is shown in Figure 4. As the figure shows, several cells have been implemented since 1988 (some of which manufactured parts/products not previously produced in the plant), and a description of these cells is given in the following sections.

3.2. Twin Lines (Cells)

Reasons for Implementation   The first "cell" implemented by SMP performed the forming and welding operations on the cases for the large products. Large products accounted for 75–80% of total product sales, 25% of all direct labor hours, and 25% of all machine hours incurred in the plant. A significant amount of WIP inventory accumulated in this area (since all 50 component parts for a large product had to be available before the welding line was started) and a large amount of storage space was required due to the configuration of the formed parts. Many worker injuries were also occurring in this area as a result of loading and unloading containers of formed parts at the press brakes and welding lines, respectively. Thus, a substantial reduction in WIP inventory, throughput time, and worker injuries was thought possible if a one-piece line flow process could be developed that eliminated some of the material handling required in the previous system.

Cell Design and Operation The cell started as two identical lines laid out side by side. These two lines were still in operation at the time of this study and will hereafter be called the Twin Lines to distinguish them from the cells that were planned for this area (see Section 4). Both lines were designed to operate concurrently in anticipation of higher demand volumes in the future. The welding equipment was laid out in the order required and a roller conveyor was used to transfer the product from welding station to welding station. This was quite similar to the previous arrangement of welding equipment on the Large Line except that several gun welders were purchased for each line to replace many of the rocker arm welders that had previously been used.

The press brakes required for the forming operations were moved and positioned at appropriate spots along the welding line so that formed parts could be fed to the correct position on the line as needed. Three additional press brakes were purchased to improve the flow of product along the lines. Short roller conveyors were used to feed the parts from the press brakes to the welding line. These roller conveyors provided a small buffer to help absorb variability in the operations between the press brake and the welding operations while still limiting the size of the queues feeding the welding lines. These conveyors also eliminated the need to put the formed parts back into containers, thus eliminating much of the twisting and bending previously required. At the same time, the plant installed a number of scissors lifts and rotational devices to allow the unit to be presented in a more ergonomically favorable position to the workers.

Containers of parts were moved to the press brakes by forklift. The press brake operator would remove an unformed part from the container, perform the required bending operation(s), and then place the formed part on the roller conveyor that fed the welding line. Welding operators took the formed part from the conveyor, placed it in the correct position on the product coming down the main conveyor line, and welded the formed part into place.

Fully welded cases were transferred by conveyor to the paint line for painting. From there, they were transferred by conveyor to the assembly area for final assembly, packaging, and shipment to a distribution center or to the final customer.

Approximately 15 different products were manufactured on these lines. Each product required slightly different processing sequences and the press brakes often processed different parts from product to product. To accommodate this, the roller conveyors that fed parts from the press brakes to the welding lines were dismantled

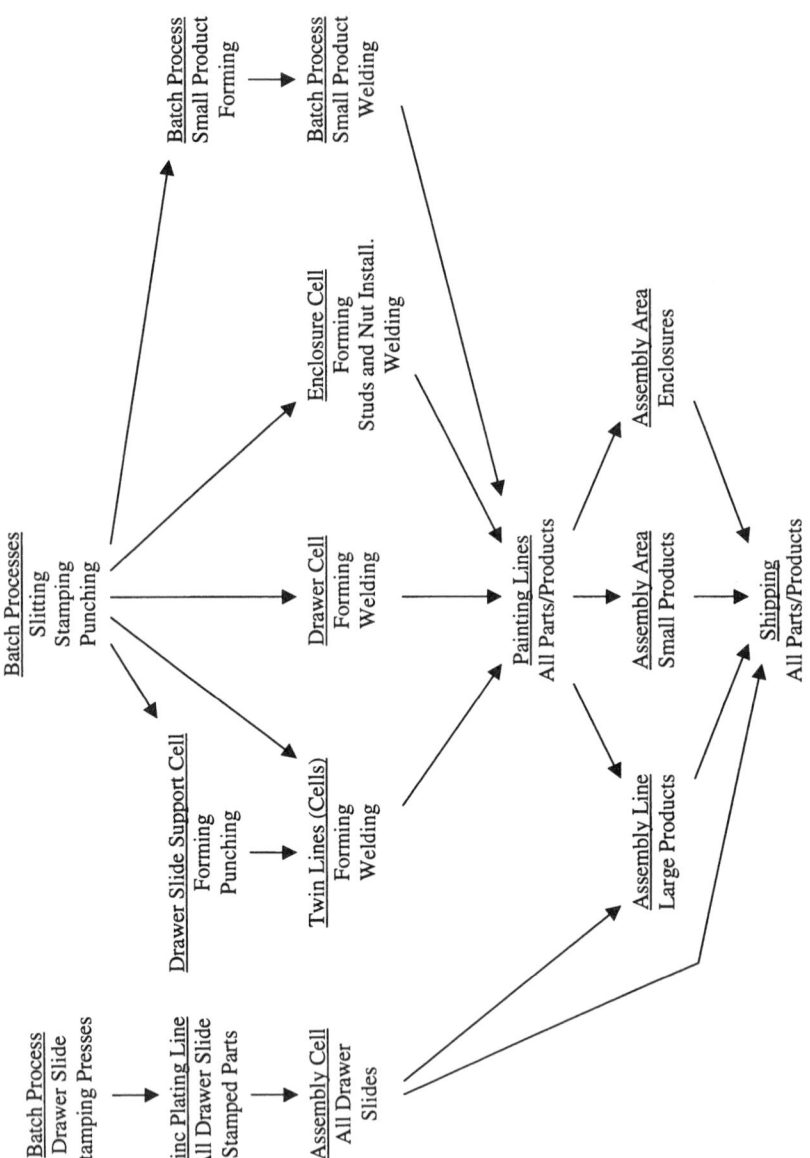

Figure 4. Part/product flow diagram for 1997 cellular system.

and repositioned from product to product (if needed) so that parts were fed to the correct positions on the line. The rocker arm and gun welders were moveable, and these were moved along the main roller conveyor as needed to accommodate the different products being produced (the main roller conveyor was seldom moved).

The high demand expected for the Twin Lines never materialized and, consequently, they were not run at the same time. Instead, one line was run while the other line was set up by lead setup people to process the next product required. When production of a product was complete on one line, the Twin Line workers immediately transferred over to the other line and began manufacturing the next product.

The traditional job classifications of press brake operator or welder were maintained when the lines were formed and different supervisors were in charge of the press brake and welding operators, respectively (thus, the job bidding procedure was not required for this cell). Welding or press brake operators rotated on a set schedule between the different welding or press brake stations, respectively, but a welding operator would not run a press brake and a press brake operator would not run a welder.

In contrast to the Twin Lines, all other cells that were implemented by SMP had cross-trained workers capable of performing all tasks in the cell and each worker was considered a cell operator. The lack of cross-trained operators caused some problems on the Twin Lines and the plant has plans to convert them to "true" cells (see Section 4). Because of this distinction, the Twin Lines are no longer considered a true cell by either the managers or the workers on the manufacturing floor.

The number of workers assigned to the Twin Lines varied with the type of product being produced and the overall level of demand. When the largest products were manufactured and/or when demand volumes were heavy, workers from the Small Line were transferred to the Twin Lines to help with production. When demand was low for the Twin Lines and/or fewer workers were needed due to the type of storage unit being manufactured, workers were transferred back to the Small Line area to build up the inventory of Small Line products. As stated by the plant manager: "The Small Line played second fiddle to the Twin Lines and what we did is carry some safety stock in our distribution center so we didn't have a service level problem."

Performance Improvements The improvements resulting from the implementation and operation of the Twin Lines were quite apparent. The average throughput time for the large products before the lines were implemented in 1988 was 47 days, the average WIP inventory level was $1.9 million, and the service level to the distribution centers was 88%. In contrast, the average throughput time in 1989 was down to 31 days, the average WIP inventory level was down to $1.4 million, and the service level to the distribution centers had increased to 91%. By the end of 1991, the average throughput time was down to 22 days, the average WIP inventory levels were down to $0.8 million, and the service level to the distribution centers had increased to 98%. During this time, the demand for products produced on the Twin Lines stayed relatively constant. As of 1996 the average throughput time had been reduced to 14.3 days and the average WIP inventory levels had increased slightly to $1.1 million. However, additional products not produced on the Twin Lines have been added since

1991, and these were included in the WIP inventory figures (the method of calculating the service level was changed in 1993 so a comparison with service levels prior to 1993 is not meaningful).

Factors Causing the Performance Improvements   A combination of factors were responsible for these improvements. First, batch sizes were reduced significantly (in both the Stamping and Punching departments and on the Twin Lines). In 1988, the average batch size was 1234 units. This was reduced to 854 units by 1989, to 683 units by the end of 1991, and to 223 units by 1996.

Second, operations overlapping was possible to a much greater extent than before. While batch processing was still used for the initial slitting, stamping, and punching operations, the transfer batch size between stations on the lines was one unit (as before, the transfer batch size through the paint and assembly areas was also one unit).

Third, a significant amount of effort was spent during this time period to reduce setup times on the equipment (both in the stamping and punching area and on the Twin Lines) so that the corresponding reduction in batch sizes did not cause the time spent on setups to grow dramatically. In fact, the total amount of time spent on setups in 1991 was virtually the same as in 1988, despite the increased number of times a machine had to be set up as a result of smaller batch sizes. In addition, when the plant quit operating the lines simultaneously, most of the equipment setup for the lines was converted to external setup time and usually no productive time was lost switching from one product to another.

Fourth, material handling time between the press brakes and the welding operations was virtually eliminated by the close proximity of the equipment and the use of roller conveyors.

Fifth, the flow-through process allowed quality problems to be spotted immediately, resulting in less rework and time spent reprocessing defective parts. As stated by the engineering manager: "The other benefit that soon became obvious was that we had a lot of work-in-process inventory in baskets. If there was a quality problem in any one of the pieces, it could have been days or even weeks before you finally discovered that putting it into a final assembly. And by shrinking that Work-In-Process, we had much more rapid feedback on potential quality problems." The engineering manager estimated that scrap and rework costs decreased by 50% after the introduction of the Twin Lines. While some of this reduction resulted from process improvements not resulting from the implementation of the Twin Lines, he estimated that half of the decrease was a result of the flow-through process and the resulting rapid feedback on quality problems.

Sixth, some capacity was added to the lines in the form of additional press brakes and gun welders. For the most part, this allowed the two lines to be independent of each other and allowed one line to be set up while the other line was running, thus eliminating unproductive operator time as a result of equipment setup. However, since the lines were not run simultaneously, it is unknown how much impact the additional equipment had on the throughput time improvement.

Finally, it is likely that the close proximity of the equipment and the ability to see what was taking place by all concerned helped improve the coordination and flow of parts from the brakes to the welding equipment, also reducing potential delays.

Due to their powerful influence on throughput time and WIP inventory levels, it is likely that the reduction in batch sizes and the use of operations overlapping with one-piece flow played a major role in the performance improvements achieved by the plant during this time period. Of course, batch size reduction of this magnitude and the use of one-piece flow would not have been possible without the elimination of material handling between operations and the reduction in equipment setup time. Thus, the factors are all tied together. In addition, the smaller batch sizes and corresponding shorter throughput time allowed the plant to react more quickly to changes in demand. This quicker reaction time resulted in an increase in the service level to the distribution center while simultaneously allowing the distribution center to lower the amount of inventory it carried.

3.3. Drawer Slide Support Cell

At about the same time that the Twin Lines were installed, SMP formed a cell to make the drawer slide supports for the large products. The drawer slide support was a panel that reinforced the side of the product where drawers were installed and supplied the holes required for mounting the drawer slides. Originally, the blanks were stamped, punched, and formed in batches at different pieces of equipment located in different areas of the plant and then transferred to the Twin Lines when needed.

The Drawer Slide Support Cell consisted of two press brakes and a straight-line punching press. A cell operator removed the blank from the container, inserted the blank into the first press brake, and formed the first half of the drawer slide support panel. The partially formed panel was placed on a short roller conveyor and pushed to the second press brake where the second half of the drawer slide support panel was formed. The completely formed drawer slide support panel was then placed on a short roller conveyor and pushed to the punching press where the required holes were punched. The completed drawer slide support panels where then put into containers and moved to the areas of the Twin Lines requiring the panels. Although no performance metrics were available for the cell, it was considered to be quite successful in reducing WIP inventory of drawer slide support parts and reducing the amount of throughput time required to manufacture them.

Over the last 7 years, a process change was made whereby the holes were punched in the drawer slide support panel as the part was stamped out, eliminating the need for the punch press operation. This change occurred gradually as the dies for punching out the drawer slide support panel were retooled. As a result of this change, the Drawer Slide Support Cell has gradually decreased in importance. The two press brakes were still used to form the drawer slide support panels, but the punching press was only used for a few low-volume drawer slide support panels whose stamping dies had not been retooled. The drawer slide supports were still put into containers and delivered to the different areas of the Twin Lines as needed. The cell was staffed as needed with a single operator.

3.4. Gauge Cell

Reasons for Implementation   When the assembly of the printed circuit boards left the plant, the assembly and inspection of hose and gauge assemblies remained for several years. This work involved cutting hose to length, crimping adapters onto the hose, attaching the hose assembly to a purchased gauge, testing the assembly for leaks, and packaging the assembly into a purchased box. Previously, the assembly of the hose and gauge assemblies had been done by electronic assemblers, and the testing and packaging of the assemblies had been done by electronic inspectors. The stations for these operations had been located in different areas of the electronics portion of the plant, and the assemblies had been produced in large quantities per batch.

When the printed circuit board assembly left the plant, this work was consolidated into a small area. The electrical assembly and inspector job classifications were eliminated, and the metal product assemblers were asked to assemble, test, and package the hose and gauge assemblies. The union thought this violated the work agreement and grievances were filed against the plant. Sheet Metal Products settled these grievances by forming a Gauge Cell Operator classification. Workers in this classification were responsible for performing all operations required to assemble, test, and package the hose and gauge assemblies. The cell operators also did their own material handling and scheduling of production and each operator was cross-trained to do all operations in the cell. This was the first cell where the work structure was changed to include all of these tasks as part of the operator's job and a new job classification was constructed to take these tasks into account. Thus, the Gauge Cell became the first true cell that SMP had in place.

Cell Design and Operation   The cell had four different assembly areas and each area was dedicated to a small set of products. The cell was staffed by three or four operators who manned the different assembly areas as needed to produce the assemblies demanded. Common fixtures were developed for each assembly area to reduce equipment setup time and material handling between operations was virtually eliminated due to their close proximity.

Performance Improvements and Factors Causing Improvement   These changes allowed batch sizes to be significantly reduced. As stated by the engineering manager: "The market only needed 50 of this size and 50 of another size, but we had the mentality at that time that we really had to build 100's and 100's of them. Well, with the cell approach, it was very easy for the operator to change over his fixture at his bench. We had to do a lot of refixturing to make setup easier, but it allowed fabrication in very small lot sizes."

The implementation of the Gauge Cell caused service levels to improve and throughput time to decrease by 90% compared to the previous functional arrangement. It is likely that the reduction in batch sizes, coupled with reductions in equipment setup time and material handling played a major role in this improvement. It is possible that other factors (e.g., changes in effective equipment capacity, worker availability, etc.) also contributed to the performance improvement. However, because of the large number of changes that occurred when the printed circuit board assembly left the plant, it is not known what these factors were or how much impact they had on the final performance of the cell.

In late 1992 or early 1993, the work performed by the cell was moved to a different plant and the cell was dismantled.

3.5. Drawer Slide Cell

Reasons for Implementation   In 1992, the corporation decided to quit purchasing and begin manufacturing the light- and medium-duty drawer slides used in some large products. This product had never been produced in the plant and none of the existing equipment had the capabilities necessary to produce and assemble the slides. New equipment was purchased for this purpose. Based on the performance improvements achieved with the previous two cells, it was decided that a cell would be the most cost-effective way to manufacture the slides, and the cell was implemented in July of 1992.

Cell Design and Operation   The equipment consisted of one large press that stamped out and formed the rails for the slides, one smaller press that stamped out and formed the ball retainers used in the slides, a zinc plating line for plating the rails and the retainers, and five assembly machines used to assemble the slides. All of this equipment was located in close proximity to each other. However, only the assembly machines and assembly operators were considered part of the cell.

The stamping presses carried a separate job classification and were run by a single operator. This operator had a lead setup person who set up the dies (i.e., assembled the dies with the correct tooling) used in the large stamping press and helped install them into position. The zinc plating line also had a separate job classification and was run by two lab technicians. While the stamping press and zinc plating line operators were not considered part of the cell in terms of job classification, they did coordinate their work with the slide assemblers to make sure that requirements for slides were met.

The operation of the stamping presses and the zinc plating line required special skills that few people in the plant had. If these operators had been considered Drawer Slide Cell operators and a layoff occurred, they potentially could have been laid off since they had lower seniority than other workers in the cell. This would shut down the production of slides and potentially idle other portions of the plant as well. By keeping the stamping presses and the zinc plating line operators as separate classifications, layoffs of these individuals would be avoided and production would not be stopped if a layoff affecting a portion of the drawer slide cell occurred.

The slides and ball retainers were first stamped out of metal on their respective machines. The output of these machines was put into small containers that were transferred to the zinc plating line by the cell operators. The slides and retainers were zinc plated and then set aside to dry for 24 hours. Once dry, the cell operators used the manufactured slides and ball retainers, along with other purchased components, to assemble the slides. Each of the five assembly machines performed a different task and every slide required an operation on each machine before it was completely assembled. Once assembled, the slides were transported by forklift to the assembly line or to shipping for transport to another plant.

Each assembly operator was considered a Drawer Slide Cell operator and was cross-trained to do all operations required to assemble the slides. This included hanging the slides and ball retainers on the plating line, removing the slides and ball retainers from the plating line, assembling the slides (seven different operations on the five machines), packing the slides, material handling within the cell, and setup of the equipment. Cell operators rotated to a different set of jobs every 1 hour and 15 minutes to help reduce repetitive motion injuries and relieve boredom.

The cell operators were also responsible for scheduling of production in the cell. A 2-month rolling schedule of slide requirements was provided to the cell, and the cell operators sequenced the orders to minimize setups while still meeting due date requirements.

The cell operated one shift per day, was staffed with eight operators, and produced five different sizes of slides in left- and right-hand configurations (for a total of 10 different slides). Batch sizes were dictated by demand for the components the slides went into and varied from 1000–10,000 slides. Since there were no prior processes to compare this cell to, performance improvement figures were not available.

Union Opposition   There was little or no opposition to this cell from the union. Slide manufacturing and assembly was seen as new work coming into the plant, and it resulted in the creation of three new job classifications. In addition, since the processes were so different from those in the rest of the plant, the union did not perceive any changes being made to the way work was structured. The job classifications were negotiated before the cell was implemented.

Startup Problems   Some startup problems were experienced by the cell. The machines initially did not work correctly and a significant amount of effort was spent getting the problems resolved and the machines running correctly. This occurred with both the stamping presses and the assembly machines. Problems with the quality of some purchased parts also caused some initial problems in the assembly area. These startup problems caused a lot of frustration and several of the initial cell assembly operators left the cell for positions elsewhere in the plant. The problems were eventually resolved, and the workers in the cell at the time of this study seemed satisfied with the way it was running.

3.6. Drawer Cell

Reasons for Implementation   At about the same time that the plant installed the Twin Lines, the two brakes used to form the drawer bodies were placed in front of the lines that welded the drawer bodies together. Seven strokes of the press brake were required to form the drawer body, and then it was placed on one of the welding lines to have the drawer fronts, drawer backs, lock clips, and so forth welded on. The drawer fronts, drawer backs, lock clips, and other items needed for the drawer were formed elsewhere in the plant and transported in containers to the welding line as needed.

The drawer area was having difficulty keeping up with demand, and the plant needed to increase its effective capacity. After some investigation, an automated piece of equipment was purchased that could perform all of the bending operations required for the drawer body faster than the operations could be done with the current press brakes. The company also purchased an automated machine that could perform all of the welding operations required to join the components of the drawer body together. The automated welding machine was connected to the automated forming machine with a roller conveyor and transfer batches of one unit were used between the machines. These two machines replaced one of the brakes and two of the welding lines for the drawers and increased the capacity of the drawer area. It also reduced the number of workers required to form and weld drawers by two-thirds, resulting in a cost savings that allowed the cost of the equipment to be recovered in less than 2 years.

Shortly after the new equipment was purchased, the brake that made the drawer fronts was positioned by the new forming and welding equipment so that drawer fronts could be fed to the line as needed. Some spot welding equipment needed to weld brackets onto the drawer front was also repositioned so that the manual brake for the drawer front fed the spot welder, which in turn fed completed drawer fronts to the line for the automated welding operation. A spot welder required to weld lock clips to the drawer back was also positioned so that drawer backs could be fed directly to the line.

Approximately 170 different drawers were manufactured, and the semiautomated drawer line could handle 85–90% of them. The real large drawers that could not be run on the line were formed separately on the brake located in front of the manual welding line and then welded on a stand-alone robot (this robot was different from the automated welding machine on the drawer line). The small drawers that were not run on the line were manufactured on the manually operated press brake and manually operated welding lines. The manually operated press brake and welding lines also manufactured some drawer subassemblies that were fed to the automated drawer line as needed.

When the automated equipment was first installed, operators from the Welding and Press Brake Departments ran welding and press brake equipment in the drawer fabrication area, respectively. Operators from the Welding Department would rotate between the automated welding equipment and any other manually operated welding equipment in the Welding Department on a daily basis. The same held true for the press brake operators. They would rotate between the automated forming equipment and any other manually operated press brakes in the Press Brake Department. Operators from the Press Brake Department would not rotate to any of the welding jobs in the drawer area and vice versa. The press brake and welding operators reported to the supervisors of their respective departments.

Early in 1993, SMP decided to convert the drawer manufacturing area into a cell. With the introduction of the automated equipment, a majority of the production in the drawer area involved individuals running the automated equipment, and the amount of manual work was significantly reduced. Productivity improvements were thought possible if the workers could be cross-trained to operate all of the equipment in the area and if the workers could transfer from machine to machine to balance the production needs as required. The equipment was already in place for the cell, but the method of work organization needed to be changed in order for the drawer fabrication area to become a true cell.

Union Opposition   The managers at SMP took the position that a cell was the correct type of work organization for manufacturing drawers, and it was something that needed to be done to stay competitive on cost. The union opposed the formation of the cell. They thought the plant was changing the terms and conditions of the working agreement and grievances were filed. From the union's viewpoint, the managers at SMP were simply shoving the changes down their throat rather than negotiating changes in the work arrangement. A series of meetings were held to settle the grievances. A new rate of pay was negotiated for the cell workers, a Drawer Cell classification was formed, and the cell was officially implemented in the middle of 1993.

A few months after the implementation of the Drawer Cell, meetings were held with the union to address the concerns relating to the formation of cells in the plant. The outcome of these meetings was a clarification memo on cell implementation. By law, the plant had a right to establish manufacturing cells, and, in this memo, the union recognized the right of SMP to do so. The memo also stated that the plant and the union would negotiate a fair wage for any cells that would be established, that cell implementation would not take place in such a way as to violate employees' contractual rights, and that ample training would be provided to assist workers in the transition to cells. In addition, both the union and the plant acknowledged that some employees might not be well suited for working in a cell type of environment and that the union and the plant would work together to resolve those situations. The memo was seen as a significant step forward for the company with respect to their efforts to implement further cells.

Volume Stability   There was some concern about the stability of demand for the cell and the potential impact it would have on the cell workers. When the cell classification was established, the workers within that classification were dedicated to the cell. The industrial engineer in charge of evaluating the feasibility of the cell was concerned about the swings in volume that could occur throughout the year. If workers were transferred in and out of the cell on a continuous basis as demand fluctuated, the cell might not be viewed in a positive manner by the union. These thoughts were also echoed by the supervisor initially in charge of the cell. As he stated: "The environment was somewhat hostile against the cell as far as the union was concerned. So it was important to me that we be able to go in and show some success right away. If I didn't have enough work, and I had to farm people out, that would be a sign that it was not working. Alternately, if I had to ask for a bunch of help, that would be a sign that it was not working." In order to level the load going through the cell, it was decided that some of the products normally produced on the Small Line would be produced in the Drawer Cell as needed.

Cell Design and Operation   The cell manufactured all drawers produced in the plant and the cell operators were required to know how to set up and operate all equipment in the cell. This included the automated forming machine, the automated welding machine, the manual press brakes, the rocker arm welders, and the robot. In addition, they were also responsible for their own material handling within the cell. Training on the manually operated welding equipment was fairly rapid, but the setup and operation of the automated forming and welding equipment and the manually operated brakes required a higher skill level and training took longer as a result.

The cell operated with 14 workers on the first shift and 10 workers on the second shift. The workers rotated between the different equipment in the cell every 1 or 2 hours.

Each day, the supervisor of the cell met with the dispatcher to determine the jobs that need to be completed that day and the order in which they should be run. The list was organized to eliminate setups as much as possible while still meeting the due dates of each job. Lot sizes ranged from 20 to 4000 and were dependent on the drawer requirements for the products being produced in the plant. A transfer batch size of one was used on the automated line and small transfer batches were used on the manual lines.

Performance Improvements   Much of the performance improvement in this area occurred when the automated equipment was purchased. The engineering manager thought that some quality and productivity improvements resulted from the cross-training of workers that occurred when the cell was actually formed but no estimates were readily available.

3.7. Enclosure Fabrication Cell

Reasons for Implementation   The fourth cell implemented by SMP was the result of a new product line coming into the plant. In 1993, the corporation purchased Company X and a decision was made to have SMP manufacture the sheet metal enclosures required for the products sold by this company since the processes required were similar to those already in place at SMP.

Component parts for the enclosures were first stamped or punched out of sheet metal. The components were then formed on press brakes and welded together using spot or wire welding technology. Some components also required the installation of threaded studs and/or nuts (to hold other component parts that were installed at Company X), and these were pressed into place with a special machine. Once forming and welding was complete, the unit was painted and then sent to a bench assembly area in the Assembly and Packaging Department for minor assembly and packaging for shipment (this bench assembly area was separate from the assembly area used for the Small Line products). The unit was then shipped to Company X where final assembly of the product occurred. Thus, with the exception of wire welding, all of the processes were similar to those already in place at SMP.

This transformation occurred over several months as Company X began to shut down its sheet metal production and SMP began to produce the enclosures. The initial production was performed on the Twin Lines and on the Small Line. This caused delivery problems for the enclosures since they were vying for the same resources required for the products normally produced in these areas. The problem became worse as Company X shut down more of its production and demand for enclosures from SMP increased. Sheet Metal Products could not run all three types of products at the same time, and the level of service provided to both the distribution centers and to Company X suffered. As stated by the engineering manager: "Some of the product was run on the Twin Lines, some was run on the Small Line, and we had our different customers saying, hey, we need these Twin Line and Small Line products. Who do you give priority to? They were vying for the same resources and they couldn't all be run at the same time." Some of the service level problems were related to a lack of capacity due to the increased load on the system. In addition, despite the fact that the enclosures required many of the same brake and welding operations, they also required some operations not usually done on the Twin Line or on the Small Line. Thus, the product did not fit real well in either of these two areas. In an effort to resolve these issues, SMP decided to form a cell dedicated to the manufacture of the enclosures. As stated by the engineering manager: "Service level was probably the over-riding reason this was done."

Union Resistance   There was some passive resistance to the formation of this cell from the union. The union saw this work as a way to add to the traditional welding and press brake classifications and did not want a cell formed. However, based on the cell clarification memo previously mentioned, the union realized the plant had the right to form the cell. An Enclosure Fabrication Cell classification was formed, a wage rate was negotiated with the union, and the cell was implemented in late 1993.

Cell Design and Operation   Most of the equipment needed to form the cell came from Company X when the enclosure production was finally shut down. The cell had five press brakes, five gun welders, seven rocker arm welders, three wire welders, three stud guns for spot welding threaded studs to sheet metal, and three machines for pressing in threaded nuts and studs. Lot sizes were fairly small, averaging 25–40 units per order. Blanks were produced by the stamping or punching presses and placed on storage racks outside the cell. When needed, the blanks for each component part were pulled from the storage racks and were formed in batches equal to the lot size of the unit. The batches were then transferred by a cell worker to the machines used to weld or press in the nuts and studs (if required). When all of the components for a unit were ready to be welded together, the cell workers assembled a roller conveyor and positioned the welding equipment by the conveyor in the order required to weld the unit together. Since different enclosures required different sequences of welding steps, the configuration of the roller conveyor, the number of pieces of welding equipment required, and the arrangement of welding equipment along the conveyor varied somewhat from product to product. The product moved along the conveyor and at each station, different welding and assembly operations were performed. When welding was complete, the unit was sent to the Painting Department for painting and then to the final assembly area for final assembly, packaging, and shipment.

Each morning, a dispatcher brought a production schedule to the cell supervisor that showed the units to be produced, the production quantity required, and the due date. If all blanks required to build a particular product were available, the cell supervisor would put the forming list for that order on a "forming" board by the press brakes, and the individuals operating the press brakes formed the parts in the order given. If a completed batch of formed parts required the installation of studs or nuts, the forming operator moved the batch to the machines that performed these operations and wrote the part number and quantity of the units requiring the operation on the "studs and nuts" board. The individuals operating the stud and nut machines then used this information to show them what needed to be done next. Containers of parts not requiring studs or nuts were staged until all components of a unit were ready to be welded together. When all components were ready, the cell supervisor wrote down the required production quantity on a "welding" board, and the individuals operating the welding equipment were responsible for welding the units in the order listed on the board. This system was very visual, allowed the operators to see what is coming in the way of production requirements, and was easily understood by all.

Approximately 15 different products were produced in the cell. The cell operated with 14 workers on the first shift and 3 workers on the second shift. Workers rotated each day to a different job according to a rotation schedule established jointly by the cell supervisor and the cell operators.

Startup Problems   This cell also experienced startup difficulties due to the cross-training required. Training on the spot welding and nut pressing equipment was fairly quick, but wire welding and the setup and operation of the press brakes required a higher skill level, and it took longer to develop competencies in these areas. In addition, some of the workers were either unwilling or unable to learn how to set up and run a press brake. On the days when these individuals were on the press brakes, little or no production was accomplished and output of the cell suffered. After the cell supervisor had several discussions with these individuals, including a group meeting involving the entire cell, the workers having problems with the press brakes left the cell for jobs elsewhere in the plant. Other workers were selected to replace those that left, and all workers in the cell were cross-trained to perform all operations required.

Performance Improvement and Factors Causing Improvement   The engineering manager estimated that throughput time in the cell was 10% less than when the enclosures were produced on the Twin Lines and on the Small Line. Most of this was probably a result of dedicating equipment to the production of just the enclosures and eliminating the conflicting priorities with the Twin Line and the Small Line product (which increased effective capacity for the enclosures). Batch sizes were already small in these areas and the company had been in the habit of keeping the product moving once the forming operation was performed. As stated by the engineering manager: "On the Enclosure Cell, we had already put in a good flow through system and were in the habit of not storing things in wiretainers and so on. We started with much less of a handicap there, so the potential improvement available to us was much more limited."

However, it should be noted that demand increased significantly since the cell was formed. Part of this was a result of the final transfer of product from Company X and part was from actual growth in demand. Thus, it is very unlikely that the Small Line area could have continued to produce this product without the additional capacity provided by the transfer of equipment from Company X and the purchase of the new equipment.

3.8. Other Manufacturing Changes

Small Line Changes   At about the same time that the Twin Lines were installed, changes were also made to the arrangement of equipment on the Small Line. The two press brakes used for the forming of these products were positioned at the head of the welding line to eliminate much of the material movement previously required for these products. However, these products were still formed in batches when workers were available to run the press brakes. They were then welded together when workers become available to operate the welding line.

Performance Measurement   Sheet Metal Products changed its performance measurement system when the Twin Lines were implemented. Previously, direct labor efficiency was one of the main measures used to assess manufacturing performance. When the Twin Line was formed, the output of the brakes was limited and dependent on the speed of the welding line. Thus, direct labor efficiency became a meaningless performance measure. The plant converted its performance measurement system to that of measuring total output of the plant and the throughput time of the Twin and Small Line products. Cost and quality (i.e., scrap and rework) continued to be measured in the same manner as before. The elimination of direct labor efficiency reporting resulted in a significant cost reduction for the Accounting Department since individuals were no longer needed to collect and analyze the data.

Instruction Sheet Development   Instruction sheets were developed for each product over the last several years that detailed how to set up the equipment to produce each component part, the dies required, how to calibrate the machine, the steps to follow, and the like. This has improved the setup process, especially for those individuals less familiar with a particular piece of equipment.

Preventative Maintenance   Sheet Metal Products instituted a preventative maintenance program and performed periodic inspection and maintenance on all equipment in the plant. This has reduced the amount of unplanned downtime and has increased the effective capacity of the equipment.

Powdered Paint System   A powdered paint system was installed in the early 1990s in order to meet customer demands for different product colors. The electropainting system was still used for the highest volume product color.

4. FUTURE CELLS PLANNED

4.1. Overview of Cell Plans

Sheet Metal Products had plans to convert the Twin Lines, the Small Line, and the assembly area to cells over the next 2 or 3 years. It also had plans to change the relative location of the Enclosure Fabrication Cell. A plant layout had been developed that showed the configuration of the individual cells and the layout of the cells relative to each other. The slitting, stamping, and punching of sheet metal would continue to operate in a batch type of mode and would not be cellularized. The paint area would also not be cellularized and would continue to operate in a fashion similar to how it operated at the time this information was collected.

These plans were driven by a need to better respond to changing market demands and at the same time, to reduce or maintain cost. According to the plant and engineering managers, customers wanted to be able to specify the types and combination of drawers, casters, locking mechanism, color, and other items that they wanted on their product. Sheet Metal Products needed to be able to respond to these demands while still keeping cost at a reasonable level if it wished to maintain or increase its market share.

Much of this customization was performed by the dealer (unless a customized order for a large number of storage units is produced in the plant to those specifications). For example, if a customer wanted a particular combination of drawers, the dealer would sell the customer a standard product and then order additional replacement drawers according to the customer's specifications. The dealer would install these drawers before the unit was delivered to the customer and then ship the drawers that originally came with the unit back to the distribution center.

This process resulted in a lot of damage that ultimately became an expense for SMP. The replacement drawers were packaged in special boxes that prevented damage to the drawer during shipment. In many cases, more money was spent on the packaging and shipment of the drawer than on the drawer itself. Despite the fact that the drawers being returned had different dimensions than the replacement drawers, the dealer usually shipped them back in the same container. As stated by the engineering manager: "The problem was that the carton you send them in was for a two inch drawer and he was trying to send you back a four inch drawer. So when he sent back the old drawers, they were scrap."

Sheet Metal Products also wanted to reduce the $8–9 million worth of inventory carried in its distribution center while still providing relatively fast delivery to the final customer. As the market moved more to a customization mentality, it would no longer be desirable to handle variations in customer demand by holding inventory since the number of different product configurations would increase drastically. In addition, the more times a product was handled, the more likely it was to incur damage. If SMP could ship product (especially large units) directly from the plant to the customer, in the configuration desired, the amount of handling required would be decreased and the number of units damaged would go down. In addition, direct shipping of customized units would eliminate the customization performed by the dealer. This, in turn, would reduce the number of replacement drawers returned, further reducing distribution center inventory and eliminating the potential damage caused by their shipment.

However, if SMP wanted to customize product and ship directly to the final customer, it had to be able to respond very quickly to a customer order. Manufacturing lead times at the time of this study for large products averaged 3–4 weeks and each type of unit was produced every 20 days. SMP wanted to reduce these manufacturing lead times so that each unit was produced every 5 days with a total manufacturing lead time of 1 week. As stated by the engineering manager: "We're looking at being able to assemble product to the customer's order, and that's reasonable if you can produce it in five days and it takes another five days for shipment. A customer will wait two weeks, he's not going to wait two months."

The managers at SMP believed that further conversion to cells provided the best means by which these objectives could be accomplished. Figure 5 shows the new cells that would be installed (as well as any cells that would remain) and the flow of production through the proposed system. The changes planned will be discussed in the following sections, while the logic behind the changes will be discussed in Sections 5 and 6.

4.2. Large Product Cells

Sheet Metal Products planned to break up the Twin Line into four cells, each dedicated to a small family of products. The dimensions and type of the product would determine the family to which each product was assigned. The welding equipment for each cell would be arranged in a line, and all four cells would be parallel to each other. The press brakes would be arranged along the lines as needed to feed components to the welding line. Some of the press brakes would feed two cells, depending on the products being produced at the time. Stamped sheet metal parts would still be made in batches by the stamping department and held in inventory. Each cell would draw from the inventory of stamped parts as needed. The output of the cells would be fed directly to the paint lines for painting and then to final assembly.

4.3. Small Line Cell

The Small Line would also be converted into a cell. Although the plans were not fully developed at the time of this study, the cell would probably be quite similar to the Small Line except that all operators would be cross-trained to perform all operations in the cell.

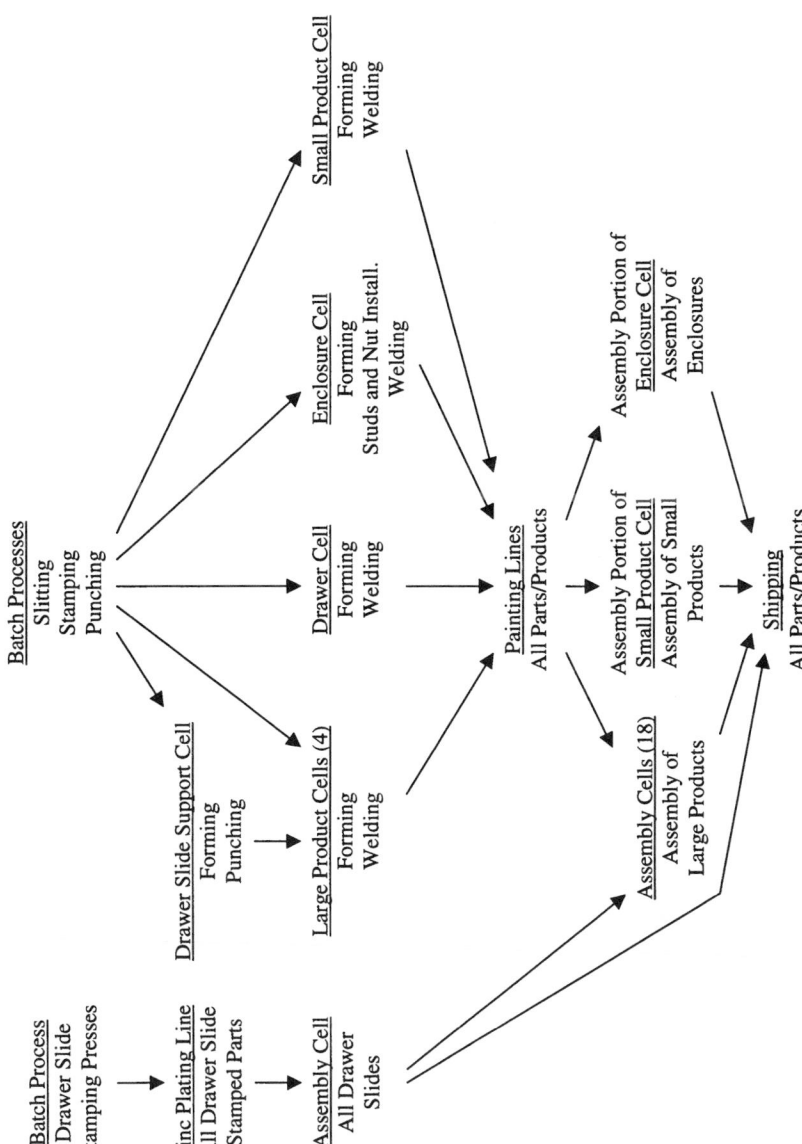

Figure 5. Part/product flow diagram for proposed cellular system.

4.4. Assembly Cells

The assembly and packing line for the Twin Line products had 30–40 stations (the number of stations varied with the product being assembled). All of the workers on the line were cross-trained to perform all of the assembly tasks, and the workers rotated every 1½–2 hours to the next station on the line in order to reduce repetitive motion injuries.

Sheet Metal Products planned to break this assembly line up into 18 cells, each staffed by two workers. Each cell would assemble a complete unit from start to finish. Component parts would be brought to the station by forklift, and the two operators in the cell would have the responsibility for complete assembly of the unit.

4.5. Other Changes to the Enclosure Fabrication, Small Line and Drawer

Sheet Metal Products also planned to relocate the Enclosure Fabrication, Small Line, and Drawer cells next to each other so that they could more easily share personnel and equipment during times of varying load. Plans were also made to have the Enclosure Fabrication and Small Line cells do their own assembly work. The plan was to bring the units back into their respective cells after painting had been completed and have the cell do the final assembly and packaging of the product.

5. CONVERTING TO CELLS VERSUS IMPROVING THE EXISTING SYSTEM

This section examines the underlying reasons why CM was able to provide the improvements desired at SMP in a more effective and efficient manner than the previous functional system.

5.1. Equipment Setup Reduction

Sheet Metal Products realized that batch size reduction required a corresponding reduction in equipment setup time, so the increased number of setups did not cause capacity problems at the machines. Accordingly, a significant amount of effort had been spent over the years to reduce setup time. While some setup reduction did occur at the press brakes as a result of forming the Twin Lines (the actual amount is difficult to determine), setup time reduction was not mentioned as a reason for installing the Twin Lines.

Interestingly, setup reduction was one of the reasons for converting the Twin Lines to a system of four cells (see considerations influencing the configuration of the proposed cell system in Section 6.5). Each cell would be dedicated to a smaller family of products than the Twin Line, and it was thought that the increased product similarity would significantly reduce setup time at the press brakes. In fact, many of the press brakes would likely need little or no changeover from product to product.

However, this dedication to product families could occur in a functional system as well. Thus, despite the plant's desire to reduce setup time through the conversion of the Twin Line to cells, cells were not needed if this was the only consideration required.

5.2. Material Handling Requirements

Sheet Metal Products also wanted to reduce the amount and cost of material movement required and the potential damage to component parts that could result. In the functional system, batches of parts were moved from the press brakes to the welding lines by forklifts. Batch sizes could have been reduced a certain amount in the functional system without increasing the number of forklift moves required, but beyond a certain point these requirements would have increased. However, it would have been virtually impossible to *decrease* the amount of material handling required as batch sizes were reduced.

For example, suppose the batch size for a large product is 1200 units and the impact of batch size reduction on a single panel for this product is examined. If 200 panels could fit in a palletized container, it would have required six containers for this batch size, and each container would have been moved individually. If batch sizes had been reduced to 200 units, the number of forklift trips would not have increased since the batch size was equal to a full container. However, if batch sizes had been reduced below 200 units, less than a full container would have been moved each time. If overall demand levels had stayed the same, more containers would have been moved each year, increasing the number of forklift trips required.

When the Twin Lines were formed, the press brakes were connected to the welding line by roller conveyors, and batch sizes were reduced without increasing the number of forklift trips between the forming and welding operations. In fact, the use of roller conveyors eliminated the need for forklifts to transfer the material between the press brake and the welding lines, resulting in a substantial cost savings. This could not have been accomplished very easily in the functional system. In addition, the elimination of forklift trips between the press brakes and the welding lines reduced the potential damage resulting from loading and unloading containers at the press brake and welding lines, respectively.

5.3. One-Piece Part Flow

Decreasing batch sizes was only part of the picture. Sheet Metal Products also wanted to use operations overlapping with one-piece flow as much as possible. There were three reasons for this. First, the use of one-piece flow would reduce the throughput time to complete a batch and the amount of WIP inventory in the system. Second, once a part was formed, it quickly began to take up more storage room, and the amount of room required increased exponentially as the formed parts were welded together. Since one-piece flow reduced the amount of WIP inventory, it also reduced the amount of storage space required. Third, if there had been a quality problem with a particular part that was not detected at the operation causing the quality problem, one-piece flow allowed this problem to be identified almost immediately at the next operation (see the discussion on the Twin Lines (cells) in Section 3.2), reducing the amount of scrap produced.

However, it was very difficult to use operations overlapping with small batch sizes, much less one-piece flow in the functional system due to the increase in the number of material moves and material tracking required. This was not a problem on the Twin Lines due to the close proximity of machines and workstations, each of which fed another station in the processing sequence.

5.4. Material Tracking and Coordination

At SMP, material handlers were used to coordinate the flow of material to the welding and the assembly lines. It was their responsibility to make sure that the welding lines were not stopped for lack of parts. Although certain areas were identified as storage areas for the departments, in reality, containers were put wherever space was available within these areas. When possible, containers for the same batch were placed together. However, sometimes this was not possible, and the material handler would have to hunt for the "last" container of a batch. As one material handler stated: "It was not unusual to spend upwards of two hours looking for that last tub or that last basket of parts because we needed to finish the order. It was not unusual, because there was no real set storage . . . , you had such quantity that you were stuffing them wherever they would fit."

However, if batch sizes were reduced without any other changes to the manufacturing system, you could still have the same number of containers on the floor, but they would have been for different orders. This would have increased the load on the material handlers coordinating the flow of material on the factory floor. Many of these material handlers were not needed when the Twin Line was formed due to the line flow process, and the number of material handlers had been reduced from 30 to 3. This would have been very hard to accomplish in the functional system unless WIP inventory had been low enough so that the inventory could have been stored right at the welding station needing the parts. As stated by the engineering manager: "I think our biggest problem in reducing lot sizes, without the cell and the flow-through approach, is that work-in-process would have been a nightmare—trying to track it, and releasing the next order before the current order is complete. And where were the parts, and did the parts belong to Model A or did they belong to Model B . . . ?"

5.5. Information System Requirements

One benefit of the Twin Lines not immediately recognized was the impact on the information system. In the functional system, a stamped part had one number, the formed part had another number, the welded unit a third part number, and so forth. For each part number, an order was released to the factory floor to start production of the batch and a receipt was recorded to the MRP system when the batch of parts was finished. If lot sizes had been reduced in the functional system and if no changes had been made to the MRP system, the number of planned order releases and order receipts would have increased, increasing the load on the information system.

For example, if batch sizes had been reduced from 1200 to 200 units and if overall demand had stayed the same, the number of order releases and order receipts for each part would have increased by a factor of 6. Thus, information requirements would have grown linearly as batch sizes were reduced.

When the Twin Lines were established, the blanks formed at the press brakes were fed directly to the welding line, eliminating the need to move a container of parts in and out of storage. In turn, this eliminated the need for order releases and order receipts for these operations. Backflushing was used to record the amount of labor and material incurred in the cells to manufacture the products produced. Thus, the bill of material was effectively collapsed as far as information requirements were concerned and the batch size reductions had less impact on the information systems requirements in the cellular system than if batch sizes had been reduced by the same amount in the functional system.

5.6. Worker Injuries

Finally, the plant needed to reduce worker injuries and the corresponding workmen's compensation payments. Many of the injuries were occurring as a result of workers removing and inserting parts into containers in the functional system. When the Twin Lines were formed, parts were still taken out of containers at the press brakes, but they were then placed on a roller conveyor system for transfer to the remaining processes. Thus, material was only lifted when initially removed from the container at the brake presses. As stated by the engineering manager, "We felt that by going to a cell that we could keep that part going. Instead of putting it back into storage in a wiretainer, we kept it on a short conveyor and right to the next operation. Whether or not it was the same person doing that next operation, at least we eliminated all that handling." This was not possible in the previous functional system.

6. FACTORS INFLUENCING CONFIGURATION OF THE CELLULAR SYSTEM

This section examines the considerations at SMP that influenced what parts, processes, and resources were combined to form cells, discusses the tradeoffs involved when these decisions were made, and presents the factors deemed important for the successful operation of the cells that were and will be formed.

6.1. Part/Product Similarity

One of the first steps in cell formation is to find part or product families that have similar processing requirements. An examination of the product population at SMP showed a high degree of similarity in the processing steps required to produce the products. The drawer slides had processing steps that were identical for all slides produced. All other products had a number of component parts that were either stamped or punched out of sheet metal on a stamping or turret press, respectively, and then bent in a press brake to form a component part. The component parts were welded together to form a semifinished unit, painted, and then assembled (if needed) to form a finished product. Most of the differences in processing steps that did occur were along product lines (i.e., the enclosures required wire welding, installation of threaded studs or nuts, etc.). These characteristics of the product population made the products at SMP good candidates for the potential use of cells and partially explain the high degree of planned cell penetration (i.e., 70% of all direct labor and 80% of all machine hours will be in cells if the proposed cells are implemented as planned). However, a high degree of similarity in the processing requirements for the products is not enough by itself to justify the dedication of equipment to form cells or to determine the configuration of the cells that should be formed. The following sections discuss other factors that determined the configuration of cells at SMP as well as the configuration of the cells planned for the next 2 or 3 years.

6.2. Processes Excluded from Cells

Sheet Metal Products had one coil slitting and one shearing machine that slit and sheared steel for all of the stamping and turret presses. Consequently, this equipment was indivisible and could not be put into cells without duplicating the equipment. This expense was not justified at the time of this study.

There were four large stamping presses, three smaller stamping presses, and four CNC turret presses, and all were excluded from the cells. Jobs were sequenced on the turret presses to avoid tooling changes in the tooling magazine as much as possible. These presses could have been dedicated to cells, but doing so would have increased the number of tooling changes required (i.e., the punching operations required for a large product side panel were more similar across products going to different cells than the punching operations for all panels on a particular product). This would have increased setup time, resulting in capacity problems at the machines and would have required the purchase of additional presses.

The stamping presses were excluded from the cells for several reasons. First, each product required numerous stamped parts that varied greatly in terms of size and not all of the panels could be produced on the same press. Thus, each stamping press fed multiple cells due to limitations on the size of the parts that each one could process. Second, some of the presses had foundations that made them expensive to move. Finally, setup times took 15–30 minutes, but a batch size of 200–250 parts could be stamped out in 8–10 minutes (cycle times were very fast once the machines were set up). If batch sizes were reduced further at the stamping presses, capacity problems would have resulted and more presses would have been required. It was more cost effective to continue producing in lot sizes equal to a 20-day supply of blanks, store the blanks, and then pull a 5-day supply of blanks from inventory when needed.

The component parts that were punched out by both the stamping and the turret presses were relatively unaffected by the type of customization planned by SMP (thus, obsolescence was not a big concern for those parts). It also did not take much space to store the parts at this stage of the production process since most were flat. Because of the indivisibility problem (i.e., each press fed multiple cells), and the long setup times, the purchase of additional stamping and turret presses would have been required if they had been dedicated to a cell. The reduction in WIP inventory (defective parts were not a big problem at these presses) that could have resulted from dedicating the presses to cells was not enough to offset the cost of duplicating the equipment.

The painting system in place at SMP had two prewash tanks for cleaning the products to be painted, a single dip tank for the electropainting system, multiple spray booths for spraying powdered paint, and several drying ovens. The system could handle the volume for the entire plant, and it was done in a separate department due to the special equipment, ventilation systems, protective clothing, and skills required. Thus, technological limitations of the process make it difficult to put into cells. Painting could have been cellularized, but the capital investment required to duplicate dip tanks, spray booths, ventilation systems, and other items for each cell would have been quite substantial and was greater than the benefits that could have been gained in the form of WIP inventory or throughput time reduction.

Zinc plating also suffered from some of the same technological limitations that painting did. In addition, the technology of this process required special skills that very few people in the plant had. These individuals had been assigned job classifications that differed from the Drawer Slide Cell operators so that the zinc plating line operators would not be affected by any bumping of personnel that could have taken place if layoffs had occurred at the plant (thus preventing the line from being shut down).

An examination of the issues discussed in this section indicates that long equipment setup times, technological limitations of some processes (and the skills required to operate them), the need for equipment foundations on some equipment, and the indivisibility of some equipment excluded certain processes and equipment from being included in cells. Their exclusion partially defined the types of cells that were formed and limited the processes to be included in cells to forming, welding, assembly, and packaging operations.

6.3. Considerations Influencing the Configuration of the Cells in Place in 1997

The development of cells at SMP followed an evolutionary pattern. Before the Twin Lines were implemented, the Welding Department had lines devoted to part and product families (i.e., large products, small products, and drawers) whose configuration was determined by processing sequence similarity and number of steps in the production process. The large products were heavy, had a large number of component parts, and required a long line with numerous stations. In contrast, the small products were significantly lighter, had a relatively small number of component parts, and required a relatively short line with few stations. Finally, the type of component parts and sequence of welding operations required for all drawers were virtually identical (and different from those required for the large and small products), and they were separated from the welding lines for the large and small products for these reasons.

When SMP began to form cells, it was natural that they should continue to evolve in the same fashion to form the Twin Line and the Drawer Cell (the Small Line had not been cellularized at the time of this study). The same logic held for the Drawer

Slide Cell and the Enclosure Fabrication Cell. Each of these were separate part or product families that really did not fit with any of the other cells or production areas in the plant due to differences in processes or process requirements. Consequently, a cell was formed for each part or product family.

The Drawer Slide Support Cell was the only cell that produced a component part that was welded into a product. All of the drawer slide supports were formed in a similar fashion and were used on both of the Twin Lines. If the cell had fed the lines directly, long conveyors would have been required to connect the cell to the lines and these conveyors would have crossed at least one aisle. Rather than do this, the drawer slide supports were produced in small batches and fed to the lines as needed. Duplicating this cell to allow direct feed to each cell would have cost more than the benefits to be obtained in the form of WIP inventory or throughput time reduction.

6.4. Justification of Installed Cells

The Twin Line did not require any new technology or changes to job classifications, but it did require a substantial capital expenditure to implement. Prior to implementation, expected expenditures included the purchase of additional gun welders, press brakes, roller conveyor lines, and electrical supplies and the cost of moving the equipment to form the cell, hook up of electrical and air lines, and so forth. The justification for these expenditures included both qualitative and quantitative forms of savings. Specifically, at 1988 demand levels, the cells were estimated to eliminate 15,972 forklift load in/load out operations annually, reduce throughput time through the form and weld operations by 50% (using batch sizes in place at that time), reduce the amount of distance a case travels during form and weld by 39%, eliminate 416 wire containers for parts storage, eliminate 9 material handlers and 2 forklifts, reduce exposure to injuries from loading individual parts into and out of wire containers by 6,388,800 occurrences per year, reduce exposure to possible part damage from loading individual parts into and out of wire containers by 6,388,800 occurrences per year, and result in 111,000 fewer exposures to welded case handling damage per year. The estimated annual dollar savings from the implementation of the Twin Line was based upon labor, inventory, scrap, and rework reduction using direct labor and variable burden rates for the press brake and large product welding line areas (estimated reductions in workmen compensation injuries were not included in the estimated savings), with an expected payback period of less than 2 years. The actual investment required to implement the Twin Line was recovered in less than 2 years through actual reductions in labor, inventory, scrap, and rework. In addition, worker injuries were significantly reduced, resulting in a dramatic drop in workmen's compensation payments. As stated by the engineering manager: "The payback was really beyond question."

Training was the main expense required to convert the drawer manufacturing area into a cell. This was treated as part of normal operating and training expense and no formal cost justification was required.

The Enclosure Fabrication Cell also did not require any formal cost justification. The majority of the equipment to form the cell came from Company X when it was purchased by the corporation that owns SMP.

6.5. Considerations Influencing the Configuration of the Proposed Cell System

Large Product Cells   There were two problems with the Twin Lines that were influencing their conversion to a system of four cells, each dedicated to a smaller family of products. First, the lack of cross-training coupled with imbalances in cycle times between the different operations caused a significant amount of idle time for the operators. This was especially prevalent between the press brake and welding operations. As stated by the engineering manager: "One of our biggest restrictions in the Twin Line area was a human resource type of problem in that we have a work classification of welder and a work classification of press brake operator. A welder isn't going to run a press brake, and a press brake operator isn't going to run a welder. There's economies from line balancing, . . . you have people sitting idle with our current situation." He went on to say: "You could balance any one product if you ran that product a month at a time, but we're into typically no more than a day for any given run. Then we're making something else the next day." If the workers on the line were cross-trained to do either press brake or welding operations (as well as their own material handling and forklift operation), the workers could be used to balance the flow of product on the line and the same amount of product could be produced with fewer workers (alternately, more product could be produced with the same amount of workers). Cross-training could also help reduce repetitive motion injuries since the workers would do a variety of different tasks throughout the day. This cross-training will be implemented when the Twin Line is broken down into four cells.

Second, long setup times on the press brakes were hindering further batch size reduction. The managers at SMP believed the most cost-effective way to reduce the amount of setup time required is to dedicate the equipment (especially the press brakes) to smaller families of products, resulting in the formation of the four Large Product cells that were planned. Within each family, the setups required for the different products on the press brakes were very similar, decreasing the amount of setup time required and allowing batch sizes to be further reduced. Forming four cells also would allow four different products to be manufactured concurrently, hopefully providing a better match between actual production and market requirements.

Some additional press brakes, spot welders, and roller conveyors were expected to be needed to form the cells. Where the component parts are quite similar between cells, a single press brake may feed two cells. This would help to reduce the amount of equipment that must be purchased. The justification for the equipment and the expenses incurred to form these cells will reside in the cost savings resulting from the reduction of inventory in the distribution center, a reduction in the amount of labor required to produce the units, a reduction in handling damage if direct shipment can be used, and a reduction in the number of drawers scrapped due to elimination of customization by the dealers.

Small Line Cell   The conversion of the Small Line into a cell was planned to improve utilization of labor. The Small Line faced many of the same type of balancing problems experienced by the Twin Lines. Also, by cross-training the workers to operate all of the equipment within the cell, worker idle time would be reduced and productivity increased. It was also planned to produce the Small Line products according to a master schedule rather than build up inventory of the product when demand for the Twin Lines is low. This would reduce the amount of inventory of the Small Line products kept as safety stock in the distribution center. This could result in some loss of volume flexibility since the Large Product Cells would no longer be able to borrow the Small Line workers as needed. If demand for the Small Line products was low at the same time the demand for the large products was high, or vice versa, workers could be temporarily transferred between the cells to help out with production as needed (workers could also be transferred between other cells to accommodate some demand variations between cells). If demand for both products was simultaneously high, overtime would be needed to meet production requirements. It was expected that the use of overtime would be cheaper than the cost of carrying inventory of the Small Line products. If demand was simultaneously low for both products and the workers were not needed elsewhere in the plant, temporary layoffs could be required. Layoffs had occurred in the past but were avoided whenever possible.

Large Product Assembly Cells   There were two main problems with the assembly lines for the large products. First, the line was not well suited for assemble-to-order production of customized products. Each station was assigned a particular set of assembly tasks for a particular product and the tasks needed to be performed in that order. This worked fine for long production runs, but it was extremely inefficient when batch sizes were small. If the assembly operations were changed to accommodate customized drawer configurations, special trim, and the like for a single unit of a product, the line had difficulty handling the change in an efficient manner (either the product had to be assembled off-line or the line reconfigured to handle the change). In addition, with the planned conversion of the Twin Lines into four cells, three or four different large products could come concurrently to the paint line. They could also arrive concurrently to the assembly and packaging area, which destroyed the idea of having one long assembly line with long production runs. As stated by the engineering manager: "It forces you to go to prekitting and some kind of smaller assembly cell."

A second problem with the line related to statistical dependencies between different stations on the line. No buffer of inventory existed between each station (due to the size of the products). The total output of the line was dependent on the output of the slowest station and any problem with any station quickly affected the rest of the line. As stated by the engineering manager: "The challenges in trying to accurately balance a 30-person work line are pretty obvious. And then, even if it were perfectly balanced, if you have any kind of a hiccup in parts supply or a defective drawer, or a purchased part that did not arrive for that model, you shut 30 people down."

Variations between tasks on the assembly line also accumulated on the assembly line, causing many of the stations to be either blocked or starved for work and the workers to become idle. For example, if station 1 completed its assembly tasks quicker than normal and station 2 took longer than normal to complete its assembly tasks, station 1 could not begin to process another unit until station 2 finished its assembly tasks. Thus, the worker at station 1 would be idle for a period of time. Similarly, if station 2 completed its assembly tasks quicker than normal and station 1 took longer than normal to complete its assembly tasks, station 2 had no work to perform and the worker would be idle for a period of time. As stated by the industrial engineer in charge of evaluating the conversion: "At the times that the assembly line was running good, I still had 50% of the people standing idle waiting for something to come to them." Each time the line stopped for worker rotation to a different station, a similar situation occurred. The line could not begin producing until all of the workers had rotated to the next station. Rotation time was usually when personal breaks were taken to use the restroom or get a drink of water, compounding the problem. As stated by the industrial engineer in charge of the cells: "In the rotation, we allow them 'x' amount of minutes to rotate. When the line shuts down, we're dependent on the first person out, and then we're dependent on the last person in. And there's 38 people and we have just wasted about three times the allotted time, several times a day."

Breaking the assembly line up into 18, two-person assembly cells would allow more efficient, assemble-to-order production. All of the component parts for each product would be brought to the cell, and the cell workers could easily install the correct components in the order required to assemble the unit to the customers specifications.

The cell method of assembly would also reduce the amount of worker idle time. If a cell experienced any problems with a particular part, the problem would only affect a single cell rather than the entire line (unless the problem affected parts going to multiple cells). In addition, the impact of statistical dependencies in the cell would be less than on the assembly line, increasing the efficiency of the assembly task. For example, suppose task A must be done before task B and the average time to complete each task was 4 minutes. If on a particular unit, task A took 3 minutes while task B took 5 minutes, the total time to complete both tasks would still be 8 minutes since task B could be performed immediately after task A (i.e., no idle time existed between the tasks). Similarly, if task A took 5 minutes while task B took 3 minutes, the total time to complete both tasks would be 8 minutes. In both cases, the variation in task times would cancel out, the total time to complete both tasks would be equal to the total average time, and no worker idle time would exist. Since the workers were already cross-trained to do all assembly tasks (and rotate through all tasks over several days), little additional training would be required, the time to perform each assembly task would likely remain the same, and the overall efficiency of the cell system would be greater than the assembly line. In addition, since each worker in the cell would perform all assembly tasks, repetitive motion injuries would be minimized without the need to rotate workers every $1\frac{1}{2}$–2 hours.

A substantial investment would be required to form these assembly cells, but SMP estimated the increased productivity, the savings in labor cost, and the reduction in scrap resulting from the reduction or elimination of product customization in the field would pay for the changes in less than 2 years.

Repositioning of Enclosure Fabrication, Small Line, and Drawer Cells

The repositioning of the Enclosure Fabrication, the Small Line, and Drawer Cells was planned so the cells could more easily share personnel and equipment when needed. Each of these cells has experienced some cyclical times where one cell was short of work while a different cell was overloaded. During those times, workers and equipment from the underloaded cell were sometimes used to help handle the demand for the overloaded cell. In addition, the cells did share some equipment that was not frequently used by all cells. Having these cells located closely together would reduce the material handling required during those times and facilitate coordination between the cells.

Enclosure and Small Product Assembly   Finally, the plans to have the Enclosure Fabrication and the Small Line do their own assembly and packaging of their products was intended to give the cells responsibility for their product. If a problem was detected at final assembly or packaging, a lot of finger pointing had occurred as to who was responsible, but little had been done to resolve the problem. By having the cell do its own assembly and packaging work, any problems found during these operations would provide feedback to the cell on how to prevent future occurrences of the problem (if the problem involved stamping or painting, the involvement of the Stamping or Painting department would be required to solve the problem).

6.6 Other Cell Design Considerations

Four other considerations should be noted with respect to the design of cells at SMP, and each of these considerations will be discussed in the following sections.

Movement of Equipment   There were very few restrictions on the movement of equipment at SMP. Some of the stamping presses had foundations that made them difficult and expensive to move, but all of the remaining punching, stamping, forming, welding, or assembly equipment were easily moved. This made it easy to rearrange equipment to form different cells as demand changed or new products were introduced. In fact, the ease with which the gun welders, rocker arm welders, and machines for pressing or welding in studs and nuts were moved permitted a great deal of flexibility with respect to the range of products that could be produced within the cell (i.e., simply rearranging the welding equipment in the Enclosure Fabrication Cell allowed different products to be welded together). This made the cells fairly robust with respect to changes in product mix that occurred as new products were introduced and old products were discontinued.

Equipment Similarity between Cells   The similarity of equipment between most cells allowed a fair amount of volume flexibility. If demand was temporarily higher than average for one cell while simultaneously being lower than average for a second cell, the equipment in the second cell may have been used to help cope with the high demand. While this could cause problems if it occurs too often (i.e., the focus

of the cell can become lost), the ability to use equipment in other cells if needed was a benefit during cases of extreme demand swings or imbalances.

Involvement of Supervisors and Operators in Cell Development and Implementation   Several managers indicated that involvement of the cell supervisor and the cell operators was especially important during the development and implementation phase since these individuals were responsible for making the cell work. Usually these individuals had a significant amount of knowledge about the processes, and their input helped foster commitment and ownership of the cell, could reduce the number of startup problems experienced, could foster a spirit of continuous improvement, and ultimately result in a better performing cell. Ironically, while interviews with both cell and noncell workers indicated that management was always willing to listen to suggestions for improvement, the vast majority of cell operators were not involved in the development of the cells to which they were assigned, in part because the operators were selected after the design was complete. While one operator mentioned that some startup problems with the cells could have been avoided if the operators were involved in the design and implementation process, it is not known to what extent the startup problems with the different cells could have been avoided through earlier involvement of the workers.

Equipment Utilization   Many of the cells formed at SMP required some duplication of equipment, and the engineering manager commented on the reduction in machine utilization and the amount of capacity cushion required for the cells in the following way: "I think in general that cells lead you toward having a little additional capacity cushion but I don't think it's a gross difference. Many areas where I've seen cells applied you have equipment sitting idle part of the time. You can also see that in some of our cells. But the equipment that's there and sitting idle is not high dollar equipment . . . a press brake is $70–90,000, and they're not cheap, but the paint system is a $14 million investment. So if we do not have continuous feed of product to our paint area, we can jeopardize utilization of $14 million worth of equipment for lack of a $90,000 press brake." He continued by saying: "You have to realize that service level is becoming our company's biggest gripe. What does the customer want? And if the customer demands the flexibility so that some days you have no Small Line product to build while you're building Enclosure Fabrication Cell product, you better have the capacity to do that. And you don't answer it with inventory because inventory obsolescence is becoming a huge factor, even in our industry."

These comments point to two issues of major importance. First, cells may result in a reduction in machine utilization, but it is important to evaluate the entire system to understand how important this reduction is. In the case of SMP, the reduction in WIP inventory, material handling, coordination, and labor costs more than compensated for the reduction in machine utilization that resulted from the formation of cells. This is not to say that machine utilization is unimportant. Rather it illustrates that machine utilization is only one factor to consider when evaluating cell adoption decisions.

Second, it illustrates the importance of considering market demands when evaluating the design of a manufacturing system and the amount of capacity cushion to have. Quick response to changing market demands was becoming an increasingly important competitive priority for SMP and was one of the main reasons for forming the Enclosure Fabrication Cell, for converting the Twin Lines to four cells, and for converting the assembly line for the large products into a series of two-person assembly and packaging cells. It also required a larger capacity cushion than before since it was no longer advisable to hold inventory as a means of meeting those market demands.

7. CELL IMPLEMENTATION AND OPERATION CONSIDERATIONS

Given the workforce resistance to the implementation of cells that occurred at SMP, it is not surprising that communication, union, and worker acceptance of the cell concept, operator selection, education and training, and the ability to work as a team arose as the dominant considerations when implementing and operating the cells. Many of the workers at SMP had been with the plant for 20 years or more. During this time, they had primarily operated in a functional context with the emphasis on the efficient completion of individual process steps. As cells were implemented, the emphasis changed to the efficient completion of a part or product and constant communication was needed to ensure that everyone understood what cells were and why the plant was adopting them. As stated by the human resource manager: "The main thing would be to really have a lot of communication right up front with the union, why you want to implement cells, what you feel it can do for your manufacturing operations, and therefore, what it can do in the long run for them as employees and union members also." He also stated that communication about cells should be an ongoing effort.

Closely related to communication, and heavily influenced by it, is the need for union and worker acceptance of the cell concept. The cell supervisor and the cell workers were the ones who ultimately determined how well the cell operated and whether it was considered a success or not. Ultimately, this comes down to how well the plant can manage the change process, the attitude of the worker toward change, and the willingness of the worker to try something different. As stated by one cell operator when asked about her perception of cells: "Until you try it you really don't know. You have to try it. It's not as bad as the operators opposed to cells think it is."

Education and training was needed on how to operate the equipment in the cell, especially with equipment that the cell operator(s) were not familiar. This training often took a considerable amount of time since the only way to become proficient at operating the equipment and performing the tasks required was to do them. This was especially problematic at the brake operations due to the number of different dies used, the complexity of the setups, and the like. It also meant that cells with more complicated processes required a longer training period than cells with less complicated processes, and this should be taken into account during cell implementation. As stated by the human resource manager: "Our normal probationary period or trial period for a new job was 30 work days or 6 weeks. Six weeks didn't come close to really giving you enough time to learn the operations in the Drawer and Enclosure Fabrication Cells, whereas it was enough time in the Gauge Cell and in the Drawer Slide Cell." This training aspect must also be considered when new members are added to the cells.

As more and more of the manufacturing system is converted to cells, the managers at SMP were concerned that they may not be able to find enough individuals who have the willingness or the capability to operate all equipment in the cell. If needed, those individuals would be put into a cell but would only be required to operate those machines that they are willing and capable of operating. It was hoped that natural attrition would eventually eliminate that problem. In addition, the ability to work in a cell environment and to operate multiple pieces of equipment will likely be one of the criteria used when hiring new workers.

Education and training was also needed on how to operate the cell according to cell concepts, and how to work as a team. As stated by one cell operator: "Do you want to know what's real important? For everybody to work together to get the product out. If you've got the teamwork, you should have everything you need." One of the cell supervisors also elaborated on the teamwork aspect by saying: "A cell is going to work and run as well as the people want it to run. And it doesn't necessarily have to be a group of people that have expertise in every single operation of that cell. It has to be a good group of people that will help each other out if needed. A group of people that won't use each other, individuals that continue to want to learn from the other people who are helping them out." I addition, job rotation within the cell was seen as an important part of this teamwork by both cell workers and cell supervisors so that everyone understood the impact of each operation on the quality of the entire product. Job rotation also eliminated the "it's not my job, I'm not going to do it" type of situation, thus reinforcing the emphasis on the efficient production of the entire product.

Finally, on a more technical note, preventative maintenance was seen as quite important to the successful operation of a cellular system. With lower levels of WIP inventory, unexpected downtime of equipment could quickly idle the cell and disrupt the flow of products to the paint lines and the assembly areas. As stated by the plant manager: "Obviously with this type of production, preventative maintenance has become very important."

8. CONCLUSION

The initial reasons for converting to CM at SMP were driven by a need to reduce batch sizes, reduce the amount of material movement required, institute one-piece part flow, reduce the amount of material tracking and coordination required, and reduce the number and severity of worker injuries in order to reduce WIP inventories, reduce throughput time, improve quality, increase service levels to the distribution center, reduce cost, and provide a safer work environment for the employees. When analyzed from a system's perspective, CM was more likely to deliver these improvements in a more effective and cost-efficient manner than would have improvements to the previous functional system.

A high degree of process sequence similarity existed for the part population at SMP, and it was anticipated that 75% of all machine hours and 80% of all direct labor hours would be spent in cells when the implementation of the proposed cells has been completed. Long equipment setup times, technological limitations of some processes, equipment foundations required by some stamping presses, and indivisibility of equipment prevented some equipment from being included in cells. The equipment excluded from cells could have been incorporated into cells, but the cost to do so would have been larger than the potential benefits that could have been gained. Excluding those processes partially defined the types of cells that were formed and limited the processes to be included in cells to forming, welding, assembly, and packaging operations.

New product introductions spurred the introduction of the Enclosure Fabrication Cell. This product did not fit well with the Twin Line or the Small Line products. New technology resulted in the formation of the Drawer Cell, and a make versus buy decision resulted in the formation of the Drawer Slide Cell.

The need to reduce equipment setup times and improve labor utilization was driving the conversion of the Twin Line to four manufacturing cells, the need to improve labor utilization and have customized assembly of large product was driving the conversion of the large product assembly line to multiple, two-person assembly cells, and the need to improve labor utilization was driving the conversion of the Small Line to a cell. Finally, placing assembly of the enclosure fabrication and the Small Line products back into their respective cells was planned to improve accountability for the product, improve quality, and foster continuous improvement.

It was anticipated that a substantial investment would be required to implement the proposed cells, and the cost to do so was justified by reductions in labor, distribution center inventory, and replacement costs for drawers that were damaged in shipment as a result of product customization by dealers.

Considerations deemed important to the implementation and operation of cells included communication, union and worker buy in to the cell concept, operator selection, education and training, the ability to work as a team, and an effective preventative maintenance program.

INDEX

A

Acceptance sampling
  appropriateness in cellular manufacturing,
    445
  defined, 307
  Dodge–Romig plans, 311
  guidelines, 308–309
  and MIL-STD-105D, 310–311
  rectifying inspections, 310
  single sampling plans for attributes, 309–
    310
  by variables, 311–312
ACD (Annual Cell Definition) module, 568–
  575
Activity network diagrams, 278
Adaptability. *See* Flexibility, cellular manu-
  facturing
Admittance conditions, 214
Admittance rules, 214
Advanced Manufacturing Technology (AMT),
  389
Affinity diagrams, 277–278, 279, 280
Algorithms. *See also* Mathematical models
  Cell Synthesis algorithm, 595, 596
  Genetic algorithms, 63–64
  Johnson's algorithm, 144
  for manufacturing cell formation, 36–66
  Production Flow Analysis, 41–44, 89–94
  Rank Order Clustering algorithm, 44–47
  similarity-based, 47–50
  simulated annealing, 59–63
  ZODIAC algorithm, 50–54
American Armory System, 457
American Dish Service Co. case study
  facilities planning using FacPlan methodol-
    ogy, 528–549

overview, 527
work cell design tasks
  cell layout design, 546–549
  infrastructure design, 540–546
  process definition, 537–540
  product selection, 530–537
work cell implementation, 551–552
work cell operation, 549–551
American National Standards Institute, 312,
  316–317
AMT (Advanced Manufacturing Technology),
  389
Analysis of variance
  experimental design, 285–286
  one-way ANOVA, 286, 288
  single-factor ANOVA, 286, 288
  tool selection example, 286–288
Annual Cell Definition (ACD) module, 568–
  575
ANOVA, 285–288
Assembly cells, 5, 742–743. *See also*
  Manufacturing cells
Assembly lines, 474, 734, 742–743. *See also*
  Job shops
Assignment model, 54–59
Autonomation, 481–483

B

Balancing workload, 254, 442–443, 476–477,
  544–545
Batch manufacturing
  alternatives to, 391–392
  versus cellular manufacturing, 180, 195
  controlling, 391
  and delivery performance, 394–397
  economics of, 390–391

Batch manufacturing (*continued*)
  simulation experiment, 199–213
Batch sizes, 239–240, 391, 714–715
Benefits, cellular manufacturing
  cost–benefit analysis, 398–399
  overview, 399–402
  predicting, 228–237
Binary matrices, 86, 88
Block-diagonal structure, 36–38, 617–618
Boundaries, cell, 115, 123–124
Brainstorming, 289, 438
Business process reengineering, 483

C

Campbell, Dudeck, and Smith (CDS) heuristic, 144
Capability. *See* Process capability
Capacity planning, 672–674
Case studies
  abrasive products manufacturer, 601–604
  American Dish Service Co., 527–552
  coil forging shop, 555–587
  color toner plant, 186–191
  control cables manufacturer, 595–601
  Erie Window Co., 604–608, 609, 610
  hydraulic pump and valve company, 406–410
  plantwide conversion to cellular manufacturing, 681–705
  sheet metal products company, 707–748
Cash flow analysis, 399
Cause-and-effect (CE) analysis, 290–291
Cell comprehensiveness flexibility, 254–255, 258, 265
Cell formation. *See also* Flexible Cell Formation (FCF) method; Machine-part matrix; Production Flow Analysis (PFA)
  capacity analysis, 620–627
  case study of alternatives, 615–658
  control cables manufacturer case study, 597–601
  elements of design, 591, 595
  flexibility considerations, 249–273
  overview, 614–615
  PFA data gathering, 82–94, 102–104
  representation of PFA results, 94–102
  scheduling case study considerations, 642–653, 654, 655–658
  stages in design process

aggregation of small cells, 106
assigning operations to machine types, 261–262
assigning part-operations to machines, 262–263
data gathering, 102–104
distribution of workcenters, 107–108
identifying candidate cells, 263–265
investment in production means, 108
modification of routings, 108–109
parts redesign, 109
search for cell cores, 105–106
suggestion of cells, 104–105
Cells. *See* Manufacturing cells
Cell Synthesis algorithm, 595, 696
Cellular Manufacturing (CM). *See also* Manufacturing cells
  advantages of, 9–10, 613
  benefits of, 399–402
  books on, 12–13
  case studies
    American Dish Service Co., 527–551
    coil forging shop, 555–587
    copier toner plant case study, 186–191
    hydraulic pump and valve company, 407–410
    sheet metal products company, 707–748
  versus conventional manufacturing, 111–112
  costs of, 402–404
  defined, 2, 35
  design and implementation problems, 31–32
  disadvantages of, 10–12, 614
  economic justification, 389–410
  evaluation framework
    dynamic evaluation, 184–186, 189, 191, 192
    static evaluation, 189–190
    stochastic evaluation, 183–184, 189, 190–191
  extracellular operations, 74–77
  factors influencing success
    average batch size, 239–240
    company size, 241–242
    complexity of manufacturing systems, 237–238
    complexity of products, 238, 239
    component variety, 239
    number of machine tools, 240–241
  financial evaluation techniques, 397–399

flexibility in, 252–261
future research issues, 33–34
gaining organizational support, 433–434
identifying potential applications, 392–393
implementation
    "Able" Company example, 362–368, 383, 384
    "Baker" Company example, 369–375, 383, 384
    "Charlie" Company example, 375–378, 383, 384
    "Davis" Company example, 379–383, 384
    list of major tasks, 421, 423
    motivating factors, 27–28
    observed impacts, 29–30
    steps in process, 348–361
    worker issues, 416–420
intercell production flow, 77–79
investment appraisal, hydraulic pump and valve company case study, 407–410
labor and machine utilization, 225–226, 254, 404–406, 745–746
list of supporting activities for introducing, 16–23
literature review, 196–198
and machine-part matrix, 36–66, 616–620
need for project management, 414–415
operating policies
    admittance policy issue, 199
    analytical modeling, 198–202
    cross training issue, 199, 202–203
    families of policies, 200, 201–202
    literature review, 196–198
    simulation experiment, 199–213
overlooked areas
    motivation, 384–385, 387, 388
    systems, 384, 385, 388
    training, 384, 386–387, 388
overview, 25–26
pitfalls, 608–611
plantwide conversion case study
    affinities, 683, 684, 695, 698
    configuration diagram, 695, 699
    data gathering, 685, 687–691, 692, 693, 694
    FacPlan levels of design, 682–683
    macro layout elements, 683–698
    manufacturing strategy development, 691, 694–695
    micro layout elements, 698–703

    model project plan, 684–685, 686, 698–699
    overview, 681–682
    space and constraints, 683, 684, 696, 699
    space planning units (SPUs), 683, 684, 695, 696, 697
quality control issues, 275–277
relationship to group technology, 2, 26, 35
sample survey questionnaire, 243–246
scheduling of work, 141–152
software for, 14–16
steps in implementation process
    developing next level, 360–361
    developing pilot project, 350–352
    developing training, 352–354
    implementing process, 354–356
    modifying process, 358–360
    reviewing process, 357–358
    selling project, 349–350
studying successful applications
    data collection and analysis, 227–228
    delivery performance factor, 232, 233
    list of factors and ratios, 227, 230–231
    overview, 226
    predicting benefits, 228–237
    sample questionnaire, 243–246
surveys of practices, 26–34, 243–246
U.S. versus other countries, 413–414
Cellular Manufacturing Systems (CMS), 2
Change
    and company culture, 322–325, 342, 348, 349
    and corporate structure, 327–328
    implementing, 348–361
    worker issues, 416–420
Change agents, 345–346
Charts. *See also* Control charts
    flowcharting, 289–290, 524
    linear responsibility charts, 431–433
    organization charts, 326
    Pareto charts, 473
    process design diagram charts, 278
CIM (Computer Integrated Manufacturing), 459, 482
CM. *See* Cellular Manufacturing (CM)
CMS (Cellular Manufacturing Systems), 2
CNC (Computer Numerical Control) machines, 404, 406–407, 440, 463
Coil forging shop case study
    beginning status
        Coil Cutting Department, 560–562

Coil forging shop case study (*continued*)
Dishwashing Machines Division, 558
Final Assembly Department, 558–560
Metal Sheets Forging Department, 560–562
overview, 555, 556
production planning and control, 564–565
sales order handling, 556–557
cellular manufacturing project
Annual Cell Definition (ACD) module, 568–575
approach, 566
constraints, 567
design phase, 565–568
impact of cellular manufacturing on production systems, 580–583
implementation phase, 578–587
objectives, 567
organization, 567–568
origin, 565
problems, 583–587
realization phase, 568–578
Weekly Items Assignment (WIA) procedure, 575–578
Color copier toner case study, 186–191
Companies. *See also* Organizations; Teams; Workers
organizational culture, 322–325, 342, 348, 349
organizational motivation, 336–339
organizational structure, 325–328
relationship with employees, 319–322
types of teams
departmental teams, 331–334
management teams, 328–331
specific teams, 334–335
Compensation. *See* Employee compensation
Comprehensiveness flexibility, 254–255, 258, 265
Computer Integrated Manufacturing (CIM), 459, 482
Computer Numerical Control (CNC) machines, 404, 406–407, 440, 463
Continuous improvement, 484, 488, 603
Control, as company culture, 324
Control charts
for attributes
fraction rejected, 302, 303–304
nonconformities, 302, 304–307
for variables

calculating control limits, 297–298
piston ring forging example, 298–301
revising control limits and centerline, 301–302
Copier toner case study, 186–191
Corporate structure, 325–328. *See also* Organizations
Cost–benefit analysis, 398–399
Costs, cellular manufacturing, 402–404
Cross training, 199, 202–203
Culture, corporate, 322–325, 342, 348, 349
Cycle time
in lean manufacturing systems, 463, 464, 466, 467, 476, 477, 486
as training example, 437–438

D

Data gathering
plantwide conversion case study, 685, 687–691, 692, 693, 694
Production Flow Analysis (PFA), 82–94, 102–104
for production systems design, 508–512
successful cellular manufacturing applications, 227–228
DCL (Dedicated Cell Loading), 200, 201–202, 203–205, 207
Decouplers, 455, 466, 484, 489, 490, 492
Dedicated Cell Loading (DCL), 200, 201–202, 203–205, 207
Delivery performance, 232, 233, 394–397
Deming, W. Edward, 470
Dies, rapid exchange, 466–470, 471, 472
Digital simulation, 461
Discounted cash flow, 399
Dodge–Romig sampling plans, 311
Downsizing, 320–321, 419
Dynamic arrivals
and sequence-dependent setup times, 147–149
and sequence-independent setup times, 146–147

E

Economic Order Quantity (EOC), 467
Employee compensation
in American Dish Service Co. case study, 546
issues determining, 443–445
Employees. *See* Workers

EOC (Economic Order Quantity), 467
Ergonomics, 493–494
Erie Window Co. case study, 604–608, 609, 610
Event graphs, 218, 219
Extracellular operations, 74–77
Eyeballing, 6–7

F

Facilities. *See* Manufacturing facilities
Facilities planning. *See* FacPlan
Facility Layout Problem (FLP), 114
FacPlan
  levels of design, 682–683
  levels of space planning, 682–683
  overview, 528–530
  for plantwide cellular conversion, 681–705
  work cell design tasks
    cell layout design, 546–549
    infrastructure design, 540–546
    process definition, 537–540
    product selection, 530–537
Factorial experiments, 288
Factory Flow Analysis (FFA), 7, 614, 615
Families of parts. *See* Part families
Family Planning (FP), 392, 393, 395–396
Fasteners, 160–161
FCF method. *See* Flexible Cell Formation (FCF) method
FFA (Factory Flow Analysis), 7, 614, 615
Financial management, 397–399
FL. *See* Functional Layout (FL)
Flexibility, cellular manufacturing
  cell design, 261–266
  versus complexity, 250
  defined, 250, 485
  labor, 197–198
  machine, 251, 254, 255, 257–258
  measurement issues
    aggregate flexibility, 259–261
    FCF example, 266–272
    machine processing capability, 257–258
    mix flexibility, 259
    overview, 255–257
    primary comprehensiveness, 258
    routing flexibility, 258
    volume flexibility, 259
  overview, 252–254
  part mix, 251, 255, 259
  routing, 252, 258

types of
  cell comprehensiveness, 254–255
  cell level, 254–255
  expansion, 252
  manufacturing system, 251–252
  market, 252
  material handling, 251
  operations, 251
  organizational, 252
  overview, 251–252
  primary comprehensiveness, 254–255
  product, 251
  production, 252
  program, 252
  shop floor, 251
  and uncertainty, 250
  volume, 252, 255, 259
Flexible Cell Formation (FCF) method
  assigning operations to machine types, 261–262
  assigning part-operations to specific machines, 262–263
  evaluating cellular configurations, 265–266
  identifying candidate manufacturing cells, 263–265
  illustration of FCF, 266–272
  improving cellular configurations, 265–266
  phases of FCF, 261–266
Flexible Manufacturing Systems (FMS), 390, 391, 408
Flowcharting, 289–290, 524
Flow control procedures, 253–254
Flow diagrams, 508–512
Flow line cells. *See also* Flow shops
  alternatives to, 393
  defined, 143
  scheduling in
    permutation schedules, 143
    static versus dynamic arrivals, 144–149
  when to use, 393
Flow shops
  background of, 457
  simulation experiment for comparing operating policies, 199–213
Flow times, 202–213
FLP (Facility Layout Problem), 114
Fluctuations, volume, 394–397
FMS (Flexible Manufacturing Systems), 390, 391, 408
FP (Family Planning), 392, 393, 395–396
Fractional factorial experiments, 288

Functional analysis
  choosing method, 512–513
  identifying system components, 513
  ranking functions, 518–519
  role in implementing production systems,
    502
  searching for component links, 513–517
  searching for solution constraints, 517
  studying solutions, 520–523
Functional Layout (FL). *See also* Batch
    manufacturing
  economics of, 390–391
  labor and machine utilization, 404–405
  number of machine tools, 240–241
  viability of rearranging machines into cells,
    390–391

G

GA. *See* Genetic algorithms; Group Analysis
    (GA)
Genetic algorithms, 63–64
Goalpost philosophy, 283
Goal programming, as machine layout design
    methodology, 117–118
Group Analysis (GA)
  classroom case study, 616–635
  defined, 7–8, 614
  example, 72–79
Group scheduling heuristics
  defined, 142
  exhaustive, 142
  for flow line cells, 143–149
  for job shop cells, 150–152
  nonexhaustive, 142
Group Technology (GT)
  as alternative to batch manufacturing, 391,
    393
  benefits of, 195
  books on, 12–13
  defined, 1–2
  examples of implementation success, 26
  overview, 25, 195–196
  relationship to cellular manufacturing, 2
  software for, 14–16
GT. *See* Group Technology (GT)

H

Hand tools, 160
Heuristics, group scheduling
  defined, 142

  exhaustive, 142
  for flow line cells, 143–149
  for job shop cells, 150–152
  nonexhaustive, 142
Hidden agendas, 331, 334
Human Integrated Manufacturing (HIM), 482
Human issues. *See* Workers

I

Information systems
  for cellular manufacturing implementation
    project, 438
  for manufacturing cell utilization, 446–447
Inspection processes
  developing for manufacturing cells, 445–
    446
  at Toyota, 472–473
Intercell production flow, 77–79
Interim manufacturing cells, 487
Interrelationship diagraphs, 278, 279, 281
Inventory. *See also* Work-in-process (WIP)
  kanban links, 442, 453, 454, 455, 477–478,
    479, 482
  and L-CMS, 455
Investment appraisal, hydraulic pump and
    valve company case study, 407–410
ISO 9000 standards family, 312, 316–317
Items file, 83–84, 103
Item-workcenter matrix, 86, 87, 89, 90, 94.
    See also Machine-part matrix

J

Japan, and quality control, 470–474
JIT (Just-In-Time), 391, 458, 481
Job satisfaction, 401–402
Job shops
  background of, 455–457
  as cell configuration, 143
  and cell scheduling, 150–152
  and labor constraints, 197
  versus manufacturing cells, 474
  routing analysis, 71–72
Johnson's algorithm, 144
Just-In-Time (JIT), 391, 458, 481

K

Kanban inventory links, 442, 453, 454, 455,
    477–478, 479, 482

L

LA. *See* Line Analysis (LA)
Labor utilization, 225–226, 404–406
Layout, factory. *See also* FacPlan; Machine
    layout design
  case studies
    American Dish Service Co., 546–549
    intercell and intracell design, 134–139
    plantwide conversion, 683–703
  cell layout design
    American Dish Service Co., 546–549
    integration of intracell and intercell
        layouts, 130–133
    intercell case studies, 134–139
    intercell example, 130–133
    intercell methodology, 124–126
    intercell overview, 122–123
    redesigning, 661–672
  machine layout design
    case studies, 134–139
    constraints and objectives, 114–117
    eight-machine test problem, 119–122
    examples, 118–122
    integration of intracell and intercell
        layouts, 130–133
    intracell example, 126–130
    intracell methodology, 123–124
    intracell overview, 122–123
    overview, 114
    six-machine test problem, 118–119
    solution methodology, 117–118
  plantwide conversion case study
    affinities, 683, 684, 695, 698
    configuration diagram, 695, 699
    macro elements, 683–698
    micro elements, 698–703
    model project plan, 698–699
    space and constraints, 683, 684, 696, 699
    space planning unit definitions, 695, 696,
        697
    space planning units (SPUs), 683, 684
L-CMS (Linked-Cell Manufacturing Sys-
    tems), 455, 458–484. *See also* Lean
    manufacturing systems
Lead times, 227, 232, 242–243, 395–396
Lean manufacturing systems
  background, 455–458
  cell design
    ergonomic issues, 493–494
    example, 487–492

list of steps, 458–484
rules for, 492–493
defined, 453, 458
evolution of, 457–458
inventory in, 453, 454, 455, 477–478, 479,
    482
list of names for, 458
steps in design
  exchanging tools and dies, 466–470
  forming U-shaped cells, 461–466
  integrating preventive maintenance,
      474–475
  integrating quality control, 470–474
  integrating suppliers, 479–481
  leveling, balancing, and synchronizing
      systems, 475–477
  linking cells, 477–478
  overview, 458–460
  reducing work-in-process, 478–479
  restructuring factory floor, 461–466
  restructuring production system, 483–
      484
  role of autonomation, 481–483
strategic advantages, 485–487
system design overview, 453–455
Lean production, defined, 458. *See also* Lean
    manufacturing systems
Leawood Group, 595
Leveling of manufacturing systems, 442, 476
Life cycles, cellular manufacturing implemen-
    tation projects
  conception phase, 421
  execution phase, 421
  list of major tasks, 421, 423
  planning phase, 421
  termination phase, 421, 449
Line Analysis (LA), 8, 614–615, 627, 636–
    640, 641, 642, 643, 644, 645, 646
Linear responsibility charts, 431–433
Linked-Cell Manufacturing System (L-CMS),
    455, 458–484
LRC. *See* Linear responsibility charts

M

Machine Batch Loading (MBL), 200, 201–
    202, 205–206, 208, 216–217
Machine-component incidence matrix. *See*
    Machine-part matrix
Machine layout design. *See also* Layout,
    factory

Machine layout design (*continued*)
case studies, 134–139
constraints and objectives, 114–117
eight-machine test problem, 119–122
examples, 118–122
integration of intracell and intercell layouts, 130–133
intracell example, 126–130
intracell methodology, 123–124
intracell overview, 122–123
overview, 114
six-machine test problem, 118–119
solution methodology, 117–118
Machine Layout Problem (MLP), 114, 123–124
Machine-part matrix
assignment model, 54–59
block-diagonal structure, 36–38, 617–618
criteria for analyzing production flow, 72–79
defined, 36
example, 36
and genetic algorithms, 63–64
and group analysis, 616–620
linear cell formation, 40
mathematical formulations, 38–40
and neural networks, 64–65
p-median formulation, 39–40
Production Flow Analysis algorithm, 41–44
and similarity coefficients, 47–50, 616–617, 618, 619
and simulated annealing techniques, 59–63
ZODIAC algorithm, 50–54
Machine requirements analysis, 672–674
Machines. *See also* Machine layout design
assigning operations to types, 261–262
assigning part-operations to, 262–263
guard removal, 162
utilization, 254, 404–406, 745–746
Machine switching conditions, 215
Machine switching rules, 215–216, 217
Manufacturing. *See* Cellular Manufacturing (CM)
Manufacturing cells. *See also* Cellular Manufacturing (CM)
advantages of, 9–10
American Dish Service Co. case study
cell implementation, 551–552
cell layout design, 546–549
cell operation, 549–551
infrastructure design, 540–546

process definition, 537–540
product selection, 530–537
versus assembly lines, 474
assessment questionnaire, 676–679
assignment model for formation, 54–59
basic components, 414
boundaries, 115, 123–124
characteristics of, 5–6
criteria for defining focus, 3, 4
defined, 2, 70
degree of automation in, 3–5
design specifics
balancing, 254, 442–443, 476–477, 544–545
determining cell capacity, 441–442
determining communication signals, 442
determining material handling methods, 442
determining number of cells, 441
determining process flow, 441–442
determining shop floor layout, 443
information and control systems, 446–447
leveling, 442, 476
selecting equipment, 441
selecting tools, jigs, and fixtures, 442
synchronizing, 443, 477
disadvantages of, 10–12
example, 70, 71
formation algorithms, 36–66
formation overview, 70–82
identifying candidates, 263–265
implementing
conceptual phase, 433–434
constructing cells, 448–449
construction phase, 448–449
develop cell operating rules and procedures, 445
direct conversion, 447
information systems, 438
inspection processes, 445–446
organizing employees for, 430–437
phased conversion, 448
phase-in strategies, 447–448
pilot conversion, 448
planning phase, 434–448
planning process, 438–448
presenting plan to management, 448
project life cycle, 421–423
project termination phase, 421, 449
quality issues, 445–446

selecting and compensating cell workers, 443–445
training for, 436–438
interim, 487
lean manufacturing systems
background, 455–457
cell design, 487–494
overview, 453, 457–458
steps in design, 458–460
steps in implementing, 460–484
strategic advantages, 485–487
system design, 453–455
and machine-part matrix, 36–66, 616–620
operating policies
defined, 196
literature review, 196–198
simulation experiment 199–213
project planning, 425–430
rack bar example, 489–492
redesigning
assessment questionnaire, 676–679
capacity planning, 672–674
cell layout, 661–672
costs of re-layout, 675
equipment changes, 675
flow patterns, 661–672
future projects, 675–676
part routings, 661–672
schedule for re-layout, 674–675
sample work breakdown structure, 426–430
and setup time, 155–178
suitable industries for, 5, 6
survey of practices, 26–34
types of cells, 3–5
U-shaped, 463, 464, 465–466, 467
Manufacturing facilities
mathematical models for machine layout design, 114–122
need for new machine layout, 113
role of cellular manufacturing, 2
types of layouts, 2
types of machine layouts, 112
Manufacturing System Design (MSD), 454–455, 484
Manufacturing systems, defined, 455, 456.
See also Lean manufacturing systems
Mass production, 457
Mathematical models. *See also* Algorithms;
Simulation
assigning operations to machine types, 261–262

assignment model, 54–59
for machine layout design, 114–122
queuing analysis, 183–184
for scheduling problems, 220
Matrices. *See also* Machine-part matrix
constructing for routings analysis, 86–94
group analysis criteria for analyzing routings, 72–79
item-workcenter matrix, 86, 87, 89, 90, 94
job shop routings example, 71–72
PFA similarities for analyzing routings, 79–82
results representation, 94–102
weighted versus binary, 88
workcenter-workcenter matrix, 86, 87, 88, 89, 94
Matrix diagrams, defined, 278
Matrix organizations, 431, 432
MBL (Machine Batch Loading), 200, 201–202, 205–206, 208, 216–217
Microsoft Project software, 644
MIL-STD-105D, 310–311
MLP (Machine Layout Problem), 114, 123–124
Motivation, organizational, 336–339
Move times
impact on policy performance, 212–213
instantaneous, 202–212
positive, 212–213
MSD (Manufacturing System Design), 454–455, 484

N

Neural networks, 64–65
NOTED (No Touch Exchange of Dies), 469, 471
Numerical control machines (CNC), 404, 406–407, 440, 463

O

Object-oriented technology, 214–216
One-Touch Exchange of Dies (OTED), 469, 471
Operating policies
comparing using simulation experiment, 199–213
conditions and rules, 214–217
framework representation, 217–219
and object-oriented model, 217–219
and object-oriented technology, 214–216

Operating policies (*continued*)
  simulating, 214–219
  tools for studying, 213–220
Optimal Scheduler program, 220
Organizational culture, 322–325, 342, 348, 349
Organizational structure, 325–328
Organization charts, 326
Organizations. *See also* Companies; Workers
  product profiling, 591, 592–594
  profile analysis, 590–591
Orthogonal arrays, 288, 291–293
OTED (One-Touch Exchange of Dies), 469, 471

P

Pareto effect, 395, 396
Part families. *See also* Parts
  advantages of cellular manufacturing, 9–10
  defined, 2
  grouping by eyeballing, 6–7
  grouping by parts classification and coding, 8–9
  grouping by production flow analysis, 7–8
  grouping methods, 440
  and manufacturing cell focus, 3
  planning groups, 439–441
  problems in grouping, 6–9
  role of subfamilies, 141–142
Part mix flexibility, 255, 259, 265
Part population analysis, 439–441
Parts. *See also* Part families
  need for redesign and reengineering, 109, 440–441
  resource-independent, 73–74
Part switching conditions, 215
Part switching rules, 215, 217
Part types, assigning to machines, 262–263
Part volume flexibility, 255, 259, 265
Period batch control, 144
Permutation schedules, defined, 143
PFA. *See* Production Flow Analysis (PFA)
Physical simulation, 461
Pilot projects, 350–352, 439, 448, 463
Planning. *See* FacPlan; Project management
Platform teams, 483
Pokayoke devices, 488
Post-it note example, 336–337
Preventive maintenance, 474–475
Primary comprehensiveness flexibility, 254–255, 258, 265

Prioritization matrices, 278
Process capability, 293–296
Process design diagram charts, 278
Processes, defining, 537–540
Production Flow Analysis (PFA)
  algorithms for, 89–94
  analysis of similarities, 79–81
  case study of cell formation alternatives, 615–658
  cell formation algorithm, 41–44
  data gathering, 82–94, 102–104
    items file, 83–84
    routings file, 85–86
    workcenter file, 84–85
  defined, 7
  matrices for, 86–102
  stages of
    Factory Flow Analysis (FFA), 614, 615
    Group Analysis (GA), 616–635
    Line Analysis (LA), 627, 636–640, 641, 642, 643, 644, 645, 646
    overview, 7–8
    Tooling Analysis (TA), 640–642
Production forecasts, 88
Production lead times, 88–89
Production systems. *See also* Manufacturing systems
  coil forging shop example, 580–583
  environment of, 499
  implementing
    list of system components, 525–526
    need for systematic method, 498–504
    operational specifications of method, 504–523
    specifying method, 500–501
    steps in method, 501–504
    and system complexity, 498–500
  versus manufacturing systems, 483
  purpose of, 499–500
  restructuring, 483–484
  role of functional analysis, 502, 512–519
  role of production operators, 500
  structure of, 498–499
Product profiling, 591, 592–594
Product realization teams, 483
Profile analysis, 590–591
Project management
  cell implementation life cycle, 421–423
  functions in process
    controlling, 424, 425
    directing, 423–424

motivating, 423–424
organizing, 423, 424
planning, 423, 424
list of major cellular manufacturing
implementation tasks, 421, 423
need for, 414–415
overview, 421–433
phases in implementing manufacturing
cells, 433–449
conceptual, 433–434
construction, 448–449
planning, 434–448
termination, 421, 449
selecting cellular manufacturing implementation teams, 434–437
work breakdown structures (WBS), 426–
430
Pull systems, 442

Q

Quality Assurance (QA), defined, 313
Quality Control (QC)
in American Dish Service Co. case study,
545
and cellular manufacturing
advantages, 276
basic issues, 276
implementation issues, 445–446
logistics, 277
defined, 313
graphical tools, 471, 473
control charts, 473
fishbone diagrams, 473
flow diagrams, 473
histograms, 473
Pareto charts, 473
run charts, 473
scatter diagrams, 473
in lean manufacturing systems, 470–474
management planning tools
activity network diagrams, 278
affinity diagrams, 277–278, 279, 280
guidelines for implementing in cellular
manufacturing, 278–281
interrelationship diagraphs, 278, 279, 281
interrelationships among, 278
matrix diagrams, 278
prioritization matrices, 278
process design diagram charts, 278
tree diagrams, 278, 279

and standards, 312–317
statistical tools
for acceptance sampling, 282, 307–310
for quality design, 282–293
for quality monitoring and control, 282,
293–307
Quality loss cost, 283–285
Quality systems. *See also* Quality Control
(QC)
defined, 313
documenting, 314–316
evaluating, 313–314
and Q9000 standards family, 316–317
QA versus QC, 313
sample manual for cellular manufacturing
facility, 314–316
Questionnaires, assessing existing manufacturing cells, 676–679
Queuing models, 198

R

Rack bar example, 489–492
Rank Order Clustering algorithm, 44–47
Redesigning manufacturing cells, 661–679
Reengineering, 320–321, 483
Research, studying
benchmark data collection, 227–228
factors influencing success, 237–242
predicting cellular manufacturing benefits,
228–237
sample survey questionnaire, 243–246
Restructuring, 320–321
Rhythm Optimal Scheduler program, 220
Routings file, 85–86, 103
Rules, operating policy, 214–217
Rush orders, 157–159

S

Sampling. *See* Acceptance sampling
Schedules, permutation, 143
Scheduling
case study considerations, 642–653, 654,
655–658
constraint programming approaches, 220
heuristics for flow line cells, 143–149
heuristics for job shop cells, 150–152
literature review, 196–198
mathematical programming models, 220
need for systems integration for parts and
workers, 447

Scheduling (*continued*)
  role of part subfamilies, 141–143
  setup time reduction project, 175
  and simulation study outcome, 219–220
  software for, 220
  static versus dynamic arrivals, 144–149
Sequence-dependent setup times
  defined, 143
  and dynamic arrivals, 147–149
  and static arrivals, 145–146
Sequence-independent setup times
  defined, 143
  and dynamic arrivals, 146–147
  and job shop cells, 150–152
  and static arrivals, 144–145
Setup time
  comparing with run time, 156–157
  deciding where to start, 165–166
  defined, 155–156
  determining costs and benefits in reduction
    efforts, 167, 169, 170
  developing strategic plan for reducing,
    157–159
  illustrating old *vs.* new methods, 167,
    168
  implementing reductions, 164–165
  importance of, 155–156
  internal versus external activities, 469
  list of activities, 171–175
  and lost capacity, 156
  management issues, 159, 167, 171
  measuring, 156–157
  need for reducing, 156–157
  need for standardized procedures, 162–164
  project schedule, 175
  reduction case studies, 175–177
  and rush orders, 157–159
  sequence-dependent, 143, 145–146, 147–
    149
  sequence-independent, 143, 144–145, 146–
    147
  tools and procedures for reducing, 160–164
Sheet metal products company case study
  benefits of converting to cells, 734–737
  cellular system
    drawer cell, 724–727
    drawer slide cell, 723–724
    drawer slide supports cell, 721–722
    enclosure fabrication cell, 727–730
    equipment setup reduction, 734–735

    factors influencing configuration, 737–
      746
    future cells planned, 731–734
    gauge cell, 722–723
    information system requirements, 736–
      737
    manufacturing changes, 730–731
    material handling requirements, 735,
      736
    overview, 716
    processes excluded from cells, 738–739
    production flow, 716, 718, 732, 733
    reasons for converting, 734–737
    twin lines, 716–721
    worker considerations, 746–747
  company description, 707–709
  factors influencing configuration, 737–746
  pre-cell environment
  manufacturing systems
    batch sizes, 714–715
    equipment, 711
    job design, 714
    material handling, 714
    plant layout, 711
    production control, 714–715
    production flow, 711–714
   organizational structure, 709, 710
   problems with, 715–716
   production processes, 709, 711
Shop layout analysis. *See* Tooling Analysis
Shortest processing time (SPT), 216–217
Similarity-based algorithm, 47–50
Similarity coefficients, 616–617, 618, 619
Simulated annealing
  algorithm for cell formation, 59–63
  as machine layout design methodology,
    117, 118
  overview, 118
Simulation
  for comparing operating policies, 199–213
  for developing operating policies, 214–219
  digital versus physical, 461
  for dynamic cellular manufacturing
    evaluation, 184–186
  and schedule development, 219–220
SMED (Single-Minute Exchange of Dies),
  466–469
Software, 14–16, 183, 220, 617–618, 644. *See
  also* FacPlan
SOH (Stock-On-Hand), 455, 489–490
Space planning. *See* FacPlan

Speed, 156. *See also* Setup time
SQC (Statistical Quality Control), 471
Standards
    ANSI/ASQC Q9000, 312, 316–317
    ISO 9000 standards family, 312, 316–317
    MIL-STD-105D, 310–311
Static arrivals
    defined, 144
    and sequence-dependent setup times, 145–146
    and sequence-independent setup times, 144–145
Statistical Quality Control (SQC), 471
Stochastic models, for cellular manufacturing evaluation, 183–184
Stock-On-Hand (SOH), 455, 489–490
STORM software, 617–618
Structure, corporate, 325–328
Subfamilies of parts, 141–142
Suppliers, in lean manufacturing systems, 479–481
Surveys, questionnaire assessing existing manufacturing cells, 676–679
Switching conditions, 214–215
Switching rules, 215
Synchronizing of manufacturing systems, 443, 477

T

TA. *See* Tooling Analysis
Takt time, 466, 476
Teams
    departmental teams, 331–334
    for implementing cellular manufacturing
        leader characteristics, 434, 436
        and linear responsibility charts, 431–433
        and matrix organization structure, 431, 432
        need for functional area representation, 434, 435, 436
        selecting members, 434–437
        training, 436–438
    level of autonomy, 32–33
    management teams, 328–331
    for production system design, 506–508
    for product realization, 483
    specific teams, 334–335
    types of, 328–335
    and work as group activity, 196
Throughput time, 454

3M Company, Post-it note example, 336–337
Tooling, rapid exchange of dies, 466–470, 471, 472
Tooling Analysis, 8, 615, 640–642
Total Quality Control (TQC), 458
Toyota Production System (TPS), 458, 475–476
Toyota Sewn Products System (TSS), 198, 200–201, 209, 210–211
TPS (Toyota Production System), 458, 475–476
TQC (Total Quality Control), 458
Training
    for cellular manufacturing implementation teams, 436–438
    and company strengths and weaknesses, 342
    cross training, 199, 202–203
    cycle time example, 437–438
    developing course details, 346–347
    developing programs, 352–353
    general versus specific, 341
    immediate versus future needs, 344–345
    management, 353–354
    and organizational culture, 342
    overview, 340–341
    resource availability, 343–344
    role of change agents, 345–346
    time constraints, 344
    types of, 341
Transfer batches, 199, 202
Transfer lines, 5, 391, 393
Traveling salesman problem, 146
Tree diagrams, 278, 279
TSS (Toyota Sewn Products System), 198, 200–201, 209, 210–211

U

U.S. Department of Defense, MIL-STD-105D, 310–311
U-shaped cells, 463, 464, 465–466, 467
Utilization, labor and equipment, 225–226, 254, 404–406, 745–746

V

Value engineering, 501, 512–513
Variance analysis, 285–288
Vendors, in lean manufacturing systems, 479–481
Visual inspection, 6–7

Volume flexibility, 255, 259, 265

W

WBA (Worker Batch Assignment), 200,
201–202, 206, 209
WBS (work breakdown structures), 426–430
WCM (World Class Manufacturing), 458
Weekly Items Assignment (WIA) procedure,
575–578
Weighted matrices, 86, 88
What-if analysis
dynamic evaluation, 191
static evaluation, 189–190
stochastic evaluation, 190–191
WIA (Weekly Items Assignment) procedure,
575–578
WIP. *See* Work-in-process (WIP)
Work breakdown structures, 426–430
Workcenter file, 84–85, 103
Workcenter-workcenter matrix, 86, 87, 88, 89,
94
Worker Batch Assignment (WBA), 200,
201–202, 206, 209
Workers. *See also* Teams
in American Dish Service Co. case study

and change, 416–420
compensation for working in manufacturing
cells, 443–445
and fear of failure, 419–420
issues in implementing cellular manufacturing, 416–420
and job loss, 418–419
and job satisfaction, 401–402
labor constraints, 196–198
and more-work-for-same-pay issue, 418
and new technology, 417–420
as participants in system implementation,
500
selecting for cells, 443–445
skill requirements and assessment, 444
Work-in-process (WIP)
in lean manufacturing systems, 478–479
reducing, 391, 399–400
rocks in the river analogy, 478–479, 480
Workload, balancing, 254, 442–443, 476–477,
544–545
World Class Manufacturing (WCM), 458

Z

ZODIAC algorithm, 50–54